Weed Control in Natural Areas
in the Western United States

UNIVERSITY OF CALIFORNIA
WEED RESEARCH & INFORMATION CENTER

2013

AUTHORS

Joseph M. DiTomaso, *Cooperative Extension Specialist, University of California, Davis, CA*

Guy B. Kyser, *Specialist, University of California, Davis, CA*

Scott R. Oneto, *Farm Advisor, University of California Cooperative Extension, Jackson, CA*

Rob G. Wilson, *Intermountain Research and Extension Center Director and Farm Advisor, University of California Cooperative Extension, Tulelake, CA*

Steve B. Orloff, *Farm Advisor, University of California Cooperative Extension, Yreka, CA*

Lars W. Anderson, *Lead Scientist (retired), USDA-ARS, University of California, Davis, CA*

Steven D. Wright, *Farm Advisor, University of California Cooperative Extension, Tulare, CA*

John A. Roncoroni, *Farm Advisor, University of California Cooperative Extension, Napa, CA*

Timothy L. Miller, *Cooperative Extension Specialist, Washington State University, Mount Vernon, WA*

Timothy S. Prather, *Professor, University of Idaho, Moscow, ID*

Corey Ransom, *Associate Professor, Utah State University, Logan, UT*

K. George Beck, *Professor and Extension Specialist, Colorado State University, Ft. Collins, CO*

Celestine Duncan, *Weed Management Services, Helena, MT*

Katherine A. Wilson, *Staff Research Associate, University of California Cooperative Extension, Tulare, CA*

J. Jeremiah Mann, *Agronomist, Yuba City, CA*

ACKNOWLEDGEMENTS

The authors would like to thank Stephen Colbert, Vanelle Peterson, Jason Giessow, and John Knapp for their helpful comments on the book. We are most grateful to Gale Pérez for her outstanding and thorough work in designing and formatting the book for publication. This project was partially funded by the 2011-2012 Renewable Resources Extension Act (RREA), Project #11-1331.

For information about ordering this publication, contact:

UC Weed Research and Information Center
Department of Plant Sciences, MS 4
University of California
One Shields Ave.
Davis, CA 95616

Telephone (530) 752-1748
FAX (530) 752-4604
E-mail: wric@ucdavis.edu

ISBN 978-0-692-01922-1
Library of Congress Control Number: 2012922754

©2013 by the UC Weed Research and Information Center
All rights reserved.

Unless indicated on the photo, nearly all photos were taken by Joseph M. DiTomaso or other authors. No part of this publication may be reproduced, stored in a retrieval system, or transmitted, in any form or by any means, electronic, mechanical, photocopying, recording, or otherwise, with the written permission of the publisher and the authors.

Printed in Canada.

Recommended citation: DiTomaso, J.M., G.B. Kyser et al. 2013. Weed Control in Natural Areas in the Western United States. Weed Research and Information Center, University of California. 544 pp.

Table of Contents

	PAGE
Areas covered in this book	4
How this book is organized	4
Abbreviations	6
Weed reports: individual plant control information	15
Susceptibility charts for additional species	433
Control options	462
Control equipment and techniques	469
Environmental concerns and safety	482
Index of common and scientific names	520

TABLE		PAGE
1	Plants covered in the book alphabetically by **FAMILY**	7
2	**Grasses and sedges** not included in the weed reports – **non-chemical** control	434
3	**Grasses and sedges** not included in the weed reports – **chemical** control	436
4	**Herbaceous forbs** not included in the weed reports – **non-chemical** control	438
5	**Herbaceous forbs** not included in the weed reports – **chemical** control	446
6	**Woody plants** not included in the weed reports – **non-chemical** control	454
7	**Woody plants** not included in the weed reports – **chemical** control	456
8	**Aquatic plants** not included in the weed reports – **non-chemical** control	458
9	**Aquatic plants** not included in the weed reports – **chemical** control	460
10	Mechanisms of herbicide degradation and half-life of selected herbicides	487
11	**Characteristics of herbicides** used for control of natural-area weeds	492
12	Rainfastness and grazing and hay harvesting restrictions for herbicides registered for terrestrial use	510
13	Characteristics of **aquatic herbicides** registered, or soon to be registered, for use in the western United States	512
14	**Biological control** programs or established agents for **non-crop weeds** in the western United States	513
15	**Biological control** programs or established agents for **aquatic weeds** in the western United States	518
16	**Grass carp** feeding preferences and effect on control	519
17	Several important **internet sites** for herbicides	519

Areas covered in this book

The weedy species covered in this book are plants that are primarily invasive or weedy in natural areas, including wildlands, rangelands, grasslands, pastures, riparian and aquatic areas. We emphasize the 13 contiguous western states (Arizona, California, Colorado, Idaho, Montana, New Mexico, Nevada, North Dakota, Oregon, South Dakota, Utah, Washington, and Wyoming). We have included species from the state noxious weed lists, as well as other non-crop weeds that are frequently problematic in natural areas of the western United States. We include most of the species from the California Invasive Species Council Inventory. In total, control options are provided in full write-ups ("weed reports") for 242 species, with a little under 100 additional species included in susceptibility charts only. Although most of species in this book are non-native, some native species are included, as they occasionally are problems in certain human use areas.

Many species covered in this book are also problems in rights-of-way, waste areas, industrial sites, urban settings, borders of cultivated fields, and utility ways, but species specifically weedy in these areas were not considered for inclusion, nor were weed control techniques and herbicides that are registered or used only in these areas.

When using any technique, whether non-chemical or chemical, care should be taken not to cause ecological damage or endanger the health of animals or humans. In particular, when using an herbicide, check to make sure that the product is registered in the state where you will be using it, and be sure to read the label on the container. The recommendations on the manufacturer's label are based on thorough testing. Following these recommendations can prevent many problems which might arise from the improper use of a chemical.

Information is supplied here with the understanding that no discrimination is intended and no endorsement by the authors or Cooperative Extension Services is implied. Trade names (brand names) of some commercial pesticides are used in portions of this book to help identify the common name used by the Weed Science Society of America (WSSA). Authors have assembled the most reliable information available at the time of publication. Due to constantly changing laws and regulations, authors can assume no liability for the recommendations. Any use of a pesticide contrary to instructions on the printed label is not legal or recommended.

How this book is organized

This book contains general information on the control and management of about 340 weeds of natural areas in the western United States. The information in the book comes from a number of sources, including personal experience of the authors, peer-reviewed literature, and non-peer reviewed literature such as conference proceedings, research progress reports, and reviews in books. In addition, we conducted extensive internet searches for credible websites that contained information on weed and invasive plant control and management. We considered all forms of control, including chemical and non-chemical. With this information, the authors summarized what we considered to be the most relevant and practical control options for each weed.

The options that are provided should not be considered recommendations by either the authors or the University of California. Rather, they are options that have been reported to be successful. Efficacy may vary from one region to another, based on climate, soils, etc. For some species there are many chemicals listed, but for others only one or a few. We recognize that many land managers will be restricted in their ability to use many of the chemical and non-chemical options provided, due to state registrations, safety and health issues, environmental risks, accessibility of infestation, geographical and topographical restraints, experience of personnel, availability of equipment or control agent, or economics. However, it is our intention to provide as many options as possible, with the hope that at least a few can achieve the desired objective and be implemented without restrictions. The choice of any option should be weighed against its desirable or undesirable impact on the ecosystem and the desired function of that system.

While the bulk of the text is dedicated to providing control options, we have also included additional information on the variety of control techniques and equipment used in natural areas, as well as safety and environmental considerations, herbicide characteristics, rainfast periods and grazing and haying restrictions for terrestrial herbicides, a list of species with biological control agents either available or under development, and helpful conversion tables. The book is not comprehensive in discussing many of these topics, as this information is available through several other resources, both published and on the internet. We have included a list of useful references and websites where additional information can be obtained.

The control options are divided into two categories. For those plants that are widespread or more significant problems throughout much of the west, we have included a full write-up ("weed report") on that plant or a cluster of closely related plants whose control options are similar. The weed report includes a brief description of the plant, its range, habitat, origin, and impact, followed by non-chemical and chemical control options. The chemical control options include the recommended rate, timing and any helpful remarks or cautions. Control information is lacking for some species, but through inference with closely related species, we have included options that we believe should be effective. About 240 species are included in these weed reports.

Second, for weeds which are less widespread, we include susceptibility charts. These charts are separated into grasses and sedges, herbaceous forbs, submerged and floating aquatic species, and woody plants. While we don't include full weed reports for these species, the susceptibility charts do have qualitative information on mechanical, cultural, and chemical control options and the best timing or growth stage for each treatment. As with the plants in the full weed reports, we have included what we consider to be likely control options, based on results or observation with related species. We also include whether or not the herbicides are considered foliar or soil active (or both) and their half-life range (or average). These susceptibility charts are a quick way to determine the potential control options for about 100 species.

Finally, weedy and invasive plants are dynamic, with new species appearing each year. Furthermore, new control techniques are being developed by researchers and field practitioners around the west. Thus, it is our goal to update and reprint the book about every three years so the information stays current. The authors recognize that there are many individuals working in the field throughout the western United States that have experience in the control of a variety of plant species. This book certainly is not complete. We ask that those using the book contact us when they are familiar with other techniques that we have not included or when the information presented here is either inaccurate or incomplete. Please contact Joe DiTomaso [jmditomaso@ucdavis.edu; University of California, Davis, Department of Plant Sciences, MS 4, One Shields Avenue, Davis, CA 95616.]

Abbreviations

a.e.	acid equivalent
a.i.	active ingredient
A or ac	acre
ACCase	acetyl Co-A carboxylase
ALS	acetolactate synthase
AMS	ammonium sulfate
CRP	Conservation Reserve Program
d	day
DF	dry flowable
DG	dispersible granules
E	excellent control
EC	emulsifiable concentrate
EPSP	5-enolpyruvylshikimate-3-phosphate
F	flowable; in tables, F = fair control
Fa	fall
fl oz	fluid ounce
FLW	flowering
ft	foot, feet
g	gram
G	granule formulation; in tables, G= good control
gal	gallon
gal/A or gal/acre	gallons per acre
GPA, gpa	gallons per acre
h	hour
ha	hectare
in	inch
lb	pounds
lb/A or lb/acre	pounds per acre
LC	liquid concentrate
LV	low volatile
M	microencapsulated
ml	milliliter
MSO	methylated seed oil
NIS	non-ionic surfactant
oz	ounce
P	pellet formulation; in tables, P = poor control
pH	measure of acidity of a solution or water; log scale for the concentration of hydrogen ions
POST	postemergence
ppb	parts per billion
ppm	parts per million
PRE	preemergence
PSI	photosystem I
PSII	photosystem II
pt	pint
pt/A or pt/acre	pints per acre
qt	quart
qt/A or qt/acre	quarts per acre
ROS	rosette
RTU	ready-to-use
SG	soluble granules
Sp	spring
Su	summer
W	winter
WDG	wettable dispersable
WG	wettable granule
WP	wettable powder
WS	water soluble
WSG	water soluble granules
WSL	water soluble liquid
WSP	water soluble powder

Table 1. Plants covered in the book alphabetically by **FAMILY**

FAMILY	SCIENTIFIC NAME	COMMON NAME	PAGE
Aizoaceae	*Carpobrotus edulis*	Hottentot fig or iceplant	83
Amaranthaceae	*Alternanthera philoxeroides*	Alligatorweed	29
Amaranthaceae	*Amaranthus* spp.	Pigweeds or amaranths	438, 446
Anacardiaceae	*Schinus* spp.	Peppertrees	364
Anacardiaceae	*Toxicodendron diversilobum*	Pacific poison-oak	403
Apiaceae	*Anthriscus caucalis*	Bur chervil	38
Apiaceae	*Anthriscus sylvestris*	Wild chervil	38
Apiaceae	*Carum carvi*	Common caraway	438, 446
Apiaceae	*Conium maculatum*	Poison-hemlock	125
Apiaceae	*Daucus carota*	Wild carrot	149
Apiaceae	*Foeniculum vulgare*	Fennel	187
Apiaceae	*Heracleum mantegazzianum*	Giant hogweed	207
Apiaceae	*Hydrocotyle ranunculoides*	Floating pennywort	224
Apiaceae	*Torilis arvensis*	Hedgeparsley	401
Apocynaceae	*Vinca major*	Big periwinkle	423
Aquifoliaceae	*Ilex aquifolium*	English holly	454, 456
Araceae	*Zantedeschia aethiopica*	Calla lily	438, 446
Araliaceae	*Hedera canariensis*	Algerian ivy	202
Araliaceae	*Hedera helix*	English ivy	202
Araliaceae	*Hedera hibernica*	Atlantic or Irish ivy	202
Arecaceae	*Phoenix canariensis*	Canary Island date palm	307
Arecaceae	*Washingtonia robusta*	Mexican fan palm	307
Asclepiadaceae	*Asclepias fascicularis*	Mexican whorled milkweed	43
Asclepiadaceae	*Asclepias speciosa*	Showy milkweed	43
Asteraceae	*Acroptilon repens*	Russian knapweed	16
Asteraceae	*Ageratina adenophora*	Croftonweed	21
Asteraceae	*Ambrosia acanthicarpa*	Annual bursage	438, 446
Asteraceae	*Ambrosia artemisiifolia*	Common ragweed	438, 446
Asteraceae	*Arctotheca calendula*	Capeweed	438, 446
Asteraceae	*Artemisia absinthium*	Absinth wormwood	438, 446
Asteraceae	*Artemisia tridentata*	Sagebrush	454, 456
Asteraceae	*Carduus acanthoides*	Plumeless thistle	79
Asteraceae	*Carduus nutans*	Musk thistle	79
Asteraceae	*Carduus pycnocephalus*	Italian thistle	79
Asteraceae	*Carduus tenuiflorus*	Slenderflower thistle	79
Asteraceae	*Carthamus lanatus*	Woolly distaff thistle	85
Asteraceae	*Centaurea calcitrapa*	Purple starthistle	87
Asteraceae	*Centaurea debeauxii*	Meadow knapweed	90
Asteraceae	*Centaurea diffusa*	Diffuse knapweed	93
Asteraceae	*Centaurea iberica*	Iberian starthistle	87
Asteraceae	*Centaurea melitensis*	Malta starthistle or tocalote	97
Asteraceae	*Centaurea solstitialis*	Yellow starthistle	100
Asteraceae	*Centaurea stoebe*	Spotted knapweed	105

Table 1 continued: Plants covered in the book alphabetically by FAMILY

FAMILY	SCIENTIFIC NAME	COMMON NAME	PAGE
Asteraceae	*Centaurea virgata* ssp. *squarrosa*	Squarrose knapweed	109
Asteraceae	*Chondrilla juncea*	Rush skeletonweed	116
Asteraceae	*Chrysanthemum coronarium*	Crowndaisy	438, 446
Asteraceae	*Chrysothamnus nauseosus*	Rabbitbrush	438, 446
Asteraceae	*Cichorium intybus*	Chicory	438, 446
Asteraceae	*Cirsium arvense*	Canada thistle	119
Asteraceae	*Cirsium vulgare*	Bull thistle	122
Asteraceae	*Conyza bonariensis*	Horseweed	438, 446
Asteraceae	*Conyza canadensis*	Hairy fleabane	438, 446
Asteraceae	*Cotula australis*	Southern brassbuttons	438, 446
Asteraceae	*Cotula coronopifolia*	Brassbuttons	438, 446
Asteraceae	*Crupina vulgaris*	Common crupina	134
Asteraceae	*Cynara cardunculus*	Artichoke thistle	138
Asteraceae	*Delairea odorata*	Cape-ivy	152
Asteraceae	*Dittrichia graveolens*	Stinkwort	157
Asteraceae	*Erechtites* spp.	Burnweeds	438, 446
Asteraceae	*Euryops multifidus*	Sweet resinbush	454, 456
Asteraceae	*Helianthus annuus*	Common sunflower	438, 446
Asteraceae	*Hemizonia fitchii*	Fitch's tarweed	205
Asteraceae	*Hemizonia parryi*	Parry's tarweed	205
Asteraceae	*Hemizonia pungens*	Spikeweed	205
Asteraceae	*Hieracium* spp.	Hawkweeds	209
Asteraceae	*Holocarpha virgata*	Virgate tarweed	214
Asteraceae	*Hypochaeris glabra*	Smooth catsear	230
Asteraceae	*Hypochaeris radicata*	Common catsear	230
Asteraceae	*Iva axillaris*	Povertyweed	438, 446
Asteraceae	*Lactuca serriola*	Prickly lettuce	438, 446
Asteraceae	*Leucanthemum vulgare*	Oxeye daisy	247
Asteraceae	*Onopordum acanthium*	Scotch thistle	292
Asteraceae	*Picris echioides*	Bristly oxtongue	313
Asteraceae	*Senecio jacobaea*	Tansy ragwort	440, 446
Asteraceae	*Silybum marianum*	Blessed milkthistle	371
Asteraceae	*Tanacetum vulgare*	Common tansy	399
Asteraceae	*Taraxacum officinale*	Dandelion	440, 446
Asteraceae	*Xanthium spinosum*	Spiny cocklebur	428
Asteraceae	*Xanthium strumarium*	Common cocklebur	428
Azollaceae	*Azolla* spp.	Mosquitoferns	52
Bignoniaceae	*Catalpa bignonioides*	Catalpa	454, 456
Boraginaceae	*Amsinckia menziesii* var. *intermedia*	Coast fiddleneck	35
Boraginaceae	*Amsinckia menziesii* var. *menziesii*	Menzies fiddleneck	35
Boraginaceae	*Cynoglossum officinale*	Houndstongue	143
Boraginaceae	*Echium plantagineum*	Vipers bugloss	159
Boraginaceae	*Echium vulgare*	Blueweed or vipers bugloss	161

FAMILY	SCIENTIFIC NAME	COMMON NAME	PAGE
Boraginaceae	*Myosotis latifolia*	Broadleaf forget-me-not	279
Brassicaceae	*Alliaria petiolata*	Garlic mustard	27
Brassicaceae	*Alyssum alyssoides*	Yellowtuft	440, 448
Brassicaceae	*Berteroa incana*	Hoary alyssum	440, 448
Brassicaceae	*Brassica nigra*	Black mustard	61
Brassicaceae	*Brassica rapa*	Birdsrape mustard	440, 448
Brassicaceae	*Brassica tournefortii*	Saharan mustard	64
Brassicaceae	*Cakile maritima*	Sea rocket	440, 448
Brassicaceae	*Capsella bursa-pastoris*	Shepherd's-purse	440, 448
Brassicaceae	*Cardaria chalepensis*	Lens-podded whitetop	76
Brassicaceae	*Cardaria draba*	Hoary cress or whitetop	76
Brassicaceae	*Cardaria pubescens*	Hairy whitetop	76
Brassicaceae	*Chorispora tenella*	Blue mustard	440, 448
Brassicaceae	*Descurainia sophia*	Flixweed or tansymustard	440, 448
Brassicaceae	*Hesperis matronalis*	Damesrocket	440, 448
Brassicaceae	*Hirschfeldia incana*	Shortpod mustard	211
Brassicaceae	*Isatis tinctoria*	Dyer's woad	236
Brassicaceae	*Lepidium latifolium*	Perennial pepperweed or tall whitetop	244
Brassicaceae	*Lepidium perfoliatum*	Clasping pepperweed	440, 448
Brassicaceae	*Lobularia maritima*	Sweet alyssum	440, 448
Brassicaceae	*Malcolmia africana*	African mustard	440, 448
Brassicaceae	*Raphanus raphanistrum*	Wild radish	331
Brassicaceae	*Raphanus sativus*	Radish	331
Brassicaceae	*Sinapis arvensis*	Wild mustard	374
Brassicaceae	*Sisymbrium altissimum*	Tumble mustard	377
Brassicaceae	*Sisymbrium irio*	London rocket	377
Brassicaceae	*Thlaspi arvense*	Field pennycress	440, 448
Buddlejaceae	*Buddleja davidii*	Butterflybush	454, 456
Butomaceae	*Butomus umbellatus*	Flowering rush	458, 460
Cabombaceae	*Brasenia schreberi*	Watershield	458, 460
Cabombaceae	*Cabomba caroliniana*	Fanwort	74
Caryophyllaceae	*Gypsophila paniculata*	Baby's-breath	198
Caryophyllaceae	*Saponaria officinalis*	Bouncingbet	440, 448
Ceratophyllaceae	*Ceratophyllum demersum*	Coontail	112
Characeae	*Chara* spp.	Chara	114
Characeae	*Nitella* spp.	Nitella	114
Chenopodiaceae	*Atriplex semibaccata*	Australian saltbush	47
Chenopodiaceae	*Bassia hyssopifolia*	Fivehook bassia	54
Chenopodiaceae	*Bassia scoparia*	Kochia	54
Chenopodiaceae	*Chenopodium album*	Common lambsquarter	440, 448
Chenopodiaceae	*Chenopodium berlandieri*	Netseed lambsquarter	440, 448
Chenopodiaceae	*Halogeton glomeratus*	Halogeton	200
Chenopodiaceae	*Salsola paulsenii*	Barbwire Russian-thistle	353

Table 1 continued: Plants covered in the book alphabetically by FAMILY

FAMILY	SCIENTIFIC NAME	COMMON NAME	PAGE
Chenopodiaceae	*Salsola tragus*	Russian-thistle or tumbleweed	353
Chenopodiaceae	*Sarcobatus vermiculatus*	Greasewood	442, 448
Clusiaceae or Hypericaceae	*Hypericum canariensis*	Canary Island hypericum	454, 456
Clusiaceae or Hypericaceae	*Hypericum perforatum*	Common St. Johnswort or klamathweed	228
Convolvulaceae	*Convolvulus arvensis*	Field bindweed	127
Cucurbitaceae	*Bryonia alba*	White bryony	73
Cuscutaceae	*Cuscuta japonica*	Japanese or giant dodder	136
Cyperaceae	*Scirpus acutus*	Hardstem bulrush or tule	434, 436
Dipsacaceae	*Dipsacus fullonum*	Common teasel	154
Dipsacaceae	*Dipsacus laciniatus*	Cutleaf teasel	154
Dipsacaceae	*Dipsacus sativus*	Fuller's teasel	154
Elaeagnaceae	*Elaeagnus angustifolia*	Russian-olive	169
Equisetaceae	*Equisetum* spp.	Horsetails or scouringrushes	175
Euphorbiaceae	*Euphorbia esula*	Leafy spurge	182
Euphorbiaceae	*Euphorbia oblongata*	Oblong spurge	182
Euphorbiaceae	*Euphorbia terracina*	Carnation spurge	182
Euphorbiaceae	*Ricinus communis*	Castorbean	333
Euphorbiaceae	*Sapium sebiferum*	Chinese tallowtree	361
Fabaceae	*Acacia melanoxylon*	Black acacia	454, 456
Fabaceae	*Alhagi pseudalhagi*	Camelthorn	25
Fabaceae	*Amorpha fruticosa*	Indigobush	33
Fabaceae	*Cytisus scoparius*	Scotch broom	147
Fabaceae	*Galega officinalis*	Goatsrue	189
Fabaceae	*Genista monspessulana*	French broom	191
Fabaceae	*Lathyrus latifolius*	Everlasting peavine	442, 448
Fabaceae	*Lotus corniculatus*	Birdsfoot trefoil	263
Fabaceae	*Lupinus* spp.	Lupines	267
Fabaceae	*Medicago polymorpha*	California burclover	442, 448
Fabaceae	*Melilotus albus*	White sweetclover	442, 448
Fabaceae	*Robinia pseudoacacia*	Black locust	335
Fabaceae	*Sesbania punicea*	Red sesbania	369
Fabaceae	*Spartium junceum*	Spanish broom	385
Fabaceae	*Sphaerophysa salsula*	Swainsonpea	388
Fabaceae	*Trifolium hirtum*	Rose clover	442, 448
Fabaceae	*Ulex europaeus*	Gorse	415
Geraniaceae	*Erodium cicutarium*	Redstem filaree	177
Geraniaceae	*Geranium dissectum*	Cutleaf geranium	193
Geraniaceae	*Geranium purpureum*	Little robin	193
Geraniaceae	*Geranium robertianum*	Herb-robert	442, 450
Haloragaceae	*Myriophyllum aquaticum*	Parrotfeather	281
Haloragaceae	*Myriophyllum spicatum*	Eurasian watermilfoil	283
Hydrocharitaceae	*Egeria densa*	Brazilian egeria	163
Hydrocharitaceae	*Elodea canadensis*	Common elodea	458, 460

FAMILY	SCIENTIFIC NAME	COMMON NAME	PAGE
Hydrocharitaceae	*Hydrilla verticillata*	Hydrilla	221
Hydrocharitaceae	*Lagarosiphon major*	Oxygenweed	240
Hydrocharitaceae	*Limnobium laevigatum*	South American spongeplant or smooth frogbit	252
Hydrocharitaceae or Najadaceae	*Najas* spp.	Naiads	286
Hypericaceae or Clusiaceae	*Hypericum canariensis*	Canary Island hypericum	454, 456
Hypericaceae or Clusiaceae	*Hypericum perforatum*	Common St. Johnswort or klamathweed	228
Iridaceae	*Crocosmia* x *crocosmiflora*	Crocosmia or Montbretia	442, 450
Iridaceae	*Iris pseudacorus*	Yellowflag iris	234
Iridaceae	*Watsonia meriana*	Watsonia	442, 450
Juncaceae	*Juncus effusus*	Soft rush	238
Juncaceae	*Juncus patens*	Spreading rush	238
Lamiaceae	*Lamium amplexicaule*	Henbit	442, 450
Lamiaceae	*Marrubium vulgare*	White horehound	273
Lamiaceae	*Mentha pulegium*	Pennyroyal	275
Lamiaceae	*Salvia aethiopis*	Mediterranean sage	356
Lamiaceae	*Trichostema lanceolatum*	Vinegarweed	410
Lemnaceae	*Lemna* spp.	Duckweeds	242
Lemnaceae	*Spirodela* spp.	Duckweeds or duckmeats	242
Lemnaceae	*Wolffia* spp.	Watermeals	242
Liliaceae	*Asparagus asparagoides*	Bridal creeper	442, 450
Liliaceae	*Asphodelus fistulosus*	Onionweed	45
Liliaceae	*Zigadenus* spp.	Deathcamas	431
Lythraceae	*Lythrum hyssopifolium*	Hyssop loosestrife	442, 450
Lythraceae	*Lythrum salicaria*	Purple loosestrife	271
Lythraceae	*Lythrum virgatum*	Wand loosestrife	442, 450
Malvaceae	*Malva neglecta*	Common mallow	442, 450
Moraceae	*Ficus carica*	Fig	185
Myoporaceae	*Myoporum laetum*	Myoporum	454, 456
Myrtaceae	*Eucalyptus globulus*	Tasmanian blue gum	180
Najadaceae or Hydrocharitaceae	*Najas* spp.	Naiads	286
Nyctaginaceae	*Mirabilis nyctaginea*	Wild four-o'clock	277
Nymphaeaceae	*Nuphar lutea*	Yellow pondlily	458, 460
Nymphaeaceae	*Nymphaea odorata*	Fragrant waterlily	458, 460
Nymphaeaceae	*Nymphoides peltata*	Yellow floatingheart	291
Oleaceae	*Ligustrum lucidum*	Glossy privet	250
Oleaceae	*Olea europaea*	Olive	454, 456
Onagraceae	*Ludwigia* spp.	Waterprimroses	265
Oxalidaceae	*Oxalis pes-caprae*	Bermuda buttercup or buttercup oxalis	295
Phytolaccaceae	*Phytolacca americana*	Common pokeweed	311
Pittosporaceae	*Pittosporum undulatum*	Victorian box	454, 456
Plantaginaceae	*Plantago lanceolata*	Buckhorn plantain	442, 450
Poaceae	*Aegilops cylindrica*	Jointed goatgrass	19
Poaceae	*Aegilops triuncialis*	Barb goatgrass	19

Table 1 continued: Plants covered in the book alphabetically by FAMILY

FAMILY	SCIENTIFIC NAME	COMMON NAME	PAGE
Poaceae	*Agrostis avenacea*	Pacific bentgrass	434, 436
Poaceae	*Agrostis stolonifera*	Creeping bentgrass	434, 436
Poaceae	*Ammophila arenaria*	European beachgrass	31
Poaceae	*Anthoxanthum odoratum*	Sweet vernalgrass	434, 436
Poaceae	*Arundo donax*	Giant reed	40
Poaceae	*Avena barbata*	Slender oat	49
Poaceae	*Avena fatua*	Wild oat	49
Poaceae	*Brachypodium distachyon*	Annual false-brome	57
Poaceae	*Brachypodium sylvaticum*	Slender false-brome	59
Poaceae	*Briza maxima*	Big quakinggrass	434, 436
Poaceae	*Bromus diandrus*	Ripgut brome	66
Poaceae	*Bromus hordeaceus*	Soft brome or soft chess	71
Poaceae	*Bromus inermis*	Smooth brome	69
Poaceae	*Bromus japonicus*	Japanese brome	71
Poaceae	*Bromus madritensis* ssp. *rubens*	Red brome	66
Poaceae	*Bromus tectorum*	Downy brome or cheatgrass	66
Poaceae	*Cenchrus* spp.	Sandburs	434, 436
Poaceae	*Cortaderia jubata*	Jubatagrass	130
Poaceae	*Cortaderia selloana*	Pampasgrass	130
Poaceae	*Cynodon dactylon*	Bermudagrass	141
Poaceae	*Cynosurus echinatus*	Hedgehog dogtailgrass	145
Poaceae	*Dactylis glomerata*	Orchardgrass	434, 436
Poaceae	*Ehrharta* spp.	Veldtgrasses	165
Poaceae	*Elytrigia repens*	Quackgrass	172
Poaceae	*Eragrostis* spp.	Lovegrasses or stinkgrasses	434, 436
Poaceae	*Festuca arundinacea*	Tall fescue	434, 436
Poaceae	*Glyceria declinata*	Waxy mannagrass	196
Poaceae	*Holcus lanatus*	Common velvetgrass	212
Poaceae	*Hordeum jubatum*	Foxtail barley	216
Poaceae	*Hordeum marinum*	Mediterranean barley	218
Poaceae	*Hordeum murinum*	Hare barley	218
Poaceae	*Lolium multiflorum*	Italian ryegrass	260
Poaceae	*Lolium perenne*	Perennial ryegrass	260
Poaceae	*Paspalum dilatatum*	Dallisgrass	434, 436
Poaceae	*Pennisetum ciliare*	Buffelgrass	297
Poaceae	*Pennisetum clandestinum*	Kikuyugrass	299
Poaceae	*Pennisetum setaceum*	Crimson fountaingrass	301
Poaceae	*Phalaris aquatica*	Hardinggrass	303
Poaceae	*Phalaris arundinacea*	Reed canarygrass	305
Poaceae	*Phragmites australis*	Common reed	309
Poaceae	*Piptatherum miliaceum*	Smilograss	315
Poaceae	*Poa bulbosa*	Bulbous bluegrass	316
Poaceae	*Poa pratensis*	Kentucky bluegrass	434, 436

FAMILY	SCIENTIFIC NAME	COMMON NAME	PAGE
Poaceae	*Polypogon monspeliensis*	Rabbitfoot polypogon	434, 436
Poaceae	*Schismus* spp.	Mediterraneangrasses	367
Poaceae	*Secale cereale*	Common rye	434, 436
Poaceae	*Setaria* spp.	Foxtails	434, 436
Poaceae	*Sorghum halepense*	Johnsongrass	380
Poaceae	*Spartina* spp.	Cordgrasses	383
Poaceae	*Sporobolus indicus*	Smutgrass	390
Poaceae	*Stipa capensis*	Mediterranean steppegrass	434, 436
Poaceae	*Taeniatherum caput-medusae*	Medusahead	392
Poaceae	*Tripidium ravennae*	Ravennagrass	412
Poaceae	*Ventenata dubia*	North African wiregrass	417
Poaceae	*Vulpia myuros*	Rattail fescue	425
Polygonaceae	*Emex spinosa*	Spiny emex	442, 450
Polygonaceae	*Polygonum arenastrum*	Prostrate knotweed	442, 450
Polygonaceae	*Polygonum convolvulus*	Wild buckwheat	444, 450
Polygonaceae	*Polygonum cuspidatum*	Japanese knotweed	318
Polygonaceae	*Polygonum lapathifolium*	Pale smartweed	320
Polygonaceae	*Polygonum persicaria*	Ladysthumb	320
Polygonaceae	*Polygonum polystachyum*	Himalayan knotweed	318
Polygonaceae	*Polygonum sachalinense*	Sakhalin or giant knotweed	318
Polygonaceae	*Polygonum x bohemicum*	Bohemian knotweed	318
Polygonaceae	*Rumex acetosella*	Red or sheep sorrel	344
Polygonaceae	*Rumex crispus*	Curly dock	346
Polygonaceae	*Rumex obtusifolius*	Broadleaf dock	346
Pontederiaceae	*Eichhornia crassipes*	Water hyacinth	167
Potamogetonaceae	*Potamogeton crispus*	Curlyleaf pondweed	323
Potamogetonaceae	*Potamogeton foliosus*	Leafy pondweed	323
Potamogetonaceae	*Potamogeton illinoensis*	Illinois pondweed	323
Potamogetonaceae	*Potamogeton natans*	Floatingleaf pondweed	326
Potamogetonaceae	*Potamogeton nodosus*	American pondweed	326
Potamogetonaceae	*Stuckenia pectinata*	Sago pondweed	323
Potamogetonaceae or Ruppiaceae	*Ruppia maritima*	Widgeongrass	349
Primulaceae	*Lysimachia vulgaris*	Garden loosestrife	269
Ranunculaceae	*Clematis vitalba*	Old-man's-beard	444, 450
Ranunculaceae	*Ranunculus acris*	Tall buttercup	444, 450
Ranunculaceae	*Ranunculus repens*	Creeping buttercup	444, 450
Ranunculaceae	*Ranunculus testiculatus*	Bur buttercup	444, 450
Rosaceae	*Cotoneaster* spp.	Cotoneasters	132
Rosaceae	*Crataegus monogyna*	English hawthorn	454, 456
Rosaceae	*Potentilla recta*	Sulfur cinquefoil	327
Rosaceae	*Prunus cerasifera*	Cherry plum	454, 456
Rosaceae	*Pyracantha* spp.	Pyracantha or firethorn	454, 456
Rosaceae	*Rosa canina*	Dog rose	338

Table 1 continued: Plants covered in the book alphabetically by **FAMILY**

FAMILY	SCIENTIFIC NAME	COMMON NAME	PAGE
Rosaceae	*Rosa eglanteria*	Sweetbriar rose	338
Rosaceae	*Rosa multiflora*	Multiflora rose	338
Rosaceae	*Rubus armeniacus*	Himalaya blackberry	341
Ruppiaceae or Potamogetonaceae	*Ruppia maritima*	Widgeongrass	349
Salicaceae	*Salix* spp.	Willows	351
Salviniaceae	*Salvinia molesta*	Giant salvinia	359
Scrophulariaceae	*Digitalis purpurea*	Foxglove	444, 452
Scrophulariaceae	*Linaria dalmatica*	Dalmatian toadflax	254
Scrophulariaceae	*Linaria vulgaris*	Yellow toadflax	257
Scrophulariaceae	*Verbascum blattaria*	Moth mullein	419
Scrophulariaceae	*Verbascum thapsus*	Common mullein	419
Scrophulariaceae	*Veronica anagallis-aquatica*	Water speedwell	422
Simaroubaceae	*Ailanthus altissima*	Tree-of-heaven	23
Solanaceae	*Hyoscyamus niger*	Black henbane	226
Solanaceae	*Nicotiana glauca*	Tree tobacco	289
Solanaceae	*Solanum elaeagnifolium*	Silverleaf nightshade	444, 452
Tamaricaceae	*Tamarix* spp.	Saltcedar or tamarisk	395
Trapaceae	*Trapa natans*	Water chestnut	405
Typhaceae	*Typha* spp.	Cattails	413
Ulmaceae	*Ulmus pumila*	Siberian elm	454, 456
Urticaceae	*Urtica dioica*	Stinging nettle	444, 452
Verbenaceae	*Verbena bonariensis*	Tall vervain	444, 452
Verbenaceae	*Verbena litoralis*	Seashore vervain	444, 452
Zannichelliaceae	*Zannichellia palustris*	Horned pondweed	458, 460
Zygophyllaceae	*Peganum harmala*	Harmal or African rue	444, 452
Zygophyllaceae	*Tribulus terrestris*	Puncturevine	407
Zygophyllaceae	*Zygophyllum fabago*	Syrian beancaper	444, 452

Weed reports
individual plant control information

Reports are organized alphabetically by SCIENTIFIC NAME

For plants that are widespread or significant problems in the west, we have included a full write-up ("weed report"). The weed report includes a brief description of the plant, its range, habitat, origin, and impact, followed by non-chemical and chemical control options. The chemical control options include recommended rates, timings, and any helpful remarks or cautions. Control information is lacking for some species, but through inference with closely related species, we have included options that we believe should be effective. Nearly 240 species are included in these weed reports.

Acroptilon repens (L.) DC.
  (= *Centaurea repens* L.)

Russian knapweed

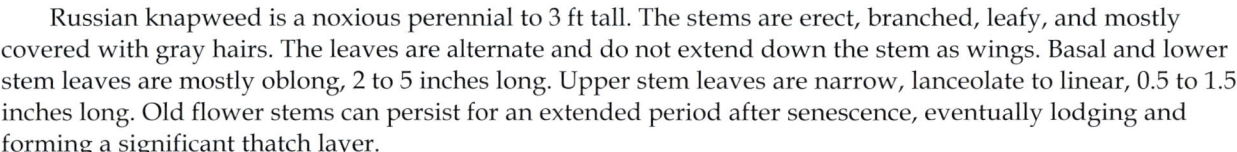

Family: Asteraceae
Range: All western and central states. Less common in eastern and southern United States.
Habitat: Fields, rangeland, cultivated sites, orchards, vineyards, roadsides, ditchbanks, and waste places. Grows on many soil types, but prefers sites that have moist soils such as drainages, riparian zones, river bottoms, irrigated fields, and runoff areas that are not excessively wet. Once established, Russian knapweed is extremely drought tolerant and favors dry sites with full sun.
Origin: Native to central Asia.
Impacts: Russian knapweed is competitive and capable of forming dense monotypic stands. It also appears to have allelopathic properties. It has been shown that the plant can take up zinc from deep in the soil profile and deposit it on the soil surface to create a toxic environment. Russian knapweed can rapidly colonize disturbed areas. Populations are often extremely long-lived due to extensive root systems. Russian knapweed is toxic to horses, but livestock usually avoid grazing it because of its bitter taste.
Western states listed as Noxious Weed: Arizona, California, Colorado, Idaho, Montana, Nevada, New Mexico, North Dakota, Oregon, South Dakota, Utah, Washington, Wyoming
California Invasive Plant Council (Cal-IPC) Inventory: Moderate Invasiveness

Russian knapweed is a noxious perennial to 3 ft tall. The stems are erect, branched, leafy, and mostly covered with gray hairs. The leaves are alternate and do not extend down the stem as wings. Basal and lower stem leaves are mostly oblong, 2 to 5 inches long. Upper stem leaves are narrow, lanceolate to linear, 0.5 to 1.5 inches long. Old flower stems can persist for an extended period after senescence, eventually lodging and forming a significant thatch layer.

The flower heads are hemispheric, in panicle-like or flat-topped clusters. They consist of about 30 white, pink, or lavender-blue disk flowers interspersed with bristles on the receptacle. The flower heads remaining on old stems can aid in identification.

Russian knapweed roots can grow several feet deep, branching frequently to form an extensive vertical and horizontal root system. The plant reproduces primarily by vegetative shoots from creeping roots. Root fragments as small as 1 inch can develop into a new plant from depths to 6 inches. Russian knapweed also produces small quantities of viable seed, which fall near the parent plant or disperse with the seedheads. However, seedlings are uncommon. Seeds appear to survive 2 to 3 years under field conditions.

NON-CHEMICAL CONTROL

Mechanical (pulling, cutting, disking)	Seedlings are easily controlled by hand-pulling or digging, but these techniques do not control established plants because shoots quickly resprout from vast root reserves.
	Multiple mowing passes during a season can suppress Russian knapweed, but mowing alone will not eliminate an infestation, and can even stimulate an increase in shoot density the following year. Removal of top vegetation stimulates shoot sprouting from roots. Cutting or mowing three times a year depletes nutrients in the roots, but unless mowing is continued, plants will recover. Mowing may not be possible in environmentally sensitive areas. Summer mowing followed by herbicide in fall can be effective.
	Root fragments resprout following tillage. Clean equipment after tillage to prevent spreading root fragments. Repeated tillage to 1 ft deep over a period of about 3 years can kill much of the root system.
Cultural	Livestock usually avoid grazing Russian knapweed because of its bitter taste. Russian knapweed is toxic to horses. Goats have been used to graze Russian knapweed.
	Burning is not effective at controlling Russian knapweed, but it is helpful at removing accumulated thatch. Thatch burns best in winter or spring under dry conditions before initiation of spring growth.
	Russian knapweed is sensitive to light competition, and crops that produce dense shade can be used to

	suppress it. Cultivation of dense competitive crops such as cereal grains, alfalfa, or perennial grasses can reduce Russian knapweed in crop fields. Reseeding with perennial grasses following control with herbicides can be effective at suppressing reinfestation in dryland areas.
Biological	*Subanguina picridis* (Russian knapweed gall nematode) and *Aceria acroptiloni* (Russian knapweed mite) have been introduced and are established in several western states. These agents may help to stress the plant but will not eliminate it.

CHEMICAL CONTROL

The following specific use information is based on published papers and reports by researchers and land managers. Other trade names may be available, and other compounds also are labeled for this weed. Directions for use may vary between brands; see label before use. Herbicides are listed by mode of action and then alphabetically. The order of herbicide listing is not reflective of the order of efficacy or preference.

GROWTH REGULATORS	
Aminocyclopyrachlor + chlorsulfuron *Perspective*	**Rate:** 4.75 to 8 oz product (*Perspective*)/acre **Timing:** Postemergence, bud stage to senescence. Although above-ground stems die back in late summer and fall, the subsurface crown buds of Russian knapweed are highly susceptible to fall applications of this herbicide. Applications can be made into winter if conditions permit. **Remarks:** *Perspective* provides broad-spectrum control of many broadleaf species. Although generally safe to grasses, it may suppress or injure certain annual and perennial grass species. Aminocyclopyrachlor provides excellent control of Russian knapweed. It also gives soil residual control 1 year after application. Do not treat in the root zone of desirable trees and shrubs. Do not apply to frozen ground. Do not apply more than 11 oz product/acre per year. At this high rate, cool-season grasses will be damaged, including bluebunch wheatgrass. Not yet labeled for grazing lands. Add an adjuvant to the spray solution. This product is not approved for use in California and some counties of Colorado (San Luis Valley).
Aminopyralid *Milestone*	**Rate:** 3 to 7 oz product/acre (0.75 to 1.75 oz a.e./acre) **Timing:** Postemergence, bud stage to senescence. Although above-ground stems die back in late summer and fall, the subsurface crown buds of Russian knapweed are highly susceptible to fall applications of this herbicide. Applications can be made into winter if conditions permit. **Remarks:** Aminopyralid is one of the most effective herbicides for this weed. It is a broadleaf herbicide like picloram, but more selective. It also has a longer soil residual activity compared to clopyralid and can provide 2 years of control. Aminopyralid is safe on most grasses, although preemergence application at high rates can greatly suppress some annual grasses, such as medusahead. Applications can decrease seed production in some annual and perennial grass species. For postemergence applications, a non-ionic surfactant (0.25 to 0.5% v/v spray solution) enhances control under adverse environmental conditions; however, this is not normally necessary. Do not apply to frozen ground. Other premix formulations of aminopyralid can also be used. These include *Opensight* (aminopyralid + metsulfuron; 2.5 to 3.3 oz product/acre) and *Forefront HL* (aminopyralid + 2,4-D; 1.5 to 2.1 pt product/acre).
Clopyralid *Transline*	**Rate:** 0.67 to 1.33 pt product/acre (4 to 8 oz a.e./acre). Use higher rate for older plants or dense stands. **Timing:** Postemergence, bud stage to senescence. Although above-ground stems die back in late summer and fall, the subsurface crown buds of Russian knapweed are highly susceptible to fall applications of this herbicide. Applications can be made into winter if conditions permit. **Remarks:** Clopyralid has shorter soil residual activity compared to aminopyralid or aminocyclopyrachlor. It controls or injures plants in the Asteraceae and Fabaceae, but is safe on most other broadleaf species and all grasses. For postemergence applications, addition of a non-ionic surfactant (0.25 to 0.5% v/v spray solution) enhances control under adverse environmental conditions; however, this is not normally necessary. Clopyralid can also be tanked mixed with aminopyralid (*Milestone*) for effective control. Do not apply to frozen ground.
Clopyralid + 2,4-D *Curtail*	**Rate:** 2 to 4 qt *Curtail*/acre **Timing:** Same as for clopyralid. **Remarks:** Add a non-ionic surfactant.

Picloram	**Rate:** 1 qt product/acre (0.5 lb a.e./acre)
Tordon 22K	**Timing:** Postemergence, bud stage to senescence. Although above-ground stems die back in late summer and fall, the subsurface crown buds of Russian knapweed are highly susceptible to fall applications of this herbicide. Applications can be made into winter if conditions permit.
	Remarks: Picloram controls a wide range of broadleaf species and has relatively long soil residual activity. Although well-developed grasses are not usually injured by labeled use rates, some applicators have noted that young grass seedlings with fewer than four leaves may be killed. Do not apply near trees, or where soil is highly permeable and where water table is high. Do not apply to frozen ground. Picloram is a restricted use herbicide. Picloram is not registered for use in California.
AROMATIC AMINO ACID INHIBITORS	
Glyphosate	**Rate:** 4 qt product (*Roundup ProMax*)/acre (4.5 lb a.e./acre). Spot treatment: 2% v/v solution.
Roundup, Accord XRT II, and others	**Timing:** Postemergence to rapidly growing plants in the bud stage.
	Remarks: Glyphosate does not control Russian knapweed as well as some other products and will not kill seeds or inhibit germination the following season. Glyphosate has no soil activity and is nonselective. It can create bare ground conditions that are susceptible to weed recruitment. In areas with desirable vegetation, use spot treatment. Glyphosate is a good control option if reseeding is planned shortly after application, as it will not injure seedlings emerging after application. Add a surfactant when using a formulation where it is not already included (e.g., *Accord, Rodeo, Aquamaster*).
BRANCHED-CHAIN AMINO ACID INHIBITORS	
Chlorsulfuron	**Rate:** 1 to 2.6 oz product/acre (0.75 to 1.95 oz a.i./acre)
Telar	**Timing:** Postemergence at flower bud to flowering stage, or fall rosette stage, or winter.
	Remarks: Always use a surfactant. Included with aminocyclopyrachlor in *Perspective*.
Imazapic	**Rate:** 12 oz product/acre (3 oz a.e./acre)
Plateau	**Timing:** Late postemergence in fall when the top 25% of the plant is necrotic, but before a hard frost. Application should be made when some green stem and foliage remains on plant. Timing should correspond to fall basal growth.
	Remarks: Some trials show this treatment to provide no or only partial control, but Russian knapweed is included on the *Plateau* label. Selective to most native grasses. Higher rates may suppress seedings of some cool-season grasses. Add ammonium sulfate and a methylated seed oil. Imazapic is not registered for use in California.
Metsulfuron	**Rate:** 1 to 2 oz product/acre (0.6 to 1.2 oz a.i./acre)
Escort	**Timing:** Postemergence at flower bud to flowering stage or to fall rosettes.
	Remarks: Always use a surfactant. Other premix formulations of metsulfuron can be used at similar application timing. These include *Cimarron Max* (metsulfuron + dicamba + 2,4-D), *Opensight* (metsulfuron + aminopyralid), and *Cimarron X-tra* (metsulfuron + chlorsulfuron). Metsulfuron is not registered for use in California.

Aegilops cylindrica Host.; jointed goatgrass
Aegilops triuncialis L.; barb goatgrass

Jointed goatgrass and barb goatgrass

Family: Poaceae
Range: Jointed goatgrass is found in all western states. Barb goatgrass is found only in California and Oregon, primarily in northern California, especially the Central Valley foothills northward to southern Oregon. Both are expanding their distribution.
Habitat: Disturbed and undisturbed grasslands, oak woodlands, fields, rangelands, pastures, and roadsides. Barb goatgrass tolerates serpentine and hard, shallow, dry, gravelly soils but usually is not found in chaparral. Jointed goatgrass also infests grain fields, especially winter wheat.
Origin: Both species are native to Mediterranean Europe and western Asia.
Impacts: Plants have high silica content, resulting in a persistent thatch that can suppress other species. Barb goatgrass is late-maturing and drought-tolerant, enabling it to occupy and form monocultures in marginal environments. Tough seedheads with long barbed awns are inconvenient for humans and can injure livestock, even fatally. Jointed goatgrass joints are difficult to separate from wheat grains, and contaminated wheat harvests are reduced in quality and value.
Western states listed as Noxious Weed: *A. cylindrica*, Arizona, California, Colorado, Idaho, New Mexico, Oregon, Washington; *A. triuncialis*, California, Oregon
California Invasive Plant Council (Cal-IPC) Inventory: *A. triuncialis*, High Invasiveness

Aegilops triuncialis

Jointed goatgrass and barb goatgrass are late-maturing winter annual grasses with spikes that resemble those of winter wheat. Both species can hybridize with wheat. Unlike wheat, goatgrass spikes break apart into hardened sections called joints.

Goatgrass plants grow up to 20 inches tall. Jointed goatgrass foliage looks similar to winter wheat, but blades, auricles, ligules, and leaf sheaths have evenly spaced, fine hairs along the margins. Barb goatgrass has gray-green foliage, spreading to erect, usually sparsely covered with fine hairs.

Goatgrass seedheads have spikelets arranged alternately along a zigzag rachis. The spikelets are large, hard, and cylindrical to cone-shaped, with long awns. Barb goatgrass, in particular, has stiff, barbed awns; mature seedheads break off whole and can work their way into fur or clothing. The goatgrasses go to seed later than most annual grasses, usually in late spring to early summer. Jointed goatgrass seedheads are 1 to 5 inches long, and barb goatgrass seedheads are 1 to 2.5 inches long. At maturity, heads turn reddish to purple and then dry to a straw color. Eventually the seedheads break apart into joints (spikelets attached to a piece of rachis). Once on the ground, the joints often can survive field burns because of their hard coat. The spikes and joints disperse by attaching to animals, humans, equipment or vehicles.

Seeds germinate in fall and winter, sometimes while still attached to the joints. In fact, joints often remain attached to the lower shoot of dug-up seedlings. Barb goatgrass seeds can remain viable for 2 years or more on the soil, while jointed goatgrass seeds can remain viable for 3 to 5 years.

NON-CHEMICAL CONTROL

Mechanical (pulling, cutting, disking)	Hand pulling or hoeing small infestations is effective, if the roots are pulled and air-dried.
	Mowing can reduce seed production, but timing is critical. Mowing should occur after flowering, but before goatgrass seeds reach the soft boot stage. Early mowing will result in new tiller growth, and late mowing will only spread viable seed.
	Tillage may be used in certain situations. In agricultural fields, sweep tillage or V-blade tillage may be used during fallow periods. Conventional deep plowing will bury goatgrass seed beyond emergence depth. However, buried seed may be viable for up to 5 years, and secondary tillage may bring goatgrass seed back up to a successful emergence depth (< 5 in).
Cultural	Heavy grazing throughout the growing season and high intensity/short duration grazing periodically during the growing season appear to increase plant density.

	Both species mature later than most rangeland annual grasses, providing a window for controlling goatgrass seed production. It is important to burn before the joints disarticulate, to ensure seed kill. Burning will not effectively control seed on the soil surface. Goatgrass germination may increase the year after burning due to increased fertility and light penetration. Therefore, a second year management strategy must be incorporated, and the population should be monitored for several years. In rangeland, burning the first year followed by herbicide and spring seeding the second year may improve barb goatgrass control. For jointed goatgrass in winter wheat, burning fields after harvest can reduce germination of joints at the surface by 90% or more.
Biological	Naturally occurring bacterial strains that infect annual brome and jointed goatgrass, but have no effect on wheat, have been isolated in Kansas and Washington. These bacteria may soon be used in a bio-herbicidal approach for jointed goatgrass control in winter wheat. However, their utility in rangelands has not been explored.

CHEMICAL CONTROL

The following specific use information is based on published papers and reports by researchers and land managers. Other trade names may be available, and other compounds also are labeled for this weed. Directions for use may vary between brands; see label before use. Herbicides are listed by mode of action and then alphabetically. The order of herbicide listing is not reflective of the order of efficacy or preference.

AROMATIC AMINO ACID INHIBITORS	
Glyphosate *Roundup, Accord XRT II*, and others	**Rate:** 1 qt product (*Roundup ProMax*)/acre for young plants < 6" tall (1.1 lb a.e./acre); 1.5 to 4 qt product (*Roundup ProMax*)/acre (1.7 to 4.5 lb a.e./acre) for larger plants or plants under stress. **Timing:** Postemergence in late winter to early spring. Apply to rapidly growing, non-stressed plants before flowering. If possible, apply before desirable perennials have emerged. According to label recommendations, seedling goatgrass can be selectively suppressed in pasture with 5 to 11 oz *Roundup ProMax* (or other trade name)/acre. **Remarks:** Glyphosate has no soil activity and is nonselective, so may kill desirable competitors. Its effectiveness is increased by addition of ammonium sulfate.
BRANCHED-CHAIN AMINO ACID INHIBITORS	
Imazapic *Plateau*	**Rate:** 4 to 6 oz product/acre (1 to 1.5 oz a.e./acre) **Timing:** Preemergence in fall or postemergence in early spring. **Remarks:** Mixed selectivity, tending to favor Asteraceae and some grasses. Safe for most native grasses, but higher rates may suppress seed of some cool-season grasses. Use lower rates for dry climates and low leaf litter and higher rates as moisture increases and/or leaf litter increases. Use methylated seed oil surfactant. Imazapic has long soil residual activity. Imazapic is not registered for use in California.
Propoxycarbazone-sodium *Canter R+P*	**Rate:** 1.2 oz product/acre (0.84 oz a.i./acre) **Timing:** Postemergence from the 2-leaf to 2-tiller stage when plants are growing rapidly. **Remarks:** Propoxycarbazone is a broad-spectrum herbicide that will control many species. It will provide only partial control of jointed goatgrass and perhaps barb goatgrass. Perennial grass species vary in tolerance. A non-ionic surfactant should be added at 0.25 to 0.5% v/v solution.
Sulfometuron *Oust* and others	**Rate:** 1.33 to 2 oz product/acre (1 to 1.5 oz a.i./acre) **Timing:** Preemergence or early postemergence in fall or in late winter after emergence but before goatgrass is 3 inches tall. **Remarks:** Sulfometuron has mixed selectivity. It is fairly safe on native perennial grasses, especially wheatgrass. Other desirable grasses may be stunted, stressed, or injured. Good for revegetation use. Should be used with a surfactant for early postemergence treatments. Sulfometuron has fairly long soil residual activity. Do not let spray drift onto sensitive crops. May move long distances in dry light windblown soils.
Sulfometuron + chlorsulfuron *Landmark XP*	**Rate:** 1.5 oz product/acre **Timing:** Preemergence, in fall or after soil thaws in spring. **Remarks:** See sulfometuron.

Ageratina adenophora (Spreng.) King & H. Robins.
Croftonweed

Family: Asteraceae
Range: In the western U.S., croftonweed is found only in California, near the coast from north of the San Francisco Bay region to the Mexican border.
Habitat: Disturbed places, coastal canyons, riparian areas, scrub, and slopes. Grows best where moisture is available year round and can tolerate both full sun and highly shaded areas.
Origin: Native to southern and central Mexico. Brought to the U.S. as an ornamental.
Impacts: Croftonweed has escaped cultivation and is especially invasive in mild coastal regions. It can cause a fatal respiratory illness in horses when ingested over a period of several months to years. Sheep and goats are unaffected, and cattle generally avoid eating the plant. Croftonweed is one of the most important invasive plants of forests in southern China. It is also a problem in forests and pastures of coastal Australia, New Zealand, South Africa, and Hawaii.
Western states listed as Noxious Weed: California, Oregon
California Invasive Plant Council (Cal-IPC) Inventory: Moderate Invasiveness

Perennial to shrub-like, to 6 ft tall, with opposite ovate-triangular leaves and a woody crown or with lower stems woody. Upper stems purplish with sparse small purplish glandular hairs. The leaves are 2 to 4 inches long and 1 to 3.5 inches wide, with three main veins from the base. The leaf margins are toothed on a petiole about 1 to 2 inches long. The lower surface of the leaves is sometimes purple and also contains the purple glandular hairs.

Plants can flower nearly year round, producing flat-topped clusters of white or sometimes pink flowerheads. The flowerheads are less than 0.5 inch wide and consist of 10 to 60 flowers in a head. The achenes (seeds) have a bristly pappus that is deciduous. Mature plants can produce abundant quantities of apomictic seed (seed develops without fertilization). Seeds disperse with wind, water, soil movement, human activities, and by clinging to the fur, feathers, and feet of animals. Seeds can germinate nearly year round under good conditions. It is unknown how long the seeds last in the soil, but it is presumed that they survive at least a couple of years. Seedlings tolerate light shade and if damaged, can regrow from the crown by 8 weeks of age. Most vegetative growth occurs in summer and fall.

NON-CHEMICAL CONTROL

Mechanical (pulling, cutting, disking)	Mechanical control is possible where plants are accessible. This can include digging plants out, but crowns must be removed. However, plants often grow on steep slopes making hand removal difficult.
	Cutting a plant may not control it, but over time it will reduce the seedbank and reduce the population.
	Disking or cultivation can also be used successfully to remove plants, but this is often impractical in infested terrain.
Cultural	Generally unpalatable to cattle; however, goats are known to eat croftonweed. The success depends on the stocking rate, weed density, and availability of other feed at the site. Goats can suppress large infestations. Because of its toxic nature, the same group of goats should be used for only one or two seasons to avoid risk of chronic health problems.
Biological	No biological control agents have been released for this species in the continental U.S. In Hawaii, South Africa, and Australia, a gall fly (*Procidochares utilis*) has been released. The fly inhibits the production of flowerheads. Its effectiveness is mixed, with no impact on the invasive plant in South Africa and Australia and some seedhead suppression in Hawaii. A leaf spot fungus (*Cercospora eupatoris*) does exert some effect, especially on seedlings. Even at best, biological control agents have never achieved high levels of control on croftonweed.

CHEMICAL CONTROL

There is very little information available for the chemical control of croftonweed. The following specific use information is based on reports by researchers and land managers. Other trade names may be available, and other compounds also are labeled for this weed. Directions for use may vary between brands; see label before use. Herbicides are listed by mode of action and then alphabetically. The order of herbicide listing is not reflective of the order of efficacy or preference.

GROWTH REGULATORS	
Fluroxypyr *Vista XRT*	**Rate:** 0.5% of concentrate for spot treatment.
	Timing: Postemergence to fully developed leaves. This is generally in late summer or autumn when the weed is growing actively.
	Remarks: Recommendations in Australia suggest a spray-to-wet application. This herbicide is not typically used alone.
Picloram *Tordon 22K*	**Rate:** 0.65% of concentrate for spot treatment.
	Timing: Postemergence to fully developed leaves. This is generally in late summer or autumn when the weed is growing actively. Application would be expected to have preemergence activity on germinating seedlings.
	Remarks: Recommendations in Australia suggest a spray-to-wet application. Can be tank mixed with triclopyr or 2,4-D. In steep or rocky areas, a low volume, high concentration application of picloram + triclopyr can give good results. Picloram is a restricted use herbicide. Picloram is not registered for use in California.
AROMATIC AMINO ACID INHIBITORS	
Glyphosate *Roundup, Accord XRT II*, and others	**Rate:** 0.5% of *Roundup ProMax Concentrate* for spot treatment.
	Timing: Postemergence to fully developed leaves. This is generally in late summer or autumn when the weed is growing actively.
	Remarks: Recommendations in Australia suggest a spray-to-wet application. Glyphosate is nonselective and can damage other non-target species it contacts.
BRANCHED-CHAIN AMINO ACID INHIBITORS	
Metsulfuron *Escort*	**Rate:** 1 oz to 50 gals (0.015%) for spot treatment.
	Timing: Postemergence to fully developed leaves. This is generally in late summer or autumn when the weed is growing actively.
	Remarks: Recommendations in Australia suggest a spray-to-wet application. Metsulfuron is not registered for use in California.

Ailanthus altissima (Mill.) Swingle
Tree-of-heaven

Family: Simaroubaceae
Range: Throughout the United States except Montana, North Dakota, South Dakota, and Wyoming.
Habitat: Disturbed places, roadsides, urban waste areas, landscaped sites, and many natural communities, including riparian areas, grassland, and woodland. Tolerates shade, many types of pollution, and harsh soil conditions, including acidic soils of mine spoils and phosphorus-poor soils. It can grow in most environments, including pavement cracks.
Origin: Native to China and introduced to North America as a landscape ornamental, as a food plant for a certain type of silkworm, and as a culturally important medicinal plant of Chinese immigrants.
Impacts: Forms dense thickets that compete with native vegetation and reduce wildlife habitat, particularly in riparian areas. Often considered a tree-fall hazard around homes and buildings, and the roots can damage pavement, roads, and foundations. May have allelopathic properties that inhibit the germination of other plants. Although not considered toxic, handling tree-of-heaven foliage can cause contact dermatitis in sensitive individuals, and the pollen is a common allergen.
Western states listed as Noxious Weed: California
California Invasive Plant Council (Cal-IPC) Inventory: Moderate Invasiveness

Tree-of-heaven is a fast-growing deciduous tree to nearly 70 ft tall, with large, pinnate-compound leaves. The leaves have an unpleasant skunky odor, especially when crushed. The tree produces a long, deep taproot with many creeping roots that sucker freely. Creeping roots may extend to about 50 ft in all directions.

Trees are dioecious (male or female), with females producing a winged fruit called a samara. Fruits mature in late summer and disperse from fall through the following spring with wind, water, and possibly birds. Most seeds probably fall close to the parent plant. Female trees typically begin to produce fruits at several years of age. One tree can produce 325,000 seeds or more annually, a large proportion of which are viable. Seeds generally survive about 1 year under field conditions and usually do not develop a persistent seed bank. Seedlings typically do not survive in shaded understory conditions. In addition to seed, plants reproduce vegetatively from creeping roots. Root sprouts tolerate shade better than seedlings, and new trees more often develop from the roots of established trees. New shoots can sprout up to 50 ft away from the parent tree and often grow more than 3 ft per year under favorable conditions. Individual trees typically have a life span of 30 to 50 years.

NON-CHEMICAL CONTROL

Mechanical (pulling, cutting, disking)	Hand pulling can remove seedlings, but once underground creeping roots have developed, this technique is generally not effective. For saplings or small trees, a weed wrench or other woody weed extractor can be used. Care must be taken to extract the entire root or stump sprouting will occur. Best results are achieved when soil is moist.
Cultural	A heavily shaded environment will reduce the establishment of tree-of-heaven.
Biological	No biological control agents have been released for the control of tree-of-heaven.

CHEMICAL CONTROL

The following specific use information is based on published papers and reports by researchers and land managers. Other trade names may be available, and other compounds also are labeled for this weed. Directions for use may vary between brands; see label before use. Herbicides are listed by mode of action and then alphabetically. The order of herbicide listing is not reflective of the order of efficacy or preference.

GROWTH REGULATORS	
Aminocyclopyrachlor + chlorsulfuron *Perspective*	**Rate:** 4.75 to 8 oz product (*Perspective*)/acre **Timing:** Postemergent to foliage of young trees in summer to early fall. **Remarks:** *Perspective* provides broad-spectrum control of many broadleaf species. Although

	generally safe to grasses, it may suppress or injure certain annual and perennial grass species. Do not treat in the root zone of desirable trees and shrubs. Do not apply more than 11 oz product/acre per year. At this high rate, cool-season grasses will be damaged, including bluebunch wheatgrass. Not yet labeled for grazing lands. Add an adjuvant to the spray solution. This product is not approved for use in California and some counties of Colorado (San Luis Valley).
Triclopyr *Garlon 3A, Garlon 4 Ultra, Pathfinder II* Aminopyralid + triclopyr *Capstone*	**Rate:** Foliar spot treatment: 1 to 2% v/v solution of *Garlon 4 Ultra* or 1 to 1.5% v/v *Capstone* and water plus 0.5% v/v non-ionic surfactant to thoroughly wet all leaves. Basal cut stump treatment: 20 to 25% *Garlon 4 Ultra* in 80% oil carrier. Cut stump treatment: 20 to 25% *Garlon 3A* in water. Basal bark treatment: 20 to 30% *Garlon 4 Ultra* in 70 to 80% oil carrier, or *Pathfinder II* as a ready-to-use formulation. Stem injection treatment: one cut per every 3 inches of stem diameter, and 1 ml of undiluted *Garlon 3A* added to each cut. For clumps, one hack per every 6 inches of total stem diameter. Treat largest stems. **Timing:** Foliar treatments best when leaves are fully expanded. Cut stump, basal cut stump, basal bark and stem injection treatments can be used anytime, but are best when used in late summer or early fall. **Remarks:** Triclopyr is a selective herbicide for broadleaf species and will not damage desirable grasses growing nearby. Foliar treatment should only be made on small trees, saplings, or seedlings. For cut stump treatments, cut stems horizontally at or near ground level, then immediately apply herbicide solution to cover the outer 20% of the stump face. Coppicing and root suckering typically occur after cutting, but the treatment should control most resprouts. Others have found that stem injection, then cut stump treatment can completely kill tree. Basal bark treatment: spray the lower trunk, including the root collar, to a height of 12 to 15 inches from the ground; the spray should thoroughly wet the lower stem but not to the point of runoff. For stem injection treatments, be sure that each cut goes well into or below the cambium layer. Trees should not be cut for at least 4 months after basal bark and stem injection treatments. A dye can be added to either product. Triclopyr can be applied as a premix with aminopyralid (*Capstone*) at 8 to 9 pt product/acre. Triclopyr can also be mixed with picloram or imazapyr for basal bark treatments (20% *Garlon 4 Ultra* + 5% *Tordon 22K* or 3% *Stalker*). Basal bark applications are made as described for triclopyr. Trees should not be cut for at least one month after treatment. Picloram is not registered for use in California.
AROMATIC AMINO ACID INHIBITORS	
Glyphosate *Roundup, Accord XRT II,* and others	**Rate:** Foliar spot treatment: 2 to 4% solution of glyphosate (*Roundup ProMax* or similar product) and water plus 0.5% v/v non-ionic surfactant to thoroughly wet all leaves. Stem injection treatment: one cut per every 3 inches of stem diameter, and 1 ml of undiluted herbicide added to each cut. For clumps, one hack per every 6 inches of total stem diameter. Treat largest stems. **Timing:** Foliar treatments best when leaves are fully expanded. For stem injection treatments, root injury is increased when applied mid-June to mid-September (fall color). **Remarks:** A nonselective systemic herbicide. Can also be mixed with dicamba to achieve good control with foliar applications. Stem injection and cut stump applications are made as described for triclopyr. Like triclopyr, there is extensive coppicing with a cut stump treatment. Others have found that stem injection with glyphosate, then cut stump treatment can completely kill tree.
BRANCHED-CHAIN AMINO ACID INHIBITORS	
Imazapyr *Arsenal, Habitat, Stalker, Chopper, Polaris*	**Rate:** Cut stump treatment: 20% *Stalker* or *Chopper* formulation v/v in 80% oil carrier or 20% *Arsenal* or *Habitat* v/v in 80% water carrier. Stem injection treatment: one cut per every 3 inches of stem diameter, and 1 ml of undiluted herbicide (*Arsenal* or *Habitat*) added to each cut. For clumps, one hack per every 6 inches of total stem diameter. Treat largest stems. Basal bark treatment: 20% *Stalker* or *Chopper* formulation v/v in 80% oil carrier. **Timing:** Best when used in late summer to early fall, but before leaf drop. **Remarks:** Soil residual herbicide. May result in bare ground around trees for some time after treatment. Applications are made as described for triclopyr. Imazapyr is the most consistent and best stem treatment for tree-of-heaven.
Metsulfuron *Escort*	**Rate:** 2 oz product/acre for foliar application (1.2 oz a.i./acre) **Timing:** Treatments best when leaves are fully expanded. **Remarks:** Can be tank mixed with glyphosate or triclopyr. Not registered for use in California.

Alhagi pseudalhagi (M. Bieb.) Desv.
  (= *A. maurorum* [Jepson Manual 2012], *A. camelorum*)

Camelthorn

Family: Fabaceae
Range: Much of the western United States, especially the desert southwest.
Habitat: Arid agricultural areas and riverbanks where roots can access water tables or other water sources during the growing season. Often grows in heavy clay soils. Tolerates some salinity.
Origin: Native to the Mediterranean region and western Asia.
Impacts: Weedy and competitive in rangeland. Because of its sharp spines, animals will not eat the foliage and plants can interfere with human activities. It can become established in croplands, where tillage may spread it.
Western states listed as Noxious Weed: Arizona, California, Colorado, Nevada, New Mexico, Oregon, Washington
California Invasive Plant Council (Cal-IPC) Inventory: Moderate Invasiveness

Camelthorn is an herbaceous perennial to shrub, 3 to 6 ft tall, with simple leaves, many thorny branches, and an extensive woody, creeping root system. One plant can rapidly colonize an area by developing new plants from the creeping roots, which grow more than 6 ft deep and as much as 30 to 40 ft in all directions. Like other legumes, the roots associate with nitrogen-fixing bacteria. Because camelthorn stores extensive reserves below ground, control efforts (including herbicide applications) must be repeated until the plant's root reserves are exhausted.

The mature plant has greenish, ridged, nearly hairless stems. Stems are highly branched, with a spiny branchlet at nearly every leaf axil. Leaves are alternate, sparse, and simple, thick and leathery, oval-shaped and 0.25 to 0.8 inch long. Upper leaf surfaces are hairless to sparsely hairy and covered with tiny red dots. Camelthorn is deciduous in cool climates. In moist habitats, thorns are smaller and fewer, and leaves larger and more numerous. The above-ground parts can be killed by hard frost.

Camelthorn flowers in summer. Each branchlet develops two to six magenta to pink pea-like flowers on short stalks. Flower production is high under hot dry conditions and low to non-existent under moist shady conditions. Reddish-brown pods with a beaked tip develop in late summer. Pods are constricted between seeds and break apart at the constrictions rather than splitting open. Plants reproduce locally from root sprouts, but can be spread greater distances by seed. Seeds disperse in water and, when eaten, can pass through the digestive tracts of herbivores. High winds may tumble clumps of branches with fruits. Seeds can survive submersion in water for at least 8 months and, like most other legumes, can remain viable for many years (> 20) in semi-arid soils. It is thought that viability decreases after 1 year in cool, moist soil conditions.

NON-CHEMICAL CONTROL

Mechanical (pulling, cutting, disking)	Mechanical removal is not effective, as it can stimulate remaining roots to spread and to develop new shoots.
	Tillage can spread plant fragments which resprout. Tilling multiple times per season may eventually deplete an infestation over several seasons but is very expensive and impractical on rangeland and in natural areas.
Cultural	When plants are less spiny, cattle and sheep can graze camelthorn. Cattle may preferentially feed on pods, but moving livestock that have browsed on fruits can disperse seeds to new locations. To avoid spread by livestock, forage should be weed-free. Restrict grazing in areas where camelthorn occurs, and quarantine livestock for seven days after they have fed on the weed.
	Burning is not an effective control method, as fire stimulates root sprouting.
Biological	Currently, no registered biocontrol agent for camelthorn is available in the United States.

CHEMICAL CONTROL

The following specific use information is based on published papers and reports by researchers and land managers. Other trade names may be available, and other compounds also are labeled for this weed. Directions for use may vary between brands; see label before use. Herbicides are listed by mode of action and then alphabetically. The order of herbicide listing is not reflective of the order of efficacy or preference.

GROWTH REGULATORS	
2,4-D Several names	**Rate:** 3.2 to 4.2 qt product/acre (3 to 4 lb a.e./acre) **Timing:** Postemergence to rapidly growing plants, particularly at the flower bud stage. **Remarks:** Experimental results indicate effective control using 3 to 4 lb a.e./acre 2,4-D twice annually for 3 years. Repeated applications at the label rate may be effective. Do not apply ester formulations when outside temperatures exceed 80°F.
2,4-D + dicamba tank mix	**Rate:** 1.6 qt product/acre of 2,4-D + 1.5 qt product/acre of dicamba **Timing:** Postemergence to rapidly growing plants. **Remarks:** One study reported 95% control using this tank mix treatment.
Aminopyralid *Milestone*	**Rate:** 5 to 7 oz product/acre (1.25 to 1.75 oz a.e./acre) **Timing:** Postemergence to rapidly growing plants before bloom. **Remarks:** Aminopyralid is a broadleaf herbicide similar to picloram, but more selective and with shorter soil residual activity. Very safe on grasses. The residual activity of aminopyralid will also provide preemergence control of germinating seeds.
Aminopyralid + triclopyr *Capstone* or tank mix	**Rate:** 4 to 8 pt product (*Capstone*)/acre **Timing:** Postemergence to rapidly growing plants before bloom. **Remarks:** Residual activity will also provide preemergence control of germinating seeds.
Picloram *Tordon 22K*	**Rate:** 2 to 4 pt product/acre (0.5 to 1 lb a.e./acre) **Timing:** Postemergence in spring or fall. **Remarks:** Picloram is one of the most effective chemical control options. It has long soil residual, so broadcast applications will also control germinating seed. Can be tank mixed with triclopyr, imazapyr, or glyphosate for improved control. Picloram is a restricted use herbicide. Not registered for use in California.
AROMATIC AMINO ACID INHIBITORS	
Glyphosate *Roundup, Accord XRT II*, and others	**Rate:** 2 qt product (*Roundup ProMax*)/acre (2.25 lb a.e./acre) **Timing:** Postemergence to rapidly growing plants. Most effective in spring or fall. **Remarks:** Glyphosate has no soil activity and is nonselective. Repeated applications will probably be necessary. This species is a good candidate for wiper applications at 33% to 50% v/v solution.
BRANCHED-CHAIN AMINO ACID INHIBITORS	
Imazapyr *Arsenal, Habitat, Chopper, Stalker, Polaris*	**Rate:** 3 to 4 pt product (*Habitat*)/acre (0.75 to 1 lb a.e./acre) broadcast, or spot treatment with 0.75 to 1.5 qt per 100 gal water (0.2 to 0.4% v/v solution). *Chopper* or *Stalker* can also be used in basal bark treatments at 20% v/v solution with methylated seed oil. **Timing:** Postemergence to rapidly growing plants. **Remarks:** Imazapyr has a fairly long soil residual and is nearly non-selective.
Metsulfuron *Escort*	**Rate:** 1 to 3 oz product/acre (0.6 to 1.8 oz a.i./acre) broadcast, or spot treatment with 1 to 2 oz per 100 gal water. **Timing:** Postemergence to rapidly growing plants at flower bud stage. **Remarks:** Metsulfuron has some soil residual activity. It is safe on most grasses. Not registered for use in California.

Alliaria petiolata (M. Bieb.) Cavara & Grande

Garlic mustard

Photo courtesy of Kelly Kearns
Washington State Noxious Weed Control Board

Family: Brassicaceae
Range: Reported in Alaska, British Columbia, Washington, Oregon, Idaho, Utah, Colorado, North Dakota, Nebraska, Kansas, and Oklahoma.
Habitat: Forest understory, parklands, rights-of-way, and riparian areas/floodplains.
Origin: Native to Eurasia and northern Africa. It was introduced into the U.S. as a food and medical plant, and also for erosion control. Apparently escaped from cultivation.
Impacts: Garlic mustard is shade tolerant and can move into stable forest understory where it rapidly outcompetes native vegetation. Once established, it can form near-monotypic stands, decreasing species diversity and function. It is considered potentially allelopathic, producing several phytotoxins that either directly affect growth of adjacent plants or interfere with mycorrhizal associations needed for normal root growth.
Western states listed as Noxious Weed: Oregon, Washington

Garlic mustard is a biennial or short-lived perennial to 3 ft tall. The basal leaves are kidney-shaped, while stem leaves are more triangular in outline. Both leaf types are alternate, petiolate, and toothed along the margin. First-year plants form a rosette, while second-year plants will bloom and set seed. All foliage produces a garlicky odor when crushed. The main taproot is distinctively S-shaped just under the soil surface.

Stems often branch near the top and bear individual flowers that alternate up the stem in a raceme pattern. Flowers bear 4 white petals, about ¼ inch long, and have 6 stamens. Seed pods are slender, 2.5-inch long siliques that split longitudinally into two sections at maturity. Garlic mustard reproduces exclusively by seed which typically fall to the ground below the parent plant. Flowers can produce up to 8,000 seeds per plant. Seeds can remain viable in the soil for 5 years.

NON-CHEMICAL CONTROL

Mechanical (pulling, cutting, disking)	Garlic mustard can be dug to remove individual plants or small stands. Remove as much root as possible, being careful to not break the root as these fragments can resprout. Hand pulling is an effective strategy on loose or coarse soil, but difficult to accomplish on heavy soil without substantial root fragmentation. When the plants have bolted in the second growing season, the stem may be easily grasped and pulled. Hand digging is often impractical for well-established or extensive infestations.
	Mowing can reduce or eliminate seed production. Plants should be cut as low as practical to limit growth of flowering stalks from below the cut. Cutting stems too soon or too infrequently will have little effect on control, while cutting too late will result in lower seedpods ripening and dispersing seed during the mowing process.
	Cultivation of first-year plants or before flowering of second-year plants can be effective.
	Mulching with plastic or fabric sheets may also suppress seed germination and plant growth.
Cultural	Garlic mustard is not known to be toxic to animals and, in fact, it is edible for humans. Grazing before flowering may reduce seeding, but could also negatively impact growth of native species that may be preferentially grazed at invaded sites. In general, most animals do not find garlic mustard to be palatable and it imparts an unpleasant odor to the taste of milk.
	Prescribed burning of large infestations for 2 consecutive years can provide effective rosette control. If fires are not hot enough, however, garlic mustard plants may not be killed and seedlings may be more likely to establish on the bare ground resulting from burning.
	In parks and other managed areas, healthy turf can be a useful tool for discouraging garlic mustard seed germination.
Biological	There are no biological control agents currently available to aid in the control of garlic mustard. However, six species have been identified that may be candidates for biological control agents in the eastern US: *Ceutorhynchus alliariae* and *C. roberti*, stem-mining weevils that attack rosettes and bolting plants; *C.*

constrictus, a seed-feeding weevil; *C. scrobicollis*, a rosette-feeding weevil; *Phyllotreta ochripes*, a root/crown mining flea beetle; and *Ophiomyia alliariae*, a leaf- and shoot-mining fly.

CHEMICAL CONTROL

The following specific use information is based on published papers and reports by researchers and land managers. Other trade names may be available, and other compounds may also be labeled for this weed. Directions for use may vary between brands; see label before use. Most herbicide applications will require multiple applications over several years to fully control garlic mustard. Herbicides are listed by mode of action and then alphabetically. The order of herbicide listing is not reflective of the order of efficacy or preference.

GROWTH REGULATORS	
2,4-D Several names	**Rate:** Broadcast foliar treatment: 1 pt product/acre (0.48 lb a.e./acre). Spot treatment: 1% v/v solution **Timing:** Postemergence to rapidly growing plants before flowering. **Remarks:** Apply 2,4-D alone or mixed with dicamba (*Banvel* or *Clarity*). Some reports have shown that 2,4-D is not particularly effective for the control of garlic mustard.
Triclopyr *Garlon 4 Ultra*	**Rate:** 8 oz product/acre (3 oz a.i./acre). Spot treatment: 1.25 to 2.5% v/v solution (*Garlon 4 Ultra*) **Timing:** Postemergence to garlic mustard foliage in spring when plants are in the rosette stage. **Remarks:** Triclopyr is selective on broadleaf species.
AROMATIC AMINO ACID INHIBITORS	
Glyphosate *Roundup, Accord XRT II*, and others	**Rate:** Broadcast foliar treatment: 2 to 4 pt product (*Roundup ProMax*)/acre (1.1 to 2.25 lb a.e./acre). Spot treatment: 1 to 3% v/v solution **Timing:** Postemergence to rosettes in late fall or early spring when plants are in the rosette stage. **Remarks:** Glyphosate is a nonselective herbicide with no soil activity. Application in winter may cause less injury to native perennial plants, but may increase spring germination of garlic mustard seed.
BRANCHED-CHAIN AMINO ACID INHIBITORS	
Imazapic *Plateau*	**Rate:** 4 to 6 oz product/acre (1 to 1.5 oz a.i./acre) **Timing:** Postemergence to garlic mustard foliage in fall or early spring. **Remarks:** Use 2 lb methylated seed oil/acre to improve herbicide uptake. Imazapic is not registered for use in California.
Metsulfuron *Escort*	**Rate:** 0.5 to 1 oz product/acre (0.3 to 0.6 oz a.i./acre) **Timing:** Postemergence to garlic mustard foliage in fall or early spring. **Remarks:** Use 0.25% v/v non-ionic surfactant to improve herbicide uptake. Metsulfuron is not registered for use in California.
Sulfometuron *Oust* and others	**Rate:** 0.5 oz product/acre (0.38 oz a.i./acre) **Timing:** Postemergence to garlic mustard foliage in fall or early spring. **Remarks:** Use 0.25% v/v non-ionic surfactant to improve herbicide uptake.
Sulfosulfuron *Outrider*	**Rate:** 2 oz product/acre (1.5 oz a.i./acre) **Timing:** Postemergence to garlic mustard foliage in fall or early spring. **Remarks:** Use 0.25% v/v non-ionic surfactant to improve herbicide uptake.

Alternanthera philoxeroides (Mart.) Griseb.

Alligatorweed

Photo courtesy of Richard Old

Family: Amaranthaceae
Range: Mainly in the southeastern states, but also in California.
Habitat: Shallow water in ditches, marshes, pond margins, and slow-moving waterways. May also be found terrestrially in wet soils. It can tolerate saline conditions up to 10% salt by volume. Alligatorweed requires a warm summer growing condition but can survive cold winters provided there are no prolonged periods of freezing temperatures.
Origin: Native to South America.
Impact: Alligatorweed forms dense floating mats on the surface of water that decrease light penetration and change the natural ecology at the site. It also competes with native vegetation, can cause flooding, and has the potential to create anoxic conditions, increase mosquito breeding, and harbor diseases.
Western states listed as Noxious Weed: Arizona, California
California Invasive Plant Council (Cal-IPC) Inventory: High Invasiveness (Alert)

Alligatorweed is an aquatic perennial that also has a terrestrial form. The aquatic form roots in shallow water and rarely grows in water deeper than 6 ft. Its stems are hollow both above and below the water. The stems are stolon-like and root at the nodes. The roots of the aquatic form are shorter and finer than the terrestrial form. The stems above water form dense interwoven floating mats that spread out over deeper water. The foliage and stems are bright green with oppositely placed obovate to narrowly lanceolate leaves. The leaves are 1 to 4 inches long and 0.5 to 1 inch wide with smooth margins and a glabrous waxy surface. The leaves are either sessile or have narrow winged petioles up to 0.5 inch long that clasp the stem at the base.

Alligatorweed produces fragrant flowers in a head inflorescence. The inflorescence can either be produced in the terminal or axillary position and is 0.5 to 1 inch in diameter with a 1 to 3 inches long stalk. The flowers and bracts are pearly white and have 5 separate sepals that are 5 to 7 mm long but lack petals. Seed set is rare and, when produced, is usually non-viable. The seed that is produced is not released from the membranous utricle. Plants typically reproduce vegetatively when pieces of the floating mats break away and colonize downstream. Any stem fragment with a node can develop into a new plant.

NON-CHEMICAL CONTROL

Mechanical (pulling, cutting, disking)	Mechanical control methods such as using a cookie cutter, flail chopper, hand removal, harvesting, hand cutter, or rotovation are good for clearing water ways, but unless all fragments of the stems are collected these management practices could exacerbate the problem. Since alligatorweed reproduces vegetatively, if any fragments move downstream they can develop into another colony and clog the waterway. Using these methods could result in an increase in the number of alligatorweed stands.
Cultural	There are no effective cultural control options for the management of alligatorweed. Drawdown of the water source will not control established infestations because the plant is also adapted to survive in moist terrestrial conditions.
Biological	Three insect species were introduced into the United States from South America between 1964 and 1971. They established in some states but populations have failed to establish in California. The alligatorweed flea beetle (*Agasicles hygrophila*) can cause considerable damage to the aquatic form of alligatorweed by eating the leaves and boring into the stem, where it pupates. Alligatorweed thrips (*Amymothrips andersoni*) produce limited damage to the stands by attacking and deforming the apical leaves. Dispersal is also limited because there is often a lack of adults with wings. The alligatorweed stem borer (*Arcola malloi* (formerly *Vogtia malloi*)) is a small moth that lays its eggs on the apical leaves. The larvae bore into the stem and work their way down the stem, resulting in wilting and drooping of the plant.

CHEMICAL CONTROL

The following specific use information is based on published papers and reports by researchers and land managers. Other trade names may be available, and other compounds also are labeled for this weed. Directions

for use may vary between brands; see label before use. Herbicides are listed by mode of action and then alphabetically. The order of herbicide listing is not reflective of the order of efficacy or preference.

GROWTH REGULATORS	
2,4-D Several names	**Rate:** 8.4 qt product/acre (8 lb a.e./acre), applied in 50 gallons of water per surface acre. **Timing:** Postemergence after plants have completely emerged, but before blooming stage. Only fair control is achieved when plants are young and succulent. **Remarks:** Because alligatorweed is a perennial, good control may require a repeat application. 2,4-D is a broadleaf-selective herbicide.
Triclopyr *Renovate*	**Rate:** 1.5 gal product/acre (4.5 lb a.e./acre) **Timing:** Summer application timing has proved to be better than spring timing. **Remarks:** Triclopyr is broadleaf-selective and safe on most grasses. It provides only partial control of alligatorweed in field trials. Add 0.25% aquatic registered non-ionic surfactant.
AROMATIC AMINO ACID INHIBITORS	
Glyphosate *Roundup, Accord XRT II* and others	**Rate:** Broadcast foliar treatment: 6 pt product (*Roundup ProMax*)/acre (3.4 lb a.e./acre) with a non-ionic surfactant at 0.75% v/v solution applied in 50 gallons of water per surface acre. Spot treatment: 1.5% v/v solution **Timing:** Postemergence when most target plants are in bloom. **Remarks:** Only partial control is achieved with a single application of glyphosate. Repeat applications will be required to maintain good control. Glyphosate is a nonselective herbicide.
BRANCHED-CHAIN AMINO ACID INHIBITORS	
Imazapyr *Habitat*	**Rate:** Broadcast foliar treatment: 1 to 4 pt *Habitat*/acre (4 to 16 oz a.e./acre) applied in 100 gal water per acre. Spot treatment: 0.5% v/v *Habitat* **Timing:** Postemergence when target plants are growing rapidly. **Remarks:** Imazapyr is nonselective. Only the formulation *Habitat* is registered for aquatic use. Apply spray-to-wet with good coverage on the rapidly growing emergent foliage. Tank mixtures with glyphosate are not recommended because they may reduce control of alligatorweed. Do not apply more than 6 pt product/acre per year.

Ammophila arenaria (L.) Link.

European beachgrass

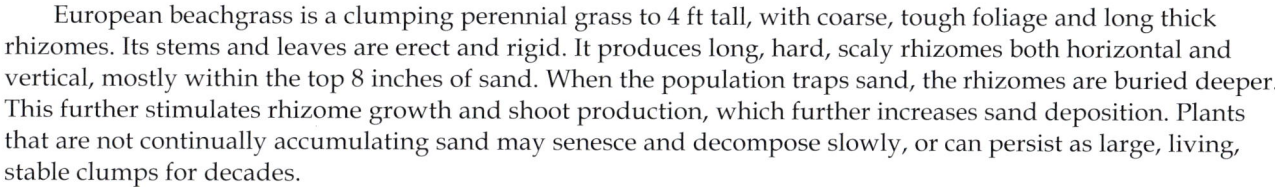

Family: Poaceae
Range: Coasts of central to northern California and the Pacific Northwest; coasts of the central eastern states.
Habitat: Coastal dune systems. Populations may extend inland for half a mile or more.
Origin: Native to northern Europe. Originally planted along the Pacific coast from the late 1800s to the mid-1900s for sand stabilization, European beachgrass has since invaded nearly every major dune system in California.
Impacts: Populations trap blowing sand, building dunes into steep slopes which do not support native coastal vegetation. Dense stands also outcompete other species and provide poor habitat for wildlife.
California Invasive Plant Council (Cal-IPC) Inventory: High Invasiveness

European beachgrass is a clumping perennial grass to 4 ft tall, with coarse, tough foliage and long thick rhizomes. Its stems and leaves are erect and rigid. It produces long, hard, scaly rhizomes both horizontal and vertical, mostly within the top 8 inches of sand. When the population traps sand, the rhizomes are buried deeper. This further stimulates rhizome growth and shoot production, which further increases sand deposition. Plants that are not continually accumulating sand may senesce and decompose slowly, or can persist as large, living, stable clumps for decades.

European beachgrass reproduces both from rhizomes and seed. Dormant rhizome fragments survive long periods of submersion in ocean water and disperse with ocean currents. It flowers in summer, producing dense, cylindrical panicles 6 to 12 inches long. Seed viability and seedling establishment are typically low and seedlings are rarely encountered.

NON-CHEMICAL CONTROL

Mechanical (pulling, cutting, disking)	Plants can be removed manually with a two-year program. Begin removal in March, as plants emerge from dormancy. Rhizomes must be dug to a depth of 8 inches, and removal must be repeated as resprouts emerge – as many as 8 removals during the first season and again during the second season. This method is expensive but allows conservation of relict native species. Removing young plants before they become well established can prevent spread and more expensive control programs.
	In suitable circumstances, heavy equipment can be used for initial removal, to be followed by manual removal. Deep ripping to 3 ft has been found to be an effective first step.
Cultural	Burning does not control beachgrass but by removing the tops may allow easier access for mechanical removal or herbicide application. One study reports effective control using a fall burn to remove beachgrass tops, followed by treating resprouts with glyphosate in the following season.
Biological	There are no biological control agents available for the management of European beachgrass.

CHEMICAL CONTROL

The following specific use information is based on published papers and reports by researchers and land managers. Other trade names may be available, and other compounds also are labeled for this weed. Directions for use may vary between brands; see label before use. Herbicides are listed by mode of action and then alphabetically. The order of herbicide listing is not reflective of the order of efficacy or preference.

AROMATIC AMINO ACID INHIBITORS	
Glyphosate
Roundup, Accord XRT II, Rodeo, and others	**Rate:** 8% to 10% v/v solution (*Roundup ProMax*) as a spot treatment, 33% to 50% v/v solution as a wiper solution
Timing: Postemergence to non-dormant plants during active growth.
Remarks: Glyphosate is a nonselective herbicide. It has no soil activity. With *Rodeo*, use a non-ionic surfactant (0.5% to 1.5% in spot treatments, 1% to 2.5% in wiper treatments). Effectiveness may be increased by addition of ammonium sulfate. Standing dead biomass may still have to be removed to allow revegetation. This treatment is only marginally effective and most land managers find better control when

	glyphosate is tank mixed with imazapyr.
BRANCHED-CHAIN AMINO ACID INHIBITORS	
Imazapyr *Habitat, Arsenal, Stalker, Chopper, Polaris*	**Rate:** 2 to 3 pt product/acre (0.5 to 0.75 lb a.e./acre) **Timing:** Best when applied pre- or postemergence in fall or spring to non-dormant plants. Applications in fall may be most effective. Some areas allow application only from September to February due to the presence of snowy plovers, an endangered species. **Remarks:** Imazapyr is a nonselective herbicide. It also has a relatively long soil residual activity and may have longer-term effects on the plant community.
Imazapyr + glyphosate + *Habitat, Arsenal,* or *Polaris* + *Roundup* and others	**Rate:** 2% *Roundup ProMax* or other trade name with similar amount of active ingredient + 1% *Habitat* v/v solution for spot treatment. **Timing:** Postemergence in fall or spring to non-dormant plants. Applications in fall may be most effective. **Remarks:** This combination appears to have improved efficacy and fewer negative effects compared to imazapyr alone. This tank mix is often used because the quicker response to glyphosate indicates that the application was successful. Success of treatment is enhanced with multiple applications per year.

Amorpha fruticosa L.
Indigobush

Family: Fabaceae
Range: Throughout the western states except Alaska, Montana, and Nevada.
Habitat: Riparian areas/floodplains, forest edges, meadows, rights-of-way, and parks.
Origin: Native to eastern and central North America. Also considered native to California. Apparently escaped from cultivation as an ornamental.
Impacts: Grows densely and almost eliminates other plants within these stands.
Western states listed as Noxious Weed: Washington

Indigobush is a woody, deciduous perennial that grows to 12 ft tall. The leaves are pinnately compound, with 13 to 25 leaflets, each 1 to 2 inches long and elliptical in outline. Leaflets are dotted with glands, covered with downy hairs, and have smooth margins. Plants produce several stems up to 3 or 4 inches in diameter. The lower stems that become buried in sediment are capable of producing new stems along their length. Old stems have a furrowed bark, while newly-grown stems are smooth, green, and lightly hairy.

Flowers are about ¼-inch long and are borne in dense clusters on upper branches in late summer. Five sepals form a cone-shaped calyx and a single dark purple to blue-violet petal wraps around 10 bright yellow stamens. The fruits are pea-like pods, ¼-inch long, oblong to curved, dark brown and dotted with glands. Each pod contains 1 or 2 seeds which likely survive in the soil for many years. Plants primarily reproduce by seed which either fall directly to the ground or move long distances by floating downstream in water.

NON-CHEMICAL CONTROL

Mechanical (pulling, cutting, disking)	Indigobush is difficult to control mechanically as it vigorously resprouts from crowns. It is not rhizomatous, however, so it can be controlled by repeated defoliation and digging and severing the root 3 to 4 inches below the crown. Repeat as necessary to control regrowth.
	Mowing and cultivation of this woody species is not usually an option, although brush hogs can be used to remove large stems. Removal of branches in midsummer can decrease seed production, and may therefore limit the spread of indigobush.
Cultural	Although it is not reported to be toxic, there are no reports that indigobush has been controlled by grazing. It is also not known how this species would respond to plastic or fabric mulches.
Biological	There are no known biological control agents to aid in the control of indigobush.

CHEMICAL CONTROL

The following specific use information is based on reports by researchers and land managers. Other trade names may be available, and other compounds may also be labeled for this weed. Directions for use may vary between brands; see label before use. Herbicides are listed by mode of action and then alphabetically. The order of herbicide listing is not reflective of the order of efficacy or preference.

GROWTH REGULATORS

Clopyralid *Transline*	**Rate:** Broadcast foliar treatment: 1.33 pt product/acre (0.5 lb a.e./acre). Cut stump treatment: up to 50% of concentrated herbicide in water
	Timing: Postemergence to rapidly growing plants. For cut stump treatment, applications can be made when plants are dormant or rapidly growing.
	Remarks: When using a cut stump treatment, cut stems as close to the soil line as practical and apply enough product to wet the cut surface. Treatments should be made as soon as possible after cutting (within one hour is recommended).

AROMATIC AMINO ACID INHIBITORS	
Glyphosate *Roundup, Accord XRT II*, and others	**Rate:** Cut stump treatment: up to 100% of concentrated herbicide **Timing:** When plants are dormant or rapidly growing. **Remarks:** Cut stems as close to the soil line as practical and apply enough full-strength solution to wet the cut surface. Treatments should be made as soon as possible after cutting (usually within an hour is recommended). Foliar treatment of indigobush may also be made, but control from 2% foliar glyphosate was rated as only fair.

Amsinckia menziesii (Lehm.) A. Nelson & J.F. Macbr. var. *menziesii*; Menzies fiddleneck
Amsinckia menziesii (Lehm.) A. Nelson & J.F. Macbr. var. *intermedia* (Fisch. & C.A. Mey.) Ganders; coast fiddleneck
(= *A. intermedia* Fisch. & C.A. Mey [Jepson Manual 2012])

Menzies and coast fiddleneck

Amsinckia menziesii var. *menziesii*

Family: Boraginaceae
Range: Widespread natives of North America, found throughout much of the western United States, except South Dakota and New Mexico. Coast fiddleneck is somewhat more common than Menzies fiddleneck.
Habitat: Both varieties are common to grasslands, pastures, roadsides, agronomic crops, orchards, vineyards, and disturbed areas. They prefer grassy and open areas below 5000 ft elevation.
Origin: Native to western North America.
Impact: Although often considered desirable natives, fiddleneck seeds and foliage are toxic to livestock due to alkaloids and sometimes high nitrate concentrations. Consuming high quantities of fiddleneck can cause irreversible liver damage in cattle, pigs and horses. Poisonings most often occur when livestock ingest contaminated grain or feed. Toxicity is fairly uncommon from grazing, as the species has poor palatability, and livestock generally avoid consuming it when other forage plants are available.

Amsinckia menziesii var. *intermedia*

There are several different native *Amsinckia* species but coast fiddleneck is the one most often reported as a weed. All of them are annuals and most are usually not problems in natural settings. While Menzies and coast fiddlenecks are the two most common types in the western United States, all *Amsinckia* species can cause poisoning in livestock. The various species differ primarily in their flowers and/or in their distribution.

Fiddleneck is a slender, erect winter annual herb that typically grows from 8 to 32 inches tall but can occasionally reach 4 ft tall. The cotyledons are very recognizable as they have a deeply lobed Y-shape. The true leaves are linear to lanceolate and up to 6 inches long. The leaves are alternate and primarily sessile (stalkless) except for the short-stalked lower leaves. The margins are usually smooth but the foliage is covered with stiff bristly hairs. Young plants are rosettes until the flower stem develops.

The most distinctive feature of fiddleneck is the flowering head that curls like the neck of a fiddle. The flowers in most fiddleneck species are yellow and often have an orange tinge. The flowers are in a coiled one-sided inflorescence up to 8 inches long. The corolla is tubular with five lobes, 7 to 11 mm long in coast fiddleneck and 4 to 7 mm long in Menzies fiddleneck. The fruits consist of four, erect, one-seeded spineless nutlets that separate after dispersal. Plants reproduce only by seed, which germinate in fall through early spring and are primarily dispersed short distances by falling from the parent plant, or occasionally longer distances when they attach to the hair or fur of animals. Seeds are expected to survive in the soil for a few years.

NON-CHEMICAL CONTROL

Mechanical (pulling, cutting, disking)	Hand pulling is effective but usually populations are too dense for this to be practical. In addition, the stiff bristly hairs on the fiddleneck plant make hand pulling without gloves unpleasant.
	Mowing before seed production can reduce seed set and will kill many of the plants.
	Tillage is effective in cultivated areas but not feasible in many range or wildland areas.
Cultural	Grazing is not an attractive control measure because the plant is not palatable and is toxic to livestock.
	Fiddleneck is a highly competitive plant in agricultural fields but it does not compete well in a dense stand of grass or other perennial plants.

	Burning kills existing fiddleneck plants. However, fiddleneck completes its life cycle early in spring so it is difficult to have the fuel load necessary to carry a fire before fiddleneck has set seed. Experimental plots that were burned had the twice the population of fiddleneck seedlings compared to unburned plots.
Biological	Because these plants are native species to North America, there are no biological control programs established.

CHEMICAL CONTROL

Chemical control is difficult due to the hairs and must be done before seed production. The following specific use information is based on reports by published papers, researchers and land managers. Other trade names may be available, and other compounds also are labeled for this weed. Directions for use may vary between brands; see label before use. Herbicides are listed by mode of action and then alphabetically. The order of herbicide listing is not reflective of the order of efficacy or preference.

GROWTH REGULATORS	
Aminocyclopyrachlor + chlorsulfuron *Perspective*	**Rate:** 3 to 4.5 oz product/acre **Timing:** Postemergence in early spring when fiddleneck plants are in the rosette stage. **Remarks:** *Perspective* provides broad-spectrum control of many broadleaf species. Although generally safe to grasses, it may suppress or injure certain annual and perennial grass species. Do not treat in the root zone of desirable trees and shrubs. Do not apply more than 11 oz product/acre per year. At this high rate, cool-season grasses will be damaged, including bluebunch wheatgrass. Not yet labeled for grazing lands. Add an adjuvant to the spray solution. This product is not approved for use in California and some counties of Colorado (San Luis Valley).
Aminopyralid *Milestone*	**Rate:** 3 to 5 oz product/acre (0.75 to 1.25 oz a.e./acre) **Timing:** Postemergence in early spring when fiddleneck plants are in the rosette stage. **Remarks:** Aminopyralid is a broadleaf herbicide similar to picloram, but more selective and very safe on grasses. Its soil residual activity will kill emerging seedlings.
AROMATIC AMINO ACID INHIBITORS	
Glyphosate *Roundup, Accord XRT II,* and others	**Rate:** For weeds < 6 in: 1 pt product (*Roundup ProMax*)/acre (0.56 lb a.e./acre); for weeds > 6 in: 22 oz product/acre (0.77 lb a.e./acre) **Timing:** Postemergence when fiddleneck plants are small and rapidly growing. **Remarks:** Glyphosate is not selective and kills or severely injures desirable vegetation. Therefore, it is ordinarily not the herbicide of choice for fiddleneck control in rangeland or natural areas with other desirable species.
BRANCHED-CHAIN AMINO ACID INHIBITORS	
Chlorsulfuron *Telar*	**Rate:** 0.25 to 0.5 oz product/acre (0.19 to 0.375 oz a.i./acre) **Timing:** Preemergence to early postemergence. **Remarks:** Chlorsulfuron has mixed selectivity and is generally safe on grasses. It is most effective preemergence for fiddleneck. Use a surfactant for postemergence applications. It has fairly long soil residual activity.
Imazapic *Plateau*	**Rate:** 4 to 6 oz product/acre (1 to 1.5 oz a.e./acre) **Timing:** Preemergence in fall to postemergence in spring. **Remarks:** Imazapic has mixed selectivity and tends to favor species in the Asteraceae, as well as some grasses. In postemergence applications, use a methylated seed oil surfactant at 0.25%. It has long soil residual activity. Imazapic is not registered for use in California.
Imazapyr *Arsenal, Habitat, Stalker, Chopper, Polaris*	**Rate:** 3 to 4 pt product (*Habitat*)/acre (12 to 16 oz a.e./acre) **Timing:** Postemergence. **Remarks:** Imazapyr is best used as a spot treatment. It is a nonselective herbicide. It also has long soil residual activity and can leave more bare ground than other treatments, even a year after application.
Metsulfuron *Escort*	**Rate:** 0.33 to 0.5 oz product/acre (0.2 to 0.3 oz a.i./acre) **Timing:** Postemergence to young, rapidly growing weeds.

	Remarks: Metsulfuron has mixed selectivity, but is generally safe on grasses. Use a surfactant. It can be tank-mixed with 2,4-D and/or dicamba, or in a premix with chlorsulfuron (*Cimarron X-tra*). Although it does not have soil activity on most species, it does have some soil residual activity. Metsulfuron is not registered for use in California.
Sulfosulfuron *Outrider*	**Rate:** 0.75 to 2 oz product/acre (0.56 to 1.5 oz a.i./acre) **Timing:** Early postemergence, winter to early spring, when desirable perennials are dormant. **Remarks:** Sulfosulfuron has mixed selectivity, but is fairly safe on native perennial grasses, especially wheatgrasses. To be most effective it may be necessary to add a non-ionic surfactant. Sulfosulfuron has fairly long soil residual activity.
PHOTOSYNTHETIC INHIBITORS	
Hexazinone *Velpar DF*	**Rate:** 0.75 to 1 lb product/acre (0.56 to 0.75 lb a.i./acre) **Timing:** Preemergence to early postemergence in late fall to early spring. **Remarks:** Apply the dry formulation of hexazinone when there is adequate moisture for activation. Hexazinone has both foliar and soil activity. Its selectivity is mixed. Use higher rates on fine soils or high organic matter soils, or when weeds are under stress. It also has fairly long soil residual activity. Hardwood trees near application site can be damaged when they absorb this chemical through the roots. High rates of hexazinone can create bare ground, so only use high rates in spot treatments.

Anthriscus caucalis M. Bieb.; bur chervil
Anthriscus sylvestris (L.) Hoffm.; wild chervil

Bur and wild chervil

A. caucalis

Family: Apiaceae
Range: Washington, Oregon, Idaho; bur chervil is also found in California, Arizona, Wyoming, and Utah.
Habitat: Bunchgrass communities, sagebrush, pastures, and meadows. Also found in Conservation Reserve Program lands.
Origin: Native to Eurasia.
Impacts: Bur chervil dominates riparian areas and within understories of hawthorn and hackberry. Wild chervil can form dense colonies in ditches, dykes, road verges, meadows, hay fields, stream banks and hedge rows.
Western states listed as Noxious Weed: *A. sylvestris*, Washington

Bur chervil is an annual forb that often germinates in fall. It is generally less than 3 ft high but can grow to over 6 ft high is some situations. Wild chervil is a biennial to short-lived perennial forb usually to 3 ft tall but up to 6 ft tall. Both plants have fern-like leaves alternating along hollow stems.

Bur chervil flowers are in the leaf axils and arranged as umbels (umbrella-like). The small white flowers have 5 notched petals, and the fruits are about 0.25 inch long with curved hairs on the seeds. Wild chervil also has white flowers with 5 petals, but these are arranged in umbels at the top of the plant instead of leaf axils. The fruits are similar in size to bur chervil but without the curved hairs.

Wild chervil appears adapted to moister areas that overlap with bur chervil along the wetter range of bur chervil distribution. Bur chervil appears well adapted to wetter communities but survives in dry communities as well. Bur chervil produces more seed in riparian areas with a tree overstory than it does in grass-shrub or grass communities. Bluebunch wheatgrass communities are drier communities than Idaho fescue communities, and bur chervil seedling mortality rates are higher in bluebunch wheatgrass versus Idaho fescue communities. Fence line comparisons of grazed and ungrazed areas have found dense bur chervil canopies in the areas not grazed.

Plants reproduce only by seed. The fruits of both species fall near the parent plant or disperse to greater distances by clinging to the fur, feathers, and feet of animals. Most seeds germinate after the first fall rains in areas with mild winters and in spring where winters are more severe. No data indicates how long the seeds survive in the soil, but it is expected that it would be a few years, based on related species.

NON-CHEMICAL CONTROL

Mechanical (pulling, cutting, disking)	Hand pulling in small infestations of bur chervil may have some effect, but wild chervil taproots would likely sprout after stems were pulled.
Cultural	Bur chervil is not dominant in areas grazed by cattle. Both chervil species are aromatic and not as palatable as grasses and other forbs. Mowing just before flowering can reduce seed production in wild chervil.
Biological	Neither chervil species currently has biological control agents released in North America.

CHEMICAL CONTROL

The following specific use information is based on reports by researchers and land managers. Other trade names may be available, and other compounds also are labeled for this weed. Directions for use may vary between brands; see label before use. Herbicides are listed by mode of action and then alphabetically. The order of herbicide listing is not reflective of the order of efficacy or preference.

GROWTH REGULATORS	
Aminocyclopyrachlor + chlorsulfuron *Perspective*	**Rate:** 3 to 4.5 lb product (*Perspective*)/acre **Timing:** Postemergence in spring. **Remarks:** *Perspective* provides broad-spectrum control of many broadleaf species. Although generally safe to grasses, it may suppress or injure certain annual and perennial grass species. Do not treat in the root zone of desirable trees and shrubs. Do not apply more than 11 oz product/acre per

	year. At this high rate, cool-season grasses will be damaged, including bluebunch wheatgrass. Not yet labeled for grazing lands. Add an adjuvant to the spray solution. This product is not approved for use in California and some counties of Colorado (San Luis Valley).
Aminopyralid + metsulfuron *Opensight*	**Rate:** 3.3 oz product/acre **Timing:** Postemergence in spring. **Remarks:** Add 0.25 to 0.5% v/v surfactant. Metsulfuron is not registered in California.
Dicamba *Banvel, Clarity*	**Rate:** 0.5 to 2 pt product/acre (0.25 to 1 lb a.e./acre) **Timing:** Postemergence in spring. **Remarks:** Rates are those listed on the label for bur chervil. Dicamba is a broadleaf-selective herbicide and may injure other non-target broadleaf species.
AROMATIC AMINO ACID INHIBITORS	
Glyphosate *Roundup, Accord XRT II,* and others	**Rate:** Broadcast foliar treatment: 2 to 4.5 pt product (*Roundup ProMax*)/acre (1.1 to 2.5 lb a.e./acre), lower rates for plants under 6 inches tall. Spray-to-wet spot treatment: 0.4% v/v solution for plants under 6 inches tall and 0.7 to 1.5% v/v solution for plants over 6 inches tall **Timing:** Postemergence in spring at the bud or early flowering stage. **Remarks:** Glyphosate is a nonselective herbicide with no soil residual activity.
BRANCHED-CHAIN AMINO ACID INHIBITOR	
Chlorsulfuron *Telar*	**Rate:** 0.5 to 1 oz product/acre (0.375 to 0.75 oz a.i./acre) **Timing:** Postemergence in fall to seedlings. **Remarks:** Add a non-ionic or silicone surfactant. The label for *Telar* specifies bur chervil, but it is expected that similar control would be obtained with wild chervil.
Imazapyr *Arsenal AC, Habitat, Stalker, Chopper, Polaris*	**Rate:** 1.5 pt *Arsenal AC* (4 lb active per gal)/acre, or 3 pt *Habitat, Stalker,* or *Chopper* (2 lb active per gal)/acre (0.75 lb a.e./acre) **Timing:** Postemergence at the bud to early flowering stage. **Remarks:** Imazapyr is a nonselective herbicide. Imazapyr has been shown to control wild chervil, but is expected to have a similar effect on bur chervil.
Metsulfuron *Escort*	**Rate:** 0.5 to 1 oz product/acre (0.3 to 0.6 oz a.i./acre) **Timing:** Postemergence in fall to seedlings. **Remarks:** Add a non-ionic or silicone surfactant. Metsulfuron is not registered for use in California.

Arundo donax L.
Giant reed

Family: Poaceae
Range: Southern region of the U.S. In the west it can be found in California, Nevada, Utah, Arizona, New Mexico, and Texas.
Habitat: Riparian areas, floodplains, ditches, typically on sites with a low slope. Occurs in a wide range of soil types, but grows best in well-drained moist soils. Tolerates some salinity and extended periods of drought. Does not survive in areas with prolonged or regular periods of freezing temperatures.
Origin: Native to the Mediterranean region and tropical Asia. In California from the late 1700s to early 1800s, giant reed was often planted for erosion control in flood channels and as wind breaks. Since then it has been cultivated as an ornamental and to produce reeds for woodwind instruments. It is now a leading candidate for cellulosic biofuel production.
Impacts: Giant reed is primarily a problem in riparian corridors. It develops dense stands which often displace native vegetation, diminish wildlife habitat, and increase flooding and siltation in natural areas. Giant reed is also adapted to a periodic fire regime. The canes are readily flammable throughout much of the year, and the presence of giant reed increases the susceptibility of riparian corridors to fire. Large stands of giant reed can increase water loss from underground aquifers in semi-arid regions due to a high evapotranspiration rate. The rate of water loss is estimated at roughly three times more than that of the native riparian vegetation. It is also an alternate host for beet western yellows virus, sugarcane mosaic virus, and maize dwarf mosaic virus.
Western states listed as Noxious Weed: California
California Invasive Plant Council (Cal-IPC) Inventory: High Invasiveness

Giant reed is a bamboo-like perennial to 25 ft tall, with thick, well-developed rhizomes. Although plants are typically terrestrial, they can tolerate periodic flooding. The canes are erect, semi-woody, and about 1 to 2 inches thick. First year green canes have unbranched stems the same diameter as older canes, but more pliable. Older canes are often branched, sometimes with leaves only on the branches. The blades are less than 3 ft long and 1 to 3 inches wide. The ligules consist of a short, even, minutely fringed membrane about 1 to 2 mm long. The auricles and collar region are distinctly pale yellowish-green. The rhizomes are creeping, thick, scaly, often forming a dense network, firm and knotty at the stem bases. Rhizome and stem fragments with a node can develop into a new plant under suitable conditions.

Inflorescences consist of large terminal plume-like panicles, 1 to 2 ft long, and silvery cream-colored to purplish or brown. Giant reed does not appear to produce viable seed in North America, although some Asian populations produce viable seed. Plants reproduce only vegetatively from rhizomes and rhizome and stem fragments; and stem and rhizome fragments generally disperse with water, mud, and human activities.

NON-CHEMICAL CONTROL

Mechanical (pulling, cutting, disking)	Minor infestations can be eradicated by manual methods, especially where sensitive native plants and wildlife might be damaged by other methods. Plants less than 6 ft in height and arising from a new stem or rhizome fragment can be hand pulled. This may be most effective in loose soils and after rains have loosened the substrate. Giant reed can also be dug using hand tools, particularly when used in combination with cutting near the base of the plant.
	Chopping, cutting or mowing (rotary brush cutter, chainsaw, or tractor-mounted mower) can also be used to reduce giant reed infestations, although the fibrous nature of giant reed makes using these techniques difficult. Such methods usually require tractor-mounted equipment, but on rough or rocky soils scythes can be used for smaller patches. These methods generally cause less soil disturbance compared to heavy equipment. However, they are nonselective and may damage other desirable species or open up new niches for weedy invasions. These methods usually require several cuttings before the underground parts exhaust their reserve food supply, and larger giant reed patches will have enough reserves to resprout even after years of treatment. The best timing for cutting is when the plants begin to flower, as this is when the reserve energy supply in the rhizomes is lowest.
	Mechanical methods using mechanized equipment (e.g., backhoe) to remove above-ground vegetation is a

	common non-chemical control method for giant reed. However, such equipment is also nonselective and can only be used on accessible terrain. Most mechanical equipment is not safe to operate on slopes over 30%. It is also of limited use where soils are highly susceptible to compaction or erosion or where excessive soil moisture is present. Site obstacles such as rocks, stumps or logs also reduce efficiency. Mechanical eradication of giant reed is extremely difficult, even with the use of a backhoe, as rhizomes buried under 3 to 10 ft of alluvium readily resprout. Regardless of the mechanical removal method employed, it is critical to remove the entire rhizome root mass. If any of the rhizome mass is left in the ground it will resprout. In addition, stems and roots should be removed, chipped or burned on site to prevent resprouting.
Cultural	Giant reed is not very palatable to cattle, but they will feed on it during the drier months. Sheep also have potential for the management of giant reed and have been shown to survive for extended periods on a strict diet of the perennial grass. However, sheep must be properly managed to prevent soil compaction problems particularly in wet areas. The most successful grazers are goats, particularly Angora and Spanish goats. Goats can have several advantages over mechanical and chemical control methods; they are less costly and can negotiate slopes too steep to manage with machines. Angoras are preferred over Spanish goats because of their smaller size and ease of transport. Since goats will trample or browse virtually any vegetation within a fenced area, any desirable trees or shrubs must be protected. A flame thrower or weed burner device can be used as a spot treatment to heat-girdle the stems at the base of giant reed plants. This technique is less costly than basal and stem herbicide treatments and is suitable for use during wet weather when the wildfire hazard is low. Its effectiveness is comparable to manual cutting. Large infestations may be burned to remove standing mature plants. This may be accomplished with or without a pre-spray of herbicides to kill and desiccate plants. When burning is used alone it will not prevent resprouting from the rhizomes. Burning is best followed by herbicide treatment of resprouting plants.
Biological	Little is known about the effects of various pathogens and insects on the growth and reproduction of *Arundo donax*. However, numerous insects are known to feed on this species. In recent work, the eurytomid wasp, *Tetramesa romana*, was evaluated as a potential biological control agent in North America. The wasp was found to be specific to *Arundo* and thus unlikely to harm native or cultivated plants in the Americas. Undoubtedly, many more years will be required before this species or any other potential biological control agents are identified and released.

CHEMICAL CONTROL

The following specific use information is based on reports by researchers and land managers. Other trade names may be available, and other compounds also are labeled for this weed. Directions for use may vary between brands; see label before use. Herbicides are listed by mode of action and then alphabetically. The order of herbicide listing is not reflective of the order of efficacy or preference.

AROMATIC AMINO ACID INHIBITORS	
Glyphosate *Roundup, Accord XRT II, Rodeo, Aquamaster,* and others	**Rate:** Broadcast foliar treatment: 2 to 4 qt product (*Roundup ProMax*)/acre (2.25 to 4.5 lb a.e./acre) or 2 to 4 qt product (*Rodeo* or *Aquamaster*)/acre (2 to 4 lb a.e./acre) around aquatic sites. Spot treatment: 2% v/v solution. However, the *Rodeo* product label allows up to an 8% v/v solution, depending on the equipment being used. **Timing:** Postemergence. Mid-summer to fall application after flowering and before dormancy is the best timing to kill plants and protect injury on many natives. Follow-up application in subsequent spring to control germinating seedlings may be necessary. **Remarks:** Glyphosate is considered the best option for control in pure stands. Two to three years of treatment are necessary. Herbicide treatment can be used after repeated mowing to reduce necessity for spring treatment to kill seedlings. Dense stands of giant reed (> 80% canopy cover) are most efficiently treated by aerial application, usually by helicopter. Helicopter application can treat at least 124 acres per day. Undiluted glyphosate can be applied as a cut stump treatment with a paint brush within 1 to 2 minutes after stem cutting. Results have shown that glyphosate used in a cut stem treatments, regardless of time of application (May, July, or September), provided excellent control with no resprouting. Another method of treatment includes cutting or burning plants followed by foliar treatment of glyphosate to cane regrowth to about 6 to 8 ft in height.

BRANCHED-CHAIN AMINO ACID INHIBITORS	
Imazapyr *Habitat*	**Rate:** 1 to 2 qt product/acre (0.5 to 1 lb a.e./acre) **Timing:** Postemergence fall application timing is most effective, similar to glyphosate. **Remarks:** Imazapyr has soil residual activity and may impact restoration efforts.
Imazapyr + glyphosate	**Rate:** 1 pt imazapyr (*Habitat*) + 1 qt glyphosate product/ acre (0.25 + 1 lb a.e./acre, respectively) **Timing:** Postemergence fall application timing is most effective. **Remarks:** The combination of the two herbicides prevents the synthesis of six amino acids, as each herbicide inhibits three amino acids. This combination is thought to provide better control at lower rates of each herbicide, thus it is more affordable compared to imazapyr alone.

Asclepias fascicularis Decne.; Mexican whorled milkweed
Asclepias speciosa Torr.; showy milkweed

Mexican whorled and showy milkweeds

Asclepias fascicularis

Family: Asclepiadaceae
Range: Mexican whorled milkweed is widespread throughout much of the western United States, including Washington, Oregon, Idaho, California, Nevada and Utah. Showy milkweed is found in all western states from Texas north to British Columbia.
Habitat: Roadsides, ditchbanks, pastures, and cultivated fields. Typically found in areas that remain moist through much of the summer, such as moist prairies and flood plains. They can grow in all soil textures from sea level to 7000 ft elevation.
Origin: Both species are native to North America.
Impact: Milkweeds are most problematic in pastures and range because in addition to being distasteful to livestock, the entire plant can be toxic to sheep, cattle, horses and domestic fowl. The toxic compound is considered to be cardenolide (cardiac glycosides). Typically, milkweeds are only eaten when forage is limited. In natural areas native milkweeds may be considered desirable plants, an important component of the plant community. The larvae of monarch butterflies feed solely on milkweed species.

Asclepias speciosa

    Mexican whorled milkweed and showy milkweed are erect perennial forbs that grow up to approximately 3 to 4 ft in height. Their sap is a milky white latex. Mexican whorled milkweed has lanceolate leaves around 6 inches long and 0.75 inch wide. The leaves are arranged in whorls of 3 to 6 and are glabrous or covered with minute hairs. In addition, the leaves are often folded upwards along the midvein. Showy milkweed has oval to oblong-shaped opposite leaves covered with soft wholly hairs. The leaves are 4 to 7 inches long on short stalks.
    Milkweeds reproduce by seed and underground roots, although the primary means of spread is by seed. They have an umbel-like inflorescence. Mexican whorled milkweed flowers are pale pink, purple or greenish-white. The flowers have 5 sepals and stamens. The petals are 5-lobed, reflexed downward, and 4 to 5 mm long. Showy milkweed has rose-purple colored petals with hairy backs and pinkish hoods that fade to yellowish.
    Mexican whorled milkweed has narrow seed pods that are 2 to 3 inches long and smooth. The seeds are light brown, oblong, flattened, and 3 to 8 mm long with a tuft of deciduous silky hairs approximately 1 inch long. Showy milkweed pods are 3 to 5 inches long and densely covered with woolly hairs. At maturity the seed pods burst and while most seeds fall close to the parent plant, some can disperse greater distances in the wind. It is not known how long the seeds survive in the soil, but it is expected that it would be several years.

NON-CHEMICAL CONTROL

Mechanical (pulling, cutting, disking)	Hand pulling is a viable method if the population size is very small.
	Mowing may reduce seed production but as the sole control measure will not kill milkweed.
	Tillage is not an effective control measure for milkweed because each root segment can give rise to a new plant.
Cultural	Grazing is not a viable control option for milkweed because it is both distasteful and toxic to livestock.
	Burning will probably top-kill milkweed, but the plants will likely recover from undamaged rhizomes. Some research has shown that burning stimulates resprouting and may also stimulate increased flowering and seed production.
Biological	Because these species are native to North America, there have been no biological control programs developed.

CHEMICAL CONTROL

The following specific use information is based on published papers or reports by researchers and land managers. Other trade names may be available, and other compounds also are labeled for this weed. Directions for use may vary between brands; see label before use. Herbicides are listed by mode of action and then alphabetically. The order of herbicide listing is not reflective of the order of efficacy or preference.

GROWTH REGULATORS	
Aminocyclopyrachlor + chlorsulfuron *Perspective*	**Rate:** 4.75 to 8 oz product/acre **Timing:** Postemergence when target plants are growing rapidly. **Remarks:** *Perspective* provides broad-spectrum control of many broadleaf species. Although generally safe to grasses, it may suppress or injure certain annual and perennial grass species. Do not treat in the root zone of desirable trees and shrubs. Do not apply more than 11 oz product/acre per year. At this high rate, cool-season grasses will be damaged, including bluebunch wheatgrass. Not yet labeled for grazing lands. Add an adjuvant to the spray solution. This product is not approved for use in California and some counties of Colorado (San Luis Valley).
Dicamba *Banvel, Clarity*	**Rate:** 4 pt product/acre (2 lb a.e./acre) **Timing:** Postemergence when the target plants are emerged and rapidly growing. **Remarks:** These specific milkweed species are not listed on the dicamba label but other milkweed species are listed. Dicamba is a broadleaf-selective herbicide often combined with other active ingredients. Several applications are likely needed for complete control. Dicamba is available mixed with diflufenzopyr in a formulation called *Overdrive*. This has been reported to be effective on some milkweed species. Diflufenzopyr is an auxin transport inhibitor which causes dicamba to accumulate in shoot and root meristems, increasing its activity. *Overdrive* is applied postemergence at 4 to 8 oz product/acre to rapidly growing plants. Higher rates should be used when treating perennial weeds. Add a non-ionic surfactant to the treatment solution at 0.25% v/v or a methylated seed oil at 1% v/v solution.
Picloram *Tordon 22K*	**Rate:** 1 qt product/acre (0.5 lb a.e./acre) **Timing:** Postemergence to rapidly growing plants at the bud to early bloom stage. **Remarks:** Picloram is sometimes used in a premix with 2,4-D (*Grazon P+D*) or tank mixed with 2,4-D at 1 lb a.e./acre plus a surfactant. Most broadleaf plants are susceptible. Although well-developed grasses are not usually injured by labeled use rates, some applicators have noted that young grass seedlings with fewer than four leaves may be killed. Do not apply near trees, or where soil is highly permeable and where water table is high. Picloram is a restricted use herbicide. It is not registered for use in California.
AROMATIC AMINO ACID INHIBITORS	
Glyphosate *Roundup, Accord XRT II, and others*	**Rate:** 2 qt product (*Roundup ProMax*)/acre (2.25 lb a.e./acre). Wiper treatment: 33 to 50% of concentrated product. **Timing:** Postemergence to rapidly growing milkweed at late bud to flowering stage. **Remarks:** Control of showy milkweed is difficult because of the pubescent leaves, and milkweed is often drought-stressed late in the season, which affects herbicide uptake and translocation. Repeated application over several seasons may be needed to provide complete control. Glyphosate is a systemic nonselective herbicide with no soil activity. Because there is often a significant height differential between milkweed and desirable pasture or range species, a wiper/rope wick applicator or some other selective application technology may be desirable. Spot treatment is also preferable when feasible.

Asphodelus fistulosus L.
Onionweed

Family: Liliaceae
Range: California, New Mexico and Texas.
Habitat: Fields, pastures, roadsides, coastal dunes, orchard and agronomic crops, and other disturbed places, especially those with sparse vegetation. Most recently, onionweed is beginning to move into grasslands and other wildland sites following burning.
Origin: Native to southern Europe.
Impacts: In pastures and on rangeland, onionweed is avoided by livestock and can develop dense populations that exclude grasses and other desirable forage species. It is a government-listed noxious weed in Australia, where it is most problematic on pastureland in the southern areas.
Western states listed as Noxious Weed: California, New Mexico, Oregon
California Invasive Plant Council (Cal-IPC) Inventory: Moderate Invasiveness (Alert)

Onionweed is an annual to short-lived perennial to 2 ft tall with thick, tuber-like stem bases and numerous glabrous, slender, grass-like basal leaves. Although the common name is onionweed, it lacks an onion or garlic scent when crushed.

Inflorescences are open, branched panicles of racemes. Flower stems are rigid, hollow, and usually several per plant. Flowers are 5 to 14 mm long, white or pale pink, with a conspicuous chocolate-brown midvein. Anthers are distinctly orange. Plants reproduce by seed, which disperse with water, mud, animals, vehicles, agricultural machinery, and other human activities. Seeds are reported to survive for many years in the soil seedbank.

NON-CHEMICAL CONTROL

Mechanical (pulling, cutting, disking)	Isolated plants can be hand-pulled or hoed before they produce viable seed. This is often too difficult with a larger population. Plants need to be cut below the soil surface to prevent resprouting. Tillage when the soil is dry can be effective since plants do not have creeping roots, true tubers or bulbs. However, tillage often leaves an ideal seedbed for new germinants, thus retreatment is generally necessary.
Cultural	Onionweed is not considered a palatable species and grazing is not generally considered an option. Prescribed burning is also not generally used for control of onionweed.
Biological	In Australia, onionweed can be affected in some seasons by a *Fusarium* fungus that destroys the roots. This occurs in moist areas. Some work is being conducted in Europe on a fungus, *Puccinia barbeyi*, as a biocontrol agent. No biological control agents are currently available in the United States.

CHEMICAL CONTROL
There is very little information available for the chemical control of onionweed.

The following specific use information is based on reports by researchers and land managers. Other trade names may be available, and other compounds also are labeled for this weed. Directions for use may vary between brands; see label before use. Herbicides are listed by mode of action and then alphabetically. The order of herbicide listing is not reflective of the order of efficacy or preference.

GROWTH REGULATORS

Picloram *Tordon 22K*	**Rate:** 2 qt product/acre (1 lb a.e./acre) **Timing:** Postemergence after the flowering stage. This is generally in late summer or autumn. Applications are expected to have preemergence activity on germinating seedlings. **Remarks:** Picloram can be tank mixed with paraquat. Picloram is a restricted use herbicide. It is not registered for use in California.

AROMATIC AMINO ACID INHIBITORS

Glyphosate

Roundup, Accord XRT II, and others

Rate: Spot treatments with 2 to 5% v/v *Roundup ProMax* or product with similar concentration of glyphosate

Timing: Plants should be fully leafed out and rapidly growing. Typically the best time to treat is in summer.

Remarks: Spray-to-wet treatments are generally used for spot treatments. Some studies show that glyphosate is not that effective for onionweed control. Glyphosate is a nonselective herbicide with no soil activity. It may be possible to use a wiper application technique to reduce drift as well as contact with non-target species, although such studies have not yet been conducted.

BRANCHED-CHAIN AMINO ACID INHIBITORS

Chlorsulfuron

Telar

Rate: 2 oz product/acre (1.5 oz a.i./acre)

Timing: In Australia, early winter treatments are recommended. Other trials in California were conducted in April on mature, rapidly growing plants and these proved successful.

Remarks: Studies from California show this to be the most effective treatment. Onionweed prefers alkaline soils, and under these conditions chlorsulfuron will remain in the soil for a longer time compared to more neutral or acid soils.

Metsulfuron

Escort

Rate: 0.1 to 1 oz product/acre (0.06 to 0.6 oz a.i./acre)

Timing: In Australia, early winter treatments are recommended.

Remarks: Metsulfuron is considered to be a bit safer on non-target species than chlorsulfuron. Metsulfuron is not registered for use in California.

Atriplex semibaccata R. Br.

Australian saltbush

Family: Chenopodiaceae
Range: Southwestern U.S. from California to Texas, also including Arizona, Nevada, Utah, and New Mexico.
Habitat: Disturbed places, roadsides, waste places, irrigation canals, sandy fields, margins of cultivated fields, grassland, scrub, shrubland, woodland, salt marsh areas on many types of soil; often grows under alkaline or saline conditions.
Origin: Native to Australia. Brought to the U.S. as a forage plant for alkaline and saline areas in the 1920s. More recently it has been promoted as a fire-resistant, drought-, salt- and alkali-tolerant plant for groundcover or erosion control and as a component of reclamation vegetation for the restoration of mined sites in the southwestern states.
Impacts: Has become regionally invasive in coastal grassland, scrub, and on the higher ground of salt marshes. Australian saltbush is susceptible to the beet western yellows virus, which can affect a variety of crop plants and is transmitted by aphids.
California Invasive Plant Council (Cal-IPC) Inventory: Moderate Invasiveness

Spreading, semi-shrubby short-lived perennial to 2 ft tall, with weakly woody stems. Leaves are alternate, mostly short-stalked and oblong to narrow-elliptic, 0.5 to 2 inches long, up to 1 inch wide, with a smooth or wavy-toothed margin. The leaves are usually gray-green and somewhat scurfy, especially the lower surface.

Plants flower between April and December. Male and female flowers develop in separate small clusters within the same plant (monoecious). Male flower clusters develop at stem tips. One or a few female flowers develop in leaf axils below the male flower clusters. They reproduce by seed with only one seed per fruit. Fruits are small, fleshy, oblong to diamond-shaped, turning reddish at maturity. Fruits persist on the parent plant, fall near it, and disperse to greater distances with water, mud, soil movement, human activities, and animals, including birds. Seeds germinate under saline and alkaline conditions. Most germination occurs in spring.

NON-CHEMICAL CONTROL

Mechanical (pulling, cutting, disking)	The plant is fairly brittle and low growing and is fairly easy to remove by hand or mechanical weeding.
Cultural	Australian saltbush is occasionally grazed by cattle, sheep, horses, and hogs and is rich in protein. It may be possible to use this as a control tool in pastures or rangelands, when it is a problem in these sites.
Biological	While some native insects can feed on the leaves, they do not appear to cause any impact on the populations.

CHEMICAL CONTROL

There is no direct information on the control of this species. As such, recommendations were used for other *Atriplex* or related Chenopodiaceae species. It is expected that the response of Australian saltbush would be similar. Directions for use may vary between brands; see label before use. Herbicides are listed by mode of action and then alphabetically. The order of herbicide listing is not reflective of the order of efficacy or preference.

GROWTH REGULATORS	
Dicamba	**Rate:** 3 to 4 pt product/acre (1.5 to 2 lb a.e./acre)
Banvel, Clarity	**Timing:** Postemergence to seedlings or to mature plants that are growing rapidly.
	Remarks: Dicamba has provided between 80 and 90% control of other *Atriplex* species in Canada. Dicamba has also been used to control *Bassia scoparia*, which is also related to Australian saltbush.
AROMATIC AMINO ACID INHIBITORS	
Glyphosate	**Rate:** 1 to 2 qt product (*Roundup ProMax*)/acre (1.1 to 2.25 lb a.e./acre). Larger plants may require higher

Roundup, Accord XRT II, and others	rates. **Timing:** Postemergence to seedlings or to mature plants that are growing rapidly. **Remarks:** Glyphosate is a nonselective herbicide that can damage other non-target species. It has been shown to give good to excellent control of other *Atriplex* species, but most of these are annuals. Glyphosate can also be mixed with 2,4-D ester or dicamba to increased efficacy. This combination should be applied during cooler weather to prevent volatilization of 2,4-D ester.
AROMATIC AMINO ACID INHIBITORS	
Chlorsulfuron *Telar*	**Rate:** 0.5 to 1 oz product/acre (0.375 to 0.75 oz a.i./acre). Apply with surfactant. **Timing:** Postemergence in late spring or early summer. **Remarks:** Chlorsulfuron has been shown to be more effective than metsulfuron in western rangelands. There is no direct information on the effect of chlorsulfuron on Australian saltbush, but it has been shown to damage a related native, Nuttall's saltbush (*Atriplex nutalllii*). Thus, it is likely that it would also provide control of Australian saltbush.

Avena fatua L.; wild oat
Avena barbata Pott ex Link; slender oat

Wild and slender oats

Family: Poaceae
Range: Wild oat is found throughout the western U.S. and slender oat is primarily found in the Pacific states. The two species have been well established here since the late 1700s.
Habitat: Grassland, crop fields, orchards, vineyards, gardens, roadsides, and other disturbed sites.
Origin: Both species are native to Eurasia.
Impact: Wild oat is a major weed throughout the small grain-growing areas of the western U.S. because it emerges throughout the cool season from autumn through spring. It causes lodging, slows harvest, clogs harvester screens, and lowers yields dramatically. When wild oats are weeds in oat crops, they cannot be controlled using herbicides because any herbicide would also cause severe damage to the cultivated crop; however, wild oats can be controlled in wheat and barley. In rangeland, wild oats make good forage as the stems are finer and more palatable than common oats.
California Invasive Plant Council (Cal-IPC) Inventory: Both species, Moderate Invasiveness

Wild and slender oats are erect, cool-season annual grasses with open-branched, nodding flower clusters. Mature plants are sturdy and can grow to about 4 ft tall. Stems are round in cross-section, hairless, or nearly so. Leaves are flat, rolled in the bud, and up to 8 inches in length. Leaves often twist counter-clockwise. Usually a few soft hairs grow from the edge of the base of the blade. The leaf sheath is open and usually has a hairless edge. Wild oat has a large membranous ligule with a rounded, jagged top. There are no auricles. Wild oat has an extensive, fibrous root system.

Wild oats are in bloom mostly from March through June. The flower head is open and branched. Spikelets hang like pendants from the flowering branches. Plants reproduce only by seed. The seeds are hairy at the base with a circular scar at the point of seed attachment. The awns on the lemmas are 1 to 2 inches long, bent once, and twisted below the bend. Depending on the biotype and the location on the panicle, a proportion of seeds can remain dormant for one or more years. Some cold-climate biotypes of wild oat can survive for 10 years or more under field conditions.

NON-CHEMICAL CONTROL

Mechanical (mowing, plowing, and cultivation)	Hand pulling can be used to remove plants in small infestations.
	Mowing can prevent seed-set in heavy to moderate infestations.
	Plants can be managed with tillage on open ground before planting when the wild oats are germinating and before seed-set.
Cultural	Burning windrows of straw immediately after cutting can reduce seed viability, as seeds are not tolerant of high temperatures. In other areas, burning after seed drop will increase wild oats in the following season.
	Some growers till once in spring to stimulate wild oat germination, then till again later to kill seedlings. This is followed by a preemergence or postemergence herbicide for control of subsequent recruitment.
	A competitive stand of perennial vegetation will discourage wild oats on pastures.
Biological	Biocontrol agents have not been developed to control wild oat. Because these species are closely related to cultivated oats and other cereals, and also because they are desirable rangeland forage species, it is unlikely that a biological control program will be established.

CHEMICAL CONTROL

The following specific use information is based on published papers and reports by researchers and land managers. Other trade names may be available, and other compounds also are labeled for this weed. Directions for use may vary between brands; see label before use. Herbicides are listed by mode of action and then alphabetically. The order of herbicide listing is not reflective of the order of efficacy or preference.

LIPID SYNTHESIS INHIBITORS	
Clethodim *Select, Envoy*	**Rate:** 0.75 to 2 pt product (*Envoy*)/acre (1.5 to 4 oz a.i./acre) **Timing:** Postemergence when target plants are between 2 and 6 inches tall. **Remarks:** Clethodim is a grass-selective herbicide. Add a non-ionic surfactant (0.25%). There are some naturally occurring populations of wild oats that are resistant to herbicides with the ACCase mode of action. Note that *Envoy* formulation is 1 lb a.i./gallon, *Select* is 2 lb a.i./gallon.
Fluazifop *Fusilade*	**Rate:** 1 to 1.5 pt product/acre (4 to 6 oz a.e./acre) **Timing:** Postemergence when target plants are between 2 and 8 inches tall and rapidly growing. **Remarks:** Fluazifop is a grass-selective herbicide. Add a non-ionic surfactant (0.06 to 0.125%) for increased control. There are some naturally occurring populations of wild oats that are resistant to herbicides with the ACCase mode of action.
Sethoxydim *Poast*	**Rate:** 1 to 2.5 pt product/acre (3 to 7.5 oz a.e./acre) **Timing:** Postemergence when wild oats are less than 12 inches tall. Label recommends 1.5 pt product/acre when seedlings are 4 inches or less. **Remarks:** Sethoxydim is a grass-selective herbicide. Add a seed oil or crop oil concentrate (0.5% to 1% v/v). Adding nitrogen to the spray solution can improve control. Do not use treated vegetation as pasture, hay, feed, or forage. Do not apply to any desired grass crop. There are some naturally occurring populations of wild oats that are resistant to herbicides with the ACCase mode of action.
AROMATIC AMINO ACID INHIBITORS	
Glyphosate *Roundup, Accord XRT II,* and others	**Rate:** 11 to 22 oz product (*Roundup ProMax*)/acre (6.2 to 12.4 oz a.e./acre) **Timing:** Postemergence when target weeds are less than 18 inches tall and rapidly growing, before planting a crop. **Remarks:** Add ammonium sulfate at a rate of 10 to 15 lb/100 gal of water. Add a non-ionic surfactant when using a formulation where it is not already included (e.g., *Accord, Rodeo, Aquamaster*).
BRANCHED-CHAIN AMINO ACID INHIBITORS	
Imazapic *Plateau*	**Rate:** 8 to 12 oz product/acre (2 to 3 oz a.e./acre) **Timing:** Postemergence when wilds oats are less than 12 inches tall. **Remarks:** Imazapic has some residual soil activity. There are some naturally occurring populations of wild oats that are resistant to herbicides with the ALS mode of action. Imazapic is not registered for use in California.
Imazapyr *Arsenal AC, Habitat, Chopper, Stalker, Polaris*	**Rate:** 2 to 3 pt product (*Arsenal AC*)/acre (1 to 1.5 lb a.e./acre) **Timing:** Preemergence or postemergence when plants are young. **Remarks:** Imazapyr is a broad-spectrum herbicide with fairly long soil residual activity. Do not apply more than 3 pt product (*Arsenal AC*)/acre per year. There are some naturally occurring populations of wild oats that are resistant to herbicides with the ALS mode of action.
Propoxycarbazone-sodium *Canter R+P*	**Rate:** 1.2 oz product/acre (0.84 oz a.i./acre) **Timing:** Postemergence from the 2-leaf to 2-tiller stage when plants are growing rapidly. **Remarks:** Propoxycarbazone is a broad-spectrum herbicide that will control many species. It will provide only partial control of wild or slender oats. Perennial grass species vary in tolerance. A non-ionic surfactant should be added at 0.25 to 0.5% v/v solution.
Rimsulfuron *Matrix*	**Rate:** 4 oz product/acre (1 oz a.i./acre) **Timing:** Preemergence or postemergence when the target plants are small. **Remarks:** Rimsulfuron controls several annual grasses and broadleaves. It provides only partial control of wild oats. Perennial grasses are tolerant to fall applications when established and grown under dryland conditions. Application to rapidly growing or irrigated perennial grasses may result in their

	injury or death. It provides soil residual control in cool climates but degrades rapidly under warm conditions. Rimsulfuron will not control summer annual weeds when applied in fall or spring. Add a surfactant when applying postemergence.
Sulfometuron *Oust* and others	**Rate:** 1 to 2 oz product/acre (0.75 to 1.5 oz a.i./acre) **Timing:** Preemergence or early postemergence when the weeds are germinating or actively growing. **Remarks:** Sulfometuron will control or injure most other plant species. To improve control, add a surfactant at the rate of 0.25% v/v or at the rate specified on the manufacturer's label. There are some naturally occurring populations of wild oats that are resistant to herbicides with the ALS mode of action.
CONTACT PHOTOSYNTHETIC INHIBITORS	
Paraquat *Gramoxone* and others	**Rate:** 1 to 2.7 pt product/acre (4 to 10.8 oz a.i./acre) **Timing:** Postemergence when target plants are less than 6 inches tall, but before planting a crop. **Remarks:** Paraquat is selective on annual species and will only cause some burndown of perennials. Spray coverage needs to be thorough. Do not make more than two applications per year. Paraquat is a restricted use herbicide.
PHOTOSYNTHETIC INHIBITORS	
Hexazinone *Velpar L*	**Rate:** 1 to 2.5 gal product/acre (2 to 5 lb a.i./acre) **Timing:** Either preemergence or postemergence when the target plants are germinating and actively growing. **Remarks:** Hexazinone is a broad-spectrum, long residual, mobile herbicide. It suppresses wild oats but does not provide control. High rates of hexazinone can create bare ground, so only use high rates in spot treatments.

Azolla spp.
Mosquitoferns

Family: Azollaceae
Range: Throughout most of the western United States, except Idaho, Wyoming, Montana and North Dakota.
Habitat: On still water or mud in ponds, small lakes, slow-moving streams and channels, ditches, rice fields, and sloughs. Often grow in eutrophic water. Do not tolerate saline water.
Origin: Native to the United States, including the western states. Mosquitoferns are sometimes sold as aquarium or pond ornamentals. In East Asia, mosquitofern is used as livestock feed and as a nitrogen source in rice fields.
Impacts: Native mosquitoferns are consumed by wildlife, especially waterfowl, and are usually a desirable component of natural aquatic communities. In addition, they provides breeding habitat for aquatic insects important to fisheries. In some human use areas, dense colonies can become a nuisance in certain situations by excluding other aquatic vegetation, encouraging the growth of algae, interfering with livestock drinking, and clogging water pumps.

Western states listed as Noxious Weed: Pinnate mosquitofern (*A. pinnata* R.Br.) is listed as a Federal Noxious Weed, but is not naturalized in the United States, although it may be cultivated as an aquarium plant. It is also a state listed Noxious Weed in California and Oregon

Mosquitofern species are small, annual to perennial, free-floating aquatic ferns that often occur in colonies. Pacific mosquitofern (*A. filiculoides* Lam.) and Mexican mosquitofern (*A. mexicana* C. Presl) are native species that occur in many western states. Upper leaf lobes are typically colonized by the nitrogen-fixing cyanobacterium *Anabaena azollae*. Stems are floating and are pinnately branched with roots suspended in the water column. Young plants are gray to green but turn red to brown with age and season.

Plants produce spores that disperse with wind, but the most common mechanism of dispersal is vegetative, with plant fragments moving in water or clinging to the feet or feathers of birds. Careless disposal of pond or aquarium contents can introduce plants to previously uninhabited areas. Colonies typically enlarge rapidly during the warmer months and diminish during the cool months.

NON-CHEMICAL CONTROL

Mechanical (floating booms, suction devices)	Small infestations of the weed in accessible areas may be removed using rakes and fine meshed nets. The disadvantage of mechanical control, however, is that under ideal conditions, the weed can double itself every 4 to 5 days. Thus, it must be repeated often. For small infestations (1 to 2 acres), floating booms can be dragged (preferably "down wind") from the shore or pushed by boats to consolidate mats of *Azolla* which can then be removed with rakes. *Azolla* provides good composting material. In large lakes, mechanical harvesters equipped with surface "skimmers" or surface suction devices can remove mats.
Cultural	The use of water-circulation devices can sometimes reduce accumulation of large biomass.
	Reducing nutrient inputs can also be helpful (e.g. divert runoff from turf or other areas that provide nutrients).
Biological	A native frond-feeding weevil, *Stenopelmus rufinasus*, has been used for control of *Azolla filiculoides* with some success outside the United States. Early research on the flea beetle *Pseudolampsis guttata* suggests it may also be useful. The triploid (sterile) grass carp (white amur) is a relatively nonselective herbivorous fish that will consume *Azolla* and other small floating plants (e.g. duckweeds). The fish do not selectively feed on "non-native" plants so careful monitoring of feeding impacts is necessary. In some Asian crop systems, use of fish and ducks are integrated to provide control of *Azolla* in rice production.

CHEMICAL CONTROL
The following specific use information is based on reports by researchers and land managers. Other trade names may be available, and other compounds also are labeled for this weed. Directions for use may vary between

brands; see label before use. Herbicides are listed by mode of action and then alphabetically. The order of herbicide listing is not reflective of the order of efficacy or preference.

AROMATIC AMINO ACID INHIBITORS	
Glyphosate *Rodeo,* *Aquamaster*	**Rate:** Spot treatment: 2% v/v solution (*Rodeo* or *Aquamaster*) for foliar spray with approved aquatic surfactant (0.5%) **Timing:** Postemergence to foliage from spring to mid-summer. **Remarks:** Glyphosate is a slow-acting, systemic herbicide. *Azolla* often forms thick mats that can prevent glyphosate (or other foliar-applied herbicides) from penetrating the canopy and therefore unexposed fronds will reestablish the population.
BRANCHED-CHAIN AMINO ACID INHIBITORS	
Bispyribac-sodium *Tradewind*	**Rate:** 8 oz product/acre (6.4 oz a.i./acre). Allow 30 days between applications and apply up to four times per year. **Timing:** Postemergence to foliage from spring to mid-summer. **Remarks:** Bispyribac-sodium is a slow-acting herbicide and may take 4 to 6 weeks to show effects.
Imazamox *Clearcast*	**Rate:** Spot treatment: 2% v/v solution as a foliar spray plus 1% methylated seed oil (MSO) **Timing:** Postemergence to foliate from spring to mid-summer. **Remarks:** Use an approved surfactant.
Penoxsulam *Galleon*	**Rate:** 5.6 to 11.2 oz product/acre (1.4 to 2.8 oz a.i./acre), but most often used at 8 oz product/acre (2 oz a.i./acre). Apply in 20 to 100 gal spray solution/acre. **Timing:** Postemergence to foliage from spring to mid-summer. **Remarks:** Penoxsulam is a slow-acting herbicide and may take 4 to 6 weeks to show effects.
PIGMENT SYNTHESIS INHIBITORS	
Fluridone *Sonar*	**Rate:** For in-water treatment: 10 to 30 ppb **Timing:** Apply directly to water from spring to mid-summer (before large biomass has developed). **Remarks:** Fluridone is a slow-acting herbicide that may take several weeks to show effects.
CONTACT PHOTOSYNTHETIC INHIBITORS	
Diquat *Reward,* *Redwing*	**Rate:** 2 to 4 pt product/surface acre (0.5 to 1 lb a.i./surface acre) **Timing:** Postemergence to foliage from spring to mid-summer. **Remarks:** Diquat is a fast-acting contact herbicide. Repeated applications may be needed. *Azolla* often forms thick mats that can prevent diquat (or other foliar-applied herbicides) from penetrating the canopy and therefore unexposed fronds will reestablish the population.
Flumioxazin *Clipper*	**Rate:** For in-water treatments: 100 to 400 ppb **Timing:** Apply directly to water from early spring to early summer, during the plant's rapid growth phase. **Remarks:** Flumioxazin is rapidly degraded and is inactive if pH exceeds 8.5. Thus, it is important to only use if pH will not exceed 8.5. It is best to apply flumioxazin in the early morning when the pH is low.

Bassia scoparia (L.) A.J. Scott; kochia
  (=*Kochia scoparia* (L.) Schrad.)
Bassia hyssopifolia (Pall.) Kuntz; fivehook bassia

Kochia and fivehook bassia

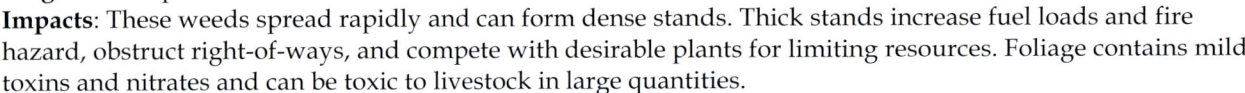

Family: Chenopodiaceae
Range: All western states, except North Dakota for *Bassia hyssopifolia*.
Habitat: Roadsides, fallow fields, crop fields, ditch margins, seasonal wetlands, and residential areas. Kochia predominately inhabits upland sites especially following soil disturbance. Fivehook bassia is a facultative wetland indicator species and often associated with alkaline areas. Both species tolerate alkaline or saline soil, drought, and frost.
Origin: Both species are native to Eurasia.
Impacts: These weeds spread rapidly and can form dense stands. Thick stands increase fuel loads and fire hazard, obstruct right-of-ways, and compete with desirable plants for limiting resources. Foliage contains mild toxins and nitrates and can be toxic to livestock in large quantities.
Western states listed as Noxious Weed: *B. scoparia*: Arizona, Oregon, Washington
California Invasive Plant Council (Cal-IPC) Inventory: Both plants are listed as Limited Invasiveness

Kochia and fivehook bassia are erect summer annuals that grow to 4 ft tall or more. In the vegetative state, these species are very difficult to distinguish from one another. The foliage is generally gray-green and covered with soft hairs. The leaves are mostly alternate, flat, linear-lanceolate to lanceolate. Kochia often appears leafier than fivehook bassia. Kochia stems sometimes appear reddish late in the season. Both species have a taproot, usually with few to several branched, fibrous lateral roots. Kochia litter appears to have allelopathic properties that affect certain plants.

Both kochia and fivehook bassia flower in late summer to early fall. They develop spikes of inconspicuous flowers that lack petals. Fivehook bassia fruit have five small 1-mm long hooked spines. Kochia fruit have five thickened or knoblike lobes, or sometimes short, horizontal wings, less than 1 mm long and wide. Fruits contain one horizontal seed and remain enclosed within the calyx. The senesced plants break off at the base and scatter seed as they tumble with the wind. Both species produce abundant seed. Seeds germinate in spring. Because seeds generally are not deeply buried, they usually survive only 1 to 2 years.

NON-CHEMICAL CONTROL

Mechanical (pulling, cutting, disking)	Small infestations can be removed by manual methods. Digging and hand-pulling are effective. When digging, sever the root below the soil surface.
	Mowing can reduce seed production. Mow before flowering to prevent seed production. On sites with high soil moisture, plants need to be removed frequently to prevent regrowth.
	Shallow tillage will control emerged plants but often stimulates recruitment. Deep tillage can reduce populations by burying seed deep enough to prevent germination. Land managers that use tillage for seedbed preparation during reseeding should prepare for a flush of seedlings when soils become saturated.
Cultural	Plants will frequently regrow following grazing. Grazing can reduce populations when small plants are grazed intensively. Kochia can provide good livestock forage in small amounts, although foliage contains mild toxins and nitrates and can be toxic to livestock in large quantities. Fivehook bassia is considered fair forage for sheep in small amounts. The foliage contains variable amounts of potassium oxalate which is toxic in large amounts and an unidentified substance that can cause digestive tract problems.
	Burning is not an effective control. In Colorado, good control of kochia was achieved using a propane flamer to sear seedlings less than 1 inch tall.
	Promoting competitive vegetation can slow spread and help prevent establishment. Perennial grass plantings have been shown to inhibit kochia establishment. Perennial grass stand density and vigor should be managed

	to minimized bare ground exposure.
Biological	No known biological controls for either species are available in the United States.

CHEMICAL CONTROL

The following specific use information is based on published papers and reports by researchers and land managers. Other trade names may be available, and other compounds also are labeled for this weed. Directions for use may vary between brands; see label before use. Herbicides are listed by mode of action and then alphabetically. The order of herbicide listing is not reflective of the order of efficacy or preference.

GROWTH REGULATORS	
Aminocyclopyrachlor + chlorsulfuron *Perspective*	**Rate:** 4.75 to 8 oz product (*Perspective*)/acre **Timing:** Both postemergence and preemergence. Postemergence applications are most effective when applied to small plants. **Remarks:** *Perspective* provides broad-spectrum control of many broadleaf species. Although generally safe to grasses, it may suppress or injure certain annual and perennial grass species. Do not treat in the root zone of desirable trees and shrubs. Do not apply more than 11 oz product/acre per year. At this high rate, cool-season grasses will be damaged, including bluebunch wheatgrass. Not yet labeled for grazing lands. Add an adjuvant to the spray solution. This product is not approved for use in California and some counties of Colorado (San Luis Valley).
Dicamba *Banvel*, *Clarity*	**Rate:** 0.5 to 2 pt product/acre (0.25 to 1 lb a.e./acre) **Timing:** Postemergence. Most effective on seedling and small plants. **Remarks:** Dicamba is a broadleaf-selective herbicide often combined with other active ingredients, particularly 2,4-D (0.5 to 1 pt dicamba + 2 pt 2,4-D/acre). Some kochia populations have developed resistance to dicamba. Where resistance is suspected, use other herbicides or combinations. Dicamba is available mixed with diflufenzopyr in a formulation called *Overdrive*. This has been reported to be effective on kochia. Diflufenzopyr is an auxin transport inhibitor which causes dicamba to accumulate in shoot and root meristems, increasing its activity. *Overdrive* is applied postemergence at 4 to 8 oz product/acre to rapidly growing plants. Higher rates should be when treating perennial weeds. Add a non-ionic surfactant to the treatment solution at 0.25% v/v or a methylated seed oil at 1% v/v solution.
Fluroxypyr *Vista XRT*	**Rate:** 6 to 22 oz product/acre (2.1 to 7.7 oz a.e./acre) **Timing:** Postemergence from seedling to bloom stage. **Remarks:** Fluroxypyr is a broadleaf-selective herbicide. It can also be effective on larger plants, but has no soil residual activity. It is recommended for herbicide resistant biotypes. Add a methylated seed oil surfactant.
AROMATIC AMINO ACID INHIBITORS	
Glyphosate *Roundup*, *Accord XRT II*, and others	**Rate:** 2 to 3 pt product (*Roundup ProMax*)/acre (1.1 to 1.7 lb a.e./acre). Spot treatment: 0.5% to 1% product v/v **Timing:** Postemergence from seedling to bloom stage. **Remarks:** Glyphosate will only provide control during the year of application; it has no soil activity and will not kill seeds or inhibit germination the following season. Glyphosate has is nonselective. It can create bare ground conditions that are susceptible to weed recruitment. In areas with desirable vegetation, use spot treatment. Glyphosate is a good control option if reseeding is planned shortly after application, as it will not injure seedlings emerging after application. Add a surfactant when it is not already included in the herbicide formulation (e.g., *Rodeo*, *Aquamaster*). Glyphosate resistant biotypes of kochia have been reported.
BRANCHED-CHAIN AMINO ACID INHIBITORS	
Chlorsulfuron *Telar*	**Rate:** 1 oz product/acre (0.75 oz a.i./acre) **Timing:** Preemergence or postemergence from seedling to bolting stage. **Remarks:** Chlorsulfuron has mixed selectivity, but is generally safe on grasses. Always use a surfactant. It can be tank-mixed with 2,4-D for quicker burndown. Chlorsulfuron is included with aminocyclopyrachlor in *Perspective*. *Telar* can be used near water, but cannot be applied to water. Some kochia populations have developed resistance to related herbicides. Where resistance is

	suspected, use other herbicides or combinations.
Imazapic *Plateau*	**Rate:** 8 to 12 oz product/acre (2 to 3 oz a.e./acre) **Timing:** Preemergence to early postemergence. **Remarks:** Imazapic is safe for most native grasses. Higher rates may suppress seedlings of some cool-season grasses. Add a methylated seed oil. Some kochia populations have developed resistance to related herbicides (ALS inhibitors). Where resistance is suspected, use other herbicides or combinations. It can be used in combination with glyphosate (premix trade name of *Journey*). Imazapic is not registered for use in California.
Imazapyr *Habitat*	**Rate:** 1 to 2 qt product/acre (0.5 to 1 lb a.e./acre) **Timing:** Can be applied preemergence or postemergence. **Remarks:** Imazapyr has soil residual activity and may impact restoration efforts. Add a spray adjuvant. Some kochia populations have developed resistance to related herbicides. Where resistance is suspected, use other herbicides or combinations.
Metsulfuron *Escort*	**Rate:** 1 to 2 oz product/acre (0.6 to 1.2 oz a.i./acre) **Timing:** Preemergence or postemergence from the rosette up until flower bud stage. **Remarks:** Always use a surfactant. Metsulfuron can be tank-mixed with 2,4-D for quicker burndown. Other premix formulations of metsulfuron can be used at similar application timing. These include *Cimarron Max* (metsulfuron + dicamba + 2,4-D) and *Cimarron X-tra* (metsulfuron + chlorsulfuron). Some kochia populations have developed resistance to related herbicides. Where resistance is suspected, use other herbicides or combinations. Metsulfuron is not registered for use in California.
Propoxycarbazone-sodium *Canter R+P*	**Rate:** 0.9 to 1.2 oz product/acre (0.63 to 0.84 oz a.i./acre) **Timing:** Postemergence to young, rapidly growing plants. **Remarks:** Propoxycarbazone is a broad-spectrum herbicide that will control many species. It will provide only partial control of kochia. Perennial grass species vary in tolerance. A non-ionic surfactant should be added at 0.25 to 0.5% v/v solution.
Rimsulfuron *Matrix*	**Rate:** 4 oz product/acre (1 oz a.i./acre) **Timing:** Preemergence or postemergence to small plants. **Remarks:** Rimsulfuron controls several annual grasses and broadleaves. Perennial grasses are tolerant to fall applications when established and grown under dryland conditions. Application to rapidly growing or irrigated perennial grasses may result in their injury or death. It provides soil residual control in cool climates but degrades rapidly under warm conditions. Rimsulfuron will not control summer annual weeds when applied in fall or spring. Add a surfactant when applying postemergence. Some kochia populations have developed resistance to related herbicides. Where resistance is suspected, use other herbicides or combinations.
PHOTOSYNTHETIC INHIBITORS	
Hexazinone *Velpar L*	**Rate:** 2 to 6 pt product/acre (0.5 to 1.5 lb a.i./acre) **Timing:** Preemergence. **Remarks:** Hexazinone has mixed selectivity and has fairly long soil residual activity. High rates of hexazinone can create bare ground, so only use high rates in spot treatments.

Brachypodium distachyon (L.) Beauv.
Annual false-brome

Family: Poaceae
Range: Common in California, but scattered throughout the other western states, including Oregon, Colorado and Texas.
Habitat: Dry slopes and fields, roadsides, disturbed grassland, margins of shrub thickets. Tolerates thin rocky soil and partial shade in oak woodlands.
Origin: Native to southern Europe. Likely an accidental introduction into North America.
Impacts: It is a poor forage grass because of its fibrous stems, little foliage, and firm spikelets with awned florets.
California Invasive Plant Council (Cal-IPC) Inventory: Moderate Invasiveness

Annual false-brome is a winter annual to 2 ft tall, with spikes that are often tinged purplish. Plants are sometimes branched at the base. The stems are erect or flat with the tips turning upward, usually lacking hairs except at the densely hairy nodes. The ligule is membranous and 2 to 3 mm long, the top irregularly jagged or fringed with short hairs, and the upper surface minutely hairy.

The inflorescence is a spike-like raceme, 1 to 3 inches long, with 1 to 6 spikelets per stem. Spikelets are nearly sessile, alternate, ascending to erect, laterally flattened, mostly 1 to 1.5 inches long, 0.25 to 0.5 inch wide. The spikelet contains numerous florets with straight awns 0.25 to 0.75 inch long. Plants reproduce only by seed. The florets fall near the parent plant and probably disperse to greater distances with animals, vehicle tires, and human activities. Germination occurs primarily in fall or early winter after the first significant rain of the season. There is no information about the longevity of the seed in the soil, but it is expected to be a couple of years.

NON-CHEMICAL CONTROL

Mechanical (pulling, cutting, disking)	Small infestations can be controlled with hand pulling, mowing, or tillage. The timing of mowing should be before viable seed production but after most soil moisture has been depleted to prevent regrowth. Shallow cultivation shortly after the main flush of germination and again a little later can eliminate most seedlings.
Cultural	Prescribed burning in early summer when plants were capable of carrying a fire was shown to significantly reduce the population of annual false-brome. Grazing has not been tested on the control of this species, but due to its poor forage quality and low palatability, such a strategy would likely require short-duration, high-intensity grazing just before seedhead production. Poor grazing practices would be expected to increase populations.
Biological	No known biological agents are available for the control of annual false-brome.

CHEMICAL CONTROL

There are no specific reports on the chemical control of annual false-brome. It is expected that control strategies will be similar to those for other annual grasses, including the various *Bromus* species. Directions for use may vary between brands; see label before use. Herbicides are listed by mode of action and then alphabetically. The order of herbicide listing is not reflective of the order of efficacy or preference.

LIPID SYNTHESIS INHIBITORS	
Fluazifop *Fusilade*	**Rate:** Broadcast foliar treatment: 1 to 1.5 pt product/acre (4 to 6 oz a.i./acre). Spot treatment: 0.5% v/v solution **Timing:** Early postemergence before boot stage. **Remarks:** Fluazifop is grass-selective and will not damage broadleaf species. It has no soil activity. To select for perennial grasses, apply before perennials emerge. Include crop oil concentrate surfactant or non-ionic surfactant.
Sethoxydim *Poast*	**Rate:** 1.5 to 2 pt product/acre (4.5 to 6 oz a.i./acre) **Timing:** Postemergence before boot stage.

	Remarks: Sethoxydim is grass-selective and will not damage broadleaf species. It has no soil activity. To select for perennial grasses, apply before perennials emerge. Include crop oil concentrate surfactant.
AROMATIC AMINO ACID INHIBITORS	
Glyphosate *Roundup, Accord XRT II*, and others	**Rate:** 1 to 2 pt product (*Roundup ProMax*)/acre (0.56 to 1.1 lb a.e./acre) **Timing:** Postemergence in late winter to early spring. Apply to rapidly growing, non-stressed plants after most seedlings have emerged. If possible, apply before desirable perennials have emerged. **Remarks:** Glyphosate is nonselective and has no soil activity.
BRANCHED-CHAIN AMINO ACID INHIBITORS	
Imazapic *Plateau*	**Rate:** 4 to 12 oz product/acre (1 to 3 oz a.e./acre) **Timing:** Preemergence in late summer or fall, or postemergence in early spring. In colder climates, spring applications after snow melt are better than fall treatments. The stage of growth of annual grasses can be critical to the effectiveness of imazapic. Late winter applications when annual grasses have 1 to 4 leaves and have not tillered can be effective, but when grasses are larger control is not as effective. **Remarks:** Imazapic has mixed selectivity. It tends to favor members of the Asteraceae. It has long soil residual activity. Imazapic can tie up in litter and its efficacy may be very much reduced under situations where there is heavy thatch on the soil surface. Imazapic is not registered for use in California.
Sulfometuron *Oust* and others	**Rate:** 0.75 to 5 oz product/acre (0.56 to 3.75 oz a.i./acre) **Timing:** Preemergence or postemergence. Fall and spring applications can both be effective, but fall applications may give full season control. In locations receiving less than 20 inches of annual precipitation, rates for control of annual bromes range from 0.75 to 2 oz product/acre; in locations receiving more than 20 inches annual precipitation, rates range from 3 to 5 oz product/acre. **Remarks:** Sulfometuron has mixed selectivity. It can cause minor damage to some native perennial grasses and has fairly long soil residual activity. Higher rates may increase control but will also give more bare ground.
Sulfosulfuron *Outrider*	**Rate:** 1.33 to 2 oz product/acre (1 to 1.5 oz a.i./acre) **Timing:** Early postemergence. **Remarks:** Sulfosulfuron has mixed selectivity and has fairly long soil residual activity. It is a newer herbicide so little is known of how it will act in many areas.
CONTACT PHOTOSYNTHETIC INHIBITORS	
Paraquat *Gramoxone*	**Rate:** 0.75 to 2 pt product/acre (3 to 8 oz a.i./acre) **Timing:** Postemergence to rapidly growing plants. More effective on smaller plants. **Remarks:** Paraquat is nonselective on annual species. It is a non-systemic herbicide that only kills contacted foliage and has no soil activity. Paraquat is a restricted use herbicide that is highly toxic to animals. The formulation has a stenching agent.

Brachypodium sylvaticum (Huds.) Beauv.

Slender false-brome

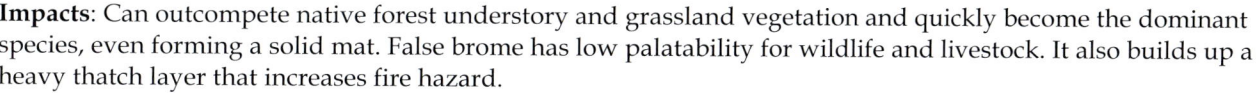

Family: Poaceae
Range: Currently found only in Oregon, Washington and California.
Habitat: Shaded woodlands, coniferous forest understories, open prairies, and roadsides. Grows well under both sun and shade conditions, in dry or moist soils.
Origin: Native to Eurasia and North Africa. Apparently escaped from cultivation as an ornamental.
Impacts: Can outcompete native forest understory and grassland vegetation and quickly become the dominant species, even forming a solid mat. False brome has low palatability for wildlife and livestock. It also builds up a heavy thatch layer that increases fire hazard.
Western states listed as Noxious Weed: Oregon
California Invasive Plant Council (Cal-IPC) Inventory: Moderate Invasiveness (Alert)

Slender false-brome is a perennial bunchgrass from 1 to 2.5 ft tall. It can remain green throughout the year under favorable conditions. The leaves are broad (up to 0.5 inch wide), flat and bright green, and the culms (stems) are soft-hairy at the nodes, as are the sheath and margins of the sheath. Ligules are membranous and about 1 to 2.5 mm long.

The spikelets are pale green and droop at the tips of the inflorescence. The number of spikelets per inflorescence is only between 5 and 10. Each spikelet is 1 to 2 inches long with 7 to 17 flowers per spikelet. The lemmas have a straight awn 0.5-0.75 inch long. Although stems can resprout from small stem or root fragments when cut, slender false-brome reproduces primarily by seed. The seedbank apparently only survives a year or so.

Brachypodium can be distinguished from *Bromus* because it has an open sheath, unlike the closed sheath of *Bromus*. In addition, the spikelets of *Brachypodium* are sessile or have a very short pedicel, whereas those of *Bromus* are on long pedicels.

NON-CHEMICAL CONTROL

Mechanical (pulling, cutting, disking)	Because the roots are fairly weak for a perennial bunchgrass, it is possible to mechanically remove the plants with a shovel, fork or other instrument when the soil is moist. Hand pulling has been successful at some sites, but generally only when patches are very small. Hand pulling in larger patches has been very labor intensive and has not resulted in lasting success. Hand pulling small patches is best in April and early May and needs to be repeated. It is important to remove the entire root system, or the plant will resprout.
	Mowing alone is unlikely to eliminate the plants, but can reduce the population provided it is repeated often. Repeated mowing should be conducted before the seeds become viable to eventually exhaust the short-lived seedbank. If native and/or rare species are present, they must be allowed to go through reproductive cycle before mowing.
Cultural	Grazing is not recommended because slender false-brome contains a fungal endophyte (*Epichloe sylvatica*) that produces an alkaloid toxic to mammals. Toxicity in sheep is known as "sheep-stagger". In Europe, however, slender false-brome was absent in heavily grazed sites, suggesting that repeated aboveground removal may eventually eliminate this species.
	Prescribed burning has not been an effective control method. In fact, slender false-brome is frequently found in recently burned sites. Any treatment, however, that can remove the above-ground biomass and deplete the seedbank will eventually reduce the population. Such treatments can also be used in conjunction with herbicide treatments to increase the efficacy of the herbicide.
Biological	European populations of the false-brome endophyte *Epichloe sylvaticum* are sexual and produce a disease called "choke" in false-brome. "Choke" can reduce or even prevent seed production. However, the North American populations of *E. sylvaticum* are an asexual strain and do not produce "choke."

CHEMICAL CONTROL

The following specific use information is based on reports by researchers and land managers. Other trade names may be available, and other compounds also are labeled for this weed. Directions for use may vary between brands; see label before use. Herbicides are listed by mode of action and then alphabetically. The order of herbicide listing is not reflective of the order of efficacy or preference.

LIPID SYNTHESIS INHIBITORS	
Fluazifop *Fusilade*	**Rate:** 8 to 16 oz product/acre (2 to 4 oz a.i./acre)
	Timing: Postemergence. Spring is best to control seedlings, but established plants can be treated in mid-summer to fall.
	Remarks: Two to three years of treatment are necessary. Fine-leaf fescues are tolerant of fluazifop. Fluazifop is a grass-specific herbicide and will not impact native forbs.
Sethoxydim *Poast*	**Rate:** 1 to 1.5 pt product/acre (3 to 4.5 oz a.i./acre)
	Timing: Postemergence. Spring is best to control seedlings, but established plants can be treated in mid-summer to fall.
	Remarks: Two to three years of treatment are necessary. Fine-leaf fescues are tolerant to sethoxydim. The activity is very similar to fluazifop. Sethoxydim is a grass-specific herbicide and will not impact native forbs.
AROMATIC AMINO ACID INHIBITORS	
Glyphosate *Roundup, Accord XRT II*, and others	**Rate:** Broadcast foliar treatment: 2 to 3.3 qt product (*Roundup ProMax*)/acre (2.25 to 3.7 lb a.e./acre). Spot treatment: 1.5% v/v solution
	Timing: Postemergence. Mid-summer to fall is the best timing to kill slender false-brome and minimize injury to natives that are dormant or have completed their life cycle. It may be necessary to make a follow-up application in the subsequent spring to control germinating seedlings.
	Remarks: Glyphosate is considered the best option for control in pure stands. Two to three years of treatment are necessary. Herbicide treatment can be used after repeated mowing to reduce the necessity for a spring treatment to kill seedlings.
PHOTOSYNTHETIC INHIBITORS	
Hexazinone *Velpar L*	**Rate:** 1 gal product/acre (2 lb a.i./acre)
	Timing: Preemergence in spring to control seedlings.
	Remarks: Hexazinone is typically recommended in combination with glyphosate. Glyphosate is used in mid-summer to fall to control existing plants, and hexazinone is used the following spring to control new germinants. High rates of hexazinone can create bare ground, so only use high rates in spot treatments.

Brassica nigra (L.) Koch

Black mustard

Family: Brassicaceae
Range: Throughout the U.S. and in all western states except Wyoming.
Habitat: Roadsides, fields, disturbed waste places, and grasslands, especially in coastal areas. Mostly inhabits areas with a mild winter climate in its native range.
Origin: Native to Europe. Introduced by the Spanish as a spice crop. The seeds of cultivars are still used to produce mustard oil.
Impact: In coastal grasslands, dense stands of black mustard outcompete native vegetation. Black mustard appears to have allelopathic properties. It is adapted to periodic fires and newly burned sites are subject to invasion. The high biomass contributes to increased fuel load and fire frequency. Black mustard contains glucosinolates, sulfur compounds that can irritate the digestive tract and cause thyroid dysfunction when consumed in large quantities over time. Toxicity problems in livestock arise when large quantities of seeds are ingested or when animals are confined to pastures that consist primarily of mustard family species.
California Invasive Plant Council (Cal-IPC) Inventory: Moderate Invasiveness

Black mustard is an erect winter annual to 6 ft tall. The basal leaves mostly have 1 to 2 pairs of distinct lateral lobes at the base, with the terminal lobe much larger than the lateral lobes. The upper stem leaves are oblong to linear, the base tapered, and the margins entire to toothed or weakly lobed.

The four petals are bright yellow, 6 to 11 mm long, and plants flower from mid-spring to mid-summer. Mature fruits are linear, 0.5 to 1 inch long, and erect, usually lying close to the stem. Plants reproduce only by seed. Most seeds fall near parent plants when fruits open at maturity. Many mustard species develop a large, persistent seedbank. Deeply buried seeds of black mustard can survive for 50 years or more. Seeds nearer to the soil surface are not as long-lived under field conditions.

NON-CHEMICAL CONTROL

Mechanical (pulling, cutting, disking)	Plants can be hand pulled or removed by other tools before they produce seed. Yearly manual removal of plants before seeds mature can eventually deplete the seedbank.
	Tillage can be used to manage black mustard in the seedling stage. Tillage should be done before black mustard has set seed. Shallow tillage is preferred over deep tillage. Deep tillage can bury weed seeds to depths where they can remain dormant for many years and become a problem at a later date.
Cultural	There is no information on the effectiveness of grazing for the control of black mustard. However, it is speculated that the plants must be readily eaten by livestock because big stands are seldom found on native ranges other than those lightly grazed. In the United States, black mustard is most common on areas protected from grazing. Reports indicate that plants are fairly palatable to sheep and cattle.
	Burning and other kinds of disturbance usually favor the increase of mustard species. Seeds on the soil during a grassland fire are not likely to be killed by the heat of the burn.
Biological	Because of the close relationship of black mustard with many important crop plants in the genus *Brassica*, there are no biological control efforts in the United States.

CHEMICAL CONTROL

The following specific use information is based on published papers and reports by researchers and land managers. Other trade names may be available, and other compounds also are labeled for this weed. Directions for use may vary between brands; see label before use. Herbicides are listed by mode of action and then alphabetically. The order of herbicide listing is not reflective of the order of efficacy or preference.

GROWTH REGULATORS	
2,4-D	**Rate:** 1 to 2 pt product/acre (0.5 to 1 lb a.e./acre)
Several names	**Timing:** Postemergence when weeds are small and rapidly growing.

	Remarks: 2,4-D is broadleaf-selective and has no soil activity.
Aminocyclopyrachlor + chlorsulfuron *Perspective*	**Rate:** 3 to 8 oz product/acre **Timing:** Preemergence or early postemergence when weeds are germinating or actively growing. **Remarks:** *Perspective* provides broad-spectrum control of many broadleaf species. Although generally safe to grasses, it may suppress or injure certain annual and perennial grass species. Do not treat in the root zone of desirable trees and shrubs. Do not apply more than 11 oz product/acre per year. At this high rate, cool-season grasses will be damaged, including bluebunch wheatgrass. Not yet labeled for grazing lands. Add an adjuvant to the spray solution. This product is not approved for use in California and some counties of Colorado (San Luis Valley).
Dicamba *Banvel, Clarity*	**Rate:** 0.5 to 1.5 pt product/acre (4 to 12 oz a.e./acre) **Timing:** Postemergence when weeds are small and rapidly growing. Use low rate for small rapidly growing weeds-higher rate for large mustards. **Remarks:** Dicamba is broadleaf-selective with little soil activity. Dicamba is available mixed with diflufenzopyr in a formulation called *Overdrive*. This has been reported to be effective on some mustards. Diflufenzopyr is an auxin transport inhibitor which causes dicamba to accumulate in shoot and root meristems, increasing its activity. *Overdrive* is applied postemergence to rapidly growing plants at 4 to 8 oz product/acre. Higher rates should be used on large annuals. Add a non-ionic surfactant to the treatment solution at 0.25% v/v or a methylated seed oil at 1% v/v.
Fluroxypyr *Vista XRT*	**Rate:** 22 oz product/acre (7.7 oz a.e./acre) **Timing:** Postemergence when weeds are small and rapidly growing. **Remarks:** Only effective when applied postemergence. Gives suppression of mustards. Can also be used in a premix with picloram (*Surmount*), but this formulation is not registered for use in California.
Triclopyr *Garlon 4 Ultra*	**Rate:** 1 to 8 qt product/acre (1 to 8 lb a.e./acre) **Timing:** Postemergence when weeds are small and rapidly growing. Higher rates are needed on more mature plants. **Remarks:** Triclopyr is broadleaf-selective and may injure other desirable species. The ester formulation (*Garlon 4 Ultra*) is more effective compared to the amine formulation. Use rate for mustard should not exceed 4 lb a.e./acre.
AROMATIC AMINO ACID INHIBITORS	
Glyphosate *Roundup, Accord XRT II,* and others	**Rate:** Spot treatment: 2% v/v solution for spot application **Timing:** Best treated postemergence when plants are small and are growing rapidly, but before flowering. **Remarks:** Some studies show that it only gives fair control on mustards. Best on seedling plants.
BRANCHED-CHAIN AMINO ACID INHIBITORS	
Chlorsulfuron *Telar*	**Rate:** 1 to 2.6 oz product/acre (0.75 to 1.95 oz a.i./acre) **Timing:** Preemergence or early postemergence, when weeds are germinating or actively growing. **Remarks:** Chlorsulfuron is primarily active on broadleaf species. It gives very effective control of most mustards, except shortpod mustard. It has fairly long residual soil activity. Do not apply more than 1.33 oz product/acre per year in pasture, range, and CRP, or 2.6 oz product/acre per year in noncrop.
Propoxycarbazone-sodium *Canter R+P*	**Rate:** 0.9 to 1.2 oz product/acre (0.63 to 0.84 oz a.i./acre) **Timing:** Postemergence to young, rapidly growing plants. **Remarks:** Propoxycarbazone is a broad-spectrum herbicide that will control many species, including black mustard. Perennial grass species vary in tolerance. A non-ionic surfactant should be added at 0.25 to 0.5% v/v solution.
Rimsulfuron *Matrix*	**Rate:** 4 oz product/acre (1 oz a.i./acre) **Timing:** Preemergence in spring or fall depending on the timing of germination. **Remarks:** Controls several annual grasses and broadleaves. Perennial grasses are tolerant to fall applications when established and grown under dryland conditions. Application to rapidly growing or irrigated perennial grasses may result in injury or death of the crop. Provides soil residual control in

	cool climates but degrades rapidly under warm conditions. Rimsulfuron will not control summer annual weeds when applied in fall or spring. Moisture is necessary for activation and the best results occur when precipitation is within 14 to 21 days of application.
Sulfometuron *Oust* and others	**Rate:** 3 to 5 oz product/acre (2.25 to 3.75 oz a.i/acre) **Timing:** Preemergence or early postemergence. **Remarks:** Sulfometuron is a broad-spectrum herbicide with long soil residual activity. Provides longer control in areas with 20 inches of annual rainfall or more.

Brassica tournefortii Gouan.

Saharan mustard

Family: Brassicaceae
Range: Throughout the southwestern United States, including California, Nevada, Arizona, New Mexico and Texas.
Habitat: Roadsides, washes, open fields, annual grasslands, coastal sage scrub, and desert shrubland. Typically grows in arid climate areas on sandy soil and where competing vegetation is sparse. Inhabits coastal and inland dunes in its native range.
Origin: Native to the Mediterranean region.
Impacts: Saharan mustard is especially problematic in the Sonoran Desert, including the Imperial Valley. It readily spreads from roadsides and other disturbed places into washes, drainages, desert shrubland, and sensitive dune areas. Saharan mustard stands contribute to increased fuel load and fire frequency. Increasing the fire frequency can lead to the type conversion of desert scrub to grassland. Because desert systems often contain rare and endangered species, Saharan mustard can be a significant threat to these species. Like other mustards, Saharan mustard can also harbor diseases and pests that attack closely related crops in the mustard family. The foliage, roots, and especially seeds of *Brassica* and many related species contain glucosinolates, which are sulfur-containing compounds that can irritate the digestive tract and cause thyroid dysfunction when consumed in large quantities over time. Toxicity problems in livestock arise when large quantities of seeds are ingested or when animals are confined to pastures that consist primarily of mustard family species. Symptoms can include colic, diarrhea, excessive salivation, and thyroid enlargement.
California Invasive Plant Council (Cal-IPC) Inventory: High Invasiveness

Erect winter annual, to 4 ft tall or more. Exists as a basal rosette until flowering stems develop at maturity. Basal leaves deeply pinnate-lobed, typically with more lobed pairs (6 to 14 pairs) than most mustard species. Lower stems have are characterized by having dense, stiff white hairs.

Inflorescences in racemes with 4-petaled pale yellow flowers (4 to 8 mm long) and long linear fruits (1.5 to 3.5 inches long). Mature fruits strongly constricted between the seeds and appearing beaded. Fruits open from the base to release seeds. Plants reproduce only by seed. Most seeds fall near parent plants when fruits open at maturity. Sometimes seeds disperse when dried plant stems break at ground level and tumble under windy conditions. Seeds become slightly sticky with mucilage when moistened with water. Like many mustards, Saharan mustard probably develops a large, persistent seedbank, with seeds that can survive for many years in the soil.

NON-CHEMICAL CONTROL

Mechanical (pulling, cutting, disking)	Hand pulling has been used successfully, but is labor intensive and must be conducted after bolting but before seeding. This leaves a narrow timing window and requires repeated monitoring and visits. A hula hoe can be effective for smaller plants in the early rosette stage. Flaming in winter has also been used for small patches. Best used during a rain event. Roadside grading with heavy equipment can also control the plant, but this too has to be conducted before seed development.
Cultural	Sheep, cattle, and goats will all graze mustard species. There are no studies to demonstrate their effectiveness. Burning is not a recommended tool since because plants are often found in desert regions and the fuel needed to carry a fire would likely be after seed production was completed.
Biological	There are no biological control agents available. However, current research efforts are underway to identify insects in Europe that feed on Saharan mustard. If suitable species are found they could eventually be imported and released to control infestations in the United States.

CHEMICAL CONTROL

The following specific use information is based on reports by researchers and land managers. Other trade names may be available, and other compounds also are labeled for this weed. Directions for use may vary between brands; see label before use. Herbicides are listed by mode of action and then alphabetically. The order of herbicide listing is not reflective of the order of efficacy or preference.

GROWTH REGULATORS	
2,4-D Several names	**Rate:** 1.5% v/v solution for spot application **Timing:** Postemergence when plants are small and are growing rapidly, but before flowering. **Remarks:** 2,4-D is broadleaf-selective with no soil activity. It is available in a premix with dicamba (trade name *Veteran 720*) and several other premix products. 2,4-D may be a restricted use herbicide in some areas.
Triclopyr *Garlon 3A, Garlon 4 Ultra*	**Rate:** 2% v/v solution for spot application **Timing:** Postemergence when plants are small (rosette stage) and growing rapidly, but before flowering. **Remarks:** Triclopyr has been shown to be a very effective control option. It is broadleaf-selective.
AROMATIC AMINO ACID INHIBITORS	
Glyphosate *Roundup, Accord XRT II*, and others	**Rate:** 2% v/v solution for spot application **Timing:** Postemergence when plants are small and growing rapidly, but before flowering. **Remarks:** Some studies show that glyphosate only gives fair control. It is best on plants in seedling stage. String trimming followed by glyphosate has been shown to be effective. Glyphosate is nonselective and has no soil activity. It can be used in combination with imazapic (premix trade name of *Journey*).
BRANCHED-CHAIN AMINO ACID INHIBITORS	
Chlorsulfuron *Telar*	**Rate:** 1 to 2 oz product/acre (0.75 to 1.5 oz a.i./acre) **Timing:** Preemergence, or postemergence to rosettes. **Remarks:** Chlorsulfuron is primarily active on broadleaf species and is very effective on Saharan mustard. It has some residual soil activity.
Imazapic *Plateau*	**Rate:** 4 to 10 oz product/acre (1 to 2.5 oz a.e./acre) **Timing:** Preemergence, or postemergence to rosettes. **Remarks:** Imazapic gives effective control with soil residual activity. It can be used in combination with glyphosate (premix trade name of *Journey*). Imazapic is not registered for use in California.
Metsulfuron *Escort*	**Rate:** 1 to 2 oz product/acre (0.6 to 1.2 oz a.i./acre) **Timing:** Preemergence, or postemergence to rosettes. **Remarks:** Metsulfuron has some soil residual activity. It is not registered for use in California.
PHOTOSYNTHETIC INHIBITORS	
Hexazinone *Velpar L*	**Rate:** 1 to 2 pt product/acre (0.25 to 0.5 lb a.i./acre) **Timing:** Typically applied preemergence. **Remarks:** There is no direct evidence of control but hexazinone is effective on other mustards, including *Descurainia* spp. High rates of hexazinone can create bare ground, so only use high rates in spot treatments.

Bromus diandrus Roth; ripgut brome
Bromus madritensis L. ssp. *rubens* (L.) Husnot; red brome
   (= *B. rubens* L.)
Bromus tectorum L.; downy brome or cheatgrass

Ripgut, red, and downy brome (cheatgrass)

Family: Poaceae
Range: Ripgut brome and red brome occur in most of the western states, except Wyoming, South and North Dakota (and Colorado for red brome). Downy brome is the most widespread of the bromes and is found throughout the U.S., particularly in the western states.
Habitat: Open disturbed areas, roadsides, fields, rangelands, agronomic crops, orchards, forestry sites, and many natural communities. They are often in areas with dry sandy soils where there is less competition with other vegetation, including desert communities. They are also common in urban waste places and can grow in most soil types. Red brome is more sensitive to winter cold climates than downy brome. Downy brome can be found as an agricultural weed in irrigated and non-irrigated cool season crops such as alfalfa and cereals.
Origin: All species are native to Eurasia.
Impact: These bromes suppress other grasses in rangeland, can infest agricultural fields, and have sharp florets that can injure grazing animals. These bromes may serve as a minor source of livestock forage early in the season, but they suppress perennial grasses that would extend the grazing season. All are fire promoters. In particular, red brome and downy brome are common in fire-sensitive arid to semiarid communities such as coastal scrub, desert shrubland, pinyon pine-juniper communities, and three-needle pine woodlands. In these ecosystems, native woody species are poorly adapted to increased fire frequencies and often decline following infestation.
Western states listed as Noxious Weed: *B. tectorum*, Colorado
California Invasive Plant Council (Cal-IPC) Inventory: *B. diandrus*, Moderate Invasiveness; *B. madritensis* and *B. tectorum*, High Invasiveness

These are cool-season annual grasses with sharp florets and straight awns. Red brome and downy brome, in particular, have a shorter life cycle than most grasses. In all these grasses, the leaves and especially the leaf sheaths are typically covered with short, soft hairs. Mature stems of ripgut brome grow to 2.5 ft tall, with leaf blades 2 to 7 mm wide and open, loose, nodding panicles 2.4 to 10 inches long. The florets have very long awns (1.4 to 2.2 inches). Red brome is shorter, to 1.5 ft tall, with leaf blades 1 to 4 mm wide; it produces much more compact, dense panicles 1 to 3 inches long, typically dark red. Mature stems of downy brome grow up to 1.3 ft tall, with leaf blades 1 to 6 mm wide and open, loose panicles 2.5 to 9 inches long, usually drooping to one side.

These weedy bromes flower in spring. They become reddish to purplish as the inflorescences mature. Red brome turns an especially dark red. Soon after maturation, the florets disperse short distances with wind and rodent activity or to greater distances by clinging to the fur, feathers, and feet of animals and to the shoes and clothing of people. Most seeds germinate the following fall after the first significant rain. Seeds typically survive in soil 2 to 3 years under field conditions, but some seeds may survive up to 5 years. Thatch accumulation or shallow burial favors establishment of germinating seeds.

NON-CHEMICAL CONTROL

Mechanical (pulling, cutting, disking)	Individual plants or small patches can be pulled by hand or hoed in early spring before seeds are ripe.
	Mowing is not usually recommended, but can reduce seed production if conducted shortly after flower initiation and before seeds mature. Plants cut earlier will regrow. Plants should be mowed to about 2 inches

	with the bolting stems removed. Repeated mowing (every 3 weeks) can eliminate seed production in areas where herbicide applications are unacceptable. Shallow cultivation shortly after the main flush of germination and again a little later can eliminate most seedlings. In agricultural fields, tilling soil with a method that deeply buries seeds in fall or early spring can help control bromes. In the Great Basin, tillage to establish summer fallows is a useful method of suppressing bromes and allowing establishment of perennial grasses.
Cultural	Overgrazing or frequent soil disturbance can increase dominance of bromes by reducing or eliminating more desirable forage species. Moderate grazing can be effective when used in combination with herbicides. Ripgut brome is susceptible to burning, if the burn is conducted before seeds mature. Burns should be conducted in late spring when most desirable vegetation is drying down but before ripgut brome heads shatter. Burning later, after seed dispersal, can increase ripgut brome densities. Red brome and downy brome mature earlier in the season, so burns usually result in an increase in these species. Burning can be effective as part of a 2 to 3 year integrated management program incorporating a spring burn, winter reseeding with native perennial grasses, and early spring application of herbicides.
Biological	There are no established biocontrol agents for the weedy bromes. Several soil fungi have been tested for their suppressive effect on downy brome. None have proven effective. A rhizobacterium native to Washington's soils, *Pseudomonas fluorescens* strain D7 (P.f. D7), has been shown to inhibit germinating cheatgrass, offering hope of managing the spread of this highly invasive species. Studies of the efficacy of this organism under a range of environmental conditions are under way to determine if this bacterium could inhibit cheatgrass across the western United States. Results are too preliminary to determine if it will be effective.

CHEMICAL CONTROL

The following specific use information is based on published papers and reports by researchers and land managers. Other trade names may be available, and other compounds also are labeled for this weed. Directions for use may vary between brands; see label before use. Herbicides are listed by mode of action and then alphabetically. The order of herbicide listing is not reflective of the order of efficacy or preference.

GROWTH REGULATORS	
Growth regulator herbicides	Although they do not generally kill annual grasses, many of the growth regulator herbicides, particularly aminopyralid and picloram, have been shown to reduce seed production in downy brome. In addition, aminopyralid and aminocyclopyrachlor have been shown in some studies to provide good preemergence control of downy brome at higher rates.
LIPID SYNTHESIS INHIBITORS	
Clethodim *Select, Envoy*	**Rate:** 6 to 8 oz product (*Select*)/acre (1.5 to 2 oz a.i./acre) for seedlings; 0.25% to 0.5% of product v/v in spot treatment **Timing:** Postemergence. Best when applications are made before plants are 6 inches tall. It is less effective if applied after a mowing. **Remarks:** Clethodim is grass-selective and safe on broadleaf species. To select for perennial grasses, apply before perennials emerge. It has no soil activity. Use a crop oil surfactant. Registered for fallow and non-crop areas, not generally for rangeland/natural areas, but has specific-use supplemental labels. Note that *Envoy* formulation is 1 lb a.i./gallon, *Select* is 2 lb a.i./gallon.
Fluazifop *Fusilade*	**Rate:** 1 to 1.5 pt product/acre (4 to 6 oz a.i./acre) for established plants, 8 oz product/acre (2 oz a.i./acre) for seedlings; 0.5% product v/v in spot treatment **Timing:** Postemergence. Best when applications are made before the boot stage. **Remarks:** Fluazifop is grass-selective and safe on broadleaf species. To select for perennial grasses, apply before perennials emerge. It has no soil activity. Use a crop oil surfactant. Registered for fallow and non-crop areas, not generally for rangeland/natural areas, but has specific-use supplemental labels.
AROMATIC AMINO ACID INHIBITORS	
Glyphosate *Roundup, Accord XRT II,* and others	**Rate:** 0.33 to 1 qt product (*Roundup ProMax*)/acre (0.375 to 1.1 lb a.e./acre) **Timing:** Postemergence in early spring to rapidly growing, non-stressed plants after most seedlings have emerged. If possible, apply before desirable perennials emerge.

	Remarks: Glyphosate is a nonselective herbicide. It has no soil activity.
BRANCHED-CHAIN AMINO ACID INHIBITORS	
Imazapic *Plateau*	**Rate:** 4 to 12 oz product/acre (1 to 3 oz a.e./acre) **Timing:** Preemergence to very early postemergence (3 leaves maximum) from fall to early spring. **Remarks:** Imazapic has long soil residual activity and mixed selectivity. It tends to favor members of the Asteraceae and some grasses. Use a spray adjuvant for postemergence applications. Effects vary depending on soil texture and soil organic matter. Heavy soils and high organic matter may require higher rates. Imazapic also can tie up in litter, and its efficacy is reduced under situations where there is thick thatch on the soil surface. Not registered for use in California.
Imazapyr *Arsenal, Habitat, Chopper, Stalker, Polaris*	**Rate:** 2 to 3 pt product/acre (8 to 12 oz a.e./acre) **Timing:** Preemergence or postemergence. **Remarks:** Imazapyr has fairly long soil residual activity. It is a nonselective herbicide.
Propoxycarbazone-sodium *Canter R+P*	**Rate:** 1.2 oz product/acre (0.84 oz a.i./acre) **Timing:** Postemergence from the 2-leaf to 2-tiller stage when plants are growing rapidly. **Remarks:** Propoxycarbazone is a broad-spectrum herbicide that will control many species, including downy brome and ripgut brome. Perennial grass species vary in tolerance. A non-ionic surfactant should be added at 0.25 to 0.5% v/v solution.
Rimsulfuron *Matrix*	**Rate:** 2 to 4 oz product/acre (0.5 to 1 oz a.i./acre) **Timing:** Preemergence in fall to early postemergence in early spring. **Remarks:** Rimsulfuron controls several annual grasses and broadleaves. Perennial grasses are tolerant to fall applications when established and grown under dryland conditions. Application to rapidly growing or irrigated perennial grasses may result in injury or death of the crop. It provides soil residual control in cool climates but degrades rapidly under warm conditions. Rimsulfuron will not control summer annual weeds when applied in fall or spring. Add a surfactant when applying postemergence.
Sulfometuron *Oust* and others	**Rate:** 0.75 to 5 oz product/acre (0.56 to 3.75 oz a.i./acre) **Timing:** Preemergence or early postemergence from fall to early spring. Most effective control is with early postemergence treatment after brome seedlings have emerged. However, *Oust* will control large downy brome (e.g., 8 to 12 inches tall). **Remarks:** Sulfometuron has mixed selectivity and is fairly safe on native perennial grasses. It is good for revegetation use. Use lower rates in arid environments and higher rates in wetter areas (> 20 inches rainfall) and on high organic matter soils. Sulfometuron has fairly long soil residual activity. At higher rates, this treatment will generally result in bare ground.
Sulfometuron + chlorsulfuron *Landmark XP*	**Rate:** 0.75 oz product/acre **Timing:** Preemergence in fall or after soil thaws in spring. **Remarks:** See sulfometuron.
Sulfosulfuron *Outrider*	**Rate:** 0.75 to 2 oz product/acre (0.56 to 1.5 oz a.i./acre) **Timing:** Early postemergence, fall to early spring, when desirable perennials are dormant. **Remarks:** Sulfosulfuron has mixed selectivity, but is fairly safe on native perennial grasses, especially wheatgrasses. It has fairly long soil residual activity. Treatments should include a non-ionic surfactant.
PHOTOSYNTHETIC INHIBITORS	
Hexazinone *Velpar L*	**Rate:** 2 to 6 pt product/acre (0.5 to 1.5 lb a.i./acre) **Timing:** Preemergence to early postemergence. **Remarks:** Hexazinone has both foliar and soil activity. In soil applications, rates will vary with soil texture and soil organic matter. Best results result when applied to moist soils. Use rates will also vary depending on the weed species to be controlled. Hardwood trees near application site can absorb this chemical through the roots and may be injured or killed. Do not spray near the root zone of desirable hardwood trees or shrubs. High rates of hexazinone can create bare ground, so only use high rates in spot treatments.

Bromus inermis Leysser

Smooth brome

Family: Poaceae
Range: Most contiguous states, except a few southeastern states; very common in the northern mountain states.
Habitat: Forests and wooded areas, edges of fields, ditches, roadsides, overgrazed pastures, prairies, rangelands, grasslands, meadows, riverbanks, and other disturbed sites. Typically not shade tolerant.
Origin: Native to Eurasia. Introduced for cultivation in California in the late 1800s. Smooth brome is cultivated for forage and has been used for revegetation of burned sites.
Impacts: Because of its vegetative spread from rhizomes, stands of smooth brome can persist for many years. Dense populations can form a thick sod and exclude other desirable species, and often lead to nitrogen deficiency. In some cases it may desirable to use smooth brome to increase C:N (decrease available N) and favor early successional species. Smooth brome is also known to significantly impact the population dynamics and movement behavior of several native arthropod species in North American prairies. It may also serve as an alternate host for viral diseases of crops. Its deep root system allows it to tolerate prolonged drought conditions. However, smooth brome is considered a desirable forage for livestock and large wild ungulates in many western locations.

Smooth brome is a cool-season perennial to 3 ft tall or more, with an extensive creeping rhizome system. Stems are glabrous and leaves can be 6 to 16 inches long. Like many other members of the genus, the sheaths are closed. Ligules are membranous and usually less than 1 mm long.

Inflorescences are erect open panicles, 2 to 8 inches long, with one to four branches per node. The spikelets are nearly cylindrical, 1 to 1.5 inches long, glabrous, and often purple-brown. Although more commonly absent, when awns are present they are less than 3 mm long. Plants are self-incompatible. Smooth brome reproduces by seed and through rhizome development, which begins 3 weeks to 6 months after germination. Individual rhizomes are reported to have 1 year longevity. Seeds remain viable for 2 to 10 years.

NON-CHEMICAL CONTROL

Mechanical (pulling, cutting, disking)	A single well-timed close mowing in the boot stage is an effective method of control, and appears to be as effective as multiple mowings, although repeated mowing is often recommended. Ideal timing is when conditions are hot and moist, followed by a dry period.
Cultural	Burning can help control the spread of smooth brome, but does not appear to eliminate it. Most studies show that fire is not very effective for the control of smooth brome. The best time for burning has been reported to be both in the early spring when plants are in the four to five leaf stage, and in the boot to early bloom stage. Earlier timing allows resources to be used by native warm-season grasses. In some cases fire can injure its competitors allowing smooth brome to increase.
Biological	There are no biological control agents for smooth brome.

CHEMICAL CONTROL

The following specific use information is based on reports in published papers or by researchers and land managers. Other trade names may be available, and other compounds also are labeled for this weed. Directions for use may vary between brands; see label before use. Herbicides are listed by mode of action and then alphabetically. The order of herbicide listing is not reflective of the order of efficacy or preference.

LIPID SYNTHESIS INHIBITORS	
Fluazifop	**Rate:** 1 to 1.5 pt product/acre (4 to 6 oz a.i./acre)
Fusilade	**Timing:** Postemergence in spring.
	Remarks: Fluazifop is grass-selective and will not damage broadleaf species. It has no soil activity. Use

	crop oil concentrate surfactant.
AROMATIC AMINO ACID INHIBITORS	
Glyphosate *Roundup, Accord XRT II*, and others	**Rate:** Broadcast foliar treatment: 1.33 to 2.67 qt product (*Roundup ProMax*)/acre (1.5 to 3 lb a.e./acre). Spot treatment: 1.5% v/v solution **Timing:** Postemergence from May to June, before stems have elongated. **Remarks:** Glyphosate is a nonselective herbicide. Its effectiveness is increased by addition of ammonium sulfate. Best where smooth brome is growing in pure patches. Glyphosate has no soil activity.
BRANCHED-CHAIN AMINO ACID INHIBITORS	
Imazapic *Plateau*	**Rate:** 8 to 16 oz product/acre (2 to 4 oz a.e./acre) **Timing:** Postemergence before stem elongation. **Remarks:** Imazapic has soil residual activity. It is a broad-spectrum herbicide that tends to favor members of the Asteraceae and some grasses. It is not as effective on smooth brome as *Landmark XP* or *Oust*. Imazapic is not registered for use in California.
Imazapyr *Arsenal, Habitat, Stalker, Chopper, Polaris*	**Rate:** 2 to 3 pt product/acre (8 to 12 oz a.e./acre) **Timing:** Preemergence or early postemergence when plants are growing rapidly. **Remarks:** Imazapyr has soil residual activity. It is a broad-spectrum to nonselective herbicide.
Sulfometuron *Oust* and others	**Rate:** 3 to 5 oz product/acre (2.25 to 3.75 oz a.i./acre) **Timing:** Preemergence or early postemergence when plants are growing rapidly. **Remarks:** Sulfometuron has long soil residual activity and may move off-site in wind-blown soils. Typically considered a nonselective herbicide combination that often gives bare ground.
Sulfometuron + chlorsulfuron *Landmark XP*	**Rate:** 1.5 oz product/acre (0.75 oz a.i./acre sulfometuron + 0.375 oz a.i./acre chlorsulfuron) **Timing:** Postemergence before stem elongation. **Remarks:** The combination gives excellent control of smooth brome, particularly on younger plants. Both herbicides have soil residual activity and may move off-site in wind-blown soils. The sulfometuron component makes *Landmark* less selective than other herbicides, and higher labeled rates may cause bare ground.
Sulfosulfuron *Outrider*	**Rate:** 1 oz product/acre (0.75 oz a.i./acre) **Timing:** Postemergence before stem elongation. **Remarks:** Sulfosulfuron provides only fair control of smooth brome and is not as effective as *Landmark XP*.

Bromus hordeaceus L.; soft brome
Bromus japonicus Thunb. ex Murr.; Japanese brome

Soft brome (soft chess) and Japanese brome

Bromus japonicus

Bromus hordeaceus

Family: Poaceae
Range: Throughout the United States and in every western state.
Habitat: Soft chess is common in grasslands, sagebrush communities, rangeland, disturbed sites, orchards and vineyards. In addition to grasslands, Japanese brome is also present in desert shrub-grasslands, sagebrush communities, pinyon-juniper communities, low elevation coniferous forests, and cropland. Both species are most common on disturbed sites.
Origin: Both species are native to Eurasia.
Impacts: These bromes compete for limited spring moisture with desirable vegetation. They can reduce desirable vegetation cover and prevent establishment of native perennials and other forage crops. In rangelands, however, they are often considered excellent livestock forage.
California Invasive Plant Council (Cal-IPC) Inventory: *Bromus hordeaceus*, Limited Invasiveness

Both of these bromes are cool-season annual grasses with stems that grow to 3 ft tall. Soft brome leaves are mostly 3 to 10 mm wide, soft-hairy on both sides. Its sheath is often densely pubescent and it has a 1 to 2 mm long ligule. Japanese brome leaves are mostly 1.5 to 6 mm wide, glabrous to soft-hairy. Its sheath is sparsely hairy to pubescent and it has ligules 2 mm long. Roots of both species are fibrous, mostly in the top 6 inches of soil. Japanese brome roots have been reported to 5 ft deep under favorable conditions.

Spikelets on both species are narrowly ovoid, slightly flattened, breaking apart above glumes and between florets, with the central axis not visible in fruit. Both species have awns 4 to 11 mm long that are slender, weak, straight or curved. Soft brome has a dense panicle, mostly 1 to 5 inches long, and the spikelets are usually covered with short hairs. Japanese brome has an open panicle, 1.5 to 10 inches long, and spikelets that are glabrous or minutely scabrous. Reproduction is only by seed. The florets disperse by falling within the vicinity of the parent plant. Longer dispersal can be through soil movement, wildlife, human activities, and as crop seed contaminants. Both species can germinate and establish in thatch or litter. Most seeds germinate in fall when a moist substrate becomes available, although some seed will germinate in spring. Seed longevity in the soil depends on environmental factors. Japanese brome seeds have been reported to remain viable for several years. Land managers should expect seeds of both species to survive somewhere from 2 to 5 years.

NON-CHEMICAL CONTROL

Mechanical (pulling, cutting, disking)	Small infestations can be removed by manual methods. Digging and hand-pulling are effective. Uproot the entire plant before flowering.
	Both species will rapidly regrow after mowing at early growth stages. Mowing in August increased Japanese brome populations in Kansas tall grass prairie. Avoid soil disturbance; avoid mowing desirable vegetation that can shade infestations during active growth, as neither species is shade tolerant.
	Tillage will control emerged plants but often stimulates germination. Deep tillage can reduce populations by burying seed deep enough to prevent germination and emergence. Land managers using tillage for seedbed preparation during reseeding should prepare for a flush of seedlings when soils become moist.
Cultural	Both species are palatable and considered desirable forage early in the season. Soft brome is palatable later in the season. Forage production is unpredictable and depends greatly on yearly precipitation. Both species are persistent on grazed lands. Studies suggest grazing can temporarily increase or decrease abundance of both species depending on site conditions and grazing timing. Both species maintain viable seed banks after several years of grazing. Intensive short duration grazing at the boot to early heading stage has reduced abundance of other winter annual grass weeds.
	Burning before seed dispersal has been shown to temporarily reduce populations. Control was dependent on

	annual precipitation and site conditions. It is recommended to burn in the ripe seed stage before seed disperse from dead plants. Sites often repopulate from the seedbank and outlying areas post-fire. Fire suppression appears to promote the increase of Japanese brome, since germination and establishment is favored by the presence of a litter layer. Promoting competitive vegetation can slow spread and help prevent establishment. Perennial grass stand density and vigor should be managed to minimize bare ground exposure.
Biological	No known biological controls are available for either species in the United States.

CHEMICAL CONTROL

The following specific use information is based on reports by researchers and land managers. Other trade names may be available, and other compounds also are labeled for this weed. Directions for use may vary between brands; see label before use. Herbicides are listed by mode of action and then alphabetically. The order of herbicide listing is not reflective of the order of efficacy or preference.

GROWTH REGULATORS	
Growth regulator herbicides	Although they do not generally kill annual grasses, many of the growth regulator herbicides, particularly aminopyralid and picloram, have been shown to reduce seed production in Japanese brome.
AROMATIC AMINO ACID INHIBITORS	
Glyphosate *Roundup, Accord XRT II*, and others	**Rate:** 1 to 3 pt (*Roundup ProMax*)/acre (0.56 to 1.7 lb a.e./acre). Spot treatment: 0.5% to 1% v/v solution **Timing:** Postemergence from seedling to boot stage. Applications at tillering often provide the best control in rangelands. **Remarks:** Glyphosate will only provide control during the year of application; it has no soil activity and will not kill seeds or inhibit germination the following season. Glyphosate is nonselective. It may damage or kill most non-target plants, creating bare ground conditions that are susceptible to weed recruitment. In areas with desirable vegetation, use spot treatment. Glyphosate is a good control option if reseeding is planned shortly after application, as it will not injure seedlings emerging after application. Add a surfactant when using a formulation where it is not already included (e.g., *Rodeo, Aquamaster*).
BRANCHED-CHAIN AMINO ACID INHIBITORS	
Imazapic *Plateau*	**Rate:** 4 to 6 oz product/acre (1 to 1.5 oz a.e./acre) **Timing:** Preemergence in fall. **Remarks:** Imazapic can be used in combination with glyphosate (premix trade name of *Journey*). Application during winter months in cold climates, when perennial species are dormant, would only control annual species. Higher rates may suppress seedings of some cool-season grasses. Add an adjuvant if applied postemergence. Imazapic is not registered for use in California.
Propoxycarbazone-sodium *Canter R+P*	**Rate:** 0.9 to 1.2 oz product/acre (0.63 to 0.84 oz a.i./acre) **Timing:** Postemergence from the 2-leaf to 2-tiller stage when plants are growing rapidly. **Remarks:** Propoxycarbazone is a broad-spectrum herbicide that will control many species, including Japanese brome. Perennial grass species vary in tolerance. A non-ionic surfactant should be added at 0.25 to 0.5% v/v solution.
Rimsulfuron *Matrix*	**Rate:** 2 to 4 oz product/acre (0.5 to 1 oz a.i./acre) **Timing:** Preemergence in fall. **Remarks:** Rimsulfuron controls several annual grasses and broadleaves. Perennial grasses are tolerant to fall applications when established and grown under dryland conditions. Application to rapidly growing or irrigated perennial grasses may result in their injury or death. It provides soil residual control in cool climates but degrades rapidly under warm conditions. Rimsulfuron will not control summer annual weeds when applied in fall or spring. Add a surfactant when applying postemergence.
Sulfometuron + chlorsulfuron *Landmark XP*	**Rate:** 1 to 1.5 oz product/acre **Timing:** Preemergence in fall. **Remarks:** Controls several annual grasses and some broadleaves. Higher rates can injure some perennial grass species.

Bryonia alba L.
White bryony

Family: Cucurbitaceae
Range: Montana, Idaho, and Utah.
Habitat: Open woodlands and brushy riparian sites.
Origin: Native to Eurasia and northern Africa. Sometimes grown as an ornamental or medicinal plant and escaped from cultivation about 1975.
Impacts: Grows up and over the top of trees, shading and sometimes killing them. Berries are toxic to humans.
Western states listed as Noxious Weed: Idaho, Oregon, Washington

White bryony is an herbaceous perennial vine that grows to 12 ft long or more, often to the tops of brush and trees. Its roots are thick, fleshy, and light yellow in color and the stems climb via tendrils that curl around other vegetation or structures; tendrils arise from leaf axils and are unbranched. The leaves are simple, with 3 to 5 lobes and broadly-toothed margins, roughly triangular or maple-like with palmate venation, and up to 5 inches long. Upper and lower surfaces bear small white glands.

Plants are monoecious, so male and female flowers are on separate plants. The flowers are borne in clusters of 8 to 10 in axils of upper leaves. Flowers are about 0.5 inch wide, with 5 or 6 greenish to cream colored petals, fused into a tube at the base. The fruits are smooth, round berries that ripen from green to black and are about 0.3 inches in diameter. Seeds are slightly ovoid, somewhat flattened, and number 3 to 6 per fruit. Birds are the primary dispersal mechanism for seeds. Birds eating the berries deposit seeds beneath other shrubs and fences which provide optimal structures for new plants to climb. There are no reports of seed longevity in the soil.

NON-CHEMICAL CONTROL

Mechanical (pulling, cutting, disking)	White bryony is difficult to control mechanically. Vines grow from a large perennial root that is extremely difficult to completely remove from soil.
	Control by digging is most effective if the root is severed at least 3 to 4 inches below the crown; the more shallow the cut, the more likely the plant will resprout. Repeat as necessary to control any regrowth.
	Mowing will not generally kill the plants. Cultivation is not usually an option as this plant generally is found growing on and over established bushes and trees that will be injured by cultivation equipment.
Cultural	Grazing is not an option for this plant, given its habit of twining over other vegetation. Apparently all parts of the white bryony plant are toxic, but most concern is about ingesting the berries or the root. Birds apparently eat the berries with no ill effect, and are the chief source of spread of the species.
	Plastic or fabric mulches placed over established crowns may aid in control, provided that vines are not allowed to grow outside the mulched area or from seams or holes.
Biological	There are no known biological control agents to aid in the control of white bryony.

CHEMICAL CONTROL

The following specific use information is based on reports by researchers and land managers. Other trade names may be available, and other compounds may also be labeled for this weed. Directions for use may vary between brands; see label before use. Herbicides are listed by mode of action and then alphabetically. The order of herbicide listing is not reflective of the order of efficacy or preference.

AROMATIC AMINO ACID INHIBITORS	
Glyphosate	**Rate:** 100% *Roundup ProMax* v/v for cut stem treatments
Roundup, Accord XRT II, and others	**Timing:** Any time of the year, including when plants are dormant or rapidly growing.
	Remarks: Cut about 3 to 4 inches below the crown and apply enough full-strength solution to wet the surface of the root. Foliar treatment of white bryony is difficult, since the plant often grows over the top of desirable vegetation that can be injured if sprayed by glyphosate. If the vine is separated from its support before foliar application, it may allow for selective foliar treatment of white bryony using glyphosate.

Cabomba caroliniana A. Gray
Fanwort

Family: Cabombaceae
Range: Western coastal states including Washington, Oregon and California.
Habitat: Grows rooted in the mud, or free-floating, in stagnant to slow flowing water, including streams, smaller rivers, lakes, ponds, sloughs, and ditches. Fanwort inhabits tropical to temperate regions and grows best in slightly acidic water.
Origin: Introduced from the eastern United States. Fanwort is commonly used as an aquarium ornamental and can be introduced into new areas when people discard their aquarium contents into a body of water. It is also considered a noxious weed in parts of Australia.
Impacts: Fanwort can form dense stands that crowd out well-established desirable plants. It can also clog ecologically, recreationally or economically important water bodies and drainage canals.
Western states listed as Noxious Weed: California, Washington

Fanwort or cabomba is a submersed rooted to free-floating perennial with short, fragile rhizomes. The erect shoots are upturned extensions of the horizontal rhizomes. Unlike coontail, fanwort has opposite leaves. without toothed margins, on short stalks. Submersed leaves are palmately dissected and fan-shaped. Sometimes a few small floating leaves develop when plants flower. Floating leaves are 0.5 to 1.5 inches long, narrowly elliptic, with the stem attached at the center (peltate) and one end shallowly forked.

Fanwort flowers are white to pink to purplish and are about 0.5 inch across. The flowers are on stalks which arise from the tips of the stems. Although fanwort produces seeds, there is no information about seed viability or soil longevity. Like many problem aquatic plants, fanwort can reproduce from small fragments. Fanwort stems become brittle in late summer, which causes the plant to break apart, facilitating its distribution and invasion of new water bodies. Although it reproduces by seed, vegetative reproduction seems to be its main method of spreading to new waters.

NON-CHEMICAL CONTROL

Mechanical (pulling, cutting, disking)	Fanwort easily fragments from disturbance, so physical control activities can actually contribute to its spread if great care is not taken to completely harvest all plant parts. Using a venturi dredge, which is like a giant vacuum cleaner, can overcome this problem. It minimizes fragmentation and also extracts the root ball.
Cultural	Fanwort is sensitive to drying out and requires permanent shallow water. Where possible, draining a water body can provide temporary control. If the base of the water storage dries out completely there is little chance of fanwort surviving, but if it remains damp there is a more than 50% chance it will recover. Extreme drying is also required to prevent regrowth from seed. With drinking water supplies, lowering water levels is particularly effective. In the southern USA, water drawdown has been used effectively to reduce fanwort growth. In addition, dewatering infested areas during periods of high temperature in the summer can suppress regrowth. Similarly dewatering during winter during periods of hard freezes can suppress growth the following spring and summer. Dewatering can also be integrated with herbicide applications.
Biological	Potential host-specific biological control agents for fanwort have been identified in Argentina and their efficacy has been investigated under laboratory conditions. However no successful field releases have yet been made. The most promising agents are the stem-boring weevil *Hydrotimetes natans* and an aquatic moth (*Paracles* spp.). The stem-boring weevils were predicted to have a larger impact on deep-water fanwort populations, while the moth larva is expected to control shallow-water populations. The triploid (sterile) grass carp (white amur) is a relatively nonselective herbivorous fish that will consume *Cabomba* spp. and most other submersed aquatic plants. However, triploid grass carp can only be used with specific permits in most U.S. states.

CHEMICAL CONTROL
The following specific use information is based on published papers and reports by researchers and land managers. Other trade names may be available, and other compounds also are labeled for this weed. Directions

for use may vary between brands; see label before use. Herbicides are listed by mode of action and then alphabetically. The order of herbicide listing is not reflective of the order of efficacy or preference.

GROWTH REGULATORS

Triclopyr
Renovate

Rate: 0.5 to 2 ppm

Timing: Apply directly to water from spring to early summer. Fall applications may also be effective when temperatures remain high.

Remarks: Triclopyr is a selective, systemic herbicide that controls broadleaf plants. Therefore, many native monocots (pondweeds and sedges) may be unaffected. Other reports indicate that 2,4-D may also be effective for control, but little information is available.

BRANCHED-CHAIN AMINO ACID INHIBITORS

Penoxsulam
Galleon

Rate: 100 to 200 ppb for 4 to 6 weeks

Timing: Apply directly to water in spring to early summer. Fall applications may also be effective if temperatures remain high.

Remarks: Penoxsulam is a slow-acting systemic herbicide that can also be used following "drawdown" (dewatering) in canals and lake shorelines. Reemerging plants will be controlled as they sprout from the sediment in the late winter to spring when water is reintroduced.

PIGMENT SYNTHESIS INHIBITORS

Fluridone
Sonar

Rate: 6 to 15 ppb for 5 to 7 weeks

Timing: Direct treatment to water is best in early spring during the rapid growth phase.

Remarks: Fluridone is a slow-acting systemic herbicide. Early symptoms typically appear within a week and include a pink to white appearance in new shoots and leaves.

CONTACT PHOTOSYNTHETIC INHIBITORS

Diquat
Reward

Rate: 0.1 to 0.25 ppm

Timing: Apply directly to water in late spring to early summer. Diquat is a fast-acting contact herbicide that can be effective in mid- to late summer, but if biomass is large, only a portion of the infested sites should be treated to minimize effects of reduced dissolved oxygen.

Remarks: Diquat is quickly bound to, and becomes inactivated on, suspended clay particles and it should not be used in moderately or highly turbid water.

Flumioxazin
Clipper

Rate: 100 to 400 ppb

Timing: Apply directly to water from spring to early summer. Fall applications may also be effective if temperatures remain high.

Remarks: Do not use flumioxazin if pH is > 8.0, or use a buffer to reduce the pH to below 7.5. To minimize effects of high pH, apply from dawn to mid-morning. Due to photosynthesis of aquatic plants and algae, pH in the water column rises from mid-day to dusk under most circumstances.

GENERAL CELL TOXICANT

Endothall
Cascade, *Teton*, *Aquathol K*

Rate: 1 to 2 ppm for 48 hours

Timing: Apply directly to water from spring to early summer.

Remarks: Endothall is a fast-acting, contact herbicide that will generally produce symptoms (flaccid tissue and collapsing plants) within 3 to 7 days.

NON-HERBICIDAL CHEMICAL

Dyes or colorants
Aquashade

Although technically not herbicides, dyes and colorants control submerged aquatic plants by absorbing light in the water column and reducing photosynthesis. Applications should be made in early spring and repeated to maintain concentration recommended on the label. Colorants are not as effective on well-established plants in mid- to late summer.

Cardaria chalepensis (L.) Hand.-Maz; lens-podded whitetop
  (= *Lepidium chalepense* L.)
Cardaria draba (L.) Desv. ; hoary cress
  (= *Lepidium draba* L.)
Cardaria pubescens (C.A. Mey.) Jarmolenko; hairy whitetop
  (= *Lepidium appelianum* Al-Shehbaz)

Cardaria draba

Hoary cress (whitetop), lens-podded and hairy whitetop

Family: Brassicaceae
Range: Nearly all western states and many central and eastern states, except southeastern states
Habitat: Disturbed open sites, ditch banks, roadsides, wetlands and riparian areas, agricultural fields including pastures, alfalfa, grain, orchards and vineyards. Often on moderately moist, alkaline to saline soils, but tolerate a wide range of soil types and moisture conditions, especially wet areas.
Origin: Hoary cress is native to Eurasia, while lens-podded and hairy whitetop are native to Central Asia.
Impacts: Hoary cress is the most common and most aggressive of the three species. However, all three species can completely displace desirable vegetation forming dense monocultures. Once established, they can be very difficult to control. They are generally considered to be unpalatable to livestock.
Western states listed as Noxious Weed: *C. draba,* Arizona, California, Colorado, Idaho, Montana, Nevada, New Mexico, Oregon, South Dakota, Utah, Washington, Wyoming; *C. chalepensis,* Arizona, California, Oregon; *C. pubescens,* Arizona, California, Oregon, Washington, Wyoming
California Invasive Plant Council (Cal-IPC) Inventory: *C. draba* and *C. chalepensis,* Moderate Invasiveness; *C. pubescens,* Limited Invasiveness

All three species of *Cardaria* are erect perennials up to 2 ft tall. The stems are generally erect and covered with short hairs. The leaves are alternate, gray-green, and variable in shape, some arrowhead shaped. The upper and especially the lower blade surface are covered with short white hairs. The basal leaves are short-stalked, and the upper leaves clasp the stem at their base. Leaves are 0.5 to 4 inches long by 0.1 to 1.5 inches wide. The basal leaves tend to be narrower but longer than stem leaves.

All three species reproduce by both seeds and through vegetative means. Numerous small, white, fragrant flowers appear in loose inflorescences in spring to summer. Flowers have four petals 2 to 4 mm long. They produce tiny pods that are heart-shaped to ovate. One plant can produce from 1,200 to 4,800 seeds. Seeds germinate in fall after the first rains. Under field conditions, seeds are short-lived. The three species also reproduce vegetatively, developing new shoots from their extensive system of vertical and horizontal roots. This is the primary method of spread. Within 3 weeks of germination the roots of seedlings can start producing buds. Root fragments also can generate new plants, but regeneration is poor in dry soils. The roots can penetrate deep into the soil, and depths well over 10 ft have been documented in some studies. Roots and rhizomes can account for 75% of the total plant biomass and store considerable amounts of carbohydrates. Carbohydrate reserves are minimal in early to mid-spring and accumulate to maximum levels by mid-summer. The foliage dies back during extended periods of freezing temperatures or drought, but the roots survive. Because of these large and deep underground systems, the three species form hard-to-control clonal colonies.

NON-CHEMICAL CONTROL

Mechanical (pulling, cutting, disking)	Hand-pulling is fairly impractical with hoary cress due to its extensive root and rhizome system. Their roots can remain alive even when the top-growth has been eliminated for a year. However, hand hoeing at intervals no longer than 4 weeks for 2 years has been effective.
	Mowing alone will not control the *Cardaria* species. A combination of mowing and competitive cropping has been used to control other *Cardaria* species and may work for hoary cress. In pasture, hoary cress has been controlled by ceasing irrigation, removing outlying plants, decreasing grazing, and generally managing for grassland health.

	Although improper cultivation can spread the *Cardaria* species by dispersing root fragments, its root systems can be exhausted through repeated cultivation. Repeat passes should be made within 10 days of weed emergence. It is important that no green leaves be allowed to form. This can eliminate colonies in 2 to 4 years. However, it is important to be aware that cultivation machinery can spread the roots of these species and increase infestations. All root fragments should be removed from machinery before it is moved to uninfested fields. Cultivation may be more successful in combination with competitive cropping, e.g., annual disking followed by planting alfalfa or grass forage crops.
Cultural	Sheep and goats will eat hoary cress and the other *Cardaria* species, especially the seedlings. Cattle tend to avoid eating them and those animals that consume it may have tainted milk. In addition, plants contain glucosinolates, which can form toxic compounds in cattle. Burning is not effective for controlling any of the *Cardaria* species. Fire will kill aboveground portions of the plant; however, plants will resprout because of their extensive root and rhizome system. Rhizomes have been found as deep as 4 ft and the root system much deeper. In addition, post-fire spread of *Cardaria*, particularly hoary cress, has been documented. Frequent searing with a weed burner at intervals less than 4 weeks apart has been found to be effective when continued for at least 2 years. Flooding an area with 6 to 10 inches of water for 2 months can be highly effective; however, short-term flooding (1 week) has no lasting impact.
Biological	Due to the taxonomic similarity to other important members of the mustard family, there are no biological control agents available for any of the *Cardaria* species.

CHEMICAL CONTROL

The following specific use information is based on reports by researchers and land managers. Much of the information is based on trials for the control of hoary cress. It is expected that these options should also provide the same results with the other two *Cardaria* species. Other trade names may be available, and other compounds also are labeled for this weed. Chemical control is often impeded by the presence of annual grass biomass intercepting foliar applications. Timing for low annual grass presence is recommended.

Directions for use may vary between brands; see label before use. Herbicides are listed by mode of action and then alphabetically. The order of herbicide listing is not reflective of the order of efficacy or preference.

GROWTH REGULATORS	
2,4-D Several names	**Rate:** 2 to 3 qt product/acre (1.9 to 2.85 lb a.e./acre) **Timing:** Postemergence early in the season before flowering, or to new growth in fall. Control is minimal after the bloom stage. **Remarks:** 2,4-D is broadleaf-selective and safe on grasses. This herbicide will most likely require repeat applications for several years. 2,4-D has little to no soil activity. It is not the most effective treatment, but widely used because of its low cost. 2,4-D is often combined with other active ingredients, e.g. clopyralid or dicamba. Do not apply the ester formulations when outside temperatures exceed 80°F.
Aminocyclopyrachlor + chlorsulfuron *Perspective*	**Rate:** 3 to 4.5 oz product (*Perspective*)/acre **Timing:** Postemergence or preemergence. Postemergence applications are most effective when applied to plants from the seedling to the mid-rosette stage. **Remarks:** *Perspective* provides broad-spectrum control of many broadleaf species. Aminocyclopyrachlor has marginal activity on mustards and control, when using *Perspective*, is largely from the chlorsulfuron component. Although generally safe to grasses, it may suppress or injure certain annual and perennial grass species. Do not treat in the root zone of desirable trees and shrubs. Do not apply more than 11 oz product/acre per year. At this high rate, cool-season grasses will be damaged, including bluebunch wheatgrass. Not yet labeled for grazing lands. Add an adjuvant to the spray solution. This product is not approved for use in California and some counties of Colorado (San Luis Valley).
Aminopyralid + metsulfuron *Opensight*	**Rate:** 3.3 oz product/acre **Timing:** Optimum timing is when the plants are in the bloom stage. **Remarks:** Follow label restrictions. Not registered for use in California.

Dicamba + 2,4-D	**Rate:** 1 pt dicamba product/acre + 3 pt 2,4-D product/acre (0.5 lb a.e. + 1.5 lb a.e./acre) **Timing:** Postemergence from bolting to early bud stage. **Remarks:** See 2,4-D.
AROMATIC AMINO ACID INHIBITORS	
Glyphosate *Roundup, Accord XRT II,* and others	**Rate:** Broadcast foliar treatment: 4 qt product (*Roundup ProMax*)/acre (4.5 lb a.e./acre). Spot treatment: 2% v/v solution **Timing:** Postemergence in the early bud stage. **Remarks:** Glyphosate is a nonselective herbicide. Spot treatment may be the best approach where feasible. Glyphosate has no soil activity. Repeat applications may be necessary. Drought stress will limit its effectiveness and its effectiveness is increased by the addition of ammonium sulfate.
BRANCHED-CHAIN AMINO ACID INHIBITORS	
Chlorsulfuron *Telar*	**Rate:** 1 oz product/acre (0.75 oz a.i./acre) **Timing:** Postemergence from bud to bloom stages, or to rosettes in fall. **Remarks:** Chlorsulfuron is one of the most effective treatments for treatments for control of the *Cardaria* species. It has mixed selectivity, but is generally safe on grasses. 2,4-D at 1 to 2 pt product/acre can be tank-mixed with chlorsulfuron for quicker burndown. Use a surfactant. Chlorsulfuron has fairly long soil residual activity. *Telar* can be used near water, but cannot be applied to water.
Imazapic *Plateau*	**Rate:** 8 to 12 oz product/acre (2 to 3 oz a.e./acre) **Timing:** Postemergence after blossoms open until plants desiccate. Fall rosettes may also be treated. **Remarks:** Imazapic is safe to apply to most native grasses. Higher rates may suppress seed of some cool-season grasses. Imazapic is not registered for use in California.
Imazapyr *Arsenal, Habitat, Stalker, Chopper, Polaris*	**Rate:** 1 to 2 pt product (*Habitat*)/acre (0.25 to 0.5 lb a.e./acre) **Timing:** Most effective when applied postemergence in spring when plants are flowering. **Remarks:** Imazapyr is a nonselective herbicide. It has long soil residual activity and leaves more bare ground than other treatments, even a year after application. Add a spray adjuvant.
Metsulfuron *Escort*	**Rate:** 1 oz product/acre (0.6 oz a.i./acre) **Timing:** Postemergence from pre-bloom to bloom stages or to rosettes in fall. **Remarks:** The effectiveness of metsulfuron is similar to chlorsulfuron. It has mixed selectivity, but is generally safe on grasses. Use a surfactant. Metsulfuron can be tank-mixed with 2,4-D for quicker burndown. Other premix formulations of metsulfuron can be used at similar application timing. These include *Cimarron Max* (metsulfuron + dicamba + 2,4-D), *Opensight* (metsulfuron + aminopyralid), and *Cimarron X-tra* (metsulfuron + chlorsulfuron). Metsulfuron typically controls hoary cress for more than one season, but its soil activity is marginal, and seedlings of Cardaria species may invade shortly after application. Metsulfuron is not registered for use in California.
Sulfometuron *Oust* and others	**Rate:** 3 to 5 oz product/acre (2.25 to 3.75 oz a.i./acre) **Timing:** Preemergence or postemergence during or just before the rainy season when the target plants are germinating and growing rapidly. **Remarks:** Add a surfactant to improve control. Sulfometuron has a long soil residual and is susceptible to off-site movement in dry light windblown soils. Use with extreme care if near crops.

Carduus acanthoides L.; plumeless thistle
Carduus nutans L.; musk thistle
Carduus pycnocephalus L.; Italian thistle
Carduus tenuiflorus Curtis; slenderflower thistle

Plumeless, musk, Italian, and slenderflower thistle

Family: Asteraceae
Range: Musk thistle is found in all western states. Plumeless thistle is found in most western states, except Oregon, Nevada, Utah and Arizona. Italian and slenderflower thistle are most common in west coast states, Washington, Oregon and California. Italian thistle is also found in Idaho.
Habitat: Disturbed open sites, roadsides, pastures, annual grasslands, and waste areas. These thistles can tolerate a relatively wide range of soil types but prefer fertile, well-drained soils.
Origin: All species are native to Europe and the Mediterranean region.
Impacts: These thistles can dominate sites and crowd out native species and forage plants. The spines inhibit grazing and discourage livestock and wildlife from entering infested areas.
Western states listed as Noxious Weed: *C. acanthoides*, Arizona, California, Colorado, Oregon, South Dakota, Washington, Wyoming; *C. nutans*, California, Colorado, Idaho, Nevada, New Mexico, North Dakota, Oregon, South Dakota, Utah, Washington, Wyoming; *C. pycnocephalus*, California, Oregon, Washington; *C. tenuiflorus*, California, Oregon, Washington
California Invasive Plant Council (Cal-IPC) Inventory: *C. acanthoides* and *C. tenuiflorus*, Limited Invasiveness; *C. nutans* and *C. pycnocephalus*, Moderate Invasiveness

The weedy *Carduus* thistles are winter annuals or biennials with prickly leaves and stems with prickly wings. Plumeless thistle and musk thistle grow to 5 ft tall. They are closely related and may hybridize. Italian thistle and slenderflower thistle are also closely related and can grow to 6.5 ft tall. All these thistles form deep taproots. Plants may grow as a biennial, germinating in winter to early spring and existing as a rosette until flower stems develop in spring to summer of the following year. After bolting, stems branch near the top. Stem leaves taper down the stem as spiny wings.

All these thistles produce purple to pink, rarely white, flowers in summer. Plumeless and musk thistle have typical hemispherical thistle flowers, while Italian and slenderflower thistle flowers are more cylindrical in shape. Musk thistle flowers are often nodding on long stalks. Reproduction in all these species is only by seed. Seeds typically fall near the parent plant or are disperse to greater distances with wind. Seeds that germinate in late summer to fall often behave as winter annuals and flower in the following summer. Seeds rarely persist in the soil seedbank for more than a few years.

NON-CHEMICAL CONTROL

Mechanical (pulling, cutting, disking)	These thistles can be cultivated or manually removed when small. To control by cutting, use a sharpened shovel at the top of the root crown. Grubbing hoes must cut the plants 2 to 4 inches below ground level to prevent resprouting from dormant axillary buds.
	Mowing the plant during flowering can greatly reduce seed production, though a single mowing is seldom sufficient due to the wide differences in the maturity of plants in a natural population. For mowing, wait till plants bolt and are about to flower. This may require repeated visits at weekly intervals over the 4 to 7 week blooming period, because not all plants bloom simultaneously. Plants will regrow if mowed before they are

	fully bolted. Plants cut 4 days after the first flowers open can produce viable seed.
Cultural	Large livestock tend to avoid grazing on thistles, although horses and cattle have been known to eat the flowerheads. Sheep will eat the rosettes. Goats like the flowerheads and are able to digest the seed.
Depending on timing, the soil seedbank, the plant community, and other factors, fire may help control thistles or may promote their invasion. Grass fires may not be hot enough to kill the root crown of thistle rosettes, but a flamer can be used to kill individual plants.
In general, thistles compete poorly with healthy established grasses and other vegetation. Establishment of selected, aggressive grasses can be an effective cultural control of thistles.
Biological	The thistle head weevil (*Rhinocyllus conicus*) is an introduced biocontrol agent that attacks *Carduus* species and several other thistles. It is established in much of the northwestern and north central United States. The weevil gives effective control of musk thistle in parts of Montana, less so in other areas. The crown weevil (*Trichosirocalus horridus*) and thistle crown fly (*Cheilosia corydon*) also are locally established. The fungus musk thistle rust (*Puccinia carduorum*) may soon be approved as a biocontrol agent.

CHEMICAL CONTROL

The following specific use information is based on published papers and reports by researchers and land managers. Other trade names may be available, and other compounds also are labeled for this weed. Directions for use may vary between brands; see label before use. Herbicides are listed by mode of action and then alphabetically. The order of herbicide listing is not reflective of the order of efficacy or preference.

GROWTH REGULATORS	
2,4-D
Several names	**Rate:** 1 to 2 qt product/acre (0.95 to 1.9 lb a.e./acre)
Timing: Postemergence in the rosette stage.
Remarks: 2,4-D is broadleaf-selective and has no soil activity. It may require repeat application. It is not the most effective treatment, but is widely used because of low cost. Do not apply ester formulations when outside temperatures exceed 80°F.
Aminocyclopyrachlor + chlorsulfuron
Perspective	**Rate:** 3 to 4.5 oz product/acre
Timing: Postemergence in spring up to flowering, or in fall rosette stage.
Remarks: *Perspective* provides broad-spectrum control of many broadleaf species. Although generally safe to grasses, it may suppress or injure certain annual and perennial grass species. Do not treat in the root zone of desirable trees and shrubs. Do not apply more than 11 oz product/acre per year. At this high rate, cool-season grasses will be damaged, including bluebunch wheatgrass. Not yet labeled for grazing lands. Add an adjuvant to the spray solution. This product is not approved for use in California and some counties of Colorado (San Luis Valley).
Aminopyralid
Milestone	**Rate:** 4 to 5 oz product/acre (1 to 1.25 oz a.e./acre)
Timing: Preemergence in winter to early spring and postemergence to seedling treatments in spring up to flower bud stage. Can be applied in fall in cold-winter areas. Apply when plants are at the late bolt through early flowering stages. For late applications starting at the late bud stage, 2,4-D at 1 lb ae/acre should be tank-mixed with *Milestone*.
Remarks: Aminopyralid is one of the most effective herbicides for thistles. It is a broadleaf herbicide like picloram, but more selective and with a shorter soil residual activity. Aminopyralid is generally safe on grasses. It has longer soil residual and higher activity than clopyralid.
Aminopyralid + 2,4-D, *Forefront HL*;
Aminopyralid + metsulfuron, *Opensight*;
Aminopyralid + triclopyr, *Capstone*	**Rate:** 1.5 to 2 pt *Forefront HL*/acre; 1 to 3 oz *Opensight*/acre; 4 to 6 pt *Capstone*/acre
Timing: Postemergence in rosette to bolting stages.
Remarks: These combinations are broadleaf-selective. *Opensight* is not registered for use in California.
Clopyralid
Transline	**Rate:** 0.25 to 1.33 pt product/acre (1.5 to 8 oz a.e./acre)
Timing: Postemergence in spring, up to the flower bud stage.
Remarks: Clopyralid is a broadleaf herbicide that is more selective to some broadleaf plants than picloram. It is very safe on grasses.

Clopyralid + 2,4-D *Curtail*	**Rate:** 2 to 4 qt product/acre (use higher rate if plants are drought-stressed) **Timing:** Postemergence to rapidly growing weeds from full rosette to early flower bud. **Remarks:** The combination is broadleaf-selective with a wide range of susceptible species.
Dicamba *Banvel, Clarity*	**Rate:** 0.5 to 2 pt product/acre (0.25 to 1 lb a.e./acre); 8 to 16 oz product/acre for rosettes, up to 2 pt product/acre for bolting plants **Timing:** Postemergence to rapidly growing plants up to bolting. Application to smaller plants gives better control. **Remarks:** Dicamba is a broadleaf-selective herbicide often combined with other active ingredients. It is effective earlier in the season than 2,4-D. It is also effective when tank-mixed with 2,4-D (0.75 lb a.e./acre 2,4-D + 0.25 lb a.e./acre dicamba). Dicamba has very limited soil residual. Avoid drift to sensitive crops. Do not apply when outside temperatures exceed 80°F. Dicamba is available mixed with diflufenzopyr in a formulation called *Overdrive*. The combination is broadleaf-selective, but safe on most grasses. This has been reported to be effective on plumeless and musk thistle, and may have a similar effect on other members of the genus *Carduus*. Diflufenzopyr is an auxin transport inhibitor which causes dicamba to accumulate in shoot and root meristems, increasing its activity. *Overdrive* is applied postemergence at 4 to 8 oz product/acre on rapidly growing plants. Higher rates should be used on large annuals and biennials or when treating perennial weeds. Add a non-ionic surfactant to the treatment solution at 0.25% v/v or a methylated seed oil at 1% v/v solution.
Fluroxypyr *Vista XRT*	**Rate:** 22 oz product/acre (7.7 oz a.e./acre) **Timing:** Postemergence to rapidly growing weeds. **Remarks:** Fluroxypyr is broadleaf-selective and safe on most grasses.
Picloram *Tordon 22K*	**Rate:** 0.67 to 1 qt product/acre (0.33 to 0.5 lb a.e./acre) **Timing:** Postemergence from flower bud stage to senescence. Most effective in fall treatments. **Remarks:** Picloram is one of the most effective herbicides for this weed and can provide 2 years of control. Most broadleaf plants are susceptible, but it is relatively safe on established grasses. It can injure young or germinating grasses. Picloram is also effective when mixed with dicamba or 2,4-D. It has long soil residual activity. Picloram can be used in a premix with 2,4-D (*Grazon P+D*) or fluroxypyr (*Surmount*) for the control of *Carduus* species. Picloram is a restricted use herbicide. Picloram and its formulations are not registered for use in California.
Triclopyr *Garlon 3A, Garlon 4 Ultra*	**Rate:** 1.33 to 2.66 qt *Garlon 3A*/acre or 1 to 2 qt *Garlon 4 Ultra*/acre (1 to 2 lb a.e./acre) **Timing:** Postemergence to rapidly growing weeds. **Remarks:** Triclopyr is broadleaf-selective and safe on most grasses. It is most effective on smaller plants. *Garlon 4 Ultra* is formulated as a low volatile ester. However, in warm temperatures, spraying onto hard surfaces such as rocks or pavement can increase the risk of volatilization and off-target damage. Rates are based on those reported for Canada thistle.
Triclopyr + 2,4-D *Crossbow*	**Rate:** 2 to 4 qt product/acre **Timing:** Postemergence during the rosette stage. **Remarks:** Include non-ionic surfactant at 1 qt/100 gal water (0.25% v/v).
AROMATIC AMINO ACID INHIBITORS	
Glyphosate *Roundup, Accord XRT II,* and others	**Rate:** 1.33 to 2.67 qt product (*Roundup ProMax*)/acre (1.5 to 3 lb a.e./acre) **Timing:** Postemergence to rapidly growing plants in bud stage. **Remarks:** Glyphosate has no soil activity and is nonselective. Repeat applications may be necessary. It is more effective with the addition of ammonium sulfate.
BRANCHED-CHAIN AMINO ACID INHIBITORS	
Chlorsulfuron *Telar*	**Rate:** 1 to 2.6 oz product/acre (0.75 to 1.95 oz a.i./acre) **Timing:** Postemergence in fall to new rosettes, or to rosettes in spring before bolting. Chlorsulfuron or metsulfuron applied from the bolting through very early bud stage eliminated viable achene development. **Remarks:** Chlorsulfuron has mixed selectivity, but is generally safe on grasses. Use a surfactant. It can be used in late season applications to reduce seed production and has fairly long soil residual activity.

	Fall application may injure desirable bromes.
Imazapic *Plateau*	**Rate:** 8 to 12 oz product/acre (2 to 3 oz a.e./acre) **Timing:** Preemergence or early postemergence. Only suppression of musk thistle when applied preemergence. **Remarks:** Imazapic has mixed selectivity and some soil residual activity. Although the label indicates that musk thistle can be controlled with imazapic, it generally tends to favor members of the Asteraceae and some grasses. Use a methylated seed oil surfactant at 0.25% v/v. Not registered for use in California.
Imazapyr *Arsenal, Habitat, Stalker, Chopper, Polaris*	**Rate:** 3 to 4 pt product/acre (0.75 to 1 lb a.e./acre) **Timing:** Preemergence or postemergence. **Remarks:** Imazapyr is a nonselective herbicide. Rates are based on those reported for bull thistle.
Metsulfuron *Escort*	**Rate:** 0.5 to 1 oz product/acre (0.3 to 0.6 oz a.i./acre) **Timing:** Postemergence to young, rapidly growing weeds in spring before flowering, or in fall to new rosettes. Chlorsulfuron or metsulfuron applied from the bolting through very early bud stage eliminated viable achene development. Metsulfuron seemed to work a bit better than chlorsulfuron. **Remarks:** Metsulfuron has mixed selectivity and is generally safe on grasses. Use a surfactant. It can be tank-mixed with aminopyralid (*Opensight*), 2,4-D and/or dicamba. Metsulfuron has some soil residual activity. It is not registered for use in California.
Metsulfuron + chlorsulfuron *Cimarron X-tra*	**Rate:** 0.5 oz product/acre **Timing:** Postemergence before flowering. **Remarks:** It is not registered for use in California.
Sulfometuron *Oust* and others	**Rate:** 6 to 8 oz product/acre (4.5 to 6 oz a.i./acre) **Timing:** Preemergence or early postemergence, during the rainy season when weeds are germinating or rapidly growing. **Remarks:** Sulfometuron has mixed selectivity, but is fairly safe on native perennial grasses, especially wheatgrass. This rate is very high and may not be the best option. Desirable grasses may be stunted, stressed, or injured. Sulfometuron has fairly long soil residual activity.
PHOTOSYNTHETIC INHIBITORS	
Hexazinone *Velpar L*	**Rate:** 4 to 6 pt product/acre (1 to 1.5 lb a.e./acre) **Timing:** Preemergence before weeds emerge or postemergence to young plants. **Remarks:** Hexazinone has both foliar and soil activity. In soil applications, rates will vary with soil texture and soil organic matter; best results when applied to moist soil. Use rates will also vary depending on the weed species to be controlled. Hardwood trees near application site can absorb this chemical through the roots and may be injured or killed. Do not spray near the root zone of desirable hardwood trees or shrubs. High rates of hexazinone can create bare ground, so only use high rates in spot treatments.

Carpobrotus edulis (L.) N.E. Br.
Iceplant (Hottentot fig)

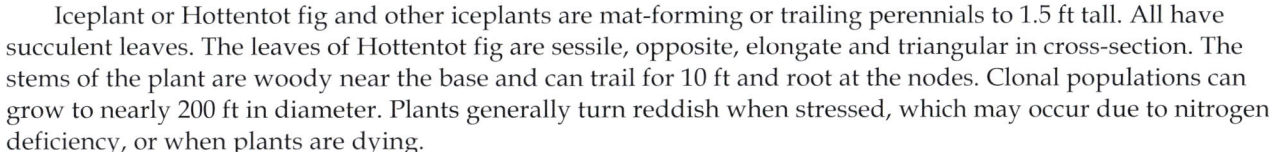

Family: Aizoaceae
Range: Primarily a weed in coastal regions of California.
Habitat: Has escaped cultivation in many coastal areas where it especially thrives in dune communities, including coastal scrub, grassland, chaparral, bluffs, dunes, and other sandy coastal sites. Does not tolerate cold winter climates.
Origin: Native to South Africa. Hottentot fig is extensively planted along highways as an ornamental and to prevent soil erosion. In addition, it is cultivated as a landscape ornamental in coastal regions and inland in mild winter areas.
Impacts: Hottentot fig displaces native dune species, and large infestations change the ecology of the community. Plants trap more sand than native dune species and generally stabilize dune communities unnaturally. Plants also die when completely buried. In time, an increased amount of organic matter in the sandy soil can promote invasion by other weed species that otherwise would not be able to inhabit dune soils.
California Invasive Plant Council (Cal-IPC) Inventory: High Invasiveness

Iceplant or Hottentot fig and other iceplants are mat-forming or trailing perennials to 1.5 ft tall. All have succulent leaves. The leaves of Hottentot fig are sessile, opposite, elongate and triangular in cross-section. The stems of the plant are woody near the base and can trail for 10 ft and root at the nodes. Clonal populations can grow to nearly 200 ft in diameter. Plants generally turn reddish when stressed, which may occur due to nitrogen deficiency, or when plants are dying.

Flowers are solitary and showy at the stem tips, 8 to 10 cm diameter, with numerous pink or yellow (aging to pink) linear petals. Plants reproduce both vegetatively by stem fragments and by seed. The capsules are berry-like, large, fleshy, and persist on plants for months, eventually turning yellow with age. There are numerous seeds per capsule, which remain in fruit until they are consumed by animals or decompose. Fruits are primarily dispersed with animals such as deer, rabbits, and rodents after they are consumed. Seeds can survive ingestion by animals, and those that pass through an animal's gut germinate more readily than seeds from intact fruits. Fruits that are not eaten become hard, forcing seeds to remain dormant until fruits decompose, usually within 3 years.

NON-CHEMICAL CONTROL

Mechanical (pulling, cutting, disking)	Mechanical removal is effective at any time of year. Hottentot fig and other iceplant species are easily removed by hand pulling. One method of removal is by tearing up the plants up by the roots. Large mats can be removed by rolling them up like a carpet but this is very labor intensive. Because the plant can grow roots and shoots from any node, all live plants and stem fragments must be removed from contact with the soil to prevent resprouting. If removal is not possible, mulching with the removed plant material is adequate to prevent most resprouting, but requires at least one follow-up visit to remove resprouts.
	Heavy equipment can also be used to remove iceplants. Earth-moving machinery such as a scoop with a brush rake attached, a skid-steer, bobcat, or tractor, can be used, but it is still necessary to remove buried stems and to mulch the soil to prevent reestablishment.
	Tarping is also a method of managing iceplant, as it blocks light and eventually kills the plant.
	Removal of iceplants can leave behind a layer of accumulated dead and decaying organic debris. The debris may contain seeds of iceplant or other weedy species. Furthermore, the carbon in the litter provides nutrients to potential invaders. It may be necessary to also remove this dead material.
Cultural	Burning is not an effective strategy for control of iceplants. While the heat of the fire will kill the seeds, the succulent foliage will not entirely be killed by fire, even if enough fuel was available to carry a burn. Grazing is also not a recommended control option, particularly on sensitive dunes.
Biological	There are currently no biological controls for *Carpobrotus edulis* or any other species of iceplant. The iceplant scale insects, *Pulvinariella mesembryanthemi* and *P. delottoi*, have a small impact on some individuals, but would likely not be useful as a control tool.

CHEMICAL CONTROL

The following specific use information is based on reports by researchers and land managers. Other trade names may be available, and other compounds also are labeled for this weed. Directions for use may vary between brands; see label before use. Herbicides are listed by mode of action and then alphabetically. The order of herbicide listing is not reflective of the order of efficacy or preference.

GROWTH REGULATORS	
Aminocyclopyrachlor + chlorsulfuron *Perspective*	**Rate:** 4.75 to 8 oz product (*Perspective*)/acre **Timing:** Postemergence and preemergence. Postemergence applications are most effective when applied to plants from the seedling to the mid-rosette stage. **Remarks:** *Perspective* provides broad-spectrum control of many broadleaf species. Although generally safe to grasses, it may suppress or injure certain annual and perennial grass species. Do not treat in the root zone of desirable trees and shrubs. Do not apply more than 11 oz product/acre per year. At this high rate, cool-season grasses will be damaged, including bluebunch wheatgrass. Not yet labeled for grazing lands. Add an adjuvant to the spray solution. This product is not approved for use in California and some counties of Colorado (San Luis Valley).
AROMATIC AMINO ACID INHIBITORS	
Glyphosate *Roundup, Accord XRT II,* and others	**Rate:** Broadcast foliar treatment: 1.3 qt product (*Roundup ProMax*)/acre (1.46 lb a.e./acre). Spot treatment: 1.5 to 2% v/v solution **Timing:** Because of the succulent nature of the plants, it is best to apply the herbicide at a time when the plant is rapidly growing. **Remarks:** Glyphosate is the only chemical option registered in California that has been shown to effectively kill *Carpobrotus edulis* or other iceplant species. The addition of 1% surfactant can increase the effectiveness of the herbicide. Since glyphosate is nonselective, it may be more appropriate to use a shielded sprayer or even a wiper application technique at 50% concentrate of the herbicide.

Carthamus lanatus L.
Woolly distaff thistle

Family: Asteraceae
Range: Primarily a problem in the central coast regions of California, but is also found in Oregon, Arizona, Texas and Oklahoma.
Habitat: Disturbed open sites, roadsides, fields, grassland, rangeland, pastures, and sometimes agricultural land, especially grain fields.
Origin: Native to Mediterranean region.

Impacts: Highly competitive with cereal crops and desirable rangeland species, and dense populations can develop. In addition, the spiny foliage and flowerheads can injure the eyes and mouths of grazing livestock.
Western states listed as Noxious Weed: California, Oregon
California Invasive Plant Council (Cal-IPC) Inventory: Moderate Invasiveness (Alert)

Erect winter annual, with rigid stems to 3 ft tall and spiny leaves. Plants exist as rosettes until flower stems develop in spring/summer. Stems usually covered with loose woolly or cobwebby hairs, as well as minute glandular hairs, especially in leaf axils and at bases of flowerheads. Stem leaves alternate, sessile, once-pinnate-lobed and prominently spine-tipped, also with glandular hairs.

Flowerheads are yellow, 1 to 2 inches long, solitary at stem tips, with spiny lobed phyllaries. Flowerheads consist of numerous disk flowers. Outer seeds (achenes) lack a pappus, whereas inner achenes have a persistent pappus, 10 to 13 mm long, of numerous narrow, unequal, brownish scales. Plants reproduce only by seed. Most achenes fall near the parent plant, but some remain in the persistent seedheads. Achenes and sometimes entire seedheads can disperse to greater distances with animals, humans, machinery such as tractors and agricultural implements, mud, and water. Most seeds germinate within the first couple of years after maturation, but some seeds can remain dormant and viable for up to 8 years under field conditions.

NON-CHEMICAL CONTROL

Mechanical (pulling, cutting, disking)	Hoeing can be effective for the control of small populations. This can be conducted in the rosette or bolting stage, but before flowering. Plants must be cut just below the soil surface to prevent resprouting.
	Mowing after bolting but just before the development of flower buds can prevent most seed production. This is generally in late spring. Mowing earlier can encourage the regrowth of flowering stems. In plants mowed after flowerheads have developed, seed can mature in cut flowerheads left on the ground. Control programs may have to be repeated two or more times throughout the season to prevent escaped plants from producing seeds.
Cultural	Heavy grazing can increase distaff thistle populations because livestock selectively graze more palatable and less spiny species, thereby reducing competition with other plants for light and nutrients.
Biological	Distaff thistle is closely related to commercial safflower (*Carthamus tinctorius* L.), which precludes the development and release of biological control agents.

CHEMICAL CONTROL
The following specific use information is based on published papers or reports by researchers and land managers. Other trade names may be available, and other compounds also are labeled for this weed. Directions for use may vary between brands; see label before use. Herbicides are listed by mode of action and then alphabetically. The order of herbicide listing is not reflective of the order of efficacy or preference.

GROWTH REGULATORS

2,4-D	**Rate:** 2 to 4 pt product/acre (0.95 to 1.9 lb a.e./acre)
Several names	**Timing:** Postemergence when plants are at seedling or small rosette stage, which is generally in late winter

	or early spring.
	Remarks: 2,4-D gives good control, but aminopyralid is considered better. It will damage other broadleaf species, but is safe on grasses. 2,4-D is a restricted use herbicide in some areas.
Aminopyralid *Milestone*	**Rate:** 3 to 7 oz product/ac (0.75 to 1.75 oz a.e./ac)
	Timing: Preemergence and postemergence. For postemergence treatment it is best applied when plants are at seedling or early rosette stage, which is generally in late winter or early spring.
	Remarks: Aminopyralid gives excellent control. It is fairly selective primarily on members of the Asteraceae and Fabaceae, and is generally safe on grasses. Aminopyralid has soil residual activity which helps to control later-germinating seedlings. A premix of aminopyralid + metsulfuron (*Opensight*) also provides excellent control at 1.5 to 2 oz product/acre.
Clopyralid *Transline*	**Rate:** 6 to 21 oz product/ac (2.25 to 8 oz a.e./ac)
	Timing: Preemergence and postemergence. For postemergence treatment, it is best applied when plants are at seedling or small rosette stage, which is generally in late winter or early spring.
	Remarks: Clopyralid gives good control, but aminopyralid is considered better. It is fairly selective primarily on members of the Asteraceae and Fabaceae, and is safe on grasses. Clopyralid has soil residual activity which helps to control later-germinating seedlings.
Dicamba *Clarity, Banvel*	**Rate:** 2 to 4 pt product/acre (1 to 2 lb a.e./acre)
	Timing: Postemergence when plants are at seedling or small rosette stage, which is generally in late winter or early spring.
	Remarks: Dicamba gives good control, but aminopyralid is considered better. It is a broadleaf-selective herbicide and is safe on grasses.
Picloram *Tordon 22K*	**Rate:** 1 to 1.5 pt product/acre (4 to 6 oz a.e./acre)
	Timing: Postemergence in spring at the rosette stage before bolting.
	Remarks: Picloram has long soil residual activity. The label indicates picloram should be mixed with 1 lb a.e./acre 2,4-D. It will damage other broadleaf species, but is generally safe on grasses. Picloram is a restricted use herbicide. It is not registered for use in California.
Triclopyr *Garlon 3A*	**Rate:** 2 qt *Garlon 3A*/acre (1.5 lb a.e./acre)
	Timing: Postemergence when plants are at seedling or small rosette stage, which is generally in late winter or early spring.
	Remarks: Triclopyr gives good control, but aminopyralid is considered better. It is a broadleaf-selective herbicide and is safe on grasses.
AROMATIC AMINO ACID INHIBITORS	
Glyphosate *Roundup, Accord XRT II*, and others	**Rate:** 1 to 2 pt product (*Roundup ProMax*)/acre (0.56 to 1.1 lb a.e./acre)
	Timing: Postemergence in late spring or early summer, but before plants begin to flower. Rosette to early bolting stage is best.
	Remarks: Glyphosate is nonselective, and it is best to apply after desirable annual grasses have dried up. If perennial grasses are present, glyphosate can cause significant damage or death.

Centaurea calcitrapa L.; purple starthistle
Centaurea iberica Spreng.; Iberian starthistle

Purple and Iberian starthistle

Centaurea calcitrapa

Family: Asteraceae
Range: Purple starthistle is found in Arizona, California, Oregon Washington, Utah, and southeastern New Mexico. All of the Oregon and Washington populations have been controlled or are currently being treated for eradication. Iberian starthistle is found in Oregon, Washington, Wyoming and California.
Habitat: Purple starthistle prefers fertile alluvial soils and forms dense stands in pasture, range, open forest, and riparian areas. Iberian starthistle often colonizes banks and watercourses and other moist areas.
Origin: Purple starthistle is native to southern Europe, and Iberian starthistle is native to southeastern Eurasia.
Impact: The rigid spines of purple and Iberian starthistle make the plants unpalatable and reduce the quality of infested hay. These plants restrict access and deter grazing by livestock and wildlife. Infestations can limit recreational opportunities, cause injuries, and degrade the quality of parks and natural areas. Purple starthistle can inhabit a wide range of environmental conditions and replace native species. Purple starthistle is known globally as an introduced weedy plant and is considered invasive or noxious in North and South America, New Zealand and Australia.
Western states listed as Noxious Weed: Both species are listed in Arizona, California, Nevada, Oregon; *C. calcitrapa* is also listed in Washington
California Invasive Plant Council (Cal-IPC) Inventory: *Centaurea calcitrapa*, Moderate Invasiveness

Centaurea iberica

Both species are annuals to perennials, to 3 ft tall. They exist as basal rosettes until they bolt. Bolting stems are erect, with highly branched flowering stems developing at maturity, usually in late spring and summer. A long taproot provides a competitive advantage over annual and perennial grasses, reducing available forage.

Both species produce spiny heads with purple flowers. The pappus on Iberian starthistle has white bristles ~1 mm long, while purple starthistle usually lacks a pappus. Plants reproduce only by seed, which disperse with the seedhead as a unit. Most fall just below the parent plant, but some can move longer distances when they attach to animals. Most seed germinates the first year, but buried seeds of both species can remain dormant for about 3 years.

The specific epithet for purple starthistle, *calcitrapa*, was derived from caltrop, a spiked weapon from the Middle Ages that was dropped on the battlefield to injure advancing troops and horses.

NON-CHEMICAL CONTROL

Mechanical (pulling, cutting, disking)	Hand pulling, grubbing, or digging can be used to control small infestations. These techniques must be repeated several times a year. Purple starthistle populations were sharply reduced after 3 years of hand grubbing.
	Mowing is not effective at killing plants but can reduce seed production if timed at full bloom. Regrowth occurs from root crowns when the tops are removed.
Cultural	Conventional grazing by sheep or cattle will not control purple starthistle and in fact can promote it, because grazing animals usually avoid this plant and selectively feed on species that would otherwise compete with it.
	Burning is not considered an effective tool for control.
	In California where purple starthistle is a common pasture weed, fertility management is occasionally used as a management tool.
Biological	While there are many biological control programs for other species of *Centaurea*, there is no biological

control program for purple or Iberian starthistle.

CHEMICAL CONTROL

The following specific use information is based on published papers or reports by researchers and land managers. These are products known to provide effective control. Those that do not provide sufficient control have been omitted from the table. Other trade names may be available, and other compounds also are labeled for this weed. For foliar applications, use low pressure and a coarse spray pattern to reduce spray drift damage to non-target species. Directions for use may vary between brands; see label before use. Herbicides are listed by mode of action and then alphabetically. The order of herbicide listing is not reflective of the order of efficacy or preference.

GROWTH REGULATORS	
2,4-D Several names	**Rate:** 1 to 2 qt product/acre (0.95 to 1.9 lb a.e./acre) **Timing:** Postemergence in spring when plants are still in rosettes but before flower stems elongate. Treat rapidly growing plants. Thoroughly covering foliage enhances control. **Remarks:** Generally requires repeat applications. It is not considered as effective as other growth regulator herbicides for season-long control. 2,4-D is broadleaf-selective and has no soil activity. Do not apply ester formulation when outside temperatures exceed 80°F. Amine forms are as effective as ester forms for small rosettes, and amine forms reduce the chance of off-target movement from volatility.
Aminocyclopyrachlor + chlorsulfuron *Perspective*	**Rate:** 4.75 to 8 oz product (*Perspective*)/acre **Timing:** Postemergence and preemergence. Postemergence applications are most effective when applied to plants from the seedling to the mid-rosette stage. **Remarks:** Aminocyclopyrachlor gives control of many members of the genus *Centaurea*, including purple starthistle. Its effect is similar to aminopyralid. *Perspective* provides broad-spectrum control of many broadleaf species. Although generally safe to grasses, it may suppress or injure certain annual and perennial grass species. Do not treat in the root zone of desirable trees and shrubs. Do not apply more than 11 oz product/acre per year. At this high rate, cool-season grasses will be damaged, including bluebunch wheatgrass. Not yet labeled for grazing lands. Add an adjuvant to the spray solution. This product is not approved for use in California and some counties of Colorado (San Luis Valley).
Aminopyralid *Milestone*	**Rate:** 4 to 7 oz product/acre (1 to 1.75 oz a.e./acre). These rates are used for control of diffuse and spotted knapweed, and are expected to provide similar control of purple and Iberian starthistle **Timing:** Postemergence to rapidly growing plants in fall, or in spring from rosette to bolting stages. Late winter to early spring applications provide residual control of germinating seedlings. **Remarks:** A non-ionic surfactant (0.25 to 0.5% v/v of spray solution) enhances control under adverse environmental conditions. Aminopyralid is a broadleaf herbicide.
Clopyralid *Transline*	**Rate:** 0.67 to 1.33 pt product/acre (4 to 8 oz a.e./acre) **Timing:** Postemergence to starthistle rosettes but before flower stem elongates. Plants should be rapidly growing at time of treatment. **Remarks:** Apply in 10 to 40 gal/acre of water.
Clopyralid + 2,4-D *Curtail*	**Rate:** 2 to 5 qt *Curtail*/acre **Timing:** Postemergence after most rosettes emerge but before flower stem elongates. **Remarks:** Use higher rates for fallow and Conservation Reserve Program (CRP) applications. With CRP applications, use in established grass only. Apply in enough total spray volume to ensure good coverage. Add a non-ionic surfactant.
Dicamba *Banvel, Clarity*	**Rate:** 2 to 4 pt product/acre (1 to 2 lb a.e./acre). Use higher rate for older plants or dense stands. **Timing:** Postemergence when plants are still in rosettes but before flower stems elongate. Plants should be rapidly growing at time of treatment. **Remarks:** Dicamba is a broadleaf-selective herbicide often combined with other active ingredients. It is not typically used alone to control starthistles.
Picloram	**Rate:** 1 to 2 pt product/acre (4 to 8 oz a.e./acre)

Tordon 22K	**Timing:** Preemergence and postemergence. With postemergence application, optimally treat at rosette to mid-bolting stage (before flowering to prevent current year seed production), or fall rosette stage. Apply when plants are growing rapidly. **Remarks:** Picloram gives a broader spectrum of control than aminopyralid, aminocyclopyrachlor, and clopyralid, and has much longer soil residual activity. Most broadleaf plants are susceptible. It will not damage perennial grasses at the suggested rate. Treatment made in bud stage may not prevent seed production in the year of application. Do not apply near trees. *Tordon 22K* is a restricted-use herbicide. Picloram is not registered for use in California.
AROMATIC AMINO ACID INHIBITORS	
Glyphosate *Roundup, Accord XRT II,* and others	**Rate:** Broadcast foliar treatment: 3 qt product (*Roundup ProMax*)/acre (3.375 lb a.e./acre). Spot treatment: 1.5% v/v solution **Timing:** Postemergence to rapidly growing plants when most plants are at bud stage. **Remarks:** Glyphosate will only provide control during the year of application; it has no soil activity and will not kill seeds or inhibit germination the following season. Glyphosate is nonselective. To achieve selectivity, it can be applied using a wiper or spot treatment to control current year's plants.
BRANCHED-CHAIN AMINO ACID INHIBITORS	
Sulfometuron *Oust* and others	**Rate:** 3 to 5 oz product/acre (2.25 to 3.75 oz a.i./acre) **Timing:** Preemergence or early postemergence, before or during the rainy season when weeds are germinating and actively growing. **Remarks:** Treated soil should be left undisturbed to reduce the potential movement of the herbicide by soil erosion due to wind or water. Treatment of powdery, dry soil or light, sandy soil when there is little likelihood of rainfall soon after treatment may result in off target movement when soil particles are moved by wind or water.
Photosynthetic inhibitors	
Hexazinone *Velpar L*	**Rate:** 1 to 2.5 gal product/acre (2 to 5 lb a.i./acre) **Timing:** Preemergence or postemergence when weeds are germinating or actively growing. **Remarks:** *Centaurea* control is only registered for uncultivated non-agricultural areas (such as rights-of-way), uncultivated agricultural areas (non-crop producing which includes uses such as farmyards and barrier strips), and industrial sites. Use lower rate on coarse-textured soils (sand to sandy loam). Use the higher rate on fine-textured soils (clay loam to clay) and on soils high in organic matter. High rates of hexazinone can create bare ground, so only use high rates in spot treatments.

Centaurea debeauxii Gren. & Godr. ssp. *thuillierii* Dostál
(= *Centaurea* x *pratensis* Thuill., *C. jacea* L. nothosubsp. *pratensis* (W.D.J. Koch) Čelak. [Jepson Manual 2012], *C. jacea* x *nigra*, *Centaurea* x *moncktonii*)

Meadow knapweed

Family: Asteraceae
Range: Occurs sporadically in California, Colorado, Idaho, Montana, Oregon, and Washington.
Habitat: Pastures and moist sites, including moist meadows, river banks, streams, irrigation ditches, and openings in forested areas tree farms, vacant lands, railroads and roadsides.
Origin: Native to Europe. Is a hybrid of black (*C. nigra*) and brown (*C. jacea*) knapweeds. Brown knapweed was introduced into North America as a forage crop and black knapweed was probably introduced in ship ballast or as an ornamental. European plants have been grown as crop or garden plants.
Impact: Meadow knapweed outcompetes grasses and other pasture species. It threatens wildlife habitat and interferes with Christmas tree plantations.
Western states listed as Noxious Weed: Colorado, Idaho, Oregon
California Invasive Plant Council (Cal-IPC) Inventory: Moderate Invasiveness (Alert)

Meadow knapweed is a taprooted perennial to 3 ft tall, generally arising from a woody crown. It resembles spotted knapweed but can be distinguished by having longer phyllaries, 15-18 mm long. Unlike squarrose, spotted and diffuse knapweeds, the leaves are simple with less lobing and are up to 6 inches long by about 1.5 inches wide.

The pink to purple flowerheads are roundish, 1 inch wide, and solitary at the tips of the branches. The phyllaries surrounding the flowers are light to dark brown, with paper-fringed margins, appearing shiny and coppery when flowers are mature. Meadow knapweed reproduces mostly by seed, but roots and root crowns can reestablish when fragmented and dispersed by cultivation or construction equipment. Achenes are brown, lack a pappus, and are much hairier than other knapweed species. Meadow knapweed seeds are primarily dispersed in rivers, streams, or irrigation water. Wildlife and birds will also spread the seed. It is not known how long seeds remain viable in the soil, but it is assumed that survival would be similar to other *Centaurea* species, 2 to 5 years, with a few seeds surviving longer.

NON-CHEMICAL CONTROL

Mechanical (pulling, cutting, disking)	It is possible to hand remove initial or small infestations. Smaller plants are much easier to remove than larger, more established plants. Established meadow knapweed plants have a large root that is hard to pull. It is best to remove as much of the root as possible before flowering. Hand pulling 2 to 4 times per year or severing plants at least 2 inches below crowns can control small infestations. This is easier when the soil is loose or wet. Plants in flower may form viable seeds even after they are pulled. As such, they may need to be placed in bags and removed from the site. Repeat visits are necessary in the following spring and summer to remove newly germinating plants or plants that have reemerged from root fragments. Monitoring of treatment sites should continue for several years.
	Mowing is not effective for most knapweed species. Plants generally resprout and flower again in the same season when mowed. Plants that are regularly mowed will persist for years and can often flower and produce seed below the level of the mower. However, mowing can improve the palatability of the foliage to grazing animals, increase the efficacy of subsequent herbicide treatments, and reduce the weed's competitiveness with desirable forage species. Furthermore, mowing at the late bud to early bloom stage can reduce seed production. Mowing after seed set can disperse seed and can stimulate stem regrowth.
	In some situations it may be feasible to cultivate infested pastures and rotate through an annual hay crop, green manure or some other cleanup crop before reseeding to the desired permanent forage species. If plans

	include using herbicides along with the cleanup crop, grasses allow more options than legumes. Meadow knapweed does not tolerate repeated cultivation but may resprout following the initial breakup of the crown and fleshy roots. Consider the seed reserve in the soil and devise methods to either deplete it (using herbicides or repeated cultivation) or to prevent its germination (by deep burial or shading of the soil surface). Irrigation may be needed for initial establishment of the new forage crop. Rototilling or plowing will eliminate knapweed. Cultivating with a disk will control young plants and seedlings, but established plants can survive if the root or root fragments remain.
Cultural	Meadow knapweed green leaves may be more palatable to livestock than other knapweed species. In pastures, good grazing practices and management of grass and forage species will greatly improve control of knapweed. There are no studies to demonstrate the effectiveness of prescribed burning, although it is likely that the response in meadow knapweed would be similar to that of spotted or diffuse knapweed.
Biological	Some biological controls introduced for other knapweeds also attack meadow knapweed. However, they have not shown much effect. Three seed-feeding insects, a moth (*Metzneria paucipunctella*), a weevil (*Larinus minutus*), and a fly (*Urophora quadrifasciata*) are established on meadow knapweed. The most promising agent is *Larinus minutus*, which may reduce weed populations if its numbers increase as they did with spotted and diffuse knapweed. *Larinus minutus* has been found attacking meadow knapweed in both Oregon and Washington, showing the most potential for damaging young plants and seedlings.

CHEMICAL CONTROL

The following specific use information is based on published papers and reports by researchers and land managers. Other trade names may be available, and other compounds also are labeled for this weed. Directions for use may vary between brands; see label before use. Herbicides are listed by mode of action and then alphabetically. The order of herbicide listing is not reflective of the order of efficacy or preference.

GROWTH REGULATORS	
2,4-D Several names	**Rate:** 1 to 2 qt product/acre (0.95 to 1.9 lb a.e./acre) **Timing:** Postemergence from rosette to beginning of bolting, or fall rosette. Optimal at early flowering stage. **Remarks:** Control with 2,4-D is only temporary and does not prevent seedling establishment the following year. Generally requires repeat applications. 2,4-D is not considered as effective as other growth regulator herbicides for season-long control. It is broadleaf-selective and has no soil activity. Do not apply ester formulation when outside temperatures exceed 80°F. Amine forms are as effective as ester forms for small rosettes, and amine forms reduce the chance of off-target movement.
Aminocyclopyrachlor + chlorsulfuron *Perspective*	**Rate:** 4.75 to 8 oz product (*Perspective*)/acre **Timing:** Postemergence and preemergence. Postemergence applications are most effective when applied to plants from the seedling to the mid-rosette stage. **Remarks:** Provides broad-spectrum control of many broadleaf species. Although generally safe to grasses, it may suppress or injure certain annual and perennial grass species. Aminocyclopyrachlor gives control of other knapweeds and is expected to also provide the same level of control on meadow knapweed. Do not treat in the root zone of desirable trees and shrubs. Do not apply more than 11 oz product/acre per year. At this high rate, cool-season grasses will be damaged, including bluebunch wheatgrass. Not yet labeled for grazing lands. Add an adjuvant to the spray solution. This product is not approved for use in California and some counties of Colorado (San Luis Valley).
Aminopyralid *Milestone*	**Rate:** 5 to 7 oz product/acre (1.25 to 1.75 oz ae/acre) **Timing:** Postemergence and preemergence. Postemergence applications are most effective when applied to plants from the rosette to the bolting stage. Effective control can also be obtained with a fall application to new regrowth. **Remarks:** Aminopyralid is one of the most effective herbicides for the control of meadow knapweed. It is safe on grasses, although preemergence application at high rates can greatly suppress invasive annual grasses, such as medusahead. Aminopyralid has a longer residual and higher activity than clopyralid. Other members of the Asteraceae and Fabaceae are very sensitive to aminopyralid. For postemergence applications, a non-ionic surfactant (0.25 to 0.5% v/v spray solution) enhances control under adverse environmental conditions; however, this is not normally necessary.

	Other premix formulations of aminopyralid can also be used for diffuse knapweed control. These include *Opensight* (aminopyralid + metsulfuron; 1.5 to 2 oz product/acre) and *Forefront HL* (aminopyralid + 2,4-D; 1.2 to 2.1 pt product/acre). Both are applied at the rosette to bolting stages.
Clopyralid *Transline*	**Rate:** 0.67 to 1.33 pt product/acre (4 to 8 oz a.e./acre). Use higher rate for older plants or dense stands. **Timing:** Preemergence (for seedling control) or postemergence (for seedling and perennial plant control). Generally optimal to apply in spring, at beginning of bolting up to the bud stage. Can also apply to fall regrowth. Results are best if applied to rapidly growing weeds. **Remarks:** While it is very safe on grasses, it will injure many members of the Asteraceae, particularly thistles, and can also injure legumes, including clovers. Most other broadleaf species and all grasses are not injured.
Clopyralid + 2,4-D *Curtail*	**Rate:** 2 to 4 qt *Curtail*/acre **Timing:** Same as for clopyralid. **Remarks:** The combination may cause increased damage to other broadleaf species. Add a non-ionic surfactant.
Dicamba *Banvel, Clarity*	**Rate:** 1 to 2 pt product/acre (0.5 to 1 lb a.e./acre). Use higher rate for older plants or dense stands. **Timing:** Postemergence from rosette to beginning of bolting, or fall rosette. Optimal at early flowering stage. **Remarks:** Dicamba is a broadleaf-selective herbicide often combined with other active ingredients. It is not typically used alone to control knapweed species. Dicamba can also be mixed with 2,4-D (1 pt dicamba + 2 pt 2,4-D/acre) or picloram (1 to 2 pt dicamba + 0.5 to 1 pt picloram/acre) for spot treatments.
Picloram *Tordon 22K*	**Rate:** 1 to 2 pt product/acre (4 to 8 oz a.e./acre). Use higher rates for older plants or dense stands. **Timing:** Preemergence and postemergence. With postemergence application, optimally treat at rosette to mid-bolting stage (before flowering to prevent current year seed production), or fall rosette stage. Apply when plants are growing rapidly. Under favorable growing conditions, application in summer can be effective if higher application volumes are used. **Remarks:** Broadleaf herbicide with a broader spectrum of control than aminocyclopyrachlor, aminopyralid, and clopyralid, and much longer soil residual activity. Lower rates may require annual spot treatments. Treatment made in bud stage may not prevent seed production in the year of application. Picloram has been shown to provide selective control of knapweeds for 3 to 4 years. Although well-developed grasses are not usually injured by labeled use rates, some applicators have noted that young grass seedlings with fewer than four leaves may be killed. Do not apply near trees. Picloram is a restricted use herbicide. It is not registered for use in California. Control with lower rates may be improved by tank mixing with dicamba or 2,4-D; picloram and dicamba (0.25 to 0.5 pt/acre + 0.125 to 0.25 pt/acre) and picloram plus 2,4-D (0.5 to 1 pt picloram + 1 to 2 pt 2,4-D/acre). A backpack sprayer or a wiper is highly recommended in small areas to minimize damage to non-target plants.
AROMATIC AMINO ACID INHIBITORS	
Glyphosate *Roundup, Accord XRT II,* and others	**Rate:** Broadcast foliar treatment: 3 qt product (*Roundup ProMax*)/acre (3.375 lb a.e./acre). Spot treatment: 1.5% v/v solution. **Timing:** Postemergence to rapidly growing knapweed when most plants are at bud stage. **Remarks:** Glyphosate will only provide control during the year of application; it has no soil activity and will not kill seeds or inhibit germination the following season. Glyphosate is nonselective. To achieve selectivity, it can be applied using a wiper or spot treatment to control current year's plants.

Centaurea diffusa Lam.
Diffuse knapweed

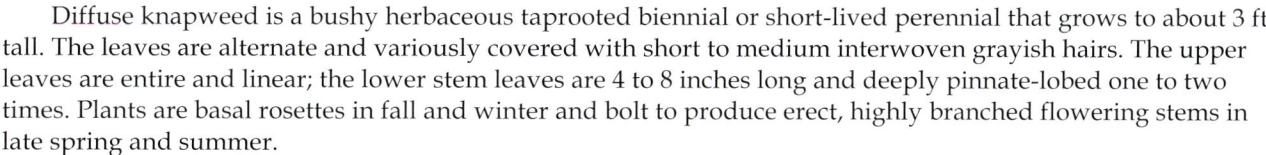

Family: Asteraceae
Range: All western states.
Habitat: Plains, rangelands, and forested benchlands, particularly on rugged terrain not well suited for cultivation; often on well drained soils. Needs less moisture than spotted knapweed; can thrive in semi-arid and arid conditions. Seldom persists in shaded places. Not common on cultivated lands or irrigated pasture because it cannot tolerate cultivation or excessive moisture.
Origin: Native to southeastern Eurasia.
Western states listed as Noxious Weed: Arizona, California, Colorado, Idaho, Montana, New Mexico, Nevada, North Dakota, Oregon, South Dakota, Utah, Washington, Wyoming
California Invasive Plant Council (Cal-IPC) Inventory: Moderate Invasiveness

Diffuse knapweed is a bushy herbaceous taprooted biennial or short-lived perennial that grows to about 3 ft tall. The leaves are alternate and variously covered with short to medium interwoven grayish hairs. The upper leaves are entire and linear; the lower stem leaves are 4 to 8 inches long and deeply pinnate-lobed one to two times. Plants are basal rosettes in fall and winter and bolt to produce erect, highly branched flowering stems in late spring and summer.

The flowerheads consist of spiny or comb-like phyllaries and white or pink to pale purple disk flowers. Unlike squarrose knapweed, the spiny phyllaries (3 mm long) do not reflex downward. The achenes either lack a pappus or have a very short bristly pappus (< 1 mm long). Plants reproduce only by seed. Diffuse knapweed inflorescences detach from the parent plant when stems break off near the ground and tumble along the ground in the wind, dispersing seed to a greater distance than most *Centaurea* species. Data shows that about 20 to 50% of plant inflorescences tumble off site. Diffuse knapweed has been shown to occasionally hybridize with spotted knapweed. It is not known how long seeds remain viable in the soil, but it is assumed that survival would be similar to other *Centaurea* species, 2 to 5 years, with a few seeds surviving longer.

NON-CHEMICAL CONTROL

Mechanical (pulling, cutting, disking)	Physical and mechanical approaches to diffuse knapweed control include hand pulling, digging, tilling, disking, and cutting or mowing. Physical removal or damage can provide some control depending on the timing and frequency of treatment, the presence of competitive, desirable vegetation, and the level of soil disturbance caused by the treatment.
	Hand pulling is practical for scattered diffuse knapweed plants, or for areas where other control methods are not feasible and sufficient labor is available. Repeated hand pulling is necessary during the season and over many years. Successful control has been reported when plants were hand removed 3 times a year (spring, summer, and late summer) over a period of 5 years. Every effort should be made to remove the entire taproot with little soil disturbance. If not possible, then cut the root 2 to 4 inches below the soil surface to remove much of the reproductive crown. Gloves should be worn when hand pulling. The best timing for hand removal is before plants produce viable seed. Hand pulling has not been effective in all areas. On dry soils, it may be difficult to remove the root crown and this can lead to rapid reestablishment.
	Mowing typically doesn't kill knapweeds; cut plants generally survive and recover to set seed. Plants mowed at the rosette stage will quickly recover, and mowing too late (after seed set) can disperse seed. However, mowing at the late bud to early bloom stage will reduce seed production. Mowing can also remove dead growth to improve herbicide coverage. A program of cutting only bolted plants, particularly at the early bloom stage, for several consecutive years can greatly suppress diffuse knapweed.
	Cultivation is effective when repeated, but diffuse knapweed typically grows in areas not conducive to tillage.
Cultural	Grazing is not an effective eradication method. Diffuse knapweed is not typically considered palatable to livestock. Furthermore, intensive grazing can create ideal seedbeds for further invasion. However, researchers have shown that cattle, sheep and goats will readily graze diffuse knapweed in early spring. Cattle grazing twice in spring decreased seed production by 50%. Sheep typically graze diffuse knapweed

	from the rosette through bud stage or when it is the only plant available. The timing of grazing may be critical to its success. Although there is no direct evidence, it is likely that the optimal timing would be similar to that of spotted knapweed. For spotted knapweed, early and late-season grazing appear to be the most effective timing with sheep: early season (spring) to reduce flower production, and late season (fall) to reduce density of young plants. In one study, two consecutive years of early and late sheep grazing reduced spotted knapweed but had little effect on the native grass community.
	Burning has been shown to give effective control of diffuse knapweed while stimulating grass regrowth. Within 2 years of burning, most diffuse knapweed rosettes were eliminated. A low-severity fire will not kill the below-ground reproductive structures of diffuse knapweed, but a severe fire can kill some of the plant crowns. Burning removes current growth but may enhance seed germination. Dry soil conditions at the time of burning can reduce germination. Burning also can remove dead growth to improve the effect of herbicide applications.
Biological	Currently, there is no single biological control agent that effectively controls diffuse knapweed populations. However, numerous biocontrol insects from diffuse knapweed's native range are established in the United States, including flies and weevils which attack seedheads. These include the banded gall fly (*Urophora affinis*), knapweed seedhead fly (*U. quadrifasciata*), knapweed peacock fly (*Chaetorellia acrolophi*), lesser knapweed flower weevil (*Larinus minutus*), and broad-nosed seedhead weevil (*Bangasternus fausti*). *Larinus minutus* in particular is an aggressive and effective biocontrol insect for diffuse and spotted knapweed.
	Root-feeding insects may have a more detrimental effect on knapweed populations than seed-feeding ones. Larvae of the diffuse knapweed root beetle (*Sphenoptera jugoslavica*) feed in the roots of diffuse knapweed. Larvae of the moths *Agapeta zoegana* and *Pterolonche inspersa* and the weevil *Cyphocleonus achates* feed in the roots of both diffuse knapweed and spotted knapweed.
	The collective stress on the plant caused by these insects sharply reduces seed production and may lead to reduced competitiveness. However, they have not been show to significantly reduce diffuse knapweed plant densities.

CHEMICAL CONTROL

The following specific use information is based on published papers and reports by researchers and land managers. Other trade names may be available, and other compounds also are labeled for this weed. Directions for use may vary between brands; see label before use. Herbicides are listed by mode of action and then alphabetically. The order of herbicide listing is not reflective of the order of efficacy or preference.

GROWTH REGULATORS	
2,4-D Several names	**Rate:** 1 to 2 qt product/acre (0.95 to 1.9 lb a.e./acre)
	Timing: Postemergence from rosette to beginning of bolting, or fall rosette. Optimal at early flowering stage.
	Remarks: Control with 2,4-D is only temporary and does not prevent seedling establishment the following year. Generally requires repeat applications. It is not considered as effective as other growth regulator herbicides for season-long control. 2,4-D is broadleaf-selective and has no soil activity. Do not apply ester formulation when outside temperatures exceed 80°F. Amine forms are as effective as ester forms for small rosettes, and amine forms reduce the chance of off-target movement.
Aminocyclopyrachlor + chlorsulfuron *Perspective*	**Rate:** 4.75 to 8 oz product (*Perspective*)/acre
	Timing: Postemergence and preemergence. Postemergence applications are most effective for plants from the seedling to the mid-rosette stage.
	Remarks: *Perspective* provides broad-spectrum control of many broadleaf species. Although generally safe to grasses, it may suppress or injure certain annual and perennial grass species. Do not treat in the root zone of desirable trees and shrubs. Do not apply more than 11 oz product/acre per year. At this high rate, cool-season grasses will be damaged, including bluebunch wheatgrass. Not yet labeled for grazing lands. Add an adjuvant to the spray solution. This product is not approved for use in California and some counties of Colorado (San Luis Valley).
Aminopyralid *Milestone*	**Rate:** 5 to 7 oz product/acre (1.25 to 1.75 oz a.e./acre)
	Timing: Postemergence and preemergence. Postemergence applications are most effective for plants from the rosette to the bolting stage. Effective control can also be obtained with a fall application to new regrowth.

	Remarks: Aminopyralid is one of the most effective herbicides for the control of diffuse knapweed. It is safe on grasses, although preemergence application at high rates can greatly suppress invasive annual grasses, such as medusahead. Aminopyralid has a longer residual and higher activity than clopyralid. Other members of the Asteraceae and Fabaceae are very sensitive to aminopyralid. For postemergence applications, adding a non-ionic surfactant (0.25 to 0.5% v/v spray solution) enhances control under adverse environmental conditions; however, this is not normally necessary. Other premix formulations of aminopyralid can also be used for diffuse knapweed control. These include *Opensight* (aminopyralid + metsulfuron; 1.5 to 3.3 oz product/acre) and *Forefront HL* (aminopyralid + 2,4-D; 1.2 to 2.1 pt product/acre), both applied at the rosette to bolting stages.
Clopyralid *Transline*	**Rate:** 0.67 to 1.33 pt product/acre (4 to 8 oz a.e./acre). Use higher rate for older plants or dense stands. **Timing:** Preemergence (for seedling control) or postemergence (for seedlings and perennial plant control). Generally optimal to apply in spring, at beginning of bolting up to the bud stage. Can also apply to fall regrowth. Results are best if applied to rapidly growing weeds. **Remarks:** While clopyralid is very safe on grasses, it will injure many members of the Asteraceae, particularly thistles, and can also injure legumes, including clovers. Most other broadleaf species and all grasses are not injured.
Clopyralid + 2,4-D *Curtail*	**Rate:** 2 to 4 qt *Curtail*/acre **Timing:** Same as for clopyralid. **Remarks:** Add a non-ionic surfactant.
Dicamba *Banvel, Clarity*	**Rate:** 1 to 2 pt product/acre (0.5 to 1 lb a.e./acre). Use higher rate for older plants or dense stands. **Timing:** Postemergence from rosette to beginning of bolting, or fall rosette. Optimal at early flowering stage. **Remarks:** Dicamba is a broadleaf-selective herbicide often combined with other active ingredients. It is not typically used alone to control diffuse knapweed. Dicamba can also be mixed with 2,4-D (1 pt dicamba + 2 pt 2,4-D/acre) or picloram (1 to 2 pt dicamba + 0.5 to 1 pt picloram/acre) for spot treatments. Dicamba is available mixed with diflufenzopyr in a formulation called *Overdrive*. This has been reported to be effective on diffuse knapweed. Diflufenzopyr is an auxin transport inhibitor which causes dicamba to accumulate in shoot and root meristems, increasing its activity. *Overdrive* is applied postemergence to rapidly growing plants at 4 to 8 oz product/acre. Higher rates should be used on large annuals and biennials or when treating perennial weeds. Add a non-ionic surfactant to the treatment solution at 0.25% v/v or a methylated seed oil at 1% v/v solution.
Picloram *Tordon 22K*	**Rate:** 1 to 2 pt product/acre (4 to 8 oz a.e./acre). Use higher rates for older plants or dense stands. **Timing:** Preemergence and postemergence. Postemergence applications are best at rosette to mid-bolting stage (before flowering to prevent current year seed production), or fall rosette stage. Apply when plants are growing rapidly. Under favorable growing conditions, summer application can be effective if higher rates are used. **Remarks:** Picloram is a broadleaf herbicide. It gives a broader spectrum of control than aminopyralid, aminocyclopyrachlor, and clopyralid, and has much longer soil residual activity. Lower rates may require annual spot treatments. Treatment made in bud stage may not prevent seed production in the year of application. Picloram has been shown to provide selective control of diffuse knapweed for 3 to 4 years. Although well-developed grasses are not usually injured by labeled use rates, some applicators have noted that young grass seedlings with fewer than four leaves may be killed. Do not apply near trees. Picloram is a restricted use herbicide. It is not registered for use in California. Control with lower rates may be improved by tank mixing with dicamba or 2,4-D; picloram and dicamba (0.25 to 0.5 pt/acre + 0.125 to 0.25 pt/acre) and picloram plus 2,4-D (0.5 to 1 pt picloram + 1 to 2 pt 2,4-D/acre). A backpack sprayer or a wiper is highly recommended in small areas to minimize damage to non-target plants.
AROMATIC AMINO ACID INHIBITORS	
Glyphosate *Roundup, Accord XRT II,* and others	**Rate:** Broadcast foliar treatment: 3 qt product (*Roundup ProMax*)/acre (3.375 lb a.e./acre). Spot treatment: 1.5% v/v solution **Timing:** Postemergence to rapidly growing knapweed when most plants are at bud stage.

	Remarks: Glyphosate will only provide control during the year of application; it has no soil activity and will not kill seeds or inhibit germination the following season. Glyphosate is nonselective. To achieve selectivity, it can be applied using a wiper or spot treatment to control current year's plants.
BRANCHED-CHAIN AMINO ACID INHIBITORS	
Imazapyr *Arsenal, Habitat, Stalker, Chopper, Polaris*	3 to 4 pt product/acre has been shown to give some level of control.

Centaurea melitensis L.

Malta starthistle or tocalote

Family: Asteraceae
Range: All southwestern states, including California, Nevada, Utah, Arizona, New Mexico and Texas, as well as Oregon and Washington.
Habitat: Open disturbed sites, open hillsides, grassland, rangeland, open woodlands, fields, pastures, roadsides, waste places. May also inhabit cultivated fields.
Origin: Native to southern Europe. Malta starthistle is thought to have been introduced into California in the late 1700s during the Spanish Missionary period.
Impacts: Not as competitive or widespread as yellow starthistle. May increase erosion and reduce water percolation, but does not survive as long as yellow starthistle so it is not likely to have the same effect on soil moisture depletion. Dense stands can displace native plants and animals, with documented negative effects on seed production in the endangered mint *Acanthiminta ilicifolia*. Malta starthistle is not known to cause chewing disease in horses and is used medicinally in Spain.
Western states listed as Noxious Weed: California, Nevada, New Mexico
California Invasive Plant Council (Cal-IPC) Inventory: Moderate Invasiveness

Malta starthistle is a simple to bushy winter annual with spiny yellow-flowered heads. Although it can resemble yellow starthistle, it is typically much shorter, growing to a maximum of 3 ft tall. In addition, Malta starthistle completes its life cycle earlier than yellow starthistle, and generally does not survive late into summer. Plants exist as basal rosettes through winter and early spring until flower stems develop in late spring. Its foliage is grayish- to bluish-green, with fine white cottony hairs. Its leaf bases form wings along the stems.

The flowerheads of Malta starthistle are solitary or in close clusters of two to three. Like yellow starthistle, the phyllaries are often sparsely covered with cobwebby hairs. The central spine of main phyllaries is half the length (5 to 12 mm long) of yellow starthistle and is generally purple- to brown-tinged. Malta starthistle produces three different types of flowerheads, including fully expanded flowers capable of cross-pollination, and two cleistogamous (self-pollinated) types, one with yellow flowers only partially protruding and the other without exerted flowers.

Malta starthistle reproduces only by seed. Malta starthistle only has one type of achene, while yellow starthistle has two. The achenes of Malta starthistle are all finely pubescent, grayish to tan, with a bristly pappus, 1 to 3 mm long. Unlike yellow starthistle, the senesced heads retain the central spines and often shed the loose receptacle and dense fuzzy gray hairs, leaving a shallow bowl of spiny phyllaries. The vast majority of seeds fall near the parent plant. Most seeds germinate after the first fall rains. Seed longevity in the soil is probably similar to yellow starthistle: few seeds survive beyond 4 years, but seeds can survive for up to about 10 years under certain conditions.

NON-CHEMICAL CONTROL

Mechanical (pulling, cutting, disking)	Removal techniques such as hand pulling, mowing, or cultivation, when used to prevent seed production over 2 to 4 years or more (the soil life of the seeds), should reduce or eliminate an infestation.
	See control techniques for yellow starthistle. These same techniques are expected to be effective on Malta starthistle.
Cultural	There have been no studies conducted on the use of grazing or prescribed burning for the control of Malta starthistle. It is possible that control with grazing would be similar to that of yellow starthistle. However, it is very unlikely that prescribed burning would be an effective tool, as the timing of the life cycle of Malta starthistle is much earlier than yellow starthistle and burning could possibly damage many desirable species. In addition, Malta starthistle does not typically occur in dense enough infestations to provide enough fuel to carry a fire.
Biological	A small seedhead-feeding beetle (*Lasioderma haemorrhoidale*) was unintentionally introduced to California from the Mediterranean region, but has had little effect in controlling Malta starthistle. It is considered a

generalist seedhead feeder and is also known to attack yellow starthistle, Sicilian starthistle, blessed milkthistle and Italian thistle. The false peacock fly and hairy weevil, introduced to control yellow starthistle, will also attack Malta starthistle, but to a lesser degree than yellow starthistle.

CHEMICAL CONTROL

The following specific use information is based on reports by researchers and land managers and much of the information is from research with *Centaurea solstitialis* (yellow starthistle). Little research has been conducted on the control of Malta starthistle. Other trade names may be available, and other compounds also are labeled for this weed. Directions for use may vary between brands; see label before use. Herbicides are listed by mode of action and then alphabetically. The order of herbicide listing is not reflective of the order of efficacy or preference.

GROWTH REGULATORS	
2,4-D Several names	**Rate:** 1 to 1.5 pt product/acre (0.48 to 0.72 lb a.e./acre) for small rosettes, 2 to 4 pt product/acre (0.95 to 1.9 lb a.e./acre) for larger plants up to bolting **Timing:** Postemergence from rosette to beginning of bolting, but before flowering. **Remarks:** 2,4-D controls larger plants well, but is not considered as effective as other growth regulator herbicides for season-long control. It is broadleaf-selective and has no soil activity. Do not apply ester formulation when outside temperatures exceed 80°F. Amine forms are as effective as ester forms for small rosettes, and amine forms reduce the chance of off-target movement.
Aminocyclopyrachlor + chlorsulfuron *Perspective*	**Rate:** 3 to 5 oz product (*Perspective*)/acre **Timing:** Postemergence and preemergence. Postemergence applications are most effective when applied to plants from the seedling to the mid-rosette stage. **Remarks:** Aminocyclopyrachlor gives control of yellow starthistle similar to aminopyralid. *Perspective* provides broad-spectrum control of many broadleaf species. Although generally safe to grasses, it may suppress or injure certain annual and perennial grass species. Do not treat in the root zone of desirable trees and shrubs. Do not apply more than 11 oz product/acre per year. At this high rate, cool-season grasses will be damaged, including bluebunch wheatgrass. Not yet labeled for grazing lands. Add an adjuvant to the spray solution. This product is not approved for use in California and some counties of Colorado (San Luis Valley).
Aminopyralid *Milestone*	**Rate:** 4 to 5 oz product/acre (1 to 1.25 oz a.e./acre). Use higher rates when weeds are larger. **Timing:** Postemergence and preemergence. Postemergence applications are most effective when applied to plants from the seedling to the mid-rosette stage. **Remarks:** Aminopyralid is one of the most effective herbicides for the control of all starthistles. It is safe on grasses, although preemergence application at high rates can greatly suppress invasive annual grasses, such as medusahead. Aminopyralid has a longer residual and higher activity than clopyralid. Other members of the Asteraceae and Fabaceae are very sensitive to aminopyralid. For postemergence applications, a non-ionic surfactant (0.25 to 0.5% v/v spray solution) enhances control under adverse environmental conditions; however, this is not normally necessary. Other premix formulations of aminopyralid can also be used for Malta starthistle control. These include *Chaparral* (aminopyralid + metsulfuron; 1.5 to 2 oz product/acre), *Opensight* (aminopyralid + metsulfuron; 1.5 to 2 oz product/acre), and *Forefront HL* (aminopyralid + 2,4-D; 1.5 to 2.1 pt product/acre), all applied at the rosette to bolting stages.
Clopyralid *Transline*	**Rate:** 0.25 to 0.67 pt product/acre (1.5 to 4 oz a.e./acre). Seedlings and rosettes can be treated at the lower rate, but bolted plants should be treated at higher rates. **Timing:** Postemergence and preemergence. For postemergence application, apply to plants from seedling to mid-bolting stage. However, optimal timing is at the later rosette stages, but before bolting. **Remarks:** Clopyralid gives excellent control of all starthistles. While it is very safe on grasses, it will injure many members of the Asteraceae, particularly thistles, and can also injure legumes, including clovers. Most other broadleaf species and all grasses are not injured. When clopyralid is used to control seedlings, a surfactant is not necessary. However, when treating older plants or plants exposed to moderate levels of drought stress, surfactants can enhance the activity of the herbicide.
Dicamba	**Rate:** 0.5 pt product/acre (0.25 lb a.e./acre) for seedlings, 1 to 1.5 pt product/acre (0.5 to 0.75 lb

Banvel, Clarity	a.e./acre) for larger plants up to bolting. **Timing:** Postemergence to plants from rosette to beginning of bolting. **Remarks:** Dicamba is a broadleaf-selective herbicide often combined with other active ingredients. It is not typically used alone to control any starthistle species.
Picloram *Tordon 22K*	**Rate:** 1 to 2 pt product/acre (4 to 8 oz a.e./acre) **Timing:** Postemergence and preemergence. Postemergence applications should be made to plants from rosette to bud formation stage. Apply when there is adequate soil moisture and weeds are growing rapidly. **Remarks:** Picloram acts much like aminopyralid, aminocyclopyrachlor, and clopyralid, but gives a broader spectrum of control and has much longer soil residual activity. It can provide about 2 to 3 years of control. Most broadleaf plants are susceptible. Although well-developed grasses are not usually injured by labeled use rates, some applicators have noted that young grass seedlings with fewer than four leaves may be killed. Do not apply near trees. *Tordon 22K* is a federally restricted use pesticide. Picloram is not registered for use in California.
Triclopyr *Garlon 3A, Garlon 4 Ultra*	**Rate:** 1 pt *Garlon 4 Ultra* or 1.33 pt *Garlon 3A*/acre (0.5 lb a.e./acre) for seedlings, up to 3 pt *Garlon 4 Ultra* or 4 pt *Garlon 3A*/acre (1.5 lb a.e./acre) for larger plants. **Timing:** Postemergence from seedling to bolting stage. **Remarks:** Triclopyr has little to no residual activity. It is broadleaf-selective and typically does not harm grasses. Formulated as both an amine and ester. *Garlon 4 Ultra* is formulated as a low volatile ester. However, in warm temperatures, spraying onto hard surfaces such as rocks or pavement can increase the risk of volatilization and off-target damage.
AROMATIC AMINO ACID INHIBITORS	
Glyphosate *Roundup, Accord XRT II,* and others	**Rate:** Broadcast foliar treatment: 1.33 to 2.67 qt product (*Roundup ProMax*)/acre (1.5 to 3 lb a.e./acre). Spot treatment: 1% to 2% v/v solution **Timing:** Postemergence from bolting to beginning of flowering. **Remarks:** Glyphosate is the most effective herbicide for late season control. Good coverage, clean water, and rapidly growing Malta starthistle plants are all essential for adequate control. It has no soil activity. Glyphosate is nonselective. To achieve selectivity, it can be applied using a wiper or spot treatment to control current year's plants. In late season treatments a surfactant should be added to the herbicide mixture.
BRANCHED-CHAIN AMINO ACID INHIBITORS	
Chlorsulfuron *Telar*	**Rate:** 1.33 to 2.6 oz product/acre (1 to 1.95 oz a.i./acre) **Timing:** Preemergence activity only. Chlorsulfuron does not have postemergence activity on yellow or Malta starthistle and must be used in combination with 2,4-D, dicamba, or triclopyr to provide effective control. **Remarks:** Chlorsulfuron has mixed selectivity on both broadleaf and grass species but is generally safe on grasses. It has a fairly long soil residual. The herbicide solution requires constant agitation during application.
Imazapyr *Arsenal, Habitat, Stalker, Chopper, Polaris*	Not often used for Malta starthistle control but has been shown to be somewhat effective at 3 to 4 pt product/acre. It has preemergence and some postemergence activity, and has a long soil residual.
Sulfometuron *Oust* and others	Not often used for Malta starthistle control but has been shown to be somewhat effective at 1 to 2 oz product/acre. It has preemergence activity only, and a long soil residual.
PHOTOSYNTHETIC INHIBITORS	
Hexazinone *Velpar L*	Not often used for Malta starthistle control but has been shown to be somewhat effective at 1 to 2.5 gal product/acre (2 to 5 lb a.i./acre). It has preemergence activity only, and a long soil residual. High rates of hexazinone can create bare ground, so only use high rates in spot treatments.

Centaurea solstitialis L.
Yellow starthistle

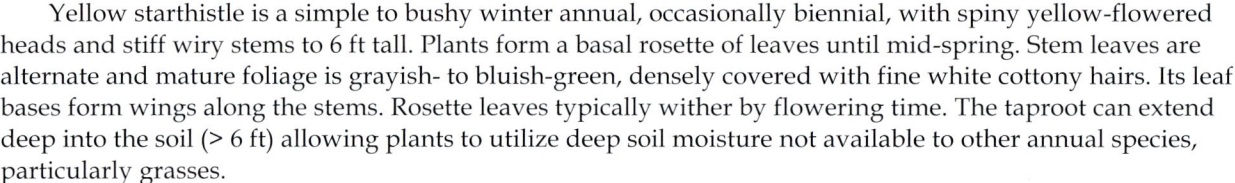

Family: Asteraceae
Range: Most contiguous states, except a few southern and northeastern states.
Habitat: Open disturbed sites, open hillsides, grassland, rangeland, open woodlands, fields, pastures, roadsides, waste places. May also inhabit cultivated fields. Does not tolerate low light areas or shading.
Origin: Southern Europe. Accidentally introduced as a seed contaminant in alfalfa. It has spread rapidly since its introduction into California in the mid-1800s.
Impacts: Plants are highly competitive and typically develop dense, impenetrable stands that displace desirable vegetation in natural areas, rangelands, roadsides and other places. Yellow starthistle is considered one of the most serious rangeland weeds in the western U.S. Yellow starthistle is sometimes problematic in grain fields, where the seeds can contaminate the grain harvest and lower its quality and value. Yellow starthistle contains an unidentified compound that causes nigropallidal encephalomalacia or chewing disease in horses.
Western states listed as Noxious Weed: Arizona, California, Colorado, Idaho, Montana, New Mexico, Nevada, North Dakota, Oregon, South Dakota, Utah, Washington
California Invasive Plant Council (Cal-IPC) Inventory: High Invasiveness

Yellow starthistle is a simple to bushy winter annual, occasionally biennial, with spiny yellow-flowered heads and stiff wiry stems to 6 ft tall. Plants form a basal rosette of leaves until mid-spring. Stem leaves are alternate and mature foliage is grayish- to bluish-green, densely covered with fine white cottony hairs. Its leaf bases form wings along the stems. Rosette leaves typically wither by flowering time. The taproot can extend deep into the soil (> 6 ft) allowing plants to utilize deep soil moisture not available to other annual species, particularly grasses.

The flowerheads are solitary on stem tips, and consist of numerous yellow disk flowers. The phyllaries are densely to sparsely covered with cottony hairs or with patches of hairs at the bases of the spines. The central spine of the main phyllaries is 10 to 25 mm long, stiff, yellowish to straw-colored throughout. Yellow starthistle reproduces only by seed and develops two types of achenes. The outer ring of achenes is a dull dark brown, often speckled with tan, lacking pappus bristles, and often remaining in heads. The inner achenes are glossy, gray or tan to mottled cream-colored and tan, with slender white pappus bristles 2 to 5 mm long. Most seeds fall near the parent plant. Some seed is viable 8 days after flower initiation. Large flushes of seeds typically germinate after the first fall rains, but smaller germination flushes can occur during winter and early spring. Seeds can survive for up to about 10 years in the field under certain environmental conditions, but it appears that few seeds survive beyond 4 years.

NON-CHEMICAL CONTROL

Mechanical (pulling, cutting, disking)	Hand removal, mowing, or cultivation, when used to prevent seed production over 2 to 3 years or more (the soil life of the seeds), can reduce or eliminate an infestation.
	Manual removal of yellow starthistle is most effective with small patches or in maintenance programs where plants are sporadically located in the grassland system. This usually occurs with a new infestation or in the third year or later in a long-term management program. These methods can also be important in steep or uneven terrain where other mechanical tools (e.g., mowing) are impossible to use. To ensure that plants do not recover it is important to detach all above-ground stem material. Leaving even a 2-inch piece of the stem can result in recovery if leaves and buds are still attached to the base of the plant. The best timing for manual removal is after plants have bolted but before they produce viable seed (i.e. early flowering). At this time, plants are easy to recognize, and some or most of the lower leaves have senesced. If hand removal is conducted after plants begin to produce seeds, it may be necessary to put pulled plants in bags and remove them from the site. Hand removal is particularly easy in areas with competing vegetation. Under this condition, yellow starthistle will develop a more erect slender stem with few basal leaves. These plants are relatively brittle and easy to remove. In addition, they usually lack leaves at the base and, consequently,

rarely recover even when a portion of the stem is left intact. Hand removal options for yellow starthistle typically include hand pulling, hoeing, or string trimming. Systematic surveys and repeated removal should be conducted every 2 to 4 weeks throughout the growing season.

Mowing is most effective when 2 to 5% of the total population of seedheads is in bloom. Mowing too early can result in higher seed production. Plants should be cut below the height of the lowest branches. It will require multiple years of continuous mowing to successfully manage yellow starthistle. Mowing is best used in an integrated approach. Since it is a late season management tool, it is best employed in the later years of a long-term management program or in a lightly infested area. Mowing is not feasible in many locations due to rocks and steep terrain. Mowing is not always successful and can decrease the reproductive efforts of insect biocontrol agents, injure late growing native forb species, and reduce fall and winter forage for wildlife and livestock.

The success of mowing depends on proper timing and the growth form of the plant. Mowing too early (before seedheads reach spiny stage) or too late (after seed set) will usually increase the yellow starthistle problem. Mowing too early in the season can remove competitive grass cover and promote vigorous yellow starthistle regrowth. If done too late, mowing scatters yellow starthistle seed. Best results were obtained by mowing once at the early flowering stage, and again 4 to 6 weeks later to cut regrowth during the floral bud stage. A dense spring canopy of desirable vegetation optimizes yellow starthistle control. Yellow starthistle plants with an erect, high-branching growth form are effectively controlled by a single mowing at the early flowering stage, while sprawling low-branching plants cannot be controlled even with repeated mowing. Despite its limitations, mowing conducted at the early flowering stage, before viable seed production, can be very effective for yellow starthistle control.

Anecdotal information also indicates that mowing the standing skeletons in fall, before the first rains, can form a mulch that blocks light and suppresses subsequent germination of yellow starthistle. A flail mower is considered best. The yellow starthistle litter layer may be less suppressive to grass germination, as it is not as light dependent as yellow starthistle.

Tillage is effective, and is occasionally used on roadsides. It is also often used in agricultural lands, which is probably why yellow starthistle is not a significant cropland weed. In wildlands and rangelands, tillage is usually not appropriate because it can damage important desirable species, increase erosion, alter soil structure, and expose the soil for rapid reinfestation if subsequent rainfall occurs. Any tillage operation that severs the roots below the soil surface can effectively control yellow starthistle. Early summer tillage, before viable seeds are set, and repeated tillage following rainfall/germination events will rapidly deplete the yellow starthistle seed bank, but may also have the same effect on the seed bank of desirable species.

Cultural

High-intensity short-duration grazing by sheep, goats, or cattle should be implemented during the period when yellow starthistle plants have bolted to just before they produce spiny heads. Cattle and sheep avoid yellow starthistle once the buds produce spines, whereas goats continue to browse plants even in the flowering stage. For this reason, goats have become a more popular method for controlling yellow starthistle in relatively small infestations.

Grazing the weed during the bolting stage can provide palatable high protein forage (8 to 14%). This can be particularly useful in late spring and early summer when other annual species have senesced. Grazing alone will not provide long-term management or eradication of yellow starthistle, but can be a valuable tool in an integrated management program. This prescription must be continued for at least 3 years in a severe infestation to reduce the yellow starthistle seed bank.

Prescribed burns can provide control if conducted at the proper timing. Burning should be timed to coincide with the very early yellow starthistle flowering stage. At this time yellow starthistle has yet to produce viable seed, whereas seeds of most desirable species have dispersed and grasses have dried to provide adequate fuel. Fire has little if any impact on seeds in the soil. Burning at other times may enhance yellow starthistle survival by removing the thatch and encouraging seed germination in fall.

The ability to use repeated burning depends on climatic and environmental conditions. In areas where resources are ample and total plant biomass is abundant, 2 or 3 consecutive years of burning may be practical. However, in other situations, fuel loads may not be sufficient to allow multiple year burns. Consequently, prescribed burning may be more appropriate as part of an integrated approach.

Air quality issues can be significant when burns are conducted adjacent to urban areas. A major risk of prescribed burning is the potential of fire escapes. This risk is greatest when burns are conducted during the summer months. In some areas, burning can lead to rapid invasion by other undesirable species with wind-dispersed seeds, particularly members of the sunflower family.

In addition to summer burning, yellow starthistle seedlings have been controlled using winter or early spring

	flaming. This technique is somewhat nonselective, and control of yellow starthistle is inconsistent. When spring drought follows a flaming treatment, control of yellow starthistle can be excellent. In contrast, a wet spring can lead to complete failure and increased yellow starthistle infestation, particularly since competing species may be dramatically suppressed.
Biological	Six insects have become established for the control of yellow starthistle in the western United States. These include three species of weevils (seed-head weevil [*Bangasternus orientalis*], flower weevil [*Larinus curtus*], and the hairy weevil [*Eustenopus villosus*]), and three species of flies (seed-head fly [*Urophora sirunaseva*], peacock fly [*Chaetorellia australis*], and the false peacock fly [*Chaetorellia succinea*]). All six insects attack the flower heads of yellow starthistle and produce larvae that develop and feed within the seedhead. Of these, only four have become well established. Of these, only two, *Eustenopus villosus* and *Chaetorellia succinea*, have any significant impact on reproduction. The combination of these two insects reduces seed production by 43 to 76%. Although this level of suppression is not sufficient to provide long-term yellow starthistle management, the use of biological control agents can be an important component of an integrated management approach. A more successful biological control program will likely require the introduction of plant pathogens or other insects which attack roots, stems, or foliage. A new potential biological control agent is a root-feeding weevil, *Ceratapion basicorne*, that has shown promise under greenhouse conditions. It has yet to be approved, but is expected to be released in the next couple of years. The most widely studied pathogen for yellow starthistle control is the Mediterranean rust fungus *Puccinia jaceae*. It can attack the leaves and stem of yellow starthistle, causing enough stress to reduce flowerhead and seed production. Although it has been released it does not seem to have much impact on yellow starthistle populations.

CHEMICAL CONTROL

The following specific use information is based on published papers and reports by researchers and land managers. Other trade names may be available, and other compounds also are labeled for this weed. Directions for use may vary between brands; see label before use. Herbicides are listed by mode of action and then alphabetically. The order of herbicide listing is not reflective of the order of efficacy or preference.

GROWTH REGULATORS	
2,4-D Several names	**Rate:** 1 to 1.5 pt product/acre (0.48 to 0.72 lb a.e./acre) for small rosettes, 2 to 4 pt product/acre (0.95 to 1.9 lb a.e./acre) for larger plants up to bolting **Timing:** Postemergence from rosette to beginning of bolting, but before flowering. **Remarks:** 2,4-D controls larger plants well, but is not considered as effective as other growth regulator herbicides for season-long control. It is broadleaf-selective and may injure other non-target species, particularly crop plants. 2,4-D has no soil activity. Do not apply ester formulation when outside temperatures exceed 80°F. Amine forms are as effective as ester forms for small rosettes, and amine forms reduce the chance of off-target movement from volatility.
Aminocyclopyrachlor + chlorsulfuron *Perspective*	**Rate:** 3 to 5 oz product (*Perspective*)/acre **Timing:** Postemergence and preemergence. Postemergence applications are most effective when applied to plants from the seedling to the mid-rosette stage. **Remarks:** Aminocyclopyrachlor gives control of yellow starthistle similar to aminopyralid. *Perspective* provides broad-spectrum control of many broadleaf species. Although generally safe to grasses, it may suppress or injure certain annual and perennial grass species. Do not treat in the root zone of desirable trees and shrubs. Do not apply more than 11 oz product/acre per year. At this high rate, cool-season grasses will be damaged, including bluebunch wheatgrass. Not yet labeled for grazing lands. Add an adjuvant to the spray solution. This product is not approved for use in California and some counties of Colorado (San Luis Valley).
Aminopyralid *Milestone*	**Rate:** 3 to 5 oz product/acre (0.75 to 1.25 oz a.e./acre). Use higher rates when weeds are larger. **Timing:** Postemergence and preemergence. Postemergence applications are most effective when applied to plants from the seedling to the mid-rosette stage. Earlier applications (i.e., in fall) may not provide full-season control, and later applications (bolting to early spiny stage) will require higher rates. **Remarks:** Aminopyralid is one of the most effective herbicides for the control of yellow starthistle. It is safe on grasses, although preemergence application at high rates can greatly suppress invasive

annual grasses, such as medusahead. Aminopyralid has a longer residual and higher activity than clopyralid. Other members of the Asteraceae and Fabaceae are very sensitive to aminopyralid. For postemergence applications, a non-ionic surfactant (0.25 to 0.5% v/v spray solution) enhances control under adverse environmental conditions; however, this is not normally necessary.

Other premix formulations of aminopyralid can also be used for yellow starthistle control. These include *Opensight* (aminopyralid + metsulfuron; 1.5 to 2 oz product/acre) and *Forefront HL* (aminopyralid + 2,4-D; 2 to 2.6 pt product/acre), both applied at the rosette to bolting stages.

Clopyralid *Transline*	**Rate:** 0.25 to 0.67 pt product/acre (1.5 to 4 oz a.e./acre). Seedlings and rosettes can be treated at the lower rate, but bolted plants should be treated at higher rates. **Timing:** Postemergence and preemergence. For postemergence application, apply to plants from seedling to mid-bolting stage. However, since clopyralid has a shorter soil residual compared to aminopyralid, optimal timing is at the later rosette stages, but before bolting. Earlier applications (i.e., in fall) may not provide full-season control, and later applications (bolting to early spiny stage) will require higher rates and may not give sufficient control. **Remarks:** Clopyralid gives excellent control of yellow starthistle. While it is very safe on grasses, it will injure many members of the Asteraceae, particularly thistles, and can also injure legumes, including clovers. Most other broadleaf species and all grasses are not injured. When clopyralid is used to control seedlings a surfactant is not necessary. However, when treating older plants or plants exposed to moderate levels of drought stress, surfactants can enhance the activity of the herbicide.
Clopyralid + 2,4-D *Curtail*	**Rate:** 2 to 4 qt *Curtail*/acre **Timing:** Same as for clopyralid. **Remarks:** Add a non-ionic surfactant.
Dicamba *Banvel, Clarity*	**Rate:** 0.5 pt product/acre (0.25 lb a.e./acre) for seedlings, 1 to 1.5 pt product/acre (0.5 to 0.75 lb a.e./acre) for larger plants up to bolting. **Timing:** Postemergence to plants from rosette to beginning of bolting. **Remarks:** Dicamba is a broadleaf-selective herbicide often combined with other active ingredients. It is not typically used alone to control yellow starthistle. Dicamba is available mixed with diflufenzopyr in a formulation called *Overdrive*. This has been reported to be effective on yellow starthistle. Diflufenzopyr is an auxin transport inhibitor which causes dicamba to accumulate in shoot and root meristems, increasing its activity. *Overdrive* is applied postemergence at 4 to 8 oz product/acre to rapidly growing plants. Higher rates should be used on large annuals. Add a non-ionic surfactant to the treatment solution at 0.25% v/v or a methylated seed oil at 1% v/v solution.
Picloram *Tordon 22K*	**Rate:** 1 to 1.5 pt product/acre (4 to 6 oz a.e./acre) **Timing:** Postemergence and preemergence. Postemergence applications should be made to plants from rosette to bud formation stage. Apply when there is adequate soil moisture and weeds are growing rapidly. **Remarks:** Picloram acts much like aminopyralid, aminocyclopyrachlor, and clopyralid, but gives a broader spectrum of control and has much longer soil residual activity. It can provide about 2 to 3 years of control. Most broadleaf plants are susceptible. Although well-developed grasses are not usually injured by labeled use rates, some applicators have noted that young grass seedlings with fewer than four leaves may be killed. Do not apply near trees. *Tordon 22K* is a federally restricted use pesticide. Picloram is not registered for use in California.
Triclopyr *Garlon 3A, Garlon 4 Ultra*	**Rate:** 1 pt *Garlon 4 Ultra* or 1.33 pt *Garlon 3A*/acre (0.5 lb a.e./acre) for seedlings, up to 3 pt *Garlon 4 Ultra* or 4 pt *Garlon 3A*/acre (1.5 lb a.e./acre) for larger plants. **Timing:** Postemergence from seedling to bolting stage. **Remarks:** Triclopyr has little to no residual activity. It is broadleaf-selective and typically does not harm grasses. *Garlon 4 Ultra* is formulated as a low volatile ester. However, in warm temperatures, spraying onto hard surfaces such as rocks or pavement can increase the risk of volatilization and off-target damage.

AROMATIC AMINO ACID INHIBITORS	
Glyphosate *Roundup, Accord XRT II,* and others	**Rate:** Broadcast foliar treatment: 1.33 to 2.67 qt product (*Roundup ProMax*)/acre (1.5 to 3 lb a.e./acre). Spot treatment: 1% to 2% v/v solution **Timing:** Postemergence to plants from bolting to beginning of flowering. **Remarks:** Glyphosate is the most effective herbicide for late season control. Good coverage, clean water, and rapidly growing yellow starthistle plants are all essential for adequate control. It has no soil activity and is nonselective. To achieve selectivity, it can be applied using a wiper or spot treatment to control current year's plants.
BRANCHED-CHAIN AMINO ACID INHIBITORS	
Chlorsulfuron *Telar*	**Rate:** 1.33 to 2.6 oz product/acre (1 to 1.95 oz a.i./acre) **Timing:** Preemergence activity only. Chlorsulfuron does not have postemergence activity on yellow starthistle and must be used in combination with 2,4-D, dicamba, or triclopyr to provide effective control. **Remarks:** Chlorsulfuron has mixed selectivity on both broadleaf and grass species but is generally safe on grasses. It has fairly long soil residual activity. Herbicide solution requires constant agitation during application.
Imazapyr *Arsenal, Habitat, Stalker, Chopper, Polaris*	Not often used for yellow starthistle control but has been shown to be somewhat effective at 3 to 4 pt product/acre. It has preemergence and some postemergence activity, and a long soil residual.
Sulfometuron *Oust* and others	Not often used for yellow starthistle control but has been shown to be somewhat effective at 1 to 2 oz product/acre. It has preemergence activity only, and a long soil residual.
PHOTOSYNTHETIC INHIBITORS	
Hexazinone *Velpar L*	Not often used for yellow starthistle control but has been shown to be somewhat effective at 1 to 2.5 gal product/acre. It has preemergence activity only, and a long soil residual. High rates of hexazinone can create bare ground, so only use high rates in spot treatments.

Centaurea stoebe L. ssp. *micranthos* (Gugler) Hayek
(= *Centaurea biebersteinii* DC., *Centaurea maculosa* Lam.)

Spotted knapweed

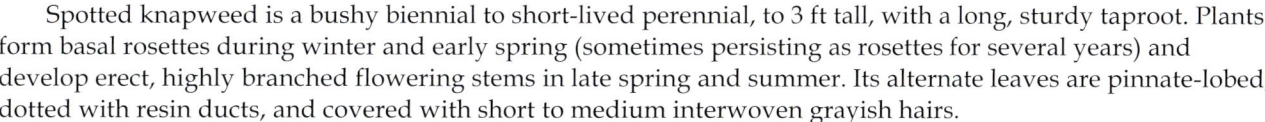

Family: Asteraceae
Range: Most contiguous states except parts of the southeast.
Habitat: Fields, roadsides, disturbed open sites, grassland, rangeland, especially degraded rangeland, logged areas. Seldom persists in shaded places. Serious infestations often occur on light, well-drained soils in areas that receive some summer rainfall.
Origin: Europe, Asia Minor; introduced into the U.S. in the 1890s.
Impact: Highly competitive with native vegetation. Forms dense stands that can exclude desirable vegetation and wildlife in natural areas.
Western states listed as Noxious Weed: Arizona, California, Colorado, Idaho, Montana, New Mexico, Nevada, North Dakota, Oregon, South Dakota, Utah, Washington, Wyoming
California Invasive Plant Council (Cal-IPC) Inventory: High Invasiveness

Spotted knapweed is a bushy biennial to short-lived perennial, to 3 ft tall, with a long, sturdy taproot. Plants form basal rosettes during winter and early spring (sometimes persisting as rosettes for several years) and develop erect, highly branched flowering stems in late spring and summer. Its alternate leaves are pinnate-lobed, dotted with resin ducts, and covered with short to medium interwoven grayish hairs.

The 30 to 40 disk flowers in each flowerhead are white, pink, or purple, and the phyllaries have comb-shaped, dark-colored tips which give the flowerheads a "spotted" appearance. After the flowerheads mature and dry out, they pop open, ejecting achenes near the parent plant. Achenes have a short (1 to 2 mm) bristly pappus on the top. Spotted knapweed can also reproduce vegetatively from lateral roots just below the soil surface. New rosettes may develop at about 3-cm intervals along lateral roots, expanding populations peripherally. Achenes can remain dormant in the soil for 8 years and have three germination patterns: non-dormant seeds that germinate with or without light exposure, dormant seeds that germinate in response to light, and dormant seeds that are not light sensitive. All germination types occur on each plant. Spotted knapweed has been shown to occasionally hybridize with diffuse knapweed.

NON-CHEMICAL CONTROL

Mechanical (pulling, cutting, disking)	Hand pulling is feasible for scattered spotted knapweed plants, or for areas where other control methods are not feasible and sufficient labor is available. Generally, this form of control is limited to small infested areas. Repeated hand pulling is necessary during the season and over many years. Successful control has been reported when plants were hand removed 3 times a year (spring, summer, and late summer) over a period of 5 years. Every effort should be made to remove the entire taproot with little soil disturbance. If not possible, then cut the root 2 to 4 inches below the soil surface to remove much of the reproductive crown. Gloves should be worn when hand pulling. The best timing for hand removal is before plant produce viable seed. Hand pulling has not been effective in all areas. When soil dries, it may be difficult to remove the root crown and this can lead to rapid reestablishment.
	Manual control methods may pose less risk to high quality waters and high value fisheries than do chemical applications. Although time- and labor-intensive, several manual control methods, including propane torching of seedlings early in the season, hand digging with small tools, mulching with black plastic, and mowing with weed eaters have proven successful for smaller populations.
	Mowing typically doesn't kill knapweeds; cut plants generally survive and recover to set seed. Plants mowed at the rosette stage will quickly recover, and mowing too late (after seed set) can disperse seed. However, mowing at the late bud to early bloom stage will reduce seed production. Mowing can also remove dead growth to improve herbicide coverage. A program of cutting only bolted plants, particularly in the early bloom stage, 2 to 4 times per year for several consecutive years can greatly suppress spotted knapweed and may shift the competitive balance in favor of desired grasses. Mowing is not possible in areas that are too rocky or steep, or with desirable shrub species.

	Spotted knapweed does not persist under annual cultivation, which is why it is not typically a cropland weed. However, tillage in wildland or rangelands can spread spotted knapweed, because tillage creates an ideal weed seed bed.
Cultural	Maintaining pasture and rangeland health by preventing overgrazing and minimizing disturbance can help limit knapweed establishment and spread.
	Grazing is not considered to be an effective eradication method. In addition, intensive grazing can create ideal seedbeds for further invasion. However, researchers have shown that cattle, sheep and goats will readily graze spotted knapweed in early spring, suppressing seed production. Sheep are the most effective. Sheep typically graze spotted knapweed from the rosette through bud stage or when it is the only plant available. The timing of grazing may be critical to its success. Early and late-season grazing appear to be the most effective control timings with sheep: early season (spring) to reduce flower production, and late season (fall) to reduce density of young plants. In one study, two consecutive years of early and late sheep grazing prevented spotted knapweed seed production, and the sheep were healthy. In addition, this grazing program had little effect on the native grass community.
	There is little information on the use of prescribed burning for control of spotted knapweed. On the one hand, burning has been shown to control diffuse knapweed while stimulating grass regrowth, and under the right conditions perhaps the same response might occur with spotted knapweed. On the other hand, spotted knapweed can be the first species to recover from a burn. A low-severity fire is not likely to kill the below-ground reproductive structures of spotted knapweed, but a severe fire may kill some of the plant crowns. Burning removes current growth but may enhance seed germination. Dry soil conditions at the time of burning can reduce germination. Another potential benefit of burning is that it can remove dead growth to improve the effect of herbicide applications.
Biological	Currently, there is no single biological control agent that effectively controls spotted knapweed populations. The banded gall fly (*Urophora affinis*), knapweed seedhead fly (*U. quadrifasciata*), lesser knapweed flower weevil (*Larinus minutus*), broad-nosed seedhead weevil (*Bangasternus fausti*) are established in the United States to date. The hairy weevil (*Eustenopus villosus*) that primarily attacks yellow starthistle has also been reported on spotted knapweed. These insects cause plants to produce fewer viable seeds and to abort flowers. *Larinus minutus* in particular is an aggressive and effective biocontrol insect for diffuse and spotted knapweed. Its larvae may destroy up to 100% of the seeds in an infested seedhead. Larvae pupate in the seedhead; the adults emerge and consume the foliage until they enter the litter and soil to overwinter.
	Three moth species (*Agapeta zoegana*, *Pelochrista medullana*, and *Pterolonche inspersa*) and a weevil (*Cyphocleonus achates*) that feed on spotted knapweed roots have also been released. Root-feeding insects may have a more detrimental effect on knapweed populations than seed-feeding ones. It is hoped that the collective stress on the plant caused by these insects will reduce seed production and lead to decreased competitiveness.

CHEMICAL CONTROL

The following specific use information is based on published papers and reports by researchers and land managers. Other trade names may be available, and other compounds also are labeled for this weed. Directions for use may vary between brands; see label before use. Herbicides are listed by mode of action and then alphabetically. The order of herbicide listing is not reflective of the order of efficacy or preference.

GROWTH REGULATORS	
2,4-D Several names	**Rate:** 1 to 2 qt product/acre (0.95 to 1.9 lb a.e./acre)
	Timing: Postemergence from rosette to beginning of bolting, or fall rosette. Optimal at early flowering stage.
	Remarks: Control with 2,4-D is only temporary and does not prevent seedling establishment the following year. Generally requires repeat applications. 2,4-D is not considered as effective as other growth regulator herbicides for season-long control. Broadleaf-selective and may injure other non-target species, particularly crop plants. 2,4-D has no soil activity. Do not apply ester formulation when outside temperatures exceed 80°F. Amine forms are as effective as ester forms at the small rosette stage, and amine forms reduce the chance of off-target movement.
Aminocyclopyrachlor + chlorsulfuron *Perspective*	**Rate:** 4.75 to 8 oz product (*Perspective*)/acre
	Timing: Postemergence and preemergence. Postemergence applications are most effective when applied to plants from the seedling to the mid-rosette stage.

	Remarks: Aminocyclopyrachlor gives control of spotted knapweed similar to aminopyralid. *Perspective* provides broad-spectrum control of many broadleaf species. Although generally safe to grasses, it may suppress or injure certain annual and perennial grass species. Do not treat in the root zone of desirable trees and shrubs. Do not apply more than 11 oz product/acre per year. At this high rate, cool-season grasses will be damaged, including bluebunch wheatgrass. Not yet labeled for grazing lands. Add an adjuvant to the spray solution. This product is not approved for use in California and some counties of Colorado (San Luis Valley).
Aminopyralid *Milestone*	**Rate:** 5 to 7 oz product/acre (1.25 to 1.75 oz a.e./acre) **Timing:** Postemergence and preemergence. Postemergence applications are most effective when applied to plants from the rosette to the bolting stage. Effective control can also be obtained with a fall application to new regrowth. **Remarks:** Aminopyralid is one of the most effective herbicides for the control of spotted knapweed. It is safe on grasses, although preemergence application at high rates can greatly suppress invasive annual grasses, such as medusahead. Aminopyralid has a longer residual and higher activity than clopyralid. Other members of the Asteraceae and Fabaceae are very sensitive to aminopyralid. For postemergence applications, a non-ionic surfactant (0.25 to 0.5% v/v spray solution) enhances control under adverse environmental conditions; however, this is not normally necessary. Other premix formulations of aminopyralid can also be used for spotted knapweed control. These include *Opensight* (aminopyralid + metsulfuron; 1.5 to 2 oz product/acre) and *Forefront HL* (aminopyralid + 2,4-D; 2 to 2.6 pt product/acre), both applied at the rosette to bolting stages.
Clopyralid *Transline*	**Rate:** 0.67 to 1.33 pt product/acre (4 to 8 oz a.e./acre). Use higher rate for older plants or dense stands. **Timing:** Applied preemergence to seedlings or postemergence to seedlings or mature plants, but generally optimal to apply postemergence in spring, at beginning of bolting up to the bud stage. Can also apply to fall regrowth. Results are best if applied to rapidly growing weeds. **Remarks:** While clopyralid is very safe on grasses, it will injure many members of the Asteraceae, particularly thistles, and can also injure legumes, including clovers. Most other broadleaf species and all grasses are not injured.
Clopyralid + 2,4-D *Curtail*	**Rate:** 2 to 4 qt *Curtail*/acre **Timing:** Same as for clopyralid **Remarks:** The addition of 2,4-D can increase the damage to other non-target broadleaf species. Add a non-ionic surfactant.
Dicamba *Banvel, Clarity*	**Rate:** 1 to 2 pt product/acre (0.5 to 1 lb a.e./acre). Use higher rate for older plants or dense stands. **Timing:** Postemergence from rosette to beginning of bolting, or fall rosette. Optimal at early flowering stage. **Remarks:** Dicamba is a broadleaf-selective herbicide often combined with other active ingredients. It is not typically used alone to control spotted knapweed. Dicamba can also be mixed with 2,4-D (1 pt dicamba + 2 pt 2,4-D/acre) or picloram (1 to 2 pt dicamba + 0.5 to 1 pt picloram/acre) for spot treatments. Dicamba is available mixed with diflufenzopyr in a formulation called *Overdrive*. This has been reported to be effective on spotted knapweed. Diflufenzopyr is an auxin transport inhibitor which causes dicamba to accumulate in shoot and root meristems, increasing its activity. *Overdrive* is applied postemergence to rapidly growing plants at 4 to 8 oz product/acre. Higher rates should be used on large annuals and biennials or when treating perennial weeds. Add a non-ionic surfactant to the treatment solution at 0.25% v/v or a methylated seed oil at 1% v/v solution.
Picloram *Tordon 22K*	**Rate:** 1 to 2 pt product/acre (4 to 8 oz a.e./acre). Use higher rates for older plants or dense stands. **Timing:** Postemergence and preemergence. Postemergence applications are best at rosette to mid-bolting stage (before flowering to prevent current year seed production), or fall rosette stage. Apply when plants are growing rapidly. Under favorable growing conditions, application in summer can be effective if higher application volumes are used. **Remarks:** Picloram is a broadleaf herbicide. It gives a broader spectrum of control than aminopyralid, aminocyclopyrachlor, and clopyralid, and has a much longer soil residual. Lower rates may require annual spot treatments. Treatment made in bud stage may not prevent seed production in the year of application. Picloram has been shown to provide selective control of spotted knapweed for 3 to 4

years. Although well-developed grasses are not usually injured by labeled use rates, some applicators have noted that young grass seedlings with fewer than four leaves may be killed. Do not apply near trees. Picloram is a restricted use herbicide. It is not registered for use in California.

Control with lower rates may be improved by tank mixing with dicamba or 2,4-D; picloram and dicamba (0.25 to 0.5 pt/acre + 0.125 to 0.25 pt/acre) and picloram plus 2,4-D (0.5 to 1 pt picloram + 1 to 2 pt 2,4-D/acre). A backpack sprayer or a wiper is recommended in small areas to minimize damage to non-target plants.

AROMATIC AMINO ACID INHIBITORS

Glyphosate

Roundup, Accord XRT II, and others

Rate: Broadcast foliar treatment: 3 qt product (*Roundup ProMax*)/acre (3.375 lb a.e./acre). Spot treatment 1.5% v/v solution

Timing: Postemergence to rapidly growing knapweed when most plants are at bud stage.

Remarks: Glyphosate will only provide control during the year of application; it has no soil activity and will not kill seeds or inhibit germination the following season. Glyphosate is nonselective. To achieve selectivity, it can be applied using a wiper or spot treatment to control current year's plants.

Centaurea virgata Lam. ssp. *squarrosa* (Boiss.) Gugler
(= *Centaurea squarrosa* Willd.; *Centaurea triumfettii* All.)

Squarrose knapweed

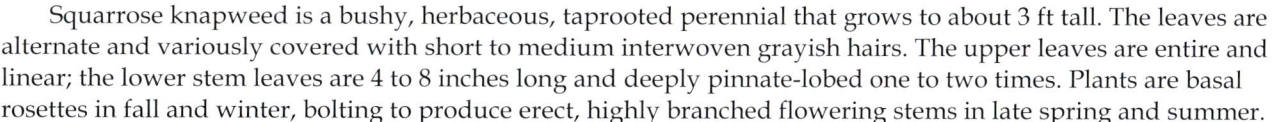

Family: Asteraceae
Range: California, Nevada, Oregon and Utah.
Habitat: Fields, roadsides, disturbed open sites, grassland, rangeland, especially degraded rangeland, and logged areas. Squarrose knapweed is more adaptable to drought and cold temperatures than spotted knapweed and diffuse knapweed. Like other knapweeds, it seldom persists in shaded places.
Origin: Native to Asia.
Impacts: Like other knapweeds, squarrose knapweed is a highly competitive, noxious weed. Dense stands can exclude desirable vegetation and wildlife in natural areas.
Western states listed as Noxious Weed: Arizona, California, Nevada, Utah
California Invasive Plant Council (Cal-IPC) Inventory: Moderate Invasiveness

Squarrose knapweed is a bushy, herbaceous, taprooted perennial that grows to about 3 ft tall. The leaves are alternate and variously covered with short to medium interwoven grayish hairs. The upper leaves are entire and linear; the lower stem leaves are 4 to 8 inches long and deeply pinnate-lobed one to two times. Plants are basal rosettes in fall and winter, bolting to produce erect, highly branched flowering stems in late spring and summer.

The flowerheads consist of spiny or comb-like phyllaries and pink to pale purple disk flowers. The spiny phyllaries usually reflex downward. This characteristic distinguishes squarrose from diffuse knapweed. The achenes either lack a pappus or have a short bristly pappus (2 to 2.5 mm long). Plants reproduce only by seed. While seeds of most *Centaurea* species fall near the parent plant, squarrose knapweed seedheads detach from the parent plant as a unit and tumble along the ground, dispersing seed to a greater distance. In addition, they can disperse by clinging to the wool, hair or fur of animals. It is not known how long seeds remain viable in the soil, but it is assumed that survival would be similar to other *Centaurea* species, 2 to 5 years, with a few seeds surviving longer.

NON-CHEMICAL CONTROL

Mechanical (pulling, cutting, disking)	Hand pulling is practical for scattered plants, or for areas where other control methods are not feasible and sufficient labor is available. Repeated hand pulling is necessary during the season and over many years. Results of hand removal are expected to be similar to diffuse knapweed, for which successful control was reported when plants were hand removed 3 times a year (spring, summer, and late summer) over a period of 5 years. Every effort should be made to remove the entire taproot with little soil disturbance. If not possible, then cut the root 2 to 4 inches below the soil surface to remove much of the reproductive crown. Gloves should be worn when hand pulling. The best timing is before plants produce viable seed. Hand pulling has not been effective in all areas. When dry soils, it may be difficult to remove the root crown and this can lead to rapid reestablishment.
	Mowing typically doesn't kill knapweeds; cut plants generally survive and recover to set seed. Plants mowed at the rosette stage will quickly recover, and mowing too late (after seed set) can disperse seed. However, mowing at the late bud to early bloom stage will reduce seed production. Mowing can also remove dead growth to improve herbicide coverage. A program of cutting only bolted plants, particularly at the early bloom stage, for several consecutive years is expected to greatly suppress squarrose knapweed.
	Cultivation is effective when repeated, but squarrose knapweed typically doesn't infest areas conducive to tillage.
Cultural	There is no direct information on management of squarrose knapweed with grazing or prescribed burning, but it is expected that this species should respond like diffuse or spotted knapweed.
Biological	Squarrose knapweed is closely related to both diffuse and spotted knapweeds. As a result, several seed-feeding biological control insects introduced for those species have attacked squarrose knapweed. In northern California, larvae of these species have reduced seed production by over 99% for many years now. The most effective appear to be the broad-nosed seedhead weevil *Bangasternus fausti*, and the two weevils

Larinus minutus and *L. obtusus*. Because squarrose knapweed plants can survive for 8 or more years, management with biological control agents will require a few more years to monitor.

CHEMICAL CONTROL

The following specific use information is based on published papers and reports by researchers and land managers. Other trade names may be available, and other compounds also are labeled for this weed. Directions for use may vary between brands; see label before use. Herbicides are listed by mode of action and then alphabetically. The order of herbicide listing is not reflective of the order of efficacy or preference.

GROWTH REGULATORS	
2,4-D Several names	**Rate:** 1 to 2 qt product/acre (0.95 to 1.9 lb a.e./acre) **Timing:** Postemergence from rosette to beginning of bolting, or fall rosette. Optimal at early flowering stage. **Remarks:** Control is only temporary and does not prevent seedling establishment the following year. Generally requires repeat applications. 2,4-D is not considered as effective as other growth regulator herbicides for season-long control. It is broadleaf-selective and may injure other non-target species, particularly crop plants. 2,4-D has no soil activity. Do not apply ester formulation when outside temperatures exceed 80°F. Amine forms are as effective as ester forms at the small rosette stage, and amine forms reduce the chance of off-target movement.
Aminocyclopyrachlor + chlorsulfuron *Perspective*	**Rate:** 4.75 to 8 oz product (*Perspective*)/acre **Timing:** Postemergence and preemergence. Postemergence applications are most effective when applied to plants from the seedling to the mid-rosette stage. **Remarks:** Aminocyclopyrachlor gives control of squarrose knapweed similar to aminopyralid. *Perspective* provides broad-spectrum control of many broadleaf species. Although generally safe to grasses, it may suppress or injure certain annual and perennial grass species. Do not treat in the root zone of desirable trees and shrubs. Do not apply more than 11 oz product/acre per year. At this high rate, cool-season grasses will be damaged, including bluebunch wheatgrass. Not yet labeled for grazing lands. Add an adjuvant to the spray solution. This product is not approved for use in California and some counties of Colorado (San Luis Valley).
Aminopyralid *Milestone*	**Rate:** 5 to 7 oz product/acre (1.25 to 1.75 oz a.e./acre) **Timing:** Postemergence and preemergence. Postemergence applications are most effective from the rosette to the bolting stage. Effective control can also be obtained with a fall application to new regrowth. **Remarks:** Aminopyralid is one of the most effective herbicides for the control of squarrose knapweed. It is safe on grasses although preemergence application at high rates can greatly suppress invasive annual grasses, such as medusahead. Aminopyralid has a longer residual and higher activity than clopyralid. Other members of the Asteraceae and Fabaceae are very sensitive to aminopyralid. For postemergence applications, a non-ionic surfactant (0.25 to 0.5% v/v spray solution) enhances control under adverse environmental conditions; however, this is not normally necessary. Other premix formulations of aminopyralid can also be used for diffuse knapweed control. These include *Opensight* (aminopyralid + metsulfuron; 1.5 to 2 oz product/acre) and *Forefront HL* (aminopyralid + 2,4-D; 2 to 2.6 pt product/acre), both applied at the rosette to bolting stages.
Clopyralid *Transline*	**Rate:** 0.67 to 1.33 pt product/acre (4 to 8 oz a.e./acre). Use higher rate for older plants or dense stands. **Timing:** Preemergence to seedlings or postemergence to seedlings or mature plants. Postemergence applications are generally best in spring, from beginning of bolting up to the bud stage. Can also apply to fall regrowth. Results are best if applied to rapidly growing weeds. **Remarks:** While clopyralid is very safe on grasses, it will injure many members of the Asteraceae, particularly thistles, and can also injure legumes, including clovers. Most other broadleaf species and all grasses are not injured.
Clopyralid + 2,4-D *Curtail*	**Rate:** 2 to 4 qt *Curtail*/acre **Timing:** Same as for clopyralid. **Remarks:** The addition of 2,4-D can increase the damage to other non-target broadleaf species. Add a non-ionic surfactant.

Dicamba *Banvel, Clarity*	**Rate:** 1 to 2 pt product/acre (0.5 to 1 lb a.e./acre). Use higher rate for older plants or dense stands. **Timing:** Postemergence from rosette to beginning of bolting, or fall rosette. Optimal at early flowering stage. **Remarks:** Dicamba is a broadleaf-selective herbicide often combined with other active ingredients. It is not typically used alone to control squarrose knapweed. Dicamba can also be mixed with 2,4-D (1 pt dicamba + 2 pt 2,4-D/acre) or picloram (1 to 2 pt dicamba + 0.5 to 1 pt picloram/acre) for spot treatments.
Picloram *Tordon 22K*	**Rate:** 1 to 2 pt product/acre (4 to 8 oz a.e./acre). Use higher rates for older plants or dense stands. **Timing:** Postemergence and preemergence. With postemergence application, optimally treat at rosette to mid-bolting stage (before flowering to prevent current year seed production), or fall rosette stage. Apply when plants are growing rapidly. Under favorable growing conditions, application in summer can be effective if higher application volumes are used. **Remarks:** Picloram is a broadleaf herbicide and gives a broader spectrum of control than aminopyralid, aminocyclopyrachlor, and clopyralid, and has much longer soil residual activity. Lower rates may require annual spot treatments. Treatment made in bud stage may not prevent seed production in the year of application. Picloram has been shown to provide selective control of squarrose knapweed for 3 to 4 years. Although well-developed grasses are not usually injured by labeled use rates, some applicators have noted that young grass seedlings with fewer than four leaves may be killed. Do not apply near trees. Picloram is a restricted use herbicide. It is not registered for use in California. Control with lower rates may be improved by tank mixing with dicamba or 2,4-D; picloram and dicamba (0.25 to 0.5 pt/acre + 0.125 to 0.25 pt/acre) and picloram plus 2,4-D (0.5 to 1 pt picloram + 1 to 2 pt 2,4-D/acre). A backpack sprayer or a wiper is highly recommended in small areas to minimize damage to non-target plants.
AROMATIC AMINO ACID INHIBITORS	
Glyphosate *Roundup, Accord XRT II,* and others	**Rate:** Broadcast foliar treatment: 3 qt product (*Roundup ProMax*)/acre (3.375 lb a.e./acre). Spot treatment: 1.5% v/v solution **Timing:** Postemergence to rapidly growing knapweed when most plants are at bud stage. **Remarks:** Glyphosate will only provide control during the year of application; it has no soil activity and will not kill seeds or inhibit germination the following season. Glyphosate is nonselective. To achieve selectivity, it can be applied using a wiper or spot treatment to control current year's plants.

Ceratophyllum demersum L.

Coontail

Family: Ceratophyllaceae
Range: Throughout the United States, including all western states.
Habitat: Ponds, slow-flowing streams, and ditches in temperate to tropical regions. Coontail tolerates low light levels and some turbidity, but not salinity. It grows best in high-nutrient water, in 8 to 16 ft of water.
Origin: Coontail is considered a native in many areas of the world, including the western United States. Plants are sometimes sold as an aquarium or pond ornamental.
Impacts: In natural areas, plants provide food and shelter for wildlife and are a desirable component of aquatic habitats. However, plants may develop dense sub-surface mats in high nutrient waters, channels, and controlled aquatic systems. Mats can inhibit water flow, block intake screens of water pumps, interfere with recreational activities, and create mosquito habitat.

Submersed annual to perennial with somewhat firm, forked bottlebrush-like leaves and stems to ~8 ft long. Plants lack roots and exist free-floating or anchored to the substrate by specialized, finely divided buried stems (rhizoid shoots). Young seedlings detach from the soil substrate when stems are about 4 inches long, and must absorb nutrients directly from the water. The stems are slender, branched, with only one branch per node. The stems usually fragment easily. The leaves are sessile, 5 to 12 whorled at each node, mostly forked 2-3 times, with margins that are conspicuously small-toothed. Turions (overwintering buds) consist of dense clusters of scale-like leaves at the stem tips.

Male and female flowers develop on the same plant (monoecious). The flowers are submerged, small, solitary and relatively inconspicuous in the leaf axils. Flowers are water-pollinated. Anthers detach and float just below the water surface before releasing pollen, which sinks down to the female flowers below. Pollination is most likely to occur in still water. Fruits are achene- or nut-like, and do not open to release the single seed. Coontail reproduces vegetatively by turions and stem fragments and also by seed. Fruits and turions sink to the bottom when separated from the parent plant. Fruits and vegetative parts disperse to greater distances with water or by clinging to the fur, feathers, or feet of animals.

NON-CHEMICAL CONTROL

Mechanical (pulling, cutting, suction dredge, bottom barriers)	Removing and destroying stem fragments from recreational equipment, such as boat propellers, docking lines, and fishing gear can help prevent the spread of coontail. Mechanical harvesting has proved sufficient to control coontail stands in some temperate areas.
	Several types of "bottom barriers" are available and are used to cover and smother specific infested areas. Materials used include polyvinyl chloride (PVC) sheets, small-mesh screens and natural fibers such as jute. Bottom barriers are best installed in spring before plants produced large biomass and exceed 20 inches tall.
Cultural	Dewatering during hot summer or exposure to hard freeze will reduce subsequent growth from fragments. However, seed can withstand severe conditions and may provide sources for reinfestation.
	Reducing nutrients in the water column can suppress growth since coontail has no true roots to obtain nutrients directly from the sediment.
Biological	The triploid (sterile) grass carp (white amur) is a relatively nonselective herbivorous fish that will consume coontail as one of their preferred plant diets. Permits are usually required in all U.S. states for use of grass carp.

CHEMICAL CONTROL
The following specific use information is based on reports by researchers and land managers. Other trade names may be available, and other compounds also are labeled for this weed. Directions for use may vary between

brands; see label before use. Herbicides are listed by mode of action and then alphabetically. The order of herbicide listing is not reflective of the order of efficacy or preference.

GENERAL CELL TOXICANT	
Acrolein *Magnacide H*	**Rate:** For in-water treatment: 1 to 15 ppm (variable, depending on target weeds, temperature and flow rates) **Timing:** Apply directly to water in late spring to fall. No more than 8 applications are allowed per year. **Remarks:** Acrolein is a very fast-acting, nonselective contact herbicide and algaecide. It is a "Restricted Use" pesticide but can be used in some irrigation canals under specific conditions, with proper permits, and may only be applied by qualified, trained applicators. Symptoms of efficacy may appear in less than an hour and include discoloration of leaves and loss of turgidity in plant shoots.
Endothall *Cascade; Aquathol*	**Rate:** 0.2 to 5.0 ppm (in-water application; e.g. flowing water in irrigation canals). Exposures must be maintained for 6 to 120 hours. Duration of contact depends on the concentration achieved. **Timing:** Apply directly to water in early spring to early summer. Endothall can be used in mid-summer, but to prevent reduction in dissolved oxygen, only partial treatments are recommended if biomass is large. **Remarks:** Endothall is a selective, contact herbicide. It affects young, rapidly growing plants and mature plants. Lower rates can be used if applied during early spring growth and when water movement is not likely to dilute or move the herbicide.
CONTACT PHOTOSYNTHETIC INHIBITOR	
Flumioxazin *Clipper*	**Rate:** For in-water treatment: 100 to 400 ppb **Timing:** Apply directly to water from early spring to early summer, during the plants rapid growth phase. **Remarks:** Flumioxazin is rapidly degraded and is inactive if pH exceeds 8.5. Thus, it is important to only use if pH will not exceed 8.5. It is best to apply flumioxazin in the early morning when the pH is low.
PIGMENT SYNTHESIS INHIBITORS	
Fluridone *Sonar*	**Rate:** 10 to 90 ppb (in-water application). Exposures must be maintained for 5 to 7 weeks for optimal control. **Timing:** Apply directly to water from early spring to early summer. **Remarks:** Fluridone is a slow-acting systemic herbicide. It affects young, rapidly growing plants. Lower rates can be used if applied during early spring growth and when water movement is not likely to dilute or move the herbicide.
INORGANIC HERBICIDES	
Chelated copper *Komeen, Cutrine-plus, Nautique*	**Rate:** 0.5 to 1 ppm elemental copper (Cu) **Timing:** Apply directly to water in spring to late summer. **Remarks:** Chelated copper is a fast-acting contact herbicide. Retreatment may be required within 3 to 5 weeks. If biomass is large, treat only one-third of infested area to minimize decrease in dissolved oxygen. Chelated copper products are less affected by high pH and "hard water" than inorganic copper products (See "Inorganic copper" below).
Inorganic copper Various granular and liquid products	**Rate:** 0.5 to 1 ppm elemental copper **Timing:** Apply directly to water in early summer when plants are short and biomass is small. **Remarks:** Copper is a fast-acting contact herbicide. Retreatment may be required within 3 to 5 weeks. If biomass is large, treat only one-third of infested area to minimize decrease in dissolved oxygen. Most inorganic copper formulations have poor efficacy in "hard water" (e.g. > 125 ppb calcium carbonate equivalent) and high pH (> 8).
NON-HERBICIDAL CHEMICALS	
Dyes or colorants *Aquashade*	Although technically not herbicides, dyes and colorants control submerged aquatic plants by absorbing light in the water column and reducing photosynthesis. Applications should be made in early spring and repeated to maintain concentration recommended on the label. Colorants are not as effective on well-established plants in mid- to late summer.

Chara spp. and *Nitella* spp.

Chara and nitella

Family: Characeae
Range: Throughout North America.
Habitat: Ponds, lakes, reservoirs, rivers, streams, bogs, canals, and rice fields. Some species inhabit brackish water. Chara often grows in hard water.
Origin: All species are native to North America.
Impacts: Chara and nitella provide food and cover for wildlife and are important components of natural aquatic ecosystems. These algae sometimes grow in rice fields and canals, but are rarely of importance as weeds. At first glance, chara and nitella are easily mistaken for vascular aquatic plants.

These submerged plant-like green algae are usually anchored to the substrate by well-developed, colorless rhizoids. There are several species that occur in various regions of the western United States. Central axes of chara and nitella are regularly jointed, solid between nodes, with whorls of branches at each node, to 12 inches long or more. These algae do not have leaves. *Chara* species are typically coarse, gray-green, sometimes encrusted with carbonates, making plants rough to touch, and often have a garlic or skunk-like odor. *Nitella* species are usually dark green, delicate, never encrusted with carbonates, and lack an unpleasant odor.

These species reproduce vegetatively from fragmentation and sexually by egg cells and motile sperm. After fertilization, the zygote (oospore) remains dormant for a period before germination occurs.

NON-CHEMICAL CONTROL

Mechanical (pulling, cutting, dredging)	Repeated mechanical harvesting can help reduce stem densities, but escaped stem fragments can drift elsewhere and develop into new plants. Removing and destroying stem fragments from recreational equipment such as boat propellers, docking lines, and fishing gear can help prevent the spread of chara and nitella.
	Several types of "bottom barriers" are available to cover and smother specific infested areas. Materials used include polyvinyl chloride (PVC) sheets, small-mesh screens and natural fibers such as jute and burlap. Bottom barriers are best installed in spring before plants produced large biomass and exceed 10 inches tall.
Cultural	Reducing nutrient inputs can help prevent invasion.
	Establishment of native pondweeds or other native submersed plants can reduce space and light needed by chara and nitella.
Biological	Triploid (sterile) grass carp is the only effective biological control agent available for these two algal species, but it is relatively nonselective and state or local permits are usually required. If other native plants are desired, careful monitoring of feeding impacts should be part of the management program so that grass carp can be removed (or added) as needed.

CHEMICAL CONTROL
The following specific use information is based on reports by researchers and land managers. Other trade names may be available, and other compounds also are labeled for this weed. Directions for use may vary between brands; see label before use. Herbicides are listed by mode of action and then alphabetically. The order of herbicide listing is not reflective of the order of efficacy or preference.

CONTACT PHOTOSYNTHETIC INHIBITORS	
Diquat	**Rate:** 0.1 to 0.25 ppm
Reward	**Timing:** Apply directly to water in late spring to early summer. Diquat is a fast-acting contact herbicide that can also be effective in mid- to late summer, but if biomass is large, only a portion of the infested sites

	should be treated to minimize effects of reduced dissolved oxygen. **Remarks:** Diquat is quickly bound to, and becomes inactivated on, suspended clay particles and it should not be used in moderately or highly turbid water.
Flumioxazin *Clipper*	**Rate:** For in-water treatment: 100 to 400 ppb **Timing:** Apply directly to water from early spring to early summer, during the plants' rapid growth phase. **Remarks:** Flumioxazin is rapidly degraded and is inactive if pH exceeds 8.5. Thus, it is important to use only if pH will not exceed 8.5. It is best to apply flumioxazin in the early morning when the pH is low.
INORGANIC HERBICIDES	
Chelated copper *Cutrine-plus,* *Nautique*	**Rate:** 0.5 to 1 ppm elemental copper **Timing:** Apply directly to water in early summer (short plants and small biomass). **Remarks:** Chelated copper is a fast-acting contact herbicide. Retreatment may be required within 3 to 5 weeks. If biomass is large, treat only one-third of infested area to minimize decrease in dissolved oxygen. Chelated copper products are less affected by high pH than inorganic copper products (see "Inorganic copper" below).
Inorganic copper Various granular and liquid products	**Rate:** 0.5 to 1 ppm elemental copper **Timing:** Apply directly to water in early summer (short plants and small biomass). **Remarks:** Copper is a fast-acting contact herbicide. Retreatment may be required within 3 to 5 weeks. If biomass is large, treat only one-third of infested area to minimize decrease in dissolved oxygen. Most inorganic copper formulations have poor efficacy in "hard water" (e.g. > 125 ppb calcium carbonate equivalent) and high pH (> 8).
NON-HERBICIDAL CHEMICALS	
Dyes or colorants *Aquashade*	Although technically not herbicides, dyes and colorants control submerged aquatic plants by absorbing light in the water column and reducing photosynthesis. Applications should be made in early spring and repeated to maintain concentration recommended on the label. Colorants are not as effective on well-established plants in mid- to late summer.

Chondrilla juncea L.
Rush skeletonweed

Family: Asteraceae
Range: Most western states; a few central, southern, and eastern states.
Habitat: Disturbed soils of roadsides, croplands (especially non-irrigated grain fields), semi-arid pastures, rangelands, and residential properties. Grows best on well-drained, sandy or gravelly soils in climates with cool winters and hot, relatively dry summers. Tolerates a wide variety of environmental conditions, including semiarid areas and cold winters. Less common on heavy clay soils.
Origin: Native to southern Europe.
Impacts: Invasive and competitive for water and nutrients. Persistent, wiry flower stems can interfere with harvest machinery.
Western states listed as Noxious Weed: Arizona, California, Colorado, Idaho, Montana, Nevada, Oregon, South Dakota, Washington
California Invasive Plant Council (Cal-IPC) Inventory: Moderate Invasiveness

Rush skeletonweed is an herbaceous biennial to perennial up to 3 ft tall with reddish basal leaves and milky sap. Its common name derives from its appearance at maturity, with wiry, branched stems with few or no stem leaves standing on a basal rosette. It forms a slender, deep, persistent taproot. Adventitious buds near the top of the taproot and on major lateral roots can generate new rosettes. Roots are easily fragmented, and pieces as small as 0.5 to 1 inch can produce new rosettes from a depth to 3 ft.

Plants exist as basal rosettes until maturity, when one or more stems develop. The upper stems are highly branched and nearly hairless, with few, greatly reduced leaves. The lower stems have dense, bristly, downward pointing hairs. Stems develop in late spring and flower until killed by frost in fall or winter. Flowerheads are yellow, like small dandelion heads, and grow on the sides or tips of stems, alone or in small clusters. Flowers are self-fertilizing, producing small pappus seeds that disperse primarily by wind. Seeds appear to survive less than 3 years. Rush skeletonweed can also reproduce vegetatively from root buds.

NON-CHEMICAL CONTROL

Mechanical (pulling, cutting, disking)	Because rush skeletonweed can resprout from root fragments, mechanical damage to established plants results in root sprouting and regrowth. Young seedlings may be controlled by cultivation. Frequent mowing may exhaust root storage, resulting in suppression.
Cultural	Under moist conditions, shallow burial of seed by hooves of grazing livestock appears to promote seed germination. In addition, moderate soil disturbance, such as grazing on a yearly basis, can increase populations by dispersing rootstocks. Continual grazing can reduce populations if seed germination is prevented.
	Burning is not effective against rush skeletonweed.
	Increasing nutrient levels on poor soils appears to discourage survival by increasing competition from other vegetation.
Biological	One of the most successful examples of classical biological control of weeds is the introduction of a rust fungus, *Puccinia chondrillina*, into Australia to control rush skeleton weed. *P. chondrillina* attacks one of three forms of the weed, the predominant type. Initially, as the population density of this susceptible type was reduced due to biocontrol, the two other types became more widespread. Therefore, additional rust strains virulent on these more resistant forms were introduced from the Mediterranean and these strains are exerting some degree of control of the resistant forms. As a result of this success, *P. chondrillina* was also introduced into the western United States to control rush skeleton weed. However, unlike in Australia, it has been only partially successful. As a result, the rust fungus is used along with two insect biocontrol agents, skeletonweed gall midge (*Cystiphora schmidtii*) and skeletonweed gall mite (*Aceria chondrillae*). All these agents are now widely established in the United States and appear to be reducing rush skeletonweed densities in California.

CHEMICAL CONTROL

The following specific use information is based on published papers and reports by researchers and land managers. Other trade names may be available, and other compounds also are labeled for this weed. Directions for use may vary between brands; see label before use. Herbicides are listed by mode of action and then alphabetically. The order of herbicide listing is not reflective of the order of efficacy or preference.

GROWTH REGULATORS	
2,4-D Several names	**Rate:** 2 qt product/acre (1.9 lb a.e./acre) **Timing:** Best applied postemergence at rosette stage. **Remarks:** 2,4-D is a broadleaf-selective herbicide with no soil activity. It is not the most effective treatment, but is widely used because of the low cost. 2,4-D will kill above-ground parts of the plant, but new rosettes will regenerate from the root system. This may require repeat applications. Do not apply the ester formulations when outside temperatures exceed 80°F. It can be used in a premix with triclopyr (*Crossbow*) at 4 qt product/acre.
Aminocyclopyrachlor + chlorsulfuron *Perspective*	**Rate:** 4.75 to 8 oz product/acre **Timing:** Postemergence in spring until flowering, or in fall to rosettes. **Remarks:** *Perspective* provides broad-spectrum control of many broadleaf species. Although generally safe to grasses, it may suppress or injure certain annual and perennial grass species. Do not treat in the root zone of desirable trees and shrubs. Do not apply more than 11 oz product/acre per year. At this high rate, cool-season grasses will be damaged, including bluebunch wheatgrass. Not yet labeled for grazing lands. Add an adjuvant to the spray solution. This product is not approved for use in California and some counties of Colorado (San Luis Valley).
Aminopyralid *Milestone*	**Rate:** 5 to 7 oz product/acre (1.25 to 1.75 oz ae/acre) **Timing:** In spring from rosette through flowering stage. In cold-winter climates, applications can be made in fall. **Remarks:** A broadleaf-selective herbicide with soil residual activity. Very safe on grasses. Longer residual and higher activity than clopyralid.
Aminopyralid + 2,4-D, *Forefront HL*; Aminopyralid + metsulfuron, *Opensight*; Aminopyralid + triclopyr, *Capstone*	**Rate:** 1.5 to 2.1 pt *Forefront HL*/acre; 2.5 to 3 oz *Opensight*/acre; 4 to 6 pt *Capstone*/acre **Timing:** Rosette to bolting stages. **Remarks:** Broadleaf-selective. Recommended rates based on those reported for similar species. *Opensight* is not registered for use in California.
Clopyralid *Transline*	**Rate:** 0.67 to 1 pt product/acre (4 to 6 oz a.e./acre) **Timing:** Postemergence to rosettes in fall, or up to bolting in spring. **Remarks:** Clopyralid is a broadleaf-selective like picloram, but more selective and with shorter soil residual activity. It is very safe on grasses. Clopyralid can be mixed with with 2,4-D (2 oz a.e./acre clopyralid + 8 oz a.e./acre 2,4-D) or dicamba (2 oz a.e./acre clopyralid + 3 oz a.e./acre dicamba).
Clopyralid + 2,4-D *Curtail*	**Rate:** 1.5 to 3 qt product/acre (use higher rate if plants are drought-stressed) **Timing:** Postemergence to rapidly growing plants from full rosette to early flower bud stage. **Remarks:** The combination is broadleaf-selective with a wide range of susceptible species. Another effective formulation is a premix of clopyralid with triclopyr (*Redeem*) at 2.5 to 4 pt product/acre.
Dicamba *Banvel, Clarity*	**Rate:** 2 to 4 lb product/acre (1 to 2 lb a.e./acre) **Timing:** Postemergence to rapidly growing plants. **Remarks:** Dicamba is a broadleaf-selective herbicide often combined with other active ingredients. It is also effective when tank-mixed with 2,4-D. Dicamba will kill above-ground parts of the plant, but new rosettes will regenerate from the root system. As result, it may require repeat applications. Do not apply when outside temperatures exceed 80°F. Dicamba is available mixed with diflufenzopyr in a formulation called *Overdrive*. The combination is broadleaf-selective, but safe on most grasses. This has been reported to be effective on rush skeletonweed. Diflufenzopyr is an auxin transport inhibitor which causes dicamba to accumulate in

	shoot and root meristems, increasing its activity. *Overdrive* is applied postemergence at 4 to 8 oz product/acre on rapidly growing plants. Higher rates should be used on large biennials or when treating perennial weeds. Add a non-ionic surfactant to the treatment solution at 0.25% v/v or a methylated seed oil at 1% v/v solution.
Picloram *Tordon 22K*	**Rate:** 2 to 4 pt product/acre (0.5 to 1 lb a.e./acre) **Timing:** Best applied postemergence to rosettes in fall or spring. **Remarks:** Picloram controls many species of broadleaf plants, but is relatively safe on established grasses. Some reports by applicators indicate that it may injure young or germinating grasses with fewer than 4 leaves. It is also effective when mixed with dicamba or 2,4-D. It has long soil residual activity. *Tordon 22K* is a federally restricted use pesticide. Not registered for use in California.
AROMATIC AMINO ACID INHIBITORS	
Glyphosate *Roundup, Accord XRT II,* *and others*	**Rate:** 2 to 4 pt product (*Roundup ProMax*)/acre (1.1 to 2.25 lb a.e./acre). Spot treatment: 1 to 2% v/v solution **Timing:** Apply to rapidly growing plants in bud stage. **Remarks:** Glyphosate is nonselective and has no soil activity. Repeat applications may be necessary. Effectiveness may be increased by addition of ammonium sulfate. Studies indicate variable results with glyphosate.
BRANCHED-CHAIN AMINO ACID INHIBITORS	
Imazapyr *Arsenal, Habitat, Stalker,* *Chopper, Polaris*	**Rate:** 3 to 4 pt product/acre (0.75 to 1 lb a.e./acre) **Timing:** Preemergence or postemergence to rapidly growing plants. **Remarks:** Nonselective herbicide with fairly long soil residual activity.

Cirsium arvense (L.) Scop.

Canada thistle

Family: Asteraceae
Range: Found throughout much of the United States, including all western states.
Habitat: Open, disturbed sites such as roadsides, gardens, pastures, hillsides, rangeland, stream banks, forest openings, and sometimes cropland such as alfalfa or grains. Prefers moist soils but will tolerate a wide range of soil types.
Origin: Native to southeastern Europe and the eastern Mediterranean area.
Impact: Competes aggressively with native plant species. It causes extensive yield loss in crops by competing for nutrients, light and water. It may also have an allelopathic effect. The productivity of pastures is significantly reduced because livestock avoid grazing Canada thistle and surrounding plants due to the spiny nature of the mature foliage. Canada thistle can also be economically damaging to ranchers by causing an increase in infections due to abrasions. Canada thistle is a host species for several agricultural insect and disease pests such as the sod-web worm, bean aphid, stalk borer, and cucumber mosaic virus.
Western states listed as a noxious weed: Arizona, California, Colorado, Idaho, Montana, Nevada, New Mexico, North Dakota, Oregon, South Dakota, Utah, Washington, Wyoming
California Invasive Plant Council (Cal-IPC) Inventory: Moderate Invasiveness

Canada thistle is an erect perennial that grows up to 3 to 5 ft tall and forms patches or clumps that are usually of a single sex. Stems ordinarily die back over winter and new shoots are formed in spring from old stem bases or root buds when the soil moisture permits. Canada thistle has an extensive creeping root system that can reach depths of 6 to 15 ft making eradication difficult. The spiny lobed leaves are 6 to 8 inches long and 1 to 1.5 inches wide. The leaves are alternate, oblong or lance-shaped and the base leaves stalkless and clasping.

Plants are dioecious (separate male and female plants) and flower heads are white to purple, borne in clusters of 1 to 5 per branch. The purplish involucre is glabrous or has white wooly hairs. Plants develop from seed and from vegetative shoots that generate from adventitious root buds. Canada thistle can produce between 1,000 and 5,000 seeds per stem. Most seeds fall near the parent plants or disperse short distances with wind. Birds and small mammals can consume and disperse some seeds. The seeds have been known to survive in the soil for up to 20 years and longevity is favored by deep burial.

NON-CHEMICAL CONTROL

Mechanical (pulling, cutting, disking)	Mowing can be used to reduce the nutrient storage in the roots and suppress flower formation. However, for mowing to be effective it must be repeated at least every 3 to 4 weeks over several growing seasons or coupled with other control practices.
	Tillage or cultivation can actually increase Canada thistle because it breaks the root system into fragments, spreading the roots through the soil and stimulating development of new plants. Small root pieces have enough stored reserves to develop new plants. Small roots can survive at least 100 days without nutrient replenishment from photosynthesis. For cultivation to be effective it must be repeated at 21 day intervals throughout the growing season.
Cultural	Neither grazing nor prescribed burning have been shown to be effective for the management of Canada thistle.
Biological	Three insects have been released as biocontrol agents. None of these species have had a significant impact on Canada thistle.
	The larvae of the Canada thistle stem weevil (*Ceutorhynchus litura*) bore into the main leaf vein and then into the crown. It is considered the most effective of the current biocontrol agents, reducing plant vigor. When present in high enough densities it can kill the plant. Larvae of the bud weevil (*Larinus planus*) feed on the bud and can reduce the potential for sexual reproduction. Larvae of the thistle stem gall fly (*Urophora cardui*) bore into the apical meristem of shoots and form a gall. They can reduce plant vigor and can prevent flower

formation depending upon the location of the gall.

A pathogenic rust (*Puccinia punctiformis*) infects Canada thistle (mix sap from infected plant with water and spray uninfected plants to infect them), but it too has not had a significant effect on its control.

CHEMICAL CONTROL

The following specific use information is based on published papers and reports by researchers and land managers. Other trade names may be available, and other compounds also are labeled for this weed. Directions for use may vary between brands; see label before use. Herbicides are listed by mode of action and then alphabetically. The order of herbicide listing is not reflective of the order of efficacy or preference.

GROWTH REGULATORS	
2,4-D Several names	**Rate:** 2 qt product/acre (1.9 lb a.e./acre) **Timing:** Postemergence in spring at the pre-bud to early bud stage. **Remarks:** Control with 2,4-D alone is only temporary; therefore, it is commonly mixed with other growth regulator herbicides. Research from Colorado showed control from a spring 2,4-D application followed by fall application with different herbicides. 2,4-D is broadleaf-selective and has no soil activity. Do not apply ester formulation when outside temperatures exceed 80°F.
Aminocyclopyrachlor + chlorsulfuron *Perspective*	**Rate:** 4.75 to 8 oz product (*Perspective*)/acre **Timing:** Postemergent to plants before they produce seed. **Remarks:** *Perspective* provides broad-spectrum control of many broadleaf species. Although generally safe to grasses, it may suppress or injure certain annual and perennial grass species. Do not treat in the root zone of desirable trees and shrubs. May need retreatment for 1 to 2 additional years. Do not apply more than 11 oz product/acre per year. At this high rate, cool-season grasses will be damaged, including bluebunch wheatgrass. Not yet labeled for grazing lands. Add an adjuvant to the spray solution. This product is not approved for use in California and some counties of Colorado (San Luis Valley).
Aminopyralid *Milestone*	**Rate:** 5 to 7 oz product/acre (1.25 to 1.75 oz a.e./acre) **Timing:** Postemergence in spring after all plants have fully emerged (some may be budding) until the oldest plants are in full flower stage. Use the higher rate when applying to flowering plants. Applications are also effective in fall before a killing frost. Use higher rates for older/dense stands or for longer residual control. **Remarks:** May need retreatment for 1 to 2 additional years. Aminopyralid is one of the most effective herbicides for the control of Canada thistle. It is safe on grasses, although preemergence application at high rates can greatly suppress invasive annual grasses, such as medusahead. Aminopyralid has a longer residual and higher activity than clopyralid. Other members of the Asteraceae and Fabaceae are very sensitive to aminopyralid. Other premix formulations of aminopyralid can also be used for Canada thistle control. These include *Opensight* (aminopyralid + metsulfuron; 2.5 to 3 oz product/acre) and *Forefront HL* (aminopyralid + 2,4-D; 1.5 to 2.1 pt product/acre), both applied at the rosette to bolting stages. The formulation with metsulfuron is not registered for use in California.
Clopyralid *Transline*	**Rate:** 0.67 to 1.33 pt product/acre (4 to 8 oz a.e./acre) **Timing:** Postemergence before the bud stage when most of the basal leaves have emerged. Fall applications are also effective. **Remarks:** One or more treatments per season may be needed for 1 to 3 consecutive years for complete control. Allow at least 20 days after application before disturbing treated areas. While clopyralid is very safe on grasses, it will injure many members of the Asteraceae, particularly thistles, and can also injure legumes, including clovers. Most other broadleaf species and all grasses are not injured. Also applied in a premix with triclopyr (*Redeem*, 2.5 to 4 pt product/acre) to rosette to bud stage Canada thistle.
Dicamba *Banvel*	**Rate:** 4 pt product/acre (2 lb a.e./acre) **Timing:** Postemergence to rosettes. Fall applications are also effective. **Remarks:** Dicamba is a broadleaf-selective herbicide often combined with other active ingredients. It is not typically used alone to control Canada thistle because it is not as effective as other herbicides

	such as aminopyralid, clopyralid or aminocyclopyrachlor.
	Dicamba is available mixed with diflufenzopyr in a formulation called *Overdrive*. This has been reported to be effective on Canada thistle. Diflufenzopyr is an auxin transport inhibitor which causes dicamba to accumulate in shoot and root meristems, increasing its activity. *Overdrive* is applied postemergence at 4 to 8 oz product/acre to rapidly growing plants. Higher rates should be used when treating perennial weeds. Add a non-ionic surfactant to the treatment solution at 0.25% v/v or a methylated seed oil at 1% v/v solution.
Picloram *Tordon 22K*	**Rate:** 2 pt product/acre (8 oz a.e./acre). **Timing:** Best when applied postemergence to rapidly growing thistle after most leaves emerge but before bud stage. Fall applications are also effective. **Remarks:** Picloram gives a broader spectrum of control than aminopyralid, aminocyclopyrachlor, and clopyralid, and has much longer soil residual activity. Most broadleaf plants are susceptible. Although well-developed grasses are not usually injured by labeled use rates, some applicators have noted that young grass seedlings with fewer than four leaves may be killed. Do not apply near trees. *Tordon 22K* is a federally restricted use pesticide. Picloram is not registered for use in California.
AROMATIC AMINO ACID INHIBITORS	
Glyphosate *Roundup, Accord XRT II,* and others	**Rate:** Broadcast foliar treatment: 2 qt product (*Roundup ProMax*)/acre (2.25 lb a.e./acre). Spot treatment: 2% v/v solution **Timing:** Postemergence to rapidly growing thistles when most plants are past the bud stage. Fall applications must be before the first killing frost. **Remarks:** Do not tank-mix other herbicides with glyphosate for thistle control. More than 1 year of treatment may be necessary for complete control. Glyphosate will only provide control during the year of application; it has no soil activity and will not kill seeds or inhibit germination the following season. Glyphosate is nonselective. To achieve selectivity, it can be applied using a wiper or spot treatment to control current year's plants.
BRANCHED-CHAIN AMINO ACID INHIBITORS	
Chlorsulfuron *Telar*	**Rate:** 1 to 1.33 oz product/acre (0.75 to 1 oz a.i./acre) **Timing:** Postemergence from bolting to bloom stages. Can also apply in fall. **Remarks:** Chlorsulfuron has mixed selectivity on both broadleaf and grass species but is generally safe on most grasses. It has fairly long soil residual activity. The herbicide solution requires constant agitation during application.
Imazapyr *Arsenal, Habitat, Stalker, Chopper, Polaris*	The herbicide label indicates that 4 to 6 pt product/acre gives some level of control, but imazapyr is not usually the herbicide of choice for the control of Canada thistle.
Sulfometuron *Oust* and others	**Rate:** 6 to 8 oz product/acre (4.5 to 6 oz a.i./acre) **Timing:** Apply preemergence or early postemergence before or during the rainy season when weeds are germinating or actively growing. **Remarks:** Sulfometuron has mixed selectivity. It can cause minor damage to some native perennial grasses and has a fairly long soil residual. Higher rates may increase control but will also give more bare ground. Requires 20 inches of annual rainfall or more for effective preemergence control.

Cirsium vulgare (Savi) Ten.
Bull thistle

Family: Asteraceae
Range: Found in every state in the U.S.
Habitat: Disturbed areas including rangeland, pastures, forest clearcuts, roadsides and waste areas. Also occurs in foothills, dry meadows and riparian areas.
Origin: Native to Europe.
Impact: Bull thistle is not palatable to livestock and reduces the forage potential of infested pasture and rangeland. Once established, it can outcompete native plants. Although common, bull thistle is generally not considered as problematic as musk or Scotch thistle.
Western states listed as Noxious Weed: California, Colorado, New Mexico, Oregon, Washington
California Invasive Plant Council (Cal-IPC) Inventory: Moderate Invasiveness

Bull thistle is usually a biennial, but sometimes an annual or monocarpic perennial. It can grow up to 7 ft in height, but 2 to 6 ft is more common. Rosettes up to 3 ft in diameter form the first year. Leaves are 3 to 12 inches long, deeply lobed with coarse prickly hairs on the top and woolly hairs underneath. Leaves have sharp spines along the midrib and at the tip of the lobes, with the tip resembling a spear. Plants can have spreading branches, and sometimes a single stem. Stem have spiny wings that run down the length of the stem. Bull thistle requires vernalization before bolting.

Plants produce solitary (or sometimes clustered) pink-magenta flowerheads at the end of each stem. They are 1.5 to 2 inches wide and 1 to 2 inches long. Large spiny bracts surround the seedheads. Bull thistle reproduces and spreads entirely from seeds. Under favorable conditions, plants can produce 100 to 300 seeds per flowerhead or more, with 1 to more than 400 flowerheads per plant. Seeds have a feathery pappus that detaches at maturity, so seeds usually do not travel great distances by wind. Most seeds fall within a few feet of the parent plant. Seeds germinate in fall or spring depending on soil moisture. Most seeds either germinate or die within the first year, but seeds buried to about 6 inches or deeper may survive for up to 3 years or more.

NON-CHEMICAL CONTROL

Mechanical (pulling, cutting, disking)	Tillage, hoeing, and hand pulling are effective as long as they are done before flowering to prevent seed production. Any mechanical or physical control measure that severs the root below the soil surface is very effective. The plant must be cut off below the soil surface and no leaves should remain attached, or the plant will recover.
	Mowing is only effective when done either immediately before flowering or when plants are just starting to flower. Mowing too early only delays flowering, while mowing too late may allow production of viable seed. Because there can be a wide variation in the maturity of plants, a single mowing is generally insufficient because some seed will still be produced. Repeated mowing throughout the growing season is a more successful approach.
Cultural	The ability of thistles to invade pastures can be changed by grazing management, primarily by changing the competitiveness of the desirable pasture species. Sheep, goats, and horses, but not cattle, will eat young plants and can have a significant effect on thistles in the early stages of an infestation. Goats tend to avoid bull thistle foliage but eat the flowerheads, which can completely prevent seed dispersal from mature plants. Light grazing by sheep may selectively reduce competition from neighboring plants, increasing seedling survival, growth, flowering and seed production in bull thistle.
	It is unclear whether fire will completely kill bull thistle. Only mature thistle plants may readily combust and their seed may already be dispersed. Fire can create conditions that favor the establishment of bull thistle, so colonization after a fire may be enhanced. Burning can be used to remove above-ground material once it dries in late summer to fall. This can facilitate subsequent herbicide applications. Burning may also encourage the seedbank to flush, providing an opportunity for seedling control.

Biological	The bull thistle gall fly (*Urophora stylata*) was released as a biocontrol agent in the Pacific Northwest, as was the thistle head weevil, *Rhinocyllus conicus*. *Urophora* is not established in California yet and has little impact elsewhere. *R. conicus* is widely established in the western United States and attacks many thistle species, including some native species. A weevil, *Trichosirocalus horridus*, was introduced to the U.S. in 1974 to control musk thistle and other thistles. Reports of its effectiveness vary.

CHEMICAL CONTROL

The following specific use information is based on published papers and reports by researchers and land managers. Other trade names may be available, and other compounds also are labeled for this weed. Directions for use may vary between brands; see label before use. Herbicides are listed by mode of action and then alphabetically. The order of herbicide listing is not reflective of the order of efficacy or preference.

GROWTH REGULATORS	
2,4-D Several names	**Rate:** 1.6 to 2.1 qt product/acre (1.5 to 2 lb a.e./acre) **Timing:** Postemergence at rosette stage. Treat seedling rosettes in fall. **Remarks:** 2,4-D is broadleaf-selective and has no soil activity. It may require repeat applications. 2,4-D is generally not the most effective treatment, but is widely used because of low cost. Use a surfactant. When using the ester formulation do not apply when outside temperatures exceed 80°F.
Aminocyclopyrachlor + chlorsulfuron *Perspective*	**Rate:** 4.75 to 8 oz product (*Perspective*)/acre **Timing:** Postemergence and preemergence. Postemergence applications are most effective when applied to plants from the seedling to the bolting stage. **Remarks:** *Perspective* provides broad-spectrum control of many broadleaf species. Although generally safe to grasses, it may suppress or injure certain annual and perennial grass species. Do not treat in the root zone of desirable trees and shrubs. Do not apply more than 11 oz product/acre per year. At this high rate, cool-season grasses will be damaged, including bluebunch wheatgrass. Not yet labeled for grazing lands. Add an adjuvant to the spray solution. This product is not approved for use in California and some counties of Colorado (San Luis Valley).
Aminopyralid *Milestone*	**Rate:** 3 to 5 oz product/acre (0.75 to 1.25 oz a.e./acre) **Timing:** Postemergence in spring to early summer when the target plants are in the rosette to bolting stage, or in fall to seedlings. **Remarks:** Aminopyralid is a broadleaf herbicide similar to picloram, but more selective and generally safe on grasses. Its soil residual activity will kill emerging seedlings. Aminopyralid has a longer soil residual and higher activity than clopyralid. Aminopyralid can also be used in a premix with 2,4-D (*Forefront HL*) at 1.2 to 1.5 pt product/acre for bull thistle control.
Clopyralid *Transline*	**Rate:** 0.67 to 1.33 pt product/acre (4 to 8 oz a.e./acre) **Timing:** Postemergence in spring up to the bud stage. Can also apply to fall regrowth. Results are best if applied to rapidly growing weeds. **Remarks:** Clopyralid is a broadleaf herbicide like picloram, but more selective. It is very safe on grasses.
Dicamba *Banvel, Clarity*	**Rate:** 1 to 2 pt product/acre (0.5 to 1 lb a.e./acre) **Timing:** Postemergence to rosettes in spring. Fall applications help control seedling rosettes. **Remarks:** Dicamba is a broadleaf-selective herbicide often combined with other active ingredients. It is also effective when tank-mixed with 2,4-D (0.75 lb a.e./acre of dicamba + 0.25 lb a.e./acre of 2,4-D). Avoid drift to sensitive crops. Do not apply when outside temperatures exceed 80°F. Dicamba is available mixed with diflufenzopyr in a formulation called *Overdrive*. This has been reported to be effective on bull thistle. Diflufenzopyr is an auxin transport inhibitor which causes dicamba to accumulate in shoot and root meristems, increasing its activity. *Overdrive* is applied postemergence at 4 to 8 oz product/acre on rapidly growing plants. Higher rates should be used on large annuals and biennials. Add a non-ionic surfactant to the treatment solution at 0.25% v/v or a methylated seed oil at 1% v/v solution.
Picloram *Tordon 22K*	**Rate:** 0.5 to 0.75 pt product/acre (2 to 3 oz a.e./acre) **Timing:** Postemergence during active growth before bud stage. **Remarks:** Picloram is one of the most effective herbicides for bull thistle control. Most broadleaf

	plants are susceptible, but relatively safe on established grasses. It is also effective when mixed with dicamba or 2,4-D. Picloram has long soil residual activity and has been reported by some to injure young or germinating grasses. Picloram can also be used in a premix with 2,4-D (*Grazon P+D*) to give control of bull thistle. Picloram products are federally restricted use pesticides. Picloram and its formulations are not registered for use in California.
Triclopyr *Garlon 3A, Garlon 4 Ultra*	**Rate:** 0.33 to 1.5 gallons *Garlon 3A*/acre or 0.25 to 1 gallons *Garlon 4 Ultra*/acre (1 to 4.5 lb a.e./acre) **Timing:** Postemergence to rapidly growing weeds, up to bud stage. **Remarks:** Triclopyr is broadleaf-selective and safe on most grasses. It is most effective on smaller plants. *Garlon 4 Ultra* is formulated as a low volatile ester. However, in warm temperatures, spraying onto hard surfaces such as rocks or pavement can increase the risk of volatilization and off-target damage. Recommended rates are based on those reported for perennial thistles. Triclopyr can also be used in a premix with 2,4-D (*Crossbow*) or clopyralid (*Transline*).
BRANCHED-CHAIN AMINO ACID INHIBITORS	
Chlorsulfuron *Telar*	**Rate:** 1 oz product/acre (0.75 oz a.i./acre) **Timing:** Postemergence to young rapidly growing weeds. **Remarks:** Chlorsulfuron provides residual control 1 year after treatment. It has mixed selectivity, but is generally safe on grasses. Always use a surfactant. 2,4-D at 1 to 2 pt product/acre can be tank-mixed with chlorsulfuron for quicker burndown.
Imazapyr *Arsenal, Habitat, Stalker, Chopper, Polaris*	**Rate:** Broadcast treatment: 4 to 6 pt product/acre (1 to 1.5 lb a.e./acre). Spot treatment: 1% v/v solution **Timing:** Postemergence at flowering. **Remarks:** Imazapyr is best used as a spot treatment. It is a nonselective herbicide. It also has long soil residual activity and can leave more bare ground than other treatments, even a year after application. Recommended rates are based on those reported for perennial thistles.
Metsulfuron *Escort*	**Rate:** 1.5 to 2 oz product/acre (0.9 to 1.2 oz a.i./acre) **Timing:** Postemergence to young, rapidly growing weeds in spring before flowering, or in fall to new rosettes. **Remarks:** Metsulfuron has mixed selectivity, but is generally safe on grasses. Use a surfactant. It can be tank-mixed with 2,4-D or aminopyralid. *Opensight* is a premix of aminopyralid and metsulfuron; use at 1 to 2.5 oz product/acre. Metsulfuron has some soil residual activity. Recommended rates are based on those reported for perennial thistles. Metsulfuron and its formulations are not registered for use in California.

Conium maculatum L.
Poison-hemlock

Family: Apiaceae
Range: Throughout the contiguous U.S., including every western state.
Habitat: Moist soil along hedgerows, along the banks of streams and rivers, roadsides and wastelands, woodlands, meadows, and pastures.
Origin: Native to Europe.
Impact: Produces piperidine alkaloids which are highly toxic to humans and animals. Domestic animals such as swine, cattle, goats, horses, and sheep can be poisoned by the toxin coniine after ingesting any portion of the plant. Poison-hemlock is the most toxic to cattle. Symptoms can include vomiting, nausea, trembling, rapid respiration, joint and movement problems, slow, weak and rapid pulse, increased salivation and urination, convulsions, paralysis, coma, and death from respiratory paralysis. Ingestion during fetal development can result in severe birth defects.
States listed as Noxious Weed: Colorado, Idaho, Nevada, New Mexico, Oregon, Utah, Washington
California Invasive Plant Council (Cal-IPC) Inventory: Moderate Invasiveness

Poison-hemlock is a tall branching biennial to 4 to 6 ft tall. The stem is erect, hollow, smooth, bright green with a distinctive mottled appearance and irregular purple splotches. The long-stalked leaves are glabrous, up to 2 ft long, bright green, alternate, and tripinnately compound (divided into leaflets which are again divided and subdivided). Its root is a long, forked, fleshy taproot, pale yellow in color, with numerous lateral roots.

The inflorescence is a compound umbel with 12 to 16 rays, and numerous small white flowers located at the terminal positions. Each flower produces two gray-brown seeds with five wavy longitudinal ridges. The seeds have the highest concentration of coniine. Poison hemlock smells like mouse urine when crushed, a characteristic of the poisonous alkaloids. Plants reproduce only by seed. Most seeds fall near the parent plant. Seed dispersal is prolonged and occurs from late summer through winter. Most seeds germinate almost immediately if conditions are favorable, but a small proportion remains dormant. Dormant seeds require a period of high summer and/or low winter temperatures before they can germinate. Seeds survive up to about 3 years under field conditions.

NON-CHEMICAL CONTROL

Mechanical (mowing, plowing, and cultivation)	Hand removal is recommended for small infestations. When pulling the plants, dig down and remove the entire taproot. Wear gloves and wash hands after working with poison-hemlock. Manual control efforts can be successful, but can cause soil disturbance encouraging further germination of seeds. Solid carpets of hemlock seedlings are not uncommon following soil disturbance.
	Cutting is ineffective; the plants send up new seed stalks in the same season the cutting occurs.
	Establishment of populations can be prevented with repeated cultivation and plowing.
Cultural	Due to the plant's toxicity, grazing is not recommended for control. Even dried plant parts are not safe as the toxins take several years to dissipate. Use certified weed-free hay to prevent the poisoning of livestock.
	Do not burn, as toxins can be released into the air through the smoke.
Biological	There are no known biological controls. Since its introduction to North America, only a few native insects have been able to overcome its toxic defenses. These attack the seedhead, but do very little damage to the rest of the plant.

CHEMICAL CONTROL
The following specific use information is based on published papers and reports by researchers and land managers. Other trade names may be available, and other compounds also are labeled for this weed. Directions for use may vary between brands; see label before use. Herbicides are listed by mode of action and then alphabetically. The order of herbicide listing is not reflective of the order of efficacy or preference.

GROWTH REGULATORS	
2,4-D Several names	**Rate:** 1 to 4 pt product/acre (0.48 to 1.9 lb a.e./acre) **Timing:** Postemergence in seedling to rosette stage. **Remarks:** Broadleaf-selective, most effective when applied soon after plants emerge. Adding a wetting agent may enhance control. Also effective tank mixed with dicamba.
Aminocyclopyrachlor + chlorsulfuron *Perspective*	**Rate:** 4.75 to 8 oz product/acre **Timing:** Postemergence in seedling to rosette stage. **Remarks:** Broad-spectrum control of many broadleaf species. Although generally safe to grasses, it may suppress or injure certain annual and perennial grass species. Do not treat in the root zone of desirable trees and shrubs. Do not apply more than 11 oz product/acre per year. At this high rate, cool-season grasses will be damaged, including bluebunch wheatgrass. Not yet labeled for grazing lands. Add an adjuvant to the spray solution. This product is not approved for use in California and some counties of Colorado (San Luis Valley).
Aminopyralid + metsulfuron *Opensight*	**Rate:** 2.5 to 3.3 oz product/acre **Timing:** Preemergence in fall, or postemergence in the seedling to rosette stage. **Remarks:** Not registered for use in California.
Triclopyr *Garlon 4 Ultra, Remedy Ultra*	**Rate:** 5 to 8 pt product/acre (2.5 to 4 lb a.e./acre) **Timing:** Postemergence in seedling to rosette stage. **Remarks:** Broadleaf-selective, safe on most grasses. Most effective on smaller plants. *Garlon 4 Ultra* and *Remedy Ultra* are low volatile esters. However, in warm temperatures, spraying onto hard surfaces such as rocks or pavement can increase the risk of volatilization and off-target damage. Also effective in a premix with 2,4-D (*Crossbow*) or tank mixed with clopyralid (*Transline*).
AROMATIC AMINO ACID INHIBITORS	
Glyphosate *Roundup, Accord XRT II,* and others	**Rate:** Broadcast treatment: 1.33 to 2.67 qt product (*Roundup ProMax*)/acre (1.5 to 3 lb a.e./acre). Spot treatment: 1 to 1.5% v/v solution **Timing:** Best when applied postemergence to rapidly growing plants before bolting. However, higher rates can control plants at the bud to full bloom stage. **Remarks:** Glyphosate is a nonselective herbicide that has no soil activity. Add a non-ionic surfactant.
BRANCHED-CHAIN AMINO ACID INHIBITORS	
Chlorsulfuron *Telar*	**Rate:** 1 to 2.6 oz product/acre (0.75 to 1.95 oz a.i./acre) **Timing:** Postemergence to rapidly growing plants. **Remarks:** Desirable grasses should be well established before application.
Imazapic *Plateau*	**Rate:** 8 to 12 oz product/acre (2 to 3 oz a.e./acre) **Timing:** Preemergence. **Remarks:** Mixed selectivity, favors members of the Asteraceae and some grasses. Some soil residual activity. In postemergence applications, use methylated seed oil at 1.5 to 2 pt/acre. Not registered for use in California.
Imazapyr *Arsenal AC, Habitat, Stalker, Chopper, Polaris*	**Rate:** 2 pt product (*Arsenal AC*)/acre (1 lb a.e./acre); 4 pt product (*Habitat*)/acre (1 lb a.e./acre) + 1 qt/acre methylated seed oil **Timing:** Preemergence or early postemergence in the rosette stage. **Remarks:** Nonselective. Long soil residual, leaves more bare ground than other treatments, even a year after application. Do not apply more than 3 qt product/acre. *Habitat* is an aquatic registered formulation for use close to water.
Metsulfuron *Escort*	**Rate:** 1 oz product/acre (0.6 oz a.i./acre) **Timing:** Postemergence to rapidly growing plants. **Remarks:** Use a non-ionic surfactant or silicone surfactant. Prevent drift to sensitive plants. Apply only to pastures, rangeland, and non-crop sites. Metsulfuron can also be used in a premix with dicamba + 2,4-D (*Cimarron Max*). Metsulfuron and its formulations are not registered in California.

Convolvulus arvensis L.

Field bindweed

Family: Convolvulaceae
Range: Found in all contiguous states and Hawaii.
Habitat: Cultivated crops, gardens, pastures, abandoned fields, and roadsides. Grows best on moist, deep fertile soils. Tolerates poor, dry gravelly soils, but seldom grows in wet soils. Inhabits regions with temperate, Mediterranean, and tropical climates. Found at elevations up to 9000 ft.
Origin: Native to Europe.
Impact: Field bindweed is considered one of the most noxious weeds in agricultural climates in the temperate zone. Plants typically form large patches that are difficult to control due to their extensive root system and long-lived seeds. It is not as important a problem in wildlands and natural areas as it is in croplands.
Western states listed as Noxious Weed: Arizona, California, Colorado, Idaho, Montana, New Mexico, North Dakota, Oregon, South Dakota, Utah, Washington, Wyoming

Field bindweed is a long-lived herbaceous perennial with vine-like stems and an extensive system of deep roots. The glabrous stems twine around other plants for support and are up to 4 ft in length. The leaves are typically dull green and arranged alternately on the stem. Leaves vary in size and shape depending on environmental factors. They are typically 1 to 2 inches in length and vary from arrowhead-shaped to almost round. The root system is an extensive network of vigorous primary and secondary taproots, horizontal creeping roots, and lateral feeder roots. The taproots can grow to a depth of 10 ft or more depending on the available soil moisture and soil depth, while most of the horizontal creeping roots develop in the top 2 ft of soil.

Plants flower from spring to the first frost. The white or pinkish flowers open for one day; they are insect pollinated and self-incompatible. The flowers are axillary, solitary or in cymes of 2 to 4, on stalks about 1 to 3 inches long. The flowers are typically 1 to 2 inches long, funnel-shaped with five fused petals with pleating that is spiraled in the bud.

Field bindweed reproduces sexually through seed and vegetatively through deep horizontal creeping roots and rhizomes. Seeds form in capsules and are dispersed only short distances. One plant can produce up to 500 seeds that can survive buried for 15 to 20 years or more. Most young plants do not produce seed in their first season.

NON-CHEMICAL CONTROL

Mechanical (pulling, mowing, tilling, solarization)	Pulling can be effective on seedlings or young adults but is not effective when the plant has developed a deep, extensive root system.
	Mowing is not effective due to the low profile of the plant.
	Intensive cultivation will control new seedlings but spreads the roots and seeds, which may spread the plant. Tilling conducted 8 to 12 days after each emergence throughout the growing season can control field bindweed, but this requires repeated treatments for 1 to 5 years.
	Deep tillage using shanks down to 3 ft with a cross bar will reduce emergence for a season. Shallow cultivation that kills all above-ground shoots can be effective if repeated several times over a couple of years.
	Solarization is an effective control method, but the black plastic or mulch must be left on the site for 3 to 5 years to eradicate field bindweed.
Cultural	Sheep and cattle have been used to graze field bindweed but this does not affect the roots of the plant and regrowth occurs quickly.
	Burning is not considered an effective control method as it only removes the aboveground biomass while

	the root system and seeds are left intact. A combination of burning with other control measures in an integrated approach is more effective.
Biological	Three biological control species have been released in the United States. *Tyta luctuosa* (European field bindweed moth) defoliates field bindweed as a caterpillar. *Chelymorpha cassidea* (tortoise beetle) is native to the United States and feeds on the leaves. *Aceria malherhae* (bindweed gall mite) is a gall mite that has established in several states and feeds on the leaves, stem, and root bud. None of these species has controlled field bindweed in most areas, although the gall mite has shown some success in Colorado.

CHEMICAL CONTROL

The following specific use information is based on published papers and reports by researchers and land managers. Other trade names may be available, and other compounds also are labeled for this weed. Directions for use may vary between brands; see label before use. Herbicides are listed by mode of action and then alphabetically. The order of herbicide listing is not reflective of the order of efficacy or preference.

GROWTH REGULATORS	
2,4-D amine Several names	**Rate:** 4 to 6 pt product/acre (1.9 to 2.85 lb a.e./acre) **Timing:** Postemergence at bud stage or in fallow in mid-summer, before bindweed is under moisture stress. **Remarks:** Use 2,4-D to help reduce bindweed stand 60 to 80% and prevent seedling establishment. 2,4-D applications must be made for several years consecutively to prevent regrowth. Avoid drift to sensitive crops.
Aminocyclopyrachlor + chlorsulfuron *Perspective*	**Rate:** 4.75 to 8 oz product (*Perspective*)/acre **Timing:** Postemergence when vegetation is fully developed. **Remarks:** *Perspective* provides broad-spectrum control of many broadleaf species. Although generally safe to grasses, it may suppress or injure certain annual and perennial grass species. Do not treat in the root zone of desirable trees and shrubs. Do not apply more than 11 oz product/acre per year. At this high rate, cool-season grasses will be damaged, including bluebunch wheatgrass. Not yet labeled for grazing lands. Add an adjuvant to the spray solution. This product is not approved for use in California and some counties of Colorado (San Luis Valley).
Dicamba *Banvel, Clarity*	**Rate:** 1 to 4 lb product/acre (0.5 to 2 lb a.e./acre) **Timing:** Postemergence when weeds are growing rapidly. Do not apply after bud break. **Remarks:** Recommended rates only suppress field bindweed. Follow-up treatments are generally necessary. Dicamba can be tank mixed with 2,4-D (0.5 to 2 lb a.e./acre) or glyphosate (3 lb a.i./acre). Dicamba is available mixed with diflufenzopyr in a formulation called *Overdrive*. This has been reported to be effective on field bindweed. Diflufenzopyr is an auxin transport inhibitor which causes dicamba to accumulate in shoot and root meristems, increasing its activity. *Overdrive* is applied postemergence at 4 to 8 oz product/acre. Higher rates should be used when treating perennial weeds. Add a non-ionic surfactant to the treatment solution at 0.25% v/v or a methylated seed oil at 1% v/v solution.
Fluroxypyr *Vista XRT*	**Rate:** 22 oz product/acre (7.7 oz a.e./acre) **Timing:** Postemergence when the target plants are growing rapidly. **Remarks:** Provides suppression and not control. Control is reduced if the plants are under stressed growth conditions.
Picloram *Tordon 22K*	**Rate:** 1 to 2 qt product/acre (0.5 to1 lb a.e./acre) **Timing:** Postemergence in the growing season when bindweed is visible. Timing is not critical, but results are most consistent if bindweed is in early bud to full bloom. **Remarks:** Apply as a coarse, low-pressure spray in sufficient volume to cover adequately. Picloram has long soil residual activity. Picloram is a restricted use herbicide. It is not registered for use in California.
Triclopyr *Garlon 3A*	**Rate:** 3 to 4 pt *Garlon 3A*/acre (1.13 to 1.5 lb a.e./acre) **Timing:** Postemergence at bud stage or at summer fallow in mid-summer. **Remarks:** Retreatment is usually necessary for effective control. Triclopyr has no soil residual activity

	and controls many broadleaf species.
AROMATIC AMINO ACID INHIBITORS	
Glyphosate *Roundup, Accord XRT II, and others*	**Rate:** 3 to 4 qt product (*Roundup ProMax*)/acre (3.4 to 4.5 lb a.e./acre) **Timing:** Postemergence when plants are growing rapidly, up to the beginning of seed production. Plants should not be under drought stress at time of application. Application in late summer is also effective. **Remarks:** Cover foliage thoroughly but avoid spray runoff. Repeat treatments may be needed for complete control. Control improves if treated area is tilled 2 to 3 weeks after treatment. Add non-ionic surfactant or 10 to 15 lb of ammonium sulfate. Glyphosate is a nonselective herbicide. It can be tank mixed with 2,4-D or dicamba.
BRANCHED-CHAIN AMINO ACID INHIBITORS	
Imazapic *Plateau*	**Rate:** 8 to 12 oz product/acre (2 to 3 oz a.e./acre) **Timing:** Postemergence, from 25% bloom through fall to rapidly growing bindweed. **Remarks:** For more effective control add 1 qt/acre methylated seed oil. Imazapic is not registered for use in California.
Imazapyr *Arsenal, Habitat, Stalker, Chopper, Polaris*	**Rate:** 1 pt product (*Arsenal*)/acre (4 oz a.e./acre) **Timing:** Preemergence or postemergence when plants are growing rapidly. **Remarks:** Imazapyr is fairly nonselective and may injure some desirable species, including grasses and broadleaves. It has fairly long soil residual activity, depending on the site.
Metsulfuron *Escort*	**Rate:** 1 to 2 oz product/acre (0.6 to 1.2 oz a.i./acre) **Timing:** Postemergence to rapidly growing bindweed in bloom stage. **Remarks:** Metsulfuron only suppresses field bindweed. Use a non-ionic or silicone surfactant to improve control. Metsulfuron is not registered for use in California.
Propoxycarbazone-sodium *Canter R+P*	**Rate:** 0.9 to 1.2 oz product/acre (0.63 to 0.84 oz a.i./acre) **Timing:** Postemergence to young, rapidly growing plants. **Remarks:** Propoxycarbazone is a broad-spectrum herbicide that will control many species. It will provide only partial control of field bindweed. Perennial grass species vary in tolerance. A non-ionic surfactant should be added at 0.25 to 0.5% v/v solution.

Cortaderia selloana (Schultes) Asch. & Graebner; pampasgrass
Cortaderia jubata (Lemoine) Stapf; jubatagrass

Pampasgrass and jubatagrass

Cortaderia selloana

Family: Poaceae
Range: Jubatagrass is found along coastal areas of California and Oregon. Pampasgrass is primarily in coastal areas of California and Oregon, but can also be found inland in both states and Utah.
Habitat: Disturbed areas, dunes, bluffs, roadsides, road-cuts, logged forests. Many coastal shrub and grasslands (including serpentine soils) and adjacent inland areas moderated by fog or other maritime influences. Pampasgrass is also found in undisturbed coastal shrubland and marshes, inland riparian areas and other interior sites where sufficient moisture is available (may tolerate standing water for prolonged periods).
Origin: Jubatagrass is native to equitable mid-elevation regions of the Andes Mountains in Ecuador, Peru, Bolivia, and northern Argentina, where the climate is similar to that of coastal California. Pampasgrass is native from the mid-elevation Andes slopes of northeastern Chile and northwestern Argentina to the low elevation subtropical grasslands and riparian areas of northern Argentina, Uruguay, and southern Brazil. Jubatagrass and pampasgrass were introduced as landscape ornamentals and for erosion control, but have since escaped cultivation and become noxious weeds in some areas of California, especially along the coast.

Cortaderia jubata

Impacts: Mature plants of both species are highly competitive with native vegetation and forestry tree seedlings. Jubatagrass produces abundant apomictic seed, and is currently more widespread than pampasgrass in California. Yet pampasgrass may be the more invasive species since it tolerates a wider range of climate variation. Both species are noxious weeds in Australia, Tasmania, and New Zealand.
Western states listed as Noxious Weed: *Cortaderia jubata*, California
California Invasive Plant Council (Cal-IPC) Inventory: Both species are High Invasiveness

Cortaderia species are large, densely tufted perennial grasses with long basal leaves and tall, showy, plume-like inflorescences. The leaves have sharply serrated margins that can easily cut the skin. Pampasgrass tussocks typically grow much larger than those of jubatagrass, and have a more erect, fountain-like appearance. The dense fibrous roots grow from shallow short lateral rhizomes. Ligules consist of a dense ring of hairs mostly 2 to 3 mm long.

Both species produce plumes that are 1 to 3 ft long, although jubatagrass often has a more purplish tinge compared to pampasgrass. Despite the similarity in appearance, the reproductive strategy of pampasgrass and jubatagrass are very different. In jubatagrass, all plants are female and develop seed without fertilization (apomixis). In pampasgrass, plants are functionally male or female. Male plumes are sometimes purplish-tinted.

Unlike jubatagrass, pampasgrass develops seed only when male and female plants are within pollination range of one another. Historically, the plumes on female plants were considered more attractive and were exclusively propagated by division within the nursery trade for ornamental purposes. In more recent years, some nursery stock has been propagated by seed, and both sexes have been widely planted as landscape ornamentals. Weedy populations spread quickly near these ornamental plantings.

Plants of both species reproduce only by seed. Each seed-bearing plume can produce up to 100,000 seeds. The seeds are very light and can disperse long distances with wind (to about 20 miles). Human activity can also disperse plants long distances. Because the seeds are so small, they do not survive long in the soil seedbank. Germination occurs in fall after the first rains, continuing through spring. Seeds typically survive for less than 6 months under field conditions, and a persistent seedbank does not accumulate.

NON-CHEMICAL CONTROL

Mechanical (pulling, cutting, disking)	Hand-pulling seedlings can help prevent the spread of either species. For removing established clumps, pulaskis, mattocks, or shovels are the safest and most effective tools. To prevent resprouting, it is important to remove the entire crown and top section of the roots. Detached plants left lying on the soil surface may take root and reestablish under moist soil conditions. Some land managers recommend turning the removed clumps upside down so the roots dry out in the air. A large chainsaw or weedeater can expose the base of the plant, allow better access for removal of the crown, and make disposal of the detached plant more manageable. Plumes can also be cut off to avoid seed dispersal. However, plants that have had plumes removed may develop more plumes during the flowering season. Mechanical removal by heavy equipment, including excavators and backhoes, can be very effective and selective. However, these methods are labor- and cost-intensive, and feasibility depends upon site accessibility, size of the infestation, funding, and availability of volunteer support
Cultural	Heavily mulching bare sites or planting desirable vegetation may prevent or reduce seedling establishment. Burning or grazing are not typically considered effective control strategies for *Cortaderia* in North America, but cattle have been shown to provide effective control for pampasgrass in commercial forests of New Zealand. Any soil disturbance that creates bare ground, including natural disturbance (fire or landslides) and human-caused disturbance, promotes invasion by jubatagrass or pampasgrass.
Biological	There are no biological control agents available for either of the *Cortaderia* species.

CHEMICAL CONTROL

The following specific use information is based on publications and reports by researchers and land managers. Other trade names may be available, and other compounds also are labeled for this weed. Directions for use may vary between brands; see label before use. Herbicides are listed by mode of action and then alphabetically. The order of herbicide listing is not reflective of the order of efficacy or preference.

LIPID SYNTHESIS INHIBITORS	
Fluazifop *Fusilade*	**Rate:** Spot treatment: 2 to 4% v/v solution (0.5 to 1% a.i.). Low volume treatment: 4% v/v solution of product. **Timing:** Postemergence. Best in late summer or fall, after flowering when translocation of herbicide to base of tillers and rhizomes is at its peak. **Remarks:** Control of jubatagrass with fluazifop was inconsistent. It has no soil residual activity. Other grass herbicides were not as effective.
AROMATIC AMINO ACID INHIBITORS	
Glyphosate *Roundup, Accord XRT II*, and others	**Rate:** Broadcast treatment: 2 to 3.3 qt product (*Roundup ProMax*)/acre (2.25 to 3.7 lb a.e./acre). High-volume spray-to-wet spot treatment: 2% v/v solution of product. Low-volume treatment: 8 to 10% v/v solution of product. Wiper treatment: 33 to 50% of concentrated product. **Timing:** Postemergence. Best in late summer or fall, after flowering when translocation of herbicide to base of tillers and rhizomes is at its peak. **Remarks:** Glyphosate provides the most consistent jubatagrass control with all plant sizes in both fall and early summer. Low volume treatment at 8% and wiper applications at 33% gave the best and most consistent control.
BRANCHED-CHAIN AMINO ACID INHIBITORS	
Imazapyr *Arsenal, Habitat, Polaris*	**Rate:** 2 to 4% v/v solution of product for spot treatment (0.45 to 0.9% a.e. solution) **Timing:** Postemergence. Best in late summer or fall, after flowering when translocation of herbicide to base of tillers and rhizomes is at its peak. **Remarks:** Results were inconsistent from site to site and year to year. Imazapyr is a slow-acting systemic herbicide and may take a year or two to achieve effective control on *Cortaderia*.

Cotoneaster franchetii Boiss.; orange cotoneaster
Cotoneaster lacteus W.W. Smith; Parney's cotoneaster
Cotoneaster pannosus Franch.; silverleaf cotoneaster

Cotoneasters

C. franchetii

C. lacteus

C. pannosus

Family: Rosaceae
Range: Coastal areas of California, Oregon, and Washington.
Habitat: Disturbed places, mixed evergreen forest, coastal scrub, and grassland, often near residential areas.
Origin: Introduced as ornamental plants from China and escaped to become invasive in coastal regions.
Impacts: On occasion, populations can become dense and crowd out native species. However, fruit likely provide a food source for some bird species.
California Invasive Plant Council (Cal-IPC) Inventory: All three species are Moderate Invasiveness

Cotoneasters are evergreen to semi-evergreen shrubs usually less than 10 ft tall, but occasionally taller. All species have simple alternate leaves and distinctive orange or red berry-like fruits. The upper surfaces of the leaves are generally dull with few or no hairs. The lower surface is usually covered with woolly hairs, but can lack hairs.

Flowers are white in both Parney's and silverleaf cotoneaster, but pink to rose in orange cotoneaster. All species have five petals. In some species, such as Parney's cotoneaster, the flower clusters are large. In others, including orange and silverleaf cotoneaster, the flower clusters are smaller and clustered more tightly. The fruit (pomes) are berry-like and red in both Parney's and silverleaf cotoneaster, but orange-red in orange cotoneaster. Plants reproduce by seed that are dispersed primarily by animals, particularly birds. Seeds require scarification and cold temperatures to germinate, so ingestion by animals is important to facilitate seed germination.

NON-CHEMICAL CONTROL

Mechanical (pulling, cutting, disking)	Seedlings and small plants can be hand pulled. Manually removing individual shrubs when discovered can help prevent the spread of cotoneaster species in natural areas. However, stumps and roots can resprout, necessitating follow-up control. Roots need to be completely removed to prevent resprouting.
Cultural	There are no known cultural control strategies developed for any species of cotoneaster.
Biological	There are many species of cultivated cotoneaster. As such, there has not been any effort to develop biological control agents for their management.

CHEMICAL CONTROL

There is little information on the control of cotoneaster species. The following specific use information is based on reports by researchers and land managers. Other trade names may be available, and other compounds also are labeled for this weed. Directions for use may vary between brands; see label before use. Herbicides are listed by mode of action and then alphabetically. The order of herbicide listing is not reflective of the order of efficacy or preference.

GROWTH REGULATORS	
Picloram	**Rate:** Undiluted concentrate for cut stump treatments
Tordon 22K	**Timing:** Treat cut stumps in late summer or fall for most effective translocation of herbicide to below-ground tissues.
	Remarks: Picloram is nonselective on most broadleaf species. Cut stump applications can provide

	selectivity, especially where plants are mixed with native species. Picloram is a restricted use herbicide. Picloram is not registered for use in California.
Triclopyr *Garlon 4 Ultra*	**Rate:** Basal bark treatment: 25% solution v/v for (*Garlon 4 Ultra*). Cut stump treatment: undiluted concentrate of either ester or amine formulation **Timing:** Treat cut stumps or basal stems in late summer or fall for most effective translocation of herbicide to below-ground tissues. **Remarks:** It is likely that foliar treatment with triclopyr would also be effective, but there is no data to demonstrate this. Triclopyr is considered the most effective chemical control option for cut stump treatment.
AROMATIC AMINO ACID INHIBITORS	
Glyphosate *Roundup, Accord XRT II*, and others	**Rate:** Spot foliar treatment: 5% v/v *Roundup ProMax* solution. Cut stump treatment: 40 to 100% of concentrate. **Timing:** Postemergence later in the season when translocation of carbohydrates is downward towards the below-ground tissues. **Remarks:** Glyphosate is a nonselective herbicide.

Crupina vulgaris Cass.
Common crupina

Photo courtesy of Richard Old

Family: Asteraceae
Range: Fairly uncommon in its distribution. Found in California, Idaho, Oregon, and Washington.
Habitat: Inhabits many moisture and temperature regimes and soil types. Often associated with disturbance. Generally found in grasslands, pastures, rangeland, forested areas, canyons, and riparian areas.
Origin: Native to the Mediterranean region of Europe.
Impacts: Primarily a noxious rangeland weed. Plants are adapted to many environmental conditions and are highly competitive for limited soil moisture. Dense populations displace desirable forage species and contaminate hay.
Western states listed as Noxious Weed: California, Colorado, Idaho, Montana, Nevada, Oregon, South Dakota, Washington
California Invasive Plant Council (Cal-IPC) Inventory: Limited Invasiveness

Common crupina is an erect cool-season annual to 2 ft tall. The rosette leaves are sessile or petioled, to 3 inches long, oblong to obovate in outline, with deep, narrow, opposite pinnate divisions, and covered with short, stiff hairs. The stems are openly branched and longitudinally ridged. Stem leaves are alternate, deeply pinnate-lobed once or twice, lobes narrow, reduced near stem tops. Plants produce dense fibrous roots.

The purple flower heads are cylindrical to ovoid or slender urn-shaped on stalks 0.5 to 1.5 inches long, with one to two fertile disk flowers in the center and two to four sterile disk-like flowers around the margin. Both flower types have a pappus of bristles. Plants produce 2 to 850 fairly large seed. Most of the seed falls near the parent plant, with some long distance dispersal via wildlife, livestock, and water. About 85% of the seeds germinate the following fall. Most germination occurs after the first significant rains of fall or early winter. Seeds can survive ingestion by most animals, except sheep, and remain viable in the soil for up to 3 years.

NON-CHEMICAL CONTROL

Mechanical (pulling, cutting, disking)	Small infestations can be removed by manual methods such as hand-pulling and digging. When digging, sever the root below the soil surface. Remove plants before seed is produced. Plants should be removed every 3 to 4 weeks during spring to ensure complete removal before seed maturity.
	Plants regrow after mowing. Mow before flowering to prevent seeds from developing in severed flowerheads. Mowing can stimulate lateral branching and increase seed production.
	Tillage will control emerged plants but often stimulates germination. Land managers using tillage for seedbed preparation for reseeding should prepare for a flush of seedlings when soils become saturated.
Cultural	Most livestock avoid grazing common crupina unless palatable forage is unavailable. However, it is highly favored by sheep and goats in the Mediterranean region. While they cannot be used to eradicate common crupina, they have been shown to suppress populations. Quarantine livestock for at least 6 days after foraging on infested rangeland to prevent introduction of common crupina to non-infested sites. Seeds often adhere to livestock.
	Fire is not an effective control method, and common crupina populations increase in response to nutrient release and more light at the soil surface following wildfire.
	Promoting competitive vegetation can slow spread and help prevent establishment. Perennial grass stand density and vigor should be managed to minimize bare ground exposure.
Biological	Potential biocontrol agents are being studied, but no biological controls are currently available in the United States.

CHEMICAL CONTROL

The following specific use information is based on published papers and reports by researchers and land managers. Other trade names may be available, and other compounds also are labeled for this weed. Directions for use may vary between brands; see label before use. Herbicides are listed by mode of action and then alphabetically. The order of herbicide listing is not reflective of the order of efficacy or preference.

GROWTH REGULATORS	
Aminocyclopyrachlor + chlorsulfuron *Perspective*	**Rate:** 4.75 to 8 oz product (*Perspective*)/acre **Timing:** Postemergence when plants are growing rapidly. **Remarks:** *Perspective* provides broad-spectrum control of many broadleaf species. Although generally safe to grasses, it may suppress or injure certain annual and perennial grass species. Do not treat in the root zone of desirable trees and shrubs. Do not apply more than 11 oz product/acre per year. At this high rate, cool-season grasses will be damaged, including bluebunch wheatgrass. Not yet labeled for grazing lands. Add an adjuvant to the spray solution. This product is not approved for use in California and some counties of Colorado (San Luis Valley).
Aminopyralid + metsulfuron *Opensight*	**Rate:** 3 to 3.3 oz product/acre **Timing:** Postemergence from the rosette to young bolting stage. **Remarks:** This combination has some soil activity. It is safe on most grasses, although preemergence application at high rates can greatly suppress some annual grasses, such as medusahead. Applications can decrease seed production in some annual and perennial grass species. For postemergence applications, add a non-ionic surfactant at 0.25 to 0.5% v/v. This combination is not registered for use in California.
Clopyralid *Transline*	**Rate:** 0.33 pt product/acre (2 oz a.e./acre) **Timing:** Apply split application at 0.33 pt in fall and spring. Both applications are postemergence. **Remarks:** Most effective for young plants. Clopyralid has a fairly short soil residual activity. It controls or injures plants in the Asteraceae and Fabaceae but is safe on most other broadleaf species and all grasses. For postemergence applications, adding a non-ionic surfactant at 0.25 to 0.5% v/v to spray solution may enhance control.
Dicamba *Banvel, Clarity*	**Rate:** 1 pt product/acre (0.5 lb a.e./acre) **Timing:** Postemergence to rapidly growing plants. **Remarks:** Dicamba is a broadleaf-selective herbicide often combined with other active ingredients. It is often mixed with 2,4-D at 0.5 to 1 lb a.e./acre.
Picloram *Tordon 22K*	**Rate:** 1 to 2 pt product/acre (4 to 8 oz a.e./acre) **Timing:** Preemergence or postemergence in fall or winter. **Remarks:** Picloram controls a wide range of broadleaf species and has relatively long soil residual activity. It is considered one of the best products to control common crupina. Although well-developed grasses are not usually injured by labeled use rates, some applicators have noted that young grass seedlings with fewer than four leaves may be killed. Do not apply near trees, or where soil is highly permeable and where water table is high. Picloram is a restricted use herbicide. Picloram is not registered for use in California.
BRANCHED-CHAIN AMINO ACID INHIBITORS	
Chlorsulfuron *Telar*	**Rate:** 1 to 2.6 oz product/acre (0.75 to 1.95 oz a.i./acre) **Timing:** Postemergence when plants are growing rapidly. **Remarks:** Always use a surfactant. Higher rates can cause grass injury.
Metsulfuron *Escort*	**Rate:** 0.5 to 1 oz product/acre (0.3 to 0.6 oz a.i./acre) **Timing:** Postemergence when plants are growing rapidly. **Remarks:** Always use a surfactant. Other premix formulations of metsulfuron can be used at similar application timing. These include *Cimarron Max* (metsulfuron + dicamba + 2,4-D) and *Cimarron X-tra* (metsulfuron + chlorsulfuron). Metsulfuron is not registered for use in California.

Cuscuta japonica Choisy.

Japanese or giant dodder

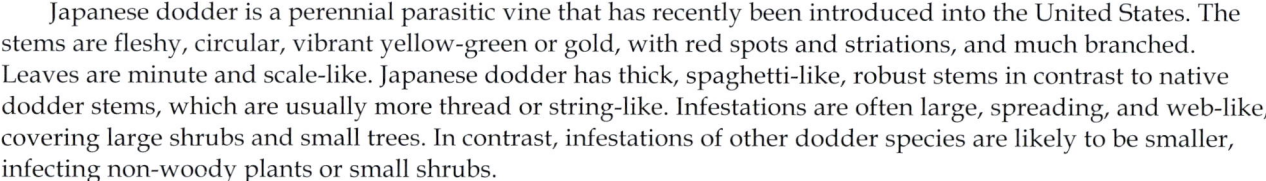

Family: Cuscutaceae
Range: South Carolina, Florida, Texas and California.
Habitat: Capable of growing in a wide range of environments and many soil types; found in fencerows, abandoned land, and residential yards, infesting trees and bushes. Can grow in semi-shade or no shade and requires moist soil.
Origin: Native to Asia and may have been intentionally, though illegally, introduced into the United States as a culturally important aphrodisiac.
Impacts: Japanese dodder is a very aggressive parasitic plant that affects ornamental plantings and agricultural crops as well as having the potential to severely alter the composition and function of riparian areas. This parasite threatens native vegetation by killing host seedlings or by making host trees more susceptible to disease. Poses a threat to crops such as alfalfa, asparagus, and tomatoes, in addition to horticultural plants. Also serves as a host for several viruses known to be detrimental to agricultural crops.
Western states listed as Noxious Weed: Arizona, California. Also on the Federal Noxious Weed list

Japanese dodder is a perennial parasitic vine that has recently been introduced into the United States. The stems are fleshy, circular, vibrant yellow-green or gold, with red spots and striations, and much branched. Leaves are minute and scale-like. Japanese dodder has thick, spaghetti-like, robust stems in contrast to native dodder stems, which are usually more thread or string-like. Infestations are often large, spreading, and web-like, covering large shrubs and small trees. In contrast, infestations of other dodder species are likely to be smaller, infecting non-woody plants or small shrubs.

Flowers are abundant, pale yellow, and sessile. While most dodders are spread both by seed and vegetatively, Japanese dodder does not produce viable seed and is therefore spread only vegetatively. It is thought that humans are the main dispersal agents, as the plants are intentionally moved around from one cultivated source to another. Japanese dodder attacks full-grown trees including natives and agricultural fruit trees.

NON-CHEMICAL CONTROL

Mechanical (pulling, cutting, disking)	Hand roguing and pruning are often used in landscape settings where the use of herbicides would damage other plants. Hand pulling is only effective if the entire vine and adjoining haustoria are removed. In severely affected host plants, the entire plant may be removed and destroyed.
Cultural	No cultural control strategies have been identified.
Biological	There are currently no biological control agents available for the control of Japanese dodder. *Melanagromyza cuscutae* and the gall-forming weevils *Smicronyx* spp. have been tested for control of other *Cuscuta* spp. Similarly the fungi *Alternaria cuscutacidae* and *Colletotrichum gloeosporioides* have shown promise in some situations but none have proved reliable enough for use in practice. On *Cuscuta japonica*, a number of fungi have been studied for their potential as biocontrol agents, including *Fusarium solani*, *Fusarium semitectum*, *Pestalotiopsis guepini* [*Pestalotia guepinii*] and *Alternaria tenuis*, but they have not yet been fully developed for use.

CHEMICAL CONTROL

The following specific use information is based on published papers and reports by researchers and land managers. Other trade names may be available, and other compounds also are labeled for this weed. Directions for use may vary between brands; see label before use. Herbicides are listed by mode of action and then alphabetically. The order of herbicide listing is not reflective of the order of efficacy or preference.

GROWTH REGULATORS

Triclopyr *Garlon 4 Ultra*	**Rate:** Broadcast treatment: 1 to 8 qt product/acre (1 to 4 lb a.e./acre). Spot treatment: 1 to 1.5% solution v/v *Garlon 4 Ultra* and water plus 0.25 to 0.5% v/v surfactant to thoroughly wet all leaves

Timing: Postemergence when plants are growing rapidly. Applications in spring provide best control.

Remarks: Triclopyr is a selective herbicide for broadleaf species. In settings where the host can be sacrificed, a systemic herbicide such as triclopyr can be used. This kills the host plant as well as the parasite, and ensures that it will not spread.

AROMATIC AMINO ACID INHIBITORS

Glyphosate

Roundup, Accord XRT II, and others

Rate: Spot treatment: 0.7% solution v/v *Roundup ProMax* (or other trade name with similar concentration of glyphosate) and water to thoroughly wet all stems

Timing: Applications should be made to young dodder plants before they are extensively attached to the host plant. Repeated applications may be necessary for complete control.

Remarks: Glyphosate is a nonselective systemic herbicide with no soil activity.

Cynara cardunculus L.
Artichoke thistle

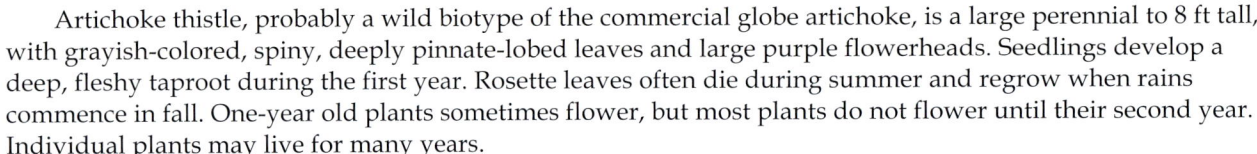

Family: Asteraceae
Range: West Coast states (Washington, Oregon and California), generally at low elevations in coastal environments.
Habitat: Disturbed, open sites in grassland, pasture, chaparral, coastal sage scrub, riparian areas, and abandoned agricultural fields, particularly in coastal areas. Often associated with areas impacted by historic or recent overgrazing. Grows best on deep clay soils. Does not tolerate heavy shade.
Origin: Native to the Mediterranean region.
Impacts: Dense colonies displace desirable vegetation through competition for space and soil moisture, and can exclude wildlife and livestock.
Western states listed as Noxious Weed: California
California Invasive Plant Council (Cal-IPC) Inventory: Moderate Invasiveness

Artichoke thistle, probably a wild biotype of the commercial globe artichoke, is a large perennial to 8 ft tall, with grayish-colored, spiny, deeply pinnate-lobed leaves and large purple flowerheads. Seedlings develop a deep, fleshy taproot during the first year. Rosette leaves often die during summer and regrow when rains commence in fall. One-year old plants sometimes flower, but most plants do not flower until their second year. Individual plants may live for many years.

Flowering occurs from spring to mid-summer. Plants send up one to several erect, thick, branched stems with spiny ribs; stems are topped with large, solitary flower heads 1 to 6 inches diameter. Stems typically die after flowering and can remain standing for several months to more than a year. Reproduction is primarily by seed, and occasionally by root fragments following mechanical disturbance. Most achenes fall near the parent plant or disperse up to about 60 ft with wind. Seeds can survive about 5 years in the soil.

NON-CHEMICAL CONTROL

Mechanical (pulling, cutting, disking)	Artichoke thistle can be cultivated or manually removed in the seedling stage. For manual control of established plants, a large portion of the taproot must be removed, otherwise the remaining root will generate new shoots.
	Cutting flower stems before maturity can reduce seed production. Some workers have found it useful to cut down dense patches with power tools or tractors, then treat the regrowth with herbicides.
	On agricultural land, repeated cultivation can eventually eliminate troublesome populations.
Cultural	Livestock usually avoid artichoke thistle because of its spiny foliage. Browsing by goats, however, can reduce seed production.
	Burning can be used to remove above-ground material once it dries in late summer to fall. This can facilitate subsequent herbicide applications. Burning may also encourage the seedbank to flush, providing an opportunity for seedling control. Because of the perennial taproot, burning alone will not kill the plant.
Biological	The artichoke fly (*Terellia fuscicornis*) was accidentally introduced into California, but is not a California Department of Food and Agriculture (CDFA) approved biocontrol agent. Preliminary studies suggest that some native thistles (*Cirsium* spp.) may be vulnerable to attack by the fly. The fly's impact on artichoke thistle populations is unknown. Larvae feed only on mature flowerheads, thus commercial artichokes are not significantly affected since they are harvested while immature.

CHEMICAL CONTROL

The following specific use information is based on published papers and reports by researchers and land managers. Other trade names may be available, and other compounds also are labeled for this weed. Directions for use may vary between brands; see label before use. Herbicides are listed by mode of action and then alphabetically. The order of herbicide listing is not reflective of the order of efficacy or preference.

GROWTH REGULATORS	
2,4-D Several names	**Rate:** 1.1 to 2.1 qt product/acre (1 to 2 lb a.e./acre) **Timing:** Postemergence at rosette stage. **Remarks:** 2,4-D is broadleaf-selective and has no soil activity. It may require repeat applications. 2,4-D is not the most effective treatment, but is widely used because of low cost. Use a surfactant. Do not apply when outside temperatures exceed 80°F.
Aminocyclopyrachlor + chlorsulfuron *Perspective*	**Rate:** 4.75 to 8 oz product/acre **Timing:** Postemergence from spring up to flowering, or fall rosette stage. **Remarks:** *Perspective* provides broad-spectrum control of many broadleaf species. Although generally safe to grasses, it may suppress or injure certain annual and perennial grass species. Do not treat in the root zone of desirable trees and shrubs. Do not apply more than 11 oz product/acre per year. At this high rate, cool-season grasses will be damaged, including bluebunch wheatgrass. Not yet labeled for grazing lands. Add an adjuvant to the spray solution. This product is not approved for use in California and some counties of Colorado (San Luis Valley).
Aminopyralid *Milestone*	**Rate:** 5 to 7 oz product/acre (1.25 to 1.75 oz a.e./acre) **Timing:** Postemergence in winter to spring, ideally before bolting. **Remarks:** Aminopyralid is a broadleaf herbicide similar to picloram, but more selective and generally safe on grasses. Its soil residual activity will kill emerging seedlings. Aminopyralid has a longer soil residual and higher activity than clopyralid.
Clopyralid *Transline*	**Rate:** 0.25 to 0.67 pt product/acre (1.5 to 4 oz a.e./acre) **Timing:** Postemergence in winter to early spring, seedling to rosette stage. **Remarks:** Clopyralid is a broadleaf herbicide like picloram, but more selective. It is very safe on grasses.
Clopyralid + 2,4-D *Curtail*	**Rate:** 1.5 to 3 qt product/acre (higher rate if plants are drought-stressed) **Timing:** Postemergence to rapidly growing weeds from full rosette to early flower bud. **Remarks:** The combination is broadleaf-selective with a wide range of susceptible species. Recommended rates are based on those reported for knapweeds and perennial thistles.
Dicamba *Banvel, Clarity*	**Rate:** 1 to 2 pt product/acre (0.5 to 1 lb a.e./acre) **Timing:** Postemergence to rapidly growing plants. Treatments are more effective on smaller plants. **Remarks:** Dicamba is a broadleaf-selective herbicide often combined with other active ingredients. It is effective earlier in the season than 2,4-D. It is also effective when tank-mixed with 2,4-D (0.75 lb a.e./acre of dicamba + 0.25 lb a.e./acre of 2,4-D). Avoid drift to sensitive crops. Do not apply when outside temperatures exceed 80°F. Recommended rates are based on those reported for knapweeds and perennial thistles.
Picloram *Tordon 22K*	**Rate:** 1 to 2 pt product/acre (4 to 8 oz a.e./acre) **Timing:** Postemergence during active growth before bud stage. **Remarks:** Picloram is one of the most effective herbicides for this artichoke thistle. Most broadleaf plants are susceptible, but relatively safe on established grasses. It is also effective when mixed with dicamba or 2,4-D. Picloram has long soil residual activity and has been reported to injure young or germinating grasses. *Tordon 22K* is a federally restricted use pesticide. Not registered for use in California.
Triclopyr *Garlon 3A, Garlon 4 Ultra*	**Rate:** 0.33 to 1.5 gallons *Garlon 3A*/acre or 0.25 to 1 gallons *Garlon 4 Ultra*/acre (1 to 4 lb a.e./acre) **Timing:** Postemergence to rapidly growing weeds, up to bud stage. **Remarks:** Triclopyr is broadleaf-selective and safe on most grasses. It is most effective on smaller plants. *Garlon 4 Ultra* is formulated as a low volatile ester. However, in warm temperatures, spraying onto hard surfaces such as rocks or pavement can increase the risk of volatilization and off-target damage. Recommended rates are based on those reported for perennial thistles.
Triclopyr + 2,4-D *Crossbow*	**Rate:** 4 qt *Crossbow*/acre **Timing:** Postemergence from rosette to early bolting stage. **Remarks:** Include non-ionic surfactant. Recommended rates are based on those reported for

	knapweeds and perennial thistles.
BRANCHED-CHAIN AMINO ACID INHIBITORS	
Imazapyr *Arsenal, Habitat, Stalker, Chopper, Polaris*	**Rate:** 4 to 6 pt product/acre (1 to 1.5 lb a.e./acre), or 1% v/v solution as a spot treatment. **Timing:** Postemergence at flowering. **Remarks:** Imazapyr is best used as a spot treatment. It is a nonselective herbicide. It also has long soil residual activity and can leave more bare ground than other treatments, even a year after application. Recommended rates are based on those reported for perennial thistles.
Metsulfuron *Escort*	**Rate:** 1.5 to 2 oz product/acre (0.9 to 1.2 oz a.i./acre) **Timing:** Postemergence to young, rapidly growing weeds in spring before flowering, or in fall to new rosettes. **Remarks:** Metsulfuron has mixed selectivity but is generally safe on grasses. Use a surfactant. It can be tank-mixed with 2,4-D. Metsulfuron has some soil residual activity. Recommended rates are based on those reported for perennial thistles. Not registered for use in California.

Cynodon dactylon (L.) Pers.
Bermudagrass

Family: Poaceae
Range: Most western states, except Wyoming, North and South Dakota.
Habitat: Disturbed sites, gardens, agronomic crops, orchards, turf, landscaped and forestry areas, on most soil types. Typically in areas that are irrigated or receive some warm-season moisture. Tolerates acidic, alkaline, or saline conditions or limited flooding. Aboveground growth does not tolerate freezing temperatures (below -1°C). Optimum growth occurs when daytime temperatures are 35 to 38°C. Grows poorly in shaded conditions.
Origin: Native to Africa.
Impact: Because of its vigorous creeping habit bermudagrass is a noxious weed in many areas where some moisture is available in the warm season. In wildland areas, it is particularly a problem in riparian sites.
Western states listed as Noxious Weed: California, Utah
California Invasive Plant Council (Cal-IPC) Inventory: Moderate Invasiveness

Bermudagrass is a warm-climate perennial with an extensive system of creeping rhizomes and stolons. Although it typically grows prostrate to the soil, it can grow to 1.5 ft tall, particularly under somewhat shady conditions. Bermudagrass is commonly grown as a turf or forage in tropical to warm temperate regions. Because of its vigorous creeping habit, it is a noxious weed in many situations where warm-season moisture is ample. Contact with plants can cause dermatitis in sensitive individuals, and the pollen is a common allergen.

Bermudagrass rhizomes and stolons are slender, tough, and scaly, producing roots at the nodes. Most rhizomes grow in the upper 2 inches of soil, but some may extend to depths of over a foot. Rhizomes survive considerable dehydration and drought, but not prolonged periods of freezing temperatures or exposure to sun. Small rhizome and stolon fragments readily generate new plants. Plants go dormant and foliage turns brown when nighttime temperatures dip below freezing, when average daytime temperature is below 10°C (50°F), or when day lengths shorten. Rhizomes and roots go dormant at soil temperatures below 18°C.

Aboveground stems are more or less erect, thin, and can be branched. Leaf blades are flat and flexible, usually less than 2.5 inches long. Usually there are long hairs around the collar region, particularly at the margins. Its inflorescence is umbel-like, with usually 4 to 8 spike-like branches mostly 1.5 to 3 inches long. Bermudagrass reproduces vegetatively from its rhizomes and stolons, and by seed. Rhizome and stolon fragments disperse with soil movement. Seeds disperse with water, soil movement, agricultural and landscape machinery, as a commercial seed impurity, in livestock feeds and bedding, and with other human activities. Seeds germinate spring through fall when temperature and moisture conditions are favorable. Some seeds survive 3 to 4 years under field conditions, but most germinate within two years.

NON-CHEMICAL CONTROL

Mechanical (pulling, cutting, disking)	Persistent manual removal of rhizomes and stolons can eliminate bermudagrass from small areas.
	A high mower setting (2 to 3 in) can suppress bermudagrass relative to other turfgrasses. Cleaning mowers and agricultural machinery after use in infested areas can prevent dispersal of rhizomes, stolons, and seeds.
	Tilling or disking as needed to expose rhizomes to sun-drying or freezing temperatures can be effective. If water (or rain) is applied during the drying process bermudagrass will regrow. Do not cultivate bermudagrass if the soil is moist, because rhizome fragments will begin to grow. Cultivating and drying will not kill seeds.
Cultural	Shading by other plants, mulches, or cloth can help to suppress bermudagrass growth.
	Using plastic to solarize moist soil for 6 weeks in summer can control small infestations.
Biological	Due to its importance as a turfgrass, there are no efforts to develop a biological control program for bermudagrass.

CHEMICAL CONTROL

The following specific use information is based on published papers and reports by researchers and land managers. Other trade names may be available, and other compounds also are labeled for this weed. Directions for use may vary between brands; see label before use. Herbicides are listed by mode of action and then alphabetically. The order of herbicide listing is not reflective of the order of efficacy or preference.

LIPID SYNTHESIS INHIBITORS	
Clethodim *Select, Envoy*	**Rate:** 12 to 32 oz product (*Envoy*)/acre (1.5 to 4 oz a.i./acre)
	Timing: Postemergence in early spring when new growth is less than 6 inches. Reapply when regrowth is less than 6 inches and repeat as necessary.
	Remarks: Clethodim is a grass-selective herbicide and will not harm broadleaf species Include crop oil concentrate surfactant. Do not apply directly to water. It has no soil activity. Note that *Envoy* formulation is 1 lb a.i./gallon, *Select* is 2 lb a.i./gallon.
Fluazifop *Fusilade*	**Rate:** 1 to 1.5 lb product/acre (4 to 6 oz a.i./acre)
	Timing: Postemergence to rapidly growing bermudagrass with 4- to 8-inch runners.
	Remarks: Fluazifop is a grass-selective herbicide and will not harm broadleaf species (will suppress filaree, *Erodium* spp.). Apply with 1% v/v crop oil concentrate or 0.25% v/v non-ionic surfactant. Fluazifop acts very slowly, taking at least 2 weeks and often 4 weeks to show effectiveness. Do not apply to stressed grasses. If weeds regrow, reapply at 3 to 6 oz a.i./acre. Fluazifop has no soil activity.
Sethoxydim *Poast*	**Rate:** 1.5 to 2.5 pt product/acre (4.5 to 7.5 oz a.i./acre)
	Timing: Postemergence in early spring when new growth is less than 6 inches. Reapply when regrowth is less than 6 inches and repeat as necessary.
	Remarks: Sethoxydim is a grass-selective herbicide and will not harm broadleaf species Include crop oil concentrate surfactant. It has no soil activity.
AROMATIC AMINO ACID INHIBITORS	
Glyphosate *Roundup, Accord XRT II*, and others	**Rate:** Broadcast foliar treatment: 1 to 2 qt (*Roundup ProMax*)/acre (1.125 to 2.25 lb a.e./acre). Spot treatment: 2% v/v solution
	Timing: Postemergence from late spring to summer when bermudagrass is growing rapidly. Some researchers recommend fall, after flowering and before dormancy.
	Remarks: Glyphosate is a nonselective herbicide. It has no soil activity. Efficacy is best when bermudagrass is not water stressed. Control can be improved by avoiding mowing for 2 to 3 weeks before application, and by cultivating 7 days after application. Effectiveness may be increased by addition of ammonium sulfate as a water conditioner, particular when water has high pH.
BRANCHED-CHAIN AMINO ACID INHIBITORS	
Imazapyr *Arsenal, Habitat, Stalker, Chopper, Polaris*	**Rate:** 4 to 6 pt product (*Habitat*)/acre (1 to 1.5 lb a.e./acre)
	Timing: Postemergence from late spring to summer when bermudagrass is growing rapidly.
	Remarks: For treatment volume, use a minimum of 75 gal/acre. Control of established stands may require repeat applications. Imazapyr is a broad-spectrum herbicide with fairly long soil residual activity. It is not the most common option for the control of bermudagrass.

Cynoglossum officinale L.
Houndstongue

Family: Boraginaceae
Range: Throughout contiguous United States, except Texas, Oklahoma, Louisiana, Mississippi, and Florida. Found in all western states.
Habitat: Woodlands, pastures, fields, rangeland, forest margins and disturbed sites such as roadsides, sand dunes, abandoned cropland, ditch banks, and urban waste areas. Often on sandy or gravelly soil; colonizes bare soil and under dripline of trees and shrubs, making control difficult.
Origin: Native to Eurasia and accidentally introduced in the late 1800s as a seed contaminant in cereal grain.
Impacts: Houndstongue can be a serious problem in rangeland, pasture and forest settings. The weed is highly invasive and can form dense monotypic stands. Foliage, especially young leaves, and fruits contain pyrrolizidine alkaloids and are liver toxins in all livestock classes, especially horses, when ingested in small amounts over time or in a single large quantity. Plants have a distinctive scent that appears to deter animals from consuming live foliage, thus most poisonings occur when animals consume hay over time.
Western states listed as Noxious Weed: Colorado, Montana, Nevada, Oregon, Utah, Washington, Wyoming
California Invasive Plant Council (Cal-IPC) Inventory: Moderate Invasiveness

Houndstongue is a biennial or short-lived perennial, with erect flower stems to 4 ft tall. The leaves can vary in size, depending on growing conditions, from 4 to 12 inches long and 1 to 3 inches wide. During its first year, the plant stores carbohydrates in a large developing taproot that becomes black and woody by the season's end.

During the second year of growth, plants develop additional leaves followed by an inflorescence, up to 4 ft tall, with reddish-purple flowers, 0.25 inch wide, often horizontal to slightly drooping. The seeds are contained within four distinctive nutlets. Each nutlet is 0.5 inch long, brown or grey-brown and covered with short, hooked prickles that cling to hair, fur, or clothing. Often referred to as "beggar's lice", these nutlets are exceptional dispersal agents. A few of the nutlets drop from the plant, but most stay attached to the persistent inflorescence many months or even years until they are picked up by a passing animal. Houndstongue reproduces solely from seed and a single plant can produce up to 2,000 seeds that can remain viable for 2 to 3 years.

NON-CHEMICAL CONTROL

Mechanical (pulling, cutting, disking)	Digging, pulling, and cutting can be effective if the root crown is severed. Cut young rosettes below the crown in fall or early spring. Clipping or mowing second-year plants close to the ground during flowering can greatly reduce seed production, even in plants which survive and regrow. Mechanical control must be done frequently to have any effect, and is only feasible for small infestations.
	Houndstongue will not withstand regular cultivation of the young rosettes.
Cultural	Grazing is not practical due to risk of poisoning. Reseeding problem areas with fast growing grasses, and not overgrazing can prevent invasion. Long-term reduction of houndstongue must involve planting competitive plant species. Many improved grass species can be seeded in late fall or winter.
Biological	A biological control program for houndstongue was initiated in 1988. The first North American releases for biological control were the root-mining flea beetle *Longitarsus quadriguttatus* and the houndstongue root-mining weevil, *Mogulones cruciger*, in British Columbia in 1997-1998. *M. cruciger* has become well-established in Alberta and has greatly reduced houndstongue there. However, this species has not been approved yet for release in the U.S. Several other insects are being evaluated, although initial results are not as promising as those of the root weevil. The native fungal pathogen that causes powdery mildew (*Golovinomyces cynoglossi*) has been reported to cause some foliar damage to houndstongue in many western states.

CHEMICAL CONTROL

The following specific use information is based on published papers and reports by researchers and land managers. Other trade names may be available, and other compounds also are labeled for this weed. Directions for use may vary between brands; see label before use. Herbicides are listed by mode of action and then alphabetically. The order of herbicide listing is not reflective of the order of efficacy or preference.

GROWTH REGULATORS	
2,4-D Several names	**Rate:** 4 pt product/acre (1.9 lb a.e./acre) **Timing:** Postemergence when plants are growing rapidly. Applications in spring provide the best control. **Remarks:** Selective herbicide for broadleaf species. In areas where desirable grasses are growing around houndstongue, 2,4-D can be used without non-target damage. Good coverage is necessary.
Aminocyclopyrachlor + chlorsulfuron *Perspective*	**Rate:** 4.75 to 8 oz product/acre plus 0.25 to 0.5% v/v surfactant **Timing:** Preemergence or postemergence. **Remarks:** *Perspective* provides broad-spectrum control of many broadleaf species. Although generally safe to grasses, it may suppress or injure certain annual and perennial grass species. Do not treat in the root zone of desirable trees and shrubs. Do not apply more than 11 oz product/acre per year. At this high rate, cool-season grasses will be damaged, including bluebunch wheatgrass. Not yet labeled for grazing lands. Add an adjuvant to the spray solution. This product is not approved for use in California and some counties of Colorado (San Luis Valley).
Aminopyralid + metsulfuron *Opensight*	**Rate:** 2.5 to 3.3 oz product/acre plus 0.25 % v/v surfactant **Timing:** Apply rosette to mid-bolt when plants are actively growing. **Remarks:** Use the higher rate on bolting plants.
AROMATIC AMINO ACID INHIBITORS	
Glyphosate *Roundup, Accord XRT II,* and others	**Rate:** Broadcast treatment: 1 to 2 pt product (*Roundup ProMax*)/acre (0.56 to 1.1 lb a.e./acre). Spot treatment: 1.5 to 2% v/v solution *Roundup* (or other trade name) and water to thoroughly wet all leaves. **Timing:** Postemergence when plants are growing rapidly. **Remarks:** Glyphosate is a nonselective systemic herbicide with no soil activity.
BRANCHED-CHAIN AMINO ACID INHIBITORS	
Chlorsulfuron *Telar*	**Rate:** 1 to 1.5 oz product/acre (0.75 to 1.125 oz a.i./acre) plus 0.25 to 0.5% v/v surfactant **Timing:** Preemergence or postemergence. Spring applications are most effective. **Remarks:** Selective herbicide effective for controlling broadleaf weeds and some grasses.
Imazapic *Plateau*	**Rate:** 8 to 12 oz product/acre (2 to 3 oz a.e./acre) plus 0.25 to 0.5% v/v surfactant **Timing:** Preemergence or early postemergence. **Remarks:** Imazapic is a selective herbicide effective for controlling broadleaf weeds and some grasses. Imazapic is not registered for use in California.
Imazapyr *Arsenal, Habitat, Stalker, Chopper, Polaris*	**Rate:** 1 pt product/acre (4 oz a.e./acre) plus 0.25 to 0.5% v/v surfactant **Timing:** Preemergence or postemergence. **Remarks:** Imazapyr is a preemergent and postemergence herbicide effective for controlling broadleaf and grass weeds.
Metsulfuron *Escort*	**Rate:** 1 oz product/acre (0.6 oz a.i./acre) plus 0.25 to 0.5% v/v surfactant **Timing:** Early postemergence. Spring applications are most effective. **Remarks:** Selective herbicide for broadleaf species. Can be used safely around desirable grasses. It can be used as a premix with aminopyralid (*Opensight*) at 2.5 to 3.3 oz product/acre. Metsulfuron is not registered for use in California.
Sulfometuron + chlorsulfuron *Landmark* XP	**Rate:** 0.75 to 2.25 oz product/acre plus 0.25 to 0.5% v/v surfactant **Timing:** Preemergence or postemergence. **Remarks:** Effective for controlling broadleaf weeds and some grasses. Long soil residual activity.

Cynosurus echinatus L.

Hedgehog dogtailgrass

Family: Poaceae
Range: One of the most invasive species of Oregon-oak woodlands in British Columbia. It is rapidly spreading in grasslands and oak woodlands in California, Oregon and Washington. Also reported in Montana.
Habitat: Roadsides, fields, grassland, chaparral, oak woodland, summer dry pasture, coastal bluffs and terraces, riverbanks, other disturbed places.
Origin: Native to Europe.
Impacts: Can increase fuel loads and create fire hazard. Heavy infestations reduce desirable forage species.
California Invasive Plant Council (Cal-IPC) Inventory: Moderate Invasiveness

Hedgehog dogtailgrass is an erect, cool-season annual to about 2 ft tall, with dense, bristly, head-like to spike-like panicles 0.5 to 2 inches long. Sheaths are open, and the ligules are membranous, translucent, 5-10 mm long. The collar generally surrounds the stem unevenly.

Plants generally flower between May and July. The dense panicles bear fertile and sterile spikelets in pairs, mostly on one side of the main axis. The fertile florets are tipped with a long awn up to about 0.5 inch long. Sterile lemmas and glumes taper to a bristle-like awn, slightly shorter than fertile lemmas including awns. Plants reproduce only by seed. Fertile florets fall near the parent plant and probably disperse to greater distances with water, mud, and by clinging to animals, vehicle tires, and human shoes and clothing. A persistent seedbank does not appear to develop.

NON-CHEMICAL CONTROL

Mechanical (pulling, cutting, disking)	Hand pulling of annual grasses such as hedgehog dogtailgrass may be effective early in spring before seed set, but this is very labor-intensive and is only used on small infestations. Minimize soil disturbance when hand pulling to minimize new seed germination. Mowing or string trimming can also to reduce hedgehog dogtailgrass if done before seed sets in the early summer. Shallow cultivation shortly after the main flush of germination and again a little later can eliminate most seedlings.
Cultural	Little is known of the effect of burning and grazing on hedgehog dogtailgrass. Burning may be effective if timed so that the plants have not yet dropped their seed, but there is enough fuel to carry a fire.
	The grass is not considered a desirable forage and it is unlikely that grazing would be very effective for its control.
Biological	No known biological agents are available.

CHEMICAL CONTROL

Much of the control information for hedgehog dogtailgrass is obtained for other similar annual grasses.

The following specific use information is based on reports by researchers and land managers. Other trade names may be available, and other compounds also are labeled for this weed. Directions for use may vary between brands; see label before use. Herbicides are listed by mode of action and then alphabetically. The order of herbicide listing is not reflective of the order of efficacy or preference.

LIPID SYNTHESIS INHIBITORS

Fluazifop *Fusilade*	**Rate:** Broadcast foliar treatment: 1 to 1.5 pt product/acre (4 to 6 oz a.i./acre). Spot treatment: 0.5% v/v solution
	Timing: Early postemergence to plant before the boot stage.
	Remarks: Fluazifop is grass-selective and will not damage broadleaf species. It has no soil activity. To select for perennial grasses, apply before perennials emerge. Include crop oil concentrate surfactant or non-ionic surfactant.

Sethoxydim *Poast*	**Rate:** 1.5 to 2 pt product/acre (4.5 to 6 oz a.i./acre)
	Timing: Early postemergence to plant before the boot stage.
	Remarks: Sethoxydim is grass-selective and will not damage broadleaf species. It has no soil activity. To select for perennial grasses, apply before perennials emerge. Include crop oil concentrate surfactant or non-ionic surfactant.
AROMATIC AMINO ACID INHIBITORS	
Glyphosate *Roundup, Accord XRT II*, and others	**Rate:** 1 to 2 pt product (*Roundup ProMax*)/acre (0.56 to 1.1 lb a.e./acre)
	Timing: Postemergence in early spring to rapidly growing, non-stressed plants after most seedlings have emerged. If possible, apply before desirable perennials have emerged.
	Remarks: Glyphosate is a nonselective herbicide. It has no soil activity.
BRANCHED-CHAIN AMINO ACID INHIBITORS	
Imazapic *Plateau*	**Rate:** 4 to 12 oz product/acre (1 to 3 oz a.e./acre)
	Timing: Preemergence in fall or postemergence in early spring. In colder climates, spring applications after snow melt are better than fall treatments.
	Remarks: Imazapic has mixed selectivity and tends to favor members of the Asteraceae. It has long soil residual activity. Imazapic can tie up in litter and its efficacy may be very much reduced under situations where there is heavy thatch on the soil surface. Imazapic is not registered for use in California.
Sulfometuron *Oust* and others	**Rate:** 2 to 6.67 oz product/acre (1.5 to 5 oz a.i./acre)
	Timing: Preemergence or early postemergence. Fall and spring applications can both be effective, but fall applications may give full season control.
	Remarks: Sulfometuron has mixed selectivity. It can cause minor damage to some native perennial grasses and has fairly long soil residual activity. Higher rates may increase control but will also give more bare ground.
CONTACT PHOTOSYNTHETIC INHIBITORS	
Paraquat *Gramoxone*	**Rate:** 0.75 to 2 pt product/acre (3 to 8 lb a.i./acre)
	Timing: Postemergence to rapidly growing plants. More effective on smaller plants.
	Remarks: Paraquat is nonselective on annual species. It is a non-systemic herbicide that only kills contacted foliage and has no soil activity. Paraquat is a restricted use herbicide that is highly toxic to animals. The formulation has a stenching agent.

Cytisus scoparius L.
Scotch broom

Family: Fabaceae
Range: The entire Atlantic and Pacific coasts from Alaska to British Columbia to California, and from Nova Scotia through Georgia. Also Idaho, Montana and Utah, as well as one Hawaiian island.
Habitat: Grasslands, shrublands, oak woodlands, forest margins, coastal habitats, riparian corridors; disturbed sites such as roadsides, pasture, gravelly floodplains, burned areas, cleared forests. Typically in mountain regions and cool coastal areas with dry summers. Grows best on sandy, high phosphorus soils with acidic to neutral pH; can tolerate high boron concentrations. Rarely grows on limestone soils.
Origin: Central and southern Europe and North Africa. Introduced to the U.S. in the 1850s as an ornamental and for erosion control.
Impacts: Grows rapidly, forming dense stands that most wildlife find impenetrable and unpalatable. Dense stems limit regeneration of most other plant species, and the accumulation of woody biomass creates a dangerous fire hazard. Broom can fix nitrogen, which increases soil fertility and gives a competitive advantage to other non-native weeds.
Western states listed as Noxious Weed: California, Idaho, Montana, Oregon, Washington
California Invasive Plant Council (Cal-IPC) Inventory: High Invasiveness

Scotch broom is a fast-growing deciduous shrub, 5 to 10 ft tall, with yellow, pea-like flowers. Stems are 5-angled or ridged, often star-shaped in cross-section. New twigs are green, erect and covered with wavy hairs, becoming smooth and woody with age. The leaves at branch bases have three leaflets alternately arranged. Upper leaves are simple, without petioles. Leaflets are < 1/3 inch long, widest at the tip and often pointed.

Plants begin flowering from 18 months to 3 years of age. The bright yellow, occasionally maroon, flowers are single or in pairs in leaf axils. Reproduction is by seed. Seeds are in small, flattened pods 0.75 to 2 inches long. Pods are dark brown or black when mature; contain 5 to 9 seeds, and have hairs along the margin. When mature, pods eject the seeds several feet from the plant. Seeds can remain viable in the soil for up to 30 years. Large soil seedbanks often accumulate making long term control difficult. Shrubs may live for up to 30 years.

NON-CHEMICAL CONTROL

Mechanical (pulling, cutting, disking)	Seedlings and small shrubs can be hand pulled. For larger established shrubs, a weed wrench or other woody weed extractor can be used. Extract the entire root or resprouting will occur. Best results are achieved when soil is moist. Disturbing the soil can stimulate the seedbank.
	Cutting broom off before it flowers will reduce seed production and will deplete the plant's energy reserves. Resprouting is common after treatment, but can be reduced by cutting broom at the beginning of the dry season. Cutting should be combined with an herbicide treatment or with multiple cuttings over a period of years. Cut shrubs at ground level with power or manual saws.
	Heavy equipment can be used to control broom in areas where soil disturbance and nonselective species removal are not important considerations. Stumps remaining following such treatment will require herbicide application to prevent regrowth.
Cultural	Grazing is not considered an effective control option. Broom flowers and seeds contain quinolizidine alkaloids and can be toxic to humans and livestock. Foliage may be mildly toxic and is unpalatable to most livestock, except goats. Goats confined to a small area can help control resprouts after a cutting or burn treatment.
	Burning alone is not effective. Although burning can remove debris, it also removes competing vegetation, releases nutrients into the soil, and stimulates germination of broom seed in the soil. It is important to employ a control strategy following a burn, otherwise the broom population may become worse. Follow-up treatments could include herbicide application, repeat burnings, and/or revegetation with desirable species.
Biological	Insects introduced as biological control agents include the Scotch broom seed beetle (*Bruchidius villosus*), the Scotch broom seed weevil (*Apion fuscirostre*), and the Scotch broom twig miner moth (*Leucoptera*

spartifoliella). The latter two species are specific to Scotch broom, while the seed beetle also attacks Portuguese broom, Spanish broom, and French broom.

CHEMICAL CONTROL

The following specific use information is based on published papers and reports by researchers and land managers. Other trade names may be available, and other compounds also are labeled for this weed. Directions for use may vary between brands; see label before use. Herbicides are listed by mode of action and then alphabetically. The order of herbicide listing is not reflective of the order of efficacy or preference.

GROWTH REGULATORS	
Picloram *Tordon 22K*	**Rate:** Broadcast foliar treatment: 2 qt product/acre (non-cropland) or 1 qt product per acre (rangeland) plus 0.25 to 0.5% v/v surfactant to thoroughly wet all leaves. **Timing:** Foliar treatments are best when plants are growing rapidly at or beyond early to full bloom stage. **Remarks:** Picloram has long soil residual activity that will control germinating broadleaf plants. Picloram is a restricted use herbicide. It is not registered for use in California.
Triclopyr *Garlon 3A, Garlon 4 Ultra, Pathfinder II* Aminopyralid + triclopyr *Capstone*	**Rate:** Broadcast treatment: 2 to 3 qt *Garlon 4 Ultra*/acre (1 to 1.5 qt a.e./acre) or 3 to 4 qt *Garlon 3A*/acre (1.125 to 1.5 qt a.e./acre). Spot treatment: 0.75 to 1.5% v/v solution of *Garlon 4 Ultra*, or 1 to 1.5% *Garlon 3A* and water plus 0.25 to 0.5% v/v surfactant to thoroughly wet all leaves. Low volume/thinline treatment: 10% v/v solution of *Garlon 4 Ultra* plus a 20% v/v ethylated crop oil in water. Basal cut stump treatment: 20% v/v *Garlon 4 Ultra* in water. Cut stump: *Garlon 3A*, undiluted or 50% in water. Basal bark treatment: 20% v/v *Garlon 4 Ultra* in 20% v/v ethylated crop oil and water, or *Pathfinder II* (ready-to-use formulation). **Timing:** Postemergence when plants are growing rapidly. Cut stump, basal cut stump, and basal bark treatments can be applied anytime as long as the ground is not frozen. **Remarks:** Selective herbicide for broadleaf species, will not injure grasses growing nearby. For cut stump treatment, cut stem horizontally at or near ground level and immediately apply herbicide solution. Roots may sucker after cutting, but the treatment should control most resprouts. For basal cut stump treatment, leave a higher stump and treat the cut surface and all the remaining bark. For basal bark treatment, spray the lower trunk, including the root collar, to a height of 12 to 15 inches; the spray should thoroughly wet the lower stem but not to the point of runoff. Plants should not be cut for at least one month after basal bark treatment. Triclopyr is also used in a premix with aminopyralid (*Capstone*) at 6 to 8 pt product/acre.
Triclopyr + 2,4-D *Crossbow*	**Rate:** Spot treatment: 0.5 to 1.5% v/v solution of *Crossbow* and water to thoroughly wet all leaves. **Timing:** Postemergence when plants are growing rapidly. **Remarks:** *Crossbow* in water forms an emulsion (not a solution).
AROMATIC AMINO ACID INHIBITORS	
Glyphosate *Roundup, Accord XRT II*, and others	**Rate:** Spot treatment: 1.5 to 2% v/v solution of *Roundup ProMax* (or other trade name with similar concentration of glyphosate) in water to thoroughly wet all leaves. Low volume/thinline treatment: 10% v/v solution of *Roundup* (or other trade name) in water. Cut stump treatment: 25% v/v *Roundup* (or other trade name) in water; 50% can reduce resprouting but may exceed label rate if stands are dense. **Timing:** Foliar treatments should be made in late summer or early fall. For cut stump treatment, apply in late summer, early fall or dormant season; treat immediately after cutting. **Remarks:** Nonselective systemic herbicide that may kill partially-sprayed plants off target. It gives good control with some resprouts. Plants should not be cut for at least 4 months after foliar treatments. Cut stump applications are made as described for triclopyr.
BRANCHED-CHAIN AMINO ACID INHIBITORS	
Imazapyr *Arsenal, Habitat, Stalker, Chopper, Polaris*	**Rate:** Spot treatment: 1 to 2% v/v solution of *Stalker* plus 0.25 to 0.5% surfactant v/v in water to thoroughly wet all leaves. Low volume/thinline treatment: 10% v/v solution of *Stalker* plus a 20% v/v ethylated crop oil in water. Cut stump treatment: 20% v/v solution of *Stalker* plus a 20% v/v ethylated crop oil in water or 20% *Habitat* v/v in 80% water carrier. Basal bark treatment: 20% v/v solution of *Stalker* plus a 20% v/v ethylated crop oil in water. **Timing:** Postemergence when plants are growing rapidly. Best when used in late summer to early fall. **Remarks:** Soil residual herbicide; may result in bare ground around plants for some time after treatment. Cut stump and basal bark applications are as described for triclopyr. Plants should not be cut for at least 4 months after basal bark treatment.

Daucus carota L.
Wild carrot

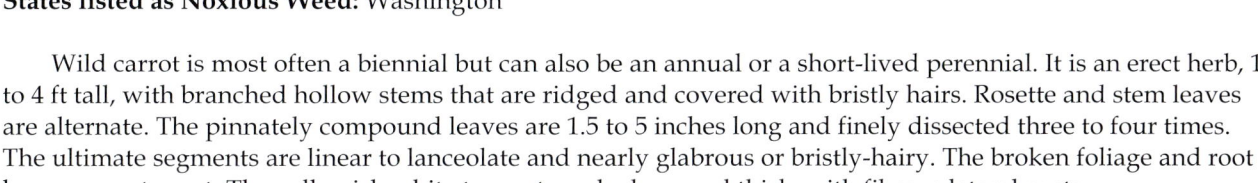

Family: Apiaceae
Range: All contiguous states, including every western state.
Habitat: Disturbed places such as roadsides, fields, pastures, orchards, and in vegetable crops. Usually grows in sandy, gravelly, or moist soils. Grows best in open areas that are not shaded.
Origin: Native to Eurasia and North Africa.
Impact: Wild carrot competes with native grassland species for resources and can result in a change in the plant community. Growing near fields of cultivated carrots, it can impact seed production because of the potential for outcrossing and hybridization, which decreases the value of the crop. Although its foliage is nutritionally similar to legumes, wild carrot makes poor rangeland feed for cattle and horses due to a slight toxicity in the foliage.
States listed as Noxious Weed: Washington

Wild carrot is most often a biennial but can also be an annual or a short-lived perennial. It is an erect herb, 1 to 4 ft tall, with branched hollow stems that are ridged and covered with bristly hairs. Rosette and stem leaves are alternate. The pinnately compound leaves are 1.5 to 5 inches long and finely dissected three to four times. The ultimate segments are linear to lanceolate and nearly glabrous or bristly-hairy. The broken foliage and root have a carrot scent. The yellowish white taproot can be long and thick, with fibrous lateral roots.

The compound umbels are 2 to 4 inches in diameter and convex to flat when in flower, becoming concave at seed maturity. The umbel contains numerous small white flowers, but the central flower of the umbel is sometimes dark red to purplish. Flowers have five petals, tipped with two unequal lobes. The bracts are finely pinnate-dissected with segments that are elongate to linear to threadlike. The fruit is an inferior two-lobed ovary, 3 to 4 mm long with five bristly, longitudinal ribs and four winged ribs lined with barb-tipped bristles. The fruit is schizocarp, separating into two mericarps, and the bristles attach well to animal fur for dispersal. Some seeds survive ingestion by pigeons, pheasants, and possibly other animals. Seeds survive several years under field conditions. Seeds submersed in water can survive up to 2 years. Buried seed can survive up to 20 years.

NON-CHEMICAL CONTROL

Mechanical (pulling, cutting, disking)	Pulling and mowing can be effective if done in the first year of the infestation before the plants have gone to seed and when the plants are 7 to 10 inches tall.
	Tilling for 2 or more years can decrease the infestation of wild carrot by depleting the seed bank.
Cultural	Wild carrot thrives in disturbed habitats. The establishment and maintenance of plant communities can reduce the numbers of wild carrot through competition.
Biological	Since wild carrot and cultivated carrots are the same species, traditional biocontrol is not an option.

CHEMICAL CONTROL
The following specific use information is based on published papers and reports by researchers and land managers. Other trade names may be available, and other compounds also are labeled for this weed. Directions for use may vary between brands; see label before use. Herbicides are listed by mode of action and then alphabetically. The order of herbicide listing is not reflective of the order of efficacy or preference.

GROWTH REGULATORS	
2,4-D	**Rate:** 2 to 4 pt product/acre (0.95 to 1.9 lb a.e./acre)
Several names	**Timing:** Postemergence when target plants are in the seedling to rosette stage.
	Remarks: 2,4-D is a broadleaf-selective herbicide. It has no soil activity and is often combined with other active ingredients, e.g., dicamba. When using the ester formulation do not apply when outside temperatures exceed 80°F. Aquatic registered formulations are available for use close to water.

Aminocyclopyrachlor + chlorsulfuron *Perspective*	**Rate:** 4.75 to 8 oz product/acre **Timing:** Postemergence when target plants are in the seedling to rosette stage. **Remarks:** *Perspective* provides broad-spectrum control of many broadleaf species. Although generally safe to grasses, it may suppress or injure certain annual and perennial grass species. Do not treat in the root zone of desirable trees and shrubs. Do not apply more than 11 oz product/acre per year. At this high rate, cool-season grasses will be damaged, including bluebunch wheatgrass. Not yet labeled for grazing lands. Add an adjuvant to the spray solution. This product is not approved for use in California and some counties of Colorado (San Luis Valley).
Aminopyralid + metsulfuron *Opensight*	**Rate:** 2.5 to 3.3 oz product/acre **Timing:** Preemergence at 3.3 oz in fall, or postemergence when target plants are in the seedling to rosette stage. **Remarks:** Not registered for use in California.
Dicamba *Banvel, Clarity*	**Rate:** 0.5 to 2 qt product/acre (0.5 to 2 lb a.e./acre) **Timing:** Postemergence when target plants are in the seedling to rosette stage. **Remarks:** Dicamba is a broadleaf-selective herbicide often combined with other active ingredients. It may injure grasses at higher rates. Do not apply when outside temperatures exceed 80°F. Dicamba is available mixed with diflufenzopyr in a formulation called *Overdrive*. This has been reported to be effective on wild carrot. Diflufenzopyr is an auxin transport inhibitor which causes dicamba to accumulate in shoot and root meristems, increasing its activity. *Overdrive* is applied postemergence at 4 to 8 oz product/acre. Higher rates should be used on biennials, such as wild carrot. Add a non-ionic surfactant to the treatment solution at 0.25% v/v or a methylated seed oil at 1% v/v solution.
Fluroxypyr *Vista XRT*	**Rate:** 22 oz product/acre (7.7 oz a.e./acre) **Timing:** Postemergence when target plants are growing rapidly. The use of a seed oil at 0.25 to 0.5% can optimize weed control. **Remarks:** Reduced control occurs if the plants are under stressed growth conditions.
Picloram + 2,4-D *Tordon 101M*	**Rate:** For early season application: 1 to 2 pt product/acre; for late season application: 3 to 4 pt product/acre **Timing:** Postemergence from early- to mid- spring when the target plants are less than 3 inches in height. Late season postemergence applications should be made in late-spring to early-summer when the target plants are 3 inches in height to early flowering. **Remarks:** Most broadleaf plants are susceptible, but it is relatively safe on established grasses. Use a non-ionic surfactant at 0.25% v/v. Picloram has a long residual activity and some potential to leach. Some reports indicate that it can injure young or germinating grasses. Do not apply near trees, or where soil is highly permeable and where water table is high. Picloram can also be used in a tank mix with fluroxypyr (*Surmount*). Picloram is a restricted use herbicide. Picloram and its formulations are not registered for use in California.
Triclopyr *Garlon 4 Ultra*	**Rate:** 0.75 to 5 qt product (*Garlon 4 Ultra*)/acre (0.75 to 5 lb a.e./acre) **Timing:** Postemergence when target plants are growing rapidly. **Remarks:** Triclopyr is a broadleaf-selective herbicide that has no soil activity. *Garlon 4 Ultra* is formulated as a low volatile ester. However, in warm temperatures, spraying onto hard surfaces such as rocks or pavement can increase the risk of volatilization and off-target damage.
BRANCHED-CHAIN AMINO ACID INHIBITORS	
Chlorsulfuron *Telar*	**Rate:** 1 to 2.6 oz product/acre (0.75 to 1.95 oz a.i./acre) **Timing:** Early postemergence to rapidly growing plants. **Remarks:** Using a non-ionic surfactant increases the effectiveness of chlorsulfuron. Apply only to pasture, range, CRP, and non-crop sites. Do not apply more than 1.33 oz product/acre per year to pasture or rangeland.
Imazapic *Plateau*	**Rate:** 8 to 12 oz product/acre (2 to 3 oz a.e./acre) **Timing:** Preemergence, or postemergence when the target plants are germinating and actively growing.

	Remarks: Imazapic has mixed selectivity and tends to favor members of the Asteraceae and some grasses. It has long soil residual activity. In postemergence applications, use a methylated seed oil surfactant at 0.25%. Imazapic is not registered for use in California.
Imazapyr *Arsenal AC, Habitat, Stalker, Chopper, Polaris*	**Rate:** 2 to 3 pt product (*Arsenal AC*)/acre (1 to 1.5 lb a.e./acre) **Timing:** Preemergence, or postemergence when the target plants are germinating and actively growing. **Remarks:** Imazapyr is a nonselective herbicide. It has long soil residual activity and leaves more bare ground than other treatments, even a year after application.
Metsulfuron *Escort*	**Rate:** 0.33 to 0.5 oz product/acre (0.2 to 0.3 oz a.i./acre) **Timing:** Early postemergence to rapidly growing plants. **Remarks:** Metsulfuron is primarily a broadleaf herbicide. The use of a non-ionic surfactant will increase the effectiveness of metsulfuron. Do not apply when the plants are under stressed growing conditions. Metsulfuron can also be used in premixes with 2,4-D + dicamba (*Cimarron Max*, 0.25 oz + 1 pt/acre) or aminopyralid (*Chaparral*, 2 to 2.5 oz product/acre). Metsulfuron and its formulations are not registered for use in California.
Sulfometuron *Oust* and others	**Rate:** 2.67 to 3 oz product/acre (2 to 2.25 oz a.i./acre) **Timing:** Preemergence, or early postemergence when the target plants are germinating and actively growing. **Remarks:** Add a surfactant at 0.25% v/v for improved control. Sulfometuron has a relatively long soil residual activity and is susceptible to off-site movement in windblown light soils.

Delairea odorata Lem.
Cape-ivy

Family: Asteraceae
Range: Mainly along the coast of California and Oregon. Also invasive in Hawaii.
Habitat: Riparian corridors, seasonal wetlands, coastal habitats, coastal bluffs and scrub, moist canyons, coastal grassland, oak woodlands, and disturbed sites such as roadsides, urban waste places, or other areas. Requires some moisture year-round. Grows in deep shade or under cloudy conditions and does not tolerate full sunlight. Tolerates serpentine soils, and established plants can survive drought conditions.
Origin: Native to the moist mountain forests of South Africa and introduced to the United States in the late 1800s as a houseplant. Also considered an invasive weed problem in Australia.
Impacts: Under favorable conditions, plants spread invasively and can develop a dense cover that outcompetes other vegetation in natural areas. Vines grow over trees and shrubs and can form dense mats that smother underlying vegetation. Such problematic infestations also reduce native species richness and seedling recruitment in the community. Cape-ivy contains pyrrolizidine alkaloids (liver toxins) and can be toxic to animals when ingested; fish can be killed when plant materials are soaking in waterways.
Western states listed as Noxious Weed: California
California Invasive Plant Council (Cal-IPC) Inventory: High Invasiveness

Cape-ivy is a fleshy perennial vine, with stems to about 30 ft long. The leaves are glossy green, glabrous, alternate, broadly deltate to "ivy-shaped", 1 to 4 inches long, 1.5 inches wide with 5 to 9 lobes. The foliage is evergreen in mild climates and the leaves and stems are deciduous elsewhere.

The flowers are yellow, grouped on terminal and axillary corymbs, with disk flowers approximately 5 mm long arranged in clusters. The fruits are achenes about 2 to 3 mm long, often with a pappus or a crown of hairs. Plants reproduce primarily vegetatively, from fragments of rhizomes, stolons, and stems. A stem fragment as small as one inch, if it has a node, can generate a new plant. Even small fragments of dying stems can resprout, although the regeneration rate is reduced by about one-third. While most seeds produced are not viable, some viable seeds develop in sites throughout California and Oregon. When viable seed are produced, they can disperse long distances by wind.

NON-CHEMICAL CONTROL

Mechanical (pulling, cutting, disking)	Manual removal of plants, including roots and rhizomes, before viable seed develops can help control infestations in areas where plants are accessible. Removing all plant material from the site will help prevent rerooting of rhizomes, stolons, or stem fragments. Follow-up removal of resprouts is essential. In some large patches, all stems can be cut at ground level and Cape-ivy rolled up like a rug. Although the below-ground reproductive tissues will resprout, this strategy makes it possible to detect and spot-treat new sprouts while avoiding contact with desirable vegetation.
	Because Cape-ivy can resprout and establish from stem fragments, mowing is not recommended.
	Cutting off Cape-ivy before it flowers will reduce seed production and deplete the plant's energy reserves. Resprouts are common after treatment. Cutting should be combined with an herbicide treatment or with multiple cuttings over a period of years. All plant parts should be bagged and properly disposed of.
Cultural	Grazing and burning are not considered effective control options. The leaves and stems can be toxic to livestock.
Biological	To date, no biological control agents have been released. However, extensive research by USDA-ARS has been ongoing since 1998. Several species of insects are being examined as potential controls, including a gall-forming fly (*Parafreutreta regalis*), a leaf-mining moth (*Acropelia* spp.), a defoliating moth (*Diota rostrata*), and a stem-boring moth (*Digitivalva delaireae*). The two most promising, the stem-boring moth and the gall-

forming fly, are going through the final stages of testing.

CHEMICAL CONTROL

The following specific use information is based on published papers and reports by researchers and land managers. Other trade names may be available, and other compounds also are labeled for this weed. Directions for use may vary between brands; see label before use. Herbicides are listed by mode of action and then alphabetically. The order of herbicide listing is not reflective of the order of efficacy or preference.

GROWTH REGULATORS	
Clopyralid *Transline*	**Rate:** Spot treatment: 0.5% v/v solution plus 0.25 v/v surfactant to thoroughly wet all leaves. **Timing:** Postemergence when plants are growing rapidly. **Remarks:** Clopyralid is a selective herbicide for broadleaf species. This compound has been shown to be successful in controlling Cape-ivy in Australia.
Triclopyr *Garlon 4 Ultra, Pathfinder II*	**Rate:** Spot treatment: 0.5 to 1% v/v solution of *Garlon 4 Ultra* and water plus 0.25 to 0.5% v/v surfactant to thoroughly wet all leaves. **Timing:** Postemergence when plants are growing rapidly. **Remarks:** Triclopyr is a selective herbicide for broadleaf species. In areas where desirable grasses are growing under or around Cape-ivy, triclopyr can be used without non-target damage.
AROMATIC AMINO ACID INHIBITORS	
Glyphosate *Roundup, Accord XRT II*, and others	**Rate:** Spot treatment: 1 to 2% v/v solution of *Roundup ProMax* (or other trade name with similar concentration of glyphosate) in water, or 1% *Roundup* (or other trade name) plus 0.5% *Garlon 4 Ultra* v/v plus silicon surfactant in water to thoroughly wet all leaves. Wiper treatment: 33 to 50% of concentrated product. **Timing:** Postemergence when plants are growing rapidly. Best results when treated in late summer or early fall. **Remarks:** Glyphosate is a nonselective systemic herbicide. It gives good control with some resprouts. In many situations, it may be more appropriate to use a wiper application to achieve selectivity. Glyphosate can be combined with triclopyr for more effective control. Use a surfactant when applying this combination.

Dipsacus fullonum L.; common teasel
Dipsacus laciniatus L.; cutleaf teasel
Dipsacus sativus (L.) Honck.; Fuller's teasel

Common, cutleaf, and Fuller's teasel

Dipsacus fullonum

Dipsacus fullonum

Family: Dipsacaceae
Range: Common teasel is present in all western states, but Fuller's teasel is only found in Oregon and California, and cutleaf teasel is only found in Oregon and Colorado within the western U.S.
Habitat: Open, sunny sites that range from wet to dry. However, these species generally grow in relatively moist situations along ditches, waterways, roads, and riparian zones. They are also found in pastures, abandoned fields, waste places, and forests, and are capable of invading healthy perennial grass stands in moist areas.
Origin: All species are native to Europe. Dried teasel heads are sometimes used in floral arrangements. Fuller's teasel heads were used in wool "fleecing" before metal carding combs were created. Common teasel is thought to be the wild ancestor of Fuller's teasel.
Impacts: Teasel species are spreading rapidly throughout western United States, especially in the Pacific Northwest. They are aggressive competitors capable of forming dense stands. Mature plants are too prickly and bitter to be eaten by most wildlife and livestock.
Western states listed as Noxious Weed: *D. fullonum*, Colorado, New Mexico, Washington (proposed); *D. laciniatus,* Colorado, Oregon
California Invasive Plant Council (Cal-IPC) Inventory: *D. fullonum* and *D. sativus*, Moderate Invasiveness

Dipsacus laciniatus

Dipsacus sativus

The teasels are biennials, annuals or short-lived perennials that can grow over 6 ft tall. The stems are coarse, straight, and ridged with stiff prickles. The branches are opposite and ascending. The rosette leaves are veined, wrinkled in appearance, with stiff prickles on the lower midrib. Stem leaves are opposite, sessile, fused at the base, lanceolate. Rosettes quickly form a strong, thick, fleshy taproot.

Teasels are easily identified by their unique flower head and bracts. The heads are terminal, egg-shaped, 1.5 to 4 inches long, and have receptacle bracts below flower heads that are often longer than the flower. Reproduction is only from seed. A plant can produce more than 2,000 seeds, of which 30% to 80% may germinate the following spring. Seed dispersal is primarily near the parent plant. Seeds can remain viable for at least 2 years.

NON-CHEMICAL CONTROL

Mechanical (pulling, cutting, disking)	Annual control treatments are usually needed for 4 to 6 years until viable seeds in the soil become sparse.
	With small infestations, digging or hand-pulling before flowering are effective controls. When digging, sever the root below the soil surface.
	Mowing is often ineffective because the root crown will resprout and flower after being cut. Repeated mowing can be effective but must be done often enough to prevent flowering regrowth.
	Tillage effectively controls emerged plants but can stimulate new germination.
Cultural	Livestock may graze rosettes, but teasel has low palatability at most growth stages.
	Fire is not an effective control and often stimulates teasel density the following season. Fire can be used to remove old growth and make teasel rosettes easier to target with herbicide applications.
	Promoting competitive vegetation can slow spread and help prevent establishment.
Biological	No known biological controls agents are available for the control of any species of *Dipsacus*, but USDA

> has recently initiated a biological control program.

CHEMICAL CONTROL

The following specific use information is based on published papers and reports by researchers and land managers. Other trade names may be available, and other compounds also are labeled for this weed. Directions for use may vary between brands; see label before use. Herbicides are listed by mode of action and then alphabetically. The order of herbicide listing is not reflective of the order of efficacy or preference.

GROWTH REGULATORS	
2,4-D Several names	**Rate:** 1 to 2 qt product/acre (0.95 to 1.9 lb a.e./acre) **Timing:** Postemergence. Most effective on small rosettes. **Remarks:** 2,4-D is often tank-mixed with chlorsulfuron or dicamba. It is available in a premix with picloram (*Grazon P+D*). Control has been inconsistent when used alone. It is broadleaf-selective and safe on most grasses. 2,4-D has minimal soil activity. Do not apply ester formulation when outside temperatures exceed 80°F. Amine forms are as effective as ester forms for small rosettes, and amine forms reduce the chance of off-target movement.
Aminocyclopyrachlor + chlorsulfuron *Perspective*	**Rate:** 4.75 to 8 oz product (*Perspective*)/acre **Timing:** Postemergence and preemergence. Postemergence applications are most effective when applied to plants from the seedling to the bolting stage. **Remarks:** Aminocyclopyrachlor provides excellent control of teasel at most growth stages. *Perspective* provides broad-spectrum control of many broadleaf species. Although generally safe to grasses, it may suppress or injure certain annual and perennial grass species. Do not treat in the root zone of desirable trees and shrubs. Do not apply more than 11 oz product/acre per year. At this high rate, cool-season grasses will be damaged, including bluebunch wheatgrass. Not yet labeled for grazing lands. Add an adjuvant to the spray solution. This product is not approved for use in California and some counties of Colorado (San Luis Valley).
Aminopyralid *Milestone*	**Rate:** 4 to 7 oz product/acre (1 to 1.75 oz a.e./acre) **Timing:** Postemergence from the rosette to young bolting stage. **Remarks:** Aminopyralid provided over 90% control when applied to rosettes in university trials. Longer soil residual activity compared to clopyralid. It is safe on most grasses, although preemergence application at high rates can greatly suppress some annual grasses, such as medusahead. Applications can decrease seed production in some annual and perennial grass species. For postemergence applications, adding a non-ionic surfactant (0.25 to 0.5% v/v spray solution) enhances control under adverse environmental conditions; however, this is not normally necessary. Other premix formulations of aminopyralid can also be used for control. These include *Opensight* (aminopyralid + metsulfuron; 2 to 3 oz product/acre) and *Forefront HL* (aminopyralid + 2,4-D; 1.5 to 2.1 pt product/acre).
Clopyralid *Transline*	**Rate:** 0.67 to 1.33 pt product/acre (4 to 8 oz a.e./acre). **Timing:** Postemergence from the rosette to young bolting stage. Results are best if applied to rapidly growing weeds. **Remarks:** Clopyralid is most effective for young plants. It has a shorter soil residual activity than aminopyralid or aminocyclopyrachlor. Trials indicate that it provides over 90% control when applied to rosettes. It controls or injures plants in the Asteraceae and Fabaceae but is safe on most other broadleaf species and all grasses. For postemergence applications, a non-ionic surfactant (0.25 to 0.5% v/v spray solution) enhances control under adverse environmental conditions; however, this is not normally necessary.
Clopyralid + 2,4-D *Curtail*	**Rate:** 2 to 4 qt *Curtail*/acre **Timing:** Same as for clopyralid. **Remarks:** Add a non-ionic surfactant.
Dicamba *Banvel, Clarity*	**Rate:** 2 pt product/acre (1 lb a.e./acre). Use higher rates for larger plants. **Timing:** Postemergence from rosette to beginning of bolting, or fall rosette stage. **Remarks:** Dicamba is a broadleaf-selective herbicide often combined with other active ingredients. In university trials, aminopyralid or clopyralid resulted in better control 1 year after treatment compared

	to dicamba used alone. Dicamba can be mixed with 2,4-D (0.5 to 1 pt dicamba + 2 pt 2,4-D/acre) from the rosette to bolting stage.
	Dicamba is available mixed with diflufenzopyr in a formulation called *Overdrive*. This has been reported to be effective on teasels. Diflufenzopyr is an auxin transport inhibitor which causes dicamba to accumulate in shoot and root meristems, increasing its activity. *Overdrive* is applied postemergence at 4 to 8 oz product/acre to rapidly growing plants. Higher rates should be used on large biennials. Add a non-ionic surfactant to the treatment solution at 0.25% v/v or a methylated seed oil at 1% v/v solution.
Fluroxypyr *Vista XRT*	**Rate:** 11 oz product/acre (3.4 oz a.e./acre) **Timing:** Postemergence from rosette to beginning of bolting, or fall rosette stage. **Remarks:** Fluroxypyr is broadleaf-selective and safe on most grasses.
Picloram *Tordon 22K*	**Rate:** 1 to 1.5 pt product/acre (4 to 6 oz a.e./acre) **Timing:** Postemergence during active growth before bud stage. **Remarks:** Most broadleaf plants are susceptible, but relatively safe on established grasses. Picloram has long soil residual activity and has been reported by some to injure young or germinating grasses. Picloram is a restricted use herbicide. Not registered for use in California.
AROMATIC AMINO ACID INHIBITORS	
Glyphosate *Roundup, Accord XRT II,* and others	**Rate:** 1 to 2 qt product (*Roundup ProMax*)/acre (1.1 to 2.25 lb a.e./acre). Spot treatment 1.5% v/v solution. **Timing:** Postemergence to rapidly growing plants from the rosette to early bolting stage. **Remarks:** Glyphosate will only provide control during the year of application; it has no soil activity and will not kill seeds or inhibit germination the following season. Glyphosate is nonselective. It can create bare ground conditions that make the area susceptible to weed recruitment. In areas with desirable vegetation, use spot treatment. Glyphosate is a good control option if reseeding is planned shortly after application, as it will not injure seedlings emerging after application. Add a surfactant if one is not already included in the herbicide formulation.
BRANCHED-CHAIN AMINO ACID INHIBITORS	
Chlorsulfuron *Telar*	**Rate:** 1 to 2.6 oz product/acre (0.75 to 1.95 oz a.i./acre) **Timing:** Postemergence from the rosette to bolting stage. **Remarks:** Always use a surfactant.
Imazapic *Plateau*	**Rate:** 8 to 12 oz product/acre (2 to 3 oz a.e./acre) **Timing:** Apply postemergence to rosettes. **Remarks:** Imazapic gives effective control with soil residual activity. It can be used in combination with glyphosate (premix trade name of *Journey*). Imazapic is safe on most native grasses. Higher rates may suppress seedings of some cool-season grasses. Add a methylated seed oil. Imazapic is not registered for use in California.
Metsulfuron *Escort*	**Rate:** 1 to 2 oz product/acre (0.6 to 1.2 oz a.i./acre) **Timing:** Postemergence from the rosette to bolting stage. **Remarks:** Metsulfuron has similar activity compared to chlorsulfuron. Metsulfuron has some soil residual activity. Always use a surfactant. Other premix formulations of metsulfuron can be used at the same application timing. These include *Chaparral* (aminopyralid + metsulfuron) at 2 to 3 oz product/acre, *Cimarron Max* (metsulfuron + dicamba + 2,4-D) at 0.5 oz + 2 pt product/acre, and *Cimarron X-tra* (metsulfuron + chlorsulfuron) at 2 oz product/acre. Metsulfuron is not registered for use in California.

Dittrichia graveolens (L.) Greuter
Stinkwort

Family: Asteraceae
Range: In the western United States, stinkwort is only found in California. It is also found in a few northeastern states.
Habitat: Disturbed soils of roadsides, wasteland, gravel areas, levees, and washes; sometimes in pastures, fields, riparian woodlands, and margins of tidal marshes or vernal pools. Grows best on well-drained, sandy or gravelly soils. Thrives in areas with hot, dry summers, but does not require them. Can grow on serpentine, saline, and metal-contaminated soils.
Origin: Native to southern Europe.
Impacts: Stinkwort is rapidly expanding its range. Often found in disturbed sites on the urban fringe, where its sticky, smelly foliage interferes with human activity. In Australia, where stinkwort has been for 150 years, it has naturalized in many habitats. It causes allergic contact dermatitis in humans; it also has been implicated in livestock deaths by enteritis owing to the barbed pappus bristles puncturing the small intestine. Its impact in natural areas is not known, but it is just beginning to invade open riparian or grazing areas.
California Invasive Plant Council (Cal-IPC) Inventory: Moderate Invasiveness (Alert)

Stinkwort is an erect, fall-flowering annual growing up to 3 ft tall. A recent introduction to the west coast, stinkwort was first reported in California in the mid-1980s. Its sticky, glandular-hairy foliage is intensely aromatic. The oils on the foliage, especially on mature plants, make it difficult to control with postemergence herbicides. Stinkwort germinates during winter but remains small until spring. During late spring and summer, it develops into a shrubby, pyramid- or sphere-shaped plant which superficially resembles Russian-thistle or kochia. Stinkwort is a member of the Inuleae tribe and is related to the cudweeds (*Gnaphalium* spp.), but more closely resembles native tarweeds (*Holocarpha*, *Hemizonia*). Its leaves are linear to lance-shaped, typically 0.5 to 1 inch long by only 1 to 3 mm wide.

Unlike most annual, and particularly winter annual weeds, it flowers from September to December, producing small yellow flowerheads around 0.25 inch diameter. The flowerheads turn reddish as they age. Its seeds are tiny, with a barbed pappus, and are readily distributed by wind, water, or by sticking to fur or clothing. The seeds reportedly have a short life in the soil, probably less than 3 years.

NON-CHEMICAL CONTROL

Mechanical (pulling, cutting, disking)	Stinkwort has a relatively shallow root system, so it can be controlled by hoeing or pulling. When doing any kind of mechanical control of stinkwort, wear appropriate protective clothing (long sleeves, long pants, gloves) to minimize exposure to the irritating oils. Once stinkwort goes into flower, plants should be bagged and taken off site, because seeds may ripen on the cut plants.
	Mowing can give partial control, but this plant often has low branches below the level of most mowers, so it may regrow. Mowing a second time, especially in mid- to late summer after the soil has dried out, may give improved control.
Cultural	There is no information available on grazing for the control of stinkwort. It is likely that the plant is not very palatable to livestock.
	Burning alone is not likely to be a good control for stinkwort. It usually grows in locations without enough grasses for dry fuel. Even so, if a burn was successful in controlling the current year's plants, this would likely just clear off the site for a subsequent flush of stinkwort the following winter. Potentially, this could be followed with herbicide treatment as part of an integrated management strategy as with yellow starthistle.
	Stinkwort establishes and thrives in open, disturbed sites. To limit its chances for establishing, minimize disturbances such as overgrazing and soil manipulation. Pastures should be managed for dense, competitive

	stands; aim for maximum ground cover in spring, when stinkwort seedlings are trying to establish.
Biological	No biocontrol agents are presently known for this species.

CHEMICAL CONTROL

The following specific use information is based on reports by researchers and land managers. Other trade names may be available, and other compounds also are labeled for this weed. Directions for use may vary between brands; see label before use. Herbicides are listed by mode of action and then alphabetically. The order of herbicide listing is not reflective of the order of efficacy or preference.

Because of stinkwort's sticky foliar oils, many sources recommend using ester formulations of growth regulator herbicides (2,4-D, MCPA, triclopyr, etc.)

GROWTH REGULATORS	
2,4-D Several names	**Rate:** 2 to 4 pt product/acre (0.95 to 1.9 lb a.e./acre) **Timing:** Postemergence to rapidly growing plants. Smaller plants are easier to control. **Remarks:** 2,4-D is a broadleaf-selective herbicide with no soil activity. Ester formulations appear better at cutting through the oils on the foliage, but do not apply ester formulations when outside temperatures exceed 80°F. A related chemical, MCPA, is commonly used for control of stinkwort in Australia at rates of 25 oz product/acre (0.75 lb a.e./acre). Timing and remarks are similar to 2,4-D.
Aminocyclopyrachlor + chlorsulfuron *Perspective*	**Rate:** 4.75 to 8 oz product/acre **Timing:** In spring either preemergence or postemergence. For postemergence applications, treat until bolting. **Remarks:** Controls many broadleaf species. Although generally safe to grasses, it may suppress some annual and perennial grass species. Do not treat in the root zone of desirable trees and shrubs. Do not apply more than 11 oz product/acre per year. At this high rate, cool-season grasses will be damaged, including bluebunch wheatgrass. Not yet labeled for grazing lands. Add an adjuvant to the spray solution. Not approved for use in California and some counties of Colorado (San Luis Valley).
Aminopyralid + 2,4-D, *Forefront HL*; Aminopyralid + metsulfuron, *Opensight*; Aminopyralid + triclopyr, *Capstone*	**Rate:** 1.5 to 2.1 pt *Forefront HL*/acre; 2.5 to 3.3 oz *Opensight*/acre; 4 to 6 pt *Capstone*/acre **Timing:** Postemergence to plants in the rosette to bolting stages. **Remarks:** These combinations are broadleaf-selective. Recommended rates based on those reported for knapweeds or perennial thistles. *Opensight* is not registered for use in California.
Dicamba *Banvel, Clarity*	**Rate:** 2 to 4 pt product/acre (1 to 2 lb a.e./acre) **Timing:** Postemergence to rapidly growing plants. **Remarks:** Broadleaf-selective herbicide often combined with other active ingredients. Also effective tank-mixed with 2,4-D. It will kill above-ground parts of the plant, but new rosettes will regenerate from the root system. Its use may require repeat applications. Avoid drift to sensitive crops. Do not apply when outside temperatures exceed 80°F. Dicamba + MCPA (3 to 5 oz dicamba product plus 14 to 20 oz MCPA product) is a standard treatment for stinkwort in Australia.
Triclopyr *Garlon 4 Ultra*	**Rate:** 2 to 4 pt product/acre (0.75 to 1.5 lb a.e./acre) **Timing:** Postemergence to rapidly growing plants before flowering. Smaller plants are easier to control. **Remarks:** Broadleaf-selective, generally safe on established grasses. It has little to no soil activity. *Garlon 4 Ultra* is formulated as a low volatile ester. However, in warm temperatures, spraying onto hard surfaces such as rocks or pavement can increase the risk of volatilization and off-target damage.
AROMATIC AMINO ACID INHIBITORS	
Glyphosate *Roundup, Accord XRT II,* and others	**Rate:** 2 to 4 pt product (*Roundup ProMax*)/acre (1.1 to 2.25 lb a.e./acre) **Timing:** Postemergence to rapidly growing plants in late spring to early summer, after desirable competitors have senesced. **Remarks:** Nonselective, no soil activity. Effectiveness may be increased by adding ammonium sulfate.

Echium plantagineum L.
Vipers bugloss

Family: Boraginaceae
Range: Currently found only in Oregon and California.
Habitat: Open woodlands, pastures, and roadsides. It is drought-tolerant and very competitive on heavy soil, but can also be found on sandier soils.
Origin: Native to Mediterranean Europe and North Africa. Apparently escaped from cultivation as an ornamental.
Impacts: Vipers bugloss outcompetes pasture and grassland vegetation and quickly becomes the dominant species. It is toxic, producing a pyrrolozidine alkaloid that can cause chronic liver damage and death, especially in cattle and horses that graze the plant. It currently infests over 80 million acres of rangeland in Australia.
Western states listed as Noxious Weed: Oregon

Vipers bugloss is also referred to as Paterson's curse or salvation jane in Australia. It is a winter annual to biennial from 1 to 6 ft tall, depending on conditions. The leaves are sessile, alternate, rough-hairy, and 2 to 7 inches long. The blades are oval in outline with distinct mid- and lateral veins and often with wavy margins. The rosettes are 4 to 14 inches wide and flat on the ground. Plants bolt in spring, with multiple scorpioid or "fiddleneck" tipped hairy stems.

The stems bear sessile flowers near the tip of the scorpion-like inflorescence. Flowers are showy, pink to blue or purple, and about 1 inch long. The corolla is trumpet-shaped with five tips, with two of the five stamens extending beyond the petal tips. Blueweed, a closely related plant (*Echium vulgare*), bears flowers in which all five stamens extend beyond the petal tips. Each flower produces four dark brown or gray teardrop-shaped "nutlets", about ¼-inch long and covered with wart-like tubercles. Plants reproduce only by seed, which disperse primarily by falling to the ground below the parent plant. The seeds are hard-coated and survive ingestion by livestock. A small proportion of the seeds can remain dormant in the soil seedbank for up to 5 years or more.

NON-CHEMICAL CONTROL

Mechanical (pulling, cutting, disking)	Vipers bugloss spreads exclusively by seed, so hand weeding must be conducted before flowering and seed set to control the species. Hand pulling is generally not appropriate for large populations. Hand control is best employed when soil is moist and plants are still in the rosette stage. It is important to remove the entire root system, or the plant will resprout.
	Mowing will not generally kill the plants, and may even be counterproductive, as this plant is often found in pastures and rangelands where too-frequent mowing will harm forage grasses.
Cultural	Grazing can be a contributing factor to the spread of vipers bugloss, since this plant invades overgrazed or poorly managed pastures and rangelands. In fact, the number of vipers bugloss seedlings was shown to decrease in the absence of grazing. Therefore, maintaining forage lands in a healthy, competitive state will deter establishment of vipers bugloss. In Australia, limiting grazing intensity in the autumn helps to improve forage competitiveness at the time when vipers bugloss seeds are germinating. Grazing the weed itself is not recommended as a management tool because vipers bugloss is toxic to cattle and horses, containing a similar alkaloid to that found in tansy ragwort (*Senecio jacobaea*).
Biological	There are no known biological control agents to control vipers bugloss in the United States. However, in Australia, six agents have been released. This biological control program became a landmark case in the history of weed biological control because conflicts occurred between ranchers and beekeepers (honey is produced from vipers bugloss). This led to a full public enquiry, and a cost: benefit analysis strongly favored controlling the plant and led to the first biological control legislation in Australia. Since then the impacts of two of the agents (*Mogulones larvatus* and *Longitarsus echii*) have been significant and sufficient for detailed economic evaluation of the project, which projected saving Australian agriculture nearly $1 billion by 2050.

CHEMICAL CONTROL

The following specific use information is based on reports by researchers and land managers. Herbicides listed below have been shown to be useful against closely the related weed species blueweed (*Echium vulgare*) and small bugloss (*Anchusa arvensis*). Other trade names may be available, and other compounds may also be labeled for this weed. Directions for use may vary between brands; see label before use. Herbicides are listed by mode of action and then alphabetically. The order of herbicide listing is not reflective of the order of efficacy or preference.

GROWTH REGULATORS	
2,-4-D Several names	**Rate:** 2.1 qt product/acre (2 lb a.e./acre) **Timing:** Postemergence. Spring is best to control seedlings, but treatments can be made in mid-summer to autumn for established plants. **Remarks:** 2,4-D is a broadleaf herbicide that, when used alone, is usually not adequate to completely control annual borage species. It is most often used in combination or sequence with other herbicides such as picloram. Picloram, however, is not registered for use in California. Use with 0.25% v/v non-ionic surfactant to improve control.
Picloram *Tordon 22K*	**Rate:** 1 to 2 pt product/acre (4 to 8 oz a.e./acre) **Timing:** Preemergence or postemergence. Treatment in autumn should provide control through spring. **Remarks:** Follow up treatment with other product may be necessary to control late-germinating seedlings. Use with 0.25% v/v non-ionic surfactant to improve postemergence control. Picloram is a restricted use herbicide. Picloram is not registered for use in California.
AROMATIC AMINO ACID INHIBITORS	
Glyphosate *Roundup, Accord XRT II*, and others	**Rate:** Broadcast treatment: 6 to 22 oz product (*Roundup ProMax*)/acre (0.21 to 0.77 lb a.e./acre). Spot treatment: 1 to 2% v/v solution. Wiper treatment: 33 to 50% of concentrated product. **Timing:** Postemergence. Broadcast applications should be applied from late spring to mid-summer to kill vipers bugloss before seeding. **Remarks:** It may be necessary to make follow-up application in subsequent years to control germinating seedlings. Glyphosate can also injure forage grasses, so this product is best applied for control of pure stands or on otherwise bare ground (such as along roadsides). Use with 0.25% v/v non-ionic surfactant to improve control. Wiper applications to bolted plants have been helpful to control the plant in Australia, but application should be made when weeds are taller than forage grasses.
BRANCHED-CHAIN AMINO ACID INHIBITORS	
Metsulfuron *Escort*	**Rate:** 1 to 2 oz product/acre (0.6 to 1.2 oz a.i./acre) **Timing:** Postemergence. Spring is best to control seedlings, but treatments can be made in mid-summer to autumn for established plants. **Remarks:** Repeat applications will likely be necessary. Use with 0.25% v/v non-ionic surfactant for optimum control. Resistance to sulfonylurea herbicides (metsulfuron and chlorsulfuron (*Telar*)) has been documented for vipers bugloss in Australia. Tank mixtures or sequential treatments with herbicides of other modes of action is recommended to delay onset of resistance. Metsulfuron is not registered for use in California.

Echium vulgare L.
Blueweed (sometimes called vipers bugloss)

Family: Boraginaceae
Range: Recorded in all western states except Nevada, Arizona, and California.
Habitat: Open woodlands, pastures, and roadsides.
Origin: Native to Europe and west-central Asia. Apparently escaped from cultivation as an ornamental.
Impacts: Outcompetes pasture and grassland vegetation and quickly becomes the dominant species.
Western states listed as Noxious Weed: Montana, Washington

Blueweed is a taprooted, upright biennial or short-lived perennial reaching 1 to 3 ft tall. The basal leaves are 3 to 10 inches long and about 1 inch wide and oblanceolate on short petioles; upper leaves become smaller and sessile on the stem. Both the leaves and stems are covered in rough hairs, and the stem is often flecked with black spots.

The inflorescences are long with many side branches, each scorpion-tailed in appearance. Flowers are borne near the uncoiling tips, and are bright blue to pink. The petals are about 1 inch long and trumpet-shaped with five unequal lobes. All five stamens usually extend beyond the petal tips, although one is usually shorter than the others. True vipers bugloss (*Echium plantagineum*) is closely related and similar in appearance, but it bears flowers with only two exerted stamens. Each blueweed flower produces four brown to tan nutlets, each about ¼-inch long, five-sided to teardrop-shaped, and covered with wart-like tubercles. Plants reproduce only by seed, which disperse primarily by falling to the ground below the parent plant. The seeds are hard-coated and survive ingestion by livestock. The seed longevity in the soil is expected to be similar to vipers bugloss (*Echium plantagineum*). In that case, a small proportion of the seeds can remain dormant in the soil seedbank for up to 5 years or more.

NON-CHEMICAL CONTROL

Mechanical (pulling, cutting, disking)	Blueweed spreads exclusively by seed, so hand weed control must be employed before flowering and seed set to control the species. Hand pulling is generally not appropriate when controlling large infestations. Hand control is best employed when soil is moist and plants are still in the rosette stage. It is important to remove the entire root system, or the plant will resprout.
	Mowing will not generally kill the plants, and may even be counterproductive, as this plant is often found in pastures and rangelands where too-frequent mowing will harm forage grasses.
Cultural	Grazing can be a contributing factor to the spread of blueweed since this plant invades over-grazed or poorly managed pastures and rangelands.
	Maintaining forage lands in a healthy, competitive state will be helpful in preventing new infestations of blueweed. Blueweed is generally not palatable to livestock, but is not reported as being toxic.
Biological	There are no known biological control agents to aid in the control of blueweed.

CHEMICAL CONTROL

The following specific use information is based on reports by researchers and land managers. Other trade names may be available, and other compounds may also be labeled for this weed. Directions for use may vary between brands; see label before use. Herbicides are listed by mode of action and then alphabetically. The order of herbicide listing is not reflective of the order of efficacy or preference.

GROWTH REGULATORS

2,-4-D Several names	**Rate:** 2.1 qt product/acre (2 lb a.e./acre)
	Timing: Postemergence. Spring is best to control seedlings, but treatments can be made in mid-summer to autumn for established plants.

Remarks: 2,4-D alone is usually not adequate to completely control borage species. It is most often used in combination or sequence with other herbicides. Use with 0.25% v/v non-ionic surfactant to improve control.

AROMATIC AMINO ACID INHIBITORS

Glyphosate
Roundup, Accord XRT II, and others

Rate: Broadcast treatment: 6 to 22 oz product (*Roundup ProMax*)/acre (0.21 to 0.77 lb a.e./acre). Spot treatment: 1 to 2% v/v solution. Wiper treatment: 33 to 50% of concentrated product.

Timing: Postemergence. Broadcast applications should be applied from late spring to mid-summer to kill blueweed before seeding.

Remarks: It may be necessary to make follow-up application in subsequent years to control germinating seedlings. Glyphosate can also injure forage grasses, so this product is best applied for control of pure stands or on otherwise bare ground (such as along roadsides). Use with 0.25% v/v non-ionic surfactant to improve control. Wiper applications to bolted plants have been helpful to control vipers bugloss in Australia, and should also work on blueweed, but application should be made when weeds are taller than forage grasses.

BRANCHED-CHAIN AMINO ACID INHIBITORS

Chlorsulfuron
Telar

Rate: 1 to 1.5 oz product/acre (0.75 to 1.125 oz a.i./acre)

Timing: Preemergence or postemergence. Treatment in autumn should provide control through spring.

Remarks: Use with 0.25% v/v non-ionic surfactant to improve postemergence control.

Metsulfuron
Escort

Rate: 1 to 2 oz product/acre (0.6 to 1.2 oz a.i./acre)

Timing: Postemergence. Spring is best to control seedlings, but treatments can be made in mid-summer to autumn for established plants.

Remarks: Repeat applications will likely be necessary. Use with 0.25% v/v non-ionic surfactant to improve control. Resistance to sulfonylurea herbicides (metsulfuron and chlorsulfuron (*Telar*)) has been documented for vipers bugloss in Australia. Tank mixtures or sequential treatments with herbicides with other modes of action are recommended to delay onset of resistance in blueweed. Metsulfuron is not registered for use in California.

Egeria densa Planch.
Brazilian egeria

Family: Hydrocharitaceae
Range: Most western states including Washington, Oregon, Idaho, Colorado, Utah, Arizona, New Mexico, and California. Most common in Washington, Oregon and California.
Habitat: Slow-flowing or still water in ditches, sloughs, canals, rivers, ponds, lakes, reservoirs; often in nutrient-rich substrates. Plants are highly susceptible to iron deficiency and grow best under low light. Does not survive prolonged periods of near freezing temperatures.
Origin: Native to eastern South America. Brazilian egeria is still commonly sold as aquarium décor under the name *Egeria* or *Anacharis*. Plants can naturalize in warm temperate to cool subtropical regions when unwanted aquarium contents are released into lakes, ponds, or waterways.
Impacts: Brazilian egeria can aggressively invade new aquatic environments, displace native aquatic vegetation by forming dense stands or large sub-surface mats, and alter the dynamics of aquatic ecosystems. Other detrimental and economic impacts from heavy infestations can include water flow impediment in waterways, increased flooding, clogged pumps and boat propellers, and reduced use of lakes and waterways for fishing and other recreational activities.
Western states listed as Noxious Weed: Oregon, Washington
California Invasive Plant Council (Cal-IPC) Inventory: High Invasiveness

Brazilian egeria is a submerged perennial aquatic weed. Its stems typically grow rooted in the substrate. The leaves are sessile, 1 to 2.5 inches long in whorls of 3-6. On lower stems, the leaves are scale-like and opposite. The leaf margins are minutely toothed, visible with low magnification.

Brazilian egeria is a dioecious species, but populations in the United States consist only of male plants. Male flowers extend 1.5 inches above the water surface on long thread-like flower tubes several inches long. The three petals are showy, glossy white and wrinkled. Because only male flowers occur in the western United States, there is no fruit production. As such, all populations of Brazilian egeria reproduce vegetatively, by stolons and stem fragments. Unlike hydrilla, Brazilian egeria does not produce tubers or turions. Plants can easily fragment into free-floating pieces that root at nodes, and these fragments can start new colonies when carried elsewhere. However, adventitious roots and lateral branches only grow from double nodes (specialized nodes separated by a shortened internode), typically spaced along stems at 6- to 12-node intervals. Only fragments with a double node develop into new plants. Vegetative parts disperse with flooding, waterfowl, and human activities, such as fishing and boating.

NON-CHEMICAL CONTROL

Mechanical (pulling, cutting, dredging)	Removing and destroying stem fragments from recreational equipment, such as boat propellers, docking lines, and fishing gear can help prevent the spread of Brazilian egeria.
	Removing dense canopies by mechanical harvesting may stimulate growth and spread viable fragments. Diver-assisted dredging is very effective in small areas (< 2 acres).
Cultural	Dewatering (drawdown) during summer months may desiccate and kill sparse populations, but large biomass tends to form large clumps that protect the interior from drying out. Brazilian egeria grows rapidly, so repeated exposure to heat or severe freezing may be necessary.
Biological	The triploid (sterile) grass carp (white amur) is a relatively nonselective herbivorous fish that will consume Brazilian egeria, and uses it as one of its most preferred diets. Some research is underway to determine the utility of a fly (*Hydrellia* sp.) whose larvae feed on Brazilian egeria, but no releases have yet been permitted in the United States.

CHEMICAL CONTROL

The following specific use information is based on published papers and reports by researchers and land managers. Other trade names may be available, and other compounds also are labeled for this weed. Directions for use may vary between brands; see label before use. Herbicides are listed by mode of action and then alphabetically. The order of herbicide listing is not reflective of the order of efficacy or preference.

BRANCHED-CHAIN AMINO ACID INHIBITORS	
Penoxsulam *Galleon*	**Rate:** For in-water treatment: 25 to 75 ppb. Treatments may need to be repeated, but not to exceed 150 ppb in an annual season. For dewatered (drawdown) treatment: 5.6 to 11.2 oz product/acre (1.4 to 2.8 oz a.i./acre) **Timing:** Apply directly to water in early spring to early summer during the period of rapid growth. For drawdown, apply during mid- to late winter before refilling. **Remarks:** Penoxsulam is a slow-acting herbicide and may take 4 to 6 weeks for effective control. For drawdown applications use 20 to 100 gal/acre of spray solution to wet the sediment.
PIGMENT SYNTHESIS INHIBITORS	
Fluridone *Sonar*	**Rate:** For in-water treatment: 5 to 20 ppb, but exposures must be maintained for 6 to 8 weeks for optimal control. Generally, two seasons of control with fluridone are optimal for management of well-established Brazilian egeria populations. **Timing:** Apply directly to water in early spring to early summer. **Remarks:** Fluridone is a systemic herbicide. It affects young, rapidly growing plants. Lower rates can be used if applied during early spring growth and when water movement is not likely to dilute or move the herbicide.
CONTACT PHOTOSYNTHETIC INHIBITORS	
Diquat *Reward*	**Rate:** For in-water treatment: 0.1 to 0.25 ppm **Timing:** Apply directly to water in late spring to early summer. **Remarks:** Diquat is a fast-acting contact herbicide that can be effective in mid- to late summer, but if biomass is large, only a portion of the infested sites should be treated to minimize effects of reduced dissolved oxygen. Diquat is quickly bound to, and becomes inactivated on, suspended clay particles and it should not be used in moderately or highly turbid water.
INORGANIC HERBICIDES	
Chelated copper *Komeen,* *Cutrine-plus*	**Rate:** For in-water treatment: 0.5 to 1 ppm elemental copper **Timing:** Apply directly to water in early summer when plants and biomass are small. **Remarks:** Chelated copper is a fast-acting contact herbicide. Retreatment may be required within 3 to 5 weeks. If biomass is large, treat only one-third of infested area to minimize decrease in dissolved oxygen. Chelated copper products are less affected by high pH than inorganic copper.
NON-HERBICIDAL CHEMICALS	
Dyes or colorants *Aquashade*	Although technically not herbicides, dyes and colorants control submerged aquatic plants by absorbing light in the water column and reducing photosynthesis. Applications should be made in early spring and repeated to maintain concentration recommended on the label. Colorants are not as effective on well-established plants in mid- to late summer.

Ehrharta spp.: including *Ehrharta calycina* Sm. (purple veldtgrass)
Ehrharta erecta Lam. (erect veldtgrass)
Ehrharta longiflora Sm. (long-flowered veldtgrass)

Veldtgrasses

Ehrharta calycina

Family: Poaceae
Range: All three species are only found in California at the current time.
Habitat: Purple veldtgrass has invaded grassland, roadsides, live oak woodlands, and coastal habitats such as dunes, scrub, and chaparral. It prefers open sunlight areas. Erect veldtgrass thrives in shade and inhabits disturbed moist places, urban areas, turf, wetlands, and possibly other moist natural communities. Long-flowered veldtgrass inhabits some coastal dune habitat areas in southern California.
Origin: All three species are native to South Africa. Purple veldtgrass was imported to California from Australia, where it was introduced in the late 1920s as a potential forage grass and was later used for erosion control. Purple and long-flowered veldtgrass are now invasive in New Zealand and Australia. Erect veldtgrass was cultivated as an experimental grass in Berkeley and Davis in the mid-1900s. It is also a weed in Europe and Australia.
Impacts: These grasses can form dense stands in coastal dunes or wetlands and forests (erect veldtgrass) and reduce native plant diversity. Purple veldtgrass is the most common of the three species. It can increase the rate of organic matter accumulation and increase fire potential in shrublands and dunes. These consequences can have a dramatic effect on native plant composition. In coastal areas, native shrub seedlings are unable to compete with purple veldtgrass during periods of rain. Eventually coastal scrub and chaparral communities are converted to grassland. Type conversion to grassland can displace native wildlife, including sensitive species such as the kangaroo rat, by eliminating the open-space shrub structure and native food plants.

Ehrharta erecta

California Invasive Plant Council (Cal-IPC) Inventory: *Ehrharta calycina*, High Invasiveness; *Ehrharta erecta*, Moderate Invasiveness; *Ehrharta longiflora*, Moderate Invasiveness (Alert)

Purple veldtgrass is a densely tufted perennial, and erect veldtgrass is a more spreading and decumbent perennial, both to about 3 to 4 ft tall. Long-flowered veldtgrass is an erect annual to about 2 ft tall. The leaves of veldtgrasses generally lack hairs. The collar is large and often violet-tinged, often with long ciliate hairs. Ligules are membranous with a ragged tip, sometimes covered with minute hairs.

The inflorescences of all the veldtgrasses are characteristically open panicles 4 to 6 inches long that only extend out on one side of the rachis. Spikelets consist of only one fertile floret and a couple of sterile florets. Sterile and fertile florets usually detach as a unit from above the glumes. Plants primarily outcross. The seeds are the main form of reproduction and they disperse by falling near the parent plant. Florets also disperse to greater distances with human activities, water, soil movement, and possibly animals. Germination can occur nearly year round in coastal areas of California that receive moisture as either rainfall or heavy fog. Seed longevity in the field is uncertain, but in South Africa, purple veldtgrass typically forms large, persistent seedbanks.

NON-CHEMICAL CONTROL

Mechanical (pulling, cutting, disking)	Manually removing mature plants, including the buried crown, may reduce plant densities, but often stimulates seed germination. All the buried plant parts must be removed on the perennial species to prevent resprouting. Repeatedly removing seedlings as they appear for a period of 2 or more years can help to control populations.
	Flaming plants during the wetter, cooler season can be effective on small infestations, but works better on seedlings.
Cultural	Purple veldtgrass appears unable to tolerate continuous or heavy grazing by sheep, especially while flowering, but it can survive moderate seasonal grazing by cattle. In general, purple veldtgrass tolerates both grazing pressure and livestock trampling more than most indigenous natives.

	Burning usually does not kill the knotty stem bases and may increase population density by damaging or killing less fire tolerant species.
	Use of mulches such as black landscape fabric and rice straw may be effective in small areas.
Biological	There are no biological control agents available for the veldtgrasses.

CHEMICAL CONTROL

The following specific use information is based on reports by researchers and land managers. Other trade names may be available, and other compounds also are labeled for this weed. Directions for use may vary between brands; see label before use. Herbicides are listed by mode of action and then alphabetically. The order of herbicide listing is not reflective of the order of efficacy or preference.

LIPID SYNTHESIS INHIBITORS	
Fluazifop	**Rate:** 1 to 1.5 pt product/acre (4 to 6 oz a.i./acre)
Fusilade	**Timing:** Some applicators report that postemergence treatment to plants over 4 inches tall is much more effective compared to treating smaller plants.
	Remarks: Fluazifop is grass-selective and will not damage broadleaf species. It has no soil activity. Not all studies have found this herbicide to be effective. Include crop oil concentrate surfactant or non-ionic surfactant.
AROMATIC AMINO ACID INHIBITORS	
Glyphosate	**Rate:** Broadcast foliar treatment: 1 to 2 pt product (*Roundup ProMax*)/acre (0.56 to 1.1 lb a.e./acre). Spot treatment: 2% v/v solution. Wiper treatment: 33 to 50% of concentrated product.
Roundup, Accord XRT II, and others	**Timing:** Postemergence in early spring, to rapidly growing, non-stressed plants after most seedlings have emerged.
	Remarks: Glyphosate is a nonselective herbicide. It has no soil activity. Broadcast applications are best used in solid stands of veldtgrass. Wiper applications or shielded sprayers can allow for selective control. Some retreatment may be necessary. Glyphosate is likely the most effective chemical option.
BRANCHED-CHAIN AMINO ACID INHIBITORS	
Imazapyr	**Rate:** 3 to 6 pt product/acre (0.75 to 1.5 lb a.e./acre)
Arsenal, Habitat, Stalker, Chopper, Polaris	**Timing:** Postemergence to rapidly growing grasses.
	Remarks: Imazapyr is fairly nonselective. While it gives effective control alone, it may give better control when combined with glyphosate.

Eichhornia crassipes (Mart.) Solms
Water hyacinth

Family: Pontederiaceae
Range: Nearly worldwide in tropical to warm temperate regions; one of the most serious aquatic weeds, especially in tropical regions. In the western U.S. it is found in Washington (perhaps eradicated), California and Arizona. Also common throughout the southern U.S. and some eastern states.
Habitat: Ponds, sloughs, channels, streams, lakes, in still or slow-moving water. Grows best in warm, high nutrient water.
Origin: Native to tropical South America (Brazil), introduced to the U.S. as an aquatic ornamental.
Impacts: Populations expand rapidly, forming dense mats that can produce high quantities of dry matter. Floating mats clog waterways, alter water oxygen levels, temperature, and pH, provide mosquito habitat, and displace native aquatic vegetation and wildlife. In some places, water hyacinth is used as green manure, to produce methane gas, or processed into a component of animal feeds, paper, particle board, activated carbon, and other products. Water hyacinth removes pollutants from water, including sewage, heavy metals, and chlorinated hydrocarbons, and is useful in waste treatment systems.
Western states listed as Noxious Weed: Arizona, California
California Invasive Plant Council (Cal-IPC) Inventory: High Invasiveness (Alert)

Water hyacinth is a floating perennial with stolons and emergent leaves to ~2 ft tall. Plants are often linked to other plants by stolons from a thick erect stem. Leaf petioles are spongy, to 1 ft long or more, usually inflated in younger plants and tapered in older plants. Water hyacinth seedlings are most often observed rooted in mud along shorelines where the water level fluctuates or on floating or beached mats of decomposing water hyacinth.

Inflorescences are spikes of funnel-shaped showy flowers, pale blue, lilac, or white. The fruit are capsules with numerous seeds. Plants reproduce by seeds and vegetatively from stolons. Vegetative parts and seeds disperse primarily with water. Vegetative reproduction is rapid under favorable conditions; plant numbers can double in ~5 days. Seeds also disperse by clinging to feet or feathers of birds. Seeds sink to the substrate upon release from capsules and then germinate in spring. Seeds are reported to survive 15 to 20 years in dried mud.

NON-CHEMICAL CONTROL

Mechanical (pulling, cutting, chopping, harvesting)	Construction of dams can create still water conditions favorable to water hyacinth establishment. Even flowing river systems are susceptible to infestations.
	Physical removal or destruction of the infestation may be achieved on a small scale by manual removal or on a large scale by harvesting equipment. Mechanical choppers and shredders leave viable fragments that can reestablish populations and may disperse in moving water or with winds. Best results are achieved when efficient systems are used to transport harvested plants to local landfill or composting sites.
	Floating booms or fixed barriers can prevent movement of water hyacinth into other areas. Booms may also be used to try and prevent movement of the weed downstream.
Cultural	Early removal of all plants is best. Dewatering in winter during hard freeze can kill mature plants, but seeds may remain viable if protected by bottom sediments. Even severe frost will not kill floating water hyacinth because the meristem is protected just below the surface of the water.
	Reducing levels of nitrogen and phosphorus in the water will reduce the growth of water hyacinth.
Biological	Two crown/petiole-boring weevils (*Neochetina bruchi* and *N. eichhorniae*) and a stem-boring moth (*Sameodes albiguttalis*) have been released as biocontrol agents in California. Only *Neochetina bruchi* established, but control has been poor. Control with these insects has been more successful in the southeast. The fungal pathogen *Cercospora rodmanii* is host specific to water hyacinth and has potential as a bioherbicide, although it is not yet registered. *C. rodmanii* has not been found on water hyacinth in the San Joaquin Delta. In August 2011, a fourth insect, *Megamelus scutellaris*, a plant hopper, was introduced at two locations in the San Joaquin Delta. Monitoring in December 2011 showed the insects

> survived the summer. Monitoring will continue to determine if these insects can survive the winter and proliferate next spring.

CHEMICAL CONTROL

The following specific use information is based on published papers and reports by researchers and land managers. Other trade names may be available, and other compounds also are labeled for this weed. Directions for use may vary between brands; see label before use. Herbicides are listed by mode of action and then alphabetically. The order of herbicide listing is not reflective of the order of efficacy or preference.

GROWTH REGULATORS	
2,4-D *Weedar 64*	**Rate:** Broadcast foliar treatment: 2 to 4 qt product/acre (1.9 to 3.8 lb a.e./acre) with a non-ionic surfactant **Timing:** Optimal from spring to early summer. Mid-summer to early fall applications can also suppress growth. **Remarks:** Relatively fast-acting, broadleaf-selective systemic herbicide. Symptoms usually appear within a few days to a week and include collapse of petioles and twisted petioles. Best results are achieved under conditions of rapid growth, high temperature and high humidity, when most plants will be killed and sink within 2 to 4 weeks. Under less favorable conditions, some plants may regrow and require repeat treatment. In any case retreatment is almost inevitably required after a few months as a result of reinfestation from incompletely sprayed plants, reinvasion from outside the sprayed area, or regrowth by seedlings.
Triclopyr *Renovate*	**Rate:** Broadcast foliar treatment: 2.67 to 5.33 pt product (*Renovate*)/acre (1 to 2 lb a.e./acre) with a non-ionic surfactant **Timing:** Postemergence to foliage from spring to early summer for optimal timing. However, mid-summer to early fall applications can also effectively suppressing growth. **Remarks:** Triclopyr is a broadleaf-selective, relatively fast-acting systemic herbicide.
AROMATIC AMINO ACID INHIBITORS	
Glyphosate *Rodeo,* *Aquamaster*	**Rate:** Broadcast treatment: 5 to 6 pt product (*Rodeo* or *Aquamaster*)/acre. Spot treatment: 1 to 2 % solution of *Rodeo* or *Aquamaster* (0.5 to 1% a.e.) plus an approved surfactant **Timing:** Postemergence when plants are growing rapidly and at or beyond the bloom stage. Optimal management is achieved with early applications and continued reapplication to new plants. **Remarks:** Slow-acting systemic herbicide. Efficacy can be reduced if plants have dust and debris on the leaves. Application after spring rains wash off the dust can increase efficacy. Glyphosate may have advantages over 2,4-D in causing a slower kill of the weed, which reduces the risks of deoxygenation during decomposition.
BRANCHED-CHAIN AMINO ACID INHIBITORS	
Bispyribac-sodium *Tradewind*	**Rate:** Broadcast foliar treatment: 1 to 2 oz/acre (0.8 to 1.6 oz a.i./acre) **Timing:** Postemergence to foliage from early spring to early summer during rapid growth. May need repeat applications. Allow 30 days between applications. Do not exceed 4 applications/year or 8 oz product/acre/year. **Remarks:** Slow-acting herbicide that may take 4 to 6 weeks to achieve effective control. It may be tank-mixed with other herbicides.
Imazamox *Clearcast*	**Rate:** Broadcast foliar treatment to emergent shoots: 16 to 32 oz/acre (2 to 4 oz a.e./acre). Spot spray-to-wet treatment: 0.25 to 5% v/v solution. For in-water treatment: 50 to 100 ppb **Timing:** Postemergence or directly to water from early spring to early summer during the period of rapid growth. **Remarks:** Use an approved surfactant. Aerial application is approved in some states.
Imazapyr *Habitat*	**Rate:** Broadcast foliar treatment to emergent shoots: 1 to 2 pt/acre (4 to 8 oz a.e./acre). Spot treatment: 0.5% v/v solution with 100 gal/acre for adequate coverage. **Timing:** Postemergence to foliage from early spring to early summer when new growth is present. **Remarks:** Use various formulations or repeated applications to achieve desired concentration for 5 to 7 weeks.
CONTACT PHOTOSYNTHETIC INHIBITORS	
Diquat *Reward*	**Rate:** Spot treatment to emergent shoot: 0.5% v/v solution (2 qt per 100 gal water) **Timing:** Postemergence to foliage from spring to early summer. Repeat treatments may be needed in mid-summer. **Remarks:** Diquat is a contact herbicide that is inactivated in turbid water. The use of clean water in spray solution is critical to its effectiveness.

Elaeagnus angustifolia L.
Russian-olive

Family: Elaeagnaceae
Range: Throughout the western U.S. and in most other states, with the exception of the southeastern U.S.
Habitat: Riparian areas, floodplains, grasslands, roadsides, fencerows, seasonally moist pastures, ditches, and other disturbed sites. Often inhabits seasonally moist areas and sites near farmlands. Grows under a wide range of environmental conditions, including clay, sandy, and fairly alkaline or saline soils. Grows best in inland areas with warm summers and cold winters. Tolerates drought, high water tables, and both freezing and hot temperatures.
Origin: Native to the temperate regions of Asia. Continues to be cultivated as a hardy landscape ornamental and windbreak tree, but has escaped cultivation in many areas of the United States.
Impacts: Russian-olive is especially invasive in seasonally wet riparian areas and may eventually replace stands of native willows (*Salix* spp.) and cottonwoods (*Populus* spp.) at some locations. Although Russian-olive fruits provide food for wildlife, trees are used to a lesser degree than the native vegetation.
Western states listed as Noxious Weed: Colorado, New Mexico
California Invasive Plant Council (Cal-IPC) Inventory: Moderate Invasiveness

Russian-olive is a fast-growing, deciduous tree to 25 ft tall, with silvery foliage. Its twigs and branches are sometimes thorny. Leaves are alternate, simple, narrowly lanceolate or elliptic, mostly 2 to 4 inches long, with smooth margins. The upper surface of the leaf is gray-green and moderately covered with silvery star-shaped hairs and scales. The twigs, leaf stalks, and lower leaf surfaces are silvery gray, and densely covered with silvery shield-shaped (peltate) scales. Cut trees typically resprout from the crown and roots. Depending on the location, roots sometimes associate with nitrogen-fixing bacteria (*Frankia*).

Russian-olive flowers are in umbel-like clusters. The flowers are highly fragrant, mostly 5 to 10 mm long and wide, consisting of a narrow, bell-shaped calyx (sepals as a unit) with four acute petal-like lobes. The fruits are drupe-like (with a fleshy outer layer covering 1 seed), ovoid, about 0.5 to 1 inch long and covered with silvery scales. Plants primarily reproduce by seed. Most fruits remain on trees until distributed by animals, especially birds. The seeds survive ingestion by animals. Seeds are dormant at maturity and require a cool moist stratification period of about 2 to 3 months. Some seeds are hard-coated and may require scarification as well as stratification. Stored seeds survive up to 3 years, but longevity in the field is undocumented.

NON-CHEMICAL CONTROL

Mechanical (pulling, cutting, disking)	Manually removing seedlings and saplings with roots before they mature is a more effective than removing mature trees. Pulling or digging out larger plants is both extremely labor-intensive and not recommended, since it can leave behind root fragments that can resprout. Ring-barking has also been used to kill older trees.
	Russian-olive plants with small diameters of 3.5 inches or less can be pulled out with a weed wrench when soils are moist. In certain situations larger trees can be removed using a bulldozer or a tractor with an attached chain. Any remaining exposed roots should be cut off below ground level and buried.
	Girdling and cutting can suppress Russian-olive but are not effective control options when used alone. Trees vigorously resprout from the roots and crown, or below the girdled or cut area, or along root lines, often resulting in even denser growth. These techniques also require frequent retreatment and cause significant soil disturbance. Cutting trees in mid-summer and then mowing the resprouts once in late summer the following year gave effective control, but was labor intensive and costly.
	Cutting trees before fruits mature can be combined with either burning the stumps or applying an herbicide in a cut stump treatment to give effective control.
	Choosing other landscape ornamentals for sites where seedlings may invade nearby natural areas can help prevent the spread of Russian-olive.
Cultural	Small seedlings of Russian-olive may be susceptible to fire, but burning alone does not adequately control larger individual plants as they vigorously resprout following fire. Stump burning of Russian-olive has been

	shown to be successful, but it is time-consuming compared to other control techniques. Prescribed burning, however, can be used as a pretreatment for another control method, particularly a subsequent herbicide treatment to the resprouts or a basal bark treatment to the stems of resprouts.
Biological	There are no efforts to develop a biological control program for Russian-olive.

CHEMICAL CONTROL

The following specific use information is based on published papers and reports by researchers and land managers. Other trade names may be available, and other compounds also are labeled for this weed. Directions for use may vary between brands; see label before use. Herbicides are listed by mode of action and then alphabetically. The order of herbicide listing is not reflective of the order of efficacy or preference.

GROWTH REGULATORS	
2,4-D Several names	**Rate:** 2 qt product/acre (1.9 lb a.e./acre) **Timing:** Postemergence when leaves are fully developed in early to mid-summer. **Remarks:** 2,4-D is a broadleaf herbicide with no soil activity. It will require two to three retreatments for effective control. 2,4-D can also be mixed with dicamba (4 lb product 2,4-D/acre + 2 lb product dicamba/acre).
Picloram Tordon 22K	**Rate:** Foliar spot treatment: 3% v/v solution provides good control of stems but vigorous suckers can develop. **Timing:** Postemergence at the end of summer to beginning of fall, but before leaf drop. **Remarks:** Picloram is broadleaf-selective and is used to control a variety of annual and perennial broadleaved herbs and woody species. High levels of picloram can give long-term soil activity for broadleaves. Picloram has long soil residual activity. Picloram is a restricted use herbicide. It is not registered for use in California. Picloram is often sold mixed with 2,4-D (*Tordon 101M*), and this formulation has also been used as a cut stump treatment to control Russian-olive or as a foliar treatment to control Russian-olive seedlings.
Triclopyr Garlon 3A, Garlon 4 Ultra, Remedy Ultra, Pathfinder II	**Rate:** Broadcast foliar treatment: 1 to 2 qt product/acre (1 to 2 lb a.e./acre); addition of 7 oz product/acre of *Milestone* can improve control of Russian-olive. Low volume foliar treatment: 5% v/v solution of triclopyr and water plus 0.5% surfactant v/v to thoroughly wet all leaves. Foliar treatment of resprouts: 25% *Garlon 4 Ultra* for the following two years. Basal cut stump treatment: 25 to 50% *Garlon 4 Ultra* in 50 to 75% oil carrier. Cut stump treatments: undiluted *Garlon 3A* or 50% *Garlon 3A* in water. Basal bark treatment: 25% *Garlon 4 Ultra* in 75% oil carrier, or *Pathfinder II* as a ready to use formulation. **Timing:** Cut stump, basal cut stump, and basal bark treatments can be applied as long as the ground is not frozen, but are best in late summer or early fall, before leaf drop. For foliar treatment, the best time to apply the herbicide is when plants are growing rapidly from May through September. **Remarks:** Triclopyr is a selective herbicide for broadleaf species and will not harm grasses growing nearby. For cut stump treatments, cut stems horizontally at or near ground level and apply herbicide solution immediately, covering the outer 20% of the cut face. Suckering from the roots typically occurs after cutting, but the treatment should control most resprouts. For basal bark treatment, spray the lower trunk, including the root collar, to a height of 12 to 15 inches from the ground; the spray should thoroughly wet the lower stem but not to the point of runoff. Trees should not be cut for 1 month to a year after basal bark treatment. *Remedy Ultra* can also be used for basal bark treatments. *Garlon 4 Ultra* at 3 qt product/acre can be tank mixed with aminopyralid (*Milestone*) at 7 oz product/acre when treating resprouts after cutting. The girdling method has also been shown to be effective for Russian-olive control. This involves making shallow, overlapping cuts into the bark around the trunk base using a hatchet or chainsaw, and then lightly spraying the entire cut surface with herbicide.
AROMATIC AMINO ACID INHIBITORS	
Glyphosate Roundup, Accord XRT II, and others	**Rate:** Foliar treatment: 1 to 1.5% v/v solution of *Roundup ProMax* (or other trade name with similar concentration of glyphosate) to thoroughly wet all leaves. Low volume spot treatment: 4 to 7% v/v solution of *Roundup* (or other trade name) to wet 50% of the leaves. Cut stump treatment: undiluted *Roundup* (or other trade name) or 50% v/v in water applied to the cambium. Frill treatment: undiluted glyphosate. **Timing:** Postemergence foliar treatments are best when leaves are fully expanded. Suckering from the

	roots might occur the following year. Cut stump and frill treatments are best applied in late summer, early fall or dormant season. Treatment should occur immediately after cutting.
	Remarks: Glyphosate is a nonselective systemic herbicide with no soil activity. It gives good control with some resprouts. Trees should not be cut for about 1 year after foliar treatment to ensure roots have been killed.
BRANCHED-CHAIN AMINO ACID INHIBITORS	
Imazapyr *Arsenal, Habitat, Stalker, Chopper, Polaris*	**Rate:** Broadcast foliar treatment: 1 to 2 qt product/acre (0.5 to 1 lb a.e./acre). Spot foliar treatment: 1 to 4% v/v solution
	Timing: Postemergence foliar treatments are best when leaves are fully expanded.
	Remarks: Imazapyr is a broad-spectrum herbicide with long soil residual activity. Add a surfactant at 0.25% v/v solution for broadcast application or 1% for spot treatment. Imazapyr has only been shown to give about 75% control of Russian-olive.
PHOTOSYNTHETIC INHIBITORS	
Tebuthiuron *Spike*	**Rate:** Individual spot treatments: 20 lb product (*Spike 20P*)/acre (4 lb a.i./acre), 7.5 lb product (*Spike 80DF*)/acre (6 lb a.i./acre); 0.5 oz product (*Spike 20P*)/1 inch of stem diameter
	Timing: Soil treatments can be applied anytime except when the soil is frozen or saturated with moisture. Applications should be made before the start of spring growth or before expected seasonal rainfall.
	Remarks: Tebuthiuron is a surface applied, soil-active product intended for total vegetation control in non-cropland. For best control, do not disturb plants for 2 years after application.

Elytrigia repens (L.) Desv. Ex Nevski
(= *Elymus repens* (L.) Gould, *Agropyron repens* (L.) P. Beauv.)

Quackgrass

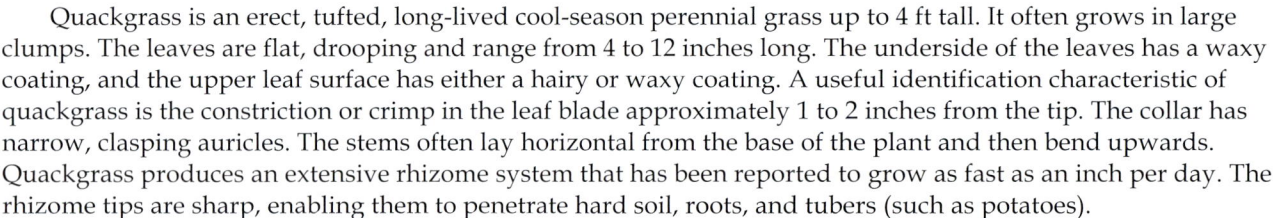

Family: Poaceae
Range: Throughout the United States and in every western state. Because quackgrass does not tolerate long, hot summers it is not present in desert environments and is more prevalent in northern states.
Habitat: Moist mountain meadows, roadsides, ditches, crop fields and other disturbed moist areas. Common in fine-textured soils with near neutral to slightly alkaline soil pH (6.5 to 8.0), however, it will tolerate sandy, acidic soils as well. Quackgrass prefers moist sites but is fairly drought and salt tolerant. Does not tolerate continuous shade.
Origin: Native to Eurasia.
Impact: Quackgrass is highly competitive and displaces native species. It is more commonly known as a noxious agricultural weed that can significantly reduce crop yield. It is a host for several cereal diseases including leaf rusts, smuts, ergot, and take-all disease. In some settings, particularly arid rangeland, quackgrass is considered desirable forage as few other grasses are able to survive in these areas.
Western states listed as Noxious Weed: Arizona, California, Colorado, Oregon, South Dakota, Utah, Wyoming

Quackgrass is an erect, tufted, long-lived cool-season perennial grass up to 4 ft tall. It often grows in large clumps. The leaves are flat, drooping and range from 4 to 12 inches long. The underside of the leaves has a waxy coating, and the upper leaf surface has either a hairy or waxy coating. A useful identification characteristic of quackgrass is the constriction or crimp in the leaf blade approximately 1 to 2 inches from the tip. The collar has narrow, clasping auricles. The stems often lay horizontal from the base of the plant and then bend upwards. Quackgrass produces an extensive rhizome system that has been reported to grow as fast as an inch per day. The rhizome tips are sharp, enabling them to penetrate hard soil, roots, and tubers (such as potatoes).

The inflorescence of quackgrass is a spike about 2 to 8 inches long with alternate stalkless (sessile) spikelets that are flattened, with the flat side facing the stem. Awns are short (< 2 mm long). Quackgrass is not a prolific seed producer and has been reported to produce as few as 25 viable seeds per plant. Seed production often gives rise to new biotypes, which explains the diversity that exists among quackgrass plants. Seed typically falls from the plant in late summer and can remain viable for 1 to 6 years. Quackgrass reproduces primarily by rhizomes but also by seed. The rhizomes can remain viable through long periods of desiccation, and fragmented rhizomes can give rise to new plants.

NON-CHEMICAL CONTROL
Cultural and mechanical practices such as burning, mowing, grazing and cultivation are more effective when done in combination with the application of an effective herbicide.

Mechanical (pulling, cutting, disking)	The extensive root and rhizome system of quackgrass makes it extremely difficult to control by mechanical or cultural means alone. Its abundant food reserves and ability to regenerate from rhizome fragments make mechanical control problematic. Hand-pulling is usually not a practical control measure, especially for larger infestations.
	Mowing has minimal effect on quackgrass.
	The effectiveness of tillage is variable and depends on the severity and frequency of the infestation. Tillage can actually stimulate new shoots to form by severing the terminal bud from the rest of the rhizome. Moldboard plowing can significantly reduce quackgrass infestations. Chisel plowing is generally not as effective and often helps spread rhizomes throughout a field. A single disking or cultivation alone is usually not very effective and may spread the rhizomes. However, disking or cultivation after plowing can be beneficial. It can help chop roots and rhizomes and bring them to the surface where they can desiccate. Vigorous tillage before or after an effective herbicide treatment can improve control.
Cultural	Grazing has minimal effect on quackgrass. It is difficult to significantly deplete the carbohydrate reserves stored in the extensive rhizome system.

	Burning is usually not very effective for quackgrass suppression. Late spring fires generally reduce cover, plant biomass and flowering, while early spring fire can have the opposite effect. Quackgrass normally recovers rapidly after a fire due to the amount of stored root reserves.
Biological	There are no biological control agents available for the management of quackgrass.

CHEMICAL CONTROL

The following specific use information is based on published papers and reports by researchers and land managers. Other trade names may be available, and other compounds also are labeled for this weed. Directions for use may vary between brands; see label before use. Herbicides are listed by mode of action and then alphabetically. The order of herbicide listing is not reflective of the order of efficacy or preference.

LIPID SYNTHESIS INHIBITORS	
Clethodim *Select*, *Envoy*	**Rate:** 9 to 18 oz product (*Envoy*)/acre (1.1 to 2.2 oz a.i./acre)
	Timing: Postemergence to rapidly growing plants before boot stage, preferably when quackgrass is 4 to 12 inches tall.
	Remarks: Clethodim is only effective on grass species and may require multiple applications. It is not sufficiently effective on sodded quackgrass. Before tank-mixing with broadleaf herbicides, read the herbicide label to avoid reduced grass control. Note that *Envoy* formulation is 1 lb a.i./gallon, *Select* is 2 lb a.i./gallon.
Fluazifop *Fusilade*	**Rate:** 1 to 1.5 pt product/acre (4 to 6 oz a.e./acre)
	Timing: Postemergence in late spring to rapidly growing quackgrass 6 to 10 inches tall.
	Remarks: Fluazifop is only effective on grass species. Repeat treatments during a single season are necessary to control established quackgrass, and additional treatments may be needed in subsequent years for complete control. Do not apply to stressed grasses.
Sethoxydim *Poast*	**Rate:** 1.5 to 2.25 pt product/acre (4.5 to 6.75 oz a.e./acre)
	Timing: Postemergence when quackgrass is 6 to 8 inches tall and rapidly growing.
	Remarks: Sethoxydim is only effective on grass species. Multiple applications may be required for improved control. Sethoxydim will not control well established sodded quackgrass. Tillage 1 to 2 weeks after application improves control but is not practical on most range sites. Some results show that sethoxydim is less effective than fluazifop to control quackgrass and is therefore not generally recommended.
AROMATIC AMINO ACID INHIBITORS	
Glyphosate *Roundup, Accord XRT II,* and others	**Rate:** Broadcast foliar treatment (non-sodded quackgrass): 1 to 2 qt product (*Roundup ProMax*)/acre (1.1 to 2.25 lb a.e./acre). Broadcast foliar treatment (sodded quackgrass): 2 to 4 qt product (*Roundup ProMax*)/acre (2.25 to 4.5 lb a.e./acre). Spot treatment: 1 to 2% v/v solution
	Timing: Postemergence to rapidly growing plants from mid-summer to fall after the target plants have reached the reproductive stage (boot stage to early flowering is best).
	Remarks: Glyphosate is the most commonly used herbicide option for control of quackgrass. It is nonselective and has no soil activity. Repeat applications are most likely necessary.
BRANCHED-CHAIN AMINO ACID INHIBITORS	
Imazapyr *Arsenal, Habitat,* *Stalker, Chopper,* *Polaris*	**Rate:** Broadcast treatment: 1 to 4 pt product/acre (4 to 16 oz a.e./acre). Spot treatment: 1% v/v solution
	Timing: Postemergence to rapidly growing plants. Use higher rates for larger plants or late-season applications.
	Remarks: Imazapyr has a fairly long soil residual and is nonselective. Rates are based on those reported for reed canarygrass.
Propoxycarbazone- sodium *Canter R+P*	**Rate:** 0.9 to 1.2 oz product/acre (0.63 to 0.84 oz a.i./acre)
	Timing: Postemergence from the 2-leaf to 2-tiller stage when plants are growing rapidly.
	Remarks: Propoxycarbazone is a broad-spectrum herbicide that will control many species. It will provide only partial control of quackgrass. Perennial grass species vary in tolerance. A non-ionic surfactant should be added at 0.25 to 0.5% v/v solution.
Rimsulfuron	**Rate:** 4 oz product/acre (1 oz a.i./acre)

Matrix	**Timing:** Preemergence in fall.
	Remarks: Rimsulfuron controls several annual grasses and broadleaves. It is only effective on seedling quackgrass. It has not been shown to be effective on established quackgrass. Perennial grasses are tolerant to fall applications when established and grown under dryland conditions. Application to rapidly growing or irrigated perennial grasses may result in their injury or death. It provides soil residual control in cool climates but degrades rapidly under warm conditions. Add a surfactant when applying postemergence.
Sulfosulfuron	**Rate:** 1.33 to 2 oz product/acre (1 to 1.5 oz a.i./acre)
Outrider	**Timing:** Postemergence in spring when the target plants are growing rapidly and in an early vegetative stage. Results are best when weeds are not disturbed by mowing or other factors for 12 days before or after application.
	Remarks: Sulfosulfuron has mixed selectivity, but is fairly safe on native perennial grasses, especially wheatgrasses. Follow-up applications should be made after suitable quackgrass regrowth and no sooner than 30 days after the previous application. Do not make applications to newly seeded perennial native grasses before the 3-leaf growth stage. Sulfosulfuron has fairly long soil residual activity. Treatments should include a non-ionic surfactant.
PHOTOSYNTHETIC INHIBITORS	
Hexazinone	**Rate:** 2.67 to 6.67 lb product/acre (2 to 5 lb a.i./acre)
Velpar DF	**Timing:** Preemergence in spring to control seedlings or postemergence when the target plants are germinating and actively growing.
	Remarks: Hexazinone is typically recommended in combination with glyphosate. Glyphosate is used in mid-summer to fall to control existing plants, and hexazinone is used the following spring to control new germinants. Requires adequate moisture for activation. Because hexazinone is costly and requires high rates to control quackgrass, it is not typically used in natural areas. High rates of hexazinone can create bare ground, so only use high rates in spot treatments.

Equisetum spp.
Horsetail and scouringrush

Family: Equisetaceae
Range: All contiguous states.
Habitat: Primarily found in moist places. In natural systems they are found in areas such as marshland, meadows, riparian zones, and on the margins of ponds and lakes. In anthropologically disturbed areas they occur in pastures, agricultural fields and orchards, and along irrigation ditches.
Origin: Native to North America, including the western states.
Impact: Many species are native to the western United States and are desirable species in some natural areas, but are considered problematic species when they enter human controlled areas such as pastures, agricultural fields, and controlled aquatic areas. The rhizomatous plants form dense colonies and are considered harmful in pasture and rangeland, because some species contain alkaloids that are toxic to livestock. These alkaloids destroy thiamine and are especially harmful to horses when ingested.
Western states listed as Noxious Weed: *Equisetum telmateia*, Oregon

The *Equisetum* genus consists of primitive rhizomatous perennials. The green photosynthetic main stems are longitudinally ribbed and hollow, except at the nodes where the branches are whorled. Species with whorls of lateral branches are referred to as horsetails, while those lacking lateral branches are called scouringrushes. The leaves of all species are small and in some species highly reduced. Leaves are united in a node-sheathing tube that is toothed along the upper margin. The rhizomatous root system is extensive and can grow to a depth of 5 ft or more. The rhizomes root at the nodes and store starch in tubers in the internodes. These tubers can detach from the parent plant and grow into a new plant.

Equisetum species reproduce by spores mainly between March and July. The strobilus is a cone-like reproductive structure at the end of the stem. The strobilus consists of numerous small flower-shaped bracts (sporophylls) on short stalks, with spore-bearing structures called sporangia on the lower surfaces. The stems in most species contain large amounts of silica, and the plants were used historically for scrubbing and cleaning.

NON-CHEMICAL CONTROL

Mechanical (pulling, cutting, disking)	Cutting and disking have short-term effectiveness, not lasting longer than a couple weeks. Cutting only destroys the top growth and delays reestablishment. Continuous tillage may result in effective control if complete destruction is achieved.
Cultural	Horsetail is top-killed by fire but regenerates rapidly. The frequency of occurrence of field horsetail is usually unchanged or increased after fire.
	Applying nitrogen fertilizer to grass crops is helpful because horsetail responds minimally to nitrogen, while grass crops respond quickly and significantly. This gives the crop a highly competitive advantage over horsetail because this weed is shade sensitive.
	Where possible, reducing the amount of water to an infested area can reduce the competitiveness of *Equisetum* species.
Biological	There are no approved biological control agents available for any of the *Equisetum* species, primarily because they are native species.

CHEMICAL CONTROL
The following specific use information is based on published papers and reports by researchers and land managers. Other trade names may be available, and other compounds also are labeled for this weed. Directions for use may vary between brands; see label before use. Herbicides are listed by mode of action and then alphabetically. The order of herbicide listing is not reflective of the order of efficacy or preference.

GROWTH REGULATORS	
2,4-D Several names	**Rate:** 1 to 4 pt product/acre (0.48 to 1.9 lb a.e./acre) **Timing:** Postemergence to rapidly growing weeds. Smaller weeds are more easily controlled. **Remarks:** 2,4-D is broadleaf-selective. It has little soil activity and is often combined with other active ingredients, e.g., dicamba. Do not apply when outside temperatures exceed 80°F. Aquatic registered formulations are available for use close to water.
Aminocyclopyrachlor + chlorsulfuron *Perspective*	**Rate:** 4.75 to 8 oz product/acre **Timing:** Postemergence to rapidly growing weeds. Smaller weeds are more easily controlled. **Remarks:** *Perspective* provides broad-spectrum control of many broadleaf species. Although generally safe to grasses, it may suppress or injure certain annual and perennial grass species. Do not treat in the root zone of desirable trees and shrubs. Do not apply more than 11 oz product/acre per year. At this high rate, cool-season grasses will be damaged, including bluebunch wheatgrass. Not yet labeled for grazing lands. Add an adjuvant to the spray solution. This product is not approved for use in California and some counties of Colorado (San Luis Valley).
Fluroxypyr *Vista XRT*	**Rate:** 22 oz product/acre (7.7 oz a.e./acre) **Timing:** Postemergence when target plants are small and rapidly growing. For optimum control add 0.25 to 0.5% v/v seed soil surfactant. **Remarks:** Provides only suppression and not complete control. Not registered for aquatic areas.
BRANCHED-CHAIN AMINO ACID INHIBITORS	
Chlorsulfuron *Telar*	**Rate:** 1 to 2.6 oz product/acre (0.75 to 1.95 oz a.i./acre) **Timing:** Preemergence or early postemergence when weeds are germinating and actively growing. **Remarks:** Do not apply more than 2.6 oz product/acre per year and only 1.33 oz product/acre per year in grazed areas. Chlorsulfuron has long soil residual activity. *Telar* can be used near water, but cannot be applied to water.
Metsulfuron *Escort*	**Rate:** 1 to 2 oz product/acre (0.6 to 1.2 oz a.i./acre) **Timing:** Early postemergence to rapidly growing plants. **Remarks:** Certain biotypes are less sensitive to metsulfuron. Not registered for aquatic areas. Metsulfuron is not registered for use in California.
Sulfometuron *Oust* and others	**Rate:** 6 to 8 oz product/acre (4.5 to 6 oz a.i./acre) for areas receiving more than 20 inches of precipitation per year **Timing:** Preemergence or early postemergence when the target plants are germinating and actively growing. **Remarks:** Add a surfactant to improve control. Sulfometuron has long soil residual activity. It is a broad-spectrum herbicide. May move long distances in dry light windblown soils. Not registered for aquatic areas.

Erodium cicutarium (L.) L'Hér. Ex Ait.

Redstem filaree

Family: Geraniaceae
Range: Throughout the United States, and in every western state.
Habitat: Roadsides, pastures, fields, grassland, rangeland, waste places, agronomic and vegetable crops, orchards, vineyards, and landscaped areas. Primarily associated with open disturbed sites.
Origin: Native to Eurasia. Redstem filaree was well established in the western United States before the arrival of the Spanish missionaries and their livestock in 1769.
Impact: Can form a dominant cover in rangelands, particularly following a burn. Does not provide good wildlife cover. Like other members of the genus, redstem filaree provides good livestock and wildlife forage before maturity.
Western states listed as Noxious Weed: Colorado
California Invasive Plant Council (Cal-IPC) Inventory: Limited Invasiveness

Redstem filaree is a common winter annual with alternate pinnate-lobed or -compound leaves 1.5 to 4 inches long, with 9 to 13 leaflets that are ovate with deeply lobed margins or dissected segments. Plants have a taproot and exist as low growing rosettes until flowering stems develop in late winter or spring. When mature they can reach 2 ft tall. The stems and leaf stalks are often reddish.

The flowers are in simple umbels, with 2 to 10 per cluster. They are pink to reddish-lavender with five separate petals. The immature fruits (schizocarps) consist of five fused ovary sections (carpels) and five elongated styles united to form a style column or beak about 1 to 2.5 inches long, the entire unit resembling a stork's head and beak. Each carpel with its associated style separates from the unit at maturity and is called a mericarp. Styles are twisted in the lower portion, sharply curved near the middle to form a right angle, with stiff upward-pointing hairs on one side. Style coils tighten under dry conditions and loosen under humid conditions to help drill seeds into the soil. Plants reproduce only by seed. Mericarps usually separate explosively and are propelled a short distance from the parent plant. Some mericarps disperse to greater distances with soil movement and especially by clinging to the fur, feathers, and feet of animals, and the shoes and clothing of people. Most seeds germinate in fall after the first significant rainfall of the season. Seeds can remain viable for many years in the soil, forming extensive seed banks.

NON-CHEMICAL CONTROL

Mechanical (pulling, cutting, disking)	Manual removal or cultivation before fruits develop can help to control filaree.
Cultural	Spring and summer burns generally increase the abundance of filaree the following year, while fall burning appears to have no effect.
Biological	No biocontrol agents have been introduced for any *Erodium* species.

CHEMICAL CONTROL

The following specific use information is based on published papers and reports by researchers and land managers. Other trade names may be available, and other compounds also are labeled for this weed. Directions for use may vary between brands; see label before use. Herbicides are listed by mode of action and then alphabetically. The order of herbicide listing is not reflective of the order of efficacy or preference.

GROWTH REGULATORS	
2,4-D Several names	**Rate:** 0.5 to 2 pt product/acre (0.21 to 0.85 lb a.e./acre)
	Timing: Postemergence to rapidly growing plants up to flowering. Smaller plants are more effectively controlled.
	Remarks: 2,4-D is broadleaf-selective and has no soil activity. It may require repeat application. 2,4-D is not the most effective treatment, but is widely used because of low cost. Do not apply ester

	formulations when outside temperatures exceed 80°F.
Aminocyclopyrachlor + chlorsulfuron *Perspective*	**Rate:** 3 to 4.5 oz product/acre **Timing:** Postemergence in spring up to flowering. **Remarks:** Very effective for the control of filaree. *Perspective* provides broad-spectrum control of many broadleaf species. Although generally safe for grasses, it may suppress or injure certain annual and perennial grass species. Do not treat in the root zone of desirable trees and shrubs. Do not apply more than 11 oz product/acre per year. At this high rate, cool-season grasses will be damaged, including bluebunch wheatgrass. Not yet labeled for grazing lands. Add an adjuvant to the spray solution. This product is not approved for use in California and some counties of Colorado (San Luis Valley).
Aminopyralid + metsulfuron *Opensight*	**Rate:** 3 to 3.3 oz product/acre **Timing:** Postemergence, in spring from rosette to flowering stages, or in fall to seedlings and rosettes. **Remarks:** Do not use in situations where forage will be harvested for hay to be moved offsite. It is expected that the combination would be effective on filaree since metsulfuron alone will provide control. Not registered for use in California.
Dicamba *Banvel, Clarity*	**Rate:** 8 to 32 oz product/acre (0.25 to 1 lb a.e./acre); 8 to 16 oz for rosettes, up to 32 oz product/acre for bolting plants **Timing:** Postemergence to rapidly growing plants up to flowering. Smaller plants are more effectively controlled. **Remarks:** Dicamba is a broadleaf-selective herbicide often combined with other active ingredients. It is effective earlier in the season than 2,4-D. It is also effective when tank-mixed with 2,4-D (0.75 lb a.e./acre dicamba + 0.25 lb a.e./acre 2,4-D). Dicamba has very limited soil residual. Do not apply when outside temperatures exceed 80°F.
AROMATIC AMINO ACID INHIBITORS	
Glyphosate *Roundup, Accord XRT II,* and others	**Rate:** 2 to 3 pt product (*Roundup ProMax*)/acre (1.1 to 1.7 lb a.e./acre) **Timing:** Postemergence to rapidly growing plants. **Remarks:** Glyphosate has no soil activity. It is a nonselective herbicide. Repeat applications may be necessary as it provides only partial control of filaree. Effectiveness is increased by addition of ammonium sulfate.
BRANCHED-CHAIN AMINO ACID INHIBITORS	
Chlorsulfuron *Telar*	**Rate:** 1 to 2 oz product/acre (0.75 to 1.5 oz a.i./acre) **Timing:** Postemergence to young, rapidly growing plants in spring before flowering, or in fall to new rosettes **Remarks:** Chlorsulfuron has mixed selectivity and is generally safe on grasses. Use a surfactant. It has a fairly long soil residual.
Imazapic *Plateau*	**Rate:** 8 to 12 oz product/acre (2 to 3 oz a.e./acre) **Timing:** Most effective postemergence. **Remarks:** Imazapic is a broad-spectrum herbicide that has some soil residual activity. Not registered for use in California.
Imazapyr *Habitat, Stalker, Arsenal, Chopper, Polaris*	**Rate:** 2 pt product (*Habitat*)/acre (0.5 lb a.e./acre) **Timing:** Preemergence or postemergence **Remarks:** Imazapyr is a nonselective herbicide. It has a relatively long soil residual activity.
Metsulfuron *Escort*	**Rate:** 0.33 to 0.5 oz product/acre (0.2 to 0.3 oz a.i./acre) **Timing:** Postemergence to young, rapidly growing plants in spring before flowering, or in fall to new rosettes **Remarks:** Metsulfuron has mixed selectivity and is generally safe on grasses. Use a surfactant. It can be tank-mixed with 2,4-D and/or dicamba and has some soil residual activity. Not registered for use in California.
Metsulfuron +	**Rate:** 0.5 oz product/acre

chlorsulfuron *Cimarron X-tra*	**Timing:** Postemergence to rapidly growing plants before flowering. **Remarks:** Safe on established grasses. Not registered for use in California.
Rimsulfuron *Matrix*	**Rate:** 4 oz product/acre (1 oz a.i./acre) **Timing:** Preemergence. **Remarks:** Rimsulfuron controls several annual grasses and broadleaves. Perennial grasses are tolerant to fall applications when established and grown under dryland conditions. Application to rapidly growing or irrigated perennial grasses may result in injury or death of the crop. It provides soil residual control in cool climates but degrades rapidly under warm conditions. Rimsulfuron will not control summer annual weeds when applied in fall or spring. Add a surfactant when applying postemergence.
Sulfometuron *Oust* and others	**Rate:** 6 to 8 oz product/acre (4.5 to 6 oz a.i./acre) **Timing:** Preemergence or early postemergence, during the rainy season when weeds are germinating or rapidly growing. **Remarks:** Sulfometuron has mixed selectivity, but is fairly safe on native perennial grasses, especially wheatgrass. Other desirable grasses may be stunted, stressed, or injured. Good for revegetation use. It has fairly long soil residual activity. Use lower rates in arid areas.
Sulfometuron + chlorsulfuron *Landmark XP*	**Rate:** 4.5 oz product/acre **Timing:** Preemergence or early postemergence. **Remarks:** See sulfometuron. Rates are based on rates reported for Carolina geranium.
PHOTOSYNTHETIC INHIBITORS	
Hexazinone *Velpar DF*	**Rate:** 2.67 to 6.67 lb product/acre (2 to 5 lb a.i./acre) **Timing:** Preemergence or early postemergence to seedlings. **Remarks:** Hexazinone is a broad-spectrum herbicide that is mobile in the soil and has long soil residual activity. It provides only suppression of filaree. It should not be used in areas with a shallow water table. Hexazinone is best used in areas with high densities of filaree, medium to shallow soils, away from watercourses, and away from trees. High rates of hexazinone can create bare ground, so only use high rates in spot treatments.

Eucalyptus globulus Labill.

Tasmanian blue gum

Family: Myrtaceae
Range: Throughout the coastal regions of California and Hawaii.
Habitat: Disturbed places, especially in riparian areas and coastal grasslands and forests. Groves can expand into intact adjacent scrub, woodlands, or grasslands. Grows best on deep, well-drained soils where roots can tap deep soil moisture or in areas that receive at least 21 inches of rain per year or moisture from additional sources, such as fog drip. Mature trees tolerate drought and short periods of temperatures as low as 17°F.
Origin: Native to southeastern Australia and Tasmania and introduced to the U.S. in the early 1850s as a landscape ornamental. Still widely planted.
Impacts: Mature Tasmanian blue gum trees create a safety hazard in public places because they tend to drop limbs. Leaves and branches decompose very slowly. Due to flammable plant compounds, dense growth of fine branches, and leaf and branch litter, groves are highly combustible and increase the risk of fire. Under drought conditions, trees tap into deep soil moisture and continue to transpire freely. The flowers are attractive to native hummingbirds, but the nectar has been implicated in clogging their beaks, causing the birds to starve. Frost dieback can exacerbate accumulation of dry, flammable leaves and branches making fire danger extremely high.
California Invasive Plant Council (Cal-IPC) Inventory: Moderate Invasiveness

Tasmanian blue gum is a fast-growing tree that can reach 150 to 180 ft tall and 4 to 7 ft in diameter. It has a straight trunk and well-developed crown with dark, rough persistent bark below and smooth, shedding, yellow-brown bark above. Leaves on older branches are 6 to 8 inches long, glossy, dark green, and leathery; they are narrowly lanceolate, often curved, alternate, and hang vertically. Juvenile leaves are opposite, sessile, broadly oblong, and covered with a gray, waxy bloom which is thicker on the bottom surface. Stems are usually square in cross-section and winged at the corners. Trees can resprout from the base when cut or damaged.

The flowers are white, sessile and solitary in the leaf axils. The fruit is a hard, woody capsule, broadly top-shaped, and often 4-angled. The fruit are 0.75 to 1 inch in diameter and 1 inch long or more, with a distinctive concave ring around the margin. Reproduction is by seed. Most seeds are released from capsules while still attached to the tree. Seeds typically fall within 300 ft from the parent plant, although some may disperse to greater distances with water, soil movement, animals, and human activities. Under favorable conditions, seeds germinate a few weeks after release from capsules, usually late fall through spring, but if conditions are dry, seeds may remain dormant for several years.

NON-CHEMICAL CONTROL

Mechanical (pulling, cutting, disking)	Hand pulling can remove seedlings and small saplings. For larger saplings and small trees, a weed wrench or other woody weed extractor can be used. Care must be taken to extract the entire root or stump sprouting will occur. Best results are achieved when soil is moist.
	Cutting a tree at ground level before it flowers will reduce seed production and deplete the plant's energy reserves. Resprouts are common after treatment. Cutting back regrowth when shoots reach 6 to 7 ft tall for 4 years or more can eventually kill the tree. Covering cut stumps with black plastic and sealing the edges with soil to exclude sunlight also gives good control. Plastic must be kept in place for at least one year. Cutting can also be combined with an herbicide treatment.
Cultural	Grazing is not considered an effective control option as animals seldom browse on seedlings.
	Burning alone is not an effective method for controlling eucalyptus. Although burning can remove debris, in many cases it can increase the population as it removes competitive vegetation, releases nutrients into the soil, and stimulates the germination of seeds left in the soil. Burning is more effective when followed by an herbicide application, subsequent burnings, and/or revegetation using desirable species. It is important to employ a control strategy following a burn, otherwise the eucalyptus population may increase in subsequent years.
Biological	No biological control agents have been released for the control of eucalyptus. In 1998, the red gum lerp psyllid

> (*Glycaspis brimblecombei*), an insect native to Australia that causes foliar damage to many eucalyptus species, was found in California. Because eucalyptus is valued as an ornamental and as a commercial forest species, a biological control program was launched for the red gum lerp psyllid. In 2000, the parasitoid *Psyllaephagus bliteus* was widely released in California to control the red gum lerp psyllid.

CHEMICAL CONTROL

The following specific use information is based on published papers and reports by researchers and land managers. Other trade names may be available, and other compounds also are labeled for this weed. Directions for use may vary between brands; see label before use. Herbicides are listed by mode of action and then alphabetically. The order of herbicide listing is not reflective of the order of efficacy or preference.

GROWTH REGULATORS	
Picloram + 2,4-D *Tordon 101M, Tordon RTU* or *Pathway*	**Rate:** Cut stump treatment: undiluted or 50% *Tordon 101M* in water or undiluted *Tordon RTU/Pathway* (ready to use). Stem injection treatment: one cut per every 3 inches of stem diameter, and 0.5 ml of undiluted or 1 ml of diluted herbicide added to each cut. **Timing:** Best when used in late summer to early fall. **Remarks:** High rates can give long-term soil activity for broadleaves. Picloram is a restricted use herbicide, not registered for use in California. Applications are as described for triclopyr.
Triclopyr *Garlon 3A, Garlon 4 Ultra, Pathfinder II*	**Rate:** Foliar spot treatment: 2% v/v solution of *Garlon 4 Ultra* and water plus 0.5% v/v non-ionic surfactant to thoroughly wet all leaves. Basal cut stump treatment (treat the cut surface and the bark on the sides of the stump): 20 to 25% *Garlon 4 Ultra* in 75 to 80% oil carrier. Cut stump treatment (apply to cut surface only): 50% *Garlon 3A* in water. Basal bark treatment: 20 to 25% *Garlon 4 Ultra* in 75-80% oil carrier, or *Pathfinder II* (ready-to-use). Stem injection treatment: one cut per every 3 inches of stem diameter, and 1 ml of undiluted *Garlon 3A* added to each cut. **Timing:** Foliar treatments best when leaves are fully expanded. Stump and stem treatments can be used any time, but are best if not used when sap is rising in the early spring. **Remarks:** Broadleaf selective; will not damage desirable grasses growing nearby. Not as effective on eucalyptus as glyphosate. Foliar treatment should only be made on small trees or seedlings. For cut stump, cut stems horizontally near ground level and immediately apply *Garlon 3A* solution, covering the outer 20% of the cut surface. Suckering may occur after cutting, but the treatment should control most resprouts. For basal cut stump, applications can be made up to 2 weeks after cutting; treat to a height of 12 to 18 inches from the ground. For basal bark, spray the lower trunk, including the root collar, to 12-15 inches from the ground; the spray should wet the lower stem but not to the point of runoff. For stem injection, be sure that each cut goes well into the cambium layer; more effective on smaller trees. Trees should not be cut for at least one month after basal bark or stem injection treatments. A dye can be added to either product.
AROMATIC AMINO ACID INHIBITORS	
Glyphosate *Roundup, Accord XRT II*, and others	**Rate:** Foliar spot treatment: 2% v/v solution (*Roundup ProMax*) glyphosate and water plus 0.5% v/v non-ionic surfactant to thoroughly wet all leaves. Cut stump treatment: undiluted or 50% *Roundup* (or other trade name) in water. Stem injection treatment: one cut per every 3 inches of stem diameter, and 1 ml of undiluted herbicide added to each cut. **Timing:** Best when used in late summer to early fall. **Remarks:** Glyphosate is a nonselective systemic herbicide. Applications are made as described for triclopyr. Glyphosate is considered the most effective herbicide for control of eucalyptus.
BRANCHED-CHAIN AMINO ACID INHIBITORS	
Imazapyr *Arsenal, Habitat, Stalker, Chopper, Polaris*	**Rate:** Low volume/thinline treatment: 20% v/v solution of *Chopper* plus a 20% v/v ethylated crop oil in water. Cut stump treatment: 20% *Stalker* or *Chopper* formulation v/v in 80% oil carrier or 20% *Arsenal* or *Habitat* v/v in 80% water carrier. Stem injection treatment: one cut per every 3 inches of stem diameter, and 1 ml of undiluted herbicide (*Arsenal* or *Habitat*) added to each cut. Basal bark treatment: 20% *Stalker* or *Chopper* formulation v/v in 80% oil carrier. **Timing:** Best when used in late summer to early fall. **Remarks:** Soil residual herbicide; may result in bare ground around trees for some time after treatment. Applications are made as described for triclopyr. Only shown to be effective on smaller eucalyptus trees.

Euphorbia esula L.; leafy spurge
   (= *E. virgata* Waldst. & Kit. [Jepson Manual 2012])
Euphorbia oblongata Griseb.; oblong spurge
Euphorbia terracina L.; carnation spurge

Leafy, oblong and carnation spurge

Euphorbia esula

Family: Euphorbiaceae
Range: Leafy spurge is found in all western states, most central states (especially in the north), and northeastern states. Oblong spurge is found in the Pacific Northwest states (Washington, Oregon and California) and is expanding its range in California. Carnation spurge, a recent introduction, is known only in southern California.
Habitat: Waste areas, disturbed sites, roadsides, fields. Leafy spurge also infests pastures, rangeland, and riparian areas, from sub-tropic to sub-arctic climates and from semi-arid to mesic conditions; it can even tolerate flooding for 4 to 5 months if shoots can grow above the water surface. Carnation spurge has been reported in disturbed places, grassland, coastal bluffs and dunes, salt marsh, riparian areas, and oak woodlands.
Origin: All species are native to southern Europe.
Impacts: These plants can form dense patches that displace desirable vegetation. Leafy spurge, in particular, is one of the most tenacious weeds in the United States, forming dense clonal colonies that suppress both native plants and forage, resulting in reduced land values. Leafy spurge is especially problematic in the north-central states and adjacent parts of Canada. It infests nearly 3 million acres of rangeland in 29 states, causing estimated economic losses of $130 million per year. The milky sap of spurges is toxic and can irritate the skin, eyes, and digestive tracts of humans and other animals. Cattle avoid foraging spurge, but goats and sheep appear tolerant to its irritant properties.

Euphorbia oblongata

Western states listed as Noxious Weed: *E. esula*, Arizona, California, Colorado, Idaho, Kansas, Montana, Nevada, New Mexico, North Dakota, Oregon, South Dakota, Utah, Washington, Wyoming
California Invasive Plant Council (Cal-IPC) Inventory: *E. esula*, High Invasiveness (Alert); *E. oblongata*, Limited Invasiveness; *E. terracina*, Moderate Invasiveness (Alert)

Euphorbia terracina

These spurges are erect perennials to nearly 3 ft tall (carnation spurge sometimes taller), with milky white sap and smooth, oblong to elliptical leaves. The leaves are alternate on the lower stems, but leaves and bracts may be whorled or opposite just under the flowering branches. Oblong and carnation spurges develop vertical taproots that can bud off new plants near the soil surface. Leafy spurge forms an extensive system of creeping roots that form adventitious root buds and generate new plants, as well as storing food reserves that enable roots to produce new shoots for many years under continuous grazing or mowing.

During summer, these spurges form umbel-like flower clusters at the stem tips, the flowers with yellow-green bracts (not petals). The flowers develop 3-chambered seed capsules with yellow-brown to grey seeds. Mature capsules rupture and eject seeds up to 16 ft from the parent plant, but some seeds disperse to greater distances with human and animal activities, water, and as hay or seed contaminants. These species generally start new infestations from seed. Leafy spurge populations also can expand vegetatively, by budding from roots or from root fragments as small as 0.5 inch. Most seeds germinate in early spring, but germination may occur throughout the growing season. Seeds can remain viable for 8 years or more in the field.

NON-CHEMICAL CONTROL

Mechanical (pulling, cutting,	Hoeing, grubbing, or hand pulling before seed production may be used for small patches. These control methods must be repeated several times over the growing season (2 to 3 week intervals), and for several

disking)	years. Use gloves when handling leafy spurge due to the irritating effects of the latex. Mowing is generally not very effective for reducing perennial spurge infestations. However, mowing every 2 to 4 weeks can reduce seed production. Mowing may result in more uniform regrowth, which is more conducive to uniform and effective herbicide applications. Two cultivations in fall to a depth of at least 4 inches will help reduce infestations. (A single cultivation may only spread sprouting root fragments.) This should be conducted for 2 to 3 years. Cultivation twice each fall for 3 consecutive years completely controlled leafy spurge in North Dakota. In other habitats, heavy cultivation every 2 weeks during the growing season and every 3 weeks during the late summer and fall for 2 or more years will reduce top growth and regenerating buds, and eventually stress the root system. Clean equipment after cultivating to avoid transporting root fragments.
Cultural	Spurges are toxic to cattle and horses, but goats and sheep have been successfully used in control programs. Graze in spring when spurges emerge. Stock sheep at 3 to 6 head/acre month, or Angora goats at 12 to 16 head/acre month. These animals will not eradicate perennial spurges but can reduce the seedbank and allow grasses to become established. Animals should be held in a pen for 3 to 5 days before moving to a new area to prevent seed dispersal. Avoid overgrazing and excessive disturbance in pastures and rangelands, and reduce cattle stocking rates in areas of known infestations. An integrated strategy of early grazing followed by herbicide application to fall regrowth has proven more effective than either strategy alone. Burning does not significantly affect roots and typically stimulates the production of new shoots from root buds. Burning before release of biocontrol insects can help these insects to become established. Burning does not appear to harm biocontrol populations once established.
Biological	Fifteen non-indigenous insect species have been approved for release in the United States for the control of leafy spurge. Five flea beetles (*Aphthona* spp.) have been established in the Great Plains and Pacific Northwest. *Aphthona nigriscutis* and *A. czwalinae/lacertosa* impact the plant by ovipositing eggs at the base of the plant; the larvae feed on the roots, increasing plant morbidity, reducing plant health and creating pathways for the introduction of plant pathogens. Several moths (*Chamaesphecia* spp. and *Hyles euphorbiae*) and a stem-boring beetle (*Obera erythrocephala*) also are being tested. *Aphthona* spp. flea beetles have produced the greatest impact on leafy spurge, reducing stem densities by as much as 80 to 90% over large areas. This has not occurred in all areas. It is hoped that *Aphthona* spp. and other insects may eventually provide long term leafy spurge suppression over much of the western United States.

CHEMICAL CONTROL

The following specific use information is based on published papers and reports by researchers and land managers. The information here is primarily from information on the management of *E. esula*. It is considered that the same effects would also occur with *E. oblongata* and *E. terracina*. Other trade names may be available, and other compounds also are labeled for this weed. Directions for use may vary between brands; see label before use. Herbicides are listed by mode of action and then alphabetically. The order of herbicide listing is not reflective of the order of efficacy or preference.

GROWTH REGULATORS	
2,4-D Several names	**Rate:** 1 to 6 qt product/acre (0.95 to 5.7 lb a.e./acre) **Timing:** Postemergence at flowering in early summer, or on fall regrowth. **Remarks:** 2,4-D is broadleaf-selective and has no soil activity. It can prevent seed formation but does not provide complete kill and will require multiple treatments. One qt product/acre will suppress seed production, 6 qt will control shoots. Do not apply when outside temperatures exceed 80°F.
Aminocyclopyrachlor + chlorsulfuron *Perspective*	**Rate:** 4.75 to 8 oz product/acre **Timing:** Postemergence in spring up to flowering, or in fall rosette stage. **Remarks:** *Perspective* provides broad-spectrum control of many broadleaf species. Although generally safe to grasses, it may suppress or injure certain annual and perennial grass species. Do not treat in the root zone of desirable trees and shrubs. Do not apply more than 11 oz product/acre per year. At this high rate, cool-season grasses will be damaged, including bluebunch wheatgrass. Not yet labeled for grazing lands. Add an adjuvant to the spray solution. This product is not approved for use in California and some counties of Colorado (San Luis Valley).
Dicamba *Banvel, Clarity*	**Rate:** 1 to 2 qt product/acre (1 to 2 lb a.e./acre) **Timing:** Postemergence in spring to early summer.

	Remarks: Dicamba is a broadleaf-selective herbicide often combined with other active ingredients. May require 3 consecutive years of treatment.
Dicamba + 2,4-D amine	**Rate:** 1 qt/acre each product **Timing:** Postemergence in spring at flower emergence and/or to fall regrowth. **Remarks:** Add non-ionic surfactant at 0.25%. Do not apply when outside temperatures will exceed 80°F.
Fluroxypyr *Vista XRT*	**Rate:** 22 oz product/acre (7.7oz a.e./acre) **Timing:** Postemergence when weeds are small and rapidly growing. **Remarks:** This rate provides only suppression of leafy spurge.
Picloram *Tordon 22K*	**Rate:** 1 to 2 qt product/acre (0.5 to 1 lb a.e./acre) **Timing:** Applications postemergence at true flower stage are most effective but it can also be applied in fall. The 1 qt product/acre rate may need to be applied annually for 3 to 4 years. *Tordon 22K* at 2 qt/acre can be applied every other year. **Remarks:** Picloram is one of the most effective herbicides for this weed. Most broadleaf plants are susceptible, but relatively safe on established grasses. Use non-ionic surfactant at 0.25%. Picloram has a long residual activity and some have reported that it may injure young or germinating grasses. Do not apply near trees. *Tordon 22K* is a federally restricted use pesticide. Not registered for use in California.
Picloram + 2,4-D	**Rate:** 1 to 1.5 pt picloram product/acre + 2 to 3 pt 2,4-D product/acre **Timing:** Postemergence in spring, at true flowering. Apply for 3 to 5 consecutive years. **Remarks:** See picloram, 2,4-D.
AROMATIC AMINO ACID INHIBITORS	
Glyphosate *Roundup, Accord XRT II,* and others	**Rate:** 1 pt product (*Roundup ProMax*)/acre (0.56 lb a.e./acre) applied three times in a growing season, or 2 pt product (*Roundup ProMax*)/acre (1.1 lb a.e./acre) applied twice in a growing season **Timing:** Postemergence in split applications (June, July, August at low rate, or June and July at high rate), or apply a high rate in fall followed by spring treatment with glyphosate or 2,4-D. **Remarks:** Glyphosate is a nonselective herbicide. It has no soil activity. Its effectiveness is increased by addition of ammonium sulfate. Glyphosate is more effective when coupled with revegetation with competitive perennial grasses.
Glyphosate + 2,4-D	**Rate:** 1 pt glyphosate product /acre + 1.5 pt/acre of 2,4-D product (glyphosate at 6 oz a.e./acre + 2,4-D at 11 oz a.e./acre) **Timing:** Postemergence in late spring just before seed set. **Remarks:** This tank mix appears to provide synergistic control. Treatment should be repeated for 3 years. See remarks for glyphosate, 2,4-D.
BRANCHED-CHAIN AMINO ACID INHIBITORS	
Imazapic *Plateau*	**Rate:** 8 to 12 oz product/acre (2 to 3 oz a.e./acre) **Timing:** Preemergence or early postemergence in fall when plants begin to grow but before hard freeze. More than one treatment will likely be necessary. **Remarks:** Imazapic has mixed selectivity and some soil residual activity. It tends to favor members of the Asteraceae and some grasses. Use a methylated seed oil surfactant at 0.25%. Not registered for use in California.
Imazapyr *Arsenal, Habitat, Stalker,* *Chopper, Polaris*	**Rate:** 1 to 1.5 pt product/acre (4 to 6 oz a.e./acre) **Timing:** Postemergence at flowering. **Remarks:** Imazapyr has a long residual activity and leaves more bare ground than other treatments, even a year after application. It is nonselective.
Other products	Quinclorac (*Paramount*) is extremely selective from 8 to16 oz product/acre applied in spring at flowering. *Overdrive* in combination with the low rate of quinclorac is also effective.

Ficus carica L.
Fig

Family: Moraceae
Range: In the western U.S., fig has only become a problem in California. It has also escaped cultivation in most southern and many eastern states.
Habitat: Riparian areas, canal banks, disturbed places, canyons, old homesteads, typically where soil moisture is available throughout the year.
Origin: Native to the Mediterranean region and introduced to California with the Spanish missionaries. Widely cultivated as an ornamental and for its edible fruits, but has escaped cultivation in some regions of California.
Impacts: In riparian areas and woodlands, fig can form dense clonal thickets which exclude native vegetation. Naturalized populations near cultivated fig can pollinate and reduce the marketability of cultivated fig harvests.
California Invasive Plant Council (Cal-IPC) Inventory: Moderate Invasiveness

Fig is a deciduous tree to 30 ft tall, with single or multiple trunks and smooth, pale grey bark. It has large, deeply lobed, palmate leaves and pear-shaped fruits. The leaves, stems, and immature fruits exude a milky white sap when cut. The sap is widely recognized to cause phytophotodermatitis, so full body covering should be used when handling plants. Trees often develop extensive shallow roots that produce new shoots, leading to dense clonal thickets.

Figs flower in several flushes from late winter through mid-summer. Male and female flowers are on separate trees, and pollination depends on an introduced species of wasp. The pear-shaped fruits mature from summer through fall. Fruits and seeds disperse primarily with birds. New plants from seed may bear fruit within 2 to 3 years. Fig also can reproduce vegetatively, from root sprouts or from stem fragments which are moved by floodwaters. Vegetative offspring may bear fruit within the first year.

NON-CHEMICAL CONTROL

Mechanical (pulling, cutting, disking)	Seedlings can be pulled or hoed. A weed wrench may be used to pull small trees, removing as much of the root as possible.
	Once trees are established, cut trees can resprout from the stump, roots, and stem fragments. Frequent repeated cutting to ground level may eventually deplete root reserves. However, this method has not been proven to successfully control naturalized populations.
Cultural	Grazing and burning are not effectively control methods for this species, as fig can resprout after damage by animals or fire.
Biological	Due to the cultivation of figs, there are no efforts to develop biological control agents for this species.

CHEMICAL CONTROL

The following specific use information is based on published papers and reports by researchers and land managers. Other trade names may be available, and other compounds also are labeled for this weed. Directions for use may vary between brands; see label before use. Herbicides are listed by mode of action and then alphabetically. The order of herbicide listing is not reflective of the order of efficacy or preference.

GROWTH REGULATORS

Triclopyr *Garlon 3A, Garlon 4 Ultra*	**Rate:** Stem injection treatment: 100% *Garlon 3A* (amine formulation). Basal bark or drizzle treatment: 10% to 25% *Garlon 4 Ultra*, mixed with crop oil concentrate or basal oil. Basal cut stump or cut stump treatment: 25% to 100% of either formulation.
	Timing: These treatments are probably most effective late in the growing season but before leaves fall. When stem injections are applied too early in the season, the sap will exude the herbicide back out the cut regions and reduce control.
	Remarks: Triclopyr is a broadleaf-selective herbicide with very low soil residual. Basal bark treatments

	have been shown effective in experimental trials. Other treatments are expected to be effective based on work with other woody species.
AROMATIC AMINO ACID INHIBITORS	
Glyphosate *Roundup, Accord XRT II*, and others	**Rate:** Stem injection treatment: 100% concentrated product. Spot foliar treatment: 10% v/v in water to resprouts. Cut stump treatment: 20 to 100% v/v in water. **Timing:** These treatments are probably most effective late in the growing season but before leaves fall. **Remarks:** Glyphosate has no soil activity and is nonselective. This use is expected to be effective based on work with other woody species.
BRANCHED-CHAIN AMINO ACID INHIBITORS	
Imazapyr *Arsenal, Habitat, Stalker, Chopper, Polaris*	**Rate:** Stem injection treatment: 100% *Habitat* or *Arsenal*. Basal bark or drizzle treatment: 13% to 25% *Chopper*, mixed with crop oil. Cut stump treatment: 20% to 100% of any formulation. **Timing:** These treatments are probably most effective late in the growing season but before leaves fall. **Remarks:** Imazapyr has fairly long soil residual activity and is nonselective. This use is expected to be effective based on work with other woody species.

Foeniculum vulgare Mill.

Fennel

Family: Apiaceae
Range: Particularly a problem in California, but found throughout many western states including Washington, Oregon, Arizona, Nevada, Utah, New Mexico and Texas.
Habitat: Fennel prefers open disturbed areas and has invaded roadsides, slopes, fields, grasslands, coastal scrub, riparian and wetland areas, and other natural communities, particularly in coastal regions.
Origin: Native to southern Europe. Birds and rodents consume the seeds, and feral pigs relish the roots.
Impacts: Established plants are competitive, and soil disturbance facilitates the development of dense stands, which can exclude native vegetation in some areas. Unlike the weedy form, cultivated varieties are seldom invasive. Fennel is also rated as a noxious weed in some regions of Australia.
California Invasive Plant Council (Cal-IPC) Inventory: High Invasiveness

Fennel is an aromatic perennial with a deep thick taproot. Plants can grow to 10 ft tall, with finely dissected leaves divided into numerous thread-like segments. Foliage and seeds have a strong licorice or anise scent, especially when crushed. Different varieties are cultivated as a spice or vegetable, for an essential oil used to flavor foods, and in some countries, for medicinal purposes.

The inflorescence consists of several flat-topped compound umbels of small yellow flowers. Fruits separate into two halves (mericarps) at maturity. Plants reproduce by seed and sometimes vegetatively from root or crown fragments. Seeds are dispersed with water, soil movement, animals, human activities, and as a seed contaminant. Seeds appear to survive several years under field conditions. Fragmentation of roots and crowns may occur during flooding events, mudslides, or agricultural operations. New shoots grow from the crown and lower portions of overwintering stems in mid-winter to early spring.

NON-CHEMICAL CONTROL

Mechanical (pulling, cutting, disking)	Hand chop small infestations (large fennel plants have a very substantial root, so this labor intensive). Slashing just before flowering may kill the plants, or repeat slashing of regrowth may be needed. Even if plants recover, slashing the stems at flowering will prevent seed set. The use of a mattock to remove the plant can also be successful, but is very labor intensive. Digging out individual plants is also possible, but also labor intensive.
	Deep cultivation will also kill the plants but is not practical in most situations.
Cultural	Grazing will not control fennel and often spreads the population. Burning is not effective, as fennel will quickly recover following the fire. However, fall burns followed by herbicide treatment the following two springs reduced fennel cover. Burning can also stimulate the seedbank to germinate, which can reduce the number of years necessary for control.
Biological	Because fennel is the same species as the cultivated fennel, there are no biological control agents available.

CHEMICAL CONTROL

The following specific use information is based on reports by researchers and land managers. Other trade names may be available, and other compounds also are labeled for this weed. Directions for use may vary between brands; see label before use. Herbicides are listed by mode of action and then alphabetically. The order of herbicide listing is not reflective of the order of efficacy or preference.

GROWTH REGULATORS	
2,4-D Several names	**Rate:** 0.25 to 0.5% of v/v solution for spot treatment
	Timing: Postemergence to fully developed leaves but before flowering.
	Remarks: 2,4-D will damage other broadleaf species and is not as effective as triclopyr.

Triclopyr *Garlon 3A, Garlon 4 Ultra*	**Rate:** Broadcast foliar treatment: 1 to 2 qt product/acre (1 to 2 lb a.e./acre). Spot treatment: 0.5 to 1% v/v solution **Timing:** Postemergence to fully developed leaves but before flowering. The best treatment timing is during the wet season from late February to early March. For spot treatment, lower rates can be used early in the season. **Remarks:** Triclopyr is a broadleaf herbicide that is the standard for fennel control. It is very effective and can also be used in combination with glyphosate at 1 lb a.e./acre each.
AROMATIC AMINO ACID INHIBITORS	
Glyphosate *Roundup, Accord XRT II*, and others	**Rate:** Broadcast foliar treatment: 5 pt product (*Roundup ProMax*)/acre (2.8 lb a.e./acre). Spot treatment: 2 to 5% v/v solution **Timing:** Postemergence to fully developed leaves but before flowering. Control is less effective once plant has bolted. **Remarks:** Glyphosate is nonselective. It gives very effective control and can also be used in combination with triclopyr at 1 lb a.e./acre each.

Galega officinalis L.
Goatsrue

Photo courtesy of Corey Ransom

Family: Fabaceae
Range: Oregon, Washington, Utah and Colorado.
Habitat: Cropland, irrigation waterways, pastures, fence lines, roadways and wet marsh areas. Prefers full sun, but will tolerate light shade.
Origin: Native to the Middle East, but it has been naturalized in Europe, western Asia, and western Pakistan. Goatsrue has been cultivated as a forage crop, an ornamental, a bee plant and as green manure. Was introduced in 1891 for evaluation as a forage crop in Utah and escaped cultivation.
Impacts: Goatsrue can form monotypic stands in wetlands, displacing native and beneficial plants and reducing food and nesting resources for wetland wildlife. Despite its sometime use as forage for livestock, the stems and leaves of goatsrue contain a poisonous alkaloid that can be fatal to humans, sheep and cattle if ingested.
Western states listed as Noxious Weed: California, Nevada, Oregon, Washington. Federally listed as a noxious weed

Goatsrue is a perennial herbaceous plant 2 to 6 ft tall. It closely resembles the native plant wild licorice (*Glycyrrhiza lepidota*), so care should be taken to accurately identify it before implementing control practices. Goatsrue forms dense crowns capable of regenerating for several seasons. It has hollow, tubular stems with alternate odd pinnately compound leaves that have 13 to 21 leaflets.

Its pea-like flowers vary from light purple to white and are clustered (20 to 50 flowers) at the tops of the stems and from leaf axils, resembling a vetch. Plants produce pods about 1 inch long with dull yellow, bean-shaped seeds about 2.5 to 3 mm long. Plants spread primarily by seed and can produce up to 15,000 seed pods per plant. Seeds disperse mainly by falling near the parent plant, but can move longer distances through the activity of humans and animals, when ingested. Seeds can remain viable in the soil for at least 5 to 10 years.

NON-CHEMICAL CONTROL

Mechanical (pulling, cutting, disking)	Small populations can be manually removing by digging or pulling up individual plants, including as much of the root system as possible, followed by frequent removal of root sprouts and seedlings. Plants will resprout without repeated effort.
	Mowing, clipping and cutting are not effective when used alone. Plants will flower and produce seeds even when cut short. Seed pods can be clipped and disposed of to help prevent the spread of seeds into uninfested areas. Mechanical measures can be followed with herbicide treatment when the plants regrow.
Cultural	Grazing is not a control option, as the plant is very toxic to livestock.
	Burning is not an effective control method, as fire stimulates root sprouting.
Biological	A rust fungus, *Uromyces galegae*, has been investigated as a biological control agent for goatsrue in Chile. Goatsrue had a high susceptibility to the rust, which decreased seed production, while other legume species were unaffected. After this discovery the fungus was distributed across Chile. The south central zone of the country had the best results due to a climate that promoted distribution and growth of the rust fungus. Currently, no registered biocontrol agents for goatsrue are available in the United States.

CHEMICAL CONTROL
The following specific use information is based on published papers and reports by researchers and land managers. Other trade names may be available, and other compounds also are labeled for this weed. Directions for use may vary between brands; see label before use. Herbicides are listed by mode of action and then alphabetically. The order of herbicide listing is not reflective of the order of efficacy or preference.

GROWTH REGULATORS	
2,4-D	**Rate:** 2 qt product/acre (1.9 lb a.e./acre)
Several names	**Timing:** Postemergence from bud stage to full flower in early to mid-summer.

	Remarks: Broadleaf herbicide with little soil residual activity. Repeated applications at the label rate may be effective. Do not apply ester formulations when outside temperatures exceed 80°F. It has been shown to be inconsistent in the control of goatsrue, which may be related to site differences.
Aminocyclopyrachlor + chlorsulfuron *Perspective*	**Rate:** 4 to 6 oz product/acre **Timing:** Postemergence when plants are in bud stage to full flower in early to mid-summer. **Remarks:** *Perspective* provides broad-spectrum control of many broadleaf species. Although generally safe to grasses, it may suppress or injure certain annual and perennial grass species. Do not treat in the root zone of desirable trees and shrubs. Do not apply more than 11 oz product/acre per year. At this high rate, cool-season grasses will be damaged, including bluebunch wheatgrass. Not yet labeled for grazing lands. Add an adjuvant to the spray solution. This product is not approved for use in California and some counties of Colorado.
Aminopyralid *Milestone*	**Rate:** 5 oz product/acre (1.25 oz a.e./acre) **Timing:** Postemergence when plants are in bud stage to full flower in early to mid-summer. **Remarks:** Broadleaf herbicide similar to picloram, but more selective and with shorter soil residual activity. It is very safe on most grasses, particularly postemergence. Broadcast applications will also provide preemergence control of germinating seeds. Very effective for the control of goatsrue.
Aminopyralid + metsulfuron *Opensight*	**Rate:** 2.5 to 3.3 oz product/acre **Timing:** Preemergence at 3.3 oz in fall or postemergence when plants are seedlings to rosettes. **Remarks:** Not registered for use in California.
Dicamba *Banvel, Clarity*	**Rate:** 2 qt product/acre (2 lb a.e./acre) **Timing:** Postemergence when plants are in bud stage to full flower in early to mid-summer. **Remarks:** Broadleaf herbicide with little soil residual activity. Very effective for the control of goatsrue. Mixtures of 2,4-D at 8 oz product/acre + dicamba at 4 oz product/acre applied twice during the growing season for 2 consecutive years will provide control of goatsrue. However, control was most effective when mechanical methods were combined with herbicide applications, by clipping plants when the initial growth reaches 2 ft tall, followed by spraying the regrowth at 2 ft.
Picloram *Tordon 22K*	**Rate:** 2 pt product/acre (0.5 lb a.e./acre) **Timing:** Postemergence when plants are in full flower in early to mid-summer. **Remarks:** One of the most effective chemical control options. Long soil residual, so broadcast applications will also control germinating seed. It is safe on most grasses. Restricted use herbicide; not registered for use in California.
Triclopyr *Garlon 3A*	**Rate:** 2 qt product/acre (1.5 lb a.e. (*Garlon 3A*)/acre) **Timing:** Postemergence when plants are in bud stage to full flower in early to mid-summer. **Remarks:** Triclopyr is a broadleaf herbicide that has no soil residual activity. It has been shown to be very effective for the control of goatsrue.
BRANCHED-CHAIN AMINO ACID INHIBITORS	
Chlorsulfuron *Telar*	**Rate:** 1 oz product/acre (0.75 oz a.i./acre) **Timing:** Postemergence when plants are in bud stage to full flower in early to mid-summer. **Remarks:** Chlorsulfuron has some soil residual activity. It is primarily a broadleaf herbicide and has been shown to be very effective for the control of goatsrue.
Imazapyr *Arsenal, Habitat, Chopper, Stalker, Polaris*	**Rate:** 2 pt product/acre (8 oz a.e./acre) **Timing:** Postemergence when plants are in bud stage to full flower in early to mid-summer. **Remarks:** Imazapyr has a fairly long soil residual and is nearly nonselective, so may kill desirable competitors. It has been shown to be effective for the control of goatsrue, but less effective compared to other compounds, particularly the ALS inhibitors and some growth regulators.
Metsulfuron *Escort*	**Rate:** 1 oz product/acre (0.6 oz a.i./acre) **Timing:** Postemergence when plants are in bud stage to full flower in early to mid-summer. **Remarks:** Some soil residual activity. Generally safe on most grasses, and shown to be very effective for the control of goatsrue. Metsulfuron is not registered for use in California.

Genista monspessulana (L.) L.A.S. Johnson

French broom

Family: Fabaceae
Range: Along the Pacific coast from southern British Columbia to southern California.
Habitat: Grasslands, shrublands, oak woodlands, forest margins, coastal habitats, riparian corridors and disturbed sites such as roadsides, pasture, burned areas, or cleared forests. Grows under varied soil moisture conditions but seems to prefer siliceous soils. Unlike other brooms in California, it grows reasonably well on alkaline soils.
Origin: Native to the Mediterranean region and Azores. Introduced to the U.S. in the 1850s as an ornamental.
Impacts: Grows rapidly and forms dense stands that most wildlife find impenetrable and unpalatable. Dense stems inhibit regeneration of most other plant species, and the accumulation of woody biomass creates a fire hazard. Broom can fix nitrogen, enabling the plant to colonize and dominate areas with poor soil. Increased soil fertility gives a competitive advantage to other non-native weeds that thrive on high nitrogen levels.
Western states listed as Noxious Weed: California, Oregon, Washington (proposed)
California Invasive Plant Council (Cal-IPC) Inventory: High Invasiveness

French broom is an upright, evergreen shrub, commonly less than 10 ft tall. Stems are green, erect, dense, and covered with silky, silvery hairs. French broom is typically leafy as compared with Scotch or Spanish broom, which have few leaves. The leaves are composed of three leaflets 0.4 to 0.8 inch long, oblong to obovate, with length about twice the width, and upper and lower surfaces sparsely to densely covered with silky, silvery hairs.

French broom produces yellow, pea-shaped flowers in dense clusters of 4 to 10 flowers on short axillary shoots. Reproduction is by seed and plants begin flowering from 18 months to 3 years of age. The seeds are produced in small, flattened pods 0.5 to 1 inch long. Pods are dark brown when mature, contain 5 to 8 seeds, and are densely covered with appressed, long silky hairs. When mature, pods eject the seeds several feet from the plant. Seeds can remain viable in the soil for up to 30 years. Large soil seedbanks often accumulate, making long term control difficult. Shrubs may live for up to 30 years.

NON-CHEMICAL CONTROL

Mechanical (pulling, cutting, disking)	Hand pulling can remove seedlings and small shrubs, but this technique is generally not effective on established shrubs. For larger established shrubs, a weed wrench or other woody weed extractor can be used. Care must be taken to extract the entire root or stump sprouting will occur. Best results are achieved when soil is moist. Disturbing the soil can stimulate the seedbank.
	Cutting broom to the ground in spring before it flowers will reduce the number of seeds and will deplete the plant's energy reserves. Resprouts are common after treatment, but can be reduced by cutting broom at the end of the dry season. Cutting should be combined with an herbicide treatment or with multiple cuttings over a period of years. Cut shrubs at ground level with power or manual saws.
	Heavy equipment can be effectively used to control broom in areas where soil disturbance and nonselective species removal are not important considerations. Stumps remaining following such treatment will require herbicide application to prevent regrowth.
Cultural	Flowers and seeds of brooms contain quinolizidine alkaloids and can be toxic to humans and livestock when ingested. Foliage may be mildly toxic and is unpalatable to most livestock, except goats. Goats confined to a small area can help control resprouting stands after a cutting or burn treatment. Goats can be trained to be quite selective at least within the vegetation structure, for example they can effectively strip flowers.
	Burning alone is not an effective method for controlling broom. Although burning can remove debris, in many cases it can increase the population as it removes competitive vegetation, releases nutrients into the soil, and stimulates the germination of broom seeds in the soil. Burning is more effective when followed by herbicide application, subsequent burnings, and/or revegetation with desirable species. It is important to employ a control strategy following a burn; otherwise the broom population in subsequent years may become worse than before.
Biological	There are no USDA-approved biocontrol agents for French broom. The native pyralid moth, *Uresiphita reversalis*, defoliates some French broom, but plants grow new leaves after the larvae metamorphose. An insect introduced

> for control of Scotch broom, the Scotch broom bruchid (*Bruchidius villosus*), also attacks French broom.

CHEMICAL CONTROL

The following specific use information is based on published papers and reports by researchers and land managers. Other trade names may be available, and other compounds also are labeled for this weed. Directions for use may vary between brands; see label before use. Herbicides are listed by mode of action and then alphabetically. The order of herbicide listing is not reflective of the order of efficacy or preference.

GROWTH REGULATOR	
Picloram *Tordon 22K*	**Rate:** Broadcast foliar treatment: 2 qt product/acre (non-cropland) or 1 qt product per acre (rangeland) plus 0.25 to 0.5% v/v surfactant to thoroughly wet all leaves. **Timing:** Postemergence foliar treatments are best when plants are growing rapidly at or beyond early to full bloom stage. **Remarks:** High levels of picloram can give long-term soil activity for broadleaves. Picloram is a restricted use herbicide. It is not registered for use in California.
Triclopyr *Garlon 3A, Garlon 4 Ultra, Pathfinder II* Aminopyralid + triclopyr *Capstone*	**Rate:** Foliar treatment: 0.75 to 1.5% v/v solution of *Garlon 4 Ultra*, or 1 to 1.5% *Garlon 3A* and water plus 0.25 to 0.5% surfactant v/v to thoroughly wet all leaves. Low volume/thinline treatment: 10% v/v solution of *Garlon 4 Ultra* plus a 20% v/v seed oil in water. Basal cut stump treatment: 20% *Garlon 4 Ultra* v/v in water. Cut stump treatment: undiluted *Garlon 3A* or 50% *Garlon 3A* in water. Basal bark treatment: 20% v/v *Garlon 4 Ultra* in 20% v/v ethylated crop oil and water, or *Pathfinder II* as a ready-to-use formulation. Use *Capstone* at 8 to 9 pints per acre. **Timing:** Postemergence when plants are growing rapidly. Cut stump and basal bark treatments can be applied anytime although they are optimal if not applied when sap is rising in the early spring. **Remarks:** Triclopyr is a selective herbicide for broadleaf species and will not damage desirable grasses growing nearby. For cut stump treatments, cut stems horizontally at or near ground level and immediately apply *Garlon 3A* solution. Suckering from the roots typically occurs after cutting, but the treatment should control most resprouts. For basal bark treatment, spray the lower trunk, including the root collar, to a height of 12-15 inches from the ground; the spray should wet the lower stem but not to the point of runoff. Plants should not be cut for at least one month after basal bark treatments.
Triclopyr + 2,4-D *Crossbow*	**Rate:** For foliar treatment: 0.5 to 1.5% v/v solution of *Crossbow* and water to thoroughly wet all leaves. **Timing:** Apply when plants are growing rapidly. **Remarks:** *Crossbow* in water forms an emulsion (not a solution).
AROMATIC AMINO ACID INHIBITORS	
Glyphosate *Roundup, Accord XRT II*, and others	**Rate:** Spot treatment: 1.5 to 2% v/v solution of *Roundup ProMax* (or other trade name with similar concentration of glyphosate) in water to thoroughly wet all leaves. Low volume/thinline treatment: 10% v/v solution of *Roundup* (or other trade name) in water. Cut stump treatment: 25% v/v *Roundup* (or other trade name) in water; 50% can reduce resprouting but may exceed label rate if stands are dense. **Timing:** Postemergence when plants are growing rapidly. Foliar treatments should be made in late summer or early fall. For cut stump treatment, application in late summer, early fall or dormant season provides best control. Stumps should be treated immediately after cutting. **Remarks:** Glyphosate is a nonselective systemic herbicide. It gives good control with some resprouts. Plants should not be cut for at least 4 months after foliar treatments. Cut stump applications are made as described for triclopyr.
BRANCHED-CHAIN AMINO ACID INHIBITORS	
Imazapyr *Arsenal, Habitat, Stalker, Chopper, Polaris*	**Rate:** Cut stump treatment: 20% v/v solution of *Stalker* plus a 20% v/v ethylated crop oil in water or 20% *Habitat* v/v in 80% water carrier. Basal bark treatment: 20% v/v solution of *Stalker* plus a 20% v/v ethylated crop oil in water. **Timing:** Best when applied in late summer to early fall, but before leaf drop. **Remarks:** Imazapyr is a soil residual herbicide and may result in bare ground around trees for some time after treatment. Cut stump and basal bark applications are made as described for triclopyr. Plants should not be cut for at least 4 months after basal bark treatment. Other ALS inhibitors, including metsulfuron, have been used effectively to control French broom in Australia.

Geranium dissectum L.; cutleaf geranium
Geranium purpureum L.; little robin

Cutleaf geranium and little robin

Family: Geraniaceae
Range: Cutleaf geranium and other common species (Carolina geranium, dovefoot geranium) are found in many western states, as well as in most other northwestern, eastern, and southern states, and in some central states. In the United States, little robin is reported only from California, where it is rapidly expanding its range.
Habitat: Roadsides, fields, pastures, orchards, vineyards, landscaped areas, waste places, turf, disturbed open woodlands, shrublands, and other plant communities, and occasionally crop fields.
Origin: Both species are native to Europe.
Impact: Once established, they may displace native herbaceous species. Little robin appears to be the most significant threat to natural areas, as it can form a carpet under an oak woodland canopy.
California Invasive Plant Council (Cal-IPC) Inventory: *G. dissectum*, Limited Invasiveness

These geraniums are small herbaceous plants, annuals to biennials, growing prostate to erect, with palmate leaves and violet-pink flowers. They are widespread but generally minor weeds of wildlands. The mature plants have forked, hairy stems, hairy foliage, and slender taproots with fibrous lateral roots. Cutleaf geranium grows to about 3 ft tall with rough stem hairs and has highly dissected, round leaves with five to seven deep lobes that are dissected into narrow segments. Cutleaf geranium flowers in spring in warm areas and all the way to October in cooler, wetter areas. Little robin is a small (to 20 inches tall) geranium recently introduced into California. Its leaves appear compound, somewhat resembling a filaree leaf, with three or five pinnate-lobed leaflets.

Flowers have five violet-pink petals and are usually two per cluster. The fruits have a cranes-bill shape like filaree, which is related, but usually are not as long as filaree. Fruits split open and the seed coils on the elongated style. Most seeds land within a short distance of the parent plant, but some disperse to greater distances with animal movement, and as a seed contaminant, especially of clover seed. The seeds of cutleaf geranium can survive in the soil for up to 10 years and it is suspected that other geraniums may have similar seed longevity.

Other widespread non-native geraniums include Carolina geranium (*G. carolinianum* L.), which closely resembles cutleaf geranium, except that the hairs on the flower stalks are not glandular; and dovefoot geranium (*G. molle* L.), which is smaller than cutleaf geranium and has lobed leaves which are not as deeply dissected. There are several other less common non-native geraniums. Herb-robert (*G. robertianum* L.) is a state listed noxious weed in Washington.

NON-CHEMICAL CONTROL

Mechanical (pulling, cutting, disking)	Geraniums can be pulled, dug, or cultivated before they produce flowers and seeds.
Cultural	Grazing is not an effective means of control, though geranium is palatable to grazers.
	Burning is probably not effective, as it tends to increase the germination of other weeds in this family.
Biological	No biocontrol agents have been introduced for any *Geranium* species, due to the importance of the family as ornamentals and to the presence of native species in the genus in the western United States.

CHEMICAL CONTROL

The following specific use information is based on published papers and reports by researchers and land managers. Other trade names may be available, and other compounds also are labeled for this weed. Directions for use may vary between brands; see label before use. Herbicides are listed by mode of action and then alphabetically. The order of herbicide listing is not reflective of the order of efficacy or preference.

GROWTH REGULATORS	
2,4-D Several names	**Rate:** 0.5 to 2 pt product/acre (0.24 to 0.95 lb a.e./acre) **Timing:** Postemergence to rapidly growing plants up to flowering, the smaller the better. **Remarks:** 2,4-D is broadleaf-selective and has no soil activity. It may require repeat application. 2,4-D is not the most effective treatment, but is widely used because of low cost. Do not apply ester formulations when outside temperatures exceed 80°F. 2,4-D can also be used in a premix with picloram (*Grazon P+D*), but this formulation is not registered for use in California.
Aminocyclopyrachlor + chlorsulfuron *Perspective*	**Rate:** 3 to 4.5 oz product/acre **Timing:** Postemergence in spring up to flowering. **Remarks:** *Perspective* provides broad-spectrum control of many broadleaf species. Although generally safe to grasses, it may suppress or injure certain annual and perennial grass species. Do not treat in the root zone of desirable trees and shrubs. Do not apply more than 11 oz product/acre per year. At this high rate, cool-season grasses will be damaged, including bluebunch wheatgrass. Not yet labeled for grazing lands. Add an adjuvant to the spray solution. This product is not approved for use in California and some counties of Colorado (San Luis Valley).
Aminopyralid *Milestone*	**Rate:** 4 to 7 oz product/acre (1 to 1.75 oz a.e./acre) **Timing:** Postemergence in spring from rosette to flowering stages, or in fall to seedlings and rosettes. **Remarks:** Aminopyralid is a broadleaf-selective herbicide with soil residual activity. It can also be used in a premix with 2,4-D (*Forefront HL*), though geranium is not included on the *Forefront HL* label.
Aminopyralid + metsulfuron *Opensight*	**Rate:** 1.5 to 2 oz product/acre **Timing:** Postemergence in spring from rosette to flowering stages, or in fall to seedlings and rosettes. **Remarks:** See label for information on use of hay. Not registered for use in California.
Dicamba *Banvel, Clarity*	**Rate:** 8 to 32 oz product/acre (0.25 to 1 lb a.e./acre); 8 to 16 oz for rosettes, up to 32 oz product/acre for bolting plants **Timing:** Postemergence to rapidly growing plants up to flowering. Smaller plants are more effectively controlled. **Remarks:** Dicamba is a broadleaf-selective herbicide often combined with other active ingredients. It is effective earlier in the season than 2,4-D. It is also effective when tank-mixed with 2,4-D (0.75 lb a.e./acre dicamba + 0.25 lb a.e./acre 2,4-D). Dicamba has very limited soil residual. Avoid drift to sensitive crops. Do not apply when outside temperatures exceed 80°F. Dicamba is available mixed with diflufenzopyr in a formulation called *Overdrive*. The combination is broadleaf-selective and safe on most grasses. This has been reported to be effective on Carolina geranium, and would be expected to have similar results on other geranium species. Diflufenzopyr is an auxin transport inhibitor which causes dicamba to accumulate in shoot and root meristems, increasing its activity. *Overdrive* is applied postemergence to rapidly growing plants at 4 to 8 oz product/acre. Higher rates should be used on large annuals and biennials or when treating perennial weeds. Add a non-ionic surfactant to the treatment solution at 0.25% v/v or a methylated seed oil at 1% v/v solution.
Fluroxypyr *Vista XRT*	**Rate:** 22 oz product/acre (7.7 oz a.e./acre) **Timing:** Postemergence to rapidly growing plants. **Remarks:** Fluroxypyr provides only suppression of geranium. It is broadleaf-selective and safe on most grasses.
AROMATIC AMINO ACID INHIBITORS	
Glyphosate *Roundup, Accord XRT II,* and others	**Rate:** 2 to 3 pt product (*Roundup ProMax*)/acre (1.1 to 1.7 lb a.e./acre) **Timing:** Postemergence to rapidly growing plants. **Remarks:** Glyphosate has no soil activity and is a nonselective herbicide. Repeat applications may be

	necessary. Effectiveness is increased by addition of ammonium sulfate.
BRANCHED-CHAIN AMINO ACID INHIBITORS	
Imazapic *Plateau*	**Rate:** 8 to 12 oz product/acre (2 to 3 oz a.e./acre) **Timing:** Most effective postemergence. **Remarks:** Not registered for use in California.
Imazapyr *Arsenal, Habitat, Stalker, Chopper, Polaris*	**Rate:** 1.5 to 2 pt product/acre (0.375 to 0.5 lb a.e./acre) **Timing:** Preemergence or postemergence **Remarks:** Imazapyr is a nonselective herbicide. It has a relatively long soil residual activity.
Metsulfuron *Escort*	**Rate:** 0.33 to 0.5 oz product/acre (0.2 to 0.3 oz a.i./acre) **Timing:** Postemergence to young, rapidly growing plants in spring before flowering, or in fall to new rosettes **Remarks:** Metsulfuron has mixed selectivity and is generally safe on grasses. Use a surfactant. It can be tank-mixed with 2,4-D and/or dicamba and has some soil residual activity. Not registered for use in California.
Metsulfuron + chlorsulfuron *Cimarron X-tra*	**Rate:** 0.5 oz product/acre **Timing:** Postemergence to rapidly growing plants before flowering. **Remarks:** Safe on established grasses. Not registered for use in California.
Sulfometuron *Oust* and others	**Rate:** 1.33 to 2 oz or 3 to 5 oz product/acre (1 to 1.5 oz or 2.25 to 3.75 oz a.i./acre). Rate depends on environmental conditions. **Timing:** Preemergence or early postemergence, during the rainy season when weeds are germinating or rapidly growing. Use the lower rate range for areas receiving less than 20 inches precipitation annually and the higher rate range for those receiving greater than 20 inches annual precipitation. **Remarks:** Sulfometuron has mixed selectivity, but is fairly safe on native perennial grasses, especially wheatgrass. Other desirable grasses may be stunted, stressed, or injured. Good for revegetation use. It has fairly long soil residual activity. Do not let spray drift onto sensitive crops. Use lower rates in arid areas.
Sulfometuron + chlorsulfuron *Landmark XP*	**Rate:** 1.5 oz product/acre **Timing:** Preemergence or early postemergence. **Remarks:** See sulfometuron. Rates are based on rates reported for Carolina geranium.

Glyceria declinata Brébiss.
Waxy mannagrass

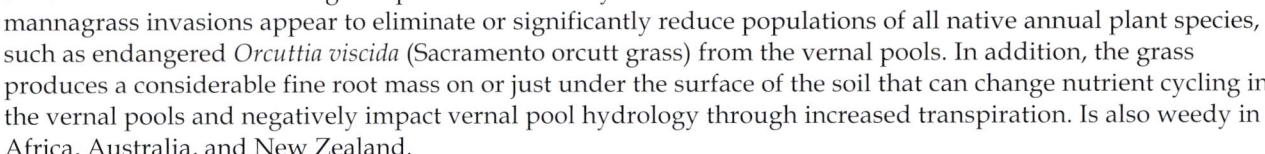
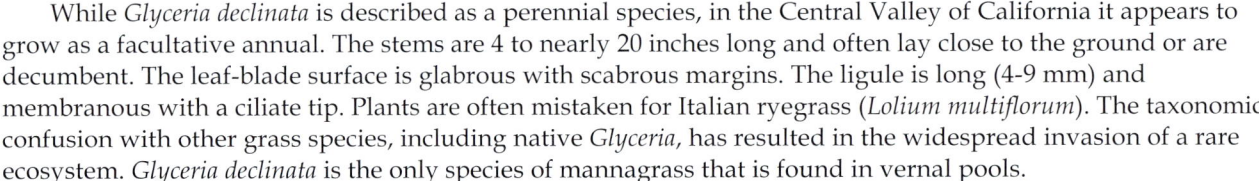

Family: Poaceae
Range: Currently only found in California.
Habitat: Moist canyons, meadows, swales, ditches, and stock ponds and, more recently, from vernal pool areas of California. Adapted to long periods of inundation.
Origin: Native to Europe, where it can also be found in vernal pools. First identified in California in 1953.
Impacts: Can compromise the integrity of vernal pools and threaten endemic and endangered plants. Dense waxy mannagrass invasions appear to eliminate or significantly reduce populations of all native annual plant species, such as endangered *Orcuttia viscida* (Sacramento orcutt grass) from the vernal pools. In addition, the grass produces a considerable fine root mass on or just under the surface of the soil that can change nutrient cycling in the vernal pools and negatively impact vernal pool hydrology through increased transpiration. Is also weedy in Africa, Australia, and New Zealand.
California Invasive Plant Council (Cal-IPC) Inventory: Moderate Invasiveness

While *Glyceria declinata* is described as a perennial species, in the Central Valley of California it appears to grow as a facultative annual. The stems are 4 to nearly 20 inches long and often lay close to the ground or are decumbent. The leaf-blade surface is glabrous with scabrous margins. The ligule is long (4-9 mm) and membranous with a ciliate tip. Plants are often mistaken for Italian ryegrass (*Lolium multiflorum*). The taxonomic confusion with other grass species, including native *Glyceria*, has resulted in the widespread invasion of a rare ecosystem. *Glyceria declinata* is the only species of mannagrass that is found in vernal pools.

The inflorescence is an open panicle, 2 to 12 inches long, and is linear or lanceolate. The spikelets can be ascending, appressed or nodding on the stem. In the Central Valley of California the spikelets mature from late April through May. Seeds are dispersed by floating on the surface of water or by becoming attached to waterfowl and grazing animals. Waterfowl, in particular, are strongly attracted to maturing plants and strip the seed from the culms with their bills. This is likely the main method of long-distance seed dispersal.

NON-CHEMICAL CONTROL

Mechanical (pulling, cutting, disking)	Because of the sensitive nature of vernal pools, great caution should be taken when implementing any control method or eradication program. To be successful in an eradication program it is critical to deplete the seedbank of waxy mannagrass. Repeated hand pulling before plants produce viable seed can be effective in vernal pools, but extreme caution should be taken to minimize disturbance or soil compaction. In addition, surrounding vernal pools and swales should also be weeded to prevent the rapid reintroduction of seed. It will probably require several years of repeated hand pulling to deplete the soil seedbank. To accomplish this, it is critical to prevent any escaped plants from reproducing. String trimmers can reduce cover of plants, as well as seed production. If only a few vernal pools on a site are severely affected and a source of water is available it may be beneficial to wet the soil very early in fall to stimulate germination, then use string trimmers to kill the young plants in their upright terrestrial stage.
Cultural	It is important to prevent the spread of seed from one area to another through animal or human activity. It is also critical that all equipment used in both natural and created vernal pools be free of plant fragments or soil that may be contaminated with seed. Artificial or stock ponds that attract waterfowl should be eliminated or replaced to reduce site desirability. Minimizing additional nutrient additions can also discourage waxy mannagrass from proliferating in sensitive areas. Because of the sensitivity of most sensitive habitats, particularly vernal pools, grazing or burning are not recommended control options.
	In some cases grazing has been shown to increase the populations due to reduced competition with other native species. However, heavy continuous grazing can reduce the cover of waxy mannagrass, but may increase nutrients and favor other species not normally found in vernal pool areas, including algae blooms. Cattle grazing should be managed in vernal pool landscapes to reduce phosphorous loading.
Biological	There are no biological control options available for *Glyceria declinata*, as there are many important native

species of *Glyceria* in the western United States.

CHEMICAL CONTROL

The following specific use information is based on reports by researchers and land managers. Other trade names may be available, and other compounds also are labeled for this weed. Directions for use may vary between brands; see label before use. Herbicides are listed by mode of action and then alphabetically. The order of herbicide listing is not reflective of the order of efficacy or preference.

LIPID SYNTHESIS INHIBITORS	
Clethodim *Select*, *Envoy*	**Rate:** 1 to 2 pt product (*Envoy*)/acre (2 to 4 oz a.i./acre) **Timing:** Postemergence before plants produce viable seeds. **Remarks:** In vernal pools, grass herbicides are only recommended when waxy mannagrass is the only grass species present. If endangered grasses such as *Orcuttia viscida* (Sacramento orcutt grass) are also present, a grass-specific herbicide should never be used. Note that *Envoy* formulation is 1 lb a.i./gallon, *Select* is 2 lb a.i./gallon.
Fluazifop *Fusilade*	**Rate:** 1 pt product/acre (4 oz a.e./acre) **Timing:** Postemergence before plants produce viable seeds. **Remarks:** In vernal pools, grass herbicides are only recommended when waxy mannagrass is the only grass species present. If endangered grasses such as *Orcuttia viscida* (Sacramento orcutt grass) are also present, a grass-specific herbicide should never be used.
Sethoxydim *Poast*	**Rate:** 1.67 pt product/acre (5 oz a.i./acre) **Timing:** Postemergence before plants produce viable seeds. **Remarks:** In vernal pools, grass herbicides are only recommended when waxy mannagrass is the only grass species present. If endangered grasses such as *Orcuttia viscida* (Sacramento orcutt grass) are also present, a grass-specific herbicide should never be used.
AROMATIC AMINO ACID INHIBITORS	
Glyphosate *Aquamaster*, *Rodeo*	**Rate:** Broadcast foliar treatment: 1 to 2 pt product (*Roundup ProMax*)/acre (0.56 to 1.1 lb a.e./acre) or 1 to 2 pt product (*Rodeo* or *Aquamaster*)/acre (0.5 to 1 lb a.e./acre) near aquatic sites. Spot treatment: 1% v/v solution. Wiper treatment: 33 to 50% of concentrated product. **Timing:** Postemergence before plants produce viable seeds. **Remarks:** In vernal pools, broadcast application of glyphosate would cause too much damage to sensitive native species. However, a wiper application to waxy mannagrass may give selectivity.

Gypsophila paniculata L.
Baby's-breath

Family: Caryophyllaceae
Range: All western states except Arizona and New Mexico.
Habitat: Disturbed sites, especially on sandy soils and in open, grassy places. It can tolerate considerable variation in temperature and moisture, including Mediterranean, semi-desert, and cold winter climates. Baby's breath is common along roadsides, ditches, and fences. Infestations are often found near homesteads, cemeteries, and residential areas where it escaped cultivation.
Origin: Native to Eurasia. Baby's breath was introduced as an ornamental for its showy white inflorescences. Dried plants were and still are used in flower arrangements.
Impacts: Baby's breath is an ornamental species that has escaped cultivation. Once established, it can form dense stands and is difficult to control. In pastures and rangeland, it competes with forage species and decreases hay forage quality.
Western states listed as Noxious Weed: California, Washington

Baby's breath is an herbaceous perennial that grows up to 3 to 4 ft tall. The stems are slender, erect to spreading, swollen at the nodes. The leaves are opposite, narrowly lanceolate and glabrous. It has a thick crown and rhizome, but the rhizome is not creeping. The new shoots grow from the crown and rhizome, but not the roots. Roots grow deep and can penetrate soil to a depth of 13 ft.

The inflorescence is open, panicle-like, and lacks bracts below the flowers. White flowers are numerous, mostly 6-8 mm wide. The seeds are black and 1.5 to 2 mm long. Plants reproduce primarily by seed. Most seed disperses near the parent, but plants can break off at ground level and tumble with the wind, dispersing seed greater distances. Plants average about 14,000 seeds. Seeds typically lack or have a short dormancy period and may survive in the soil for 1 to 2 years.

NON-CHEMICAL CONTROL

Mechanical (pulling, cutting, disking)	Suppressing the root system and preventing seed production is the management goal for established plants. Hand pulling mature baby's breath plants is difficult. The root system must be severed below the thickened crown and rhizome. This may require digging to a depth of 6 inches to 1 ft. Regrowth is usually rare if the complete crown and rhizome are removed.
	Plants regrow following mowing. Frequent mowing along roadsides has not resulted in a noticeable decrease in populations in northeastern California.
	Plants can tolerate shallow infrequent tillage, but plants rarely persist in fields that undergo frequent deep tillage.
Cultural	Baby's breath has low nutritional value and medium palatability for grazing. Populations often increase in lightly grazed systems.
	Plants left to grow in small patches for their ornamental and horticultural attraction often spread over time and are the source of most infestations. Promoting competitive vegetation can slow spread and help prevent establishment.
Biological	No biological controls are available for the control of baby's breath.

CHEMICAL CONTROL
The following specific use information is based on published papers and reports by researchers and land managers. Other trade names may be available, and other compounds also are labeled for this weed. Directions for use may vary between brands; see label before use. Herbicides are listed by mode of action and then alphabetically. The order of herbicide listing is not reflective of the order of efficacy or preference.

GROWTH REGULATORS	
2,4-D	**Rate:** 1 to 2 qt product/acre (0.95 to 1.9 lb a.e./acre)

Several names	**Timing:** Postemergence to spring rosettes or to bolting plants with green basal leaves. **Remarks:** 2,4-D provides only suppression of baby's breath. The best control is achieved at high rates applied to bolting plants. 2,4-D can be tank-mixed with chlorsulfuron for quicker burndown. It is broadleaf-selective and is safe on most grasses. 2,4-D has minimal soil activity. Do not apply ester formulation when outside temperatures exceed 80°F.
Aminopyralid + metsulfuron *Opensight*	**Rate:** 2.5 to 3 oz product/acre **Timing:** Preemergence at 3.3 oz in fall, or postemergence when target plants are in the seedling to rosette stage. **Remarks:** Not registered for use in California.
Dicamba *Banvel, Clarity*	**Rate:** 1.5 to 3 pt product/acre (0.75 to 1.5 lb a.e./acre). **Timing:** Postemergence to spring growth or to bolting plants with green basal leaves. **Remarks:** Dicamba is most effective at high rates. It is a broadleaf-selective herbicide and is safe on most grasses. Dicamba has very little soil activity.
AROMATIC AMINO ACID INHIBITORS	
Glyphosate *Roundup, Accord XRT II*, and others	**Rate:** 2 qt product (*Roundup ProMax*)/acre (2.25 lb a.e./acre). Spot treatment 1.5 % v/v solution. **Timing:** Postemergence to spring growth or to bolting plants with green basal leaves. **Remarks:** Glyphosate has no soil activity. It is nonselective and also kills most non-target plants, creating bare ground conditions that are susceptible to weed recruitment. In areas with desirable vegetation, use spot treatment. Glyphosate is a good control option if reseeding is planned shortly after application, as it will not injure seedlings emerging after application. Add a surfactant when it is not already included in the herbicide formulation (e.g., *Rodeo, Aquamaster*).
BRANCHED-CHAIN AMINO ACID INHIBITORS	
Chlorsulfuron *Telar*	**Rate:** 1 to 2.6 oz product/acre (0.75 to 1.95 oz a.i./acre) **Timing:** Postemergence to spring growth or to bolting plants with green basal leaves. **Remarks:** Chlorsulfuron is an effective treatment. At higher rates (2.6 oz product/acre) it will cause grass injury. Always use a surfactant.
Imazapic *Plateau*	**Rate:** 8 to 12 oz product/acre (2 to 3 oz a.e./acre) **Timing:** Postemergence to spring growth or to bolting plants with green basal leaves. **Remarks:** Imazapic gives effective control with soil residual activity. The addition of 2,4-D at 1 to 2 pt product/acre can aid in burndown. It is selective to most native grasses. Higher rates may suppress seedings of some cool-season grasses. Add a methylated seed oil. Imazapic can be used in combination with glyphosate (premix trade name of *Journey*). Imazapic is not registered for use in California.
Metsulfuron *Escort*	**Rate:** 1 to 2 oz product/acre (0.6 to 1.2 oz a.i./acre) **Timing:** Postemergence to spring growth or to bolting plants with green basal leaves. **Remarks:** Use a non-ionic or silicone surfactant to improve control. Metsulfuron is not registered for use in California.

Halogeton glomeratus (M. Bieb.) C.A. Meyer

Halogeton

Family: Chenopodiaceae
Range: Throughout the dry arid regions of the western U.S.
Habitat: Disturbed open sites, dry lakebeds, shrublands, roadsides, typically where native vegetation is sparse. Inhabits arid and semi-arid regions, especially where winters are cold. Primarily adapted to alkaline and saline soils.
Origin: Native to the cold desert regions of Eurasia.
Impacts: Plant tissues accumulate salts from lower soil horizons. The salts leach from dead plant material, increasing topsoil salinity and favoring halogeton seed germination and seedling establishment. The foliage contains variable amounts of soluble sodium oxalates and can be fatally toxic to livestock, especially sheep, when ingested in quantity. Impacts grazing capacity as animals generally avoid consuming the bitter-tasting foliage if more palatable forage is available. Caution should be exercised, however, when unloading hungry livestock onto halogeton-infested rangeland.
Western states listed as Noxious Weed: Arizona, California, Colorado, New Mexico, Oregon
California Invasive Plant Council (Cal-IPC) Inventory: Moderate Invasiveness

Halogeton is an erect winter or summer annual to 1.5 ft tall, with small fleshy leaves. Stems are usually tinged reddish or purple and the leaves are alternate, sessile, dull green to bluish-green, with a stiff bristle tip.

The flower clusters are numerous and dense in most leaf axils. Flowers lack petals and produce one-seeded fruits (utricles) with the sepals forming a fan-shaped structure concealing the black or brown seeds. Plants reproduce only by seed, which are dispersed by seed-gathering ants, animals, and when dry plants break off at ground level and tumble with the wind. Many seeds survive ingestion by animals, including sheep and rabbits. Black seeds can imbibe water and germinate in less than 1 hour. Because seed form small coiled embryos in fruit, they do not persist long in the soil.

NON-CHEMICAL CONTROL

Mechanical (pulling, cutting, disking)	Because halogeton is a simple shallow-rooted annual, it can be controlled effectively by tillage or pulling. Plants are easiest to control as seedlings or in early vegetative growth. Plants not controlled until after flowering begins may contain seeds and should be removed and destroyed to prevent reseeding. Periodic mowing close to the soil surface can significantly reduce but not completely prevent seed production. Surviving branches below the reach of mower blades will continue to produce viable seeds. It is best to avoid increasing disturbance unless successful restoration of perennials is highly probable.
Cultural	Disturbances such as overgrazing and fire typically reduce desirable vegetation and increase open sites with bare soil. This can encourage invasion and establishment of halogeton. Any cultural control strategy should increase perennial vegetation, as halogeton has been shown to compete poorly with established perennial species.
	Grazing alone is not a control option because of the toxicity of the plant. However, timely grazing of desired vegetation has been shown to reduce halogeton spread. For example, halogeton in Nevada decreased under late spring to early summer (mid-April to mid-June) grazing at moderate intensity, compared to high intensity grazing in early spring (March to April).
	While fire can kill standing halogeton plants, fire disturbance often enhances seed germination and favors the growth of dense stands when the burns are not hot enough. In most cases, halogeton is one of the first plants to reestablish following wildfire on infested rangeland.
Biological	No biological control agents are available for the control of halogeton. A stem-boring moth (*Coleophora parthenica*) from Pakistan was released for halogeton control in the U.S. However, it failed to establish. Other potential biological control agents have been identified in Central Asia, but they have not yet been developed and tested.

CHEMICAL CONTROL

The following specific use information is based on reports by researchers and land managers. Other trade names may be available, and other compounds also are labeled for this weed. Directions for use may vary between brands; see label before use. Herbicides are listed by mode of action and then alphabetically. The order of herbicide listing is not reflective of the order of efficacy or preference.

GROWTH REGULATORS	
2,4-D Several names	**Rate:** 2 to 6 pt product/acre for ester formulation (0.95 to 2.85 lb a.e./acre) **Timing:** Postemergence in early spring when plants are growing rapidly before bloom stage. **Remarks:** 2,4-D gives good, but not excellent, control and may damage desirable broadleaf natives, particularly at high rates. Reapplications are required to control subsequent germinants. Ester formulations are considered more effective than amine formulations. Apply with a crop oil concentrate for consistent control. Because of injury to native shrubs and a lack of desirable forage species adapted to alkali conditions, the use of 2,4-D for halogeton control has declined in recent years.
Aminocyclopyrachlor + chlorsulfuron *Perspective*	**Rate:** 3 to 4.5 oz product/acre **Timing:** Postemergence in early spring when plants are growing rapidly before bloom stage. **Remarks:** *Perspective* provides broad-spectrum control of many broadleaf species. Although generally safe to grasses, it may suppress or injure certain annual and perennial grass species. Little is known of the herbicide for halogeton control as its registration is relatively new. However, it has been shown to be very effective in herbicide trials. Do not treat in the root zone of desirable trees and shrubs. Do not apply more than 11 oz product/acre per year. At this high rate, cool-season grasses will be damaged, including bluebunch wheatgrass. Not yet labeled for grazing lands. Add an adjuvant to the spray solution. This product is not approved for use in California and some counties of Colorado (San Luis Valley).
AROMATIC AMINO ACID INHIBITORS	
Glyphosate *Roundup, Accord XRT II*, and others	**Rate:** Spot treatment, 2% v/v solution *Roundup ProMax* **Timing:** Postemergence in summer before plants bloom. **Remarks:** Use as a spot treatment on small infestations. Repeat treatments are necessary to control flushes emerging later in the season.
BRANCHED-CHAIN AMINO ACID INHIBITORS	
Chlorsulfuron *Telar*	**Rate:** Label recommends 0.5 to 1 oz product/acre (0.375 to 0.75 oz a.i./acre), but field results have shown 0.2 to 0.5 oz product/acre (0.15 to 0.375 oz a.i./acre) to be effective. Apply with surfactant. **Timing:** Postemergence in late spring or early summer when plants are only 1 to 3 inches tall. **Remarks:** Chlorsulfuron has been shown to be more effective than metsulfuron in western rangelands. Chlorsulfuron can damage some native shrubs, including Nuttall's saltbush (*Atriplex nutalllii*). Chlorsulfuron is considered the most effective herbicide for control of halogeton in rangelands.
Imazapic *Plateau*	**Rate:** 4 to 6 oz product/acre (1 to 1.5 oz a.e./acre) **Timing:** Preemergence or early postemergence to seedlings 1 to 3 inches tall. **Remarks:** For postemergence application, add a surfactant at about 1.5 oz/acre. Imazapic is selective to most native grasses, but will injure some species. Higher rates may suppress seedlings of some cool-season grasses. Imazapic is not registered for use in California.
Metsulfuron *Escort*	**Rate:** 0.5 to 1 oz product/acre (0.3 to 0.6 oz a.i./acre). Apply with surfactant. **Timing:** Postemergence in the late spring or early summer when seedlings have emerged and are growing rapidly, generally 1 to 3 inches tall. **Remarks:** Metsulfuron does not cause injury to grasses and this may be a desirable feature in areas with crested wheatgrass or other forage grasses. Metsulfuron is not registered for use in California.
PHOTOSYNTHETIC INHIBITORS	
Tebuthiuron *Spike 20P*	Tebuthiuron is a pelleted formulation that provides total vegetation control for several years and may be desirable for use on railroad ballast and oil field locations, where halogeton is often found. It has a very long soil residual activity and will provide total vegetation control for 3 to 5 years.

Hedera helix L.; English ivy
Hedera canariensis Willd.; Algerian ivy
  (= *H. helix* L. ssp. *canariensis* (Willd.) Cout.)
Hedera hibernica (G. Kirchn.) Bean; Irish or Atlantic ivy

English, Algerian and Atlantic ivy

Family: Araliaceae
Range: Many western states, including Washington, Oregon, California, Idaho, Arizona, and Utah.
Habitat: Riparian corridors, moist woodlands, forest margins, coastal habitats, and disturbed sites such as cleared forests, urban waste places, and old homesteads. Requires some moisture year-round. Tolerates deep shade, but thrives where plants receive some summer shade and direct winter sun.
Origin: Native to Europe and introduced to the United States as an ornamental. English ivy is still a common landscape ornamental of which there are numerous cultivars.
Impacts: Under favorable conditions, plants spread invasively and can develop a dense cover that outcompetes other vegetation in natural areas. Infestations around old homesteads have been present for many years and serve as nursery sites for further spread. It has escaped from cultivation in many places, especially near the coast and along riparian corridors. English ivy grows over the natural vegetation in an area, including trees, and eventually kills most resident plants by shading them out with its dense canopy of foliage. It thrives in deciduous trees, which allow plants to receive more light and to continue upward growth during winter months. Trees covered with ivy are more susceptible to wind damage from the extra weight. English ivy berries and leaves can be toxic to humans and cattle when ingested in quantity, and the sap can cause contact dermatitis in sensitive individuals, which includes about 10% of the population.
Western states listed as Noxious Weed: *H. helix*, Oregon, Washington; *H. canariensis* and *H. hibernica*, Washington
California Invasive Plant Council (Cal-IPC) Inventory: High Invasiveness

English ivy and other closely related *Hedera* species are fast growing, perennial, evergreen vines that vigorously climb other vegetation and on structures. Plants have two growth forms. The juvenile form has viny stems to about 12 inches long with leaves that are usually three-lobed. The adult reproductive form has erect shrubby stems with ovate to diamond-shaped leaves. Juvenile stems are vine-like, growing both on the ground and vertically into canopies. Juvenile stems develop adventitious roots along the ground and aerial root-like structures that enable stems to cling to objects such as trees and buildings. Juvenile leaves are palmately three to five lobed and vary in size, up to 12 inches long. Adult reproductive stems are erect, shrubby, lack aerial roots, and are non-climbing. Adult leaves are ovate to diamond-shaped and up to 6 inches long. Leaves of both forms have smooth upper surfaces, often slightly glossy, and usually dark green. Leaf stalks and lower leaf surfaces are sometimes glabrous but usually covered with grayish star-shaped hairs.

Unlike most plants in the region that flower in spring or summer, ivy flowers in fall. The shrubby adult form develops flowers in racemes or panicles of simple umbels. The juvenile stage may last for 10 years or more before reproducing by seed. Fruits are berrylike, dark blue to black, about 4 to 8 mm wide. Fruits mature in spring with an individual plant producing tens of thousands of fruit each year. Fruits are consumed and dispersed primarily by birds. English ivy also reproduces vegetatively from juvenile stems. Stem fragments of juvenile and adult plants left in contact with moist soil can regenerate into a new plant. Plants can live 100 years or more.

NON-CHEMICAL CONTROL

Mechanical (pulling, cutting, disking)	When the plant carpets the forest floor, individual stems can be readily pulled off the ground; however, it is essential to remove all runners. If off-site removal is not possible, all plant parts must be placed off the ground in such a way that they can dry out. Repeated removal efforts over multiple years may allow desirable vegetation to colonize the area. Because ivy can resprout and establish from stem fragments, mowing or cutting is not recommended. Small or young ivy plants can be pulled off supporting structures or trees, and roots dug out. The roots of young plants can be easily dug out, particularly when the soil is moist, from the ground around the base of infested trees. Older individuals generally do not resprout. Gloves should

	be worn as many people are sensitive to the dermatitis-causing agents in the plant. Cutting ivy off before it flowers will reduce seed production and deplete the plant's energy reserves. Resprouts are common after treatment. Cutting should be combined with an herbicide treatment or with multiple cuttings over a period of years. Cut ivy at ground level with power or manual saws, and then pry the vines from the tree or structure. Once the vines are cut they will eventually die and fall from the tree, usually after the first extended hot and dry period. Occasionally vines will be embedded in the trunk of the tree, which makes control by both hand and chemicals very difficult.
Cultural	Grazing and burning are not considered effective control options. The leaves and fruit can be toxic to livestock. Deer have been shown to feed on ivy in its native range. Although prescribed burning is not an effective control option, the use of a blowtorch can be successful. To be successful, plants and resprouts must be repeatedly burned until the plant's resources are exhausted.
Biological	Because *Hedera* species are still widely used as ground covers and ornamentals, there is no biological control program established for their management.

CHEMICAL CONTROL

The following specific use information is based on published papers and reports by researchers and land managers. Other trade names may be available, and other compounds also are labeled for this weed. Directions for use may vary between brands; see label before use. Herbicides are listed by mode of action and then alphabetically. The order of herbicide listing is not reflective of the order of efficacy or preference.

GROWTH REGULATORS	
Picloram *Tordon 22K*	**Rate:** Broadcast foliar treatment: 3 to 4 pt product/acre (0.75 to 1 lb a.e./acre) plus 0.25 to 0.5% v/v surfactant to thoroughly wet all leaves **Timing:** Postemergence foliar treatments are best when plants are growing rapidly at or beyond early to full bloom stage. **Remarks:** High levels of picloram can give long-term soil activity for broadleaves. Picloram has proved successful in Australia. Picloram is a restricted use herbicide. It is not registered for use in California. Do not apply near trees, or damage may occur through root uptake.
Triclopyr *Garlon 3A, Garlon 4 Ultra, Pathfinder II*	**Rate:** Spot treatment: 2 to 5% v/v solution of *Garlon 4 Ultra* and water plus 0.25 to 0.5% v/v surfactant to thoroughly wet all leaves. Low volume/thinline treatment: 10% v/v solution of *Garlon 4 Ultra* plus a 20% v/v basal oil concentrate in water. Basal cut stump treatment: 20% v/v *Garlon 4 Ultra* in water. Cut stump treatment: undiluted *Garlon 3A* or 33% *Garlon 3A* in water. Stem injection treatment: drill and fill the stem of large mature plants that are climbing up other trees with 100% *Garlon 3A* or *4 Ultra*. Basal bark treatment: 20% v/v *Garlon 4 Ultra* in 20% v/v basal oil and water, or *Pathfinder II* as a ready-to-use formulation. **Timing:** Postemergence when plants are growing rapidly. Cut stump and basal bark treatments can be applied anytime as long as the ground is not frozen. **Remarks:** Triclopyr is a selective herbicide for broadleaf species. In areas where desirable grasses are growing under or around ivy, triclopyr can be used without non-target damage. For cut stump treatments, cut stems horizontally at or near ground level. Apply herbicide solution immediately after the stump is cut. Suckering from the roots typically occurs after cutting, but the treatment should control most resprouts. Basal bark treatment: spray the lower trunk, including the root collar, to a height of 12 to 15 inches from the ground; the spray should thoroughly wet the lower stem but not to the point of runoff. When making bark treatments, be careful not to get the spray solution on the bark of desirable trees. Plants should not be cut for at least one month after basal bark treatments. Spraying triclopyr immediately after the removal of most leaves and young shoots with a string trimmer has also proved successful.
AROMATIC AMINO ACID INHIBITORS	
Glyphosate *Roundup, Accord XRT II,* and others	**Rate:** Spot treatment: 2 to 4% v/v solution of *Roundup ProMax* (or other trade name with similar concentration of glyphosate) in water to thoroughly wet all leaves. Low volume/thinline treatment: 10% v/v solution of *Roundup* (or other trade name) in water. Cut stump treatment: 25% v/v *Roundup* (or other trade name) in water. **Timing:** Postemergence when plants are growing rapidly. Foliar treatments should be made in late summer or early fall. For cut stump treatment, application in late summer, early fall or

	dormant season provides best control. Treatment should occur immediately after cutting.
	Remarks: Glyphosate is a nonselective systemic herbicide with no soil activity. It gives good control with some resprouts. Plants should not be cut for at least 4 months after foliar treatments. Cut stump applications are made as described for triclopyr. Glyphosate has also proved successful in Australia.
BRANCHED-CHAIN AMINO ACID INHIBITORS	
Imazapyr *Arsenal, Habitat, Stalker, Chopper, Polaris*	**Rate:** Spot treatment: 1 to 2% v/v solution of *Stalker* plus 0.25 to 0.5% surfactant v/v in water to thoroughly wet all leaves. Low volume/thinline treatment: 10% v/v solution of *Stalker* plus a 20% v/v ethylated crop oil in water. Cut stump treatment: 20% v/v solution of *Stalker* plus a 20% v/v ethylated crop oil in water or 20% *Habitat* v/v in 80% water carrier. Basal bark treatment: 20% v/v solution of *Stalker* plus a 20% v/v ethylated crop oil in water. **Timing:** Postemergence when plants are growing rapidly. Best when used in late summer to early fall. **Remarks:** Imazapyr is a soil residual herbicide and may result in bare ground around plants for some time after treatment. Cut stump and basal bark applications are made as described for triclopyr. Plants should not be cut for at least 4 months after basal bark treatment. Another ALS inhibitor, metsulfuron, has proved successful in Australia.

Hemizonia pungens (Hook. & Arn.) Torr. & A. Gray; spikeweed
 (= *Centromadia pungens* (Hook. & Arn.) Greene)
Hemizonia fitchii A. Gray ; Fitch's tarweed
 (= *Centromadia fitchii* (A. Gray) Greene)
Hemizonia parryi Greene; Parry's tarweed
 (= *Centromadia parryi* (Greene) Greene)

Spikeweed, Fitch's and Parry's tarweed

Family: Asteraceae
Range: All three species are most common in grasslands of California, but spikeweed can also be found in Washington, Oregon, Idaho, Nevada and Arizona, and Fitch's tarweed can be found in Oregon.
Habitat: Dry grasslands, seasonal wetlands, waste areas, woodlands, pastures, rangeland, and along roadsides. Often encountered in areas where soil has been disturbed.
Origin: All are native to California and other western states.
Impacts: Spikeweed and the tarweeds can be a problem in range, pasture, and grain fields (primarily poor growth areas). The plants are spiny, can form dense stands and are avoided by livestock. Populations can increase to an undesirable density in grazed pastures and rangeland. In natural areas, these species are not be a problem and may be a desirable component of the ecosystem.
Western states listed as Noxious Weed: Although native to the west, *Hemizonia pungens* is listed as a noxious weed in Oregon and Washington

These three species are very similar in appearance and can be separated by seedhead characteristics. They are late season annuals with rigid, bristly, branching stems. Although they normally grow from 1.5 to 3 ft tall, they are occasionally smaller or larger, ranging from 4 inches to 4 ft tall. The basal leaves are yellowish green, linear-lanceolate, stiff, and 2 to 6 inches long with narrow lobes. The stem leaves are alternate, approximately 0.5 inch long and spine-tipped. Often there are dwarf stems in the axils. The rosettes bolt in late spring and early summer, and plants typically flower from July to September. Plants are covered with sticky, glandular hairs that may tie up foliar-applied herbicides.

The flower heads are yellow and consist of ray and disk flowers. They are located at the tips of short lateral stems. Bracts at the base of the flower head are partially covered by the upper leaves. The bracts are also spine-tipped and have short stiff hairs. These species reproduce by seeds, which fall near the parent plant and are dispersed only short distances by wind. The achenes (seeds) are approximately 2 mm long. There is little evidence to show how long the seeds survive in the soil, but it is expected that they can survive for at least 3 to 5 years.

NON-CHEMICAL CONTROL

Mechanical (pulling, cutting, disking)	Hand-pulling may be adequate for small, recently established infestations, but is not practical for larger infestations. Hand-pulling is best done in spring when the plants are still green and somewhat soft. Gloves still must be worn as more mature plants will be spiny.
	Mowing just before flowering is probably effective but there is little information available on this topic. Tillage in late spring when the soil has dried is believed to be effective but it is not practical in many or most range or natural areas.
Cultural	Livestock will consume spikeweed/tarweed in winter and early spring when plants are young and succulent. However, when plants become more spiny and tough, livestock will generally avoid these plants and as a result select for these species. There is some evidence that sheep grazing may be effective at reducing populations before plants become too spiny for them to eat.
	Burning does not appear to be an effective control measure. Fire may reduce the population the year of the

	burn but there does not appear to be an extended effect into subsequent years, and competing vegetation may be detrimentally affected.
Biological	Because these plants are native to the western United States, there are no biological control programs designed to management any species of *Hemizonia*.

CHEMICAL CONTROL

The following specific use information is based on reports by researchers and land managers. Other trade names may be available, and other compounds also are labeled for this weed. Directions for use may vary between brands; see label before use. Herbicides are listed by mode of action and then alphabetically. The order of herbicide listing is not reflective of the order of efficacy or preference.

GROWTH REGULATORS	
2,4-D Several names	**Rate:** 1.5 qt product/acre (1.4 lb a.e./acre) **Timing:** Postemergence when plants are in rosette stage in winter or early spring. Application during cool weather allows for the use of ester formulations of 2,4-D, which may have better absorption into the glandular leaves. **Remarks:** 2,4-D is a broadleaf herbicide and will not damage grasses. It can be a restricted use herbicide in some areas. The ester formulation should not be applied when temperatures are above 80°F.
Aminocyclopyrachlor + chlorsulfuron *Perspective*	**Rate:** 1.75 to 2.75 ounces product/acre **Timing:** Preemergence or early postemergence before bolting. **Remarks:** *Perspective* provides broad-spectrum control of many broadleaf species. Although generally safe to grasses, it may suppress or injure certain annual and perennial grass species. Do not treat in the root zone of desirable trees and shrubs. Do not apply more than 11 oz product/acre per year. At this high rate, cool-season grasses will be damaged, including bluebunch wheatgrass. Not yet labeled for grazing lands. Add an adjuvant to the spray solution. This product is not approved for use in California and some counties of Colorado (San Luis Valley).
Dicamba *Banvel, Clarity*	**Rate:** 1 to 2 pt product/acre (0.5 to 1 lb a.e./acre) **Timing:** Postemergence when target plants are small and rapidly growing. **Remarks:** Dicamba is a broadleaf-selective herbicide often combined with other active ingredients such as 2,4-D. Do not apply when outside temperatures exceed 80°F.
BRANCHED-CHAIN AMINO ACID INHIBITORS	
Chlorsulfuron *Telar*	**Rate:** 0.5 to 1 oz product/acre (0.375 to 0.75 oz a.i./acre) **Timing:** Preemergence or postemergence to plants in rosette stage. **Remarks:** Chlorsulfuron has mixed selectivity and is generally safe on grasses. It is most effective preemergence. Use a surfactant for postemergence applications. It has fairly long soil residual activity.

Heracleum mantegazzianum Sommier & Levier

Giant hogweed

Photo courtesy of Kings County Noxious Weed Program

Family: Apiaceae
Range: Western Washington and isolated locations in northwestern Oregon. Also found in the northeastern U.S. and is expanding its range elsewhere.
Habitat: Riparian areas, disturbed sites, roadsides, waste places. Often grows in wet places.
Origin: Native to southwestern Asia.
Impact: Giant hogweed can develop a dense canopy that will crowd out native and other species. This loss of understory vegetation increases stream bank erosion. While cattle and pigs can consume giant hogweed without any apparent problems, skin contact with sap can cause severe photosensitizing dermatitis on most people and animals.
Western states listed as Noxious Weed: California, Oregon, Washington

Robust biennial or perennial to 16 ft tall with large three-part compound leaves. Once cultivated as an unusual garden ornamental, giant hogweed has escaped cultivation to become an ecological and health problem. Infestations are nearly always associated with garden escapes. Unlike other emergent aquatics, giant hogweed cannot tolerate prolonged root submergence in water.

Plants reproduce by seed and vegetatively by forming new crowns from the tuberous rootstocks. Giant hogweed can produce abundant seed that primarily disperse with water. Seeds survive 7 years or more under field conditions. Individual plants appear to flower once and die, but new plants grow from crowns developed from rootstock during the previous year.

NON-CHEMICAL CONTROL

Mechanical (pulling, cutting, disking)	Root cutting is usually performed with an ordinary spade with a sharpened blade. It is best undertaken in early spring, and should be repeated in summer if regrowth appears. The main tap root should be cut 3 to 4 inches below the soil level. The method is very effective but labor intensive (therefore potentially costly) and thus suitable only for single plants or small stands. Removal of umbel inflorescences can temporarily prevent seed set, but the inflorescence may regrow and set seed after a single cutting. Therefore, umbel removal requires several visits over the flowering season. Umbel removal is most effective when terminal umbels just start to flower. It is important to wear protective clothing and avoid getting the sap on skin.
	Mechanical mowing, e.g., using a flail mower, has been shown to be useful for clearing large areas of giant hogweed. Smaller stands can be trimmed or scythed. However, plants usually regrow from the rootstalk. Mowing may control plants if done persistently, two to three times during the growing season, to starve the rootstalk. Mowing multiple times also hinders resprouting plants from flowering and setting seeds. Cutting in late spring (May to June), when plants are taller and have used more root reserves, is more effective than cutting in early spring (March). It appears particularly effective to repeatedly mow plants in the mid-flowering stage. This timing prevents seed production and depletes the underground reserves, and should eradicate a population within a few years as the seedbank declines.
	Digging or plowing to destroy the crown (below 4 inches soil depth) can completely kill the plant. It is possible that large infestations may be controlled by deep cultivation (plowing) although this has not been tested and is generally impractical on river banks.
Cultural	Giant hogweed is not controlled by light grazing. Intensive grazing by sheep and rooting by pigs has been shown to be effective. In Denmark, the population of the weed was much reduced after two years of sheep grazing and completely eliminated after five years, when no viable seeds were found to remain in the soil. The weed may be slightly less palatable to cattle, but grazing by cattle as well as pigs is recommended in Ireland.
	Sheep and goats often seek out young plants of giant hogweed. Recommended grazing timing is in mid-spring. Herds should preferably include individuals already familiar with the weed to reduce the risk of over-eating and poisoning. Dark-skinned, thick-pelted animals are less likely to acquire dermatitis.
	Burning is not practical or effective for the management of giant hogweed.

| Biological | There are no biocontrol insects available for giant hogweed at this time. |

CHEMICAL CONTROL

The following specific use information is based on published papers and reports by researchers and land managers. Other trade names may be available, and other compounds also are labeled for this weed. Directions for use may vary between brands; see label before use. Herbicides are listed by mode of action and then alphabetically. The order of herbicide listing is not reflective of the order of efficacy or preference.

GROWTH REGULATORS	
Triclopyr *Garlon 3A*, *Garlon 4 Ultra*, *Renovate*	**Rate:** Spot treatment: 1% v/v solution *Garlon 3A* or *Garlon 4 Ultra* **Timing:** Postemergence in summer months during bud stage and while the plant is rapidly growing. **Remarks:** Triclopyr is a broadleaf herbicide. It has very little soil residual activity. *Garlon 4 Ultra* is formulated as a low volatile ester. However, in warm temperatures, spraying onto hard surfaces such as rocks or pavement can increase the risk of volatilization and off-target damage.
AROMATIC AMINO ACID INHIBITORS	
Glyphosate *Roundup, Rodeo, Aquamaster, Accord XRT II*, and others	**Rate:** Broadcast application with *Rodeo* or *Aquamaster*: 3 to 5 pt product/acre (1.5 to 2.5 lb a.e./acre) **Timing:** Optimal treatment timing for postemergence application is during summer months when plants are at the bud stage and rapidly growing. **Remarks:** Glyphosate is the most widely used compound for the control of giant hogweed. It is important to cover the leaf surfaces thoroughly (spray-to-wet), but do not spray to the point at which liquid is dripping off the leaves. Do not cut or dig up the plant until the top growth has died back. If the leaves remain green two weeks after initial treatment, spray again with glyphosate. Control methods can also include cutting to the ground (see above), cutting and spraying regrowth with glyphosate in spring and summer, or treating uncut plants with glyphosate (sometimes with more than one application). Glyphosate is a nonselective herbicide with no soil activity. Glyphosate can also be used in an injection technique for individual plant control. Using a hand-held injector, inject 5 ml of a 5% v/v solution into one leaf cane per plant, 12 inches above the root crown. Treated canes should be marked to avoid unnecessary retreatment. It is important to note that for treatments that use high concentration, such as injection, glyphosate cannot exceed 8.5 qt product/acre of *Roundup Pro Concentrate* (or other trade name). This volume of product would treat about 32,000 stems/acre.
BRANCHED-CHAIN AMINO ACID INHIBITORS	
Imazapic *Plateau*	**Rate:** 12 oz product/acre (3 oz a.i./acre) **Timing:** Postemergence in spring during the bolting stage. **Remarks:** Avoid physically contacting plant when applying. Imazapic has some soil residual activity. Add an appropriate adjuvant to spray mix. Imazapic is not registered for use in California.
Imazapyr *Habitat*	**Rate:** Up to 6 pt *Habitat*/acre for aquatic use **Timing:** Postemergence early in the season (March to May) for best effect. **Remarks:** Imazapyr has a long residual effect in the soil that prevents further germination but may also impact non-target species.

Hieracium caespitosum Dumort.; meadow hawkweed
Hieracium aurantiacum L.; orange hawkweed
Hieracium glomeratum Froeland; queendevil hawkweed
Hieracium piloselloides Vill.; kingdevil hawkweed
Hieracium pilosella L.; mouse-ear hawkweed

Hawkweeds

Family: Asteraceae
Range: Orange hawkweed is widely distributed across western provinces of Canada and the western states of Alaska, Washington, Oregon, California, Idaho, Montana, Wyoming, and Colorado. Meadow hawkweed is found in British Columbia, Washington, Oregon, Idaho, Montana and Wyoming. Tall hawkweed is found in Washington, Idaho and Montana. Yellow devil hawkweed has been found in two counties of Idaho.
Habitat: Peat bogs, forests with open canopies, mesic bunchgrass, sagebrush, and meadows. Also problematic in pastures and Conservation Reserve Program lands.
Origin: All non-native hawkweeds are native to Eurasia.
Impacts: Hawkweeds can dominate grasslands to the near exclusion of other species. Their low growth habit yields little usable forage for livestock or wildlife. Currently the species having the greatest impact include orange and meadow hawkweed.
Western states listed as Noxious Weed: *H. caespitosum,* Idaho, Montana, Oregon, Washington; *H. aurantiacum,* Colorado, Idaho, Montana, Oregon, Washington; *H. piloselloides,* Montana, Oregon, Washington (proposed); *H. pilosella,* Oregon, Washington; *H. atratum, H. glomeratum,* and *H. laevigatum,* Washington

Hawkweeds are perennial forbs in a genus that is divided into subgenera. Until recently, all invasive hawkweeds were in the subgenus *Pilosella*, a group with stolons and more-or-less leafless flowering stems. Hawkweeds native to North America are in the subgenus *Hieracium* or *Chionoracium* (and all *Chionoracium* are from the western hemisphere). The species in the *Hieracium* subgenus do not have stolons or leafless stems, so it formerly was easy to identify non-native hawkweeds. Unfortunately, other non-native hawkweeds in the subgenus *Hieracium* have established in North America, including *H. atratum* Fries., *H. lachenalii* C.C. Gmel., *H. laevigatum* Willd., *H. maculatum* Schrank, *H. murorum* L., and *H. sabaudum* L. If dense patches of hawkweeds have leaves along the flowering stems and do not have stolons, they may be one of the above species within the *Hieracium* subgenus. (However, native hawkweeds can also occur in dense clumps depending on site conditions; for example, the native *H. albertinum* can form dense stands in old burn piles.)

The five species listed at the top are all in the *Pilosella* subgenus, with stolons and leafless flowering stems. Orange hawkweed has orange flowers, and the others are yellow-flowered. Nearly all the weedy species have flowering stems arranged so that all flowerheads are at the same height, or all branches arise from the same point. Hair shape can also help to identify *Hieracium* species. Some species have star-like hairs and others simple hairs. Plants with mostly simple hairs, without hairs, and with few to no star-shaped hairs include *H. floribundum, H. bauhini* and *H. pilloselloides*.

Most hawkweeds are nearly obligate with respect to mycorrhizal relationships, and their competitiveness with grasses appears linked to their utilization of mycorrhizal networks. Dense, well-fertilized perennial grass stands will have lower levels of mycorrhizal fungi and more resilience to invasion from hawkweeds.

All weedy species reproduce vegetatively from stolons, rhizomes, and by seeds. Stolon fragments can generate new plants. Seed reproduction is generally less important than vegetative reproduction in a given locality. Most seeds fall near the parent plant but can disperse long distances with wind. Polyploid populations generally produce asexual seed (apomixis). Seeds appear to survive up to about 7 years under field conditions.

NON-CHEMICAL CONTROL

Mechanical (pulling, cutting, disking)	Hand pulling has limited success for the control of hawkweeds, since disturbing the stolons and rhizomes may only help the plant to spread. In addition, due to their mat-forming growth, stoloniferous *Hieracium* spp. can successfully escape mowing. Although mowing prevents seed production by removing flowering stems, repeated mowing encourages faster vegetative spread. Tillage may spread stolons and root buds, so only repeated cultivation would have a chance of success.
Cultural	Competitive grass communities can delay hawkweed reinfestations for several years, perhaps allowing herbicide maintenance applications to be made only every 4 to 7 years. Maintaining fertilization at adequate levels for perennial grasses can also limit hawkweed invasion.
Biological	Biological control has been pursued for meadow hawkweed. No agents have, thus far, been registered for field release. A wasp *Aulacidea subterminalis* forms galls on stolons and has the potential to reduce stolon size and vegetative reproduction. Two flies are considered potential biological control agents. One feeds externally on the root system (*Cheilosia urbana*) and the other (*C. psilophthalma*) feeds on rosettes.

CHEMICAL CONTROL

The following specific use information is based on published papers and reports by researchers and land managers. Other trade names may be available, and other compounds also are labeled for this weed. Directions for use may vary between brands; see label before use. Herbicides are listed by mode of action and then alphabetically. The order of herbicide listing is not reflective of the order of efficacy or preference.

GROWTH REGULATORS	
Aminopyralid *Milestone*	**Rate:** 4 to 7 oz product /acre (1 to 1.75 oz a.e./acre) **Timing:** Postemergence to rosettes or bolting plants. Can also be effective when plants are flowering. **Remarks:** Not effective in fall. Earlier applications allow for release of suppressed perennial grasses.
Aminopyralid + metsulfuron *Opensight*	**Rate:** 2.5 to 3.3 oz product/acre **Timing:** Postemergence to rosettes or bolting plants. Also effective when applied during flowering. **Remarks:** Not effective in fall; earlier applications allow for release of perennial grasses. The combination will control many broadleaf species. This product is not registered for use in California.
Aminopyralid + 2,4-D *Forefront HL*	**Rate:** 1.2 to 2.1 pt product/acre **Timing:** Postemergence to rosettes or bolting plants. Also effective when applied during flowering. **Remarks:** Not effective in fall. Earlier applications allow for release of suppressed perennial grasses.
Clopyralid *Transline*	**Rate:** 0.67 to 1.33 pt product/acre (4 to 8 oz a.e./acre) **Timing:** Postemergence in spring to rosettes and bolted plants. **Remarks:** Not effective in fall. Earlier applications allow for release of suppressed perennial grasses.
Clopyralid + 2,4-D *Curtail*	**Rate:** 2 qt product/acre **Timing:** Postemergence in spring to rosettes. **Remarks:** Not effective in fall. The combination will damage other broadleaf species.
Dicamba *Banvel, Clarity*	**Rate:** 1 to 1.5 pt product/acre (0.5 to 0.75 lb a.e./acre) **Timing:** Postemergence to plants from rosette to beginning of bolting. **Remarks:** Broadleaf-selective herbicide often combined with other active ingredients. It is not typically used alone to control hawkweeds.
Picloram *Tordon 22K*	**Rate:** 0.5 pt product/acre (2 oz a.e./acre) **Timing:** Postemergence in spring to seedlings or bolting plants. **Remarks:** Broadleaf-selective herbicide with long soil residual. Also formulated as a premix with 2,4-D (*Grazon P+D*) to increase its effectiveness. Restricted use herbicide; not registered for use in California.
PHOTOSYNTHETIC INHIBITORS	
Hexazinone *Velpar L*	Not often used for hawkweed control, but hawkweed is included on the *Velpar* label. It has preemergence activity only and a long soil residual. High rates of hexazinone can create bare ground, so only use high rates in spot treatments.

Hirschfeldia incana (L.) Lagr.-Fossat

Shortpod mustard

Family: Brassicaceae
Range: California, Nevada and Oregon.
Habitat: Disturbed places, roadsides, fields, pastures, agronomic crops, orchards, ditch banks, vineyards, and dry washes.
Origin: Native to the Mediterranean region.
Impact: Spreads into natural areas displacing natives. Becoming more problematic in wildland areas of southern California.
California Invasive Plant Council (Cal-IPC) Inventory: Moderate Invasiveness

Shortpod mustard is an erect yellow-flowered mustard to 3 to 4 ft tall. It is a biennial or short-lived perennial, occasionally a winter annual. The lower leaves of mature plants are obovate, irregularly pinnate-lobed and toothed, with the terminal lobe larger than lateral lobes on a long stalk. The upper leaves do not clasp the stem. The stem bases are moderately to densely covered with stiff, downward-directed hairs. The basal leaves usually form a rosette, and the leaves are moderately to densely covered with stiff grayish hairs.

The flowers are pale yellow, not as bright as most other mustard species. They form in an elongated raceme. The mature fruit is 8 to 15 mm long – shorter than in other common yellow-flowered mustards –and appressed to the stem. Shortpod mustard reproduces only by seed, although plants can resprout from the base when damaged. Seed production is high. Most seeds disperse by falling close to the parent plant. Like most other mustards, the seeds likely survive in the soil for several years.

NON-CHEMICAL CONTROL

Mechanical (pulling, cutting, disking)	Manual removal or cultivating before seeds develop, particularly during the seedling stage, can control populations. Manual removal and other control methods implemented over a period of years will eventually exhaust the seedbank.
Cultural	Neither grazing nor prescribed burning have been tested for the management of shortpod mustard, but it is not expected that these would be effective.
Biological	Because of its close relationship to important cultivated members of the Brassicaceae, there is no biological control program developed for the management of shortpod mustard.

CHEMICAL CONTROL

Few herbicides give effective control of shortpod mustard. Many herbicides used on annual and biennial mustards are less effective on this perennial. Research is ongoing. Herbicides are listed by mode of action and then alphabetically. The order of herbicide listing is not reflective of the order of efficacy or preference.

AROMATIC AMINO ACID INHIBITORS	
Glyphosate *Roundup, Accord XRT II*, and others	**Rate:** 1 to 2 pt product (*Roundup ProMax*)/acre (0.56 to 1.1 lb a.e./acre) **Timing:** Early postemergence to small plants. **Remarks:** Glyphosate provides suppression of shortpod mustard. It has no soil activity and is nonselective. Its effectiveness is increased by addition of ammonium sulfate.
BRANCHED-CHAIN AMINO ACID INHIBITORS	
Chlorsulfuron *Telar*	**Rate:** 0.25 to 0.5 oz product/acre (0.19 to 0.375 oz a.i./acre) **Timing:** Preemergence to early postemergence. **Remarks:** Mixed selectivity, generally safe on grasses. Most effective preemergence. Use a surfactant for postemergence applications. Fairly long soil residual. Seems to be less effective in arid environments.
Sulfometuron *Oust* and others	**Rate:** 0.5 oz product/acre (0.38 oz a.i./acre) **Timing:** Preemergence to early postemergence to small plants. **Remarks:** Use 0.25% v/v non-ionic surfactant to improve herbicide uptake.

Holcus lanatus L.
Common velvetgrass

Family: Poaceae
Range: Most of the western U.S. except Wyoming and South Dakota.
Habitat: Roadsides, disturbed grassland, cultivated fields, and orchards. Often found on soils with low fertility. Some biotypes of common velvetgrass tolerate high salt concentrations. Grows best under moist conditions, but established plants tolerate moderate drought. A facultative wetland indicator species in California and some other western states. It does not survive a period of severe frost.
Origin: Native to Europe.
Impact: Dense populations of common velvetgrass have been shown to reduce the establishment of native species and the growth of tree seedlings. This species rapidly colonizes disturbed areas, where it out competes native species for soil moisture and nutrients, especially in nutrient-limited substrates. The accumulation of litter can prevent the germination of native grasses and increase risk of fire. In California, common velvetgrass is particularly invasive in coastal grasslands and wetlands in the northern part of the state. Common velvetgrass provides food for game birds, deer, elk, and insects.
California Invasive Plant Council (Cal-IPC) Inventory: Moderate Invasiveness

Common velvetgrass is a tufted perennial typically 2 to 3 ft tall, with soft pubescent, grey-green foliage. The stems are ascending to erect and slightly flattened in cross-section. The foliage, including internodes, are moderately to densely covered with long grayish hairs. The roots are fibrous and deep, especially in low-nitrogen soil. Individual clumps enlarge rapidly by adding shoots and roots at the nodes. The ligules are membranous, 1 to 2 mm long.

The inflorescence consists of an open to contracted panicle, 2 to 6 inches long. It is covered with short, velvety, gray hairs, and often is purplish-tinged. Plants reproduce by seed, most of which fall below the parent plant, but which can be dispersed long distances by water. Seed production is typically high. Seeds can germinate immediately, or they can build up a large soil seedbank; deeply buried seeds can survive for up to 10 years.

NON-CHEMICAL CONTROL

Mechanical (pulling, cutting, disking)	Hand pulling of plants can reduce populations, and removing inflorescences can contain population expansion, but common velvetgrass can resprout from basal shoots following the removal of the above-ground growth. It is important to remove the entire plant. Common velvetgrass is easier to control with hand pulling compared to more rhizomatous perennial species.
	Mowing and tillage can reduce infestations of common velvetgrass, but are generally impractical in areas where infestations occur. In some cases, mowing and cutting can stimulate regrowth and flower production and increase the reproductive potential of common velvetgrass. Seeds are easily spread by mowing equipment.
Cultural	Intensive grazing may reduce infestations of common velvetgrass, but occasional low intensity grazing may enhance establishment and spread.
	Burning can decrease populations, but is not practical in most situations.
	Long-term flooding can eradicate common velvetgrass, while elimination of irrigation can reduce its abundance.
Biological	There are no biological control agents available for the control of common velvetgrass.

CHEMICAL CONTROL
The following specific use information is based on published papers and reports by researchers and land managers. Other trade names may be available, and other compounds also are labeled for this weed. Directions

for use may vary between brands; see label before use. Herbicides are listed by mode of action and then alphabetically. The order of herbicide listing is not reflective of the order of efficacy or preference.

LIPID SYNTHESIS INHIBITORS	
Fluazifop *Fusilade*	**Rate:** 1 to 1.5 pt product/acre (4 to 6 oz a.e/acre) **Timing:** Postemergence in early spring when grass is rapidly growing. **Remarks:** Fluazifop is a grass herbicide and will not injure broadleaf species. Repeat applications may be needed to control well-established populations. Add 1% crop oil concentrate or 0.25% non-ionic surfactant.
Sethoxydim *Poast*	**Rate:** 1 to 2.5 pt product/acre (3 to7.5 oz a.e./acre) **Timing:** Apply in early spring when grass is rapidly growing. Applying in mid- to late spring may be less effective. **Remarks:** Sethoxydim is a grass herbicide and will not injure broadleaf species. Repeat applications may be needed to control well-established grass. Add 1% crop oil concentrate or 0.25% non-ionic surfactant. It is possible that clethodim may also provide good control, although there is no evidence for this yet.
PHOTOSYNTHETIC INHIBITORS	
Hexazinone *Velpar L*	**Rate:** 2 to 6 pt product/acre (0.5 to 1.5 lb a.i./acre) **Timing:** Preemergence or postemergence to rapidly growing plants. **Remarks:** Hexazinone is a broad-spectrum herbicide that is mobile in the soil and has long soil residual activity. It should not be used in areas with a shallow water table. Because common velvetgrass is typically found in wet areas, hexazinone is not the best choice under most circumstances. High rates of hexazinone can create bare ground, so only use high rates in spot treatments.

Holocarpha virgata (Gray) Keck

Virgate tarweed

Family: Asteraceae
Range: Very common grassland species in California.
Habitat: Grasslands, woodlands, fields, pastures, rangeland, and roadsides.
Origin: Native to California.
Impacts: Virgate tarweed is usually not considered a weed in natural areas. Birds and mammals consume the seeds, and the pollen is an important food source for bees. However, mature virgate tarweed is unpalatable to livestock, and populations can increase to an undesirable density in grazed pastures and rangeland. Its population density may increase following late spring rainfall after annual grasses have matured.

Virgate tarweed is a 3 to 4 ft tall, very aromatic native annual covered with a sticky resin. It is a winter annual that lasts into late summer. Plants germinate in fall and winter, and overwinter as rosettes until flower stems develop in spring. Stem leaves are tipped with a very distinctive sessile open pit resin gland. Main stems are branched well above the base. Lower stem leaves are opposite or alternate, whereas upper stem leaves can appear whorled and lay close to the stem.

Virgate tarweed flowers in late summer. The flowerheads are yellow, about 0.25 inch long, and appear sessile on the stems. Flowers consist of 9 to 25 yellow disk flowers with black anthers surrounded by 3 to 7 yellow ray flowers. All flowers and seeds lack a pappus. Plants reproduce only by seed (achenes). Most achenes fall near the parent plant or disperse short distances with wind, rain, and animals. Little is known of the seed longevity, but it is expected that the seeds can persist for a few years in the soil.

NON-CHEMICAL CONTROL

Mechanical (pulling, cutting, disking)	Mowing pasture to a height of 4 inches just before flowering in July or August can greatly reduce the population that year and the density of virgate tarweed the following year. Tillage was also shown to be effective in late spring when the soil was dry.
Cultural	Livestock will use tarweed as forage in winter and early spring while plants are young and succulent. However, plants become less palatable as they increase in height and resin content. By late season, livestock rarely use virgate tarweed as forage, and grazing would not be an effective control strategy.
	It would be expected that a late season (summer) prescribed burn before the plant develops viable seed would be effective for control, but would have to be repeated a couple of years in a row.
	In one study, fall application of nitrogen fertilizer (ammonium sulfate) reduced tarweed density the following season. This was thought to be due to an increase in soil moisture usage by other plants in the following spring, increasing their competitive effect on virgate tarweed.
Biological	Virgate tarweed is a native species so a biological control effort is not desirable.

CHEMICAL CONTROL

There is little information on the control of virgate tarweed, as it is a native species that is only a problem in rangelands.

The following specific use information is based on reports by researchers and land managers. Other trade names may be available, and other compounds also are labeled for this weed. Directions for use may vary between brands; see label before use. Herbicides are listed by mode of action and then alphabetically. The order of herbicide listing is not reflective of the order of efficacy or preference.

GROWTH REGULATORS

2,4-D Several names	**Rate:** 3 pt product/acre (1.4 lb a.e./acre) **Timing:** Postemergence, when plants are in rosette stage in winter or early spring (before late April). Application during cool weather allows for the use of ester formulations of 2,4-D, which may have better absorption into the glandular leaves. Early applications are much more effective compared to

	applications made after bolting. **Remarks:** 2,4-D is a broadleaf-selective herbicide and has no soil activity. It can be a restricted use herbicide in some areas. This treatment has been shown to give the most effective control in the absence of other studies. Although no information is provided, a relative of virgate tarweed, *Hemizonia pungens*, is listed on the *Clarity* (dicamba) label as susceptible to that herbicide. It is not known if *Clarity* will also control virgate tarweed.
Aminocyclopyrachlor + chlorsulfuron *Perspective*	**Rate:** 1.25 to 1.75 oz product/acre **Timing:** Preemergence or early postemergence before bolting. **Remarks:** Trials have shown this product to give very good control of virgate tarweed. *Perspective* provides broad-spectrum control of many broadleaf species. Although generally safe to grasses, it may suppress or injure certain annual and perennial grass species. Do not treat in the root zone of desirable trees and shrubs. Do not apply more than 11 oz product/acre per year. At this high rate, cool-season grasses will be damaged, including bluebunch wheatgrass. Not yet labeled for grazing lands. Add an adjuvant to the spray solution. This product is not approved for use in California and some counties of Colorado (San Luis Valley).
Aminopyralid *Milestone*	**Rate:** 5 to 7 oz product/acre (1.25 to 1.75 oz a.e./acre) **Timing:** Preemergence or early postemergence before bolting. **Remarks:** Although there are few experimental results indicating that aminopyralid can control virgate tarweed, anecdotal information suggests that it will be effective when applied to young plants.
BRANCHED-CHAIN AMINO ACID INHIBITORS	
Chlorsulfuron *Telar*	**Rate:** 1 to 2.6 oz product/acre (0.75 to 1.95 oz a.i./acre) **Timing:** Preemergence or postemergence to plants in rosette stage. **Remarks:** Although there are few experimental results indicating that chlorsulfuron can control virgate tarweed, anecdotal information suggests that it will be effective when applied to young plants. In addition, it is listed on the label for control of another tarweed, *Hemizonia pungens*. Use a surfactant for postemergence applications.

Hordeum jubatum L.
Foxtail barley

Family: Poaceae
Range: Indigenous to the western U.S.; now throughout most of the country, except for southern Atlantic and Gulf Coast states.
Habitat: Disturbed areas, meadows, basins, ditchbanks and roadsides. Thrives in moist areas that are saline or alkaline but will also grow under non-saline conditions. Occasionally weedy on agricultural sites such as grain and hay fields, especially those with poor drainage. It is most abundant on the edge of sloughs and salt marshes and rapidly invades areas exposed by a receding water table.
Origin: Native to western North America.
Impact: In disturbed areas, foxtail barley forms monotypic stands that displace favorable vegetation. The barbed awns can injure the mouths, nose, eyes and digestive tract of grazing animals. It is also a host for several viruses.

Foxtail barley is a non-rhizomatous, short-lived, cool-season perennial grass. It has erect stems and grows in dense bunches or tufts usually 1 to 2 ft tall. The leaves are gray-green, sometimes pubescent, and up to 9 mm wide. The sheath margins have numerous soft hairs and the ligules are short, membranous and lack auricles.

Each culm terminates in a nodding spike about 2.5 to 4 inches long. The spikes are light green with reddish or purplish tints. They appear silky and glisten in the sunlight. When mature, the spikes fade to a light tan color and are very brittle. The awns are 1 to 2.25 inches long and sharp, with backward-pointing barbs. Foxtail barley reproduces sexually by seed and can reproduce vegetatively by tillering. Foxtail barley initiates growth in April or May, and flowers and sets seed from June to late July or August. The seeds appear to survive less than 7 years under field conditions.

NON-CHEMICAL CONTROL

Mechanical (pulling, cutting, disking)	Foxtail barley is fairly tolerant of cutting or mowing. Mowing can help interrupt seed production, but the plants recover, often grow more prostrate, and continue to flower.
	This species usually can be controlled by plowing in fall followed by cultivation such as harrowing in spring. However, spring tillage alone is significantly less effective.
Cultural	Foxtail barley is most sensitive to a late-spring burn that coincides with its active growing period. Moderate fires will likely top-kill the plant, and hot fires may kill the root system as well.
	Foxtail barley can be managed somewhat by controlling water levels and managing drainage. When water is allowed to stand in low, poorly drained areas, it may kill off competing vegetation, allowing foxtail barley to establish. Low spots should be managed to prevent standing water.
Biological	Because foxtail barley is native to the western United States, there is no biological control program for its management.

CHEMICAL CONTROL

The following specific use information is based on published papers or reports by researchers and land managers. Other trade names may be available, and other compounds also are labeled for this weed. Directions for use may vary between brands; see label before use. Herbicides are listed by mode of action and then alphabetically. The order of herbicide listing is not reflective of the order of efficacy or preference.

LIPID SYNTHESIS INHIBITORS	
Clethodim *Select, Envoy*	**Rate:** 12 to 32 oz product (*Envoy*)/acre (1.5 to 4 oz a.i./acre)
	Timing: Postemergence when the target plants are growing rapidly and between 2 and 6 inches in height.
	Remarks: Clethodim is a grass-selective herbicide. Add a non-ionic surfactant containing at least 80% a.i. at the rate of 1 pt/50 gal (0.25% v/v). Note that *Envoy* formulation is 1 lb a.i./gallon, *Select* is 2 lb a.i./gallon.

BRANCHED-CHAIN AMINO ACID INHIBITORS	
Imazapic *Plateau*	**Rate:** 8 to 12 oz product /acre (2 to 3 oz a.e./acre) **Timing:** Postemergence when target plants are growing rapidly. **Remarks:** Splitting the 12 ounce application into two applications of 6 oz each in May and late June has improved control and seedhead suppression. Imazapic is not registered for use in California.
Propoxycarbazone-sodium *Canter R+P*	**Rate:** 0.9 to 1.2 oz product/acre (0.63 to 0.84 oz a.i./acre) **Timing:** Postemergence from the 2-leaf to 2-tiller stage when plants are growing rapidly. **Remarks:** Propoxycarbazone is a broad-spectrum herbicide that will control many species. It will provide only partial control of foxtail barley. Perennial grass species vary in tolerance. A non-ionic surfactant should be added at 0.25 to 0.5% v/v solution.
Sulfometuron *Oust*	**Rate:** 1.33 to 2 oz product/acre (1 to 1.5 oz ai/acre); or 3 to 5 oz product/acre (2.25 to 3.75 oz a.i./acre). Rate depends on environmental conditions. **Timing:** Preemergence or postemergence, just before or during the rainy season when the target plants are growing rapidly and germinating. Use the lower rate range for areas receiving less than 20 inches precipitation annually and the higher rate range for those receiving greater than 20 inches annual precipitation. **Remarks:** Sulfometuron is a broad-spectrum herbicide that will control many species. It also has long soil residual activity and is susceptible to off-site movement in light wind-blown soils. Add a surfactant to improve control.
PHOTOSYNTHETIC INHIBITORS	
Hexazinone *Velpar L*	**Rate:** 2.75 to 4.5 pt product/acre (0.7 to 1.1 lb a.i./acre) **Timing:** Early postemergence when the target plants are in the seedling stage. **Remarks:** Hexazinone has minimal selectivity and will kill most other vegetation. It is also mobile in the soil and should not be used in areas with a shallow water table. High rates of hexazinone can create bare ground, so only use high rates in spot treatments.

Hordeum marinum Huds. ssp. *gussonianum* (Parl.) Thell.; Mediterranean barley
Hordeum murinum L. ssp. *leporinum* (Link) Arcang.; hare barley

Mediterranean and hare barley

Hordeum murinum

Family: Poaceae
Range: Mediterranean barley occurs in most western states, except North and South Dakota, Wyoming, Colorado and New Mexico. Hare barley also occurs in most western states, except North and South Dakota, and Colorado.
Habitat: Roadsides, annual grassland, open hillsides, managed forestlands, agronomic crops (especially small grains and alfalfa), orchards, vineyards, waste areas and other disturbed sites. Mediterranean barley is more common in moister areas, including some wetlands.
Origin: These species are indigenous to the Mediterranean region of Europe.
Impact: The weedy annual barleys can persist and become dominant in continuously disturbed areas with a Mediterranean climate (wet winters and dry summers). Both species compete with desirable vegetation for limited spring moisture. They can reduce desirable vegetation cover and prevent establishment of native perennial species.
California Invasive Plant Council (Cal-IPC) Inventory: Both species are listed as Moderate Invasiveness

Hordeum marinum

Mediterranean and hare barley are cool-season annual grasses. Mature plants can reach over 3 ft tall but 1 to 2 ft is much more common. Hare barley is typically taller than Mediterranean barley. Mediterranean barley lacks auricles or they are short, less than 2 mm long. In contrast, hare barley has well-developed auricles that clasp the stem. The leaves are flat and narrow, about 3 to 8 mm wide and typically hairy. Stems are round in cross-section, grow erect to somewhat spreading, and often bend abruptly at the base.

Both species produce a bristly thick spike, 1 to 3 inches long, in April through June. The central axis of the spike breaks apart at the nodes at maturity. Reproduction is only by seed. Most spikelets fall close to the parent plant, but some can also be distributed longer distances by clinging to humans or the wool or fur of animals. Seeds have a short-term dormancy after maturity, which is typically lost after a few weeks with the onset of lower fall temperatures. Most of the seeds germinate in fall the year they are produced, so large persistent seedbanks are unlikely for these grasses. Seed longevity in the soil is expected to be at least a couple of years.

NON-CHEMICAL CONTROL

Mechanical (pulling, cutting, disking)	Small infestations can be removed by digging and hand-pulling. The entire plant should be uprooted before spike formation.
	Mowing can somewhat suppress Mediterranean and hare barley but results have been variable. The weed typically recovers if mowed before maturity and senescence. However, fewer seedheads are produced when the plant regrows. Mowing and subsequent removal of the biomass can reduce germination the following year.
	Tillage is effective at controlling emerged plants but can stimulate subsequent emergence.
Cultural	These barley species are sometimes grazed during the immature growth stages before seedhead formation. The awns of mature plants can injure the mouth, eyes, nasal passages, feet and skin of grazing animals. Populations can increase in moderately grazed pastures but in some cases heavy grazing has diminished populations. Overall, grazing is marginally effective, with variable results depending on the degree of grazing and on residual soil moisture status.
	Thick mulches can reduce germination.
	Fire is an effective control measure for these *Hordeum* species. In one study, burning alone as a control

	measure reduced the cover of annual barley from 90% to less than 5% cover for approximately 3 years. The timing of burning is in the ripe seed stage before seed disperses from dead plants.
Biological	Because of the close relationship to cultivated barley, there are no known biological controls for either species of weedy annual barley.

CHEMICAL CONTROL

The following specific use information is based on published papers and reports by researchers and land managers. Other trade names may be available, and other compounds also are labeled for this weed. Directions for use may vary between brands; see label before use. Herbicides are listed by mode of action and then alphabetically. The order of herbicide listing is not reflective of the order of efficacy or preference.

ACCASE INHIBITORS	
Clethodim *Select, Envoy*	**Rate:** Broadcast foliar treatment: 6 to 8 oz product (*Select*)/acre (1.5 to 2 oz a.i./acre) for seedlings. Spot treatment: 0.25% to 0.5% v/v solution
	Timing: Postemergence. Best when applications are made before plants are 6 inches tall. It is less effective if applied after a mowing.
	Remarks: Clethodim is grass-selective and safe on broadleaf species. To select for perennial grasses, apply before perennials emerge. It has no soil activity. Use a crop oil surfactant. Registered for fallow and non-crop areas, not generally for rangeland/natural areas, but has specific-use supplemental labels. Note that *Envoy* formulation is 1 lb a.i./gallon, *Select* is 2 lb a.i./gallon.
Fluazifop *Fusilade*	**Rate:** Broadcast foliar treatment: 1 to 1.5 pt product/acre (4 to 6 oz a.i./acre) for established plants, 8 oz product/acre (2 oz a.i./acre) for seedlings. Spot treatment: 0.5% v/v solution
	Timing: Postemergence. Best when applications are made before the boot stage.
	Remarks: Fluazifop is grass-selective and safe on broadleaf species. To select for perennial grasses, apply before perennials emerge. It has no soil activity. Use a crop oil surfactant. Registered for fallow and non-crop areas, not generally for rangeland/natural areas, but has specific-use supplemental labels.
AROMATIC AMINO ACID INHIBITORS	
Glyphosate *Roundup, Accord XRT II*, and others	**Rate:** For foliar broadcast treatment: 1 to 3 pt product (*Roundup ProMax*)/acre (0.56 to 1.7 lb a.e./acre). Spot treatment: 0.5% to 1% v/v solution with the rate depending on weed size.
	Timing: Postemergence when the plants are small and rapidly growing, from seedling to boot stage. Applications at tillering often provide the best control in rangelands.
	Remarks: Glyphosate has no soil activity and will only provide control the year of application. It is nonselective and kills most non-target plants creating bare ground conditions that leave the area susceptible to weed encroachment. In areas with desirable vegetation, use spot treatment. Glyphosate is an effective option when reseeding is planned shortly after application, as it will not injure seedlings emerging after application. Add a surfactant when using a formulation where it is not already included (e.g., *Rodeo, Aquamaster*).
BRANCHED-CHAIN AMINO ACID INHIBITORS	
Imazapyr *Arsenal, Habitat, Chopper, Stalker, Polaris*	**Rate:** 2 to 3 pt product/acre (8 to 12 oz a.e./acre)
	Timing: Preemergence or postemergence.
	Remarks: Imazapyr has fairly long soil residual activity. It is a nonselective herbicide.
Rimsulfuron *Matrix*	**Rate:** 2 to 3 oz product/acre (0.5 to 0.75 oz a.i./acre)
	Timing: Preemergence in fall to early postemergence in early spring. The higher rate is generally needed in spring.
	Remarks: Rimsulfuron controls several annual grasses and broadleaves. Limited data suggests it will control these *Hordeum* species. Perennial grasses are tolerant to fall applications when established and grown under dryland conditions. Application to rapidly growing or irrigated perennial grasses may result in injury or death of the crop. It provides soil residual control in cool climates but degrades rapidly under warm conditions. Rimsulfuron will not control summer annual weeds when applied in fall or spring. Add a surfactant when applying postemergence.
Sulfometuron	**Rate:** 0.75 to 5 oz product/acre (0.56 to 3.75 oz a.i./acre)
	Timing: Preemergence or early postemergence from fall to early spring. Most effective control is with early

Oust and others	postemergence treatment after seedlings have emerged.
	Remarks: Sulfometuron has mixed selectivity and is fairly safe on native perennial grasses. It is good for revegetation use. Use lower rates in arid environments and higher rates in wetter areas (> 20" rainfall) and on high organic matter soils. Sulfometuron has fairly long soil residual activity. Higher rates generally result in bare ground.
Sulfometuron + chlorsulfuron *Landmark XP*	**Rate:** 2.25 oz product/acre **Timing:** Preemergence in fall or after soil thaws in spring. **Remarks:** See sulfometuron.
Sulfosulfuron *Outrider*	**Rate:** 0.75 to 2 oz product/acre (0.56 to 1.5 oz a.i./acre) **Timing:** Early postemergence, fall to early spring, when desirable perennials are dormant. **Remarks:** Sulfosulfuron has mixed selectivity, but is fairly safe on native perennial grasses, especially wheatgrasses. It has fairly long soil residual activity. Treatments should include a non-ionic surfactant.
PHOTOSYNTHETIC INHIBITORS	
Hexazinone *Velpar L*	**Rate:** 2 to 6 pt product/acre (0.5 to 1.5 lb a.i./acre) **Timing:** Preemergence to early postemergence. **Remarks:** Hexazinone has both foliar and soil activity. In soil applications, rates will vary with soil texture and soil organic matter. Best results when applied to moist soils. Use rates will also vary depending on the situation. Hardwood trees near application site can absorb this chemical through the roots. High rates of hexazinone can create bare ground, so only use high rates in spot treatments.

Hydrilla verticillata (L.f.) Royle

Hydrilla

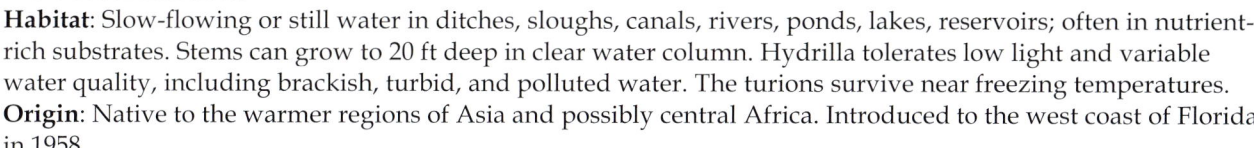

Family: Hydrocharitaceae
Range: Tropical to temperate regions on all continents except Antarctica. In the western U.S., it has been found in Washington, California, and Arizona, but is far more common in the southern states.
Habitat: Slow-flowing or still water in ditches, sloughs, canals, rivers, ponds, lakes, reservoirs; often in nutrient-rich substrates. Stems can grow to 20 ft deep in clear water column. Hydrilla tolerates low light and variable water quality, including brackish, turbid, and polluted water. The turions survive near freezing temperatures.
Origin: Native to the warmer regions of Asia and possibly central Africa. Introduced to the west coast of Florida in 1958.
Impacts: Hydrilla can aggressively invade new aquatic environments, displace native aquatic vegetation by forming dense stands or large sub-surface mats, and alter the dynamics of aquatic ecosystems. Other detrimental and economic impacts from heavy infestations can include water flow impediment in waterways, increased flooding, clogged pumps and boat propellers, and reduced use of lakes and waterways for fishing and other recreational activities.
Western states listed as Noxious Weed: Arizona, California, Colorado, New Mexico, Nevada, Oregon, Washington. Hydrilla is also on the U.S. Federal Noxious Weed List.
California Invasive Plant Council (Cal-IPC) Inventory: High Invasiveness (Alert)

Hydrilla is a submerged aquatic perennial. Its stems typically grow rooted in the substrate. There are many biotypes, including monoecious and dioecious types. Usually monoecious biotypes spread horizontally first, then extend vertically; the dioecious biotypes produce tall vertical shoots first, then spread laterally. The leaves are sessile, whorled, often scale-like and opposite on lower stems. Middle and upper leaves are often 5 to 8-whorled, up to 1 inch long, 1 to 4 mm wide. Leaf margins are minutely toothed, visible with low magnification. Certain nodes on stems and stolons develop adventitious roots, typically at nodes of dormant axillary buds or branches. During brief periods in spring and for longer periods in fall, hydrilla plants often direct some newly formed shoots (from the root crown) straight down into the sediment, and subterranean turions, more commonly called "tubers", form at these shoot tips.

Female flowers extend to the water's surface on thread-like flower tubes several inches long. The sepals and petals are translucent, white to reddish, 3 to 5 mm long. Fruit production and seed set is typically low in hydrilla and seedlings are seldom encountered. As such, nearly all populations of hydrilla in the United States reproduce vegetatively. Stems can easily fragment into free-floating pieces that root at nodes. In addition, plants develop overwintering tubers in the soil substrate, and stem shoots form specialized perennating buds enclosed by tough leaf scales called turions. The stem turions separate from the parent stem in late fall or at maturity. Turions usually germinate readily and survive up to 4 to 5 years under normal aquatic conditions. Tubers can only survive from a few days to weeks under dry conditions. All these vegetative reproductive parts disperse with water and also with human activities, such as fishing and boating.

NON-CHEMICAL CONTROL

Mechanical (pulling, cutting, dredging, excavating)	Removing and destroying stem fragments from recreational equipment, such as boat propellers, docking lines, and fishing gear can help prevent the spread of hydrilla.
	Removing dense canopies by mechanical harvesting may stimulate growth and spread thousands of viable fragments that can disperse, lodge along the bottom or shoreline, and start new populations of hydrilla.
	Diver-assisted dredging can be very effective in small areas (< 2 acres). Since populations over a year old will have already produced a very large "tuber bank" in the sediment, dredging or excavation to remove at least 18 inches of sediment is recommended if eradication is the goal.
Cultural	Preventing introduction and spread is critical to avoiding long-term impacts and costs of controlling hydrilla. Imported benign plants can harbor hydrilla fragments and even tubers and turions. Careful inspection is critical.
	Dewatering (drawdown) during mid-summer can reduce the potential for production of a viable tuber

	bank. Dewatering in winter alone has little effect since tubers usually form 0.5 to 1.5 ft below the sediment. Dewatering coupled with sediment-applied herbicides can be effective (see chemical control options).
Biological	The most effective biological control agent is the triploid (sterile) grass carp (white amur). Grass carp is a relatively nonselective herbivorous fish that will consume several species of submersed plants. It has been used for management and as part of hydrilla eradication programs in California since 1985. Several host-specific herbivorous insects have been released in the United States, including the flies *Hydrellia pakistanae* and *Hydrellia balciunasi*, a moth *Paraponyx dimunutalis*, and the tuber feeding weevil *Bagous affinis*. These insects have shown variable efficacy but should be considered if long-term management (not eradication) is the goal. The native fungal pathogen *Mycoleptodiscus terrestris* ("MT") has been investigated for 20 years but to date no efficacious product has been developed.

CHEMICAL CONTROL

The following specific use information is based on published papers and reports by researchers and land managers. Other trade names may be available, and other compounds also are labeled for this weed. Directions for use may vary between brands; see label before use. Herbicides are listed by mode of action and then alphabetically. The order of herbicide listing is not reflective of the order of efficacy or preference.

BRANCHED-CHAIN AMINO ACID INHIBITORS	
Bispyribac-sodium *Tradewind*	**Rate:** For in-water treatment: 20 to 45 ppb
	Timing: Apply directly to water from early spring to early summer during the period of rapid growth. Repeat applications, no less than 14 days after last application, may be needed to achieve target concentration.
	Remarks: Maximum of four applications per year. Slow activity, requires 2 to 6 weeks to get full effect. Can increase some natives due to selective effects. Residual in water is 2 to 3 months.
Imazamox *Clearcast*	**Rate:** For in-water treatment: 150 to 200 ppb; 50 to 75 ppb can be effective if applied in early spring, but repeat application is generally necessary at these lower rates. For drawdown (dewatered) treatment: 64 oz product/acre (0.5 lb a.e./acre); first flush of water in canals must NOT be used for irrigation
	Timing: Apply directly to water from early spring to early summer during the period of rapid growth. For dewatered treatments, apply in late winter at least 14 days before water will be reintroduced.
	Remarks: Use an approved surfactant. Aerial application is approved in some states.
Penoxsulam *Galleon*	**Rate:** For in-water treatment: 25 to 75 ppb. Repeat treatments may be required, but not to exceed 150 ppb in an annual season. For drawdown (dewatered) treatment: 5.6 to 11.2 oz product/acre (1.4 to 2.8 oz a.i./acre)
	Timing: Apply directly to water from early spring to early summer during the period of rapid growth. For drawdown treatment, apply during mid- to late winter before refilling.
	Remarks: Penoxsulam is a slow-acting herbicide and may take 4 to 6 week for effective control. For drawdown applications use 20 to 100 gal/acre to wet the sediment.
PIGMENT SYNTHESIS INHIBITORS	
Fluridone *Sonar*	**Rate:** For in-water treatment: 5 to 20 ppb, but exposures must be maintained for 6 to 8 weeks for optimal control.
	Timing: Apply directly to water from early spring to early summer.
	Remarks: Fluridone is a systemic herbicide. It affects young, rapidly growing plants. Lower rates can be used if applied during early spring growth and when water movement is not likely to dilute or move the herbicide. Due to the development of resistance to fluridone in some hydrilla populations, careful monitoring of efficacy is required, and alternative active ingredients should be used periodically as part of an integrated management program.
CONTACT PHOTOSYNTHETIC INHIBITORS	
Diquat *Reward*	**Rate:** For in-water treatment: 0.1 to 0.25 ppm
	Timing: Apply directly to water from late spring to early summer.
	Remarks: Diquat is a fast-acting contact herbicide that can be effective in mid- to late summer, but if biomass is large, only a portion of the infested sites should be treated to minimize effects of reduced dissolved oxygen. Diquat is quickly bound to, and becomes inactivated on, suspended clay particles and it should not be used in moderately or highly turbid water.

Flumioxazin *Clipper*	**Rate:** For in-water treatment: 100 to 400 ppb **Timing:** Apply directly to water from early spring to early summer. Fall applications may also be effective if temperatures remain high. **Remarks:** Do not use flumioxazin if pH is > 8.0. If pH is high, use a buffer to reduce it below 8. To minimize effects of high pH, apply from dawn to mid-morning. Due to photosynthesis of aquatic plants and algae, pH in the water column rises from mid-day to dusk under most circumstances.
GENERAL CELL TOXICANT	
Acrolein *Magnacide H*	**Rate:** For in-water treatment: 1 to 15 ppm. The recommended rate is variable and depends on target weeds, temperature and flow rates **Timing:** Apply directly to water in late spring to fall and use no more than 8 applications per year. **Remarks:** Acrolein is a very fast-acting, nonselective contact herbicide and algaecide. It is a "Restricted Use" pesticide but can be used in some irrigation canals under specific conditions, with proper permits, and may only be applied by qualified, trained applicators with proper protective clothing. Symptoms of efficacy may appear in less than an hour and include discoloration of leaves and loss of turgidity.
Endothall *Cascade; Teton; Aquathol K*	**Rate:** For in-water treatment: 0.2 to 5.0 ppm (e.g., flowing water in irrigation canals). For effective control, exposures must be maintained for 6 to 120 hours. Duration of contact depends on the concentration achieved. **Timing:** Apply directly to water from early spring to early summer. Endothall can be used in mid-summer, but if biomass is heavy, partial treatments are recommended in order to prevent large reduction in dissolved oxygen. **Remarks:** Endothall is a selective, contact herbicide. It affects young, rapidly growing plants and mature plants. Lower rates can be used if applied during early spring growth and when water movement is not likely to dilute or move the herbicide.
INORGANIC HERBICIDES	
Chelated copper *Komeen, Cutrine-Plus*	**Rate:** For in-water treatment: 0.5 to 1 ppm elemental copper **Timing:** Apply directly to water in early summer when plants and biomass are small. **Remarks:** Chelated copper is a fast-acting contact herbicide. Retreatment may be required within 3 to 5 weeks. If biomass is large, treat only one-third of infested area to minimize decrease in dissolved oxygen. Chelated copper products are less affected by high pH compared to inorganic copper.
NON-HERBICIDAL CHEMICAL	
Dyes or colorants *Aquashade*	Although technically not herbicides, dyes and colorants control submerged aquatic plants by absorbing light in the water column and reducing photosynthesis. Applications should be made in early spring and repeated to maintain concentration recommended on the label. Colorants are not as effective on well-established plants in mid- to late summer.

Hydrocotyle ranunculoides L.f.
Floating pennywort

Family: Apiaceae
Range: Oregon, Washington, Arizona, and California. Also very common in the southern United States.
Habitat: Ponds and lake margins, marshes, low swamps, slow streams, irrigation and drainage ditches.
Origin: Floating pennywort is a widespread native of North America. Plants are sometimes sold as aquatic or pond ornamentals and have escaped cultivation in some regions. In Britain, floating pennywort has become a problematic weed of natural aquatic habitats, and in southern and western Australia, it is a government listed noxious weed.
Impacts: In natural areas, colonies are usually considered a desirable component of aquatic ecosystems. Because of its creeping habit, floating pennywort can be a nuisance in irrigation and drainage ditches. It has become a more prevalent problem in some areas where water hyacinth has been controlled.

Floating pennywort is an aquatic or terrestrial perennial with branched creeping stems that root at the nodes. Plants grow in dense, low-growing mats in shallow water or on wet soil near water. Occasionally small colonies are free-floating. The foliage is glabrous and somewhat fleshy. Stems are easily fragmented. The leaves are alternate and round or kidney-shaped.

The inflorescence is a simple umbel with ~5 to 10 flowers on stalks usually shorter than the leaves. Flowers consist of five greenish- or yellowish-white to purplish petals. Fruits (schizocarps) separate into halves at maturity. Plants reproduce by seed and/or vegetatively from creeping stems and stem fragments. Seeds and stem fragments primarily disperse with water. In Britain, floating pennywort plants rooted in the substrate produce seed, while floating colonies primarily reproduce vegetatively. There are no data to indicate seed longevity in the soil, but it is expected to be around 3 years based on other members of the Apiaceae.

NON-CHEMICAL CONTROL

Mechanical (pulling, cutting, chopping)	Construction of dams or narrow restrictions in rivers and streams can create still water conditions that are favorable to establishment of pennywort. Even high-flowing river systems are susceptible to infestations in quiescent areas near turns and within riparian vegetation such as cattails and bulrushes.
	Mechanical harvesting can be effective. Best results are achieved when closed systems are used to transport harvested plants to local landfill or composting sites. Mechanical removal followed by hand picking four times a year during the growing season is now the accepted practice in the UK. Mechanical choppers and shredders leave viable pieces of pennywort rhizomes that can easily reestablish populations, and also provide an opportunity for dispersal by moving water or wind.
Cultural	Although pennywort is a native plant, it can create large mats when there are sufficient nutrients in the water and when roots can access sediment-borne nutrients (e.g. shallow water and shorelines). Some suppression in mat size and density can be achieved if there is adequate shoreline shade from trees.
	Dewatering (drawdown) small lakes or ponds alone will not control pennywort, but can be done to allow scraping and removal with large equipment. This may also provide the opportunity to use foliar-applied herbicides without overspraying water.
Biological	There is some interest in developing biological control agents in the UK, and a weevil native to South America, *Listronotos elongates*, has been identified as a candidate. The triploid (sterile) grass carp (white amur) is a relatively nonselective herbivorous fish that will consume pennywort if its more favored submersed plants are absent. This may be an effective method in small ponds or lakes that do not support many plants. Permits are required for use of grass carp in most U.S. states.

CHEMICAL CONTROL
The following specific use information is based on reports by researchers and land managers. Other trade names may be available, and other compounds also are labeled for this weed. Directions for use may vary between

brands; see label before use. Herbicides are listed by mode of action and then alphabetically. The order of herbicide listing is not reflective of the order of efficacy or preference.

GROWTH REGULATORS	
2,4-D *Weedar 64*	**Rate:** 1 to 4 pt product/acre (0.48 to 1.9 lb a.e./acre) with a non-ionic surfactant **Timing:** Postemergence in spring to early summer. However, mid-summer to early fall applications can also be effective in suppressing growth. **Remarks:** 2,4-D is a relatively fast-acting, selective systemic herbicide. Repeated application of 2,4-D amine, at intervals of 3 weeks, delays regrowth of the plant by up to 12 weeks.
Triclopyr *Renovate*	**Rate:** 4 to 16 pt product/acre (1.5 to 6 lb a.e./acre) with an approved non-ionic surfactant **Timing:** Postemergence in early spring to early summer. Applications in mid-summer can also suppress growth, but may give rapid dieback that can result in depressed dissolved oxygen. **Remarks:** Triclopyr is a selective, systemic herbicide. Lower rates can be used if applied during early spring growth when plants are small.
BRANCHED-CHAIN AMINO ACID INHIBITORS	
Bispyribac-sodium *Tradewind*	**Rate:** 1 to 2 oz product/acre (0.8 to 1.6 oz a.i./acre) **Timing:** Postemergence from early spring to early summer during the period of rapid growth. Treatments may need to be repeated, but allow 30 days between applications and do not exceed four applications per year or 8 oz product/acre/year. **Remarks:** Bispyribac-sodium is a slow-acting herbicide that may take 4 to 6 weeks to achieve control. It can also be tank mixed with other herbicides.
Imazamox *Clearcast*	**Rate:** Broadcast treatment to emergent foliage: 2 to 4 pt product/acre (4 to 8 oz a.e./acre). Spot treatment: 0.25 to 5% v/v solution **Timing:** Postemergence from early spring to early summer during the period of rapid growth. **Remarks:** Use an approved surfactant. Aerial application is approved in some states.
Imazapyr *Habitat*	**Rate:** Broadcast treatment to emergent foliage: 1 to 2 pt/acre (4 to 8 oz a.e./acre). Spot treatment: 0.5% v/v solution using 100 gal/acre spray solution. **Timing:** Postemergence in early spring to early summer when new growth is present. **Remarks:** Imazapyr is a relatively slow-acting systemic herbicide.
CONTACT PHOTOSYNTHETIC INHIBITORS	
Diquat *Reward*	**Rate:** Spot treatment to emergent shoots: 0.5% v/v solution or 2 qt per 100 gal water **Timing:** Postemergence in spring to early summer. Repeat treatments may be needed in mid-summer. **Remarks:** Diquat is a contact herbicide that is inactivated in turbid water. Use only clean water to mix and spray herbicide. Repeat applications are required to achieve eradication.

Hyoscyamus niger L.

Black henbane

Family: Solanaceae
Range: Most western states with the exception of Arizona and California.
Habitat: Disturbed open sites, roadsides, fields, waste places, abandoned gardens and other non-crop areas. Grows best in sandy or well-drained loam soils with moderate fertility.
Origin: Native to Eurasia and introduced to eastern North America in the early 1600s as a medicinal herb.
Impacts: All plant parts contain the tropane alkaloids hyoscyamine, scopolamine, and atropine and are toxic to humans and animals when ingested. Seeds have the highest concentration of alkaloids. Livestock rarely consume foliage because of the unpleasant odor and bitter taste. Most toxicity problems occur in humans that ingest seeds, particularly children and people who abuse black henbane for its neurological effects.
Western states listed as Noxious Weed: California, Colorado, Idaho, Nevada, New Mexico, Utah, Washington

Black henbane is an erect summer annual or biennial generally to about 3 ft, but it can grow to 6 ft tall under some conditions. The coarse foliage has sticky glandular hairs and a foul odor. The leaves are alternate, gray-green, oblong to lanceolate, 2 to 8 inches long, with coarsely toothed to acutely pinnate-lobed margins.

The inflorescence is a terminal raceme that is one-sided and somewhat coiled at the tip. Petals are fused and funnel-shaped with five unequal lobes. Flowers are 1 to 1.5 inches long and pale yellow to greenish with conspicuous purple veins and a purple throat. The calyx is persistent and urn-shaped, five-lobed and densely covered with long glandular hairs at the base. The fruit are capsules from 0.5 to 0.75 inch long and contain numerous seeds. Plants reproduce by seed only. Seeds disperse by falling at the base of the parent plants. Under field conditions, seeds appear to remain viable for up to about 4 years.

NON-CHEMICAL CONTROL

Mechanical (pulling, cutting, disking)	Hand removal has been shown to offer some level of control, as has mowing and cultivation. Gloves should be worn for any hand removal as the plant is poisonous. Taproots must be removed to 2 inches below ground to ensure that resprouting does not occur. Mechanical methods should be repeated annually to exhaust the soil seed reserve.
Cultural	Plants with mature fruits can be burned to kill seed. Plants are poisonous, thus they are not recommended in a grazing control program.
Biological	There are no biological control agents available for black henbane.

CHEMICAL CONTROL

The following specific use information is based on published papers or reports by researchers and land managers. Other trade names may be available, and other compounds also are labeled for this weed. Directions for use may vary between brands; see label before use. Herbicides are listed by mode of action and then alphabetically. The order of herbicide listing is not reflective of the order of efficacy or preference.

GROWTH REGULATORS

2,4-D Several names	**Rate:** 2 to 4 pt product/acre (0.95 to 1.9 lb a.e./acre)
	Timing: Postemergence before flowering to prevent seed production and dispersal. Best applied to young plants.
	Remarks: 2,4-D is a restricted use herbicide in some areas. It will damage most broadleaf species.
Dicamba Banvel, Clarity	**Rate:** 1 to 2 pt product/acre (8 to 16 oz a.e./acre)
	Timing: Postemergence before flowering to prevent seed production and dispersal. Best applied to young plants from rosette to bolting stage.
	Remarks: Dicamba is a broadleaf herbicide with little soil activity.

Fluroxypyr *Vista XRT*	**Rate:** 15 to 22 oz product/acre (5.3 to 7.7 oz a.e./acre) **Timing:** Postemergence before flowering to prevent seed production and dispersal. Best applied to young plants from rosette to bolting stage. **Remarks:** Fluroxypyr is a broadleaf herbicide with little soil activity.
Picloram *Tordon 22K*	**Rate:** 1 to 2 pt product/acre (4 to 8 oz a.e./acre) **Timing:** Preemergence or postemergence in spring when plants are growing rapidly, but before bloom. Treatments can also be made in late summer for preemergence activity. Picloram can be used in a premix with 2,4-D (*Grazon P+D*) or tank mixed with 2,4-D at 1 lb a.e./acre. **Remarks:** Picloram is a restricted use herbicide. It is not registered for use in California.
AROMATIC AMINO ACID INHIBITORS	
Glyphosate *Roundup, Accord XRT II*, and others	**Rate:** 2 to 4 qt product (*Roundup ProMax*)/acre (2.25 to 4.5 lb a.e./acre) **Timing:** Postemergence before flowering to prevent seed production and dispersal. Best applied to young plants. **Remarks:** Glyphosate provides effective control. It is nonselective and has no soil activity. Wiper applications for small patches can provide selectivity.
BRANCHED-CHAIN AMINO ACID INHIBITORS	
Chlorsulfuron *Telar*	**Rate:** 0.5 to 1 oz product/acre (0.375 to 0.75 oz a.i./acre) **Timing:** Postemergence to rapidly growing plants from bolting to early flowering stage. **Remarks:** Chlorsulfuron is a very effective control option, but has a broad spectrum of susceptible species. Its residual soil activity gives effective control one year later. Chlorsulfuron can be applied in combination with metsulfuron in the premix trade name *Cimarron Plus*. This combination is not registered for use in California.
Metsulfuron *Escort*	**Rate:** 1 to 2 oz product/acre (0.6 to 1.2 oz a.i./acre) **Timing:** Postemergence to rapidly growing plants from bolting to early flowering stage. **Remarks:** Metsulfuron has some residual control activity. Use with a non-ionic or silicone-based surfactant. It can be applied in combination with chlorsulfuron in the premix trade name *Cimarron X-tra*. Metsulfuron is not registered for use in California.

Hypericum perforatum L.
Common St. Johnswort (Klamathweed)

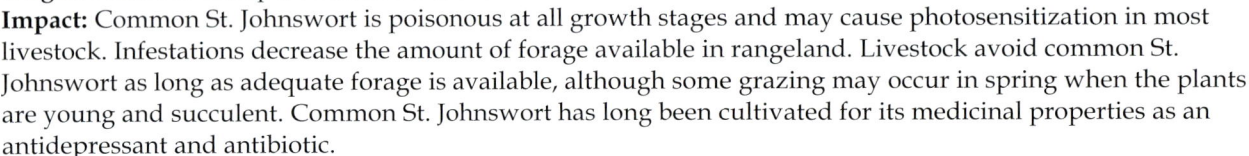

Family: Clusiaceae or Hypericaceae
Range: Nearly all of the contiguous states, including most western states except for Arizona. Also found in Hawaii.
Habitat: Forest, woodland, rangeland, and prairie communities. Less commonly occurs in riparian areas. Often associated with disturbances caused by roads, logging, grazing, and fire.
Origin: Native to Europe, western Asia, and North Africa.
Impact: Common St. Johnswort is poisonous at all growth stages and may cause photosensitization in most livestock. Infestations decrease the amount of forage available in rangeland. Livestock avoid common St. Johnswort as long as adequate forage is available, although some grazing may occur in spring when the plants are young and succulent. Common St. Johnswort has long been cultivated for its medicinal properties as an antidepressant and antibiotic.
Western states listed as Noxious Weed: California, Colorado, Montana, Nevada, Oregon, South Dakota, Utah, Washington, Wyoming.
California Invasive Plant Council (Cal-IPC) Inventory: Moderate Invasiveness

Common St. Johnswort is an erect, herbaceous, tap-rooted perennial forb that is typically around 2 to 3 ft tall but under ideal conditions can grow up to 5 ft. One to 30 stems arise from a woody root crown. The stems are woody at the base and branched and leafy in the upper half. The leaves are linear and about 1 inch long by about 0.4 inch wide. They are opposite and lack a petiole. The foliage is dotted with tiny translucent and black oil glands that are most noticeable when held up to the light.

The flowers are yellow and about 0.8 inch wide with 5 petals and numerous stamens. Flowers occur in terminal clusters of 25 to 100 at the stem tips. The fruit is a sticky, many-seeded, 3-celled capsule that is 5 to 10 mm long. Common St. Johnswort spreads by seeds and rhizomes, but seeds are the primary mechanism of reproduction. Estimates of seed production vary, but range from about 15,000 to 34,000 seeds per plant. Common St. Johnswort seed generally fall below the parent plant, but can be dispersed longer distances by water. Seeds of *Hypericum* species may remain viable in the soil for periods longer than 50 years.

NON-CHEMICAL CONTROL

Mechanical (pulling, cutting, disking)	Given common St. Johnswort's deep taproot and ability to regenerate, hand-pulling or digging is only practical for very small isolated infestations.
	Mowing is ineffective because plants can resprout from underground root reserves. However, mowing can postpone or reduce seed drop and repeated mowing may deplete the underground root reserves. Results from Australia showed that mowing at two week intervals was effective especially when done in spring.
	Common St. Johnswort is not a problem in cultivated crops, indicating it does not tolerate tillage. However, tillage is not practical in most natural areas.
Cultural	Grazing for control of common St. Johnswort is generally not recommended because of the potential for livestock poisoning. Intensive grazing by goats, which are less susceptible to the poison, may help to keep common St. Johnswort density at low levels.
	Burning may kill top growth but may not damage the crown and root system and may stimulate common St. Johnswort to resprout. There is also evidence that fire stimulates common St. Johnswort seeds to germinate. In general, burning is not a recommended control measure as it generally encourages establishment of the plant.
	Some research has shown that mulching—cutting the plants down to 2 inches and covering them with tarpaper—has been effective.
Biological	Common St. Johnswort has been regarded as one of the success stories for biological control of weeds. The overall impact of common St. Johnswort has been reduced through biological control agents. However, they have not yet managed to reduce common St. Johnswort infestations to levels that do not cause unacceptable

economic or environmental damage in some areas, and the weed has continued to spread. This may be due to climatic issues where the beetles are unable to reach high enough densities to result in an adequate level of control. In some areas where the beetles are well adapted, common St. Johnswort populations have been reduced 97 to 99%. In some cases, densities may be so low that the population of biocontrol agents cannot respond when the weed populations increase, and reintroduction of biocontrol agents may be necessary. The primary biological control agents include St. Johnswort or Klamathweed beetles (*Chrysolina hyperici* and *quadrigemina*). These leaf-feeding flea beetles provide excellent control in some areas. In addition, St. Johnswort root borer (*Agrilus hyperici*) also provides excellent control, whereas the St. Johnswort inchworm (*Aplocera plagiata*) and the St. Johnswort gall midge (*Zeuxidiplosis giardia*) provide only fair control.

CHEMICAL CONTROL

The following specific use information is based on published papers and reports by researchers and land managers. Other trade names may be available, and other compounds also are labeled for this weed. Directions for use may vary between brands; see label before use. Herbicides are listed by mode of action and then alphabetically. The order of herbicide listing is not reflective of the order of efficacy or preference.

GROWTH REGULATORS	
2,4-D Several names	**Rate:** 2 to 4 pt product/acre (0.95 to 1.9 lb a.e./acre) **Timing:** Application postemergence to new seedlings is most effective, but if not possible apply before any blossoms open. **Remarks:** 2,4-D is broadleaf-selective and has no soil activity. It may require repeat applications. It is often tank mixed with other herbicides. It should be used with a surfactant. Do not apply ester formulations when outside temperatures exceed 80°F.
Aminopyralid *Milestone*	**Rate:** 5 to 7 oz product/acre (1.25 to 1.75 oz a.e./acre) **Timing:** Postemergence to rapidly growing plants before bloom. **Remarks:** Aminopyralid is broadleaf-selective and is safe on grasses, although preemergence application at high rates can greatly suppress invasive annual grasses, such as medusahead. For postemergence applications, a non-ionic surfactant (0.25 to 0.5% v/v spray solution) enhances control under adverse environmental conditions; however, this is not normally necessary. Other premix formulations of aminopyralid can also be used for common St. Johnswort control. These include *Opensight* (aminopyralid + metsulfuron; 2.5 to 3.3 oz product/acre) and *Forefront HL* (aminopyralid + 2,4-D; 2 pt product/acre), both applied at the rosette to bolting stages.
Picloram *Tordon 22K*	**Rate:** 1 to 2 qt product/acre (0.5 to 1 lb a.e./acre) **Timing:** Postemergence when the target plants are small and rapidly growing, preferably before bloom. **Remarks:** Most broadleaf plants are susceptible to picloram. Control using a lower rate may be improved by tank mixing with 0.5 to 1 lb a.e./acre 2,4-D. Although well-developed grasses are not usually injured by labeled use rates, some applicators have noted that young grass seedlings with fewer than four leaves may be killed. Do not apply near trees. *Tordon 22K* is a federally restricted use pesticide. Picloram is not registered for use in California.
AROMATIC AMINO ACID INHIBITORS	
Glyphosate *Roundup, Accord XRT II*, and others	**Rate:** 1 to 2 qt product (*Roundup ProMax*)/acre (1.1 to 2.25 lb. a.e./acre). Wiper treatment: 33 to 50% of concentrated product. **Timing:** Postemergence to rapidly growing plants at late bud to flowering stage. **Remarks:** Glyphosate is nonselective and has no soil activity. Long-term management will likely require additional treatments to control newly germinating plants the following season. It is generally used as part of a revegetation program or for spot treatment. Rope wick or sponge wick (wiper) applications have also been effective for common St. Johnswort control, while reducing damage to desirable vegetation.
BRANCHED-CHAIN AMINO ACID INHIBITORS	
Metsulfuron *Escort*	**Rate:** 1 oz product/acre (0.6 oz a.i./acre) **Timing:** Postemergence when the target plants are small and rapidly growing. **Remarks:** Metsulfuron has mixed selectivity but is generally safe on grasses. Use a non-ionic surfactant at 0.5% by volume surfactant. It can be tank-mixed with 2,4-D or aminopyralid (*Opensight*). Metsulfuron has some soil residual activity. It is not registered for use in California.

Hypochaeris glabra L.; smooth catsear
Hypochaeris radicata L.; common catsear

Smooth and common catsear

Family: Asteraceae
Range: Smooth catsear is found throughout California, Oregon, and Washington, except for the deserts; also in many southern states and a few eastern states. Common catsear is found in most western states, except North and South Dakota, Wyoming and Arizona.

Habitat: Disturbed places, fields, grassland, pastures, roadsides, orchards, vineyards, landscaped areas, and gardens. Smooth catsear also invades agricultural fields and can grow in serpentine soil. Common catsear is more drought-tolerant and can be found in turf.
Origin: Both species are native to Europe.
Impacts: The catsears move into overgrazed pastures and rangeland, crowding out palatable forage species.
Western states listed as Noxious Weed: *H. radicata*, Washington
California Invasive Plant Council (Cal-IPC) Inventory: *H. glabra*, Limited Invasiveness; *H. radicata*, Moderate Invasiveness

The catsears superficially resemble dandelion, with milky juice, a basal rosette of leaves, and yellow dandelion-like flowerheads. However, the catsears have branched flowering stems, unlike dandelion, and dandelion has "rocket"-shaped leaves with sharp lobes pointed back toward the base.

Smooth catsear is an annual with flower stems to 16 inches tall, with a slender taproot. Its leaves are hairless, with smooth to shallow-lobed margins. Common catsear is a perennial with flower stems to 32 inches tall, with a fibrous root system often with several deep roots. It has toothed to lobed leaves covered with rough, yellowish hairs. Both plants grow as prostrate rosettes on open ground, sending up leafless, usually branched flowering stems.

Both catsears produce plumed seeds which are distributed by wind or by clinging to the fur, feathers, and feet of animals. Seeds generally do not persist in the soil seedbank. In heavily grazed areas or mowed turf, common catsear can reproduce vegetatively by offsets from the crown, and diffuse clonal patches can develop. However, root fragments don't regenerate when detached from the crown.

NON-CHEMICAL CONTROL

Mechanical (pulling, cutting, disking)	Hand removal and cultivation can control both species. Common catsear is more difficult, as it has a taproot which can resprout. If the entire taproot can be removed, as with a shovel, to several inches below the root crown, common catsear will not grow back.
	Mowing is not effective in controlling catsear.
Cultural	Light to moderate grazing usually facilitates survival of these species.
	Burning can stimulate germination of common catsear and probably smooth catsear seed.
Biological	No known biocontrol agents are available for either species of *Hypochaeris*.

CHEMICAL CONTROL

The following specific use information is based on published papers and reports by researchers and land managers. Other trade names may be available, and other compounds also are labeled for this weed. Directions for use may vary between brands; see label before use. Herbicides are listed by mode of action and then alphabetically. The order of herbicide listing is not reflective of the order of efficacy or preference.

GROWTH REGULATORS	
2,4-D Several names	**Rate:** 1 to 2 qt product/acre (0.95 to 1.9 lb a.e./acre) **Timing:** Postemergence to rosettes; most effective on smaller plants. **Remarks:** Broadleaf-selective, no soil activity. Common catsear may require repeat application. Not the most effective treatment, but widely used because of low cost. Do not apply ester formulations when outside temperatures exceed 80°F. Recommended rates are based on those reported for similar species.
Aminocyclopyrachlor + chlorsulfuron *Perspective*	**Rate:** 3 to 4.5 oz product/acre **Timing:** Postemergence in spring up to flowering, or at the fall rosette stage. **Remarks:** Provides broad-spectrum control of many broadleaf species. Although generally safe to grasses, it may suppress or injure certain annual and perennial grass species. Do not treat in the root zone of desirable trees and shrubs. Do not apply more than 11 oz product/acre per year. At this high rate, cool-season grasses will be damaged, including bluebunch wheatgrass. Not yet labeled for grazing lands. Add an adjuvant to the spray solution. Recommended rates are based on those reported for similar species. This product is not approved for use in California and some counties of Colorado (San Luis Valley).
Aminopyralid *Milestone*	**Rate:** 5 to 7 oz product/acre (1.25 to 1.75 oz a.e./acre) **Timing:** Apply in winter to early spring for preemergence and seedling treatments, or in spring up to flower bud stage. Can be applied in fall in cold-winter areas. **Remarks:** Broadleaf herbicide like picloram, but more selective. Safe on grasses. Longer residual and higher activity than clopyralid. Recommended rates are based on those reported for several perennial thistle species.
Aminopyralid + 2,4-D, *Forefront HL*; Aminopyralid + metsulfuron, *Opensight*; Aminopyralid + triclopyr, *Capstone*	**Rate:** 1.2 to 2 pt *Forefront HL*/acre; 2.5 to 3.3 oz *Opensight*/acre; 4 to 6 pt *Capstone*/acre **Timing:** Postemergence from rosette to bolting stages. *Opensight* may also be applied in fall to seedlings and rosettes. **Remarks:** Rates for *Forefront HL* and *Capstone* are based on label rates for similar species. *Opensight* is not registered for use in California.
Clopyralid *Transline*	**Rate:** 0.67 to 1.33 pt product/acre (4 to 8 oz ae/acre) **Timing:** Postemergence in spring, up to the flower bud stage. **Remarks:** A broadleaf herbicide like picloram, but more selective. Very safe on grasses. Relatively long residual. Recommended rates are based on those reported for similar species.
Clopyralid + 2,4-D *Curtail*	**Rate:** 2 to 3 qt *Curtail*/acre (use higher rate if plants are drought-stressed) **Timing:** Postemergence to rapidly growing weeds from full rosette to early flower bud. **Remarks:** This mix is broadleaf-selective with a wide range of susceptible species. Recommended rates are based on those reported for similar species.
Dicamba *Banvel, Clarity*	**Rate:** 1 to 2 pt product/acre (0.5 to 1 lb a.e./acre) **Timing:** Postemergence to rapidly growing plants in the rosette stage. Smaller plants are more easily controlled. **Remarks:** Broadleaf-selective herbicide often combined with other active ingredients. It is effective earlier in the season than 2,4-D. It is also effective when tank-mixed with 2,4-D (0.75 lb a.e./acre dicamba + 0.25 lb a.e./acre 2,4-D). Dicamba has very limited soil residual. Avoid drift to sensitive crops. Do not apply when outside temperatures exceed 80°F. It can kill or injure legumes. Recommended rates are based on those reported for similar species. Dicamba is available mixed with diflufenzopyr in a formulation called *Overdrive*. Diflufenzopyr is an auxin transport inhibitor which causes dicamba to accumulate in shoot and root meristems, increasing its activity. *Overdrive* is applied postemergence at 4 to 8 oz product/acre to rapidly growing weeds. Higher rates should be used on large annuals or on perennial weeds. Add a non-ionic surfactant to the treatment solution at 0.25% v/v or a methylated seed oil at 1% v/v solution.
Fluroxypyr *Vista XRT*	**Rate:** 22 oz product/acre (7.7 oz a.e./acre) **Timing:** Postemergence to rapidly growing plants.

	Remarks: Fluroxypyr is broadleaf-selective and safe on most grasses.
Picloram *Tordon 22K*	**Rate:** 1 to 2 pt product/acre (4 to 8 oz a.e./acre) **Timing:** Postemergence in the rosette to flower bud stage in spring, or to new rosettes in fall. **Remarks:** Long soil residual. Most broadleaf plants are susceptible; relatively safe on established grasses but can injure young or germinating grasses. Also effective mixed with dicamba or 2,4-D. Recommended rates are based on those reported for several perennial thistle species. Picloram is a restricted use herbicide. Not registered for use in California.
Triclopyr *Garlon 3A, Garlon 4 Ultra*	**Rate:** 2 pt product/acre (0.75 lb a.e./acre *Garlon 3A*, 1 lb a.e./acre *Garlon 4 Ultra*) **Timing:** Postemergence to rapidly growing plants. **Remarks:** Triclopyr is broadleaf-selective and safe on most grasses. It is most effective on smaller plants. *Garlon 4 Ultra* is formulated as a low volatile ester. However, in warm temperatures, spraying onto hard surfaces such as rocks or pavement can increase the risk of volatilization and off-target damage. Recommended rates are based on those reported for similar species.
Triclopyr + 2,4-D *Crossbow*	**Rate:** 2 to 4 qt product/acre **Timing:** Postemergence in the rosette stage. **Remarks:** Include non-ionic surfactant. Recommended rates are based on those reported for similar species.
AROMATIC AMINO ACID INHIBITORS	
Glyphosate *Roundup, Accord XRT II,* and others	**Rate:** 1.33 to 2.67 qt product (*Roundup ProMax*)/acre (1.5 to 3 lb a.e./acre) **Timing:** Postemergence to rapidly growing plants from rosette to bud stage. **Remarks:** Glyphosate has no soil activity and is nonselective. Repeat applications may be necessary. Effectiveness is increased by addition of ammonium sulfate.
BRANCHED-CHAIN AMINO ACID INHIBITORS	
Chlorsulfuron *Telar*	**Rate:** 1 to 2.6 oz product/acre (0.75 to 1.95 oz a.i./acre) **Timing:** Postemergence in fall to new rosettes, or to rosettes in spring before bolting. **Remarks:** Mixed selectivity, generally safe on grasses, but fall application may injure bromes. It has fairly long soil residual activity. Use a surfactant. Can be used in late season applications to reduce seed production. Recommended rates are based on those reported for similar species.
Imazapyr *Arsenal, Habitat, Stalker, Chopper, Polaris*	**Rate:** 2 to 3 pt product/acre (0.5 to 0.75 lb a.e./acre) **Timing:** Preemergence or postemergence. **Remarks:** Nonselective. Recommended rates are based on those reported for similar species.
Metsulfuron *Escort*	**Rate:** 0.5 to 1 oz product/acre (0.3 to 0.6 oz a.i./acre) **Timing:** Postemergence to young, rapidly growing plants in spring before flowering, or in fall to new rosettes. **Remarks:** Mixed selectivity, generally safe on grasses. Metsulfuron has some soil residual activity. Use a surfactant. It can be tank-mixed with 2,4-D and/or dicamba. Recommended rates are based on those reported for several perennial thistle species. Not registered for use in California.
Metsulfuron + chlorsulfuron *Cimarron X-tra*	**Rate:** 0.5 to 1 oz product/acre **Timing:** Postemergence before flowering. **Remarks:** Recommended rates are based on those reported for similar species. Not registered for use in California.
Sulfometuron *Oust* and others	**Rate:** 3 to 5 oz product/acre (2.25 to 3.75 oz a.i./acre) **Timing:** Preemergence or early postemergence when weeds are germinating or rapidly growing. **Remarks:** Mixed selectivity, fairly long soil residual. It is fairly safe on native perennial grasses, especially wheatgrass. It is also safe on smooth brome. Other desirable grasses may be stunted, stressed, or injured. Good for revegetation use. Do not let spray drift onto sensitive crops. Recommended rates are based on those reported for similar species.
PHOTOSYNTHETIC INHIBITORS	
Hexazinone	**Rate:** 4 to 6 pt product/acre (1 to 1.5 lb a.i./acre)

Velpar L	**Timing:** Preemergence before weeds emerge or postemergence to young plants.
	Remarks: Both foliar and soil activity. In soil applications, rates will vary with soil texture and soil organic matter; best results if applied when soil is moist. Hardwood trees near application site can absorb this chemical through the roots and may be injured or killed. Do not spray near the root zone of desirable hardwood trees or shrubs. High rates of hexazinone can create bare ground, so only use high rates in spot treatments.

Iris pseudacorus L.
Yellowflag iris

Family: Iridaceae
Range: Many western states, except North and South Dakota, Wyoming, Colorado, Arizona and New Mexico.
Habitat: Moist soils near pond margins, irrigation ditches, and wetland sites.
Origin: Native to Europe. Was introduced and is still widely used as a pond ornamental.
Impacts: Yellowflag iris is toxic to humans and animals when a certain amount of the foliage or rhizomes is ingested. It can form dense stands at the edge of lakes, streams and ponds and reduce native plant biodiversity. Yellowflag iris may also reduce habitat needed by waterfowl and fish.
Western states listed as Noxious Weed: Montana, Oregon, Washington
California Invasive Plant Council (Cal-IPC) Inventory: Limited Invasiveness

A perennial with thick rhizomes (1.5 to 2 inches in diameter), sword-like leaves, and several yellow to cream-colored flowers per stem. Plants are typically 3 ft tall, but can grow to 5 ft tall. There are numerous native irises throughout the western United States. These play an important role in plant communities. Yellowflag iris is one of the few that is found near aquatic areas and has bright yellow flowers.

Plants take 3 years to mature before flowering. The beautiful yellow flowers produce large seed capsules to 3 inches long containing many seeds per capsule (up to 120). Plants reproduce both vegetatively and by seed. Yellowflag iris seeds are dispersed by water. Seeds have a hard seed coat beneath which there is a gas space. This allows seeds to float on the water surface in fall and early spring and germinate along shorelines when water recedes. It is not clear how long the seeds persist in the soil seed bank.

NON-CHEMICAL CONTROL

Mechanical (pulling, cutting, disking)	Mechanical removal of yellowflag iris in sensitive aquatic areas may cause extensive disturbance that facilitates the establishment of other weedy plants. Nevertheless, physical and mechanical methods may be effective in controlling small populations of yellowflag iris. However, it is necessary to remove the entire plant and rhizome system. Repeated mowing is not often considered effective for iris control, but may eventually weaken the plant.
Cultural	Plastic tarps have been used to control yellowflag iris in small patches. Woven plastic and landscape fabric proved to be the best materials.
Biological	There are no biological control agents for yellowflag iris. Because the genus has many important natives, as well as ornamentals, it is not likely that a biological control program will be initiated.

CHEMICAL CONTROL
The following specific use information is based on published papers or reports by researchers and land managers. Other trade names may be available, and other compounds also are labeled for this weed. Directions for use may vary between brands; see label before use. Herbicides are listed by mode of action and then alphabetically. The order of herbicide listing is not reflective of the order of efficacy or preference.

GROWTH REGULATORS		
2,4-D	**Rate:** 5 lb a.e. in 100 gal water	
Several names	**Timing:** Postemergence at early bloom stage. Recommendation for terrestrial species is for a spray-to-wet treatment.	
	Remarks: Terrestrial irises have been controlled with 2,4-D ester formulations, but these are not registered for use in aquatic areas. It is possible that the aquatic amine formulation of 2,4-D may be effective on yellowflag iris, but no results have confirmed this. Rates are those used for terrestrial iris species using the ester formulations. Imazapyr or glyphosate are the preferred control methods.	

AROMATIC AMINO ACID INHIBITORS	
Glyphosate *Rodeo,* *Aquamaster*	**Rate:** 4% v/v solution of *Rodeo* or *Aquamaster* (2% a.e.) for spot treatment **Timing:** Postemergence to foliage when plants are growing rapidly, but before flowering in late spring or early summer. Can also apply in fall. **Remarks:** Use a non-ionic surfactant registered for use in aquatic areas. Glyphosate is nonselective. In some cases reapplication may be necessary. Application with a drizzle gun gives good results and is far easier to treat compared to a conventional spray boom. In addition, it greatly reduces the risk of drift. Glyphosate (2% v/v solution) can be tank mixed with imazapyr (1% v/v solution). Application with a drizzle gun was not as effective with glyphosate as it was with imazapyr.
BRANCHED-CHAIN AMINO ACID INHIBITORS	
Imazapyr *Habitat*	**Rate:** Spot treatment: 1 to 3% v/v solution. Drizzle application: 20% *Habitat* in 30% aquatic registered modified vegetable oil (e.g., *Competitor*) at 5 gallons per acre. **Timing:** Postemergence to plants at pre-bloom stage or to late season plants in fall. **Remarks:** Use a non-ionic surfactant registered for use in aquatic areas. Application with a drizzle gun gives excellent results and is far easier to treat compared to a conventional spray boom. In addition, it greatly reduces the risk of drift. This technique with imazapyr is the most effective for iris control.

Isatis tinctoria L.
Dyer's woad

Family: Brassicaceae
Range: Throughout western U.S., except Arizona, North and South Dakota.
Habitat: Disturbed and undisturbed sites, roadsides, railroad rights-of-way, fields, pastures, grain and alfalfa fields, forest and rangeland. Often grows on dry, rocky or sandy soils. Most problematic on rangeland, in disturbed non-crop sites, and in undisturbed natural areas in the intermountain west.
Origin: Native to Europe. Was cultivated for centuries in Europe as a medicinal herb and source of blue dye, and was also cultivated by the early settlers of the eastern states.
Impact: Highly competitive and can grow in large, dense colonies that displace desirable rangeland species, crop plants, and native vegetation. It has a deep taproot (3 to 5 ft long), which makes it extremely competitive, especially on gravelly or sandy soil. The foliage contains compounds that appear to have insecticidal and fungicidal properties. The rotting seed pods are believed to be allelopathic.
Western states listed as Noxious Weed: Arizona, California, Colorado, Idaho, Montana, New Mexico, Nevada, Oregon, Utah, Washington, Wyoming
California Invasive Plant Council (Cal-IPC) Inventory: Moderate Invasiveness

Dyer's woad is an erect biennial, sometimes winter annual or short-lived perennial that grows to heights of 1 to 3 ft, occasionally 4 ft under ideal conditions. The leaves are bluish-green, with a pale midvein, and covered with a powdery white film. Rosette leaves are long and narrow, mostly 1.5 to 7 inches long, 0.4 to 1.5 inches wide, with weakly toothed to wavy margins. The stem leaves are broad to narrowly arrowhead-shaped, alternate and clasping the stem. Plants exist as basal rosettes until flower stems develop at maturity.

Dyer's woad reproduces only by seed. It flowers in spring, producing umbrella-shaped panicles with small, bright yellow, four-petaled flowers. These develop into hanging (pendant) blue/blackish fruits 8 to 18 mm long and 2.5 to 7 mm wide. After senescence, dried plants with a few fruits may persist well into winter. Most fruits fall near the parent plants, but some disperse short distances with wind and to greater distances with water, and as a seed and hay contaminant. New seedlings emerge in fall and early spring. Anecdotal evidence suggests the seedbank may persist for several years. Therefore, several consecutive years of control are generally necessary.

NON-CHEMICAL CONTROL

Mechanical (pulling, cutting, disking)	Hand pulling may be very effective provided the crown is removed. Hand pulling is easiest after the plants have bolted but should be done before seed set. It is important to visit the site 2 to 3 weeks later to rogue plants that have resprouted or were missed the first time through. It is necessary to follow up for several years to prevent reinfestation.
	Mowing is not effective due to resprouting from the crown, but mowing multiple times can reduce root reserves and seed production. Dyer's woad populations can be reduced if seed production can be prevented for a few years by cutting off the seed stalks and removing them from the field. Close clipping (2 inches from the soil surface) is more effective. This should be done as soon as possible after flowering to minimize resprouting and prevent seed production. Multiple visits to the field may be necessary to minimize seed production on resprouting plants.
	Spring cultivation can control infestations in crop fields but is not practical in most range settings.
Cultural	Livestock generally avoid eating dyer's woad. However, significant reductions in dyer's woad have been observed when livestock are forced under heavy grazing pressure to consume dyer's woad. The plant is more palatable before bolting and grazing should be done before flowering to minimize seed production.
	The effect of fire on dyer's woad is not well known, but it is likely that even if fire kills the above-ground part of the plant, it will regenerate from root buds on the crown.
Biological	Insects for biological control of Dyer's woad are not available in the U.S. but are being evaluated. A native rust fungus (*Puccinia thlaspeos*) causes systemic infection in dyer's woad. Infected plants may appear chlorotic, stunted, malformed and have reduced seed production. Even though the disease is systemic, plants

derived from the seed of infected plants did not show symptoms of infection. The effectiveness of this fungus is not known at this time.

CHEMICAL CONTROL

The following specific use information is based on published papers and reports by researchers and land managers. Other trade names may be available, and other compounds also are labeled for this weed. Directions for use may vary between brands; see label before use. Herbicides are listed by mode of action and then alphabetically. The order of herbicide listing is not reflective of the order of efficacy or preference.

GROWTH REGULATORS	
2,4-D Several names	**Rate:** Broadcast treatment: 2 to 3 qt/acre (1.9 to 2.85 lb a.e./acre). Spot treatment: 1% v/v solution
	Timing: Postemergence from seedling to rosette stages.
	Remarks: 2,4-D is broadleaf-selective and has no soil activity. It may require repeat application. It is an effective treatment for seedlings that may emerge following mechanical treatments. Do not apply ester formulations when outside temperatures exceed 80°F.
Aminocyclopyrachlor + chlorsulfuron *Perspective*	**Rate:** 4.75 oz product (*Perspective*)/acre
	Timing: Postemergence and preemergence. Postemergence applications are most effective when applied to plants from the seedling to the mid-rosette stage.
	Remarks: *Perspective* provides broad-spectrum control of many broadleaf species. Although generally safe to grasses, it may suppress or injure certain annual and perennial grass species. Do not treat in the root zone of desirable trees and shrubs. Do not apply more than 11 oz product/acre per year. At this high rate, cool-season grasses will be damaged, including bluebunch wheatgrass. Not yet labeled for grazing lands. Add an adjuvant to the spray solution. This product is not approved for use in California and some counties of Colorado (San Luis Valley).
BRANCHED-CHAIN AMINO ACID INHIBITORS	
Chlorsulfuron *Telar*	**Rate:** 1 to 1.33 oz product/acre (0.75 to 1 oz a.i./acre)
	Timing: Preemergence or postemergence to seedlings and rosettes.
	Remarks: Chlorsulfuron has mixed selectivity, but is generally safe on grasses. Use a surfactant for postemergence applications. Chlorsulfuron is not recommended for late-season applications when the chances for rainfall for incorporation are low. The herbicide has long soil residual activity. *Telar* can be used near water, but cannot be applied to water.
Imazapic *Plateau*	**Rate:** 8 to 12 oz product/acre (2 to 3 oz a.e./acre)
	Timing: Postemergence to rosettes or bolting plants.
	Remarks: Imazapic is will not injure most native grasses. It has mixed selectivity and tends to favor members of the Asteraceae and some grasses. It has some soil residual activity. Use a methylated seed oil surfactant at 0.25%. Imazapic is not registered for use in California.
Metsulfuron *Escort*	**Rate:** 0.5 to 1 oz product/acre (0.3 to 0.6 oz a.i./acre)
	Timing: Postemergence to rapidly growing plants up to early flowering.
	Remarks: Metsulfuron has mixed selectivity, but is generally safe on grasses. Tolerance to metsulfuron increases as the flowering stage progresses. This herbicide applied at any stage has been found to reduce seed development. It can be tank mixed with 2,4-D. Use a surfactant. Metsulfuron is not registered for use in California.
	Dyer's woad is also on the *Opensight* label (metsulfuron + aminopyralid) at 3.3 oz product/acre. However, this will provide only suppression and a second treatment may be necessary. The addition of 0.5 lb a.e./acre 2,4-D may improve control.

Juncus spp., primarily *Juncus patens* E. Mey
(spreading rush) and *Juncus effusus* L. (soft rush)

Rushes

Juncus patens

Family: Juncaceae
Range: Rush species vary in their range but they occur throughout the U.S. *J. patens* is found in California, Oregon and Washington. *J. effusus* is native to nearly all western states except South Dakota, Wyoming and Utah.

Juncus effusus

Habitat: Pastures, meadows, roadsides, ditches, and disturbed areas. They are often observed along the margins around lakes, ponds, streams, and canals. They can grow in full or partial sun but require moist to wet conditions.
Origin: Both species are native to North America, including the western United States.
Impact: Rushes are often considered beneficial in wildland ecosystems. They provide food for birds and small mammals and cover for waterfowl, songbirds and several small mammals. Because of their extensive root system they help prevent soil erosion. However, rushes are usually considered detrimental in pastures and grazed rangeland because cattle avoid grazing rushes.

Rushes can be both annuals and perennials but most are perennials. The two species that are most commonly considered weedy in California and other western states are soft rush (*J. effusus*) and spreading rush (*J. patens*). They are both clump-forming erect perennial plants with pale-green stems; they grow 2 to 5 ft tall but 3 ft is more common. Their stems are cylindrical and about 1.5 to 3.5 mm in diameter at the base. The stems of soft rush are weakly grooved, whereas those of spreading rush are distinctively grooved. The plant has only basal leaves that wrap the stems at the base. The sheath is brown with a threadlike blade at the tip. Both species have branched rhizomes with thick scaly roots.

These rushes flower from May through August. The inflorescence is open and branched and appears to be coming out of the side of the stem rather than the end. Each branch has 30 to 100 small greenish-brown flowers. These rushes spread both by seed and vegetatively by rhizomes. Most seed fall to the ground below the parent plant, but can also disperse longer distances with water.

NON-CHEMICAL CONTROL

Mechanical (pulling, cutting, disking)	Hand pulling or digging is not practical to control rushes except for extremely small populations.
	Mowing is usually only a temporary solution because rushes recover quickly. Repeated mowing may be more effective, but repeated mowing can be difficult or impossible to achieve because most of the sites where rushes are present are wet for most of the year.
	Disking or other tillage operations are not generally feasible at sites infested with rushes because they are typically too wet. Cultivation followed by reseeding can be effective provided excessively wet conditions are ameliorated.
Cultural	Grazing is not usually an effective control measure. Most rushes actually increase in abundance when livestock preferentially graze other species, which allows rush plants to proliferate.
	Burning is not an effective control measure for rushes. The plant is only top-killed and rapidly regenerates by rhizomes. Some rushes have been observed to have a faster growth rate and grow taller on burned than unburned sites.
	Fertilization and good water management help maximize the competitive ability of desirable grasses in pastures and help shift the competitive advantage away from rushes.
Biological	There are no established biological control efforts because rushes are native species and considered beneficial in most settings.

CHEMICAL CONTROL

The following specific use information is based on reports by researchers and land managers. Other trade names may be available, and other compounds also are labeled for this weed. Directions for use may vary between brands; see label before use. Herbicides are listed by mode of action and then alphabetically. The order of herbicide listing is not reflective of the order of efficacy or preference.

GROWTH REGULATORS	
2,4-D Several names	**Rate:** 4 pt product/acre (1.9 lb a.e./acre) **Timing:** Postemergence to rush plants in mid- to late spring, generally April to May. **Remarks:** 2,4-D is broadleaf-selective and has no soil activity. It is preferable not to graze for 6 months after herbicide application to enhance competitiveness of desirable grasses and to prevent regrowth of existing rush plants or germination of rush seeds. Effective control may require repeat applications. Use a surfactant. Do not apply ester formulations when outside temperatures exceed 80°F. 2,4-D can be mixed with various other compounds (e.g., dicamba or triclopyr), either in tank mixes or in commercial combinations.

BRANCHED-CHAIN AMINO ACID INHIBITORS	
Imazapic *Plateau*	**Rate:** 8 to 12 oz product/acre (2 to 3 oz a.e./acre) **Timing:** Preemergence or postemergence. **Remarks:** Imazapic typically only suppresses rushes, and at the rates required for suppression can injure desirable grasses as well. Imazapic has mixed selectivity and tends to favor species in the Asteraceae, as well as some grasses. In postemergence applications, use a methylated seed oil surfactant at 0.25%. It has some soil residual activity. Imazapic is not registered for use in California.

Lagarosiphon major (Ridley) Moss

Oxygenweed

Family: Hydrocharitaceae
Range: Not yet present in the United States.
Habitat: Lakes, rivers, streams and ponds, including high mountain streams and ponds. Prefers cooler waters of the temperate zone, particularly clear, still or slow-moving freshwater with silty or sandy soils. Growth is greatest in sheltered areas protected from wind, waves and currents. It also tolerates relatively alkaline conditions.
Origin: Native to southern Africa. Initially introduced as an aquarium plant.
Impacts: Should the plant become established it is believed that the impacts would be similar to those of hydrilla. Problematic in New Zealand and Europe. *Lagarosiphon major* can form dense floating mats in deep-water reservoirs and other water bodies, and it can block the intakes of hydroelectric systems. Populations can grow to 23 ft in depth and reduce light levels to about 1% within 2 ft of the water surface. This dramatic reduction in light penetration can eliminate growth of native water plants and reduce associated populations of aquatic invertebrates. Despite its common name, dense infestations of oxygenweed can dramatically reduce dissolved oxygen levels and impact fish populations. In addition, it can also restrict the passage of boats and limit recreational activities like swimming and angling. Storms can tear weed mats loose and deposit large masses of rotting vegetation on beaches, spoiling their amenity value.
Western states listed as Noxious Weed: While it is not yet in the United States, oxygenweed is listed as a noxious weed in California, Oregon, and Washington. Also on the U.S. Federal Noxious Weed List

Lagarosiphon major is a rhizomatous, perennial, submerged aquatic plant. Its stems are long (to 20 ft) and brittle with many small narrow leaves that are curved downward. Plants are related to and resemble *Egeria densa* and *Hydrilla verticillata*. The stems grow fast, and when they reach the water surface they spread to form thick mats. Shallow lakes to 10 ft deep can be completely covered.

Flowers are tiny, transparent to white or pinkish. Plants are dioecious (sexes on different plants), but only female plants are known outside its native range. Female flowers reach the surface on long thread-like tubes (to 10 inches long). In the native range where male plants are present, male flowers separate from the plant and pollinate the female flowers by chance when they bump into them. Outside the native range, the lack of male flowers restricts reproduction to vegetative fragmentation or local growth by rhizomatous spread. Dispersal is through movement of vegetative fragments by water or from boating, fishing, weed harvesters and float planes, and possibly some species of bird. In New Zealand, spread is significantly associated with boating and fishing activities.

NON-CHEMICAL CONTROL

Mechanical (pulling, cutting, disking)	Hand removal, mechanical harvesting and diver-assisted suction removal and dredging can be effective, but these actions also produce viable fragments that can disperse, lodge on the bottom or shoreline, and start new populations.
	Bottom barriers (synthetic or natural fibers) can control small areas (< 2 acres). Jute (similar to burlap) has been used successfully in Ireland to control *Lagarosiphon*. This also resulted in releasing native plants that emerged through the jute. Some jute material is treated with chemicals to increase its preservation; only untreated jute or other natural matting should be used. Suppliers and sources should be carefully evaluated..
Cultural	Dewatering (drawdown) during summer can reduce some subsequent growth, but unless the bottom is well-dried for several weeks, plants will recover.
	Reduced nutrient inputs can also help, but *Lagarosiphon* derives most of its nutrients from the sediment.
Biological	The most effective biological control agent is the triploid (sterile) grass carp (white amur), though it is a relatively nonselective herbivorous fish that will consume several species of submersed plant species. *Lagarosiphon* is a moderately preferred food source. Careful monitoring of consumption and impacts on native plants should be considered.

CHEMICAL CONTROL

The following specific use information is based on reports by researchers and land managers. Other trade names may be available, and other compounds also are labeled for this weed. Directions for use may vary between brands; see label before use. Herbicides are listed by mode of action and then alphabetically. The order of herbicide listing is not reflective of the order of efficacy or preference.

_	_
CONTACT PHOTOSYNTHETIC INHIBITORS	
Diquat *Reward*	**Rate:** For in-water treatment: 0.1 to 0.25 ppm **Timing:** Apply directly to water in late spring to early summer. **Remarks:** Diquat is a fast-acting contact herbicide that can also be effective in mid- to late summer. If biomass is large, only a portion of the infested sites should be treated to minimize effects of reduced dissolved oxygen. Diquat is quickly bound to, and becomes inactivated on, suspended clay particles and it should not be used in moderately or highly turbid water.
GENERAL CELL TOXICANTS	
Acrolein *Magnacide H*	**Rate:** For in-water treatment: 1 to 15 ppm. The recommended rate is variable and depends on target weeds, temperature and flow rates. **Timing:** Apply directly to water in late spring to fall and use no more than 8 applications per year. **Remarks:** Acrolein is a very fast-acting, nonselective contact herbicide and algaecide. It is a "Restricted Use" pesticide but can be used in some irrigation canals under specific conditions, with proper permits, and may only be applied by qualified, trained applicators. Symptoms of efficacy may appear in less than an hour and include discoloration of leaves and loss of turgidity. Acrolein is toxic to all organisms.
Endothall *Cascade;* *Teton;* *Aquathol K*	**Rate:** For in-water treatment: 1 to 3 ppm; exposures must be maintained for 24 to 48 hours for optimal control. **Timing:** Apply directly to water in early spring to early summer. Endothall can also be used in mid-summer, but partial treatments are recommended if biomass is large to prevent dramatic reduction in dissolved oxygen. **Remarks:** Endothall is a selective, contact herbicide. It affects young, rapidly growing plants and mature plants. Lower rates can be used if applied during early spring growth and when water movement is not likely to dilute or move the herbicide.
INORGANIC HERBICIDES	
Chelated copper *Komeen,* *Cutrine-Plus*	**Rate:** 0.5 to 1 ppm elemental copper. **Timing:** Apply directly to water in early summer when plants and biomass are small. **Remarks:** Chelated copper is a fast-acting contact herbicide and retreatment may be required within 3 to 5 weeks. If biomass is large, treat only one-third of the infested area to minimize decrease in dissolved oxygen. Chelated copper products are less affected by high pH compared to non-chelated copper.
NON-HERBICIDAL CHEMICALS	
Dyes or colorants *Aquashade*	Although technically not herbicides, dyes and colorants control submerged aquatic plants by absorbing light in the water column and reducing photosynthesis. Applications should be made in early spring and repeated to maintain concentration recommended on the label. Colorants are not as effective on well-established plants in mid- to late summer.

Lemna spp.; duckweeds
Spirodela spp.; duckweeds and duckmeats
Wolffia spp.; watermeals

Duckweeds, duckmeats, and watermeals

Family: Lemnaceae
Range: Throughout the United States.
Habitat: Ponds, lakes, slow-moving streams, ditches, and canals, sometimes on mud. Inhabit freshwater with intermediate to high nutrient levels. Tolerate a broad pH range and heavy shade. Grow best under warm conditions.
Origin: All species are native in the western states, and most are widespread in other areas of the United States and elsewhere. Common duckweed, duckmeat, and a few other species occur nearly worldwide.
Impacts: Common duckweed is used as animal feed in Asia and in bioassays to detect water pollution levels, because it can sequester certain pollutants. Duckweeds are common in the aquarium and ornamental pond trade. All duckweed, duckmeat, and watermeal species are a valuable food source for wildlife, especially waterfowl. In irrigation canals they can form dense colonies and clog filters, and in ponds with high organic matter duckweeds can grow rapidly and cover the surface in a very short time.

The Lemnaceae includes three genera of similar plants that often grow together. These are duckweed (*Lemna* spp.), duckmeat or giant duckweed (*Spirodela* spp.), and watermeal (*Wolffia* spp.). All are very small floating perennials, often growing in clusters and forming dense colonies. The plant bodies are stems, often called fronds, typically in pairs or trios and slightly obovate and succulent. Daughter fronds bud from a marginal pouch on each side at the base of the parent frond. Roots vary with the genus. *Wolffia* has no roots, *Lemna* has one root and *Spirodela* has more than two roots.

Flowers are minute and are rarely even seen. Fruits are very small and are not generally important in reproduction of the Lemnaceae. Plants primarily reproduce vegetatively by budding. During the warm season the rate of budding can be very high, with each plant producing a daughter plant about every 3 days. Plants and seeds typically disperse with water or by clinging to the feet, fur, and feathers of animals. Some members of the Lemnaceae, particularly *Spirodela* and *Wolffia* species, develop minute winter buds (turions) that separate from the mother fronds and sink to the sediment. As the water warms in spring, the buds float to the surface.

NON-CHEMICAL CONTROL

Mechanical (floating booms, suction devices)	Manual or mechanical removal can help control troublesome colonies. For small infestations (1 to 2 acres), floating booms can be dragged (preferably "down wind") from the shore or pushed by boats to consolidate mats of duckweed that can then be removed with rakes. These duckweeds provide good composting material. In large lakes, mechanical harvesters equipped with surface "skimmers" or surface suction devices can remove mats.
Cultural	Use of water-circulation devices can sometimes reduce accumulation of large biomass. Reducing nutrient inputs can also be helpful (e.g. divert runoff from turf or other areas that provide nutrients).
Biological	Biological control organisms for the control of duckweeds include ducks, fish, turtles and crustaceans (water shrimp, crayfish, ostracods, freshwater prawns, daphnia, amphipods, etc.). The triploid (sterile) grass carp (white amur) is a relatively nonselective herbivorous fish that will consume duckweeds and other small floating plants (e.g. *Azolla*). The fish do not selectively feed on "non-native" plants so careful monitoring of feeding impacts is necessary. There are a number of other species of freshwater fish that eat duckweeds to supplement their diets.

CHEMICAL CONTROL

The following specific use information is based on reports by researchers and land managers. Other trade names may be available, and other compounds also are labeled for this weed. Directions for use may vary between brands; see label before use. Herbicides are listed by mode of action and then alphabetically. The order of herbicide listing is not reflective of the order of efficacy or preference.

AROMATIC AMINO ACID INHIBITORS	
Glyphosate *Rodeo,* *Aquamaster*	**Rate:** 2% v/v solution *Rodeo* or *Aquamaster* for foliar treatment, with approved surfactant at 0.5% **Timing:** Postemergence from spring to mid-summer. **Remarks:** Glyphosate is a slow-acting, systemic herbicide. The *Lemna* species can often form thick mats which can prevent glyphosate (or other foliar-applied herbicides) from penetrating the canopy, and unexposed plants will reestablish the population.

BRANCHED-CHAIN AMINO ACID INHIBITORS	
Bispyribac-sodium *Tradewind*	**Rate:** 8 oz product/acre (6.4 oz a.i./acre). Repeated applications may be necessary, but allow 30 days between applications and apply up to 4 times per year. **Timing:** Postemergence from spring to mid-summer. **Remarks:** Slow-acting herbicide: may take 4 to 6 weeks for control.
Imazapyr *Habitat*	**Rate:** Broadcast treatment to emergent shoots: 2 to 3 pt product/acre (8 to 16 oz a.e./acre). Spot treatment: 1% v/v solution using 100 gal/acre for adequate coverage. **Timing:** Postemergence in early spring to early summer when new growth is present. **Remarks:** May require repeated applications to achieve desired effect.
Penoxsulam *Galleon*	**Rate:** 5.6 to 11.2 oz product/acre (1.4 to 2.8 oz a.i./acre). Apply in 20 to 100 gal spray solution/acre. **Timing:** Postemergence from spring to mid-summer. **Remarks:** Penoxsulam is a slow-acting herbicide and may take 4 to 6 weeks to achieve effective control.

PIGMENT SYNTHESIS INHIBITORS	
Fluridone *Sonar*	**Rate:** For in-water treatment: 10 to 30 ppb **Timing:** Apply directly to water from spring to mid-summer before large biomass has developed. **Remarks:** Fluridone is a slow-acting herbicide and may take several weeks to control. It needs a long contact period to achieve maximum effectiveness.

CONTACT PHOTOSYNTHETIC INHIBITORS	
Diquat *Reward,* *Redwing*	**Rate:** 2 to 4 lb product/surface acre (0.5 to 1 lb a.i./surface acre) **Timing:** Postemergence from spring to mid-summer. **Remarks:** Diquat is a fast-acting contact herbicide; repeated applications may be needed. *Lemna* species often form thick mats which can prevent diquat (or other foliar-applied herbicides) from penetrating the canopy, and unexposed plants will reestablish the population.
Flumioxazin *Clipper*	**Rate:** For in-water treatment: 100 to 400 ppb **Timing:** Apply directly to water from early spring to early summer, during the plants' rapid growth phase. **Remarks:** Flumioxazin is rapidly degraded and is inactive if pH exceeds 8.5. Thus, it is important to only use if pH will not exceed 8.5. It is best to apply flumioxazin in the early morning when the pH is low.

Lepidium latifolium L.
Perennial pepperweed (tall whitetop)

Family: Brassicaceae
Range: All western states, except North and South Dakota.
Habitat: Many different areas and habitats, including wetlands, riparian areas, meadows, vernal pools, salt marshes, flood plains, sand dunes, roadsides, irrigation ditches, ornamental plantings, and agronomic crops, including alfalfa, orchards, vineyards, and irrigated pastures. Most typically found on moist or seasonally wet sites in the west, and most problematic in riparian or wetland areas. Tolerates saline and alkaline conditions.
Origin: Native to Eurasia.
Impacts: Perennial pepperweed can rapidly form large, dense stands that displace desirable vegetation and wildlife. Populations easily spread along waterways and can infest entire stream corridors, riparian areas, or irrigation structures. Roots do not hold soil together well, allowing erosion of river, stream, or ditch banks. Flooded streams often wash away roots growing along the streambank, and new infestations develop downstream. Once established, perennial pepperweed is persistent and difficult to control in crops, natural areas, and ornamental plantings. Perennial pepperweed reduces forage quality in hay and pasture. Perennial pepperweed plants extract salts from deep soil and deposit them on the soil surface, inhibiting the germination and growth of other species that are sensitive to salinity.
Western states listed as Noxious Weed: California, Colorado, Idaho, Montana, New Mexico, Oregon, South Dakota, Utah, Washington, Wyoming
California Invasive Plant Council (Cal-IPC) Inventory: High Invasiveness

Perennial pepperweed is an erect perennial to 6 ft tall. The crown and lower stems are weakly woody. The foliage lacks hairs and is green to gray-green, often dusted with powdery white caused by a rust fungus. The basal leaves are larger and wider than stem leaves, to 1 ft long and 4 inches wide, with serrate margins. The aboveground parts typically die in late fall and winter, leaving dead stems and thatch which can persist for several years. The roots are long, thick, minimally branched, and vigorously creeping. Most grow in the top 2 ft of soil, but some can penetrate to a depth of 10 ft or more.

The inflorescences are rounded to pyramidal and consist of numerous small white flowers. The flowers have four petals, producing small pods (about 2 mm long) with tiny reddish-brown seeds (about 1 mm long). Perennial pepperweed is a prolific seed producer. Laboratory tests suggest seeds germinate readily with fluctuating temperatures and adequate moisture; however, seeds do not appear to remain viable in the soil for extended periods. As a result, perennial pepperweed reproduces primarily vegetatively from roots and root fragments. Large root fragments can survive desiccation on the soil surface for extended periods, and fragments as small as 0.5 to 1 inch long and 2 to 8 mm in diameter can develop into new plants. Root fragments and seeds disperse with flooding, soil movement, and human and animal activities.

NON-CHEMICAL CONTROL

Mechanical (pulling, cutting, disking)	Seedlings are easily controlled by hand-pulling or tillage, but these techniques do not control established plants because shoots quickly resprout from vast root reserves. In addition, seedlings are not often encountered. Root segments as small as 1 inch are capable of producing new shoots. Cultivation and tillage typically increase infestations by dispersing root fragments. Clean equipment after tillage to prevent spreading root fragments.
	Mowing stimulates perennial pepperweed plants to resprout and produce new growth, but mowing is helpful for removing accumulated thatch. Mowing breaks old stems into small fragments and helps prevent shading of favorable species. Combining mowing with herbicides has been shown to be an effective control strategy. For best results, mow plants at the bolting or flower bud stage and apply herbicides to resprouting shoots once they have reached the flower bud stage.
Cultural	Cattle, sheep, and goats will graze perennial pepperweed, especially rosettes in early spring. When stands

	are dense it becomes difficult for most animals to graze. Goats appear to tolerate heavy consumption of fresh plants. Sheep and goats permanently maintained in a pasture suppress growth of perennial pepperweed. However, once livestock are removed, plants quickly resprout. Burning is not effective at reducing perennial pepperweed stands, but it is helpful at removing accumulated thatch. Perennial pepperweed thatch burns best in winter or spring under dry conditions before initiation of spring growth. Seasonal flooding for an extended period during the growing season can significantly reduce populations. It is not known how long perennial roots can survive flooded conditions, but anecdotal information indicates that 6 months of submergence are required. Establishing desirable vegetation in disturbed areas can suppress perennial pepperweed and slow reinvasion after control. Because perennial pepperweed is very competitive, seed or transplant desirable vegetation after dense perennial pepperweed stands are controlled. Choose vigorous, fast-growing plant species that are adapted to the site. Perennial grasses are a good choice for natural areas and pastures. Grasses are tolerant of broadleaf-selective herbicides, and over time grasses form a thick sod that prevents future weed establishment. In pastures, promote grass expansion and vigor with fertilization and grazing management.
Biological	Biological control agents are being evaluated for use on perennial pepperweed in the United States, but currently no organisms are available.

CHEMICAL CONTROL

The following specific use information is based on reports by researchers and land managers. Other trade names may be available, and other compounds also are labeled for this weed. Directions for use may vary between brands; see label before use. Herbicides are listed by mode of action and then alphabetically. The order of herbicide listing is not reflective of the order of efficacy or preference.

GROWTH REGULATORS	
2,4-D Several names	**Rate:** 2 qt product/acre (1.9 lb a.e./acre) **Timing:** Postemergence at rosette to flowering stages. Most effective at flower bud or flowering stage. **Remarks:** 2,4-D often requires annual application for multiple years. It is broadleaf-selective and safe on most grasses. 2,4-D has minimal soil activity. Do not apply ester formulations when outside temperatures exceed 80°F. Amine forms are as effective as ester forms, and amine forms reduce the chance of off-target movement.
AROMATIC AMINO ACID INHIBITORS	
Glyphosate *Roundup, Accord XRT II*, and others	**Rate:** 2 to 4 qt product (*Roundup ProMax*)/acre (2.25 to 4.5 lb a.e./acre). Spot treatment: 2 % product v/v **Timing:** Postemergence from seedling to bloom stage. Most effective at flower bud or flowering stage. **Remarks:** Glyphosate will not kill seeds or inhibit germination the following season. Glyphosate is nonselective and has no soil activity. It can create bare ground conditions that are susceptible to weed recruitment. In areas with desirable vegetation, use spot treatment. Glyphosate is a good control option if reseeding is planned shortly after application, as it will not injure seedlings emerging after application. Add a surfactant when using a formulation where it is not already included (e.g., *Rodeo, Aquamaster*).
BRANCHED-CHAIN AMINO ACID INHIBITORS	
Chlorsulfuron *Telar*	**Rate:** 1 to 2.6 oz product/acre (0.75 to 1.95 oz a.i./acre) **Timing:** Postemergence from seedling to flowering stage. Most effective at flower bud or flowering stage. **Remarks:** Chlorsulfuron has long soil residual activity. It has mixed selectivity, but is generally safe on grasses. Always use a surfactant. 2,4-D at 1 to 2 pt/acre can be tank-mixed with chlorsulfuron for quicker burndown. *Telar* can be used near water, but cannot be applied to water. Chlorsulfuron is included with aminocyclopyrachlor in *Perspective*. This combination (i.e., formulation) is expected to provide good control, but has not been reported.
Imazapic *Plateau*	**Rate:** 8 to 12 oz product/acre (2 to 3 oz a.e./acre) **Timing:** Postemergence from seedling to flowering stage. Most effective from the bud to the late flowering stage. **Remarks:** Imazapic gives effective control with some soil residual activity. It can be used in combination

	with glyphosate (premix trade name of *Journey*). Selective to most native grasses. Higher rates may suppress seedings of some cool-season grasses. Add a methylated seed oil. Imazapic is not registered for use in California.
Imazapyr *Habitat*	**Rate:** 1 to 2 qt product/acre (0.5 to 1 lb a.e./acre) **Timing:** Postemergence from seedling to flowering stage. Most effective from the bud to the late flowering stage. **Remarks:** *Habitat* is registered for aquatic use. Also effective following early season mowing and/or disking. It is nonselective, has long soil residual activity, and leaves more bare ground than other treatments, even a year after application. Add a spray adjuvant.
Metsulfuron *Escort*	**Rate:** 1 to 2 oz product/acre (0.6 to 1.2 oz a.i./acre) **Timing:** Postemergence from seedling to flowering stage. Most effective at flower bud and flowering stage. **Remarks:** Metsulfuron has similar activity compared to chlorsulfuron. Metsulfuron has some soil residual activity. Always use a surfactant. Metsulfuron can be tank-mixed with 2,4-D for quicker burndown. Other premix formulations of metsulfuron can be used at similar application timing. These include *Cimarron Max* (metsulfuron + dicamba + 2,4-D) and *Cimarron X-tra* (metsulfuron + chlorsulfuron). Metsulfuron is not registered for use in California.
Propoxycarbazone-sodium *Canter R+P*	**Rate:** 0.9 to 1.2 oz product/acre (0.63 to 0.84 oz a.i./acre) **Timing:** Postemergence to young, rapidly growing plants. **Remarks:** Propoxycarbazone is a broad-spectrum herbicide that will control many species. It will provide only partial control of perennial pepperweed. Perennial grass species vary in tolerance. A non-ionic surfactant should be added at 0.25 to 0.5% v/v solution.

Leucanthemum vulgare Lam.
Oxeye daisy

Family: Asteraceae
Range: Throughout the United States, including all western states.
Habitat: Disturbed places, roadsides, pastures, grassland, coastal scrub. Often grows on poor soil, but thrives on moist clay soils.
Origin: Native to Europe. It is widely cultivated as an ornamental but has escaped cultivation in all contiguous states.
Impacts: It forms spreading clumps which crowd out other plants, including native plants and ornamentals. Livestock generally avoid grazing the foliage, and milk from dairy cattle that have consumed the plant can have an unpleasant flavor. Oxeye daisy can host the yellow dwarf potato virus.
Western states listed as Noxious Weed: Colorado, Montana, Washington, Utah, Wyoming
California Invasive Plant Council (Cal-IPC) Inventory: Moderate Invasiveness

A clumping perennial to 3.5 ft tall, with white daisy flowerheads. Its main roots are creeping, shallow, and extensive on favorable sites, developing into dense colonies. Roots seasonally develop new shoots, and root fragments can regenerate into new plants. The foliage is more or less hairless. The lower stems have alternate oval-shaped leaves on stalks as long as the leaf blades. The stalks plus leaves are up to 6 inches long. The leaf blades have coarse, rounded teeth and are lobed near the stalk. Upper stem leaves are shorter and without stalks.

The flowerheads are daisy-like with white ray flowers and yellow centers, mostly 1 to 3 inches in diameter, solitary on a long stalk. The flowers produce abundant seed, especially when moisture is available. Seeds have no pappus but primarily disperse with water, soil, and human activities—including distribution in some commercial "wildflower" packets. Seeds can survive ingestion by animals and can remain viable for 38 years or more under field conditions. The plants also reproduce vegetatively from creeping roots or root fragments, which can be dispersed by machinery.

NON-CHEMICAL CONTROL

Mechanical (pulling, cutting, disking)	Small patches may be removed with hand tools or by repeated hand-pulling.
	Mowing before bloom can reduce seed set but will not control the plant. Mowing during or after flowering will disperse seeds.
	Because of its shallow root system, oxeye daisy can be controlled with cultivation. A single shallow cultivation may spread root fragments and enlarge the population. However, cultivation at a depth of about 6 inches in summer followed by repeated shallow cultivations can help control patches. Equipment should be cleaned after use, and the site should be monitored for resprouts.
Cultural	Oxeye daisy is palatable to most livestock, but grazing alone does not control it. Cattle tend to avoid it, and it can give an off-flavor to milk.
	Burning does not appear to control this species.
	Plants establish most readily on bare soil. Undisturbed vegetative cover discourages oxeye daisy establishment and reproduction. Fertilization to favor grasses can help to reduce daisy infestations.
Biological	No known biocontrol agents are available for control of oxeye daisy, primarily because the plant is still widely cultivated and also related to many other cultivated chrysanthemums.

CHEMICAL CONTROL
The following specific use information is based on published papers and reports by researchers and land managers. Other trade names may be available, and other compounds also are labeled for this weed. Directions for use may vary between brands; see label before use. Herbicides are listed by mode of action and then alphabetically. The order of herbicide listing is not reflective of the order of efficacy or preference.

GROWTH REGULATORS
Aminocyclopyrachlor +   **Rate:** 3 to 4.5 oz product/acre

chlorsulfuron *Perspective*	**Timing:** In spring up to flowering, or in the fall rosette stage. **Remarks:** *Perspective* provides broad-spectrum control of many broadleaf species. Although generally safe to grasses, it may suppress or injure certain annual and perennial grass species. Do not treat in the root zone of desirable trees and shrubs. Do not apply more than 11 oz product/acre per year. At this high rate, cool-season grasses will be damaged, including bluebunch wheatgrass. Not yet labeled for grazing lands. Add an adjuvant to the spray solution. Recommended rates are based on those reported for similar species. This product is not approved for use in California and some counties of Colorado (San Luis Valley).
Aminopyralid *Milestone*	**Rate:** 4 to 7 oz product/acre (1 to 1.75 oz a.e./acre) **Timing:** In winter to early spring for preemergence and seedling treatments; in spring up to flower bud stage. Can be applied in fall in cold-winter areas. **Remarks:** A broadleaf herbicide like picloram, but more selective. Very safe on grasses. Longer residual and higher activity than clopyralid. Will kill most legumes.
Aminopyralid + 2,4-D, *Forefront HL*; Aminopyralid + metsulfuron, *Opensight*; Aminopyralid + triclopyr, *Capstone*	**Rate:** 1.2 to 1.5 pt *Forefront HL*/acre; 2.5 to 3.3 oz *Opensight*/acre; 5 to 8 pt *Capstone*/acre **Timing:** From rosette to bolting stages. *Opensight* may also be applied in fall to seedlings and rosettes. **Remarks:** *Opensight* is not registered for use in California.
Clopyralid *Transline*	**Rate:** 0.67 to 1.33 pt product/acre (4 to 8 oz ae/acre) **Timing:** In spring, up to the flower bud stage. **Remarks:** A broadleaf herbicide like picloram, but more selective. Very safe on grasses. Will kill most legumes.
Clopyralid + 2,4-D *Curtail*	**Rate:** 2 to 3 qt *Curtail*/acre (use higher rate if plants are drought-stressed) **Timing:** Apply to rapidly growing weeds from full rosette to early flower bud. **Remarks:** This mix is broadleaf-selective with a wide range of susceptible species. Recommended rates are based on those reported for similar species.
Dicamba *Banvel, Clarity*	**Rate:** 1 to 2 pt product/acre of (0.5 to 1 lb a.e./acre) **Timing:** Apply to rapidly growing plants in the rosette stage. Smaller plants are more effectively controlled. **Remarks:** Dicamba is a broadleaf-selective herbicide often combined with other active ingredients. It is effective earlier in the season than 2,4-D. It is also effective when tank-mixed with 2,4-D (0.75 lb a.e./acre dicamba + 0.25 lb a.e./acre 2,4-D). Dicamba has very limited soil residual. Avoid drift to sensitive crops. Do not apply when outside temperatures exceed 80°F. It will kill or injure legumes. Recommended rates are based on those reported for similar species. Dicamba is available mixed with diflufenzopyr in a formulation called *Overdrive*. This has been reported to be effective on oxeye daisy. Diflufenzopyr is an auxin transport inhibitor which causes dicamba to accumulate in shoot and root meristems, increasing its activity. *Overdrive* is applied postemergence at 4 to 8 oz product/acre to rapidly growing plants. Higher rates should be used on large annuals and biennials or when treating perennial weeds. Add a non-ionic surfactant to the treatment solution at 0.25% v/v or a methylated seed oil at 1% v/v solution.
Picloram *Tordon 22K*	**Rate:** 1.5 to 2 pt product/acre (6 to 8 oz a.e./acre) **Timing:** Apply at rosette to flower bud stage in spring, or to new rosettes in fall. **Remarks:** Most broadleaf plants are susceptible to picloram, but it is relatively safe on established grasses. Picloram is also effective mixed with dicamba or 2,4-D. It has long soil residual activity and some applicators note that it can injure young or germinating grasses. Picloram is a restricted use herbicide. Not registered for use in California.
Triclopyr *Garlon 3A, Garlon 4 Ultra*	**Rate:** 2 pt product/acre (0.75 lb a.e./acre *Garlon 3A*, 1 lb a.e./acre *Garlon 4 Ultra*) **Timing:** Postemergence to rapidly growing plants. **Remarks:** Broadleaf-selective and safe on most grasses. Most effective on smaller plants. *Garlon 4 Ultra* is formulated as a low volatile ester. However, in warm temperatures, spraying onto hard surfaces such as rocks or pavement can increase the risk of volatilization and off-target damage.

	Recommended rates are based on those reported for similar species.
Triclopyr + 2,4-D *Crossbow*	**Rate:** 2 to 4 qt product/acre **Timing:** Rosette stage. **Remarks:** Include non-ionic surfactant. Recommended rates are based on those reported for similar species.
AROMATIC AMINO ACID INHIBITORS	
Glyphosate *Roundup, Accord XRT II,* and others	**Rate:** 1.33 to 2.67 qt product (*Roundup ProMax*)/acre (1.5 to 3 lb a.e./acre) **Timing:** Apply to rapidly growing plants from rosette to bud stage. **Remarks:** Glyphosate has no soil activity and is a nonselective herbicide. Repeat applications may be necessary. Effectiveness is increased by addition of ammonium sulfate.
BRANCHED-CHAIN AMINO ACID INHIBITORS	
Chlorsulfuron *Telar*	**Rate:** 1 to 2.6 oz product/acre (0.75 to 1.95 oz a.i./acre) **Timing:** In fall to new rosettes, or to rosettes in spring before bolting. **Remarks:** Mixed selectivity; generally safe on grasses, but fall application may injure bromes. Use a surfactant. It can be used in late season applications to reduce seed production and has fairly long soil residual activity. Recommended rates are based on those reported for similar species.
Imazapyr *Arsenal, Habitat, Stalker,* *Chopper, Polaris*	**Rate:** 2 to 3 pt product/acre (0.5 to 0.75 lb a.e./acre) **Timing:** Preemergence or postemergence. **Remarks:** Imazapyr is a nonselective herbicide.
Metsulfuron *Escort*	**Rate:** 0.5 to 1 oz product/acre (0.3 to 0.6 oz a.i./acre) **Timing:** Apply to young, rapidly growing weeds in spring before flowering, or in fall to new rosettes. The best control in Colorado studies with metsulfuron occurred when applied during flowering. **Remarks:** Mixed selectivity, generally safe on grasses. Metsulfuron has some soil residual activity. Use a surfactant. *Opensight* is a premix of aminopyralid + metsulfuron and can be applied at 2.5 to 3.3 oz product/acre. Metsulfuron can also be tank-mixed with 2,4-D and/or dicamba. Not registered for use in California.
Metsulfuron + chlorsulfuron *Cimarron X-tra*	**Rate:** 0.5 to 1 oz product/acre **Timing:** Most recommendations on timing indicate to treat before flowering. However, the best control in Colorado studies with metsulfuron occurred when applied during flowering. **Remarks:** Recommended rates are based on those reported for similar species. Not registered for use in California.
Sulfometuron *Oust* and others	**Rate:** 3 to 5 oz product/acre (2.25 to 3.75 oz a.i./acre) **Timing:** Preemergence or early postemergence, when weeds are germinating or rapidly growing. **Remarks:** Sulfometuron has mixed selectivity. It is fairly safe on native perennial grasses, especially wheatgrass. Other desirable grasses may be stunted, stressed, or injured. Good for revegetation use, but with a fairly long soil residual.
PHOTOSYNTHETIC INHIBITORS	
Hexazinone *Velpar L*	**Rate:** 4 to 6 pt product/acre (1 to 1.5 lb a.e./acre) **Timing:** Apply before weeds emerge or to young plants. **Remarks:** Mixed selectivity, fairly long soil residual. It can be effective in both foliar and soil applications. In soil applications, rates will vary with soil texture and soil organic matter; best results if applied when soil is moist. Use rates will also vary depending on the weed species to be controlled. Hardwood trees near application site can absorb this chemical through the roots and may be injured or killed. Do not spray near the root zone of desirable hardwood trees or shrubs. High rates of hexazinone can create bare ground, so only use high rates in spot treatments.

Ligustrum lucidum Ait.

Glossy privet

Family: Oleaceae
Range: California Coast Ranges and some adjacent interior counties. Also found in most southern states from Texas to North Carolina.
Habitat: Disturbed areas, particularly in riparian forest; usually found in urban fringe. The seedlings are tolerant of shade, various soil types, and a range of temperature and moisture regimes.
Origin: Native to China, Korea, and Japan.
Impacts: In Australia, privet establishes dense thickets with shallow, fibrous root systems. These stands occupy riparian areas and exclude native species. Monospecific stands have not yet been reported in California, but the plant is widely naturalized. The fruit and bark contain compounds poisonous to livestock. Some people are allergic to the pollen.

Glossy privet is a popular ornamental tree, 20 to 25 ft tall, that sometimes escapes cultivation. Like its relative, olive, privet is an evergreen tree with opposite leaves. However, glossy privet has larger leaves, leathery and smooth, more than 1 inch wide and 2.5 to 6 inches long, dark glossy green above and pale green below.

Both glossy privet and olive produce pyramid-shaped panicles of cream-colored flowers in early summer. Under moist conditions, a large tree may produce 3 million seeds or more. Seeds have a high rate of viability but usually survive less than 12 months in the soil. Glossy privet reproduces by seed, but can also resprout from the base if cut. Birds disperse the small purplish- to bluish-black, berry-like fruits.

NON-CHEMICAL CONTROL

Mechanical (pulling, cutting, disking)	Privet can be effectively controlled by hand pulling of young plants. Plants should be pulled as soon as they are large enough to grasp but before they produce seeds. Plants are best pulled after a rain when the soil is loose. Larger stems, up to 2.5 inches wide, can be removed using a weed wrench or similar uprooting tool. The entire root must be removed since broken fragments may resprout.
	Mowing or cutting can be used for small initial populations or environmentally sensitive areas where herbicides cannot be used. Stems should be cut at least once per growing season, as close to ground level as possible. Repeated mowing or cutting will control the spread of privet, but will not eradicate it.
Cultural	This species does not carry a burn well. If a burn can be supplemented with other fuels, then privet may be top-killed. This treatment, repeated annually, can eliminate privet over time.
Biological	No biological control agents are available for management of glossy privet, as it is still a widespread ornamental species.

CHEMICAL CONTROL

The following specific use information is based on published papers and reports by researchers and land managers. Other trade names may be available, and other compounds also are labeled for this weed. Directions for use may vary between brands; see label before use. Herbicides are listed by mode of action and then alphabetically. The order of herbicide listing is not reflective of the order of efficacy or preference.

GROWTH REGULATORS

Triclopyr *Garlon 3A, Garlon 4 Ultra*	**Rate:** Spot treatment: 2% product v/v in water, plus 0.5% non-ionic surfactant. Stem injection treatment: 50-100% *Garlon 3A* (amine formulation). Basal bark or drizzle treatment: 10% to 25% v/v *Garlon 4 Ultra* (ester formulation), mixed with crop oil. Cut stump treatment: 20% to 25% v/v solution of any formulation.
	Timing: Foliar treatments are effective in spring or fall, less effective in summer. Stem and stump treatments are probably most effective late in the growing season.
	Remarks: Triclopyr is a broadleaf-selective herbicide with very short soil residual. Basal bark treatments have been shown effective in experimental trials. These stem treatments are expected to be effective based on work with other woody species.

Triclopyr + fluroxypyr *Pasturegard HL*	**Rate:** Broadcast treatment: 2 qt product/acre foliar application. Basal bark, drizzle, or cut stump treatment: 50% in oil carrier. **Timing:** These treatments are probably most effective late in the growing season. **Remarks:** Not registered for use in California.
AROMATIC AMINO ACID INHIBITORS	
Glyphosate *Roundup, Accord XRT II*, and others	**Rate:** Spot treatment: 2% to 3% v/v *Roundup ProMax* in water as spray-to-wet foliar treatment. Cut stump treatment: 25% v/v in water. Stem injection treatment: undiluted product. **Timing:** Foliar treatments are effective in spring or fall, less effective in summer. Stem and stump treatments are probably most effective late in the growing season. **Remarks:** Glyphosate is nonselective and has no soil activity. Stem and stump treatments are not labeled but are expected to be effective based on work with other woody species.
BRANCHED-CHAIN AMINO ACID INHIBITORS	
Imazapyr *Arsenal, Habitat, Chopper, Stalker, Polaris*	**Rate:** Broadcast treatment: 4 to 6 pt product/acre (1 to 1.5 lb a.e./acre) as foliar application. Spot treatment: 1% v/v solution for spray-to-wet applications. Stem injection treatment: 20 to 100% *Habitat* or *Arsenal* (aqueous formulation). Basal bark or drizzle treatment: 25% *Chopper* (emulsifiable formulation), mixed with crop oil. Cut stump treatment: 10% to 25% of any formulation. **Timing:** These treatments are probably most effective late in the growing season. **Remarks:** Imazapyr has a fairly long soil residual. It is a nonselective herbicide.
Metsulfuron *Escort*	**Rate:** Spot treatment: 0.035% to 0.05% product plus 0.1% surfactant (e.g., 4.5 to 6.5 oz product plus 13 oz surfactant in 100 gal water) for spray-to-wet applications. Stem injection or cut stump treatment: 0.06% to 0.12% product v/v in water. **Timing:** These treatments are probably most effective late in the growing season. **Remarks:** Metsulfuron has mixed selectivity, but is generally safe on grasses. Use a surfactant. It can be tank-mixed with 2,4-D and/or dicamba and has some soil residual activity. Not registered for use in California.
PHOTOSYNTHETIC INHIBITORS	
Hexazinone *Velpar L*	**Rate:** Broadcast treatment: 2 to 8 qt product/acre (1 to 4 lb a.i./acre) as foliar application. Cut-stump treatment: 10% v/v in water. **Timing:** Apply in spring, from bud break until new growth hardens off. **Remarks:** Hexazinone is active with both foliar and soil treatments. In soil applications, rates will vary with soil texture and soil organic matter; best results if applied when soil is moist. The use rates will also vary depending on the situation. Hardwood trees near application site can absorb this chemical through the roots. High rates of hexazinone can create bare ground, so only use high rates in spot treatments.
Tebuthiuron *Spike*	**Rate:** 7.5 lb *Spike 80DF*/acre (6 lb a.i./acre) in soil-applied spot treatments; 0.375 to 0.75 oz *Spike 20P* per 100 sq ft for control of individual plants. **Timing:** Apply in spring before rapid seasonal growth. **Remarks:** Tebuthiuron has no foliar activity; root uptake only. Do not apply in the vicinity of desirable plants. Nearly all vegetation in the application area will be killed, and this chemical has long soil residual activity. Results may vary with soil type. Tebuthiuron will leach in sandy soils.

Limnobium laevigatum (Humb. & Bonpl. ex Willd.) Heine
(= *L. spongia* (Bosc) Rich. ex Steud. [Jepson Manual 2012])

Smooth frogbit or South American spongeplant

Photos courtesy of D.W. Kratville

Family: Hydrocharitaceae
Range: Only found in California to date.
Habitat: Found growing wild in lakes, ponds, and slow rivers.
Origin: Introduced from tropical to sub-tropical Central and South America. In California, smooth frogbit has escaped cultivation as a pond ornamental
Impacts: Can form dense stands in aquatic systems and impede the flow of water. Also weedy in Canada and South America.
Western states listed as Noxious Weed: California
California Invasive Plant Council (Cal-IPC) Inventory: High Invasiveness (Alert)

Smooth frogbit or spongeplant is a floating to rooted stoloniferous perennial with foliage that may be confused with that of water hyacinth. The floating rosettes send runners out into the water, the ends of which form juvenile plants. The juvenile form has thick, spongy, floating ovate to spatula-shaped leaves, usually with rounded tips and on an inflated stalk. Juvenile rosettes gradually develop into mature clumps to about 2 ft tall, with leathery, emergent, broadly elliptic leaves. Unlike water hyacinth, smooth frogbit typically has juvenile leaves and sometimes mature leaves with a patch or disc of honeycomb-like spongy tissue (aerenchyma) on the lower surfaces.

Plants are monoecious and produce small solitary or paired unisexual white flowers (~0.5 inches in diameter) with inferior ovaries on stalks up to ~1/3 the height of the leaves. Capsules are fleshy, berry-like, containing seeds that are covered with short hairs. Dispersal is by seed and stem fragments and by attaching to watercraft. A large mat of runners and adult plants can develop very quickly, shading plants growing below. Smooth frogbit seeds germinate rapidly to produce extremely small, floating seedlings that can resemble duckweed (e.g. *Lemna* spp.) and are easily dispersed by wind, currents, tidal action and also on waterfowl, boats and even trapped on water hyacinth plants. Seeds appear to survive for at least 4 years.

NON-CHEMICAL CONTROL

Mechanical (pulling, cutting, harvesting)	Construction of dams can create still water conditions that are favorable to *Limnobium* establishment. Even flowing river systems are susceptible to infestations. Best results are achieved when efficient systems are used to transport harvested plant to local landfill or composting sites.
	Mechanical choppers and shredders leave viable pieces of frogbit that can easily reestablish populations and also may disperse with moving water or wind. An added concern with mechanical removal is that it can dislodge and spread very small (ca. 0.1 to 2 cm) frogbit seedlings that then float to adjacent areas. These can also be easily transported by the workboats and harvesting equipment. Where possible, containment nets or curtains should be deployed to minimize off-site dispersal of seedling and small plants. When populations are discovered early and before seeds are produced, physical removal can be a very effective tool.
Cultural	Since frogbit is a relatively new invasive (since 2001 and with limited distribution), prevention of further movement and import is critical. Any horticultural shipments of floating plants should be inspected for the presence of frogbit. The similarity between water hyacinth and frogbit at young stages of growth make the presence of frogbit easily overlooked.
Biological	There are no host-specific biological control agents available for *Limnobium*. The triploid grass carp may be useful in small infestations but it is a nonselective herbivore that prefers submersed plants, which it is likely to consume before it feeds on frogbit.

CHEMICAL CONTROL

The following specific use information is based on reports by researchers and land managers. Other trade names may be available, and other compounds also are labeled for this weed. Directions for use may vary between brands; see label before use. Herbicides are listed by mode of action and then alphabetically. The order of herbicide listing is not reflective of the order of efficacy or preference.

GROWTH REGULATORS

2,4-D
Weedar 64

Rate: 1 to 2 pt product/acre (0.48 to 0.95 lb a.e./acre) with a non-ionic surfactant

Timing: Postemergence in spring to early summer is optimal. However, mid-summer to early fall applications can also be effective in suppressing growth.

Remarks: 2,4-D is a relatively fast-acting, selective systemic herbicide. Effects (symptoms) usually appear within a few days to a week and include collapse of petioles and twisted petioles.

AROMATIC AMINO ACID INHIBITORS

Glyphosate
Rodeo,
Aquamaster

Rate: 1 to 2 % v/v solution *Rodeo* or *Aquamaster* (1% a.e.) plus an approved surfactant

Timing: Postemergence in early spring to fall. Optimal management is achieved with early applications and continued reapplication to new plants.

Remarks: Glyphosate is a slow-acting systemic herbicide. Efficacy can be reduced if plants have dust and fine debris on the petioles (leaves). Therefore, applications made 24 hr after rains wash off the dust can often have increased efficacy.

BRANCHED-CHAIN AMINO ACID INHIBITORS

Imazamox
Clearcast

Rate: Broadcast treatment to emergent shoots: 1 to 2 pt product/acre (2 to 4 oz a.e./acre). Spot spray-to-wet treatment: 0.25 to 5% v/v solution. For in-water treatment: 50 to 100 ppb

Timing: Postemergence from early spring to early summer during the rapid growth phase.

Remarks: Use an approved surfactant. Aerial application is approved in some states.

Imazapyr
Habitat

Rate: Broadcast treatment to emergent shoots: 1 to 2 pt product/acre (4 to 8 oz a.e./acre). Spot treatment: 0.5% v/v solution applied at 100 gal/acre for adequate coverage.

Timing: Postemergence from early spring to early summer when new growth is present.

Remarks: May require repeated applications to maintain desired concentration for 5 to 7 weeks.

Penoxsulam
Galleon

Rate: Foliar treatment: 2 to 5.6 oz product/acre (0.5 to 1.4 oz a.i./acre) plus approved surfactant.

Timing: Postemergence to foliage from spring to mid-summer.

Remarks: Penoxsulam is a slow-acting herbicide and may take 4 to 6 weeks to show effects.

CONTACT PHOTOSYNTHETIC INHIBITORS

Diquat
Reward

Rate: 0.5% v/v solution or 2 qt/100 gal water

Timing: Postemergence applications in spring to early summer are optimal. Repeat treatments may be needed in mid-summer.

Remarks: Diquat is a contact herbicide that is inactivated in turbid water. Only clean water should be used to mix and spray the herbicide.

Flumioxazin
Clipper

Rate: 6 to 12 oz product/acre (3 to 6 oz a.i./acre). It is critical to buffer the spray solution to achieve pH values below 7.

Timing: Postemergence from spring to early summer. Fall applications may also be effective if temperatures remain high.

Remarks: If infestation is dense, treat in sections and wait 10 to 14 days before treating the next section. Do not treat the same section of water within 28 days. Flumioxazin may be tank mixed with 2,4-D.

Linaria dalmatica (L.) Mill. ssp. *dalmatica*
 (= *L. genistifolia* (L.) Mill. ssp. *dalmatica* (L.) Maire & Petitm.)

Dalmatian toadflax

Family: Scrophulariaceae
Range: Throughout much of North America, including all western states.
Habitat: Open fields, pastures, riparian areas, rangeland and disturbed sites such as roadsides, forest clearings, and agricultural fields. Grows in most environments and can tolerate many soil types. Grows best in cool, semiarid climates and on dry, coarse soils at neutral to slightly alkaline pH. Although it often invades disturbed areas, it has been shown to move into relatively undisturbed prairies and riparian habitats. Tolerates sub-arctic conditions. It is one of the few weeds that have invaded the shortgrass steppe in eastern Colorado.
Origin: Native to Europe and the Mediterranean region and brought to North America as a garden ornamental in the mid to late 1800s.
Impacts: Dalmatian toadflax is a persistent, aggressive invader capable of forming dense colonies through adventitious buds from creeping root systems. These colonies can push out native grasses and other perennials, thereby altering the species composition of natural communities. The plant decreases forage for domestic livestock and some big game species and decreases habitat for associated animal communities. Dalmatian toadflax contains quinazoline alkaloids that could possibly pose toxicity problems to livestock if ingested in sufficient quantity, but intoxications of livestock have not been reported. Goats and sheep have been known to graze the plants with little effect.

Western states listed as Noxious Weed: Arizona, California, Colorado, Idaho, Montana, Nevada, New Mexico, Oregon, South Dakota, Utah, Washington, Wyoming
California Invasive Plant Council (Cal-IPC) Inventory: Moderate Invasiveness

Dalmatian toadflax is an herbaceous perennial that can reach a height of 3 ft or more, with individual plants producing up to 25 stems in the first year of growth. Stems are rough and woody at the base becoming smooth, waxy and herbaceous toward the top. The leaves are waxy with a bluish green color, ovate to heart-shaped but sometimes lanceolate, 0.5 to 2.3 inches long, with smooth margins. The leaves are alternate and clasping on the upper portion of the stem. Dalmatian toadflax produces both taproots and creeping roots, with adventitious buds forming new individuals. Roots can grow 4 to 10 ft deep and can extend 10 ft from the parent plant.

The flowers resemble snapdragons with petals ranging from 0.75 to 1.5 inches long. Flowers are two-lipped, yellow, often with an orange, bearded throat and a long spur. They mature from the lower part of the stem upwards, and various stages of flowering and fruiting can be present on an inflorescence. The fruits are two-celled capsules with many irregularly-shaped, sharply angular, slightly winged, black seeds. Reproduction is by seed and vegetatively from creeping lateral roots. Most seed falls near the parent plant. Seed production and viability is highly variable, depending on out-crossing and presence of pollinators. A single plant can produce as many as 500,000 seeds per year. Seeds germinate in spring and fall when conditions become favorable. Seedlings compete poorly with established vegetation for soil moisture. Plants can rapidly colonize a site by vegetative reproduction from creeping roots. It is not known how long the seed survive in the soil, but because they are so small, it is likely that they do not survive for more than a couple of years.

NON-CHEMICAL CONTROL

Mechanical (pulling, cutting, disking)	Hand pulling is only effective on seedlings before plants become established and the extensive creeping root system develops.
	Mowing can prevent the plant from going to seed, but mowing also stimulates vegetative reproduction from the lateral roots and rhizomes.
	Tilling on arable lands can be effective but tilling needs to be done every 7 to 10 days over the course of the

	season and repeated yearly for several years in order to eradicate resprouting root fragments.
Cultural	In some cases, grazing is not considered an effective control option. However, in Colorado, goat grazing has worked well, particularly when in combination with the biocontrol insect *Mecinus*. Overgrazing can reduce competition and increase site disturbance, creating an ideal environment for toadflax establishment. The plant is not preferred by grazing livestock and contains quinazoline alkaloids that are moderately toxic. Fire is not effective because the underground root system is not damaged and will resprout. Reseeding with competitive annual and perennial grasses reduces survival and helps prevent further spread.
Biological	Eight insects have been introduced and approved by the USDA-APHIS for release as biocontrol agents for both Dalmatian and yellow toadflax in the United States with varying success. *Brachypterolus pulicarius*, a shoot and flower feeding beetle, can reduce seed set on attacked plants by 74%. *Gymnaetron antirrhini* and *G. netum*, both seed-capsule feeding weevils, have been shown to impact seed production in these species. *Calophasia lunula*, a moth introduced from Eurasia, has been shown to defoliate up to 20% of a plant. Establishment in California is uncertain for these species. Another agent, the toadflax stem-mining weevil (*Mecinus janthiniformis*) was misidentified as *Mecinus janthinus* when first introduced in 1995. European research showed that the *Mecinus* attacking Dalmatian toadflax was actually *M. janthiniformis*, and that *M. janthinus* was attacking yellow toadflax. It has had substantial and dramatic impacts on Dalmatian toadflax populations in many western states and is even becoming established in California. The weevil damages the foliage but also destroys flower production. In Colorado, it has reduced Dalmatian toadflax populations 4 to 5 years following release.

CHEMICAL CONTROL

The following specific use information is based on published papers and reports by researchers and land managers. Other trade names may be available, and other compounds also are labeled for this weed. Directions for use may vary between brands; see label before use. Herbicides are listed by mode of action and then alphabetically. The order of herbicide listing is not reflective of the order of efficacy or preference.

GROWTH REGULATORS	
2,4-D Several names	**Rate:** 2 to 4 pt product/acre (0.95 to 1.9 lb a.e./acre). **Timing:** Postemergence when plants are growing rapidly. Applications in spring provide best control. **Remarks:** 2,4-D is a selective herbicide for broadleaf species. It was found to provide only fair control of Dalmatian toadflax in a California study. Good coverage is necessary. Efficacy is improved when tank-mixed with picloram, chlorsulfuron, or metsulfuron.
Aminocyclopyrachlor + chlorsulfuron *Perspective*	**Rate:** 4 to 6 oz product/acre plus 0.25 to 0.5% v/v surfactant **Timing:** Postemergence when plants are in the rosette stage or in mid-fall when plants are dormant. **Remarks:** *Perspective* provides broad-spectrum control of many broadleaf species. It gave excellent control in a California study and appears to be the best product for Dalmatian toadflax. Although generally safe to grasses, it may suppress or injure certain annual and perennial grass species. Do not treat in the root zone of desirable trees and shrubs. Do not apply more than 11 oz product/acre per year. At this high rate, cool-season grasses will be damaged, including bluebunch wheatgrass. Not yet labeled for grazing lands. Add an adjuvant to the spray solution. This product is not approved for use in California and some counties of Colorado (San Luis Valley).
Dicamba *Banvel, Clarity*	**Rate:** 4 qt product/acre (4 lb a.e./acre) and water plus 0.25 to 0.5% v/v surfactant **Timing:** Early postemergence in spring before toadflax reaches bloom stage. **Remarks:** Dicamba is a selective herbicide for broadleaf species. In a California study, 2 lb a.e./acre gave partial control of Dalmatian toadflax at the rosette stage, and poor control when applied at the bolting or dormant stage. It has a short soil residual activity. Repeated applications may be necessary for better control.
Picloram *Tordon 22K*	**Rate:** 1 to 2 qt product/acre (0.5 to 1 lb a.e./acre) plus 0.25 to 0.5% v/v surfactant **Timing:** Postemergence when plants are growing rapidly in spring before full bloom, or in late summer to early fall. **Remarks:** At 8 oz a.e./acre in a California study, picloram gave only partial control (80%) when applied at the dormant stage in mid-fall, and poor control at the rosette and bolting stages (< 60% control). Higher rates may be necessary in some areas. High levels of picloram can give long-term soil activity for broadleaves. *Tordon 22K* is a federally restricted use pesticide. It is not registered for use

	in California.
Picloram + 2,4-D *Tordon 101M*	**Rate:** 2 qt product/acre plus 0.25 to 0.5% v/v surfactant **Timing:** Postemergence when plants are growing rapidly in spring before full bloom. **Remarks:** May require annual treatment for 2 to 3 years. High levels of picloram can give long-term soil activity for broadleaves. It is not registered for use in California.
Picloram + chlorsulfuron *Tordon 22K + Telar*	**Rate:** 1 qt product/acre *Tordon 22K* + 1.25 oz product/acre *Telar* plus 0.25 to 0.5% v/v surfactant **Timing:** Postemergence when plants are growing rapidly, from bloom through fall. Fall treatments give best control. **Remarks:** High levels of picloram can give long-term soil activity for broadleaves. Retreatment may be necessary. *Tordon 22K* is a federally restricted use pesticide. It is not registered for use in California.
AROMATIC AMINO ACID INHIBITORS	
Glyphosate *Roundup, Accord XRT II,* and others	**Rate:** Broadcast treatment: 1 to 2 qt product (*Roundup ProMax*)/acre (1.1 to 2.25 lb a.e./acre). Spot treatment: 1.5 to 2% solution v/v *Roundup* (or other trade name) and water to thoroughly wet all leaves **Timing:** Postemergence when plants are growing rapidly. Applications in early spring provide best control. **Remarks:** Glyphosate is a nonselective systemic herbicide that may kill non-target partially-sprayed plants. Repeated applications may be necessary for complete control.
BRANCHED-CHAIN AMINO ACID INHIBITORS	
Chlorsulfuron *Telar*	**Rate:** 2 to 2.6 oz product/acre (1.5 to 1.95 oz a.i./acre) plus 0.25 to 0.5% v/v surfactant **Timing:** While most recommendations note that applications should be made postemergence when plants are growing rapidly in the bud to bloom stage, others have found that the best timing is when plants are in the rosette stage or when they are dormant in mid-fall. **Remarks:** Chlorsulfuron is a selective herbicide effective for controlling mainly broadleaves, but also some grasses. A California study showed best results with a rosette or dormant stage application at the highest rate (1.95 oz a.i./acre). Chlorsulfuron can be tank mixed with picloram or 2,4-D.
Imazapic *Plateau*	**Rate:** 12 oz product/acre (3 oz a.e./acre) plus 1 qt/acre methylated seed oil in the spray mix **Timing:** Postemergence in fall when the top 25% of the plant is necrotic. This typically occurs after a hard frost. **Remarks:** Imazapic is a selective postemergence herbicide effective for controlling broadleaf weeds and some grasses. In one study, it was not found to be very effective for the control of Dalmatian toadflax at 12 oz product/acre. It is not registered for use in California.
Imazapyr *Arsenal, Habitat, Stalker, Chopper, Polaris*	**Rate:** 3 pt product/acre (0.75 lb a.e./acre) plus 0.25 to 0.5% v/v surfactant **Timing:** Some reports note that the best timing is postemergence when plants are growing rapidly, whereas others show that a dormant application in mid-fall was the best timing. **Remarks:** Imazapyr is a preemergent and postemergence herbicide effective for controlling broadleaf weeds and grasses. In a California study using 3 pt product/acre, excellent control was only achieved with a mid-fall application to dormant plants. It has fairly long soil residual activity.
Metsulfuron *Escort*	**Rate:** 1.5 to 2 oz product/acre (0.9 to 1.2 oz a.i./acre) plus 0.25 to 0.5% v/v surfactant; efficacy is improved with the addition 2,4-D at a rate of 1 qt product/acre **Timing:** Early postemergence. **Remarks:** Metsulfuron is a selective herbicide for broadleaf species. In areas where desirable grasses are growing around toadflax, metsulfuron can be used without non-target damage. In one study, however, metsulfuron at 2 oz product/acre gave poor control regardless of timing from the rosette to the dormant stage. It is not registered for use in California.

Linaria vulgaris Miller
Yellow toadflax

Family: Scrophulariaceae
Range: Throughout North America and every state except Hawaii.
Habitat: Fields, pastures, riparian areas, rangeland and disturbed sites such as roadsides, forest clearings, and agricultural fields. Grows in most environments and can tolerate many soil types. Often inhabits moist, coarse soils, particularly gravelly or sandy soils. Although it often invades disturbed areas, it has been shown to move into relatively undisturbed prairies and riparian habitats. Tolerates sub-arctic conditions.
Origin: Native to Europe and the Mediterranean region and brought to North America as a garden ornamental in the mid-1600s.
Impacts: Yellow toadflax is highly competitive for soil moisture with winter annuals and shallow-rooted perennials. Large colonies that displace desirable vegetation can develop in natural areas. The plant decreases forage for domestic livestock and some big game species and decreases habitat for associated animal communities. Yellow toadflax contains quinazoline alkaloids that could possibly pose toxicity problems to livestock if ingested in sufficient quantity, but intoxications of livestock have not been reported. Goats and sheep have been known to graze the plants with little effect.
Western states listed as Noxious Weed: Idaho, Montana, Nevada, New Mexico, Oregon, South Dakota, Utah, Washington, Wyoming

Yellow toadflax is an herbaceous creeping rooted perennial that can reach a height of about 3 ft, but is generally about 1 ft tall. The stems are erect, glabrous to glandular-hairy near the top, but with a woody base. The leaves are pale green, 1 to 2 inches long, linear to narrow, alternate, and sessile.

The flowers are bilateral, resembling snapdragon flowers, and on dense racemes of 15 to 20 flowers in the axils of the upper portion of the stem. The flowers are about 1 inch long, yellow to pale yellow with an orange bearded throat and yellow spur in which nectar collects. The fruits are brown capsules, 0.25 to 0.5 inch long, ovate, and contain multiple flat, dark brown seeds that have a papery wing to aid dispersal. Reproduction is by seed and vegetatively from creeping lateral roots. Most seed falls near the parent plant. Seeds germinate in spring and fall when conditions become favorable. Seedlings compete poorly for soil moisture with established vegetation. Plants can rapidly colonize a site by vegetative reproduction from creeping roots. It is not known how long the seed survive in the soil, but because they are so small, it is likely that they do not survive for more than a couple of years. Yellow and Dalmatian toadflax readily cross to produce a very vigorous intermediate.

NON-CHEMICAL CONTROL

Mechanical (pulling, cutting, disking)	Hand pulling is only effective on seedlings before plants become established and the extensive creeping root system develops.
	Mowing can prevent the plant from going to seed, but mowing also stimulates vegetative reproduction from the lateral roots and rhizomes.
	Tilling on arable lands can be effective but tilling needs to be done every 7 to 10 days over the course of the season and repeated yearly for several years to eradicate resprouting root fragments.
Cultural	Grazing is not considered an effective control option. Overgrazing can reduce competition and increase site disturbance, creating an ideal environment for toadflax establishment. The plant is not preferred by grazing livestock and contains quinazoline alkaloids that are moderately toxic.
	Fire is not effective because the underground root system is not damaged and will resprout.
	Reseeding with competitive annual and perennial grasses reduces survival and helps prevent further spread.
Biological	Two insects active on yellow toadflax were accidentally introduced into the United States in the early 1900s. The toadflax flower feeding beetle (*Brachypterolus pulicarius*) and the toadflax capsule weevil (*Gymnetron antirrhini*) are well established in the Pacific Northwest. Both significantly reduce seed production, but do not have a significant impact on populations. However, the most promising biocontrol agent is the toadflax stem-

mining weevil (*Mecinus janthinus*). It is too early to know how successful this insect will be.

CHEMICAL CONTROL

The following specific use information is based on published papers and reports by researchers and land managers. Other trade names may be available, and other compounds also are labeled for this weed. Directions for use may vary between brands; see label before use. Herbicides are listed by mode of action and then alphabetically. The order of herbicide listing is not reflective of the order of efficacy or preference.

GROWTH REGULATORS	
2,4-D Several names	**Rate:** 2 to 4 pt product/acre (0.95 to 1.9 lb a.e./acre) **Timing:** Postemergence when plants are growing rapidly. Applications in spring provide best control. **Remarks:** 2,4-D is a selective herbicide for broadleaf species. In areas where desirable grasses are growing around toadflax, 2,4-D can be used without non-target damage. Good coverage is necessary. Efficacy is improved when tank-mixed with picloram, chlorsulfuron, or metsulfuron.
Aminocyclopyrachlor + chlorsulfuron *Perspective*	**Rate:** 4 to 6 oz product/acre plus 0.25 to 0.5% v/v surfactant **Timing:** Postemergence when plants are growing rapidly in the bud to bloom stage. **Remarks:** *Perspective* provides broad-spectrum control of many broadleaf species. Although generally safe to grasses, it may suppress or injure certain annual and perennial grass species. Do not treat in the root zone of desirable trees and shrubs. Do not apply more than 11 oz product/acre per year. At this high rate, cool-season grasses will be damaged, including bluebunch wheatgrass. Not yet labeled for grazing lands. Add an adjuvant to the spray solution. This product is not approved for use in California and some counties of Colorado (San Luis Valley).
Picloram *Tordon 22K*	**Rate:** 2 qt product/acre (1 lb a.e./acre) plus 0.25 to 0.5% v/v surfactant **Timing:** Postemergence when plants are growing rapidly in spring before full bloom, or in late summer to early fall. **Remarks:** High levels of picloram can give long-term soil activity for broadleaves. Picloram at 2 pt product/acre plus 6 oz *Overdrive*/acre controlled yellow toadflax better (98% control) than picloram at 2 qt product/acre (70% control) 2 years after treatment. *Tordon 22K* is a federally restricted use pesticide. It is not registered for use in California.
Picloram + 2,4-D *Tordon 101M*	**Rate:** 2 qt product/acre plus 0.25 to 0.5% v/v surfactant **Timing:** Postemergence when plants are growing rapidly in spring before full bloom. **Remarks:** May require annual treatment for 2 to 3 years. High levels of picloram can give long-term soil activity for broadleaves. *Tordon 101M* is a federally restricted use pesticide. It is not registered for use in California.
Picloram + chlorsulfuron *Tordon 22K + Telar*	**Rate:** 1 qt product/acre *Tordon 22K* + 1.25 oz product/acre *Telar* plus 0.25 to 0.5% v/v surfactant **Timing:** Postemergence when plants are growing rapidly from bloom through fall. Fall treatments give best control. **Remarks:** High levels of picloram can give long-term soil activity for broadleaves. Retreatment may be necessary. *Tordon 22K* is a federally restricted use pesticide. It is not registered for use in California.
AROMATIC AMINO ACID INHIBITORS	
Glyphosate *Roundup, Accord XRT II*, and others	**Rate:** Broadcast treatment: 1 to 2 qt product (*Roundup ProMax*)/acre (1.1 to 2.25 lb a.e./acre). Spot treatment: 1.5 to 2% solution v/v *Roundup* (or other trade name) and water to thoroughly wet all leaves **Timing:** Postemergence when plants are growing rapidly. Applications in early spring provide best control. **Remarks:** Glyphosate is a nonselective systemic herbicide with no soil activity. Repeated applications may be necessary for complete control.
BRANCHED-CHAIN AMINO ACID INHIBITORS	
Chlorsulfuron *Telar*	**Rate:** 2 to 2.6 oz product/acre (1.5 to 1.95 oz a.i./acre) plus 0.25 to 0.5% v/v surfactant **Timing:** Postemergence when plants are growing rapidly in the bud to bloom stage. **Remarks:** Chlorsulfuron is a selective herbicide effective for controlling broadleaves and grasses.

	While it is often stated that *Telar* provides only suppression of yellow toadflax and is often tank mixed with picloram or 2,4-D, this depends on the timing of the application. *Telar* will control yellow toadflax, but timing is of critical importance. In a research report, 1.75 oz product/acre *Telar* gave 76 to 98% control 2 years after treatment. This treatment was applied when most of the shoots had already flowered so that adventitious root bud activity had begun.
Imazapic *Plateau*	**Rate:** 12 oz product/acre (3 oz a.e./acre) plus 1 qt/acre methylated seed oil in the spray mix **Timing:** Postemergence in fall when top 25% of the plant is necrotic, usually after a hard frost. **Remarks:** Imazapic is a selective postemergence herbicide effective for controlling broadleaf weeds and some grasses. It is not registered for use in California.
Imazapyr *Arsenal, Habitat, Stalker, Chopper, Polaris*	**Rate:** 3 pt product/acre (12 oz a.e./acre) plus 0.25 to 0.5% v/v surfactant **Timing:** Postemergence when plants are growing rapidly. **Remarks:** Imazapyr is a preemergent and postemergence herbicide effective for controlling broadleaf weeds and grasses. It has fairly long soil residual activity and at the high rates needed, it will often leave bare ground. Best used in spot treatments.
Metsulfuron *Escort*	**Rate:** 1.5 to 2 oz product/acre (0.9 to 1.2 oz a.i./acre) plus 0.25 to 0.5% v/v surfactant; efficacy is improved with the addition of 2,4-D at a rate of 1 qt product/acre **Timing:** Early postemergence. **Remarks:** Metsulfuron is a selective herbicide for broadleaf species. It provides only suppression of yellow toadflax. In areas where desirable grasses are growing around toadflax, metsulfuron can be used without non-target damage. It is not registered for use in California.

Lolium multiflorum Lam.; Italian ryegrass
 (= *L. perenne* L. ssp. *multiflorum* (Lam.) Husnot)
Lolium perenne L.; perennial ryegrass
(=*Festuca perennis* (L.) Columbus & J.P. Sm. [Jepson Manual 2012])

Italian and perennial ryegrass

Lolium multiflorum

Lolium multiflorum

Family: Poaceae
Range: Throughout the U.S., including all western states.
Habitat: Roadsides, fields, pastures, agronomic crops, orchards, and vineyards. They grow best in fertile, well-drained soils and tolerate saturated soil, but do not thrive during periods of drought.
Origin: Native to Europe. Italian ryegrass is thought to have originated as an early European agricultural cultivar of perennial ryegrass. The ryegrasses are cultivated as pasture forage and turf, and Italian ryegrass is sometimes cultivated as a cover crop.
Impact: Both ryegrasses have developed resistance to glyphosate and other herbicide classes. In coastal areas of California ryegrass is occasionally infected with the fungus that causes ryegrass staggers in livestock, which may reduce milk production in ewes after giving birth. In humid regions of North America, other fungi infecting ryegrass can cause intoxication or photosensitization in livestock. The ryegrasses can impact sensitive wildland areas, particularly vernal pools.
California Invasive Plant Council (Cal-IPC) Inventory: Moderate Invasiveness

Lolium perenne

Either species can be an annual, biennial, or short-lived perennial under the right conditions. Annual ryegrass typically does not have shoots at the base. Perennial ryegrass usually has shoots at its base when in flower. They can reach 3 ft tall and the foliage of both species is usually glabrous and glossy.

Both species have spike-like inflorescences. Italian ryegrass is usually distinguished from perennial ryegrass by having lemma awns. However, they readily hybridize with one another. The offspring are highly fertile and can intrograde with either parent, resulting in populations of plants that are difficult to categorize as either species, with a continuity of characteristics. Because of this, taxonomists often classify them as the same species (*L. perenne*). Plants reproduce by abundant seed production. Seed germination is usually during fall, but can occur year-round under favorable conditions. Most seeds fall below the parent plant. Some seeds can survive in the soil for many years, but typically the seed bank appears to be short-lived. These cool-season grasses grow most vigorously during spring and fall.

NON-CHEMICAL CONTROL

Mechanical (pulling, cutting, disking)	Ryegrasses tolerate trampling, mowing and grazing. However, small patches can be controlled by hand pulling before they reproduce.
	Mowing is not considered an effective tool for the control of ryegrass, as it will readily recover with any soil moisture remaining.
	Tillage is not usually practical for the control of ryegrasses. It is primarily effective when plants are young, but at this time the soil is usually moist and either new plants will germinate or tilled plants will recover. By the time soils dry out in late spring or early summer, ryegrass has generally produced seed or has completed its life cycle and tillage is not effective.
Cultural	The ryegrasses are a preferred forage species, and grazing has not been shown to be an effective control method without causing serious damage to the ecosystem through overgrazing.
	Because the ryegrasses disperse their seed relatively early in the season, burning often selects for ryegrass and is not an effective control option.
	Ryegrass does not persist on infertile soil or where there is competition from other grasses.
Biological	Several fungal pathogens attack ryegrass seedheads, but there is little reduction in seed production and

> there is no active biological control program.

CHEMICAL CONTROL

The following specific use information is based on published papers and reports by researchers and land managers. Other trade names may be available, and other compounds also are labeled for this weed. Directions for use may vary between brands; see label before use. Herbicides are listed by mode of action and then alphabetically. The order of herbicide listing is not reflective of the order of efficacy or preference.

GROWTH REGULATORS	
Aminocyclopyrachlor + chlorsulfuron *Perspective*	**Rate:** 3 to 8 oz product/acre **Timing:** Early postemergence when weeds are germinating or actively growing. **Remarks:** Only gives suppression. *Perspective* provides broad-spectrum control of many broadleaf species. Although generally safe to grasses, it may suppress or injure certain annual and perennial grass species, including ryegrass. It is the chlorsulfuron in the mix that causes the suppression on ryegrass. Do not treat in the root zone of desirable trees and shrubs. Do not apply more than 11 oz product/acre per year. At this high rate, cool-season grasses will be damaged, including bluebunch wheatgrass. Not yet labeled for grazing lands. Add an adjuvant to the spray solution. This product is not approved for use in California and some counties of Colorado (San Luis Valley).
LIPID SYNTHESIS INHIBITORS	
Clethodim *Select, Envoy*	**Rate:** 0.75 to 2 pt product (*Envoy*)/acre (1.5 to 4 oz a.i./acre) **Timing:** Postemergence when target plants are between 2 and 6 inches in height. **Remarks:** Clethodim is a grass-selective herbicide. Add a non-ionic surfactant (0.25%) or a crop oil concentrate (1%). Note that *Envoy* formulation is 1 lb a.i./gallon, *Select* is 2 lb a.i./gallon.
Fluazifop *Fusilade*	**Rate:** 1 to 1.5 pt product/acre (4 to 6 oz a.e./acre) **Timing:** Postemergence when target plants are between 2 and 4 inches tall and rapidly growing. **Remarks:** Fluazifop is a grass-selective herbicide. Add either a crop oil concentrate at 0.5 to 1% v/v or a non-ionic surfactant at 0.25 to 0.5% v/v spray volume.
Sethoxydim *Poast*	**Rate:** Broadcast foliar treatment: 1.5 pt product/acre (4.5 oz a.e./acre). Spot treatment: 1 to 1.5% v/v solution **Timing:** Postemergence when plants are less than 8 inches tall. **Remarks:** Sethoxydim is a grass-selective herbicide. Use a crop oil concentrate or methylated seed oil surfactant. Do not apply to stressed grasses.
AROMATIC AMINO ACID INHIBITORS	
Glyphosate *Roundup, Accord XRT II,* and others	**Rate:** For foliar broadcast treatment: 1 to 3 pt product (*Roundup ProMax*)/acre (0.56 to 1.7 lb a.e./acre). Spot treatment: 0.5% to 1% v/v solution with the rate depending on weed size. **Timing:** Early postemergence for Italian ryegrass and early head formation for perennial ryegrass. **Remarks:** Glyphosate has no soil activity and will only provide control in the year of application. It is nonselective and can result in bare ground conditions that leave the area susceptible to weed encroachment. In areas with desirable vegetation, use spot treatment. Ryegrass in many areas of the west has developed resistance to glyphosate. Add a surfactant when using a formulation where it is not already included (e.g., *Rodeo, Aquamaster*).
BRANCHED-CHAIN AMINO ACID INHIBITORS	
Imazapic *Plateau*	**Rate:** 8 to 12 oz product/acre (2 to 3 oz a.e./acre) **Timing:** Preemergence or postemergence for control of Italian ryegrass, and postemergence for control of perennial ryegrass. **Remarks:** Imazapic is a broad-spectrum herbicide, but is considered most effective on annual grasses. It has some residual soil activity. Imazapic is not registered for use in California.
Imazapyr *Arsenal, Habitat, Stalker, Chopper, Polaris*	**Rate:** 1 to 3 pt product/acre (4 to 12 oz a.e./acre) **Timing:** Preemergence or early postemergence. **Remarks:** Imazapyr is a broad-spectrum herbicide with fairly long soil residual activity.
Sulfometuron	**Rate:** 3 to 5 oz product/acre (2.25 to 3.75 oz a.i./acre)

Oust and others	**Timing:** Preemergence or early postemergence when the weeds are germinating or actively growing.
	Remarks: Sulfometuron has mixed selectivity and is fairly safe on native perennial grasses. It is good for revegetation use. Use lower rates in arid environments and higher rates in wetter areas (> 20" rainfall) and on high organic matter soils. Sulfometuron has fairly long soil residual activity. At higher rates, this treatment will generally result in bare ground. A related herbicide, rimsulfuron (*Matrix*), has also been reported to give some control used preemergence or early postemergence.
PHOTOSYNTHETIC INHIBITORS	
Hexazinone *Velpar L*	**Rate:** 2 to 6 pt product/acre (0.5 to 1.5 lb a.i./acre)
	Timing: Preemergence to early postemergence when weeds are growing rapidly.
	Remarks: Hexazinone is a broad-spectrum, long residual, mobile herbicide. It has both foliar and soil activity. In soil applications, rates will vary with soil texture and soil organic matter. Best results when applied to moist soils. Use rates will also vary depending on the weed species to be controlled. Hardwood trees near application site can absorb this chemical through the roots and may be injured or killed. Do not spray near the root zone of desirable hardwood trees or shrubs. High rates of hexazinone can create bare ground, so only use high rates in spot treatments.

Lotus corniculatus L.
Birdsfoot trefoil

Family: Fabaceae
Range: Throughout the United States, including nearly all western states.
Habitat: Turf, pastures, roadsides, crop fields, ditches, orchards, vineyards, managed forests, disturbed grassland, wetland and riparian sites; often found notably in swales inland of coastal dunes. Grows year-round in warm-climate areas, winter-dormant in cold winter regions. Tolerates drought and infertile, dry, wet, saline, acidic, or limestone-based soils.
Origin: Native to Eurasia. Cultivars of birdsfoot trefoil are used for pasture forage and hay, especially on poorly drained, low-fertility soils, and to control erosion in some areas. It generally provides excellent green forage for livestock and wildlife, but has escaped cultivation on numerous occasions.
Impacts: Can form dense mats and choke out native vegetation. On rare occasions, birdsfoot trefoil has been implicated with cyanide poisoning of livestock in other areas of the world.

Birdsfoot trefoil is a trailing, mat-forming perennial with stems about 3 ft long and yellow flowers in head-like umbels. Foliage of mature plants is a distinct blue-green color, glabrous to sparsely hairy. Leaves have 5 leaflets, 3 at the tip of the leaf axis and 2 as 'wings' toward the base. The leaflets are narrow, with a dented tip. The foliage dies back in fall. Birdsfoot trefoil has a well-developed, woody taproot to ~3 ft deep, with fibrous lateral roots that form a mat near the soil surface. New shoots can grow from the roots in spring or when the crown is damaged. Roots usually develop nitrogen-fixing nodules.

In spring to summer, birdsfoot trefoil develops bright yellow pea-like flowers in umbels of 3 to 8 flowers on long stalks. These develop into clusters of small pods that twist into spirals at maturity, ejecting the seeds. Seeds may disperse with water and soil, animals, human activities, or as a seed and feed contaminant. Seeds germinate primarily in spring, but some germination can occur in fall. Some seeds are hard-coated and survive for several years under field conditions. Seeds may also survive ingestion by animals. Seedlings grow slowly and compete poorly with other vegetation. Birdsfoot trefoil also can reproduce vegetatively from roots and stems. Under favorable conditions, root fragments can develop into new plants, and mature stems can root at the nodes.

NON-CHEMICAL CONTROL

Mechanical (pulling, cutting, disking)	Hand pulling can be used on small infestations or isolated plants. It is critical to remove all of the below-ground tissues to prevent resprouting. Repeated clipping or mowing near the ground can prevent seed production and weaken the root system, but plants can survive and even thrive under mowing. Tillage is probably ineffective unless conducted repeatedly and over several years.
Cultural	Prescribed burns can facilitate seed germination and establishment, ultimately increasing the population. Close grazing that takes all stem growth reduces regrowth and stand life. Delay spring grazing until plants are at least 8 inches tall.
Biological	There are no biological control agents available.

CHEMICAL CONTROL
The following specific use information is based on reports by researchers and land managers. Other trade names may be available, and other compounds also are labeled for this weed. Directions for use may vary between brands; see label before use. Herbicides are listed by mode of action and then alphabetically. The order of herbicide listing is not reflective of the order of efficacy or preference.

GROWTH REGULATORS	
Aminopyralid *Milestone*	**Rate:** 7 oz product/acre (1.75 oz a.e./acre) **Timing:** Postemergence in spring before flowering. **Remarks:** Aminopyralid selectivity is similar to clopyralid, but has shown greater activity than clopyralid on other species tested. It is safe on grasses. Aminopyralid has longer soil residual activity than clopyralid.

Clopyralid *Transline*	**Rate:** 0.25 to 1.33 pt/acre (1.5 to 8 oz a.e./acre)
	Timing: Postemergence in spring before flowering.
	Remarks: Clopyralid selectively controls certain broadleaf species, particularly members of the Asteraceae and Fabaceae. It is safe on grasses. Clopyralid has some soil residual activity, but not as long as aminopyralid. Repeat treatments will be necessary until seedbank is depleted, possibly several years.
Dicamba *Banvel, Clarity*	**Rate:** 4 to 6 pt/acre (2 to 3 lb a.e./acre)
	Timing: Postemergence in early spring when plants are growing rapidly.
	Remarks: Dicamba is broadleaf-selective and has no soil activity. Do not apply when outside temperatures exceed 80°F.
Triclopyr *Garlon 3A, Garlon 4 Ultra*	**Rate:** 1 to 1.5 pt *Garlon 4 Ultra* /acre (0.375 to 0.75 lb a.e./acre) or 1.5 to 2 pt *Garlon 3A* (0.56 to 0.75 a.e./acre)
	Timing: Postemergence to rapidly growing plants.
	Remarks: Triclopyr is a growth regulator herbicide with little or no soil residual activity. It is broadleaf-selective and typically does not harm grasses. *Garlon 4 Ultra* is formulated as a low volatile ester. However, in warm temperatures, spraying onto hard surfaces such as rocks or pavement can increase the risk of volatilization and off-target damage.
AROMATIC AMINO ACID INHIBITORS	
Glyphosate *Roundup, Accord XRT II*, and others	**Rate:** 3 to 5 qt product (*Roundup ProMax*)/acre (3.375 to 5.625 lb a.e./acre)
	Timing: Postemergence in spring before flowering.
	Remarks: Glyphosate has no soil activity and is a nonselective herbicide.

Ludwigia spp.

Waterprimroses

Ludwigia peploides

Family: Onagraceae
Range: Primarily in the coastal states, Washington, Oregon and California; creeping waterprimrose is also found in Arizona and New Mexico.
Habitat: Slow-flowing rivers, lake and reservoir margins, and in the shallow waters of canals and floodplains.
Origin: Most species are native to South America. *L. peploides* ssp. *peploides* is native to California, Arizona, New Mexico, Texas, and Louisiana; ssp. *glabrescens* (Kuntze) Raven is native to the central and eastern U.S.; and ssp. *montevidensis* (Spreng.) Raven is introduced from southern South America. *L. peploides* is sometimes sold as an aquarium or pond ornamental.
Impacts: Dense stands degrade natural communities, reduce water quality and floodwater retention, and prevent effective mosquito control. Plants can develop a tangled mat of stems that can reduce water flow in irrigation channels and drainage ditches.
Western states listed as Noxious Weed: *L. grandiflora,* Washington
California Invasive Plant Council (Cal-IPC) Inventory: *L. hexapetala,* High Invasiveness (Alert); *L. peploides,* High Invasiveness

Waterprimroses are floating to emergent perennials with stems to 10 ft long. Stems and leaf veins are often reddish. Leaves are alternate with smooth margins. Species, or even subspecies or varieties, differ in hairiness. Plants expand by creeping rhizomes.

The taxonomy of *Ludwigia* is still very confusing. Two or three species are problematic, including creeping waterprimrose (*L. peploides* (Kunth) Raven) and Uruguay waterprimrose (*L. grandiflora* (Michx.) Greuter & Burdet; = *L. hexapetala* (Hook. & Arn.) Zardini, Gu & Raven and *L. uruguayensis* (Camb.) Hara var. *major* (Hassler) Munz). Recent evidence suggests that *L. grandiflora* and *L. hexapetala* are two distinct species.

Flowering stems are usually creeping and floating to ascending. Flowers are solitary in upper leaf axils, trumpet-shaped with a long slender tube (inferior ovary). Flowers have five petals, generally bright yellow and showy. Plants reproduce by seed and vegetatively from creeping stems and stem fragments, and to some degree from rhizomes. The fruits are hard, narrowly cylindrical capsules, 4 to 5-chambered, 1 to 2 inches long, typically bent downward. The fruits contain numerous small seeds, float in water and are easily dispersed by currents. Seeds do not individually disperse from capsules. Despite the production of numerous seeds, seedlings are rarely encountered. The plants also produce creeping submerged stems that root at nodes and produce aerial shoots. Floating vegetative mats or shoot fragments readily break off and are carried away by flowing water.

NON-CHEMICAL CONTROL

Mechanical (pulling, cutting, roguing)	Ideally, mechanical control measures should include removal of plant material that contains viable propagules (fruit, rhizomes, seed). Equipment that can dig up, plow or "rogue" the stands of *Ludwigia* spp. usually is capable of lifting and depositing it on trucks.
	Mowing devices typically leave fragmented pieces that can reinfest or disperse downstream. However, mowing may be used as part of an integrated program if done before seed set and in conjunction with properly applied herbicides.
Cultural	Preventing accumulation of nutrients and sediment can reduce the spread of *Ludwigia* spp., but this usually requires significant reduction in existing nutrient sources.
	Managed flood/dry conditions can be used in conjunction with both mechanical removal and approved herbicides.
Biological	The native flea beetles *Lysathia flavipes* and *L. ludoviciana* can defoliate some *Ludwigia* species. The chrysomelid *Altica cyanea* has been investigated for use in China. The USDA-ARS Exotic and Invasive Research

> Lab at UC Davis has begun a search for potential agents in South America. The grass carp (white amur, *Ctenopharyngodon idella*) is a relatively nonselective herbivorous fish that will consume some *Ludwigia* species, particularly those producing prostrate, floating mats. However, since the grass carp prefers submersed plants, its use must be weighed against potential impacts to native submersed plants.

CHEMICAL CONTROL

The following specific use information is based on published papers and reports by researchers and land managers. Other trade names may be available, and other compounds also are labeled for this weed. Directions for use may vary between brands; see label before use. Herbicides are listed by mode of action and then alphabetically. The order of herbicide listing is not reflective of the order of efficacy or preference.

GROWTH REGULATORS

2,4-D
Weedar 64

Rate: Broadcast foliar treatment: 1 to 2 pt product/acre (0.48 to 0.95 lb a.e./acre) with a non-ionic surfactant

Timing: Optimal timing is to apply 2,4-D postemergence from spring to early summer. However, applications from mid-summer to early fall can also be effective in suppressing growth.

Remarks: 2,4-D is a relatively fast-acting, selective systemic herbicide.

Dicamba + diflufenzopyr
Overdrive

Rate: 4 to 8 oz product/acre

Timing: Postemergence to rapidly growing plants.

Remarks: Reported effective on some waterprimrose species. Diflufenzopyr is an auxin transport inhibitor which causes dicamba to accumulate in shoot and root meristems, increasing its activity. Higher rates should be used when treating perennial weeds. Add a non-ionic surfactant to the treatment solution at 0.25% v/v or a methylated seed oil at 1% v/v solution. This product does not have an aquatic registration and cannot be used near water.

Triclopyr
Renovate

Rate: Broadleaf foliar treatment: 2.67 to 5.33 pt product/acre (1 to 2 lb a.e./acre) with a non-ionic surfactant

Timing: Postemergence, spring to early summer, is optimal. However, mid-summer applications can also be effective in suppressing growth. Late summer to fall applications can reduce subsequent spring regrowth.

Remarks: Triclopyr is a selective, relatively fast-acting systemic herbicide.

AROMATIC AMINO ACID INHIBITORS

Glyphosate
Rodeo, Aquamaster

Rate: Spot foliar treatment: 1 to 2% v/v solution (*Rodeo* or *Aquamaster*) with approved surfactants.

Timing: Postemergence from spring through fall.

Remarks: Nonselective, slow-acting systemic herbicide. Efficacy can be reduced if plants have dust and debris on the petioles (leaves). Applications made after rains remove the dust can often increase efficacy.

BRANCHED-CHAIN AMINO ACID INHIBITORS

Imazamox
Clearcast

Rate: Broadcast treatment to emergent shoots: 2 pt product/acre (4 oz a.e./acre). Spot spray-to-wet treatment: 0.25 to 5% v/v solution. Direct in-water treatment: 50 to 100 ppb. Efficacy may be improved by adding 1 qt/acre glyphosate (*Rodeo* or *Aquamaster*).

Timing: All applications (in-water or foliar) should be made from early spring to early summer during the period of rapid growth.

Remarks: Use an approved surfactant. Aerial application is approved in some states.

Imazapyr
Habitat

Rate: Broadcast treatment to emergent shoots: 4 to 6 pt product/acre (1 to 1.5 lb a.e./acre). Spot treatment: 1.5% v/v solution in 100 gal/acre for adequate coverage.

Timing: Early spring to early summer (when new growth is present)

Remarks: Use repeated applications to achieve desired concentration for 5 to 7 weeks. Do not tank mix with glyphosate for *Ludwigia* control.

CONTACT PHOTOSYNTHETIC INHIBITORS

Diquat
Reward

Rate: Spot (emergent shoot) treatment: 0.5% v/v solution (2 qt/100 gal water)

Timing: Postemergence foliar treatment from spring to early summer is optimal. Repeat treatments may be needed in mid-summer.

Remarks: Contact herbicide that is inactivated in turbid water; use only clean water to mix and spray.

Lupinus spp.

Lupines

Lupinus arboreus

Family: Fabaceae
Range: Several species present in all western states.
Habitat: Lupines are adapted to dry, relatively infertile soils. They often require ecological disturbance to persist and are common in pasture, rangeland, recent burns, sand dunes, forests, sagebrush communities, and grasslands.
Origin: Nearly all species are native plants in the western United States.
Impacts: Native lupines are a desirable component in natural communities. Lupines and their ability to fix nitrogen play an important role in colonization of sites following disturbance, and in most cases, lupine control is unnecessary. Lupine control is generally limited to small pastures or rangeland with the major goal of preventing livestock poisoning. Many lupine species contain poisonous alkaloids throughout the growing season that can potentially poison livestock and/or cause birth defects (crooked calf syndrome). Lupine control is also justified to prevent unwanted changes in native plant communities. Yellow bush lupine (*L. arboreus* Sims) has expanded its range in California and can negatively impact coastal dune communities by changing vegetation structure and soil nitrogen levels.

Lupine species are mostly herbaceous perennials, but some can be annuals. They range in size from small plants shorter than 1 ft to large shrubs taller than 8 ft. The leaves are palmate-compound with 5 to 28 leaflets. Roots typically are associated with nitrogen-fixing bacteria.

The flowers are produced in dense or open whorls, racemes or on an erect spike. Each flower is 0.5 to 1 inch long, with a typical pea-flower shape. Flowers can range in color from white to yellow to purple. Due to the flower shape, several species are known as bluebonnets or Quaker bonnets. The fruit is an exploding pod containing several seeds. Seeds are generally dispersed close to the parent plant. Viability of seeds is considered to be similar to most other legumes, i.e., seeds probably survive in the soil for several years.

NON-CHEMICAL CONTROL

Mechanical (pulling, cutting, disking)	Hand pulling, tillage, and digging are effective for controlling established plants, but the disturbance from these methods can promote new recruitment. The root system should be severed below the thickened crown. Larger species can be removed with a weed wrench.
	Most lupine species will quickly regrow following mowing. Mowing is not effective unless done frequently enough to prevent seed production and reduce vigor of established plants.
Cultural	Grazing is not an effective control method. Some lupine species can be intermittently grazed without problems, but livestock producers should determine the toxicity of the lupine species before grazing. Avoid grazing during times of year that livestock are susceptible to poisoning or birth defects. Populations often increase in heavily grazed systems.
	Like many other legumes, lupines usually respond positively to fire. Germination of seeds and resprouting of established plants is common post-fire for many lupine species. Several species typically occur in habitats subject to fairly frequent fire.
	Promoting competitive vegetation can slow spread and help prevent establishment.
Biological	Several native insects such as beetles, butterflies, and moths feed on lupine species. No introduced biological controls are known or are likely to occur with this group of natives.

CHEMICAL CONTROL

The following specific use information is based on published papers and reports by researchers and land managers. Other trade names may be available, and other compounds also are labeled for this weed. Directions for use may vary between brands; see label before use. Herbicides are listed by mode of action and then alphabetically. The order of herbicide listing is not reflective of the order of efficacy or preference.

GROWTH REGULATORS	
2,4-D Several names	**Rate:** 1 to 2 qt product/acre (0.95 to 1.9 lb a.e./acre)
	Timing: Postemergence to rapidly growing plants before flowering. In grazed areas, allow enough time for complete burndown before grazing.
	Remarks: 2,4-D is broadleaf-selective and safe on most grasses. 2,4-D has minimal soil activity. Often tank-mixed with chlorsulfuron or dicamba. Do not apply ester formulation when outside temperatures exceed 80°F.
Dicamba *Banvel, Clarity*	**Rate:** 0.5 to 2 pt product/acre (0.25 to 1 lb a.e./acre). Use higher rates for large established plants.
	Timing: Postemergence to rapidly growing plants before flowering.
	Remarks: Dicamba is a broadleaf-selective herbicide. Dicamba is often mixed with 2,4-D (0.5 to 1 pt dicamba + 2 pt 2,4-D/acre).
AROMATIC AMINO ACID INHIBITORS	
Glyphosate *Roundup, Accord XRT II*, and others	**Rate:** 1 to 2 qt product (*Roundup ProMax*)/acre (1.1 to 2.25 lb a.e./acre). Spot treatment: 1.5% v/v solution.
	Timing: Postemergence to rapidly growing plants before flowering. In grazed areas, allow enough time for complete burndown before grazing.
	Remarks: Glyphosate will not kill seeds or inhibit germination the following season. Glyphosate has no soil activity and is nonselective. It can create bare ground conditions that are susceptible to weed recruitment. In areas with desirable vegetation, use spot treatment. Glyphosate is a good control option if reseeding is planned shortly after application, as it will not injure seedlings emerging after application. Add a surfactant when using a formulation where it is not already included (e.g., *Rodeo, Aquamaster*).
BRANCHED-CHAIN AMINO ACID INHIBITORS	
Chlorsulfuron *Telar*	**Rate:** 1 to 2.6 oz product/acre (0.75 to 1.95 oz a.i./acre)
	Timing: Postemergence to rapidly growing plants before flowering. In grazed areas, allow enough time for complete burndown before grazing.
	Remarks: Always use a surfactant. Chlorsulfuron can be tank-mixed with 2,4-D for quicker burndown. Included with aminocyclopyrachlor in *Perspective*.
Metsulfuron *Escort*	**Rate:** 1 to 2 oz product/acre (0.6 to 1.2 oz a.i./acre)
	Timing: Postemergence to rapidly growing plants before flowering. In grazed areas, allow enough time for complete burndown before grazing.
	Remarks: Similar activity compared to chlorsulfuron. Always use a surfactant. Metsulfuron can be tank-mixed with 2,4-D for quicker burndown. Other premix formulations of metsulfuron can be used at similar application timing. These include *Cimarron Max* (metsulfuron + dicamba + 2,4-D) and *Cimarron X-tra* (metsulfuron + chlorsulfuron). Metsulfuron is not registered for use in California.

Lysimachia vulgaris L.

Garden loosestrife

Family: Primulaceae
Range: Reported in Washington, Oregon, Montana, and Colorado.
Habitat: Wetlands, riparian areas, and shores of lakes and ponds.
Origin: Native to Eurasia and northern Africa. Escaped from cultivation as an ornamental.
Impacts: Forms dense stands in wetlands and outcompetes native vegetation.
Western states listed as Noxious Weed: Washington

Garden loosestrife is a rhizomatous perennial to 3 to 6 ft tall. The leaves are opposite or in whorls of 3 to 5, lanceolate in shape, to 5 inches long and about an inch wide, and generally sessile on the stem. Leaf margins are not toothed, and blades are softly hairy and dotted with black or orange glands. The stems are usually branched near the top, bearing multi-flowered clusters along these branches and in upper leaf axils. Rhizomes, or sometimes stolons, can measure as long as 15 ft from a single stem.

Flowers are 0.5 inch wide with 5 yellow petals that are fused and often orange near the base. Sepals also number 5 and are distinctly orange-tipped. Fruits are dry capsules about ¼-in wide that split apart at maturity to spill the small seeds. Garden loosestrife spreads both by seed and from rhizomes. However, new seedlings are not as important to its spread compared to vegetative expansion. Although the seed have dormancy, they do not appear to survive for more than 3 years in the soil. Garden loosestrife has successfully invaded cattail (*Typha latifolia*) dominated sites in Washington, and is reported to outcompete purple loosestrife (*Lythrum salicaria*). Another species also called garden or yellow loosestrife (*Lysimachia punctata*) is sometimes found in similar habitats. Of the two species, *Lysimachia vulgaris* is more likely to be found in wetland areas, and bears flowers on branches and clusters at the top of the plant. *Lysimachia punctata* is generally on drier sites and bears flowers in axillary clusters on mostly unbranched stems.

NON-CHEMICAL CONTROL

Mechanical (pulling, cutting, disking)	Digging of garden loosestrife is feasible for individual plants or small stands. Remove as much root and rhizome as possible, as broken root and rhizome sections will resprout from fragments. Hand pulling is not an effective strategy, as it is rarely possible to remove roots and rhizomes without substantial breaking. Hand digging is often impractical for well-established and extensive infestations.
	Mowing is difficult in wetland sites, and unless applied repeatedly, mowing will not generally control this perennial species. However, timely mowing can prevent seed production.
	Mulching with plastic or fabric sheets may suppress growth of garden loosestrife. Shoots emerging from around edges or holes in the mulch should be controlled as they appear.
Cultural	Draining water from infested areas may slow growth of garden loosestrife, or allow other control options to be employed. Conversely, inundation with deep water for an extended period of time may result in some control of the species.
Biological	There are no known biological control agents to aid in the control of garden loosestrife.

CHEMICAL CONTROL

The following specific use information is based on reports by researchers and land managers. Other trade names may be available, and other compounds may also be labeled for this weed. Directions for use may vary between brands; see label before use. Most herbicide applications will require multiple applications to fully control garden loosestrife. Because it usually is found growing in or near standing water, only aquatic herbicide formulations are recommended for use. Additionally, most states require specific aquatic endorsements for applicators of aquatic herbicides. Herbicides are listed by mode of action and then alphabetically. The order of herbicide listing is not reflective of the order of efficacy or preference.

GROWTH REGULATORS	
Triclopyr *Renovate*	**Rate:** Broadcast treatment: 6 to 8 qt product/acre (4.5 to 6 lb a.e./acre). Spot treatment: 1 to 1.5% v/v solution **Timing:** Postemergence to rapidly growing plants at mid- to full-bloom. **Remarks:** Use up to 1% non-ionic surfactant approved for aquatic use to improve herbicide uptake. Triclopyr is relatively safe on monocot plants such as grasses, rushes, sedges, and cattails.
AROMATIC AMINO ACID INHIBITORS	
Glyphosate *Rodeo* or *Aquamaster*	**Rate:** Broadcast treatment: 4 pt product (*Rodeo* or *Aquamaster*)/acre (2 lb a.e./acre). Spot treatment: 1 to 1.5% v/v solution **Timing:** Postemergence to rapidly growing plants at full to late flowering, or as an autumn application. **Remarks:** Use up to 1% non-ionic surfactant approved for aquatic use to improve herbicide uptake. Glyphosate overspray will injure or kill other plants that it contacts.
BRANCHED-CHAIN AMINO ACID INHIBITORS	
Imazapyr *Habitat*	**Rate:** 1 pt product/acre (0.25 lb a.e./acre) **Timing:** Postemergence to rapidly growing plants in summer. **Remarks:** Use up to 1% non-ionic surfactant approved for aquatic use to improve herbicide uptake. Imazapyr overspray will injure or kill other plants that it contacts.

Lythrum salicaria L.
Purple loosestrife

Family: Lythraceae
Range: Throughout the western United States, except Arizona. It is more prevalent in the Midwestern and northeastern states.
Habitat: Freshwater and brackish wetlands including marshes, riparian areas, lakeshores, floodplains, seasonally wet areas, intermittent streams, ditches and canals. It sometimes invades upland sites as well. Moisture is critical for establishment, but once established, purple loosestrife can survive for years at dry sites. Somewhat shade-tolerant and grows in most soil types, including infertile soils, but prefers slightly acid to neutral soils.
Origin: Considered to be native to Eurasia.
Impact: An extremely aggressive colonizer which disrupts the ecology of wetland sites by displacing native vegetation and wildlife. An isolated colony of purple loosestrife plants can spread to cover wetland sites in a single season under optimum growing conditions. Purple loosestrife can clog irrigation systems causing significant economic losses. The species has poor palatability and impacts the value of meadows and wetland pastures for grazing. Purple loosestrife is not considered a threat to cultivated crops.
Western states listed as Noxious Weed: Arizona, California, Colorado, Idaho, Montana, New Mexico, Nevada, North Dakota, Oregon, South Dakota, Utah, Washington, Wyoming
California Invasive Plant Council (Cal-IPC) Inventory: High Invasiveness

Purple loosestrife is a perennial, aquatic herb that grows 3 to 7 ft tall, but can reach 10 ft under ideal conditions. It has a persistent taproot and spreading root stock with a dense bushy growth pattern. Plants start producing multiple stems from a single rootstock as early as the second year and can have more than 50 stems per plant. Established plants can have a crown 5 ft wide. Stems are erect and green to purple in color, with few branches, and are either 4- or 6-sided. The stalkless leaves are 0.75 to 4 inches long, 0.2 to 0.5 inch wide, sometimes with fine hairs. They can be opposite or whorled. The leaves are lance-shaped, rounded or heart-shaped at the base, clasping and with smooth margins. Shoots die each fall and new shoots develop the following spring from buds located at the top of the root crown.

Purple loosestrife has showy purple to magenta flowers, clustered on a spike a few inches to 3 ft long. The seed capsule has two cells with numerous tiny reddish-brown seeds (1 mm long or less). Seed production per plant ranges from around 100,000 seeds for young plants to over 2.7 million seeds for established plants. Seeds are dispersed by water and to a lesser degree by wind. Seeds can germinate underwater, and their longevity is at least 3 years. While sexual reproduction by seed is most important, purple loosestrife can also spread vegetatively from stem cuttings. Buried stems have adventitious buds that can produce new shoots or roots.

NON-CHEMICAL CONTROL

Mechanical (pulling, cutting, disking)	Manually digging or hand-pulling is effective for early infestations to help prevent the establishment of dense colonies. Early detection and removal is essential because established plants are too large and deep-rooted to remove easily. All plant material, especially the root crown, should be removed to prevent resprouting. Pulled plants should be dried or burned. Repeat visits are important to ensure there is no regrowth.
	Mowing purple loosestrife can be impractical due to the sites it occupies. Mowing or cutting stems can help reduce seed bank accumulation. Late-season cutting was found to reduce shoot production to a greater degree than mid-summer cutting. Mowing or cutting may be more effective when used as part of an integrated approach with herbicides.
	Tillage is probably not an effective control measure for purple loosestrife. Wetland sites where it grows are not conducive to tillage operations. In addition, any disturbance that fragments live stem or root tissue is likely to spread purple loosestrife, and its extensive soil seedbank is likely to reinvade open areas created with tillage. Three to four consecutive years of tillage might be effective.
Cultural	Purple loosestrife grows in wet areas that are not usually grazed and it has poor palatability.

	There is little information on the effectiveness of fire, but some sources indicate that purple loosestrife does not burn well and it would be difficult to get a fire to carry through an infested area. Additionally, it is doubtful that burn temperatures in wetland areas would get high enough to kill the massive crown. However, burning removes biomass and may improve the effectiveness of herbicides. Continuous flooding has been somewhat effective for large infestations where the water level can be controlled. The duration of the flooding appears to be more important than the depth of flooding. The precise parameters for maximum effectiveness need further study. Black plastic mulch was found to be marginally effective, but while it can reduce growth and seed production, it does not kill the roots of mature plants.
Biological	Several non-native insects have been released in the United States to control purple loosestrife. These include the black-margined and golden loosestrife beetles (*Galerucella calmariensis* and *G. pusilla*), whose larvae and adults feed on the foliage and flowers, reducing seed production. In addition, the larvae and adults of the root weevil (*Hylobius transversovittatus*) feed on the root. Larvae of the loosestrife flower weevil (*Nanophyes marmoratus* and *N. brevis*) feed on flowers, and the adults feed on foliage and flowers. The two *Galerucella* spp. have been the most successful of these biocontrol agents. These beetles were released between 1992 and 1994 and have become established in some states. They are not present in California yet but have shown promising results in Oregon. The root weevil and the flower weevil have been released at test sites in California with mediocre results. There is ongoing research on the use of pathogenic fungi as biocontrol agents for purple loosestrife.

CHEMICAL CONTROL

The following specific use information is based on published papers and reports by researchers and land managers. Other trade names may be available, and other compounds also are labeled for this weed. Directions for use may vary between brands; see label before use. Herbicides are listed by mode of action and then alphabetically. The order of herbicide listing is not reflective of the order of efficacy or preference.

GROWTH REGULATORS	
Triclopyr *Garlon 3A*	**Rate:** Broadcast foliar treatment: 6 to 8 qt *Garlon 3A*/acre (4.5 to 6 lb a.e./acre). Spot treatment: 1.5 to 2% v/v solution **Timing:** Postemergence at bud to mid-flowering stage. **Remarks:** Thorough coverage and a minimum of 50 gal/acre spray solution is recommended. Follow-up applications should be made to regrowth the following year. Triclopyr is broadleaf-selective and safe on most grasses. It is most effective on smaller plants and has little or no residual activity. *Garlon 3A* and other amine formulations are registered for aquatic use. Triclopyr (1 to 2 qt *Garlon 3A*/acre) can be mixed with aminopyralid (*Milestone*) at 7 oz product/acre or with 1 to 2 pt product/acre 2,4-D.
AROMATIC AMINO ACID INHIBITORS	
Glyphosate *Roundup, Rodeo, Aquamaster,* and others	**Rate:** Broadcast foliar treatment: 1 to 2 pt product (*Roundup ProMax*)/acre (0.56 to 1.1 lb a.e./acre). Spot treatment: 1% v/v solution **Timing:** Postemergence to rapidly growing plants in the full to late flowering stage. Seedlings may be treated in spring following a fall treatment. **Remarks:** Nonselective, no soil activity. Effectiveness is increased by addition of ammonium sulfate. Aquatic registered formulations, e.g., *Rodeo* and *Aquamaster*, are available for use close to water.
BRANCHED-CHAIN AMINO ACID INHIBITORS	
Imazapyr *Arsenal, Habitat, Stalker, Chopper, Polaris*	**Rate:** 1 to 2 pt product/acre (4 to 8 oz a.e./acre) **Timing:** Postemergence to rapidly growing loosestrife after mid-bloom until killing frost. **Remarks:** Nonselective, long soil residual activity. Leaves more bare ground than other treatments, even a year after application. *Habitat* is an aquatic registered formulation available for use close to water.
Metsulfuron *Escort*	**Rate:** 1 to 2 oz product/acre (0.6 to 1.2 oz a.i./acre) **Timing:** Postemergence from seedling to flowering stage. Most effective at flower-bud and flowering stage. **Remarks:** Primarily active on broadleaf species. Always use a surfactant. It can be tank-mixed with 2,4-D for quicker burndown. Other premix formulations of metsulfuron can be used at similar application timing. These include *Cimarron Max* (metsulfuron + dicamba + 2,4-D) and *Cimarron X-tra* (metsulfuron + chlorsulfuron). Metsulfuron is not registered for use in California.

Marrubium vulgare L.
White horehound

Family: Lamiaceae
Range: Nearly all states, including all western states, except perhaps North Dakota. Found almost worldwide.
Habitat: Pastures, fields, roadsides, rangeland, disturbed natural areas, waste places, ditches, and other disturbed places. Most often in dry places. White horehound is especially common in overgrazed areas.
Origin: Native to Eurasia. White horehound was once cultivated as a medicinal herb and for flavoring candy.
Impacts: Expands range during drought conditions, especially under heavy grazing. Once established, it can outcompete native vegetation and form dense stands in annual grasslands. Livestock avoid consuming the bitter-tasting foliage, and the plant thrives in the absence of competition with other vegetation.
California Invasive Plant Council (Cal-IPC) Inventory: Limited Invasiveness

White horehound is a cool-season perennial to 2 ft tall, often looking like a low shrub. It has densely hairy white-woolly stems, thick and square in cross-section, which mostly branch near the base of the plant. Its leaves are aromatic, opposite on the stems, ovate to nearly round, 0.5 to 2.5 inches long, with round-toothed margins. Both upper and lower leaf surfaces are hairy, and the veins are depressed so the leaves look puckered. White horehound forms a branched woody taproot with numerous fibrous laterals.

In spring to summer, white horehound produces head-like whorls of small white flowers in the upper leaf axils, spaced along the stems. After senescence in late fall, dead stems with persistent fruits can persist through winter. The plant reproduces only by seed. Fruits disperse primarily by falling to the ground beneath the parent plants, but long distance dispersal can occur when seeds cling to the fur, feathers, and feet of animals or to the shoes and clothing of people. Anecdotal reports suggest that the seed of white horehound can survive in soil for 7 to 10 years.

NON-CHEMICAL CONTROL

Mechanical (pulling, cutting, disking)	In small patches, manual removal (pulling or hoeing) is effective and should be repeated as needed when new seedlings emerge.
	Mowing before seed production can help to restrict growth of a patch.
	Deep cultivation, especially when the soil is dry, can control horehound, although partially buried plants may survive. However, this technique is not practical in most natural areas.
Cultural	Sheep will feed on horehound, but it is not preferred forage. Intensive grazing may open up the ground for the plant to spread. Ensure that livestock are quarantined or clean of burs before entry onto clean land. In some trials in pasture, the use of broadleaf-selective chemicals followed by heavy grazing has been shown to reduce horehound abundance.
	Burning can kill mature plants but usually stimulates seed germination the following season. Burning can be effective when followed by other control measures, such as chemical control of seedlings.
Biological	Some moth species have been released for horehound control in Australia but have not been made available in North America. In Australia, the combination of biocontrol insects and herbicides such as 2,4-D appears promising.

CHEMICAL CONTROL
Fall is generally the best time for controlling mature white horehound. It is very important to include surfactant in foliar treatments of this plant because of the hairy leaves. White horehound is slow to show herbicide symptoms.

The following specific use information is based on published papers and reports by researchers and land managers. Other trade names may be available, and other compounds also are labeled for this weed. Directions

for use may vary between brands; see label before use. Herbicides are listed by mode of action and then alphabetically. The order of herbicide listing is not reflective of the order of efficacy or preference.

GROWTH REGULATORS	
2,4-D Several names	**Rate:** 1 to 4 pt product/acre (0.48 to 1.9 lb a.e./acre) **Timing:** Apply to young, rapidly growing plants. **Remarks:** 2,4-D is broadleaf-selective and has no soil activity. Use higher rates for larger, more established plants. Reports on control of established plants are mixed. Do not apply when outside temperatures exceed 80°F.
Dicamba *Banvel, Clarity*	**Rate:** 1 to 4 pt product/acre (0.5 to 2 lb a.e./acre) **Timing:** Apply to young, rapidly growing plants. **Remarks:** Dicamba is a broadleaf-selective herbicide often combined with other active ingredients. Higher rates are needed for control of mature plants, but it can injure grasses at higher rates.
Picloram + 2,4-D *Grazon P+D, Gunslinger*	**Rate:** 2 to 4 pt product/acre **Timing:** Apply during active growth. **Remarks:** Both compounds are broadleaf-selective and relatively safe on grasses. Not registered for use in California.
Triclopyr *Garlon 3A, Garlon 4 Ultra*	**Rate:** 3.33 pt *Garlon 3A*/acre, 2.5 pt *Garlon 4 Ultra*/acre (1.25 lb a.e./acre) **Timing:** Postemergence to rapidly growing weeds. **Remarks:** Triclopyr is broadleaf-selective and is safe on most grasses. It is most effective on smaller plants. *Garlon 4 Ultra* is formulated as a low volatile ester. However, in warm temperatures, spraying onto hard surfaces such as rocks or pavement can increase the risk of volatilization and off-target damage.
BRANCHED-CHAIN AMINO ACID INHIBITORS	
Metsulfuron *Escort*	**Rate:** 0.2 to 1 oz product/acre (0.12 to 0.6 oz a.i/.acre) **Timing:** Apply to young, rapidly growing weeds in spring before flowering, or in fall to new rosettes. **Remarks:** Metsulfuron has mixed selectivity and is generally safe on grasses. Use a surfactant. It can be tank-mixed with 2,4-D and/or dicamba and has some soil residual activity. Not registered for use in California.

Mentha pulegium L.
Pennyroyal

Family: Lamiaceae
Range: California, Oregon and Washington.
Habitat: Disturbed moist places, ditches, roadsides, pastures, seasonally flooded sites, seeps, vernal pools, marsh margins, stream and pond margins. Grows best in clay or silty soils where moisture is plentiful. Often grows in the partial shade of other vegetation. Tolerates some alkalinity and seasonal drought. In addition to pastures, roadsides, and various disturbed sites, pennyroyal is invading wetland habitats, including sensitive vernal pools and other wetland areas.
Origin: Native to Europe. Pennyroyal is cultivated as a garden ornamental and medicinal plant, but has escaped cultivation and appears to be rapidly spreading in some areas of the western states.
Impacts: Livestock generally avoid consuming pennyroyal, which can favor the expansion of pennyroyal in pastures and on rangeland, reducing the grazing capacity of these areas. Pennyroyal is sometimes used medicinally, but the foliage contains an essential oil that, when ingested as an extract, can be fatally toxic to humans. Handling the plants or the oil can cause contact dermatitis in sensitive individuals. The foliage has been used as an insect repellent.
California Invasive Plant Council (Cal-IPC) Inventory: Moderate Invasiveness

Pennyroyal is a low-growing, aromatic perennial to 2 ft tall. Rhizomes are short, and sometimes stolons are present. Plants have square stems, opposite leaves, and are covered with short, white hairs and glandular dots.

The flowers are lavender to violet, in clusters that are head-like in whorls around stems (verticillate). Each cluster sits just above a pair of small down-turned leaves or leaf-like bracts. The fruits consist of 4 nutlets enclosed by the calyx. Plants reproduce by seed and vegetatively from rhizomes and stolons. The nutlets and calyx disperse as a unit primarily by water or by clinging to the fur, feathers, and feet of animals. Seeds can germinate under water, and the seedlings continue to grow during an extended period of shallow inundation. Like other members of the mint family, it is expected that the seeds of pennyroyal are long-lived in the soil.

NON-CHEMICAL CONTROL

Mechanical (pulling, cutting, disking)	Pennyroyal infestations can be suppressed by manual removal of individual plants and small patches before flowering, including the rhizomes and stolons, followed by the removal of seedlings as soon as discovered. Below-ground reproductive tissues should be severed approximately 3 inches below the soil surface when the plants are beginning to bolt. This can be difficult, however, because pennyroyal has brittle stems that make it hard to remove below-ground reproductive tissues.
	Late spring or early summer mowing, repeated over several years, may weaken plants by depleting photosynthetic reserves. However, it is often difficult to get mowing equipment into infested areas because these are generally in wetland sites. Cutting will generally result in crown resprouting. In addition, plants in the rosette stage and early stages of growth often have a prostrate growth habit that is not conducive to management by mowing.
	Tillage can be an effective control strategy for rosettes and bolting plants, but tillage is generally not an option in typical wetland infestations. Larger colonies on dry soils can be controlled by repeated cultivation. It is important to maintain a loose soil, because weeds are likely to survive in large clods and reestablish. Best results are obtained with offset discs or a rotary hoe. One-way discs and tined implements often lead to recovery of fragments that can increase the size of the colony.
Cultural	Prescribed burning has been noted as a possible control method. However, it is unlikely that fire will control below-ground rhizomes, and high soil moisture levels in most pennyroyal habitats may limit the effectiveness of burning.
	Control by livestock grazing is unlikely. Pennyroyal is unpalatable as forage for cows or sheep and sometimes causes gastrointestinal irritation when ingested.

Biological	There are no biological control agents available for the management of *Mentha pulegium*. This is likely due to the importance of mints as a crop in some areas of the west, as well as the number of native mints that potentially could be affected by bioagents.

CHEMICAL CONTROL

There is little information on the chemical control of *Mentha pulegium*.

The following specific use information is based on reports by researchers and land managers. Other trade names may be available, and other compounds also are labeled for this weed. Directions for use may vary between brands; see label before use. Herbicides are listed by mode of action and then alphabetically. The order of herbicide listing is not reflective of the order of efficacy or preference.

GROWTH REGULATORS	
2,4-D Several names	**Rate:** 4 pt product/acre (1.9 lb a.e./acre) for amine formulation **Timing:** Postemergence when plants are mature; best when they have bolted but before seed production. **Remarks:** In Australia they recommend the ester formulation of 2,4-D, but this cannot be used when pennyroyal infests wetland areas. Only the amine formulation is registered for aquatic areas. Results from New Zealand show that 2,4-D provides only good control, not excellent control. 2,4-D is a broadleaf-selective herbicide.
Triclopyr *Garlon 3A*	**Rate:** 2 to 4 qt product/acre (1.5 to 3 lb a.e./acre) **Timing:** Postemergence when plants are mature; best when they have bolted but before seed production. **Remarks:** Only the amine formulation should be used in wetland areas. Triclopyr is a broadleaf-selective herbicide.
AROMATIC AMINO ACID INHIBITORS	
Glyphosate *Rodeo,* *Aquamaster*	**Rate:** Broadcast treatment: 2 to 4 qt product (*Roundup ProMax*)/acre (2.25 to 4.5 lb a.e./acre) or 2 to 4 qt product (*Rodeo* or *Aquamaster*)/acre (2 to 4 lb a.e./acre) near aquatic sites. Spot treatment: 1 to 2% v/v solution **Timing:** Postemergence when plants are mature; best when they have bolted but before seed production. **Remarks:** Glyphosate is a nonselective herbicide with no soil activity. Thus, it is best used in monotypic stands or as a spot treatment. Only aquatic formulations should be used in wetland areas. This will require the addition of a surfactant registered for use in aquatic sites.
BRANCHED-CHAIN AMINO ACID INHIBITORS	
Metsulfuron *Escort*	Some evidence indicates that metsulfuron can be effective for control of pennyroyal. Although this herbicide is registered for use in aquatic systems in other countries, e.g. Australia, it is not registered for use in wetland habitats in the United States. On dry sites it can be effective. It is not registered for use in California.

Mirabilis nyctaginea (Michx.) MacMill.
Wild four-o'clock

Family: Nyctaginaceae
Range: All western states except Arizona and Oregon.
Habitat: Found in a wide range of habitats, including perennial crops such as orchards and alfalfa fields, waste areas and along roadsides, railroad lines, woodlands, pastures, riparian areas, and dry meadows and rangelands. It is often found on sandy or rocky soil, but can also grow on clay soils or along waterways. It rarely establishes in annually cultivated ground.
Origin: Native east of the Rocky Mountains, from Montana to Mexico, and east to Wisconsin and Alabama.
Impacts: Wild four-o'clock can spread from small infestations to hundreds of acres very quickly. It colonizes both perennial agriculture and rangelands. It can outcompete pasture and grassland plants. It can compete in the same habitat as Macfarlane's four-o'clock (*Mirabilis macfarlanei*), a rare species that is considered threatened in Idaho and endangered in Oregon.
Western states listed as Noxious Weed: Washington

Wild four-o'clock is a taprooted perennial to 4 ft tall. Plants are deeply rooted with thick, black roots, and sometimes producing a semi-woody crown. The leaves are opposite, to 4 inches long and 3 inches wide, and are heart- or egg-shaped. Leaves are smooth and waxy, with lower and middle leaves larger and borne on petioles and upper leaves smaller and more typically sessile. The stems are oppositely branching and usually smooth with a bluish to whitish waxy bloom on their surfaces.

The flowers are borne in clusters of three to five on short hairy stalks near the top of the plant. Flowers are about 10 mm in diameter but have no petals. Instead, flowers consist of five showy pink to red or lavender sepals with a whorl of bracts at the base. The name "four-o'clock" refers to the flowers, which open late in the day and wither early the next morning. Fruits are prominently five-ribbed, warty, somewhat hairy, grayish brown in color and from 3 to 6 mm long. Dispersal is only by seed. Seed spread primarily by falling to the ground below the parent plant; long-distance dispersal can occur when the umbel-like inflorescences catch on and are transported by vehicles and equipment. The seed is a hard, elongated nutlet. Seeds do undergo dormancy, but the longevity in the soil is not well understood. It is expected, however, that the seeds would remain in the soil for a few years.

NON-CHEMICAL CONTROL

Mechanical (pulling, cutting, disking)	Wild four-o'clock spreads primarily by seed, so hand weeding must be employed before flowering and seed set. However, hand pulling is generally not appropriate for this species, as roots rarely can be removed without breaking and subsequent resprouting.
	Unless applied repeatedly, mowing will not generally control this perennial species. Timely mowing can prevent seed production, however.
	While some reports indicate that cultivation alone can control wild four-o'clock, this method was not effective in Washington. However, if employed more than once per season, cultivation may be more effective. Hand hoeing following initial cultivation can be effective.
Cultural	Wild four-o'clock was reported to produce a mildly toxic alkaloid, but it is not widely considered to be a toxic species. In addition, pigs are fond of the fleshy roots. Selective grazing has not been shown to be an effective control strategy.
Biological	There are no known biological control agents to aid in the control of wild four-o'clock. This is because the species is native to the eastern United States. It is, however, susceptible to Mirabilis mosaic caulimovirus, which could potentially reduce growth and seed production.

CHEMICAL CONTROL

Wild four-o'clock is reported as being very tolerant of 2,4-D, but suppressed by dicamba (*Banvel*); it is not known if non-crop sulfonylurea herbicides such as metsulfuron (*Escort*) or chlorsulfuron (*Telar*) also suppress wild four-o'clock.

The following specific use information is based on reports by researchers and land managers. Other trade names may be available, and other compounds may also be labeled for this weed. Directions for use may vary between brands; see label before use.

AROMATIC AMINO ACID INHIBITORS	
Glyphosate *Roundup, Accord XRT II*, and others	**Rate:** 1.5 qt product (*Roundup ProMax*)/acre (1.7 lb a.e./acre) **Timing:** Postemergence to early bloom. **Remarks:** Usually one application at low rate is adequate in non-crop areas or glyphosate-resistant crops.

Myosotis latifolia Poir.
Broadleaf forget-me-not

Family: Boraginaceae
Range: Primarily in coastal ranges from Washington to California.
Habitat: Shady moist disturbed places, riparian areas, coastal forests and woodlands, moist meadows and fields, roadsides, old gardens and other areas near human habitation.
Origin: Native to northwestern Africa and introduced as a garden ornamental.
Impacts: Dense stands can reduce biodiversity in riparian or woodland communities. May be toxic to livestock and other animals.
California Invasive Plant Council (Cal-IPC) Inventory: Limited Invasiveness

Erect to nearly prostrate perennial to 2 ft tall. Forget-me-not has creeping roots and coiled racemes of pale blue to pink funnel-shaped flowers. The stems are woody at the base and the leaves are broadly elliptic to narrowly oblanceolate, 1 to 7 inches long. The lower leaves taper to a narrowly winged stalk and the upper leaves are sessile.

Flowering racemes are initially coiled, later elongate and open. The pale blue or pink flowers are funnel-shaped and about 0.25 to 0.5 inch wide, with five spreading lobes and five pale crests that nearly close the top of the tube. The fruits consist of four erect nutlets surrounded by the calyx. Plants reproduce by seed and vegetatively from creeping roots. Like other members of the family, it is expected that the seeds can survive many years in the soil seedbank.

NON-CHEMICAL CONTROL

Mechanical (pulling, cutting, disking)	Roots are easily uprooted, and individual plants or small populations can be removed by hand pulling. All roots need to be removed, however, to prevent resprouting.
Cultural	Areas typically infested with broadleaf forget-me-not are not conducive to the use of grazing or burning as a control strategy.
Biological	Because there are so many cultivated and native species of *Myosotis*, there are no biological control agents available.

CHEMICAL CONTROL

There is very little information available on the control of *Myosotis*. Many of the chemical control options listed below were obtained from recommendations for the control of a related species, blue heliotrope (*Heliotropium amplexicaule*). It is expected that similar recommendations would also be effective on broadleaf forget-me-not.

The following specific use information is based on reports by researchers and land managers. Other trade names may be available, and other compounds also are labeled for this weed. Directions for use may vary between brands; see label before use. Herbicides are listed by mode of action and then alphabetically. The order of herbicide listing is not reflective of the order of efficacy or preference.

GROWTH REGULATORS	
Dicamba *Banvel, Clarity*	**Rate:** Broadcast foliar treatment: 2 to 4 pt product/acre (1 to 2 lb a.e./acre). Spot treatment: 0.6% v/v solution **Timing:** Postemergence to young rapidly growing plants. **Remarks:** Dicamba will injure other broadleaf species. Recommendation is based on control of blue heliotrope, a closely related species.
Fluroxypyr *Vista XRT*	**Rate:** Broadcast foliar treatment: 6 to 22 oz product/acre (2.1 to 7.7 oz a.e./acre). Spot treatment: 1% v/v solution **Timing:** Postemergence during flowering. **Remarks:** Recommendation is based on control of blue heliotrope, a closely related species.

Picloram *Tordon 22K*	**Rate:** Spot application rates not to exceed 2 qt product/acre
	Timing: Postemergence on rapidly growing plants up to the flowering stage.
	Remarks: Picloram is a broadleaf herbicide. For control of blue heliotrope, recommendations suggest tank mixing with 2,4-D or triclopyr. While this treatment is used in Australia, there are few situations where it should be used in the western U.S. Picloram is a restricted use herbicide. Picloram is not registered for use in California.
AROMATIC AMINO ACID INHIBITORS	
Glyphosate *Roundup, Accord XRT II*, and others	**Rate:** Broadcast foliar treatment: 1 to 2 qt product (*Roundup ProMax*)/acre (1.1 to 2.25 lb a.e./acre); larger plants may require higher rates. Spot treatment: 1% v/v solution. Wiper treatment: 33 to 50% of concentrated product.
	Timing: Postemergence to seedlings or to mature plants that are growing rapidly.
	Remarks: Glyphosate is a nonselective herbicide. The control information is based on that of other forget-me-not species. Compared to other chemical options, glyphosate in a wiper application method would be the best control option for most situations.
BRANCHED-CHAIN AMINO ACID INHIBITORS	
Metsulfuron *Escort*	**Rate:** Broadcast treatment: 1 to 2 oz product/acre (0.6 to 1.2 oz a.i./acre). Spot treatment: 0.1% w/v solution
	Timing: Postemergence when plants are growing rapidly in spring to fall.
	Remarks: Recommendation is based on control of other forget-me-not species.

Myriophyllum aquaticum (Vell. Conc.) Verdc.
Parrotfeather

Family: Haloragaceae
Range: New Mexico, Arizona, California, Oregon, Idaho, Montana, and Washington. Also common in the southern and eastern United States.
Habitat: Ponds, lakes, rivers, streams, canals, ditches. Usually in still or slow-moving water, but occasionally found in faster moving water of streams and rivers. Grows best in tropical regions, can survive freezing conditions by becoming dormant. Does not tolerate brackish water; requires high light conditions.
Origin: Introduced from South America as an aquarium plant and pond ornamental in the late 1800s or early 1900s.
Impacts: Parrotfeather can develop colonies that form large sub-surface or surface mats. Mats impede water flow, interfere with boat traffic and recreational activities, create mosquito habitat, and displace native aquatic vegetation.
Western states listed as Noxious Weed: Washington
California Invasive Plant Council (Cal-IPC) Inventory: High Invasiveness (Alert)

Parrotfeather is a perennial with creeping rhizomes. It has emergent stems and is sometimes semi-terrestrial on mud banks. The total stem length can be up to 16 ft long, though the majority is submerged. Stems fragment easily and root at lower nodes. The submersed leaves are arranged in whorls of three to six leaves per node, pinnately dissected into linear lobes. The emergent leaves resemble submersed leaves, but are slightly thicker and not as finely dissected. In addition, the emerged leaves are typically light gray-green, mostly 5 to 6-whorled, 1 to 2 inches long, with a flattened midrib that is broader than the lobes.

Flowers are dioecious (male and female flowers develop on separate plants) and inconspicuous in the leaf axils. Most plants in the introduced range are female. Only populations within the native range have been observed to develop seed. In the introduced range, reproduction is only vegetatively by rhizomes, stem fragments, and axillary buds. Stem fragments form new roots and shoots and disperse primarily with water, or by clinging to the feet or feathers of water birds, and with human activities such as boating, mechanical harvesting, and the dumping of unwanted pond or aquarium contents. Mats sometimes detach and float to infest new areas.

NON-CHEMICAL CONTROL

Mechanical (pulling, cutting, dredging)	Repeated mechanical harvesting can help reduce stem densities, but escaped stem fragments can drift elsewhere and develop into new plants. More effective harvesting systems that remove the biomass and nutrient reserves accumulated in the emergent tissues may be an effective control measure. Removing and destroying stem fragments from recreational equipment, such as boat propellers, docking lines, and fishing gear can help prevent the spread of non-native watermilfoils. Since nearly all spread and reproduction is via shoot and rhizomes, physically removing 6 to 10 inches of infested sediment should eliminate regrowth. For small ponds or small infestations in lakes, this may be a cost-effective approach if it is coupled with stopping any further introductions.
Cultural	Dewatering (draining) can be effective if the exposed sediment is subject to hard freezes. Dewatering coupled with excavation to removed sediment-borne roots and rhizomes will also control parrotfeather.
Biological	Herbivorous insects from Argentina have been investigated and were released for control of parrotfeather in South Africa, including *Listronotus marginicollis* (stem-miner) and *Lysathia* spp. (leaf-feeders). Neither has been released in the U.S. The triploid grass carp can provide some suppression, but the fish prefers other submersed plants (native and non-native) and will consume those first if present. Use of grass carp usually requires a permit.

CHEMICAL CONTROL

The following specific use information is based published papers and reports by researchers and land managers. Other trade names may be available, and other compounds also are labeled for this weed. Directions for use may vary between brands; see label before use. Herbicides are listed by mode of action and then alphabetically. The order of herbicide listing is not reflective of the order of efficacy or preference.

GROWTH REGULATORS	
2,4-D *Weedar 64*	**Rate:** For emergent shoots: 1 to 2 pt product/acre (0.48 to 0.95 lb a.e./acre) with a non-ionic surfactant **Timing:** Spring to early summer is optimal; however mid-summer applications can be effective in suppressing growth. **Remarks:** Emergent shoots of parrotfeather are difficult to "wet" due to dense waxy cuticle. The use of a surfactant is highly recommended.
Triclopyr *Renovate*	**Rate:** For emergent shoots: 2.67 to 5.33 pt product/acre (1 to 2 lb a.e./acre) with a non-ionic surfactant. Submersed shoot stage (in-water application): 1 to 2.5 ppm. **Timing:** Spring to early summer is optimal; however mid-summer applications can be effective in suppressing growth. **Remarks:** Emergent shoots of parrotfeather are difficult to "wet" due to dense waxy cuticle. The use of a surfactant is highly recommended.
BRANCHED-CHAIN AMINO ACID INHIBITORS	
Bispyribac-sodium *Tradewind*	**Rate:** Foliar treatment to emergent shoots: 1 to 2 oz product/acre (0.8 to 1.6 oz a.i./acre); no more than 4 treatments per year. **Timing:** Postemergence to foliage in early spring to early summer (during rapid growth). **Remarks:** Bispyribac-sodium can be tank-mixed with 2,4-D.
Imazamox *Clearcast*	**Rate:** Foliar treatment to emergent shoots: 64 oz product/acre (8 oz a.e./acre). Spot treatment (spray-to-wet): 0.25 to 5% v/v solution. **Timing:** Postemergence to foliage in early spring to early summer (rapid growth) **Remarks:** Use an approved surfactant. Aerial application is approved in some states.
Imazapyr *Habitat*	**Rate:** Foliar treatment to emergent shoots: 2 to 4 pt/acre (0.5 to 1 lb a.e./acre). Spot treatment: 1% v/v solution. **Timing:** Postemergence to foliage in early spring to early summer (when new growth is present). **Remarks:** Imazapyr is a slow-acting systemic herbicide.
Penoxsulam *Galleon*	**Rate:** Foliar treatment to emergent shoots: 2 to 5.6 oz/acre (0.5 to 1.4 oz a.i./acre) with approved surfactant. **Timing:** Early spring to early summer **Remarks:** Provides partial control and suppression. May be tank-mixed with endothall or other herbicides.
PIGMENT SYNTHESIS INHIBITORS	
Fluridone *Sonar*	**Rate:** For in-water treatment: 10 to 30 ppb **Timing:** Apply to water in early spring to early summer (when new growth is present). **Remarks:** Use various formulations (variable release-rates) or repeated applications to achieve desired concentration for 5 to 7 weeks.
CONTACT PHOTOSYNTHETIC INHIBITORS	
Diquat *Reward*	**Rate:** Spot treatment of emergent shoots: 0.5% v/v solution (2 qt/ 100 gal water) **Timing:** Spring to early summer is optimal. Repeat treatments may be needed in mid-summer. **Remarks:** Use only clean water to mix and spray as diquat is inactivated in turbid water. Since diquat is a contact herbicide, repeat treatment will be necessary at 3 to 5 week intervals.

Myriophyllum spicatum L.
Eurasian watermilfoil

Family: Haloragaceae
Range: Primarily temperate regions in the Northern Hemisphere, but also in some sub-tropical to tropical areas. It is present in nearly every state in the United States, with the possible exception of Wyoming.
Habitat: Ponds, lakes, rivers, streams, canals, ditches. Usually in still or slow-moving water, but occasionally found in faster moving water of streams and rivers.
Origin: Native to Eurasia and first documented near Maryland around 1942, possibly introduced through the aquarium trade.
Impacts: Eurasian watermilfoil can develop colonies that form large sub-surface or surface mats. Mats impede water flow, interfere with boat traffic and recreational activities, create mosquito habitat, and displace native aquatic vegetation.
Western states listed as Noxious Weed: Colorado, Idaho, Montana, New Mexico, Nevada, Oregon, South Dakota, Washington
California Invasive Plant Council (Cal-IPC) Inventory: High Invasiveness

Eurasian watermilfoil is a common submerged perennial with creeping rhizomes and finely dissected whorled submersed leaves. The stems can be 12 to 20 ft long, becoming emergent only while flowering or after stream or canal drawdown when moisture is still present. Submersed leaves are 3 to 6-whorled per node, pinnately dissected into linear lobes.

Inflorescences are terminal emerged spikes about 2 to 4 inches long. Flowers are pinkish, whorled and monoecious (male and female flowers develop separately on same plant). Fruits break apart into four one-seeded nutlets. At maturity, fruits detach and float for a period before sinking. In the field, reproduction by seed appears to be insignificant. Some populations produce many seeds, but seedlings are rarely observed. Seeds can survive at least 7 years under dry conditions. Seeds are consumed by waterfowl and may disperse to great distances with migrating birds. Reproduction is primarily vegetatively by rhizomes, stem fragments, and axillary buds. Stem fragments easily detach and root at lower nodes. They disperse with water, by clinging to the feet or feathers of water birds, and with human activities such as boating, mechanical harvesting, and the dumping of unwanted pond or aquarium contents.

NON-CHEMICAL CONTROL

Mechanical (pulling, cutting, rotovating, bottom barriers)	Repeated mechanical removal or harvesting can help reduce stem densities, but escaped stem fragments can drift elsewhere and develop into new plants. This is mainly used to control small infestations. Removing and destroying stem fragments from recreational equipment, such as boat propellers, docking lines, and fishing gear can help prevent the spread of non-native watermilfoils. Several types of "bottom barriers" are available and are used to cover and smother specific infested areas. Materials used include polyvinyl chloride (PVC) sheets, small-mesh screens, and natural fibers such as jute. Bottom barriers are best installed in spring before plants produce large biomass and exceed 10 inches tall. Unlike parrotfeather, Eurasian watermilfoil is more easily controlled by several aquatic herbicides.
Cultural	Dewatering infested areas during periods of high temperature in summer can suppress regrowth. Similarly, dewatering during winter during periods of hard freezes can suppress growth the following spring and summer. Dewatering can also be integrated with a subsequent herbicide application.
Biological	There are three herbivorous insects that have been used to control Eurasian watermilfoil with varying success: milfoil weevil (*Euhrychiopsis lecontei*), a midge (*Critotopus myriophylli*), and a moth (*Acentria ephemerella*). The milfoil weevil is commercially available. The (sterile) triploid grass carp (white amur, *Ctenopharyngodon idella*) is a relatively nonselective herbivorous fish that will consume Eurasian watermilfoil, but usually only after it first consumes its preferred submersed plants such as native pondweeds and hydrilla.

CHEMICAL CONTROL

The following specific use information is based on published papers and reports by researchers and land managers. Other trade names may be available, and other compounds also are labeled for this weed. Directions for use may vary between brands; see label before use. Herbicides are listed by mode of action and then alphabetically. The order of herbicide listing is not reflective of the order of efficacy or preference.

GROWTH REGULATORS	
2,4-D *Weedar 64*	**Rate:** For in-water applications: 0.5 to 1.5 ppm; exposures must be maintained for 5 to 10 days for optimal control **Timing:** Apply to water in early spring to early summer. **Remarks:** 2,4-D is a selective, systemic herbicide. It affects young, rapidly growing plants. Lower rates can be used if applied during early spring growth and when water movement is not likely to dilute or move the herbicide.
Triclopyr *Renovate*	**Rate:** For in-water applications: 1 to 2.5 ppm **Timing:** Apply to water in early spring to early summer. Applications in mid-summer can suppress biomass, but may result in rapid die-back that results in depressed dissolved oxygen. If mid-summer applications are made to infestations with high biomass, then only treat 1/3 to 1/2 of the area and wait 10 to 14 days before treating the remainder. **Remarks:** Triclopyr is a selective, systemic, fast-acting herbicide. Lower rates can be used if applied during early spring growth and when water movement is not likely to dilute or move the herbicide.
BRANCHED-CHAIN AMINO ACID INHIBITORS	
Bispyribac-sodium *Tradewind*	**Rate:** For in-water applications: 20 to 45 ppb; may be repeated to maintain concentration but not within 14 days of initial treatment; do not exceed 4 applications per year **Timing:** Apply to water in early spring to early summer (rapid growth). For drawdown treatment, apply during mid- to late-winter before refilling. **Remarks:** Bispyribac-sodium is a slow-acting herbicide and may take 4 to 6 weeks for control. For drawdown applications use 20 to 100 gal/acre of spray solution to wet the sediment. Bispyribac-sodium may be tanked mixed with other herbicides.
Penoxsulam *Galleon*	**Rate:** For in-water applications: 25 to 75 ppb; may be repeated but not to exceed 150 ppb in an annual season. For dewatered (drawdown) treatment: 5.6 to 11.2 oz product/acre (1.4 to 2.8 oz a.i./acre) **Timing:** Apply to water in early spring to early summer (rapid growth). For drawdown, apply during mid-late winter before refilling. **Remarks:** Penoxsulam is a slow-acting herbicide and may take 4 to 6 weeks for control. For drawdown applications use 20 to 100 gal/acre of spray solution to wet the sediment.
PIGMENT SYNTHESIS INHIBITORS	
Fluridone *Sonar*	**Rate:** For in-water applications: 5 to 20 ppb; exposures must be maintained for 5 to 7 weeks for optimal control. For dewatered (drawdown) treatment: 4 pt product (*Sonar*)/acre (2 lb a.i./acre) **Timing:** Apply to water in early spring to early summer. For dewatered treatment, apply to drained canal or exposed shorelines in fall or late winter before water is reintroduced. **Remarks:** Fluridone is a systemic herbicide that is slow-acting. It affects young, rapidly growing plants. Lower rates can be used if applied during early spring growth and when water movement is not likely to dilute or move the herbicide.
CONTACT PHOTOSYNTHETIC INHIBITORS	
Diquat *Reward*	**Rate:** For in-water applications: 0.1 to 0.25 ppm **Timing:** Apply to water in late spring to early summer. Diquat is a fast-acting contact herbicide that can be effective in mid- to late summer, but if biomass is large, only a portion of the infested sites should be treated to minimize effects of reduced dissolved oxygen. **Remarks:** Diquat is quickly bound to, and becomes inactivated on, suspended clay particles and it should not be used in moderately or highly turbid water.

Flumioxazin *Clipper*	**Rate:** For in-water applications: 100 to 400 ppb **Timing:** Apply to water in early spring to early summer (rapid growth). **Remarks:** Flumioxazin is rapidly degraded and is inactive if pH exceeds 8.5; use only if pH will not exceed 8.5; best to check water pH and apply in the early morning when pH is low.
GENERALL CELL TOXICANTS	
Endothall *Cascade, Teton* and *Aquathol K*	**Rate:** For in-water applications: 1 to 3 ppm; exposures must be maintained for 24 to 48 hours or more for optimal control (see label for specific rate and duration of exposure). **Timing:** Apply to water in early spring to early summer. Can be used in mid-summer, but if biomass is large, partial treatments are recommended to prevent large reduction in dissolved oxygen. **Remarks:** Endothall is a selective, contact herbicide. It affects young, rapidly growing plants and mature plants. Lower rates can be used if applied during early spring growth and when water movement is not likely to dilute or move the herbicide.
INORGANIC HERBICIDES	
Chelated copper *Komeen, Cutrine-Plus*	**Rate:** For in-water applications: 0.5 to 1 ppm elemental copper **Timing:** Apply to water in early summer (short plants and small biomass). **Remarks:** Chelated copper is a fast-acting contact herbicide. Retreatment may be required within 3 to 5 weeks. If biomass is large, treat only one-third of infested area to minimize decrease in dissolved oxygen (e.g. > 125 ppb calcium carbonate equivalent) and high pH (> 8). Chelated copper products are less affected by hard water and high pH compared to inorganic copper. Copper can accumulate in the environment.
Inorganic copper Various granular and liquid products	**Rate:** For in-water applications: 0.5 to 1 ppm elemental copper **Timing:** Apply to water in early summer (short plants and small biomass). **Remarks:** Copper is a fast-acting contact herbicide. Retreatment may be required within 3 to 5 weeks. If biomass is large, treat only one-third of infested area to minimize decrease in dissolved oxygen. Most inorganic copper formulations have poor efficacy in "hard water" (e.g. > 125 ppb calcium carbonate equivalent) and high pH (> 8). Copper can accumulate in the environment.
NON-HERBICIDAL CHEMICALS	
Dyes or colorants *Aquashade*	Although technically not herbicides, dyes and colorants control submerged aquatic plants by absorbing light in the water column and reducing photosynthesis. Applications should be made in early spring and repeated to maintain concentration recommended on the label. Colorants are not as effective on well-established plants in mid- to late summer.

Najas guadalupensis (Spreng.) Magnus; southern naiad
Najas marina L.; hollyleaf naiad

Naiads

Family: Hydrocharitaceae or Najadaceae
Range: Southern naiad is found throughout the United States. Hollyleaf naiad is found in California, Arizona, Nevada, Utah and New Mexico.
Habitat: Southern naiad inhabits still or slow-moving water in a broad range of substrates, including ponds, lakes, reservoirs, canals, rice fields, and irrigation ditches. Grows at water depths of 3 to 15 ft and tolerates polluted water or slightly brackish water. Hollyleaf naiad inhabits fresh to brackish water marshes, ponds, lakes, slow-moving streams, canals, and irrigation systems
Origin: Southern naiad is a common widespread native of North and South America. Hollyleaf naiad is native to the southwestern United States.
Impacts: Both naiads are usually not considered weedy in natural habitats. The foliage and seeds are an important food source for wildlife, especially shorebirds and waterfowl. However, they can become troublesome in ditches, human-made ponds, and disturbed or controlled aquatic systems where populations can become locally dominant, forming dense submersed mats of vegetation.

A number of naiad species can be problems, but the two most common are southern naiad (*N. guadalupensis*) and hollyleaf naiad (*N. marina*). Both are submersed annuals with fibrous root systems and stolons, but the stolons do not persist. Detached parts can root in substrate or can survive freely suspended in the water. Southern naiad can have stems to 2 ft long and opposite to sub-opposite leaves. The stems are slender, highly branched, and rooting at the lower nodes. Unlike other *Najas* species in North America, hollyleaf naiad has stiff, conspicuously prickle-toothed leaf blades 1 to 3 mm wide, and stem internodes covered with minute prickles.

While southern naiad is monoecious (male and female flowers separate, but on the same plant), hollyleaf naiad is dioecious (male and female flowers on separate plants). Flowers are submerged, inconspicuous and typically solitary in the leaf axils. Flowers are water-pollinated. Plants produce abundant hard-seeded achenes that are dispersed by water and persist for a long time in the seedbank, though the actual length of viability in the seedbank is unknown.

NON-CHEMICAL CONTROL

Mechanical (pulling, cutting, dredging)	Repeated mechanical harvesting can help reduce stem densities, but escaped stem fragments can drift elsewhere and develop into new plants. Removing and destroying stem fragments from recreational equipment, such as boat propellers, docking lines, and fishing gear can help prevent the spread of naiads. However, *N. marina* has robust spines that can irritate and scratch the skin of workers who may be physically removing and handling this plant. Several types of "bottom barriers" are available to cover and smother specific infested areas. Materials used include polyvinyl chloride (pvc) sheets, small-mesh screens and natural fibers such as jute and burlap. Bottom barriers are best installed in spring before plants produce large biomass and exceed 20 inches tall.
Cultural	Draining or dewatering during summer can suppress regrowth for a while, but seed bank and rhizomes may reestablish populations within a few weeks to months. Mid-winter dewatering may reduce subsequent spring growth, but only if a hard freeze occurs, and even then some propagules can survive.
Biological	Triploid (sterile) grass carp is the only effective biological control agent available for naiads, but it is relatively nonselective and state or local permits are usually required. If other native plants are desired, careful monitoring of feeding impacts should be part of the management program so that grass carp can be removed (or added) as needed.

CHEMICAL CONTROL

The following specific use information is based on published papers and reports by researchers and land managers. Other trade names may be available, and other compounds also are labeled for this weed. Directions for use may vary between brands; see label before use. Herbicides are listed by mode of action and then alphabetically. The order of herbicide listing is not reflective of the order of efficacy or preference.

BRANCHED-CHAIN AMINO ACID INHIBITORS	
Penoxsulam *Galleon*	**Rate:** For in-water application: 100 to 200 ppb for 4 to 6 weeks. For dewatered (drawdown) applications: 5.6 to 11.2 oz/acre (1.4 to 2.8 oz a.i./acre)
	Timing: Apply to water in spring to early summer. Fall applications may be effective when temperatures remain high. For drawdown treatment, apply in mid- to late winter.
	Remarks: Penoxsulam provides partial control and suppression. It is a slow-acting systemic herbicide.
PIGMENT SYNTHESIS INHIBITORS	
Fluridone *Sonar* (various formulations)	**Rate:** For in-water application: 45 to 90 ppb; exposures must be maintained for 5 to 7 weeks for optimal control. For dewatered sediments (drawdown) application: 4 pt product (*Sonar*)/acre (2 lb a.e./acre); use 30 to 100 gal/ acre spray volume
	Timing: Apply to water in early spring to early summer. For dewatered sites, apply in late fall or late spring before water is introduced.
	Remarks: Systemic, slow-acting herbicide. It affects young, rapidly growing plants. Lower rates can be used during early spring growth and when water movement is not likely to dilute or move the herbicide.
CONTACT PHOTOSYNTHETIC INHIBITORS	
Diquat *Reward*	**Rate:** For in-water application: 0.1 to 0.25 ppm
	Timing: Apply to water in late spring to early summer. Diquat is a fast-acting contact herbicide that can be effective in mid to late summer, but if biomass is large, only a portion of the infested sites should be treated to minimize effects of reduced dissolved oxygen.
	Remarks: Diquat is quickly bound to, and becomes inactivated on suspended clay particles and it should not be used in moderately or highly turbid water. It is a contact herbicide that acts rapidly.
Flumioxazin *Clipper*	**Rate:** For in-water application: 100 to 400 ppb
	Timing: Apply to water in spring to early summer. Fall applications may be effective when temperatures remain high.
	Remarks: Do not use if pH is > 8.0; or use a buffer to reduce the pH to below 8.0. To minimize effects of high pH, apply from dawn to mid-morning. Due to photosynthesis of aquatic plants and algae, pH in the water column rises from mid-day to dusk under most circumstances.
GENERAL CELL TOXICANT	
Endothall *Cascade; Aquathol K*	**Rate:** For in-water application: 1 to 3 ppm; exposures must be maintained for 24 to 48 hours or more for optimal control (see label for specific rates and duration of exposures needed).
	Timing: Apply to water in early spring to early summer. Can be used in mid-summer but partial treatments are recommended if biomass is large in order to prevent large reduction in dissolved oxygen.
	Remarks: Endothall is a selective, contact herbicide. It affects young, rapidly growing plants and mature plants. Lower rates can be used if applied during early spring growth and when water movement is not likely to dilute or move the herbicide.
INORGANIC HERBICIDES	
Chelated copper *Komeen, Cutrine-Plus, Nautique*	**Rate:** For in-water application: 0.5 to 1 ppm elemental copper.
	Timing: Apply to water in early summer (short plants and small biomass).
	Remarks: Chelated copper is a fast-acting contact herbicide. Retreatment may be required within 3 to 5 weeks. If biomass is large, treat only one-third of infested area to minimize decrease in dissolved oxygen. Chelated copper products are less affected by hard water and high pH than inorganic copper. Copper can accumulate in the environment.
Inorganic copper Various granular and liquid products	**Rate:** For in-water application: 0.1 to 0.25 ppm
	Timing: Apply to water in late spring to early summer (short plants and small biomass).
	Remarks: Copper is a rapid acting contact herbicide. Retreatment may be required within 3 to 5 weeks. If

	biomass is large, treat only one-third of infested area to minimize decrease in dissolved oxygen. Most inorganic copper formulations have poor efficacy in "hard water" (e.g. > 125 ppb calcium carbonate equivalent) and high pH (> 8). Copper can accumulate in the environment.
NON-HERBICIDAL CHEMICALS	
Dyes or colorants *Aquashade*	Although technically not herbicides, dyes and colorants control submerged aquatic plants by absorbing light in the water column and reducing photosynthesis. Applications should be made in early spring and repeated to maintain concentration recommended on the label. Colorants are not as effective on well-established plants in mid- to late summer.

Nicotiana glauca Graham
Tree tobacco

Family: Solanaceae
Range: Much of the southwestern United States, including California, Nevada, Arizona, and New Mexico.
Habitat: Disturbed places, roadsides, urban waste areas, gravel quarries, landscaped sites, and many natural communities, including riparian areas, grassland, and woodland. Drought-tolerant and can survive under a wide range of growing conditions, but is most commonly found in sandy or gravelly soils along riparian areas, near cultivated areas, around old dwellings and ditch banks.
Origin: Native to South America and introduced to the United States in the early 1800s as a landscape ornamental.
Impacts: Tree tobacco grows rapidly and forms dense stands. It displaces native vegetation used by wildlife and contributes to bank erosion and flooding. Large infestations can decrease water flow and reduce recreational uses. Tree tobacco is toxic to humans and animals. Unlike other members of the *Nicotiana* genus, tree tobacco does not contain the alkaloid nicotine. However, it produces a similar compound called anabasine, which is highly toxic to humans and animals. Anabasine is more toxic than nicotine and can cause fetal deformities in livestock when the mother ingests small amounts of plant material during early pregnancy.
California Invasive Plant Council (Cal-IPC) Inventory: Moderate Invasiveness

Tree tobacco is a slender, erect, straggling shrub or small tree growing 6 to 20 ft tall. It has ovate, bluish-grey leaves, 2 to 8 inches long, with entire margins. The plants lack hairs on the foliage.

Tree tobacco produces sprays of nodding, tubular bright yellow flowers and is a prolific seed producer. Plants reproduce only by seed. An individual tree can produce 10,000 to 1,000,000 seeds per year with viability approaching 100%. The fruit are capsules about 7 to 15 mm long. The minute seeds (0.6 mm long) are chiefly spread by water movement; however, animals also serve as dispersal agents. The reproductive biology of this species is poorly documented, but the small size of the seed suggests that they do not survive more than a year or two in the soil.

NON-CHEMICAL CONTROL

Mechanical (pulling, cutting, disking)	Hand pulling can remove seedlings and small saplings. For larger established shrubs, a weed wrench or other woody weed extractor can be used. Care must be taken to extract the entire root or stump sprouting will occur. Best results are achieved when soil is moist.
	Cutting tree tobacco off before it flowers will reduce seed production and deplete the plant's energy reserves. Resprouts are common after treatment. Cutting at the end of the dry season can help reduce resprouting from the root crown. Cutting should be combined with an herbicide treatment or with multiple cuttings over a period of years. Cut trees at ground level with power or manual saws.
Cultural	Grazing is not considered an effective control option. Foliage contains anabasine and can be toxic to livestock when ingested.
	Burning is also not considered to be an effective control method, as plants will resprout from the base.
Biological	No biological control agents have been released for the control of tree tobacco. Tree tobacco is susceptible to tobacco mosaic virus, but this only has a minor effect on its population.

CHEMICAL CONTROL

The following specific use information is based on published papers and reports by researchers and land managers. Other trade names may be available, and other compounds also are labeled for this weed. Directions for use may vary between brands; see label before use. Herbicides are listed by mode of action and then alphabetically. The order of herbicide listing is not reflective of the order of efficacy or preference.

GROWTH REGULATORS

Triclopyr	**Rate:** Spot treatment: 0.5 to 2% v/v solution of *Garlon 4 Ultra*, or 1 to 1.5% *Garlon 3A* and water plus 0.25 to

Garlon 3A, Garlon 4 Ultra, Pathfinder II	0.5% surfactant to thoroughly wet all leaves. Low volume/thinline treatment: 20% v/v solution of *Garlon 4 Ultra* plus a 20% v/v basal oil in water. Cut stump treatment: 50% v/v *Garlon 4 Ultra* in 20% v/v ethylated crop oil and water, undiluted *Garlon 3A* or 50% *Garlon 3A* in water. Basal bark treatment: 20% v/v *Garlon 4 Ultra* in 20% v/v basal oil and water, or *Pathfinder II* as a ready-to-use formulation. **Timing:** Postemergence when plants are growing rapidly. Cut stump and basal bark treatments can be applied anytime, except when the soil is saturated or frozen. **Remarks:** Triclopyr is a selective herbicide for broadleaf species. In areas where desirable grasses are growing under or around tree tobacco, triclopyr can be used without non-target damage. For cut stump treatments, cut stems horizontally at or near ground level. Apply herbicide solution immediately after the stump is cut. For basal bark treatment, spray the lower trunk, including the root collar, to a height of 12 to 15 inches from the ground; the spray should thoroughly wet the lower stem but not to the point of runoff. Plants should not be cut for at least 1 month after basal bark treatments.
AROMATIC AMINO ACID INHIBITORS	
Glyphosate *Roundup, Accord XRT II*, and others	**Rate:** Spot treatment: 1 to 2% v/v solution of *Roundup ProMax* (or other trade name with a similar concentration of glyphosate) in water to thoroughly wet all leaves. Low volume/thinline treatment: 10% v/v solution of *Roundup* (or other trade name) in water. Cut stump treatment: 50% v/v *Roundup* (or other trade name) in water. **Timing:** Postemergence when plants are growing rapidly. **Remarks:** Glyphosate is a nonselective systemic herbicide with no soil activity. Plants should not be cut for at least 4 months after foliar treatments.
BRANCHED-CHAIN AMINO ACID INHIBITORS	
Imazapyr *Arsenal, Habitat, Stalker, Chopper, Polaris*	**Rate:** Spot treatment: 0.5 to 1.5% v/v solution of *Stalker* plus 0.25 to 0.5% surfactant v/v to thoroughly wet all leaves. Low volume/thinline treatment: 10% v/v solution of *Stalker* plus a 20% v/v ethylated crop oil in water. Cut stump treatment: 50% v/v solution of *Stalker* in 20% v/v ethylated crop oil and water. Basal bark treatment: 20% v/v *Stalker* in 20% v/v ethylated crop oil and water. **Timing:** Postemergence when plants are growing rapidly. For both foliar and stem treatments, best results when used in late summer to early fall. **Remarks:** Imazapyr is a soil residual herbicide and may result in bare ground around trees for some time after treatment. Cut stump and basal bark applications are made as described for triclopyr. Trees should not be cut for at least 4 months after basal bark treatments.

Nymphoides peltata (J.G. Gmel.) Kuntze
Yellow floatingheart

Family: Nymphaeaceae
Range: Washington, California and Arizona.
Habitat: Lakes, reservoirs and ponds, and slow moving rivers.
Origin: Introduced from Eurasia and the Mediterranean region as well as Japan, China, and India. Yellow floatingheart is cultivated as a pond ornamental, but has been released into certain natural lakes where it has become a nuisance weed.
Impacts: Yellow floatingheart often develops dense mat-like patches that displace desirable vegetation. Dense mats can also reduce recreational activities and create stagnant low-oxygen conditions in the water below.
Western states listed as Noxious Weed: Oregon, Washington

Photo courtesy of Robin Breckenridge

Yellow floatingheart is a submersed perennial water lily-like plant with creeping rhizomes and stolons and floating rounded heart-shaped leaves 2 to 5 inches in diameter that may be confused with those of the water lilies. The flowering stems have opposite leaves.

The inflorescence is a simple umbel of showy yellow flowers with five ciliate-margined petals. The flowers are on long stalks that rise a few inches above the water. Yellow floatingheart reproduces by seed and vegetatively from rhizomes, stolons, rhizome and stolon fragments, and separated leaves. The seeds are water-dispersed individually or in chain-like floating rafts. Seeds can also be dispersed by waterfowl. Seeds readily germinate, but there is no information on seed longevity in the soil. Fragmented leaves with part of a stem still attached will also form new plants, and vegetative fragments can also be dispersed by water. Plants can survive exposure on wet mud.

NON-CHEMICAL CONTROL

Mechanical (pulling, cutting, dredging)	Mechanical control of *Nymphoides peltata* is very difficult due to its ability to propagate vegetatively through fragments, and through underwater roots and rhizomes. Mechanical harvesting may create abundant plant fragments, potentially aiding in dispersal to new locations. Leaf petioles cut by mechanical harvesting will eventually form new leaves, requiring one or two cuts each spring and summer to maintain controlled areas. Nevertheless, these plants are sometimes controlled by cutting, harvesting, and covering with bottom barrier materials (synthetic and natural fibers). In severe infestations, excavation may be necessary to remove plants, rhizomes and seed in the sediment. However, both roots and rhizomes are also able to withstand mechanical removal by dredging. Hand raking can be effective in very small, localized areas where fishing or navigation lanes need to be created.
Cultural	Use alternative native floating plants and keep contained within pots.
	Dewatering is usually not sufficient to control this plant because the below-ground propagules (rhizomes, stolons) often survive.
Biological	The (sterile) triploid grass carp (white amur) is a relatively nonselective herbivorous fish that may partially consume the seedlings and young, tender parts of floatingheart, but usually only after it first consumes its preferred submersed plants such as native pondweeds. Grass carp do not eat water lilies in Washington and it is not known if they would readily eat yellow floating heart.

CHEMICAL CONTROL
The following specific use information is based on reports by researchers and land managers. Other trade names may be available, and other compounds also are labeled for this weed. Directions for use may vary between brands; see label before use. Other herbicides may be effective, but few tests have been conducted to demonstrate which products control yellow floatingheart.

AROMATIC AMINO ACID INHIBITORS	
Glyphosate *Rodeo*, *Aquamaster*	**Rate:** Use a 2% v/v *Rodeo* or *Aquamaster* solution (1% a.e.) with an approved surfactant and spray to thoroughly wet the floating leaf surface.
	Timing: Postemergence in late spring to mid-late summer.
	Remarks: Repeated applications are generally necessary.

Onopordum acanthium L.
Scotch thistle

Family: Asteraceae
Range: All western states.
Habitat: Disturbed areas such as river and stream corridors, roadsides, right-of-ways, trails, rangeland, pasture, forest clearings, and abandoned cropland. Best suited to areas with high soil moisture during germination periods. Often associated with degraded annual plant communities and areas with high rodent activity. Not common on annually cultivated lands.
Origin: Native to Eurasia.
Impacts: Scotch thistle populations spread rapidly and can form dense stands over large acreage. Seeds contain a water-soluble germination inhibitor causing a high percentage of seeds to remain dormant in the soil. This makes control difficult and a long-term commitment. Populations often expand rapidly on infested land during wet years due to the large amount of seed that break dormancy. Sharp spines on leaves, stems, and seedheads deter livestock and wildlife from grazing. Dense stands create a natural barrier that prevents movement by livestock, wildlife, and humans.
Western states listed as Noxious Weed: Arizona, California, Colorado, Idaho, Nevada, New Mexico, Oregon, Utah, Washington, Wyoming
California Invasive Plant Council (Cal-IPC) Inventory: High Invasiveness

Scotch thistle is a biennial with large, green, spiny leaves that are covered with fine, cottony hairs. Rosettes can often grow up to 1 to 2 ft in diameter and mature plants are generally 4 to 6 ft tall. Scotch thistle's tall, spiny stature made it useful as a natural fence in Europe centuries ago.

Flower heads are large (1 to 3 inches diameter), spherical to hemispherical, and either solitary of in clusters of 2 to 7. The heads consist of numerous spine-tipped phyllaries in many overlapping rows. The numerous disk flowers are purple (sometimes white) and showy. The pappus is composed of bristles (7 to 9 mm long) that form a detachable ring at the base. Scotch thistle reproduces only by seeds. It produces large numbers of achenes that remain viable in the soil for 7 to 39 years. Seeds contain a water-soluble germination inhibitor, and greater than 80% of the seeds display innate dormancy at maturity. Germination often occurs in spring and fall during saturated soil conditions. Seed dispersal is primarily by wind, but can also occur with water, rodents, livestock, or vehicles.

NON-CHEMICAL CONTROL

Mechanical (pulling, cutting, disking)	Small infestations can be removed by manual methods. Digging is effective and the preferred manual removal method. When digging, sever the root below the soil surface.
	Mowing in the late bolting or bud stage can reduce seed production. Mow before flowering to prevent seeds from developing in severed flowerheads. On sites with high soil moisture, plants may need to be remowed to prevent flowering on regrowth. When mowing is conducted too early, it only delays flowering. Similarly, when plants are cut too late in the flowering process, viable seed may still develop in the seedhead. Because there can be wide variation in plant maturity, a single mowing is unlikely to provide satisfactory control, but repeated mowing throughout the entire growing season can be successful. For total kill, plants must be cut off below the soil surface with no leaves remaining.
	Tillage will control emerged plants but often stimulates germination. Land managers using tillage for seedbed preparation for reseeding efforts should prepare for a flush of seedlings when soils become saturated.
Cultural	Sheep, goats and horses, but not cattle, have a significant effect on thistles in the early stages of infestation when they eat young thistle plants. Sheep may graze small rosettes. Goats ignored the leaves of Scotch thistle, but they ate all the flower heads and prevented seed production and dispersal. Cattle avoid grazing Scotch thistle. Overgrazing promotes Scotch thistle.

	Fire is not an effective control.
	Promoting competitive vegetation can slow spread and help prevent establishment. Perennial grass plantings have been shown to inhibit Scotch thistle seedling establishment and can reduce Scotch thistle populations. Perennial grass stand density and vigor should be managed to minimized bare ground exposure.
Biological	No biological controls are currently available in the United States. Australia has released biocontrol insects, but some of them have failed host specificity tests in the U.S. Insects are being evaluated by USDA for release in the U.S.

CHEMICAL CONTROL

The following specific use information is based on published papers and reports by researchers and land managers. Other trade names may be available, and other compounds also are labeled for this weed. Directions for use may vary between brands; see label before use. Herbicides are listed by mode of action and then alphabetically. The order of herbicide listing is not reflective of the order of efficacy or preference.

GROWTH REGULATORS	
2,4-D Several names	**Rate:** 1 to 2 qt product/acre (0.95 to 1.9 lb a.e./acre) **Timing:** Postemergence from rosette to beginning of bolting, or fall rosette. Most effective on small rosettes. **Remarks:** Often tank-mixed with chlorsulfuron or dicamba for quicker burndown. Does not control large bolting plants. Broadleaf-selective and safe on most grasses. 2,4-D has minimal soil activity. Do not apply ester formulation when outside temperatures exceed 80°F. Amine forms are as effective as ester forms for small rosettes, and amine forms reduce the chance of off-target movement.
Aminocyclopyrachlor + chlorsulfuron *Perspective*	**Rate:** 4.75 to 8 oz product (*Perspective*)/acre **Timing:** Postemergence and preemergence. Postemergence applications are most effective when applied to plants from the seedling to the bolting stage. **Remarks:** Aminocyclopyrachlor provides excellent control of Scotch thistle at most growth stages. One of the few herbicides that provides soil residual control 1 year after application. *Perspective* provides broad-spectrum control of many broadleaf species. Although generally safe to grasses, it may suppress or injure certain annual and perennial grass species. Do not treat in the root zone of desirable trees and shrubs. Do not apply more than 11 oz product/acre per year. At this high rate, cool-season grasses will be damaged, including bluebunch wheatgrass. Not yet labeled for grazing lands. Add an adjuvant to the spray solution. This product is not approved for use in California and some counties of Colorado (San Luis Valley).
Aminopyralid *Milestone*	**Rate:** 5 to 7 ounces product/acre (1.25 to 1.75 oz a.e./acre) **Timing:** Postemergence from the rosette to young bolting stage. **Remarks:** Longer soil residual than clopyralid. Safe on most grasses, although preemergence application at high rates can greatly suppress some annual grasses, such as medusahead. Applications can decrease seed production in some annual and perennial grass species. Provides over 90% control when applied to rosettes. For postemergence applications, adding a non-ionic surfactant (0.25 to 0.5% v/v spray solution) enhances control under adverse environmental conditions; however, this is not normally necessary. Other premix formulations of aminopyralid can also be used for Scotch thistle control. These include *Opensight* (aminopyralid + metsulfuron; 1.5 to 2.5 oz product/acre) and *Forefront HL* (aminopyralid + 2,4-D; 1.5 to 2.1 pt product/acre); apply at the rosette to bolting stage.
Clopyralid *Transline*	**Rate:** 0.67 to 1.33 pt product/acre (4 to 8 oz a.e./acre). Use higher rate for older plants or dense stands. **Timing:** Postemergence from the rosette to young bolting stage. Results are best if applied to rapidly growing weeds. **Remarks:** Most effective for young plants. Shorter soil residual than aminopyralid or aminocyclopyrachlor. Controls or injures plants in the Asteraceae and Fabaceae but safe on most other broadleaf species and all grasses. For postemergence applications, a non-ionic surfactant (0.25 to 0.5% v/v spray solution) enhances control under adverse environmental conditions; however, this is not normally necessary.
Clopyralid + 2,4-D	**Rate:** 2 to 4 qt *Curtail*/acre

Curtail	**Timing:** Same as for clopyralid. **Remarks:** Add a non-ionic surfactant.
Dicamba *Banvel, Clarity*	**Rate:** 0.5 to 2 pt product/acre (0.25 to 1 lb a.e./acre). Use higher rates for large rosettes and bolting plants. **Timing:** Postemergence from rosette to beginning of bolting, or fall rosette. **Remarks:** A broadleaf-selective herbicide often combined with other active ingredients. Not typically used alone to control Scotch thistle. Dicamba can also be mixed with 2,4-D (0.5 to 1 pt dicamba + 2 pt 2,4-D/acre) from rosette to bolting stage.
Picloram *Tordon 22K*	**Rate:** 0.5 to 0.75 pt product/acre (2 to 3 oz a.e./acre). **Timing:** Preemergence and postemergence. With postemergence application, best time is rosette to early bolting stage, when plants are growing rapidly. **Remarks:** Picloram controls a wide range of broadleaf species and has relatively long soil residual activity. Lower rates in rate range may require annual spot treatments. Although well-developed grasses are not usually injured by labeled use rates, some applicators have noted that young grass seedlings with fewer than four leaves may be killed. Picloram is a restricted use herbicide. Picloram is not registered for use in California. Do not apply near trees, or where soil is highly permeable and where water table is high. Control with lower rates may be improved by tank mixing with dicamba or 2,4-D: picloram + dicamba (0.25 to 0.5 pt picloram + 0.125 to 0.25 pt dicamba/acre), or picloram + 2,4-D (0.5 to 1 pt picloram + 1 to 2 pt 2,4-D/acre).
AROMATIC AMINO ACID INHIBITORS	
Glyphosate *Roundup, Accord XRT II,* and others	**Rate:** 1 to 2 qt product (*Roundup ProMax*)/acre (1.1 to 2.25 lb a.e./acre). Spot treatment: 1.5% v/v solution. **Timing:** Postemergence to rapidly growing plants from the rosette to early bolting stage. **Remarks:** Glyphosate will only provide control in the year of application, and will not kill seeds or inhibit germination the following season. Glyphosate has no soil activity and is nonselective. It can create bare ground conditions that can make an area susceptible to weed recruitment. In areas with desirable vegetation, use spot treatment. Glyphosate is a good control option if reseeding is planned shortly after application, as it will not injure seedlings emerging after application. Add a surfactant when using a formulation where it is not already included (e.g., *Rodeo, Aquamaster*).
BRANCHED-CHAIN AMINO ACID INHIBITORS	
Chlorsulfuron *Telar*	**Rate:** 1 to 2.6 oz product/acre (0.75 to 1.95 oz a.i./acre) **Timing:** Postemergence from the rosette to flower bud stage. **Remarks:** One of the more effective treatments for large bolting plants. Always use a surfactant. Can be tank-mixed with 2,4-D for quicker burndown.
Metsulfuron *Escort*	**Rate:** 1 to 2 oz product/acre (0.6 to 1.2 oz a.i./acre) **Timing:** Postemergence from the rosette up until flower-bud stage. **Remarks:** Similar activity compared to chlorsulfuron. Always use a surfactant. Metsulfuron can be tank-mixed with 2,4-D for quicker burndown. Other premix formulations of metsulfuron can be used at similar application timing. These include *Cimarron Max* (metsulfuron + dicamba + 2,4-D), *Opensight* (metsulfuron + aminopyralid), and *Cimarron X-tra* (metsulfuron + chlorsulfuron). Metsulfuron and premixes containing metsulfuron are not labeled or use in California.

Oxalis pes-caprae L.
Buttercup oxalis (Bermuda buttercup)

Family: Oxalidaceae
Range: In the western U.S., it occurs in California and Arizona.
Habitat: Coastal dunes, scrub, grasslands, oak woodlands, gardens, turf, urban areas, orchards, vineyards and agricultural fields. Grows in most environments and can tolerate many soil types. Grows in full sun in cool coastal areas, but inland it grows primarily in semi-shaded sites.
Origin: Native to South Africa and brought to North America as a garden ornamental.
Impacts: Buttercup oxalis is a major problem in field-grown flowers and in the home landscape, especially in groundcovers. In the last 10 years, this plant has spread extensively throughout California invading native coastal dunes and natural areas along the coast. Due to its extensive occurrence in yards and gardens, buttercup oxalis has the potential to rapidly spread via the production of bulbs and the movement of contaminated soils into adjacent natural areas. Plants contain variable quantities of soluble oxalates and can be lethally toxic to livestock when ingested in quantity.
California Invasive Plant Council (Cal-IPC) Inventory: Moderate Invasiveness

Buttercup oxalis is a low-growing perennial with clover-like leaves and yellow flowers. Plants grow from bulbs and produce a loose basal rosette of leaves to 14 inches tall. Stems are mostly below ground. Leaf stalks are 5 inches long with a trifoliate leaf, green to dark purple-tinged, 6 to 10 mm long, and 5 to 24 mm wide. Leaflets are glabrous to sparsely pubescent, broadly heart-shaped, often pubescent below, and typically folding downward at midday and at night. Small, whitish-brown bulblets develop on the stem at the base of the rosette of leaves, and new bulbs form underground along the rhizome. A plant forms about a dozen small bulbs per year, each less than 1 inch long. Slender white rhizomes are about 4 inches long with true roots growing upward from the mature bulb apex. The leaves and flowers develop from the top of the rhizome. A threadlike rhizome grows downward from the mature bulb base and produces a tuberous root with many fibrous roots below. Small bulbils develop in the leaf scale axils along the length of the threadlike rhizome. Bulbs and bulblets readily detach from rhizomes.

The flowers are bright yellow, 0.75 to 1.5 inches wide, and are borne on top of a leafless stalk rising 6 to 12 inches tall. Viable seed never has been documented in the United States, and rarely has it been seen anywhere else in the world. The foliage dies and the bulbs become dormant when temperatures rise in late summer. Plants reproduce vegetatively by bulbs and spread with cultivation, soil movement, intentional planting, and disposal of garden refuse and nursery soil.

NON-CHEMICAL CONTROL

Mechanical (pulling, cutting, disking)	Hand pulling can provide control but care must be taken to remove the entire plant, including underground rhizome and bulbs. Repeated pulling of the tops will deplete the bulb's carbohydrate reserves, but these efforts take years to be successful.
	Cultivation can provide control on new infestations. Repeated tillage is required to effectively control the bulbs.
Cultural	Grazing is not considered an effective control option. Plants contain variable quantities of soluble oxalates and can be lethally toxic to livestock when ingested in quantity.
	Burning is also not considered to be an effective control option.
Biological	There is currently no biocontrol agent available for the control of *O. pes-caprae* in North America. A potential biocontrol agent is *Klugeana philoxalis*, a larval feeder on shoots of *O. pes-caprae*, but no other information on this species is available.

CHEMICAL CONTROL

The following specific use information is based on published papers and reports by researchers and land managers. Other trade names may be available, and other compounds also are labeled for this weed. Directions for use may vary between brands; see label before use. Herbicides are listed by mode of action and then alphabetically. The order of herbicide listing is not reflective of the order of efficacy or preference.

GROWTH REGULATORS	
Fluroxypyr *Vista XRT*	**Rate:** 15 to 22 oz product/acre (5.3 to 7.7 oz a.e./acre) to thoroughly wet all leaves **Timing:** Early postemergence when plants are growing rapidly. **Remarks:** Fluroxypyr provides selective postemergent control of many annual and perennial broadleaf weeds. It has no soil activity.
Triclopyr *Garlon 4 Ultra*	**Rate:** 2 to 4 qt product/acre (2 to 4 lb a.e./acre) plus 0.25 to 0.5% v/v surfactant to thoroughly wet all leaves **Timing:** Postemergence when plants are growing rapidly. **Remarks:** Triclopyr is a selective herbicide for broadleaf species. It has no soil activity.
AROMATIC AMINO ACID INHIBITORS	
Glyphosate *Roundup, Accord XRT II*, and others	**Rate:** Spot treatment: 2% v/v solution *Roundup ProMax* (or other trade name with similar concentration of glyphosate) and water to thoroughly wet all leaves **Timing:** Postemergence when plants are growing rapidly. Applications in early spring provide best control. **Remarks:** Glyphosate is a nonselective systemic herbicide with no soil activity. Repeated applications may be necessary for complete control.
BRANCHED-CHAIN AMINO ACID INHIBITORS	
Imazapyr *Arsenal, Habitat, Stalker, Chopper, Polaris*	**Rate:** 3 pt product/acre (12 oz a.e./acre) plus 0.25 to 0.5% v/v surfactant to thoroughly wet all leaves **Timing:** Postemergence when plants are growing rapidly. **Remarks:** Imazapyr is a preemergent and postemergence herbicide effective for controlling broadleaf weeds and grasses. It has fairly long soil residual activity.
PHOTOSYNTHETIC INHIBITORS	
Hexazinone *Velpar L*	**Rate:** 2.75 to 4.5 pt product/acre (0.7 to 1.1 lb a.i./acre) to thoroughly wet all leaves **Timing:** Early postemergence when plants are growing rapidly. **Remarks:** Hexazinone provides both contact and residual control of many broadleaf and grasses. It has a long soil activity, is considered mobile, and should not be used in areas where the water table is shallow. High rates of hexazinone can create bare ground, so only use high rates in spot treatments.

Pennisetum ciliare (L.) Link
 (= *Cenchrus ciliaris* L.)

Buffelgrass

Family: Poaceae
Range: While it is found in California, the most significant infestations are in Arizona and New Mexico.
Habitat: Mostly disturbed sites and fields; has greatly expanded its range into the Sonoran Desert.
Origin: Native to Africa, India and western Asia. It was introduced into Texas in the 1940s to stabilize overgrazed rangelands and provide livestock forage. It was introduced into Arizona in the 1930s and 1940s to control erosion.
Impacts: Buffelgrass is becoming especially problematic in southwestern desert regions that receive summer rain, where it and other exotic grasses are contributing to the type conversion of desert shrubland to grassland by making the fire-intolerant shrubland more susceptible to periodic burning. It is also problematic in Hawaii and Australia. Buffelgrass is highly palatable and nutritious to livestock.
Western states listed as Noxious Weed: Arizona

Buffelgrass is a tufted, or sometimes rhizomatous or stoloniferous, perennial to 4 ft tall. The ligules are hairy and the leaf blades contain small stiff hairs along the blade from the stem to the tip of the leaf. Plants can survive with only a one- or two-month rainy season.

The inflorescence of buffelgrass is a bristly spike 1 to 5 inches long, with each spikelet consisting of about 30 to 50 plumose bristles that are typically purplish. The spikes somewhat resemble those of crimson fountaingrass (*Pennisetum setaceum*). Unlike crimson fountaingrass, buffelgrass spikes are erect, dense, and mostly 0.5 to 1 inch wide, and the bristles are stiff, wavy, and mostly less than 0.5 inch long. Plants reproduce primarily by seed, although rhizome fragments can also spread the plant. Plants are apomictic. Seeds are dispersed long distances on animal skin and fur, but wind can also blow seed moderate distances away from parent plants. There is no information on the longevity of the seed in the soil seedbank. Germination of new seedlings is episodic and occurs only in high rainfall years.

NON-CHEMICAL CONTROL

Mechanical (pulling, cutting, disking)	Hand tools such as shovels, digging bars, rock picks, Caliche bars, and Pulaskis are commonly used to uproot plants. However, all roots must be removed to prevent resprouting. While the technique can be used year-round on most sites, the optimal time for manual removal is at least 4 days after 0.5 to 1 inch of precipitation. At this time, the soil is moist and removal is much easier. Two years are required to successfully remove mature buffelgrass from a site. Seedlings will need to be monitored and removed thereafter. Hand removal can be time consuming and labor-intensive and is best used for small infestations.
	Mowing or cutting are not effective methods for buffelgrass control and can even increase growth rates. However, cutting or mowing can be used to decrease standing biomass before herbicide applications. In this way, less herbicide can be used and the herbicide may be more effective. Mowing or cutting work best in large, relatively flat, dry areas.
	Repeated cultivation (tilling) can also eventually eliminate a buffelgrass infestation, but the applicability of such measures is limited in natural areas. In addition, if all vegetative reproductive parts are not removed, it can spread the infestation. Disking can also stimulate germination of the seedbank.
Cultural	Grazing is not a practical option for the management of buffelgrass, as it has been shown, on some occasions, to withstand continuous grazing. In other situations, continual heavy grazing was shown to reduce deeper root development.
	Buffelgrass can tolerate burning better than most long-lived perennials in the Sonoran Desert. Frequent fires allow buffelgrass to persist and spread to the detriment of native species, and thus, prescribed burning is not a good control option. Burning is particularly ineffective when soil water availability is high, as is often the case during winter burns. However, research indicates that when burned at the peak of live biomass production, buffelgrass production was reduced by almost 50%, even 4 years post-treatment.

Biological	There are no biological control agents available for buffelgrass. In Mexico, spittlebugs (*Aeneolamia albofasciata*) killed more than 50% of buffelgrass aboveground biomass. Leaf blight (also called rice blast or buffelgrass blight), caused by the fungus *Pyricularia grisea*, causes lesions in leaves and can damage up to 90% of individual plants, but some buffelgrass cultivars are resistant to this fungus. Since there is concern that this fungus may affect agricultural crops and because buffelgrass is considered a valuable pasture grass in Texas, the fungus is not being developed as a potential biocontrol agent. In the United States, buffelgrass is largely free of insect pests.

CHEMICAL CONTROL

The following specific use information is based on published papers and reports by researchers and land managers. Other trade names may be available, and other compounds also are labeled for this weed. Directions for use may vary between brands; see label before use. Herbicides are listed by mode of action and then alphabetically. The order of herbicide listing is not reflective of the order of efficacy or preference.

LIPID SYNTHESIS INHIBITORS	
Fluazifop *Fusilade*	**Rate:** 1 to 1.5 pt product/acre (4 to 6 oz a.e./acre) **Timing:** Postemergence to rapidly growing plants before boot stage. **Remarks:** Treatments may need to be repeated during a single season. Fluazifop is only effective on grass species. It is possible that fluazifop and other grass-selective herbicide, including clethodim and sethoxydim, will be effective for the control of buffelgrass, but no information is available. Fluazifop does not currently have a rangeland label.
AROMATIC AMINO ACID INHIBITORS	
Glyphosate *Roundup, Accord XRT II*, and others	**Rate:** Broadcast foliar treatment: 1 to 4 qt product (*Roundup ProMax*)/acre (1.1 to 4.5 lb a.e./acre). Spot treatment: 2% v/v solution, thoroughly wetting leaves. **Timing:** Postemergence to rapidly growing plants from mid-summer to fall. Plants should be at least 50% green when applications are made. This usually occurs during the monsoon rains, but can also occur in winter if climatic conditions are right. **Remarks:** Glyphosate is the standard herbicide option for control of buffelgrass. It is a nonselective herbicide. In many cases, it is best to add a dye to the chemical solution to avoid spraying non-target species. Adding ammonium sulfate to spray solution will enhance control. This application may need to be repeated several times to kill all the underground portions of buffelgrass. Glyphosate can be tank mixed with 1 to 2 oz a.e./acre imazapic (*Plateau*). However, this treatment is more expensive compared to glyphosate alone.
BRANCHED-CHAIN AMINO ACID INHIBITORS	
Imazapyr *Arsenal, Habitat, Stalker, Chopper, Polaris*	**Rate:** 4 to 6 pt product/acre (1 to 1.5 lb a.e./acre) **Timing:** Preemergence or postemergence to rapidly growing plants. **Remarks:** Imazapyr is a broad-spectrum herbicide with fairly long soil residual activity. It is expected to be effective on buffelgrass, though no direct evidence is available. The higher rates should be used in heavy or well-established infestations.
PHOTOSYNTHETIC INHIBITORS	
Hexazinone *Velpar L*	**Rate:** 1.25 to 2.5 gal *Velpar L*/acre (2.5 to 5 lb a.i./acre). Rates will vary depending on formulation used. **Timing:** Preemergence or postemergence to rapidly growing plants. **Remarks:** Hexazinone is a broad-spectrum herbicide that is mobile in the soil and has long soil residual activity. At high rates it may damage most vegetation. It should not be used in areas with a shallow water table. Hexazinone was effective in reducing growth of older buffelgrass plants (older than 45 days) and also controlled seedlings. High rates of hexazinone can create bare ground, so only use high rates in spot treatments.
Tebuthiuron *Spike 80DF*	**Rate:** 5 lb product/acre (4 lb a.i./acre) **Timing:** Preemergence in fall before rains. **Remarks:** At this high rate, tebuthiuron is not very selective and has long soil residual activity. It has been shown to provide some control of buffelgrass when applied preemergence. Tebuthiuron was effective in decreasing growth of older buffelgrass individuals (older than 45 days) and also controlled seedlings.

Pennisetum clandestinum Hochst. ex Chiov.

Kikuyugrass

Family: Poaceae
Range: California, especially the central coast, and Hawaii.
Habitat: Disturbed sites, roadsides, urban places, gardens, landscaped areas, orchards, cropland, turf, forested sites, and occasionally wetland areas. Grows best in areas with mild winters that receive some summer moisture. Tolerates periods of drought, light shade, and most soil types, but does not survive prolonged periods of freezing temperatures.
Origin: Native to tropical Africa. Kikuyugrass was introduced to California as an erosion-controlling ground cover but now is a noxious weed in many agricultural and landscape situations.
Impacts: May be able to act as a cover crop and smother native vegetation. Under certain conditions, kikuyugrass can accumulate high levels of nitrates and soluble oxalates that are toxic to livestock when ingested in quantity.
Western states listed as Noxious Weed: California, Oregon, Utah; also on Federal Noxious Weed list
California Invasive Plant Council (Cal-IPC) Inventory: Limited Invasiveness

Kikuyugrass is a warm-climate perennial grass which has escaped cultivation in the mild coastal and warm arid areas of California. It is tough, low-growing to 20 inches tall, with an extensive network of coarse, creeping stolons and rhizomes and flowerheads that remain mostly hidden within the upper leaf sheaths. The branched rhizomes develop an extensive network within the top 10 cm of soil. Kikuyugrass has flat or folded leaves, smooth to sparsely hairy, with a pronounced midvein on the underside. The ligules consist of a fringe of white hairs, and the collar margins have long white hairs. Kikuyugrass usually goes dormant and turns brown during the cool season where nighttime frosts are common.

Flowers are on short racemes of 2 to 4 spikelets which are hidden within the upper leaf sheaths, usually with only the long white stigmas protruding. Spikelets are 0.5 to 1 inch long. This plant can reproduce by seed, but most reproduction is vegetative by rhizomes and stolons. Seeds are probably only a minor mechanism of dispersal compared to rhizome and stolon fragments, which can readily root and generate new plants. Rhizomes, stolon fragments and seeds disperse with landscape maintenance and agricultural machinery, hand tools, soil movement, water, and other human activities.

NON-CHEMICAL CONTROL

Mechanical (pulling, cutting, disking)	Small patches can be pulled by hand. Avoid disking or cultivating kikuyugrass, as this will spread stem fragments.
Cultural	Burning is not effective in controlling this species.
	Before bringing in soil, sod, or planting stock, make sure they are clean of kikuyugrass.
Biological	There are no known biocontrol agents available for kikuyugrass.

CHEMICAL CONTROL
The following specific use information is based on published papers and reports by researchers and land managers. Other trade names may be available, and other compounds also are labeled for this weed. Directions for use may vary between brands; see label before use. Herbicides are listed by mode of action and then alphabetically. The order of herbicide listing is not reflective of the order of efficacy or preference.

GROWTH REGULATORS	
Triclopyr *Garlon 3A*, *Garlon 4 Ultra*	**Rate:** 0.67 to 1.33 qt *Garlon 3A* /acre or 0.5 to 1 qt *Garlon 4 Ultra*/acre of (0.5 to 1 lb a.e./acre) **Timing:** Apply to rapidly growing plants. **Remarks:** Triclopyr is safe on most grasses, but appears to have some activity on kikuyugrass. It has very little soil residual activity. This is a suppression treatment; complete control will require repeat treatments at 1-month intervals.

LIPID SYNTHESIS INHIBITORS	
Clethodim *Select*, *Envoy*	**Rate:** 1 pt product (*Select*)/acre (4 oz a.i./acre); 0.25% to 0.5% product v/v in spot treatment. **Timing:** Postemergence before runners are 6 inches long. It is less effective if applied after mowing and may require repeat treatments. **Remarks:** Clethodim is grass-selective and is safe on broadleaf species. To select in favor of other perennial grasses, apply before they emerge. It has no soil activity. Use a crop oil surfactant. Registered for fallow and non-crop areas, not generally for rangeland/natural areas, but has specific-use supplemental labels. Rates are based on those reported for bermudagrass. Note that *Envoy* formulation is 1 lb a.i./gallon, *Select* is 2 lb a.i./gallon.
Fluazifop *Fusilade*	**Rate:** 1 to 1.5 pt product/acre (4 to 6 oz a.i./acre) **Timing:** Postemergence when runners are less than 8 inches long. For good control, repeat treatments may be necessary. **Remarks:** Fluazifop is grass-selective and is safe on broadleaf species. To select in favor of other perennial grasses, apply before they emerge. It has no soil activity. Use a crop oil surfactant. This chemical is registered for fallow and non-crop areas, not for rangeland/natural areas.
AROMATIC AMINO ACID INHIBITORS	
Glyphosate *Roundup*, *Accord XRT II*, and others	**Rate:** 1.5 to 2 qt product (*Roundup ProMax*)/acre (1.7 to 2.25 lb a.e./acre); 1.5% v/v solution as a spot treatment. **Timing:** Apply to rapidly growing, non-stressed plants after most seedlings have emerged. **Remarks:** Glyphosate has no soil activity and is a nonselective herbicide.
BRANCHED-CHAIN AMINO ACID INHIBITORS	
Imazapyr *Arsenal*, *Habitat*, *Chopper*, *Stalker*, *Polaris*	**Rate:** 4 to 6 pt product/acre (1 to 1.5 lb a.e./acre) **Timing:** Preemergence or postemergence. **Remarks:** Imazapyr has a fairly long soil residual. It is a nonselective herbicide. Rates are based on those reported for bermudagrass.

Pennisetum setaceum (Forssk.) Chiov.

Crimson fountaingrass

Family: Poaceae
Range: Oregon, California, Arizona, New Mexico, and Colorado.
Habitat: Disturbed sites, roadsides, urban places, but also does well on undisturbed coastal dunes, coastal sage scrub, warm desert shrubland and canyons. Grows best in areas with mild winters that receive some summer moisture. Plants tolerate periods of drought, light shade, and most soil types, but do not survive prolonged periods of freezing temperatures. Crimson fountaingrass can grow in rock crevices and pavement cracks, but does not tolerate saline conditions.
Origin: Native to northeastern Africa and western Asia. Crimson fountaingrass is a common landscape ornamental and is often cultivated as an annual in cold winter areas. However, some cultivars, such as 'Rubrum' and 'Eaton Canyon,' usually do not produce viable seed.
Impacts: In some desert regions, crimson fountaingrass is contributing to the type conversion of shrubland to grassland by facilitating the occurrence of periodic fire, which usually kills the shrubs. Crimson fountaingrass is also a major noxious weed in Hawaii.
California Invasive Plant Council (Cal-IPC) Inventory: Moderate Invasiveness

Crimson fountaingrass is a fire-adapted warm-climate tufted perennial to 4 ft tall. The leaves are narrow, 8 to 13 inches long, folded or flat, glabrous to sparsely short-hairy, with a pronounced midvein on the underside. The ligules are 1 mm long and consist of a fringe of white hairs. The collar margins are also ciliate with long white hairs.

The inflorescences are showy, purplish, spike-like panicles that are lax or slightly drooping. The bristly panicles are 4 to 12 inches long, mostly 1.5 to 2 inches wide, with spikelets 4 to 7 mm long. The bristles are purplish, numerous, unequal, straight, and 0.75 to 1.5 inches long. Plants reproduce by seed, which are developed asexually by apomixis and, to a smaller degree, sexually by pollination. Spikelets with bristles disperse in late spring with wind, or more likely by clinging to the fur, feathers, and feet of animals. In Hawaii, seeds appear to survive for about 6 years in the soil seedbank. Individual plants may live up to 20 years or more.

NON-CHEMICAL CONTROL

Mechanical (pulling, cutting, disking)	Hand removal may be an effective method of controlling crimson fountaingrass because of the bunchgrass nature of the plant. Small infestations can be removed by uprooting or cutting with weed eaters. A heavy tool such as a pick, shovel or mattock may be needed to uproot large plants with a basal diameter over 6 inches. If inflorescences are present, they should be cut and placed in plastic bags, then destroyed to prevent spread of seeds. Removal by hand may need to be repeated several times a year at one to 2 month intervals. Seedlings will need to be monitored and removed thereafter.
	Mowing is not an effective method for crimson fountaingrass control.
	Tillage is not practical in most areas where crimson fountaingrass grows and is not likely to be successful as a control option.
Cultural	Susceptibility of *Pennisetum setaceum* to grazing/browsing damage is typically low, but in Hawaii cattle were shown to eat crimson fountaingrass when no other grasses were available.
	Crimson fountaingrass recovers quickly after fire, and the population may increase in density after a burn.
Biological	There are no biological control agents available for crimson fountaingrass, as the species is still a widely planted ornamental.

CHEMICAL CONTROL

The following specific use information is based on published papers and reports by researchers and land managers. Other trade names may be available, and other compounds also are labeled for this weed. Directions

for use may vary between brands; see label before use. Herbicides are listed by mode of action and then alphabetically. The order of herbicide listing is not reflective of the order of efficacy or preference.

LIPID SYNTHESIS INHIBITORS	
Fluazifop *Fusilade*	**Rate:** 1 to 1.5 oz product/acre (4 to 6 oz a.e./acre) **Timing:** Postemergence to rapidly growing plants. **Remarks:** Treatments may need to be repeated during a single season. Fluazifop is only effective on grass species, both annual and perennial. Use a non-ionic surfactant or crop oil to enhance activity. Do not apply to water-stressed plants.
Sethoxydim *Poast*	**Rate:** 1 to 2.5 pt product/acre (3 to 7.5 oz a.i./acre) **Timing:** Postemergence to rapidly growing plants. Control is fastest during warm weather and when weeds are small. **Remarks:** Treatments may need to be repeated during a single season. Sethoxydim is only effective on grass species. It is also possible that other grass-selective herbicides, including clethodim, will be effective for the control of crimson fountaingrass, but no information is available. Use of a crop oil concentrate enhances activity.
AROMATIC AMINO ACID INHIBITORS	
Glyphosate *Roundup, Accord XRT II*, and others	**Rate:** Broadcast foliar treatment: 1 to 4 qt product (*Roundup ProMax*)/acre (1.1 to 4.5 lb a.e./acre). Spot treatment: 1 to 2% v/v solution. Wiper treatment: 33 to 50% of concentrated product. **Timing:** Postemergence to rapidly growing plants from mid-summer to fall. Best kill of rhizomes occurs when applications are made at the flowering stage. **Remarks:** Glyphosate is a nonselective herbicide and has no soil residual activity. A wiper applicator can provide a higher level of selectivity. Control can be enhanced with the addition of ammonium sulfate. Glyphosate is not as consistent as other herbicides for control of crimson fountaingrass.
BRANCHED-CHAIN AMINO ACID INHIBITORS	
Imazapyr *Arsenal, Habitat, Stalker, Chopper, Polaris*	**Rate:** 4 to 6 pt product/acre (1 to 1.5 lb a.e./acre) **Timing:** Preemergence or postemergence to rapidly growing plants. **Remarks:** Imazapyr is a broad-spectrum herbicide with fairly long soil residual activity. Treatment recommendations on the label are for *Pennisetum villosum*, but it is expected that similar results should occur with *Pennisetum setaceum*. The higher rates should be used for heavy or well-established infestations.
PHOTOSYNTHETIC INHIBITORS	
Hexazinone *Velpar L*	**Rate:** 1.25 to 2.5 gal product/acre (2.5 to 5 lb a.i./acre) **Timing:** Preemergence or postemergence to rapidly growing plants. **Remarks:** Hexazinone is a broad-spectrum herbicide that is mobile in the soil and has long soil residual activity. It should not be used in areas with a shallow water table. Hexazinone is best used in areas with high densities of fountaingrass, medium to shallow soils, away from watercourses, and away from trees. High rates of hexazinone can create bare ground, so only use high rates in spot treatments.

Phalaris aquatica L.

Hardinggrass

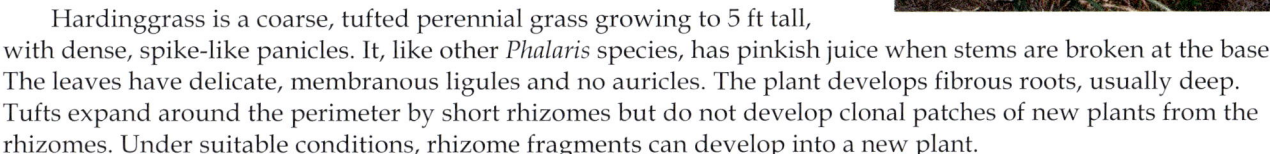

Family: Poaceae
Range: Arizona, California, Montana, Oregon, Texas, and some southeastern states. It is generally more invasive in coastal regions.
Habitat: Riparian areas, ditch banks, fields. Tolerates frost and drought.
Origin: Native to Mediterranean Europe. Hardinggrass was introduced to Australia and the United States to extend the forage season on pastures and rangeland, but has escaped cultivation in many locations.
Impacts: Once established, robust clumps are competitive for water and space, displacing native species. Under drought conditions, hardinggrass may develop toxic levels of alkaloids.
California Invasive Plant Council (Cal-IPC) Inventory: Moderate Invasiveness

Hardinggrass is a coarse, tufted perennial grass growing to 5 ft tall, with dense, spike-like panicles. It, like other *Phalaris* species, has pinkish juice when stems are broken at the base. The leaves have delicate, membranous ligules and no auricles. The plant develops fibrous roots, usually deep. Tufts expand around the perimeter by short rhizomes but do not develop clonal patches of new plants from the rhizomes. Under suitable conditions, rhizome fragments can develop into a new plant.

Hardinggrass usually flowers from late spring to the end of summer, producing dense, cylindrical spikes like other canarygrass species. The spikes are 0.5 to 1 inch in diameter and up to 4.5 inches long. Spikes typically remain intact for a period after senescence, but most florets are shed at maturity. Most reproduction is by seeds, which typically fall near the parent plant. The seed soil life is short, generally less than 2 years.

NON-CHEMICAL CONTROL

Mechanical (pulling, cutting, disking)	Hand-pulling is practical only for small stands and requires a large time commitment. It can be effective if done over the entire population 2 to 3 times per year for 5 years.
	Close mowing late in the season, when plants are still green, can reduce the plants' vigor. Mowing can be used to remove excess biomass, thus enhancing the effectiveness of follow-up herbicide applications. Plants should be allowed to regrow before treating.
	Cultivation of hardinggrass is usually not effective, because the plant can regenerate both from seed and from pieces of rhizome. Cultivation may be used to control seedlings, and repeated cultivations may eventually exhaust established stands.
Cultural	Grazing by livestock or geese can be used, similar to mowing, to remove biomass and stimulate new growth that can be treated with an herbicide. Grazing alone will not eradicate hardinggrass, but intensive grazing may help to suppress it.
	Burning in early spring, when there are large numbers of new shoots, can suppress this species. Burning alone is not an effective control but may facilitate later herbicide application. Plants can be burned first, then the regrowth treated with herbicide. In denser stands, plants can be treated with herbicide first so that their dead foliage provides fuel for a following burn.
Biological	No effective biocontrol agents are known for hardinggrass.

CHEMICAL CONTROL

The following specific use information is based on published papers and reports by researchers and land managers. Other trade names may be available, and other compounds also are labeled for this weed. Directions for use may vary between brands; see label before use. Herbicides are listed by mode of action and then alphabetically. The order of herbicide listing is not reflective of the order of efficacy or preference.

LIPID SYNTHESIS INHIBITORS	
Clethodim	**Rate:** 16 oz product (*Select*)/acre (4 oz a.i./acre) for seedlings; 0.5% of product v/v in spot treatment.
Select, *Envoy*	**Timing:** Postemergence; best before 6 inches tall. Less effective if applied after a mowing.

	Remarks: Clethodim is grass-selective and is safe on broadleaf species. To select in favor of other perennial grasses, apply before they emerge. It has no soil activity. Use a crop oil surfactant. The first treatment may provide only suppression of established plants; retreat as needed. Registered for fallow and non-crop areas, not generally for rangeland/natural areas, but has specific-use supplemental labels. Rates are based on high-end rates reported for annual canarygrass. Note *Envoy* is 1 lb a.i./gallon, *Select* is 2 lb a.i./gallon.
Fluazifop *Fusilade*	**Rate:** 1 to 1.5 pt product/acre (4 to 6 oz a.i./acre); 0.5% product v/v in spot treatment. **Timing:** Postemergence to rapidly growing plants. Best before boot stage. **Remarks:** Fluazifop is grass-selective and is safe on broadleaf species. It has no soil activity. To select in favor of other perennial grasses, apply before they emerge. Use a crop oil surfactant. The first treatment may provide only suppression of established plants; retreat as needed. Registered for fallow and non-crop areas, not for rangeland/natural areas. Rates are based on those reported for reed canarygrass and other perennial grasses.
AROMATIC AMINO ACID INHIBITORS	
Glyphosate *Roundup, Rodeo, Aquamaster*, and others	**Rate:** 2 to 3 qt product (*Roundup ProMax*)/acre (2.25 to 3.375 lb a.e./acre); 2% to 5% product v/v in water for spot treatment; 33% to 50% product v/v in water for wiper applications. **Timing:** For selective use, apply in early spring when hardinggrass is just sprouting and before other species germinate. More generally, application to rapidly growing flowering plants provides the best control. **Remarks:** Glyphosate is a nonselective herbicide. It has no soil activity. In addition to foliar sprays it can be applied using a rope wiper. Its effectiveness is increased by addition of ammonium sulfate. Also effective following removal of dead residue by burning, mowing, or grazing. Some formulations, e.g. *Rodeo* and *Aquamaster*, are registered for use in or near water. Rates are based on those reported for reed canarygrass.
Glyphosate + imazapyr *Rodeo + Habitat*	**Rate:** 1 qt *Rodeo* + 1 pt *Habitat*/acre **Timing:** Apply in spring to young growth. **Remarks:** Other formulations of each chemical are available; these brands are both registered for aquatic use. Rates are based on those reported for reed canarygrass.
BRANCHED-CHAIN AMINO ACID INHIBITORS	
Imazapic *Plateau*	**Rate:** 8 to 12 oz product/acre (2 to 3 oz a.e./acre) **Timing:** Preemergence in fall. **Remarks:** Imazapic has mixed selectivity and tends to favor members of the Asteraceae, as well as some grasses. It is safe for most native grasses, but higher rates may suppress seed of some cool-season grasses. Use methylated seed oil surfactant. Imazapic has some residual activity. Rates are based on those reported for reed canarygrass. Not registered for use in California.
Imazapyr *Arsenal, Habitat, Chopper, Stalker, Polaris*	**Rate:** 1.5 to 4 pt product/acre (6 to 16 oz a.e./acre) broadcast, or spot treatment with 1% product v/v in water. **Timing:** Apply to rapidly growing plants. Use higher rates for larger plants or late-season applications. **Remarks:** Imazapyr has a fairly long soil residual and is nonselective, so may kill desirable competitors. *Habitat* is registered for aquatic use. Rates are based on those reported for reed canarygrass.
Sulfometuron *Oust* and others	**Rate:** 1.33 to 2 oz product/acre (1 to 1.5 oz a.i./acre) for areas receiving 20 inches or less annual precipitation; 3 to 5 oz product/acre (2.25 to 3.75 oz a.i./acre) for areas receiving more than 20 inches precipitation. **Timing:** Preemergence or early postemergence, or apply to soil before the beginning of seasonal growth. **Remarks:** Sulfometuron has mixed selectivity. Do not apply to frozen ground. Add non-ionic surfactant for postemergence applications. It has fairly long soil residual activity.
PHOTOSYNTHETIC INHIBITORS	
Hexazinone *Velpar L*	**Rate:** 1.5 to 3 gal product/acre (3 to 6 lb a.i./acre) **Timing:** Spot apply before hardinggrass begins seasonal growth. **Remarks:** Results of efficacy on hardinggrass are from a trial conducted in New Zealand. High rates of hexazinone can create bare ground, so only use high rates in spot treatments.

Phalaris arundinacea L.
Reed canarygrass

Family: Poaceae
Range: Most of the United States except Hawaii and some southeastern states; Canada and worldwide.
Habitat: Wet sites along streams, in grassland, and woodlands throughout California and the West, except deserts and Great Basin.
Origin: Some biotypes are native to North America. Other invasive biotypes originated in Europe.
Impacts: Creeping rhizomes allow reed canarygrass to establish dense stands that outcompete native vegetation, particularly in wetland areas. It is sometimes planted for livestock forage, for streambank stabilization and erosion control, and for revegetation of mine spoils. In natural areas, reed canarygrass is an important component of the ecosystem and provides food for seed-eating birds. However, it may sometimes develop colonial patches in ditches, irrigation channels, and other controlled aquatic systems. Given time, it can overgrow small watercourses and alter soil hydrology.
Western states listed as Noxious Weed: Washington

Reed canarygrass is a cool-season, coarse, perennial grass to 5 ft tall, with creeping rhizomes. This species and other *Phalaris* species have pinkish juice when stems are broken at the base. The creeping rhizomes allow reed canarygrass to form dense monocultures that spread radially. The rhizomes can tolerate prolonged flooding. Because reed canarygrass is tolerant of freezing temperatures and begins to grow early in spring, it can outcompete other native species in wetland and riparian areas. Reed canarygrass establishment is promoted by disturbance, e.g., ditching, channelizing of streams, overgrazing, flooding, or sedimentation.

Panicles are green to purplish and often interrupted near the base. In addition to rhizomes, reed canarygrass also produces abundant seeds that are highly viable and can help the plant to disperse over greater distances. Seeds buried below the soil surface can survive up to about 20 years.

NON-CHEMICAL CONTROL

Mechanical (pulling, cutting, disking)	Hand-pulling is practical only for small stands and requires a large time commitment. It can be effective if done over the entire population 2 to 3 times per year for 5 years.
	Repeated mowing, e.g., five times per season, can effectively control reed canarygrass. Mowing can be used to remove excess biomass, thus enhancing the effectiveness of follow-up herbicide applications. Plants should be allowed to regrow before treating.
	Reed canarygrass is sensitive to disking or plowing. Pre-treating the plants with glyphosate early in the season kills many of the rhizomes, allowing them to deteriorate and enhancing the effect of disking. Following disking, later applications of herbicide may be necessary to control seedlings.
Cultural	Grazing can suppress reed canarygrass, but the palatability of this plant decreases in late season and following grazing. Grazing may be inappropriate in wetland settings.
	In wetlands, fire can suppress reed canarygrass and increase relative competitiveness of other wetland species. Fire is best suited for sites containing a healthy seedbank of fire-adapted native species that will colonize the area after a burn. Prescribed fire may be required for 5 or 6 years, or in 2 to 3 year rotations. Late spring or late fall burns are most effective, though late spring burns are most likely to damage desirable plant species. Burning has also been used to control resprouts and new germination 2 to 3 weeks after glyphosate application. Burning also can be combined with herbicides. Plants can be burned first, then the regrowth treated with herbicide. Early-season burning, in particular, stimulates shoot production. In denser stands, plants can be treated with herbicide first so that their dead foliage provides fuel for burning a few weeks later.
Biological	No biocontrol agents are known for reed canarygrass.

CHEMICAL CONTROL

The following specific use information is based on published papers and reports by researchers and land managers. Other trade names may be available, and other compounds also are labeled for this weed. Directions for use may vary between brands; see label before use. Herbicides are listed by mode of action and then alphabetically. The order of herbicide listing is not reflective of the order of efficacy or preference.

LIPID SYNTHESIS INHIBITORS	
Clethodim *Select, Envoy*	**Rate:** 16 oz product (*Select*)/acre (4 oz a.i./acre) for seedlings; 0.5% product v/v in spot treatment. **Timing:** Postemergence; best before 6 inches tall. Less effective if applied after a mowing. **Remarks:** Clethodim is grass-selective and safe on broadleaf species. To select in favor of other perennial grasses, apply before they emerge. It has no soil activity. Use a crop oil surfactant. The first treatment may provide only suppression of established plants. Retreatment may be necessary. Registered for fallow and non-crop areas, not generally for rangeland/natural areas, but has specific-use supplemental labels. Rates are based on high-end rates reported for annual canarygrass. Note that *Envoy* formulation is 1 lb a.i./gallon, *Select* is 2 lb a.i./gallon.
Fluazifop *Fusilade*	**Rate:** 1 to 1.5 pt product/acre (4 to 6 oz a.i./acre); 0.5% product v/v in water for spot treatment. **Timing:** Postemergence to rapidly growing plants; best before boot stage. **Remarks:** Fluazifop is grass-selective and safe on broadleaf species. It has no soil activity. To select in favor of other perennial grasses, apply before they emerge. Use a crop oil concentrate. The first treatment may provide only suppression of established plants, but retreat as needed. Registered for fallow and non-crop areas, not for rangeland/natural areas.
AROMATIC AMINO ACID INHIBITORS	
Glyphosate *Roundup, Rodeo, Aquamaster*, and others	**Rate:** 2 to 3 qt product (*Rodeo* or *Aquamaster*)/acre (2 to 3 lb a.e.); 2% to 5% product v/v solution in water for spot treatment; 33% to 50% product v/v solution in water for wiper applications. **Timing:** For selective use, apply in early spring when reed canarygrass is just sprouting and before other species germinate or emerge. More generally, apply to rapidly growing plants. **Remarks:** Glyphosate is a nonselective herbicide. It has no soil activity. Can be applied using a wiper (e.g., rope wick). Its effectiveness is increased by addition of ammonium sulfate. Also effective following removal of dead residue by burning, mowing, or grazing. Some formulations are registered for use in or near water (e.g., *Rodeo, Aquamaster*).
Glyphosate + imazapyr *Rodeo + Habitat*	**Rate:** 1 qt *Rodeo* + 1 pt *Habitat*/acre **Timing:** Postemergence in spring to young growth. **Remarks:** Use aquatic formulations near water. These herbicides are nonselective.
BRANCHED-CHAIN AMINO ACID INHIBITORS	
Imazapic *Plateau*	**Rate:** 8 to 12 oz product/acre (2 to 3 oz a.e./acre) **Timing:** Preemergence in fall. **Remarks:** Imazapic has mixed selectivity and tends to favor species in the Asteraceae and some grasses. It is safe for most native grasses, but higher rates may suppress seed of some cool-season grasses. Use with a methylated seed oil for maximum activity. It has long soil residual activity. Not registered for use in California.
Imazapyr *Arsenal, Habitat, Chopper, Stalker, Polaris*	**Rate:** 1 to 4 pt product/acre (4 to 16 oz a.e./acre) broadcast, or spot treatment with 1% v/v solution in water. **Timing:** Postemergence to rapidly growing plants. Use higher rates for larger plants or late-season applications. **Remarks:** Imazapyr has fairly long soil residual activity. It is a nonselective herbicide. *Habitat* is registered for aquatic use.
Sulfometuron *Oust* and others	**Rate:** 3 to 5 oz product/acre (2.25 to 3.75 oz a.i./acre) **Timing:** Preemergence to early postemergence. **Remarks:** Sulfometuron has mixed selectivity. Do not apply to frozen ground. Add non-ionic surfactant for postemergence applications. It has fairly long soil residual activity. *Oust* cannot be used adjacent to water where reed canarygrass usually grows.

Phoenix canariensis Chabaud; Canary Island date palm
Washingtonia robusta H. Wendl.; Mexican fan palm

Date and fan palm

Family: Arecaceae
Range: These are wildland weeds in southern California, but plants have also escaped along roadsides and urban areas in other states.
Habitat: Landscaped areas, urban places, riparian streams, particularly near rural areas, and orchards.
Origin: Canary Island date palm is native to the Canary Islands. Mexican fan palm is native to central Mexico, but not the northern mountain deserts. It has also invaded northwestern Mexico. Both are commonly cultivated as landscape ornamentals.
Impacts: Expansion of both species in riparian stream and river corridors can threaten native biodiversity. Populations are densest downstream from the sources of invasion, which are typically residential areas.
California Invasive Plant Council (Cal-IPC) Inventory: *Phoenix canariensis*, Limited Invasiveness; *Washingtonia robusta*, Moderate Invasiveness (Alert)

Both palms are perennials with single woody-like trunks. Mature plants can grow to 100 ft tall, but invasive populations are typically much shorter. Canary Island date palm has pinnately-compound leaves, up to 20 ft long and persistent on the trunk. Mexican fan also has persistent leaves, palmately veined to compound when torn, that are 3 to 6 ft wide.

Flowers develop within the leaf crowns. In Canary Island date palm, plants are dioecious with male and female plants. Fruit are edible and brown when mature. Mexican fan palms have black fruit that are not edible. Both species only reproduce by seed. Birds ingest fruits and disperse the seeds with their droppings. Seeds are large and readily carried by winter rains from landscaped areas down storm drains into nearby creeks and rivers.

NON-CHEMICAL CONTROL

Mechanical (pulling, cutting, disking)	Young palm plants can be hand pulled. Older plants can be cut at the base with a chainsaw. This can be a very effective control strategy for taller plants, where herbicide treatment to the foliage can lead to significant drift.
Cultural	Fire and grazing are not appropriate strategies for the control of palms. Reports indicate that cut stumps can be covered with clear plastic for solarization treatment.
Biological	There are no biological control agents available for palms.

CHEMICAL CONTROL

The following specific use information is based on reports by researchers and land managers. Other trade names may be available, and other compounds also are labeled for this weed. Directions for use may vary between brands; see label before use. Herbicides are listed by mode of action and then alphabetically. The order of herbicide listing is not reflective of the order of efficacy or preference.

GROWTH REGULATORS

Triclopyr *Garlon 3A, Garlon 4 Ultra, Pathfinder II*	**Rate:** Spot treatment: 10% of *Garlon 4 Ultra* concentrate (v/v) or 50% of *Garlon 3A* concentrate (v/v), or *Pathfinder II* undiluted as a ready to use herbicide **Timing:** Postemergence into apical buds of smaller plants. **Remarks:** Spot treatments should be made into the centers of smaller plants. This will reduce damage to other species adjacent to invasive palms.

AROMATIC AMINO ACID INHIBITORS	
Glyphosate	**Rate:** 50% v/v for spot treatment into drill holes. Undiluted for cut stump treatments.
Roundup, Accord XRT II, and others	**Timing:** Apply directly into stems either in drilled holes or after cutting.
	Remarks: Plants can be trimmed with a chainsaw, then drilled with a 5/16-inch construction drill bit. A 50% dilution of concentrated glyphosate (about 0.5 oz) is added to the drill hole. Fan palms can usually be killed with a single drill hole. Date palms are harder to kill because they have three vascular bundles, so more holes will need to be drilled. Undiluted glyphosate can be used to treat cut stumps.
BRANCHED-CHAIN AMINO ACID INHIBITORS	
Imazapyr	**Rate:** 1% v/v solution for spot treatment
Arsenal, Habitat, Stalker, Chopper, Polaris	**Timing:** Postemergence to fully developed leaves of smaller plants.
	Remarks: Imazapyr is fairly nonselective. When not near aquatic areas, the ester formulation (*Stalker* or *Chopper*) would be expected to be more effective.

Phragmites australis (Cav.) Trin. ex Steud.

Common reed

Family: Poaceae
Range: A western biotype is native throughout the western United States and adjacent Canada. A Gulf Coast biotype is found from Florida across to southern California, although it is not known if this biotype is native or introduced from Mexico and Central America. A non-native biotype has been found throughout the contiguous U.S. and adjacent Canada.
Habitat: Wetlands, riparian areas, shores of lakes and ponds.
Origin: The non-native biotype was introduced from Europe, apparently via ships' ballast. There are also native biotypes of *Phragmites australis*.
Impacts: Forms dense stands in wetlands and reduces native plant biodiversity.
Western states listed as Noxious Weed: Oregon, Washington

Common reed is a rhizomatous perennial grass to 15 ft tall, usually found growing in water or along the shores of streams, ponds, and lakes. The leaves are typical for a grass, but large: up to 1.5 inches wide and 8 to 16 inches long. The ligule (at the junction of the leaf blade and the stem) is tipped by a fringe of hairs, while the blades and sheaths are smooth. Leaf sheaths of the introduced biotype adhere tightly to the stem into winter, while in the native biotype the sheaths fall away as the leaves die in the autumn. Stems ("culms") are sturdy, up to 0.5 inch thick at the base, and slightly ridged or rough to the touch under the leaf sheath.

The inflorescence is 6 to 16 inches long, tawny brown or purplish, and feathery in appearance. The glumes (bracts below the flowering spikelet) are shorter than the lemmas (bracts at the base of individual florets) and are hairless. Plants reproduce by seeds, rhizomes, and stem fragments. The florets often have many hairs, allowing seed to blow with the wind or in the airstreams of vehicles, or to float in the water. The seeds are short-lived (likely < 2 years) under field conditions, and a persistent seedbank does not accumulate. Common reed rhizomes form a dense network under the colony; each rhizome can grow 10 ft or more in a single growing season.

There are several biotypes of common reed, some native and some introduced. The following table offers a good comparison between the native biotype and the major European biotype. The authors of the table caution that these characters may not distinguish the Gulf Coast type of common reed, which has been introduced into portions of southern California and Arizona in the west.

SUMMARY OF MORPHOLOGICAL CHARACTERS THAT DISTINGUISH NATIVE AND INTRODUCED *PHRAGMITES AUSTRALIS* IN THE UNITED STATES.

Character	Native	Introduced
Ligule length	> 1.0 mm	< 1.0 mm
Lower glume length	3.0 to 6.5 mm Most > 4.0 mm	2.5 to 5.0 mm Most < 4.0 mm
Upper glume length	5.5 to 11.0 mm Most > 6.0 mm	4.5 to 7.5 mm Most < 6.0 mm
Adherence of dead leaf sheaths	Loose, drop off easily	Tight, remain on dead stems
Growth form (stem density)	Typically in mixed communities. Stem density may be low to high, dead stems less likely to persist to the next growing season.	Often grows as a monoculture. Stem density is high, dead stems often persist to the next growing season.
Culm texture	Smooth, shiny	Dull or flat color, slightly ridged
Culm color	May be dark red at nodes and internodes, where exposed to UV. May be green as well.	Typically green, occasionally with some red color at the lower nodes
Spots on culms	May be present	Not present; mildew may be present
Leaf color	Lighter, yellow green to dark green	Typically darker green, but may be lighter in saline areas

From: Phragmites Field Guide: Distinguishing Native and Exotic Forms of Common Reed (*Phragmites australis*) in the United States, by Jil Swearingen and Kristin Saltonstall (2010). (http://www.nps.gov/plants/alien/fact/pdf/phau1-powerpoint.pdf)

NON-CHEMICAL CONTROL

Mechanical (pulling, cutting, disking)	Digging and removal of common reed is usually not feasible, given its dense root and rhizome system and its tendency to grow on rocky or rough ground or in standing water. If attempted, remove as much root and rhizome as possible, as broken root and rhizome sections will resprout from fragments. Hand pulling is not an effective strategy, as it rarely is possible to remove roots and rhizomes without breaking and fragmenting these tissues.
	Mowing is difficult in wetland sites, and unless applied repeatedly, mowing will not generally control this perennial species. Timely mowing can prevent seed production, however.
	Mulching with plastic or fabric sheets has not been shown to be effective, given that shoots from rhizomes are sharp-tipped as they emerge from the soil.
Cultural	Dredging and draining of water where common reed is found may reduce the vigor of common reed colonies. However, draining and dredging are not appropriate for use on most wetland preserves where the weed is often found in abundance.
	Prescribed burning is sometimes used for controlling this species, primarily by removing old growth of common reed and allowing seeds of other species to germinate and perhaps establish. Burning is sometimes used to remove old growth in preparation for herbicide application.
Biological	There are currently no biological control agents to aid in the control of common reed. Literature and field surveys in the northeastern United States and eastern Canada indicate that at least 26 native herbivores attack common reed in North America. There have been no deliberate releases made of European insects known to feed on the introduced biotype, but at least 21 species have been accidentally introduced to North America.

CHEMICAL CONTROL

The following specific use information is based on published papers or reports by researchers and land managers. Other trade names may be available, and other compounds may also be labeled for this weed. Directions for use may vary between brands; see label before use. Most herbicide applications will require multiple applications to fully control common reed. Because it usually is found growing in or near standing water, only aquatic herbicide formulations are recommended for use. Additionally, most states require specific aquatic endorsements for applicators of aquatic herbicides. Herbicides are listed by mode of action and then alphabetically. The order of herbicide listing is not reflective of the order of efficacy or preference.

AROMATIC AMINO ACID INHIBITORS	
Glyphosate *Rodeo,* *Aquamaster*	**Rate:** Broadcast treatment: 4 to 6 pt product (*Rodeo* or *Aquamaster*)/acre (2 to 3 lb a.e./acre). Spot treatment: 0.75% v/v solution. Wiper treatment: 33 to 50% of concentrated product.
	Timing: Postemergence to plants in full bloom in late summer or autumn.
	Remarks: Use up to 1% non-ionic surfactant approved for aquatic use to improve herbicide uptake. Removal of old stalks and foliage by mowing or burning in spring may be necessary for the herbicide application to adequately cover the foliage and for the treatment to be effective. Glyphosate overspray will injure or kill other plants that it contacts. Wiper application can also be used to apply glyphosate to common reed. Glyphosate can be combined with imazapyr for more effective control under some circumstances.

BRANCHED-CHAIN AMINO ACID INHIBITORS	
Imazapyr *Habitat*	**Rate:** 4 to 6 pt product/acre (1 to 1.5 lb a.e./acre)
	Timing: Postemergence to plants fully leafed out in summer.
	Remarks: Use up to 1% non-ionic surfactant approved for aquatic use to improve herbicide uptake. Removal of old stalks and foliage by mowing or burning in spring may be necessary for the herbicide application to adequately cover the foliage and for the treatment to be effective. Imazapyr overspray will injure or kill other plants that it contacts. Imazapyr can be combined with glyphosate for more effective control under some circumstances.

Phytolacca americana L.

Common pokeweed

Family: Phytolaccaceae
Range: Throughout the contiguous U.S., but less common in the western states. In the west, it is found in Washington, Oregon, California, Arizona, and New Mexico.
Habitat: Woodlands, pastures, fields, forest margins and disturbed sites such as roadsides, ornamental landscapes, agricultural fields, and urban waste areas. Survives in most environments and can tolerate many soil types.
Origin: Native to the eastern United States. It is sometimes cultivated as an ornamental or garden vegetable.
Impacts: Pokeweed is an occasional weed throughout much of the United States and is rapidly increasing in abundance in some areas. All plant parts, especially the root, contain numerous saponins and oxalates and can be fatally toxic to humans and livestock when ingested raw or with improper preparation. Severe digestive tract irritation is the primary symptom. Birds are reported to eat the berries without ill-effect and may occasionally become intoxicated following ingestion.

Pokeweed is a large herbaceous perennial shrub, 2 to 8 ft tall, with large leaves and showy purple-black berries. It has a smooth, stout, purplish stem that branches extensively. The bright green, elliptic leaves are simple, alternate on the stem, and have a strong unpleasant scent when crushed. Pokeweed's above-ground growth dies back every winter, but it has a large white fleshy rootstock that allows it to survive and regenerate each spring.

The flowers form in elongated clusters that hang from the branches. Flowers are white to magenta and give way to distinct deep purple berries with crimson juice by mid-summer to fall. The purple berries hanging from the bright green leaves in late summer are the most distinguishing characteristic of pokeweed. Reproduction is only by seed and a single plant can produce numerous seeds. The seeds are large, lens-shaped, glossy, and black. Birds eat the berries and scatter the seeds. This probably accounts for single, isolated plants in areas where pokeweed has never been noticed before. The seeds occasionally are found as impurities in garden and vegetable seeds. It is not known how long the seeds survive in the soil.

NON-CHEMICAL CONTROL

Mechanical (pulling, cutting, disking)	Hand pulling is effective on small plants. Once plants are established and develop an extensive root system, hand removal is difficult.
	Cultivation or cutting before fruits mature can control common pokeweed. Because plants produce numerous viable seed, cutting after fruit production will only result in numerous seedlings the following year. Cutting well below the root crown prevents regrowth.
Cultural	Grazing is not considered an effective control option. Seeds and foliage contain numerous saponins and oxalates and can be fatally toxic to livestock when ingested.
Biological	There are no biological control agents available for the management of pokeweed.

CHEMICAL CONTROL

The following specific use information is based on published papers and reports by researchers and land managers. Other trade names may be available, and other compounds also are labeled for this weed. Directions for use may vary between brands; see label before use. Herbicides are listed by mode of action and then alphabetically. The order of herbicide listing is not reflective of the order of efficacy or preference.

GROWTH REGULATORS

2,4-D Several names	**Rate:** Broadcast treatment: 1 to 2 qt product (3.8 lb a.e./gal formulation)/acre (0.95 to 1.9 lb a.e./acre). Spot treatment of seedlings: 2% v/v with the 6 lb a.e./gal ester formulation, plus 5% methylated seed oil
	Timing: Postemergence when plants are growing rapidly. Applications in spring provide best control.
	Remarks: 2,4-D is a selective herbicide for broadleaf species; will not damage desirable grasses growing nearby. Good coverage is necessary. Some reports indicate that using a premix combination of 2% 2,4-D

	amine with 2% fluroxypyr (*Vista XRT*) can be very effective as a foliar mix.
Dicamba *Banvel, Clarity*	**Rate:** Broadcast treatment: 1 to 2 pt product/acre (0.5 to 1 lb a.e./acre) and water plus 0.25 to 0.5% v/v surfactant to thoroughly wet all leaves. **Timing:** Early postemergence when plants are small and rapidly growing. **Remarks:** Dicamba is a selective herbicide for broadleaf species and will not damage desirable grasses growing nearby. Dicamba is available mixed with diflufenzopyr in a formulation called *Overdrive*. This has been reported to be effective on common pokeweed. Diflufenzopyr is an auxin transport inhibitor which causes dicamba to accumulate in shoot and root meristems, increasing its activity. *Overdrive* is applied postemergence at 4 to 8 oz product/acre to rapidly growing plants. Higher rates should be used when treating perennial weeds. Add a non-ionic surfactant to the treatment solution at 0.25% v/v, or a methylated seed oil at 1% v/v solution.
Triclopyr *Garlon 4 Ultra*	**Rate:** Basal bark or drizzle treatment: 5% v/v *Garlon 4 Ultra* (ester formulation), mixed with 95% crop oil **Timing:** Foliar treatments using drizzle technique are effective in spring or fall, less effective in summer. Stem and stump treatments are probably most effective late in the growing season. **Remarks:** Triclopyr is a broadleaf-selective herbicide with very short soil residual. Basal bark treatments have been shown effective in experimental trials. Treat basal stems to 18 inches in height and thoroughly soak the root collar. Older plants have a large root mass, and it is necessary to get effective downward translocation to control pokeweed. Downward translocation is most active late in the growing season.
AROMATIC AMINO ACID INHIBITORS	
Glyphosate *Roundup, Accord XRT II*, and others	**Rate:** Broadcast treatment: 2 qt product (*Roundup ProMax*)/acre (2.25 lb a.e./acre). Spot treatment: 3 to 5% solution v/v *Roundup* (or other trade name) and water to thoroughly wet all leaves. Higher rates may be necessary with less concentrated formulations. **Timing:** Postemergence when plants are growing rapidly. Applications in mid- to late-summer provide best control. New seedlings can be controlled in spring. **Remarks:** Glyphosate is a nonselective herbicide with no soil activity.
BRANCHED-CHAIN AMINO ACID INHIBITORS	
Imazapyr *Arsenal, Habitat, Stalker, Chopper, Polaris*	**Rate:** Broadcast treatment: 3 to 4 pt product/acre (0.75 to 1 lb a.e./acre) plus 0.25 to 0.5% v/v surfactant to thoroughly wet all leaves. **Timing:** Preemergence or postemergence. **Remarks:** Imazapyr is a preemergence and postemergence herbicide effective for controlling broadleaf weeds and grasses. Its success on pokeweed is variable; it may not always provide effective control in preventing new seedling germination.

Picris echioides L.
(= *Helminthotheca echioides* (L.) Holub [Jepson Manual 2012])

Bristly oxtongue

Family: Asteraceae
Range: Primarily in California, but also found in Nevada, Oregon, Washington, Montana, and North Dakota.
Habitat: Roadsides, waste places, fields, pastures, crop fields, orchards, vineyards, landscaped areas, gardens, and other disturbed open places. Most common in seasonally wet places.
Origin: Native to the Mediterranean regions of Europe.
Impacts: Can form dense stands in rangelands and other areas near coastal grasslands.
California Invasive Plant Council (Cal-IPC) Inventory: Limited Invasiveness

Bristly oxtongue is an erect winter, or sometimes summer, annual or biennial to nearly 3 ft tall. It has milky sap, stiff-bristly foliage, and yellow dandelion-like flowerheads. Young plants overwinter as rosettes before bolting in late spring. The leaves are alternate and covered with stiff, coarse, papilla-based hairs that are minutely branched at the tips.

The flowerheads are both terminal and axillary, mostly 1 to 2 inches wide, and consist only of yellow ligulate flowers. The achenes have a white bristly to plumose pappus on a stalk. Plants reproduce only by seed. Seeds probably disperse short distances with wind. Some seeds disperse greater distances by clinging to tools, vehicle tires, and landscaping and agricultural machinery. No studies have determined the seed longevity in the soil, but seeds would be expected to persist for a couple of years.

NON-CHEMICAL CONTROL

Mechanical (pulling, cutting, disking)	Control can be achieved by hand pulling, string trimming, or hoeing when soil is moist. Roots should be removed to 2 inches below the soil surface. Mowing repeatedly will suppress plants, but basal leaves may result in some recovery.
Cultural	It is not known whether plants are palatable to livestock.
	Burning may be an effective control option, but there are no studies to support this. However, bristly oxtongue often occurs in areas with annual grasses, and it is a late season plant. As such, there is likely a window of opportunity for burning after grasses have dried to provide fuel, but before bristly oxtongue has produced viable seed.
Biological	There are no biological control agents available for the control of bristly oxtongue.

CHEMICAL CONTROL

There is very little information available for the control of bristly oxtongue, but control measures for other members of the Asteraceae are expected to be effective. In particular, the chemical control options for yellow starthistle are likely to also be effective on bristly oxtongue.

The following specific use information is based on reports by researchers and land managers. Other trade names may be available, and other compounds also are labeled for this weed. Directions for use may vary between brands; see label before use. Herbicides are listed by mode of action and then alphabetically. The order of herbicide listing is not reflective of the order of efficacy or preference.

GROWTH REGULATORS	
2,4-D Several names	**Rate:** 1 to 4 pt product/acre (0.48 to 1.9 lb a.e./acre) **Timing:** Postemergence to seedlings or plants no later than the bolting stage. **Remarks:** 2-4-D is a broadleaf herbicide with no soil activity. Older plants are expected to require a higher rate compared to seedlings or plants in the early rosette stage.
Aminocyclopyrachlor + chlorsulfuron *Perspective*	**Rate:** 3 to 4.5 oz product/acre **Timing:** Postemergence in spring and early summer to rosettes or bolting plants, or in fall to seedlings and rosettes before the ground freezes. **Remarks:** Higher rates are necessary after plants bolt. Aminocyclopyrachlor has similar activity as aminopyralid and is expected to provide similar control of bristly oxtongue. *Perspective* provides broad-spectrum control of many broadleaf species. Although generally safe to grasses, it may suppress or injure certain annual and perennial grass species. Do not treat in the root zone of desirable trees and shrubs. Do not apply more than 11 oz product/acre per year. At this high rate, cool-season grasses will be damaged, including bluebunch wheatgrass. Not yet labeled for grazing lands. Add an adjuvant to the spray solution. This product is not approved for use in California and some counties of Colorado (San Luis Valley).
Aminopyralid *Milestone*	**Rate:** 5 to 7 oz product/acre (1.25 to 1.75 oz a.e./acre) **Timing:** Postemergence in spring and early summer to rosettes or bolting plants, or in fall to seedlings and rosettes before the ground freezes. **Remarks:** Higher rates are necessary after plants bolt.
Clopyralid *Transline*	**Rate:** 6 to 10 oz product/acre (2.25 to 3.75 oz a.e./acre) **Timing:** Postemergence in spring and early summer to rosettes or bolting plants, or in fall to seedlings and rosettes before the ground freezes. **Remarks:** Higher rates are necessary after plants bolt. See recommendations for yellow starthistle; control is expected to be similar for bristly oxtongue.
Picloram *Tordon 22K*	**Rate:** 1 pt product/acre (4 oz a.e./acre) **Timing:** Preemergence in winter, or early postemergence in late fall or spring. **Remarks:** Picloram has a very long soil residual activity and should provide 2 years of control. It is selective on broadleaf species and does not generally injure grasses. *Tordon 22K* is a federally restricted use pesticide. Picloram is not registered for use in California.
AROMATIC AMINO ACID INHIBITORS	
Glyphosate *Roundup, Accord XRT II,* and others	**Rate:** Broadcast foliar treatment: 2 to 4 pt product (*Roundup ProMax*)/acre (1.1 to 2.25 lb a.e./acre). Spot treatment: 1% v/v solution **Timing:** Postemergence to seedlings or plants no later than the bolting stage. **Remarks:** Glyphosate is a nonselective herbicide. Studies with yellow starthistle show good control with glyphosate, and it is expected that similar results would occur with bristly oxtongue.

Piptatherum miliaceum (L.) Coss.

Smilograss

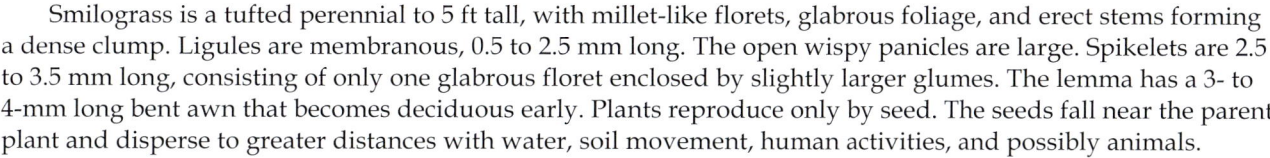

Family: Poaceae
Range: California, Nevada, Arizona, Utah and Idaho.
Habitat: Riparian areas, canyons, roadsides, fields, waste places, and other disturbed sites.
Origin: Native to Eurasia. Introduced as a livestock forage. Also used for heavy metal uptake in mine tailings.
Impacts: Smilograss has escaped cultivation in some areas of the western U.S. and appears to be increasing in riparian areas, ditches along roadsides, and canyons, especially in southern California, where it can threaten native plant diversity and ecosystem function.
California Invasive Plant Council (Cal-IPC) Inventory: Limited Invasiveness

Smilograss is a tufted perennial to 5 ft tall, with millet-like florets, glabrous foliage, and erect stems forming a dense clump. Ligules are membranous, 0.5 to 2.5 mm long. The open wispy panicles are large. Spikelets are 2.5 to 3.5 mm long, consisting of only one glabrous floret enclosed by slightly larger glumes. The lemma has a 3- to 4-mm long bent awn that becomes deciduous early. Plants reproduce only by seed. The seeds fall near the parent plant and disperse to greater distances with water, soil movement, human activities, and possibly animals.

NON-CHEMICAL CONTROL

Mechanical (pulling, cutting, disking)	Hand removal is possible because of the bunchgrass nature of the plant. However, all roots must be removed to prevent resprouting. Tillage may fragment the root system and could spread the plant.
Cultural	Smilograss was introduced as a forage species and may be used in a grazing program. However, there is no information on the effectiveness of grazing for control of smilograss. Fire is not likely to control this perennial bunchgrass.
Biological	There are no biological control agents available for smilograss.

CHEMICAL CONTROL

There is very little information available for the control of smilograss.

The following specific use information is based on reports by researchers and land managers. Other trade names may be available, and other compounds also are labeled for this weed. Directions for use may vary between brands; see label before use. Herbicides are listed by mode of action and then alphabetically. The order of herbicide listing is not reflective of the order of efficacy or preference.

LIPID SYNTHESIS INHIBITORS		
Clethodim *Select, Envoy*	**Rate:**	9 to 18 oz product (*Envoy*)/acre (1.1 to 2.2 oz a.i./acre)
	Timing:	Postemergence to rapidly growing plants before boot stage.
	Remarks:	Treatments may need to be repeated during a single season. Clethodim is only effective on grass species. It is possible that other grass-selective herbicides, including sethoxydim, will be effective for the control of smilograss, but no information is available. Note *Envoy* is 1 lb a.i./gallon, *Select* is 2 lb a.i./gallon.
Fluazifop *Fusilade*	**Rate:**	1 to 1.5 pt product/acre (4 to 6 oz a.e./acre)
	Timing:	Postemergence to rapidly growing plants before boot stage.
	Remarks:	Treatments may need to be repeated during a single season. Fluazifop is only effective on grass species. It is also possible that other grass-selective herbicides, including sethoxydim, will effectively control smilograss, but no information is available.
AROMATIC AMINO ACID INHIBITORS		
Glyphosate *Roundup, Accord XRT II*, and others	**Rate:**	Broadcast foliar treatment: 1 to 4 qt product (*Roundup ProMax*)/acre (1.1 to 4.5 lb a.e./acre). Spot treatment: 1 to 2% v/v solution
	Timing:	Postemergence to rapidly growing plants from mid-summer to fall.
	Remarks:	The standard herbicide option for control of smilograss. Nonselective, no soil activity.

Poa bulbosa L.
Bulbous bluegrass

Family: Poaceae
Range: Most prevalent in northeastern California, but found in all other western states and nearly all states in the U.S.
Habitat: Disturbed sites, roadsides, abandoned fields, alfalfa and grass hay fields, and rangelands. Sometimes found in pastures and grain fields. Tolerates a wide range of environmental conditions but best adapted to disturbed shallow soils that are moist during winter and early spring. Typically found in areas with 12 to 40 inches annual precipitation. It is well adapted to shallow soils that only receive rain in winter and early spring, but not to continually moist areas.
Origin: Native to Asia, Europe, and northern Africa.
Impact: Bulbous bluegrass produces little biomass for grazing and competes with more desirable vegetation in range sites. It can also invade crop and hay fields.

Bulbous bluegrass is a densely tufted cool-season perennial from 6 inches to 2 ft tall. Because it grows new roots each season and has a relatively short life span, it has been described as an annual with "perennial tendencies". Stems and culms are erect. The lower stems are flattened, while the upper stems are wiry and round in cross-section. The stems are thickened and bulblike at the base, a distinguishing characteristic. Leaves are narrow (1 to 3 mm wide) and 2 to 6 inches long. The leaf blades are flat or loosely rolled and have membranous ligules about 3 mm long without auricles. The leaves are also keeled, most conspicuously near the base.

Bulbous bluegrass produces 2- to 5-inch long panicles. The flowers typically develop into leafy bulblets with a dark purple-colored base. There are 4 to 6 bulblets per spikelet. The panicle is usually dense, with a plume-like appearance. The plant senesces soon after bulblets mature, typically around early May. The bulblets develop asexually and germinate immediately without a period of dormancy. Thus, the bulblets likely do not survive long in the soil. Bulbous bluegrass reproduces primarily through asexual means in the United States, but in its native Europe it reproduces sexually by seed.

NON-CHEMICAL CONTROL

Mechanical (pulling, cutting, disking)	Hand-pulling can control bulbous bluegrass but it is difficult to remove all of the bulbs.
	Mowing is not considered an effective control method.
	Bulbous bluegrass can be effectively controlled with early-season cultivation or tillage.
Cultural	Intensive grazing for several growing seasons can reduce bulbous bluegrass infestations.
	Bulbous bluegrass is likely killed or top-killed by fire, but little information is available. Survival of basal bulbs and bulblets depends on fire intensity. Buried bulblets may survive fire. Effects of fire are not well known, but bulbous bluegrass is often present on burned sites.
	Bulbous bluegrass is not competitive in dense stands of perennial crops like alfalfa or pasture.
Biological	No biological controls are currently available for the management of bulbous bluegrass.

CHEMICAL CONTROL

The following specific use information is based on reports by researchers and land managers. Other trade names may be available, and other compounds also are labeled for this weed. Directions for use may vary between brands; see label before use. Herbicides are listed by mode of action and then alphabetically. The order of herbicide listing is not reflective of the order of efficacy or preference.

LIPID SYNTHESIS INHIBITORS	
Clethodim *Select*, Envoy	**Rate:** Broadcast foliar treatment: 6 to 8 oz product (*Select*)/acre (1.5 to 2 oz a.i./acre) for seedlings. Spot treatment: 0.25% to 0.5% v/v solution
	Timing: Postemergence. Best when applications are made before plants are 6 inches tall. It is less effective if applied after a mowing.

	Remarks: Clethodim is grass-selective and safe on broadleaf species. To select for perennial grasses, apply before perennials emerge. It has no soil activity. Use a crop oil surfactant. Registered for fallow and non-crop areas, not generally for rangeland/natural areas, but has specific-use supplemental labels. Note that *Envoy* formulation is 1 lb a.i./gallon, *Select* is 2 lb a.i./gallon.
Aromatic amino acid inhibitors	
Glyphosate *Roundup, Accord XRT II*, and others	**Rate:** 0.33 to 1 qt product (*Roundup ProMax*)/acre (0.37 to 1.1 lb a.e./acre) **Timing:** Postemergence in early spring, to rapidly growing, non-stressed plants after most seedlings have emerged. If possible, apply before desirable perennials emerge. **Remarks:** Glyphosate is a nonselective herbicide. It has no soil activity.
BRANCHED-CHAIN AMINO ACID INHIBITORS	
Imazapic *Plateau*	**Rate:** 4 to 12 oz product/acre (1 to 3 oz a.e./acre) **Timing:** Preemergence to early postemergence from fall to early spring. **Remarks:** Long soil residual activity, mixed selectivity. Tends to favor members of the Asteraceae and some grasses. Use a spray adjuvant for postemergence applications. Effects vary depending on soil texture and organic matter. Heavy soils and high organic matter may require higher rates. Can tie up in litter; efficacy is reduced in situations where there is thick thatch on the soil surface. Not registered for use in California.
Imazapyr *Arsenal, Habitat, Chopper, Stalker, Polaris*	**Rate:** 2 to 3 pt product/acre (8 to 12 oz a.e./acre) **Timing:** Preemergence or postemergence. **Remarks:** Imazapyr has long soil residual activity. It is a nonselective herbicide.
Rimsulfuron *Matrix*	**Rate:** 2 to 4 oz product/acre (0.5 to 1 oz a.i./acre) **Timing:** Preemergence in fall to early postemergence in early spring. **Remarks:** Controls several annual grasses and broadleaves. Perennial grasses are tolerant to fall applications when established and grown under dryland conditions. Application to rapidly growing or irrigated perennial grasses may result in their injury or death. It provides soil residual control in cool climates but degrades rapidly under warm conditions. Will not control summer annual weeds when applied in fall or spring. Add a surfactant when applying postemergence.
Sulfometuron *Oust* and others	**Rate:** 1 oz product/acre (0.75 oz a.i./acre) **Timing:** Preemergence or early postemergence from fall to early spring. Most effective control is with early postemergence treatment after bluegrass seedlings have emerged. **Remarks:** Mixed selectivity, fairly safe on native perennial grasses. Good for revegetation use. Use lower rates in arid environments and higher rates in wetter areas (> 20" rainfall) and on high organic matter soils. Fairly long soil residual activity. At higher rates, this treatment will generally result in bare ground.
Sulfometuron + chlorsulfuron *Landmark XP*	**Rate:** 0.75 oz product/acre **Timing:** Preemergence in fall or after soil thaws in spring. **Remarks:** See sulfometuron.
Sulfosulfuron *Outrider*	**Rate:** 0.75 to 2 oz product/acre (0.56 to 1.5 oz a.i./acre) **Timing:** Early postemergence, fall to early spring, when desirable perennials are dormant and target plants are growing rapidly. **Remarks:** Mixed selectivity; fairly safe on perennial grasses, especially wheatgrasses. Fairly long soil residual. Treatments should include non-ionic surfactant. Sequential applications can be made (minimum of 21 days between applications) as long as the total rate does not exceed 2.66 oz product/acre per year.
PHOTOSYNTHETIC INHIBITORS	
Hexazinone *Velpar L*	**Rate:** 2 to 6 pt product/acre (0.5 to 1.5 lb a.i./acre) **Timing:** Preemergence to early postemergence. **Remarks:** Both foliar and soil activity. In soil applications, rates will vary with soil texture and soil organic matter. Best results when applied to moist soils. Hardwood trees can absorb this chemical through the roots and may be injured or killed. Do not spray near the root zone of desirable hardwood trees or shrubs. High rates of hexazinone can create bare ground, so only use high rates in spot treatments.

Polygonum cuspidatum Siebold & Zucc.; Japanese knotweed
  (= *Fallopia japonica* (Houtt.) Ronse Decr.)
Polygonum sachalinense F. Schmidt ex Maxim.; Sakhalin or giant knotweed
  (= *Fallopia sachalinensis* (F. Schmidt ex Maxim.) Ronse Decr.)
Polygonum × bohemicum (J. Chrtek & Chrtkova) Zika & Jacobson [*cuspidatum* × *sachalinense*]; bohemian knotweed
  (= *Fallopia × bohemica* (Chrtek and Chrtkova) J. Bailey)
Polygonum polystachyum Wall. ex Meisn.; Himalayan knotweed
  (= *Persicaria wallichii* Greuter & Burdet)

Japanese, Sakhalin, Bohemian, and Himalayan knotweeds

P. cuspidatum

P. sachalinense

P. x bohemicum

P. polystachyum

Family: Polygonaceae
Range: Alaska, British Columbia, Washington, Oregon, Idaho, Montana, California, Utah, Colorado, South Dakota, Nebraska, Kansas, and Oklahoma.
Habitat: Riparian areas/floodplains, forest edges, meadows, rights-of-way, and parks.
Origin: Native to Japan, China, and eastern Asia. Apparently escaped from cultivation.
Impacts: All species are partially shade tolerant and frequently found along river corridors. They form dense colonies, almost eliminating other plants within these stands.
Western states listed as Noxious Weed: *P. cuspidatum*, California, Colorado, Oregon, Washington; *P. sachalinense*, California, Colorado, Oregon, Washington; *P. x bohemicum*, Colorado, Washington; *P. polystachyum*, California, Colorado, Oregon, Washington
California Invasive Plant Council (Cal-IPC) Inventory: *P. cuspidatum* and *P. sachalinense*, Moderate Invasiveness (Alert)

These species are all herbaceous rhizomatous perennials to 12 ft tall or more. Leaf shape and size vary by species. Japanese knotweed leaves are thick and leathery, about 4 to 5 inches long, bearing few hairs, square across the base, and with an abruptly tapered tip. Sakhalin knotweed leaves are heart-shaped in outline, up to 12 inches across and 18 inches long, and bearing a few hairs along the margin and on the underside along the veins. Leaves of the hybrid Bohemian knotweed are mid-range between Japanese and Sakhalin knotweeds, usually with more or less square bases, and a few rough hairs on the undersurface and along the margins, but usually only bluntly tipped. Himalayan knotweed leaves are slender, up to 12 inches long but only 3 inches wide, bearing hairs along the margins and on the main veins. The stems of all species are smooth, somewhat waxy, and hollow. They superficially resemble bamboo, with enlarged nodes. Sakhalin knotweed stems may be as much as 2 inches thick at the base, with Bohemian, Japanese, and Himalayan knotweed stems progressively thinner to about 0.5 inch thick. Stems of Japanese and Bohemian knotweed tend to be spotted with reddish-purple.

Flowers of all species are white, about 3 mm across, each bearing 3 to 5 petals and borne in racemes from 2 to 4 inches long that arise in the upper leaf axils. In Japanese knotweed, male and female flowers are on separate plants, but Himalayan and Sakhalin knotweeds have perfect (male/female combined) flowers. Female and perfect flowers may bear teardrop-shaped fruits, each with a 2- to 3-mm reddish brown to black, prism-shaped seed. All species spread primarily, or perhaps exclusively, by woody rhizomes, or from fragmented stem sections that can root at the nodes. They can move long distances in streams and other water sources.

NON-CHEMICAL CONTROL

Mechanical (pulling, cutting, disking)	These knotweed species are nearly impossible to control by digging. Their extensive rhizomes are difficult to remove intact, and fragments easily resprout.
	Mowing can reduce growth, but will seldom, if ever, eliminate these plants. Repeated cutting at least every 4 weeks and at least 7 weeks before senescence can give effective suppression.
	Cultivation appears to be effective if conducted with enough frequency and for a sufficient number of years.
	Mulching with plastic or fabric sheets for 2 years or more may result in some control. Fabric mulches have been reported to produce better results when applied loosely over the colony and stems walked on or otherwise crushed when their growth pushes up the fabric.
Cultural	None of these knotweeds are known to be toxic to animals and, in fact, Japanese knotweed is edible to humans. Grazing by livestock may provide some growth reduction, but has not been shown to eliminate colonies.
Biological	There are no biological control agents currently available to aid in the control of these knotweeds. However, a species of sap sucking psyllid (*Aphalara itadori*) has been released on Japanese knotweed in Europe, and at some future date may be cleared for release in the United States and Canada. There are also other biological control candidates under investigation, including a leaf spot pathogen in the genus *Mycosphaerella*.

CHEMICAL CONTROL

The following specific use information is based on published papers and reports by researchers and land managers. Other trade names may be available, and other compounds may also be labeled for this weed. Directions for use may vary between brands; see label before use. Herbicides are listed by mode of action and then alphabetically. The order of herbicide listing is not reflective of the order of efficacy or preference.

GROWTH REGULATORS	
Aminopyralid *Milestone*	**Rate:** 7 oz product/acre (1.75 oz a.e./acre) or spot treatment at 14 oz product/acre
	Timing: Postemergence to foliage in mid-summer to autumn, when plants are fully leafed. Optimum results when plants are 3 to 5 ft tall in early to mid-summer. Good, thorough coverage of foliage is necessary for control.
	Remarks: Repeat applications will be necessary.

AROMATIC AMINO ACID INHIBITORS	
Glyphosate *Aquamaster, Rodeo, Roundup,* and others	**Rate:** Broadcast foliar treatment: 4 qt product (*Rodeo* or *Aquamaster*)/acre (4 lb a.e./acre). Spot treatment: 2.5 to 8% v/v solution. Individual stem injection treatment: 5 ml undiluted product/stem up to a maximum 2 gallons product per acre (8 lb a.e./acre)
	Timing: Postemergence to foliage in mid-summer to autumn, when plants are fully leafed out. For injection treatment, inject full-strength product into the stem about 3 nodes above the soil line.
	Remarks: Use 0.25% non-ionic surfactant to improve herbicide uptake. Repeat applications will be necessary. Glyphosate is nonselective and will injure or kill other plants growing near treated knotweed that are oversprayed or drifted on. Glyphosate applied by injection technique to plants growing in sandy or gravelly soil will sometimes leak out into soil and be taken up by roots of adjacent vegetation.

BRANCHED-CHAIN AMINO ACID INHIBITORS	
Imazapyr *Habitat*	**Rate:** Broadcast foliar treatment: 3 to 4 pt product/acre (0.75 to 1 lb a.e./acre). Spot treatment: 1% v/v solution
	Timing: Postemergence to foliage in mid-summer to autumn, when plants are fully leafed out.
	Remarks: Imazapyr is considered the most effective treatment method for control of the large knotweeds. Use of 0.25% non-ionic surfactant or 1 qt/acre methylated seed oil may improve herbicide uptake. Repeat applications will be necessary. Imazapyr is nonselective and will injure or kill other plants it contacts.

Polygonum persicaria L.; ladysthumb
Polygonum lapathifolium L.; pale smartweed
   (= *Persicaria lapathifolia* (L.) Gray [Jepson Manual 2012])

Ladysthumb and pale smartweed

Polygonum persicaria

Polygonum lapathifolium

Family: Polygonaceae
Range: Throughout the United States, including the western states.
Habitat: Edges of ponds, shallow lakes, marshes, and streams; wet fields, areas subject to seasonal flooding, irrigation ditches, pastures, orchards, rice fields, grain fields, irrigated crops. Plants typically grow in wet to moist soils but tolerate periods of dryness. Terrestrial plants grow best in disturbed places with fertile soils and minimal competition. Plants do not tolerate highly acidic soils.
Origin: Pale smartweed is a widespread native of North America. Ladysthumb is native to Europe.
Impacts: Can invade rice fields, pastures, orchards, and irrigated crops; stands of emergent plants can impede the flow of water in irrigation ditches, canals, and drainage areas. In natural areas they can be a desirable component of the flora. However, these species can invade protected wetlands and agricultural systems. Their seeds are an important food source for many species of songbirds, waterfowl, and mammals, and the foliage provides cover for wildlife. The two species can hybridize.

Pale smartweed and ladysthumb are coarse, emergent aquatic to terrestrial herbs. Both are summer annuals, growing to 5 ft tall (pale smartweed) or 3 ft (ladysthumb). Mature plants have coarse stems which are swollen at the nodes, with papery sheaths (ocrea) around the nodes. The stems are openly branched and often reddish. Leaves are alternate, lanceolate to elliptic, with stalks 0.25 to 1 inch long. The leaves are 1 to 8 inches long, 0.25 to 2 inches wide, with tapered bases. Both species have minute glands on the lower leaf surfaces; pale smartweed's are sunken, while ladysthumb's are raised. Ladysthumb also typically has a dark purplish central spot on the leaves (as may some other smartweed species). Lower stems typically root at the nodes. These species have shallow, branched taproots, with fibrous secondary roots. Under certain conditions, fragmented stems can regenerate into new plants.

Both species produce dense flowering spikes from early summer into early fall. Pale smartweed spikes are 1 to 3 inches long, cylindrical, and drooping. Ladysthumb spikes are erect, oblong, and 0.5 to 1.5 inches long. The seeds are glossy, brown to black, about 2 mm long. Seeds fall near and/or remain on the parent plant, or disperse to greater distances generally with water or as contaminants of crop seed. The foliage is killed by the first frost, turning brown to reddish, and does not persist through winter. Seed longevity is not well understood, but seeds are expected to remain dormant and viable for a few years after dispersal.

NON-CHEMICAL CONTROL

Mechanical (pulling, cutting, disking)	Cultivation or hand removal can control these weeds.
	Mowing before flowering can reduce seed set.
	Tillage is not generally an option in wet areas.
Cultural	These plants grow in wet areas which are not usually grazed.
	These plants grow in wet locations, so burning is not a feasible control method.
	Improving drainage can help to control smartweed populations.
Biological	There are no biological control agents available for the management of any of the *Polygonum* species.

CHEMICAL CONTROL

The following specific use information is based on published papers and reports by researchers and land managers. Other trade names may be available, and other compounds also are labeled for this weed. Directions for use may vary between brands; see label before use. Herbicides are listed by mode of action and then alphabetically. The order of herbicide listing is not reflective of the order of efficacy or preference.

GROWTH REGULATORS	
2,4-D Several names	**Rate:** 0.5 to 2 pt product/acre (0.24 to 0.95 lb a.e./acre) **Timing:** Postemergence to rapidly growing weeds, the smaller the better. **Remarks:** 2,4-D is broadleaf-selective. It has no soil activity and is often combined with other active ingredients, e.g., dicamba. Do not apply when outside temperatures exceed 80°F. Aquatic registered formulations are available for use close to water. 2,4-D and dicamba can be used together in a premix with metsulfuron (*Cimarron Max*), but not near water.
Aminopyralid *Milestone*	**Rate:** 3 to 5 oz product/acre (0.75 to 1.25 oz a.e./acre) **Timing:** Apply in spring to rapidly growing plants, or treat soil shortly before plants emerge. **Remarks:** Aminopyralid is a broadleaf herbicide like picloram, but more selective. It is safe on grasses but will kill most legumes and other members of the Asteraceae. It has longer residual and higher activity than clopyralid and is registered for application up to water's edge.
Aminopyralid + 2,4-D, *Forefront HL*; Aminopyralid + metsulfuron, *Opensight*; Aminopyralid + triclopyr, *Capstone*	**Rate:** 1.5 to 2.1 pt *Forefront HL*/acre; 1.5 to 2 oz *Opensight*/acre; 4 to 6 pt *Capstone*/acre **Timing:** Postemergence to rapidly growing plants. **Remarks:** These premixes (and equivalent tank mixes using the same active ingredients) are all broadleaf-selective. *Opensight* is not registered for use in California.
Clopyralid *Transline*	**Rate:** 0.67 to 1.33 pt product/acre (4 to 8 oz a.e./acre) **Timing:** Postemergence to rapidly growing weeds. **Remarks:** A broadleaf herbicide like picloram, but more selective. Very safe on grasses but will kill most legumes and other members of the Asteraceae. It has some soil residual activity. May provide only suppression of smartweeds.
Clopyralid + 2,4-D *Curtail*	**Rate:** 1 to 2 qt product/acre **Timing:** Postemergence to rapidly growing weeds. **Remarks:** The mixture is broadleaf-selective with a wide range of susceptible species.
Dicamba *Banvel, Clarity*	**Rate:** 0.25 to 1 pt product/acre (0.125 to 0.5 lb a.e./acre) **Timing:** Postemergence to rapidly growing weeds, the smaller the better. Use higher rates for larger plants. **Remarks:** Dicamba is a broadleaf-selective herbicide often combined with other active ingredients. It may injure grasses at higher rates. Do not apply when outside temperatures exceed 80°F. Dicamba and 2,4-D can be used together in a premix with metsulfuron (*Cimarron Max*), but not near water. Dicamba is available mixed with diflufenzopyr in a formulation called *Overdrive*. This combination is broadleaf-selective and safe on most grasses. This has been reported to be effective on ladysthumb and other smartweeds. Diflufenzopyr is an auxin transport inhibitor which causes dicamba to accumulate in shoot and root meristems, increasing its activity. *Overdrive* is applied postemergence at 4 to 8 oz product/acre to rapidly growing plants. Higher rates should be used on large annuals and biennials or when treating perennial weeds. Add a non-ionic surfactant to the treatment solution at 0.25% v/v or a methylated seed oil at 1% v/v solution.
Triclopyr *Garlon 3A, Garlon 4 Ultra*	**Rate:** 1.33 to 6 qt *Garlon 3A*/acre (1 to 4.5 lb a.e./acre), 1 to 4 qt *Garlon 4 Ultra*/acre (1 to 4 lb a.e./acre) **Timing:** Postemergence to rapidly growing plants. **Remarks:** Triclopyr is broadleaf-selective and safe on most grasses. It is most effective on smaller plants and has little or no residual activity. *Garlon 3A* and other amine formulations are registered for aquatic use. *Garlon 4 Ultra* is formulated as a low volatile ester. However, in warm temperatures, spraying onto hard surfaces such as rocks or pavement can increase the risk of volatilization and off-target damage.

AROMATIC AMINO ACID INHIBITORS	
Glyphosate *Roundup, Rodeo, Aquamaster*, and others	**Rate:** 1 to 2 pt product (*Roundup ProMax*)/acre (0.56 to 1.1 lb a.e./acre). Use 1 to 2 pt *Rodeo* or *Aquamaster*/acre (0.5 to 1 lb a.e./acre) around aquatic areas. **Timing:** Postemergence to rapidly growing plants. **Remarks:** Glyphosate is a nonselective herbicide. It has no soil activity and its effectiveness is increased by addition of ammonium sulfate. Aquatic registered formulations, e.g., *Rodeo* and *Aquamaster*, are available for use close to water.
BRANCHED-CHAIN AMINO ACID INHIBITORS	
Imazapic *Plateau*	**Rate:** 4 to 6 oz product/acre (1 to 1.5 oz a.e./acre) **Timing:** Preemergence or postemergence. **Remarks:** Imazapic has mixed selectivity and tends to favor members of the Asteraceae and some grasses. It has some soil residual activity. In postemergence applications, use a methylated seed oil surfactant at 0.25%. It is not registered for use in California.
Imazapyr *Arsenal, Habitat, Stalker, Chopper, Polaris*	**Rate:** 2 pt product/acre (8 oz a.e./acre) **Timing:** Shortly after emergence. **Remarks:** Imazapyr is a nonselective herbicide. It has long soil residual activity and leaves more bare ground than other treatments, even a year after application. *Habitat* is an aquatic registered formulation available for use close to water.

Potamogeton spp.; several pondweeds
Stuckenia pectinata (L.) Börner; sago pondweed

Pondweeds (submerged)

Potamogeton foliosus

Potamogeton crispus

Family: Potamogetonaceae
Range: Throughout the U.S., including all western states.
Habitat: Ponds, lakes, streams, rivers, reservoirs, irrigation ditches, marshy areas. Most species commonly grow in shallow water, but can grow to depths of ~20 ft or more in clear water. Leafy and sago pondweed also grow in brackish water.
Origin: All pondweed species are native to the western United States except curlyleaf pondweed (*P. crispus* L.), which was introduced from Eurasia.
Impacts: In natural areas, most pondweeds are a desirable component of the aquatic community. They provide habitat and are an important food source for wildlife. However, colonies can be troublesome in drainage canals, irrigation ditches, and other controlled aquatic systems. Curlyleaf pondweed is very invasive in aquatic systems and is considered the most significant problem among the submerged pondweed species.
Western states listed as Noxious Weed: *P. crispus*, Washington
California Invasive Plant Council (Cal-IPC) Inventory: *P. crispus*, Moderate Invasiveness

Stuckenia pectinata

The genus Potamogeton is comprised of many widespread, highly variable species that are often difficult to distinguish, resulting in much taxonomic confusion. The most common submerged species in the western United States are curlyleaf pondweed (*P. crispus*), leafy pondweed (*P. foliosus* Raf. var. *foliosus*), Illinois pondweed (*P. illinoensis* Morong), and sago pondweed (*S. pectinata* (L.) Börner).

The submerged pondweeds are all perennials, most with rhizomes. Most species hybridize with one or more other species. The leaves are generally alternate, but sometimes nearly opposite. Leaves of most species have a prominent midvein. Leaves vary in size and shape among species. Curlyleaf pondweed has a very wavy (undulate) leaf which makes it easy to distinguish.

Inflorescences consist of cylindrical spikes that are above the surface of the water, or floating on the surface in the case of sago pondweed. The flowers are greenish and inconspicuous. Plants are wind- and water-pollinated. Fruits are achene- or nutlet-like structures. Despite the fact that all species, except hybrids, produce seeds, seedlings are seldom encountered. When seeds are produced, they float and disperse with water, are ingested by wildlife, or cling to the feet, fur, or feathers of animals. Seeds surviving ingestion by birds germinate readily. Illinois pondweed seeds may survive up to ~5 years under moist conditions. Sago pondweed seeds survive up to ~1.5 years under dry conditions. Seed longevity of other species is poorly documented.

Most plants reproduce vegetatively from rhizomes or stem fragments. Curlyleaf pondweed develops turions (specialized stem buds that survive unfavorable conditions) in the leaf axils and/or at the tips of short axillary branches before dormancy. Turions are composed of few to several reduced, overlapping leaves. Curlyleaf pondweed turions resemble brown pinecones. Sago pondweed reproduces vegetatively by tubers formed at the tips of the rhizomes.

NON-CHEMICAL CONTROL

Mechanical (pulling, cutting, disking)	Repeated mechanical harvesting can help reduce stem densities, but escaped stem fragments can drift elsewhere and develop into new plants. Removing and destroying stem fragments from recreational equipment such as boat propellers, docking lines, and fishing gear can help prevent the spread of pondweeds.
	Several types of "bottom barriers" are available and are used to cover and smother specific infested areas. Materials used include polyvinyl chloride (pvc) sheets, small-mesh screens, and natural fibers such as jute.

	Bottom barriers are best installed in spring before plants produced large biomass and exceed 20 inches tall. In canals, backhoes and telescoop devices can be used to remove infestations, but these operations usually remove sediment as well. Because sago pondweed forms tubers in the sediment, excavation depths should reach at least 12 inches.
Cultural	Dewatering (draining) canals or lakeshores in mid-summer may suppress subsequent growth, but plants can easily resprout from rhizomes if bottom sediments remain moist and cool. Winter drawdown alone will not impact growth of sago pondweed since it produces vegetative propagules (tubers) protected within the sediment.
Biological	The (sterile) triploid grass carp (white amur) is a relatively nonselective herbivorous fish that will consume most pondweed species, including sago pondweed and curlyleaf pondweed. The fish do not selectively feed on "non-native" plants so careful monitoring of feeding impacts is necessary. In many canal systems non-selectivity is not a problem, as few or no plants are desired since they can interfere with efficient water delivery.

CHEMICAL CONTROL

The following specific use information is based on published papers and reports by researchers and land managers. Other trade names may be available, and other compounds also are labeled for this weed. Directions for use may vary between brands; see label before use. Herbicides are listed by mode of action and then alphabetically. The order of herbicide listing is not reflective of the order of efficacy or preference.

BRANCHED-CHAIN AMINO ACID INHIBITORS	
Penoxsulam *Galleon*	**Rate:** For in-water applications: 25 to 75 ppb; may be repeated but not to exceed 150 ppb in an annual season. For dewatered (drawdown) applications: 5.6 to 11.2 oz product/acre (1.4 to 2.8 oz a.i./acre) to canal or shorelines where plants grow
	Timing: Apply to water in early spring to early summer (rapid growth). For drawdown treatments, apply during mid to late winter before refilling.
	Remarks: Penoxsulam is a slow-acting herbicide and may take 4 to 6 weeks to achieve control. For drawdown applications use 20 to 100 gal/acre of spray solution to wet the sediment.
Imazamox *Clearcast*	**Rate:** For in-water applications: 50 to 100 ppb. For dewatered (drawdown) applications: 64 oz product/acre (8 oz a.e./acre); first flush of water in canals must NOT be used for irrigation
	Timing: Apply to water in early spring to early summer (rapid growth). Dewatered applications should be made in late winter at least 14 days before water will be reintroduced.
	Remarks: Use an approved surfactant.
PIGMENT SYNTHESIS INHIBITORS	
Fluridone *Sonar*	**Rate:** For in-water applications: 5 to 10 ppb; exposures must be maintained for 5 to 7 weeks for optimal control. For dewatered (drawdown) applications to dry canals or shorelines: 4 lb product (*Sonar*)/acre (2 lb a.e./acre), applied in 30 to 100 gal water/acre
	Timing: Apply to water in early spring to early summer. Dewatered applications should be made from fall to late winter, optimally when there is no standing water.
	Remarks: Fluridone is a slow-acting systemic herbicide that affects young, rapidly growing plants. Lower rates can be used if applied during early spring growth and when water movement is not likely to dilute or move the herbicide.
CONTACT PHOTOSYNTHETIC INHIBITORS	
Diquat *Reward*	**Rate:** For in-water applications: 0.1 to 0.25 ppm
	Timing: Apply to water in late spring to early summer. Diquat is a fast-acting contact herbicide that can be effective in mid- to late-summer, but if biomass is large, only a portion of the infested sites should be treated to minimize effects of reduced dissolved oxygen.
	Remarks: Diquat is quickly bound to, and becomes inactivated on, suspended clay particles and it should not be used in moderately or highly turbid water.
Flumioxazin *Clipper*	**Rate:** For in-water applications: 100 to 400 ppb
	Timing: Apply to water in early spring to early summer during rapid growth.
	Remarks: Flumioxazin is rapidly degraded and is inactive if pH exceeds 8.5. Use only if pH will not exceed 8.5. It

is best to apply in the early morning when pH is low.

GENERAL CELL TOXICANTS

Acrolein *Magnacide H*	**Rate:** For in-water applications: 1 to 15 ppm. Rate is variable and depends on target weeds, temperature and flow rates **Timing:** Apply to water in late spring to fall. No more than 8 applications are allowed per year. **Remarks:** Acrolein is a very fast-acting, nonselective contact herbicide and algaecide. It is a "Restricted Use" pesticide but can be used in some irrigation canals under specific conditions, with proper permits, and may only be applied by qualified, trained applicators. Symptoms of efficacy may appear in less than an hour and include discoloration of leaves and loss of turgidity.
Endothall *Cascade*; *Teton*; *Aquathol K*	**Rate:** For in-water applications: 1 to 3 ppm; exposures must be maintained for 24 to 48 hours or more for optimal control. The duration of contact depends on the concentration achieved (see label for specific rate and duration of exposure). **Timing:** Apply to water in early spring to early summer. Endothall can be used in mid-summer, but partial treatments are recommended if biomass is large to prevent large reduction in dissolved oxygen. **Remarks:** Endothall is a selective, contact herbicide. It affects young, rapidly growing plants and mature plants. Lower rates can be used if applied during early spring growth and when water movement is not likely to dilute or move the herbicide. Needs at least 4 hours of exposure time. Can mix with 0.5 ppm of copper (*Cleargate*) to also give good control. In moving water, an additional 'bump' treatment may be needed 3 to 4 miles from the main site of application to maintain a high enough concentration.

NON-HERBICIDAL CHEMICALS

Dyes or colorants *Aquashade*	Although technically not herbicides, dyes and colorants control submerged aquatic plants by absorbing light in the water column and reducing photosynthesis. Applications should be made in early spring and repeated to maintain the concentration recommended on the label. Colorants are not as effective on well-established plants in mid- to late summer.

Potamogeton natans L.; floatingleaf pondweed
Potamogeton nodosus Poir.; American pondweed

Floatingleaf and American pondweed (floating leaves)

Family: Potamogetonaceae
Range: Throughout much of North America, including all western states.
Habitat: Ponds, lakes, streams, rivers, reservoirs, irrigation ditches, marshy areas. They commonly grow in shallow water, but can grow to depths of ~20 ft or more in clear water. Floatingleaf pondweed also grows in brackish water.
Origin: Both species are widespread natives of North America and elsewhere.
Impacts: Both of these pondweeds provide habitat, and are an important food source, for wildlife. In natural areas, they are a desirable component of the aquatic community. However, colonies can be troublesome in drainage canals, irrigation ditches, and other controlled aquatic systems.

Floatingleaf and American pondweeds are creeping rhizomatous perennials with both submerged and floating leaves. The floating leaves are glabrous, waxy-shiny, elliptic to ovate, stalked, and leathery. Floatingleaf pondweed has a broader leaf compared to American pondweed.

Inflorescences consist of cylindrical spikes that are above the surface of the water. The flowers are greenish and inconspicuous. Plants are wind- and water-pollinated. Fruits are achene- or nutlet-like structures. Despite the prolific production of seeds, seedlings are seldom encountered. When seeds are produced, they float and disperse with water, are ingested by wildlife, or cling to the feet, fur, or feathers of animals. Seeds surviving ingestion by birds germinate readily. Most plants reproduce vegetatively from rhizomes or stem fragments. In late summer to fall, many pondweed species (e.g., *P. nodosus*, *P. gramineus*) form dozens of distinct vegetative propagules called "winter buds" from a few inches to more than a foot deep in the sediments at the tips of their rhizomes. These are usually dormant until the following spring. Seed longevity of other species is poorly documented, but may be similar to those of sago pondweed, which survive up to 1.5 years under dry conditions.

NON-CHEMICAL CONTROL

Mechanical (pulling, cutting, excavating)	Repeated mechanical harvesting can help reduce stem densities, but escaped stem fragments can drift elsewhere and develop into new plants. Removing and destroying stem fragments from recreational equipment, such as boat propellers, docking lines, and fishing gear can help prevent the spread of pondweeds. Several types of "bottom barriers' are available and are used to cover and smother specific infested areas. Materials used include polyvinyl chloride (pvc) sheets, small-mesh screens, and natural fibers such as jute. Bottom barriers are best installed in spring before plants produced large biomass and exceed 10 inches tall. In canals, backhoes and telescoop devices can be used to remove infestations, but these operations usually remove sediment as well. Because plants form winter buds, excavation depths should reach at least 12 inches.
Cultural	Dewatering (draining) canals or lakeshores in mid-summer may suppress subsequent growth, but plants can easily resprout from rhizomes if bottom sediments remain moist and cool. Winter drawdown alone will not impact growth since vegetative propagules are protected within the sediments.
Biological	The (sterile) triploid grass carp (white amur) is a relatively nonselective herbivorous fish that will consume most pondweed species readily. The fish do not selectively feed on "non-native" plants so careful monitoring of feeding impacts is necessary. In many canal systems non-selectivity is not a problem, as few or no plants are desired since they can interfere with efficient water delivery.

CHEMICAL CONTROL

The following specific use information is based on published papers and reports by researchers and land managers. Other trade names may be available, and other compounds also are labeled for this weed. Directions for use may vary between brands; see label before use. Herbicides are listed by mode of action and then alphabetically. The order of herbicide listing is not reflective of the order of efficacy or preference.

BRANCHED-CHAIN AMINO ACID INHIBITORS

Penoxsulam
Galleon

Rate: For in-water applications: 25 to 75 ppb; may be repeated but not to exceed 150 ppb in an annual season. For dewatered (drawdown) applications: 5.6 to11.2 oz product/acre (1.4 to 2.8 oz a.i./acre) to canal or shorelines where plants grow

Timing: Apply to water in early spring to early summer (rapid growth). For dewatered treatment, apply during mid- to late-winter before refilling.

Remarks: Penoxsulam is a slow-acting herbicide and may take 4 to 6 weeks to achieve control. For drawdown applications use 20 to 100 gal/acre of spray solution to wet the sediment.

Imazamox
Clearcast

Rate: For in-water applications: 50 to 100 ppb. For dewatered (drawdown) applications: 64 oz product/acre (8 oz a.e./acre); first flush of water in canals must not be used for irrigation

Timing: Apply to water in early spring to early summer (rapid growth). For dewatered treatment, apply in late winter at least 14 days before water will be reintroduced.

Remarks: Use an approved surfactant.

PIGMENT SYNTHESIS INHIBITORS

Fluridone
Sonar

Rate: For in-water applications: 5 to 10 ppb; exposures must be maintained for 5 to 7 weeks for optimal control. For dewatered (drawdown) applications to dry canals or shorelines: 4 pt product (*Sonar*)/acre (2 lb a.e./acre) applied with 30 to 100 gal water/acre

Timing: Apply to water in early spring to early summer. For dewatered treatment, apply from fall to late winter, optimally when there is no standing water.

Remarks: Fluridone is a slow-acting systemic herbicide that affects young, rapidly growing plants. Lower rates can be used if applied during early spring growth and when water movement is not likely to dilute or move the herbicide.

CONTACT PHOTOSYNTHETIC INHIBITORS

Flumioxazin
Clipper

Rate: For in-water applications: 100 to 400 ppb

Timing: Apply to water in early spring to early summer, during rapid growth.

Remarks: Flumioxazin is rapidly degraded and is inactive if pH exceeds 8.5. Use only if pH will not exceed 8.5. It is best to apply in the early morning when pH is low.

GENERAL CELL TOXICANTS

Acrolein
Magnacide H

Rate: For in-water applications: 1 to 15 ppm; rate is variable and depends on target weeds, temperature and flow rates

Timing: Apply to water in late spring to fall. No more than 8 applications are allowed per year.

Remarks: Acrolein is a very fast-acting, nonselective contact herbicide and algaecide. It is a "Restricted Use" pesticide but can be used in some irrigation canals under specific conditions, with proper permits, and may only be applied by qualified, trained applicators. Symptoms of efficacy may appear in less than an hour and include discoloration of leaves and loss of turgidity.

Endothall
Cascade;
Aquathol K

Rate: For in-water applications: 1 to 3 ppm; exposures must be maintained for 24 to 48 hours or more for optimal control (see labels for specific duration of contact and concentration needed).

Timing: Apply to water in early spring to early summer. Endothall can be used in mid-summer, but partial treatments are recommended if biomass is large in order to prevent large reduction in dissolved oxygen.

Remarks: Endothall is a selective, contact herbicide. It affects young, rapidly growing and mature plants. Lower rates can be used if applied during early spring growth and when water movement is not likely to dilute or move the herbicide.

Potentilla recta L.
Sulfur cinquefoil

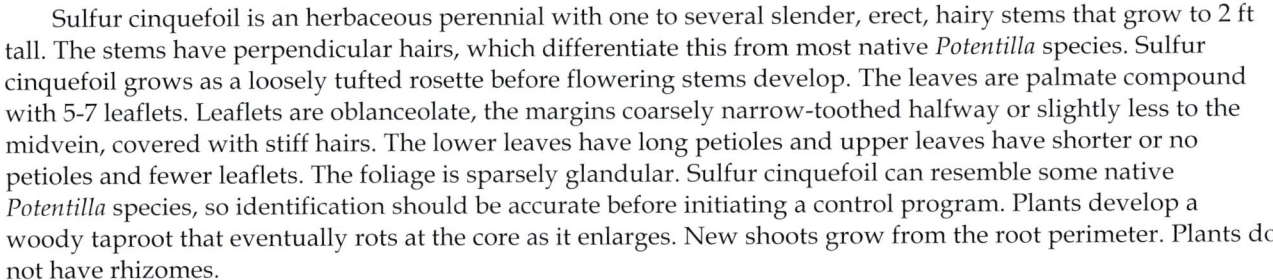

Family: Rosaceae
Range: Most western states except Arizona, New Mexico, and Utah.
Habitat: Open canopy forests, grasslands, and shrubby areas. Commonly associated with disturbed areas, although populations are also found in healthy plant communities.
Origin: Native to the eastern Mediterranean region of Eurasia.
Impacts: Sulfur cinquefoil is primarily a problem in pastures, rangeland, and wildlands. Little information is available on the ability of sulfur cinquefoil to compete with other plant species. Reports from Montana suggest it can become a significant component of the plant community and has become dominant on some sites. Sulfur cinquefoil canopy cover on most sites in Montana ranged from 5 to 15%. It is not readily grazed by livestock and wildlife.
Western states listed as Noxious Weed: Colorado, Montana, Nevada, Oregon, Utah, Washington

Sulfur cinquefoil is an herbaceous perennial with one to several slender, erect, hairy stems that grow to 2 ft tall. The stems have perpendicular hairs, which differentiate this from most native *Potentilla* species. Sulfur cinquefoil grows as a loosely tufted rosette before flowering stems develop. The leaves are palmate compound with 5-7 leaflets. Leaflets are oblanceolate, the margins coarsely narrow-toothed halfway or slightly less to the midvein, covered with stiff hairs. The lower leaves have long petioles and upper leaves have shorter or no petioles and fewer leaflets. The foliage is sparsely glandular. Sulfur cinquefoil can resemble some native *Potentilla* species, so identification should be accurate before initiating a control program. Plants develop a woody taproot that eventually rots at the core as it enlarges. New shoots grow from the root perimeter. Plants do not have rhizomes.

Sulfur cinquefoil bears many pale to sulfur-yellow flowers in open branched, flat-topped inflorescences. Flowers are about an inch in diameter with five petals. Plants may produce over 1,500 seeds. Most mature seed is dispersed near the parent plant. In a lab experiment, seeds remained viable after 28 months of burial at 3 inches deep in the soil. Other reports suggest seed longevity may exceed 3 to 4 years.

NON-CHEMICAL CONTROL

Mechanical (pulling, cutting, disking)	Small infestations can be removed by manual methods such as hand-pulling and digging. Remove the entire root crown to prevent resprouting.
	Plants regrow after mowing. Mowing is not recommended as it can stimulate lateral branching and root growth.
	Plowing and deep disking can control emerged plants but can stimulate recruitment. These types of tillage are not typically practical in most natural areas. Reseeding with other species following tillage resulted in complete control in one study.
Cultural	Sulfur cinquefoil has a tannin content of 17 to 22% dry weight and has low palatability. Utilization by cattle was less than 1% on 83 of 85 sites studied in Montana. There are no reports of toxicity.
	Little information was found on fire or burning effects. In one study, fire did not result in mortality of large plants. The study also suggested fire may enhance recruitment.
	Sulfur cinquefoil is intolerant of shade, thus land management strategies should promote competitive vegetation and shading.
Biological	No biological controls are available for the control of sulfur cinquefoil in the United States. This is likely due to the high number of native species within the *Potentilla* genus. Several insects and fungi associate with sulfur cinquefoil, including root and crown boring insects and fungi that cause crown rot.

CHEMICAL CONTROL
The following specific use information is based on published papers and reports by researchers and land managers. Other trade names may be available, and other compounds also are labeled for this weed. Directions

for use may vary between brands; see label before use. Herbicides are listed by mode of action and then alphabetically. The order of herbicide listing is not reflective of the order of efficacy or preference.

GROWTH REGULATORS	
2,4-D Several names	**Rate:** 1 to 2 qt product/acre (0.95 to 1.9 lb a.e./acre) **Timing:** Postemergence when plants are in the pre-bud stage. **Remarks:** 2,4-D is broadleaf-selective and safe on most grasses. It has minimal soil activity. Repeat application is usually required. Do not apply ester formulation when outside temperatures exceed 80°F.
Aminocyclopyrachlor + chlorsulfuron *Perspective* Aminocyclopyrachlor + metsulfuron *Streamline*	**Rate:** 4.75 to 8 oz product (*Perspective*)/acre or 4.75 to 9 oz product (*Streamline*)/acre **Timing:** Postemergence when plants are in the pre-bud stage. **Remarks:** These products provide broad-spectrum control of many broadleaf species. Although generally safe to grasses, they may suppress or injure certain annual and perennial grass species. Do not treat in the root zone of desirable trees and shrubs. Do not apply more than 11 oz product/acre per year. At this high rate, cool-season grasses will be damaged, including bluebunch wheatgrass. Not yet labeled for grazing lands. Add an adjuvant to the spray solution. These products are not approved for use in California and some counties of Colorado (San Luis Valley).
Aminopyralid *Milestone*	**Rate:** 4 to 7 oz product/acre (1 to 1.75 oz a.e./acre) **Timing:** Postemergence when plants are in spring rosette to pre-bud stage. **Remarks:** Aminopyralid is safe on most grasses, although preemergence application at high rates can greatly suppress some annual grasses, such as medusahead. Applications can decrease seed production in some annual and perennial grass species. For postemergence applications, adding a non-ionic surfactant (0.25 to 0.5% v/v spray solution) enhances control under adverse environmental conditions or when weeds are heavily pubescent. Other premix formulations of aminopyralid can be used. These include *Opensight* (aminopyralid + metsulfuron) at 2 to 2.5 oz product/acre and *Forefront HL* (aminopyralid + 2,4-D) at 1.2 to 1.5 pt product/acre. *Opensight* is not registered for use in California.
Picloram *Tordon 22K*	**Rate:** 1 pt product/acre (4 oz a.e./acre) **Timing:** Postemergence when plants are in the pre-bud stage or to fall regrowth. **Remarks:** Picloram controls a wide range of broadleaf species and has relatively long soil residual activity. Although well-developed grasses are not usually injured by labeled use rates, some applicators have noted that young grass seedlings with fewer than four leaves may be killed. Do not apply near trees, or where soil is highly permeable and where water table is high. Picloram is a restricted use herbicide. Picloram is not registered for use in California.
Triclopyr *Garlon 4 Ultra*	**Rate:** 1 to 2 pt *Garlon 4 Ultra* product/acre (0.5 to 1 lb a.e./acre) **Timing:** Postemergence when plants are in the rosette stage. **Remarks:** Triclopyr controls several woody plants and broadleaf forbs, but is safe on most grass species. Add 0.25 to 0.5% non-ionic surfactant.
AROMATIC AMINO ACID INHIBITORS	
Glyphosate *Roundup, Accord XRT II,* and others	**Rate:** Broadcast treatment: 1 to 2 qt product (*Roundup ProMax*)/acre (1.1 to 2.25 lb a.e./acre). Spot treatment: 1.5% v/v solution **Timing:** Postemergence when plants are in the pre-bud stage. **Remarks:** Glyphosate will not injure or kill germinating seedlings because it has no soil activity. It is, however, nonselective; overspray will kill most non-target plants, creating bare ground conditions that are susceptible to weed recruitment. In areas with desirable vegetation, use spot treatment. Glyphosate is a good control option if reseeding is planned shortly after application, as it will not injure seedlings emerging after application. Add a surfactant when using a formulation where it is not already included (e.g., *Accord XRT II, Rodeo, Aquamaster*).
BRANCHED-CHAIN AMINO ACID INHIBITORS	
Chlorsulfuron *Telar*	**Rate:** 1 to 2.6 oz product/acre (0.75 to 1.95 oz a.i./acre) **Timing:** Postemergence in the rosette stage. **Remarks:** Always use a surfactant. Chlorsulfuron is effective on other species of *Potentilla*, and

	although sulfur cinquefoil is not on the label, it is expected to have activity.
Metsulfuron *Escort*	**Rate:** 1 to 2 oz product/acre (0.6 to 1.2 oz a.i./acre) **Timing:** Postemergence in the rosette stage. **Remarks:** Metsulfuron has mixed selectivity, but is considered fairly safe on most grasses. Metsulfuron has some soil residual activity. Always use a surfactant. Other premix formulations of metsulfuron can be used at similar application timing. These include *Cimarron Max* (metsulfuron + dicamba + 2,4-D) and *Cimarron X-tra* (metsulfuron + chlorsulfuron). Metsulfuron is not registered for use in California.
PHOTOSYNTHETIC INHIBITORS	
Hexazinone *Velpar L*	**Rate:** 4 to 7 qt product/acre (2 to 3.5 lb ai/acre) **Timing:** Preemergence in fall. **Remarks:** Hexazinone is listed on the label as an effective control option for *Potentilla* species in blueberry, and it is expected to also control sulfur cinquefoil in natural areas. Can give total vegetation control, so generally should be used in a spot treatment at the base of target plants. Mobile in the soil. High rates of hexazinone can create bare ground, so only use high rates in spot treatments.

Raphanus sativus L.; radish
Raphanus raphanistrum L.; wild radish

Radish and wild radish

Family: Brassicaceae
Range: Most western states, with the exception of desert regions.
Habitat: Although they are typically weeds of cultivated crops, orchards, vineyards, and neglected gardens, they can also be common in parks, roadsides, and disturbed locations in natural areas.
Origin: Both species are native to eastern Europe and Asia.
Impact: The weedy radishes can cause significant yield losses in crops, due to their quick establishment, fast growth rate, allelopathy, and strong competitiveness. In non-crop areas they can be toxic to livestock if consumed in large amounts.
California Invasive Plant Council (Cal-IPC) Inventory: *R. sativus*, Limited Invasiveness

Wild radishes are winter or summer annuals, occasionally biennials. Mature plants are erect and up to 2 ft in height. Their cotyledons are distinctively kidney-shaped and 0.5 to 1 inch long. Plants form rosettes until the flower stem develops at maturity. The lower leaves are alternate and vary in size and shape from being deeply lobed and ovate to having leaflets, but are usually covered with stiff, flattened hairs. The lobed leaves are usually 2.3 to 8 inches long and have irregularly round edges or sharp toothed edges. The taproot can be more than 3 ft deep and provides the plant with access to water during dry periods as well as an underground energy reserve.

At maturity, several flowering stems develop. Flowering occurs from spring through summer. The flowers have four petals and can be white, yellow, pink, or purple. Radishes reproduce only by seed, producing yellow-brown pods that each contains 1 to 10 seeds. Seeds are not dispersed from the fruit and primarily fall to the base of the parent plant. The seeds have a long dormancy and can stay viable in the soil for several years.

NON-CHEMICAL CONTROL

Mechanical (pulling, cutting, disking)	Hand-pull, removing most of the root system, before plants produce seed. Hand weeding may need to be repeated to control later developing plants.
	Mowing can help reduce seed production but does not harm the basal leaves, thus allowing plants to regrow. Repeated mowing is required to prevent seed set. This is not an effective means of control.
	Tillage is a common and effective method of control in agricultural areas and would also be effective, if practical, in natural areas or other non-crop sites.
Cultural	Maintain competitive grasses and avoid overgrazing.
	Radish is a very early season flowering plant, so burning is not practical for its control.
Biological	Redlegged earth mite, cabbage moth, thrips, Rutherglen bug, and white Italian snail attack wild radish but do not provide sufficient control.

CHEMICAL CONTROL

The following specific use information is based on published papers and reports by researchers and land managers. Other trade names may be available, and other compounds also are labeled for this weed. Directions for use may vary between brands; see label before use. Herbicides are listed by mode of action and then alphabetically. The order of herbicide listing is not reflective of the order of efficacy or preference.

GROWTH REGULATORS	
2,4-D Several names	**Rate:** 1 to 2 pt product/acre (0.48 to 0.9 lb a.e./acre) **Timing:** Postemergence. Apply before budding when plants are small and rapidly growing. **Remarks:** 2,4-D is a broadleaf herbicide with no soil activity.
Dicamba *Banvel, Clarity*	**Rate:** 0.25 to 1 pt product/acre (2 to 8 oz a.e./acre) **Timing:** Postemergence. Apply before budding when plants are small and rapidly growing. **Remarks:** Dicamba is a broadleaf-selective herbicide often combined with other active ingredients. *Overdrive*, a premix of dicamba with diflufenzopyr, has been reported to be effective on wild radish. Diflufenzopyr is an auxin transport inhibitor which causes dicamba to accumulate in shoot and root meristems, increasing its activity. *Overdrive* is applied at 4 to 8 oz product/acre, postemergence to rapidly growing plants. Higher rates should be used on large annuals. Add a non-ionic surfactant to the treatment solution at 0.25% v/v or a methylated seed oil at 1% v/v solution.
BRANCHED-CHAIN AMINO ACID INHIBITORS	
Imazapic *Plateau*	**Rate:** 4 to 6 oz product/acre (1 to 1.5 oz a.i./acre) **Timing:** Either preemergence or postemergence before budding when plants are small and rapidly growing. **Remarks:** Applying preemergence provides suppression, while applying postemergence at a rate of 4 oz product/acre provides good control. Imazapic is not registered for use in California.
Rimsulfuron *Matrix*	**Rate:** 2 to 4 oz product/acre (0.5 to 1 oz a.i./acre) **Timing:** Preemergence or postemergence when the target weeds are 1 to 3 inches in height. **Remarks:** Rimsulfuron controls several annual grasses and broadleaves. Perennial grasses are tolerant to fall applications when established and grown under dryland conditions. Application to rapidly growing or irrigated perennial grasses may result in their injury or death. It provides soil residual control in cool climates but degrades rapidly under warm conditions. Rimsulfuron will not control summer annual weeds when applied in fall or spring. Add a surfactant when applying postemergence. For preemergence activity, the herbicide must be activated by rainfall or irrigation of at least half an inch. For the best results, rain should occur within 2 to 3 weeks of application and during cooler temperatures. Do not apply more than 4 oz product/acre per year.
PHOTOSYNTHETIC INHIBITORS	
Hexazinone *Velpar L*	**Rate:** 2 to 4 pt product/acre (0.5 to 1 lb a.i./acre) **Timing:** Either preemergence or postemergence. **Remarks:** Hexazinone is used as a nonselective herbicide in non-cropland areas and as a selective herbicide in reforestation practices. High rates of hexazinone can create bare ground, so only use high rates in spot treatments.

Ricinus communis L.

Castorbean

Family: Euphorbiaceae
Range: Although more common in the southeastern U.S., it can be problematic in California, Arizona and Utah.
Habitat: Roadsides, fields, riparian areas, and disturbed waste places. Grows best when moisture is plentiful, but tolerates considerable drought. Plants are highly susceptible to frost damage and grow as an annual in cold winter areas.
Origin: Native to tropical Africa and Eurasia. Some varieties of castorbean are cultivated as ornamentals and for the oil contained in the seeds.
Impacts: Seeds and, to a lesser degree, foliage are highly toxic to humans and animals when ingested. Plant material contains an extremely toxic protein, ricin. Ingestion of 4 to 8 seeds can kill an adult, and fewer can kill a child. Seeds must be broken or chewed to release the toxin. Castor oil derived from the seeds does not contain the water-soluble toxin. Crushed foliage has a disagreeable odor and is usually avoided by animals. Livestock poisonings most often occur when feed is contaminated with castorbean seeds. Handling foliage and seeds can cause severe contact dermatitis in sensitive individuals.
California Invasive Plant Council (Cal-IPC) Inventory: Limited Invasiveness

Castorbean is a summer annual, perennial, shrub, or small tree to 10 ft tall or more. The leaves are alternate and large, shield-like, with 5 to 11 palmate lobes. The stems are hollow, often tinged red. The plant lacks hairs.

Male and female flowers develop on the same plant (monoecious). Flower clusters are terminal and panicle-like, with female flowers above the male flowers. All flowers lack petals. The fruit are capsules covered with soft spines, usually gray-green, but occasionally purplish or red. The seeds are oblong, 10 to 14 mm long, smooth, glossy, and mottled silver and brown. Plants reproduce by seed that disperse short distances when capsules snap open at maturity. On the soil surface seeds can be moved a short distance by ants. They can also disperse to greater distances in water or by human activities. Castorbean does not form persistent seedbanks except when buried by tillage.

NON-CHEMICAL CONTROL

Mechanical (pulling, cutting, disking)	For individual plants, it is possible to remove by digging. Smaller plants can be hand pulled, although it is important to wear gloves, as castor bean is poisonous. Even larger plants can be removed by hand pulling in wet sandy soils. It is important to remove the bulk of the root crown to prevent regeneration. Weed wrenches can be a very effective tool for small to medium-sized plants with a single stem. This is much more effective in moist soils. In dry soils the plant can break off and leave the root system. Removed larger plants can be piled and burned on site.
	With larger colonies, where cultivation is practical, slashing followed by cultivation gives effective control. It is essential to keep the cultivation shallow to prevent deep burial of seeds. Repeat cultivation as needed to kill seedlings and any regrowth from old crowns.
Cultural	Neither burning nor grazing is recommended for castor bean control. Burning in coastal areas can lead to ideal conditions for habitat conversion. Seeds in or on the soil readily germinate after fire, and seedlings grow so rapidly that they outcompete other species, dominating the area and driving out desirable natives. Animals do not find castor bean palatable and, in fact, the plant is poisonous.
Biological	Castor bean is a crop in some locations in the United States. As such, there are no efforts to develop a biological control program. However, a large number of diseases and pests are known to impact castor bean crops. Mung moth, pink bollworm, scab, wilt, leaf spot, seedling blight, inflorescence rot, pod rot, rust spot, graymold, crown rot, stem canker, leaf blight, bacterial wilt, and angular leaf spots can impact castor bean crops, but are rarely seen in riparian wildland plants.

CHEMICAL CONTROL

The following specific use information is based on published papers and reports by researchers and land managers. Other trade names may be available, and other compounds also are labeled for this weed. Directions for use may vary between brands; see label before use. Herbicides are listed by mode of action and then alphabetically. The order of herbicide listing is not reflective of the order of efficacy or preference.

GROWTH REGULATORS	
2,4-D Several names	**Rate:** 2 qt product/acre (1.9 lb a.e./acre) **Timing:** Postemergence when plants are growing actively, thoroughly wetting leaves and stems. **Remarks:** 2,4-D is a broadleaf-selective herbicide. It has no soil activity. When using the ester formulation do not apply when outside temperatures exceed 80°F.
Dicamba *Banvel, Clarity*	While not typically used in the United States, cut stump treatments with dicamba have been reported to be effective in Australia.
Picloram *Tordon 22K*	Picloram is not typically used for castorbean control, but can be used in a cut stump treatment when combined with 2,4-D or triclopyr. This technique is sometimes used effectively with large plants in Australia. Picloram is a restricted use herbicide. It is not registered for use in California.
Triclopyr *Garlon 3A,* *Garlon 4 Ultra,* *Pathfinder II*	**Rate:** Foliar spot treatment: 1% v/v solution of *Garlon 3A* or *4*. Basal bark treatment: 20% v/v solution of *Garlon 4 Ultra* or undiluted *Pathfinder II*. Cut stump treatment: 100% v/v solution *Garlon 3A*, or 20% *Garlon 4 Ultra* with 80% basal oil. Basal cut stump treatment: 20% Garlon *4 Ultra* with 80% basal oil. **Timing:** Postemergence, to seedlings or to fully developed leaves of mature plants. Where castorbean grows as a perennial, cut stump or basal bark treatments should be made in the late season when plants are translocating carbohydrates to below-ground tissues. For cut stump treatments, applications must be made immediately after cutting. **Remarks:** For basal bark treatments, apply with a backpack sprayer using low pressure and a solid cone or flat fan nozzle. Spray the lower stems, including the root collar, to a height of 12 to 15 inches from the ground; the spray should thoroughly wet the lower stem but not to the point of runoff. Cut stump treatments should be used for larger diameter plants. Treated sites must be revisited several times after the initial herbicide treatment to hand pull seedlings that sprout. Triclopyr is a broadleaf herbicide and will kill other sensitive broadleaf species it contacts. *Garlon 4 Ultra* is a low volatile ester. However, in warm temperatures, spraying onto hard surfaces such as rocks or pavement can increase the risk of volatilization and off-target damage.
AROMATIC AMINO ACID INHIBITORS	
Glyphosate *Roundup, Accord XRT II*, and others	**Rate:** Foliar broadcast treatment to seedlings: 1.5 pt product (*Roundup ProMax*)/acre (0.85 lb a.e./acre). Spot foliar treatment: 2% v/v solution for spray-to-wet applications. Cut stump treatment: 25% v/v solution. **Timing:** Postemergence, to seedlings or to fully developed leaves of mature plants. **Remarks:** Glyphosate is a nonselective herbicide. Stem injection and cut stump applications can be made to large trees and shrubs. These techniques can reduce risk of non-target damage. Cutting can be accomplished with loppers or small saws.
PHOTOSYNTHETIC INHIBITORS	
Hexazinone *Velpar L*	**Rate:** 4 to 7 qt product/acre (2 to 3.5 lb a.i./acre) **Timing:** Preemergence to base of plant while plants are growing rapidly and when subsequent precipitation can move chemical into root zone. **Remarks:** In Australia, castorbean has been successfully controlled with a preemergence treatment of hexazinone. It can give total vegetation control, so generally should be used in a spot treatment at the base of target plants. Hexazinone is mobile in the soil. High rates of hexazinone can create bare ground, so only use high rates in spot treatments.

Robinia pseudoacacia L.
Black locust

Family: Fabaceae
Range: Most contiguous states, except Arizona.
Habitat: Disturbed places, sites near old habitations, roadsides, landscaped sites, and many natural communities, including riparian areas, canyon slopes, mixed conifer forests, floodplain forests and woodlands. Grows best on well-drained, neutral to mildly acidic, limestone soils. Tolerates some drought, fire, and temperatures as low as -33°F. Does not tolerate shade, salinity, or extended periods of excessive soil moisture.
Origin: Native to the eastern United States. Black locust was widely planted in the United States as a landscape ornamental and for its hard, rot-resistant wood. In many areas of the country, black locust has escaped cultivation and become invasive.
Impacts: Black locust produces numerous suckers from the roots, and forms dense clonal colonies that exclude native vegetation. Foliage, seeds, and especially bark are toxic to humans and livestock when ingested, but rarely fatal. Flowers are reported as non-toxic to slightly narcotic. Black locust flowers attract bees, but the seeds and foliage are poorly used by wildlife.
California Invasive Plant Council (Cal-IPC) Inventory: Limited Invasiveness

Black locust is a fast-growing deciduous tree to nearly 80 ft tall, with pinnate-compound leaves, highly fragrant, white, pea-like flowers, and creeping roots that sucker readily. The bark is light gray-brown, rough, and deeply furrowed. Some trees may have a pair of stout spines 0.5 to 1 inch long at the base of the leaf. The leaflets are ovate to oblong with a smooth margin and a minute bristle at the tip. Some cultivars lack spines on the twigs or have pinkish flowers.

Trees reproduce vegetatively from root sprouts and by seed. Reproduction by root suckers is more prevalent in natural areas. Suckers usually appear from both stump and roots in the fourth or fifth year. Suckering can be extensive especially after being cut or damaged. Seeds are produced in flat pods, light brown to nearly black, 2 to 4 inches long. The hard-coated seed can remain dormant for 10 years or more. Seeds typically germinate in spring and do not tolerate shade or compete well with established vegetation. Seedlings that establish on favorable sites can grow up to 3 ft per year. Saplings generally begin to fruit at 6 years of age.

NON-CHEMICAL CONTROL

Mechanical (pulling, cutting, disking)	There are no mechanical strategies effective for the control of mature black locust. Hand pulling can remove seedlings, but once underground creeping rhizomes have developed, this technique is generally not effective.
	Cutting or girdling will result in prolific root suckering. Repeated cutting of sprouts over multiple years may provide some control.
Cultural	A heavily shaded environment will reduce the establishment of black locust.
Biological	Black locust suffers considerable damage from insects, particularly the black locust borer, *Megacylline robinine*. However, no USDA biological control program for black locust has been attempted, and no USDA-approved biocontrol agents exist for this species.

CHEMICAL CONTROL

The following specific use information is based on published papers and reports by researchers and land managers. Other trade names may be available, and other compounds also are labeled for this weed. Directions for use may vary between brands; see label before use. Herbicides are listed by mode of action and then alphabetically. The order of herbicide listing is not reflective of the order of efficacy or preference.

GROWTH REGULATORS	
Aminopyralid *Milestone*	**Rate:** Broadcast foliar treatment: 7 oz product/acre (1.75 oz a.e./acre). Foliar spot treatment: 0.05 to 0.8% v/v solution. Total herbicide should not exceed 7 oz product/acre or 14 oz product/acre on half of the target area. Use higher rates for low volume treatments and lower rates when thoroughly

	wetting leaves. Cut stump treatment: 10% v/v *Milestone* in water. Basal bark treatment: 0.5 to 5% v/v *Milestone* in 95-99.5% oil carrier. Stem injection treatment: one cut per every 3 inches of stem diameter, and 1 ml of 10% v/v *Milestone* in water added to each cut. For clumps, one hack per every 6 inches of total stem diameter. Treat the largest stems. **Timing:** Postemergence foliar treatments are best when leaves are fully expanded. For cut stump, basal bark, and stem injection treatments, applications can be made at most times of the year, except when ground is frozen, but are best used in late summer or early fall, before leaf drop. **Remarks:** Aminopyralid provides preemergent and postemergent control of susceptible weeds, and will not damage desirable grasses growing nearby. Cut stump and basal bark applications are made as described for triclopyr. For stem injection treatments, be sure that each cut goes well into or below the cambium layer. Trees should not be cut for at least 1 month after basal bark and stem injection treatments. Aminopyralid can be used in a premix combination with triclopyr (*Capstone*) at 6 to 9 pt product/acre.
Clopyralid *Transline*	**Rate:** 1 to 1.3 pt product/acre (6 to 8 oz a.e./acre) **Timing:** Postemergence foliar treatments are best when leaves are fully expanded. **Remarks:** A broadleaf herbicide like picloram, but more selective. Very safe on grasses but will kill most legumes and many members of the Asteraceae. It has a relatively short residual activity.
Picloram *Tordon 22K*	**Rate:** Foliar spot treatment: 3% v/v solution provides good control of stems but vigorous suckers can develop. Efficacy is improved with the addition of 3 to 5 qt of *Garlon 4 Ultra* or 4 to 8 qt of *Garlon 3A*/acre. **Timing:** Postemergence at the end of summer to beginning of fall, but before leaf drop. **Remarks:** High rates of picloram can give long-term soil residual control for broadleaves. Picloram is a restricted use herbicide. It is not registered for use in California.
Triclopyr *Garlon 3A, Garlon 4 Ultra, Pathfinder II*	**Rate:** Low volume foliar treatment: 5% v/v solution of triclopyr and water plus 0.5% surfactant v/v to thoroughly wet all leaves. Cut stump treatment: 25% *Garlon 4 Ultra* in 75% oil carrier, or undiluted *Garlon 3A* or 50% *Garlon 3A* in water. Basal bark treatment: 20% *Garlon 4 Ultra* in 80% oil carrier, or *Pathfinder II* as a ready to use formulation. Basal cut stump treatment: 25% *Garlon 4 Ultra* in 75% oil carrier. **Timing:** Cut stump and basal bark treatments can be applied as long as the ground is not frozen, but are most effective in late summer or early fall, before leaf drop. **Remarks:** Triclopyr is a selective herbicide for broadleaf species and will not damage desirable grasses growing nearby. For cut stump treatments, cut stems horizontally at or near ground level, and immediately apply herbicide solution to cover the outer 20% of the stump face. Suckering from the roots typically occurs after cutting, but the treatment should control most resprouts. For basal bark treatment, spray the lower trunk, including the root collar, to a height of 12 to 15 inches from the ground; the spray should thoroughly wet the lower stem but not to the point of runoff. Trees should not be cut for at least 1 month following basal bark treatment.
AROMATIC AMINO ACID INHIBITORS	
Glyphosate *Roundup, Accord XRT II*, and others	**Rate:** Foliar treatment: 1 to 1.5% v/v solution of *Roundup ProMax* (or other trade name with similar concentration of glyphosate) to thoroughly wet all leaves. Low volume spot treatment: 4 to 7% v/v solution of *Roundup* (or other trade name) to wet 50% of the leaves. Cut stump treatment: undiluted *Roundup* (or other trade name) or 50% v/v in water. **Timing:** Postemergence foliar treatments are best when leaves are fully expanded. Suckering from the roots might occur the following year. For cut stump treatment, apply immediately after cutting; application in late summer, early fall or dormant season provides best control. **Remarks:** Glyphosate is a nonselective systemic herbicide. It gives good control with some resprouts. Trees should not be cut for at least 4 months after foliar treatments.
BRANCHED-CHAIN AMINO ACID INHIBITORS	
Imazapyr + glyphosate *Arsenal, Habitat* or *Polaris* + *Roundup* or other trade name Imazapyr + metsulfuron	**Rate:** Low volume foliar treatment: 0.5 to 1% *Arsenal* + 2 to 3% *Rodeo* or 2.3 g/gallon *Escort* (0.08 oz/gal or 0.06% solution) **Timing:** Postemergence. Best when used in late summer to early fall, but before leaf drop. **Remarks:** Imazapyr is a soil residual herbicide and may result in bare ground around trees for some time after treatment. Metsulfuron is not registered for use in California.

Arsenal, Habitat or *Polaris + Escort*	
PHOTOSYNTHETIC INHIBITORS	
Hexazinone *Velpar L*	**Rate:** Broadcast or basal-soil (to the soil beneath the target plant) treatment: 5 to 10 qt product/acre (1.25 to 2.5 lb a.i./acre) **Timing:** Apply to the soil below the target plants from late winter through summer. **Remarks:** Hexazinone is a residual herbicide applied as a broadcast or basal-soil treatment for brush control. High rates of hexazinone can create bare ground, so only use high rates in spot treatments.
Tebuthiuron *Spike 80DF*	**Rate:** Broadcast treatment: 2.5 lb product (*Spike 80DF*)/acre (2 lb a.i./acre) **Timing:** Soil treatments can be applied anytime except when the soil is frozen or saturated with moisture. Applications should be made before the start of spring growth or before expected seasonal rainfall. **Remarks:** Tebuthiuron is a surface applied, soil-active product intended for total vegetation control in non-cropland. For best control, do not disturb plants for 2 years after application.

Rosa eglanteria L.; sweetbriar rose
  (=*Rosa rubiginosa* L. [Jepson Manual 2012])
Rosa canina L.; dog rose
Rosa multiflora Thun. ex Murr.; multiflora rose

Sweetbriar, dog and multiflora roses

Photo courtesy of Richard Old

Family: Rosaceae
Range: Sweetbriar rose is reported from all western states except Nevada, New Mexico and Arizona. Dog rose is reported in British Columbia, Washington, Idaho, Oregon, Utah and California. Multiflora rose has been reported in British Columbia, Washington, Oregon, California and New Mexico. The only known location in Idaho was eradicated.
Habitat: Sweetbriar rose is present in grass and shrublands, while dog rose is found in riparian areas within grasslands and shrublands, within the inland northwestern United States. Multiflora rose occurs in successional fields, pastures, and roadsides. It also may occur in dense forests, particularly near disturbances, such as treefall gaps.
Origin: Sweetbriar and dog roses are both native to Europe, northern Africa and western Asia. Multiflora rose was introduced from Japan, Korea, and eastern China in 1886 as rootstock for ornamental roses. In the 1930s it was widely promoted as a "living fence" for soil conservation and in wildlife programs.
Impacts: Sweetbriar rose can form dense stands that exclude livestock and larger grazing animals from grasslands and shrublands because of their curved thorns. Dog rose can also form dense thickets in grassland riparian areas. Multiflora rose forms dense thickets on disturbed sites, roadsides and improperly grazed pastures.
Western states listed as Noxious Weed: *Rosa multiflora*, South Dakota

Sweetbriar rose develops distinct shrubs up to 9 ft tall, even when in dense thickets. Dog rose can have less distinct individuals in riparian areas and can grow to 9, and occasionally 12 ft, tall. Sweetbriar and dog rose both have large thorns that curve towards the base of the stem. Both also have odd-pinnate leaves, with 5 to 9 leaflets for sweetbriar and 5 to 7 leaflets for dog rose. Unlike the other weedy roses, sweetbriar rose has gland-tipped hairs on the underside of leaves.

The flowers of sweetbriar and dog rose are white to pink and have 5 petals in a single whorl. The stamens and pistils are numerous. The sepals of sweetbriar rose have glandular hairs, but dog rose either has no hairs or only a few glandular hairs on its sepals. The flowers of both species can be solitary or in clusters on the branches.

Multiflora rose grows to 9 ft tall and can form dense thickets. It also has curved thorns that make the thickets impenetrable. The leaves are odd-pinnate with an average of 9 leaflets (range of 5 to 11) that lack hairs on the top of the leaf but are pubescent on the underside. The flowers of multiflora rose are white to pink, about 1 inch across, and arranged in clusters. The sepals on the flowers are pinnately lobed and the lobes are fringe-like.

Roses reproduce by seed and vegetatively from stems that root at the tips when in contact with the soil. Hips remain on plants through winter and into the following spring. Animals sometimes consume the fruits and disperse the achenes. Unconsumed fruits become leathery and eventually fall off. Ingestion by animals probably enhances germination. It is not known how long the seedbank can survive.

NON-CHEMICAL CONTROL

Mechanical (pulling, cutting, disking)	A weed wrench can be used to remove smaller plants. Removal should be conducted when soil is moist and care should be taken because of the sharp spines (technically prickles).
	A single mowing or cutting to the base will not control any of these rose species, as they all are capable of resprouting from the base. If 3 to 6 mowings are done each year for 2 to 4 years, control may be possible. After stems have died from an herbicide application, mowing will allow site access to livestock or wildlife.
	Repeated cultivation effectively controls these species.
Cultural	Burning may remove canes but is not considered an effective control option as shrubs will resprout from the base following the burn.
	Competitive grass communities may slow establishment of new shrubs but likely will not prevent increasing shrub densities.
Biological	Because of the commercial importance of roses and the numerous native rose species in the western United States, there are no biological control agents available.

CHEMICAL CONTROL

The following specific use information is based on published papers and reports by researchers and land managers. Most of the information was obtained for the control of multiflora rose, but it is expected that these methods would also be effective on the other weedy roses. These are the products that provide effective control. Those that do not provide sufficient control have been omitted from the table. Other trade names may be available, and other compounds also are labeled for this weed. Directions for use may vary between brands; see label before use. Herbicides are listed by mode of action and then alphabetically. The order of herbicide listing is not reflective of the order of efficacy or preference.

GROWTH REGULATORS	
Aminopyralid *Milestone*	**Rate:** 7 oz product/acre (1.75 oz a.e./acre) or with 2,4-D as 1.5 pt/acre of *Forefront HL* herbicide with 1 to 2 pt/acre of *Garlon 4 Ultra/Remedy Ultra* herbicide
	Timing: Postemergence, to plants after full leave expansion, generally around the flowering period. Preemergence control can also be achieved with seedlings.
	Remarks: Aminopyralid is a broadleaf-selective herbicide and will injure or kill most members of the Asteraceae and Fabaceae.
Aminopyralid + metsulfuron *Opensight*	**Rate:** 3.3 oz product/acre
	Timing: Applications can be made from spring leaf development through plant senescence in fall; optimum timing is during flowering.
	Remarks: This herbicide combination is not registered for use in California.
Dicamba *Banvel, Clarity*	**Rate:** Foliar spot treatment: 2% v/v solution with a 0.5% non-ionic surfactant, thoroughly wet leaves
	Timing: Postemergence, to rapidly growing leaves and before killing frost in fall. Best applied when plants are flowering.
	Remarks: Dicamba is selective against broadleaf plants and should not be used if desirable broadleaf vegetation is present.
Picloram *Tordon 22K*	**Rate:** 1 qt/acre (0.5 lb a.e./acre). Picloram can also be mixed with 4 pt 2,4-D (2 lb a.e./acre), use 0.5% v/v non-ionic surfactant
	Timing: Postemergence, to fully expanded leaves in late spring to summer.
	Remarks: Picloram can also be used alone but is best applied in combination with 2,4-D or fluroxypyr (*Surmount*). Picloram and the combination can injure or kill a wide range of species. It is well suited for control of dense thickets with few non-target species. Picloram is a restricted use herbicide. Picloram and its formulations are not registered for use in California.
Triclopyr *Garlon 3A, Garlon 4 Ultra*	**Rate:** Spot treatment (spray-to-wet): 2% v/v solution plus 0.5% non-ionic surfactant. Basal bark treatment: 25% *Garlon 4 Ultra* in a weed or basal oil. Cut stem treatment: 25% *Garlon 3A* in water; apply to cut surfaces immediately after cutting.
	Timing: Treatments can be made to foliage, stems, or cut stems. For foliar treatments, the best timing is when leaves are fully expanded. For basal bark application, treatments should be made to dormant plants in fall after frost. Cut stem applications are best in fall.
	Remarks: Triclopyr is a selective herbicide for broadleaf species and will not damage desirable grasses

growing nearby. Fall applications during the dormant period can reduce effects on non-target species. Triclopyr can also be used in a premix with fluroxypyr (*PastureGard*) at 1.5 to 4 pt product/acre or in a tank mix with aminopyralid (*Milestone*) at 7 oz product/acre.

AROMATIC AMINO ACID INHIBITORS

Glyphosate
Roundup, Accord XRT II, and others

Rate: Spot treatment: 1 to 2% v/v solution of *Roundup ProMax* or similar product plus 0.5% non-ionic surfactant; thoroughly wet all leaves. Cut stem treatment: 25 to 50% v/v product in water; apply to cut surfaces immediately after cutting.

Timing: Both foliar and cut stem treatments are best applied when leaves are fully expanded, especially when the flowers are in full bloom.

Remarks: Glyphosate can also be mixed with dicamba to achieve good control with foliar applications. Glyphosate is a nonselective systemic herbicide with no soil activity. Use a low pressure and coarse spray pattern to reduce spray drift damage to non-target species. Application to cut stems may minimize non-target effects.

BRANCHED-CHAIN AMINO ACID INHIBITORS

Imazapyr
Arsenal, Habitat, Stalker, Chopper, Polaris

Rate: 2.5% v/v product

Timing: Postemergence in mid- to late summer. Applications should be made before a killing frost occurs.

Remarks: Imazapyr requires special mixing and a specialized hand applicator or applicator mounted on a spray boom. Apply to leaves but stop application before product runs off foliage. A low volume application without runoff reduces the non-target effect of imazapyr.

Metsulfuron
Escort

Rate: Foliar treatment: 1 to 3 oz/acre (0.6 to 1.8 oz ai/acre) broadcast to dense infestations, or 1 to 2 oz per 100 gal of water as a spray-to-wet treatment.

Timing: Apply to fully expanded leaves before a killing frost occurs.

Remarks: Can injure or kill herbaceous plants and other shrubs, so care should be taken to minimize non-target injury. Metsulfuron can also be used in premixes with aminopyralid (*Opensight*) or 2,4-D + dicamba (*Cimarron Max*). Metsulfuron and its formulations are not registered for use in California.

PHOTOSYNTHETIC INHIBITORS

Hexazinone
Velpar L

Rate: 2 to 4 gal product/acre (4 to 8 lb a.i./acre)

Timing: Preemergence in late winter through summer before bud break until new growth hardens off.

Remarks: Hexazinone is soil mobile and should not be used in areas with a shallow water table. It has a fairly long soil residual. High rates of hexazinone can create bare ground, so only use high rates in spot treatments.

Tebuthiuron
Spike

Rate: 10 to 20 lb product (*Spike 20P*)/acre (2 to 4 lb a.i./acre)

Timing: Preemergence anytime except when the soil is frozen or is saturated with moisture. For optimum control, apply just before the active season of growth in spring.

Remarks: Tebuthiuron is applied as a pellet at the base of the target plant. It has a very long soil residual and will control most plants around the base of the target species. Rainfall is necessary to move the herbicide into the root zone of the plant.

Rubus armeniacus Focke
Himalaya blackberry

Family: Rosaceae
Range: Common throughout the western United States, except in Wyoming, North and South Dakota.
Habitat: Disturbed, open, moist sites such as canals, ditch banks, fencerows, roadsides, open fields, and riparian zones, in a variety of plant communities. It can also tolerate periodic flooding with brackish water.
Origin: A cultivar introduced from Eurasia, originating from Armenia, quickly spread throughout Europe and the rest of the world.
Impact: Himalaya blackberry is a highly competitive plant with a growth form that allows it to quickly crowd out native species. Its thickets have dense canopies allowing little light penetration and reducing the growth of understory plants. In riparian areas it can prevent access to water sources for livestock and wildlife.
Western states listed as Noxious Weed: California, Oregon
California Invasive Plant Council (Cal-IPC) Inventory: High Invasiveness

Himalaya blackberry is an evergreen erect shrub that grows up to 10 ft tall and is climbing, mounded, or trailing. The aboveground canes are usually biennial while the roots are perennial. The roots are found in the top 20 inches of the soil but may grow down to a depth of 7 ft in loose soil. The roots can sprout new shoots from root buds, and in good conditions root fragments may sprout a new plant. The stems are green to purplish-red, woody, strongly angled, and are protected against predation by straight or curved pickles with a thick base. The leaves are pinnately compound with 3 to 5 leaflets that are dark green with a white underside covered with dense short hairs. The leaflets are broadest above the middle, toothed and sometimes shallowly lobed.

The flowers of Himalaya blackberry are white to pinkish and numerous in non-glandular panicles. They are self-fertile with 5 petals, and numerous stamens and pistils. The fruit are edible and an aggregate of drupelets that adhere to a fleshy receptacle. The mature berries are ovoid to oblong, black, 0.75 inch long, glossy, and glabrous or slightly pubescent. They typically ripen later in the season than the native berries. Seeds are dispersed primarily by birds. In addition to seeds, plants reproduce by root sprouts and stem tip rooting. Seeds likely only survive a few years in the soil.

NON-CHEMICAL CONTROL

Mechanical (pulling, cutting, disking)	Hand pulling can be an effective control method for small populations. To successfully control populations with mechanical removal, it is important to remove the canes, roots and the root crowns to prevent resprouting. A Pulaski, mattock or similar device can be used to remove plants. Bulldozing may cause resprouting and can spread the weed by fragmenting roots and stems.
	Cutting and removing only the aboveground biomass will result in the stimulated growth of root sprouts. The root sprouts must be controlled and repeated cutting of the above-ground biomass during flowering time will exhaust the root stores.
	Tillage can be effective if the canes are raked and removed from the site. However, this will cause significant soil disturbance and is unsuitable in riparian areas.
Cultural	Goats will readily consume Himalaya blackberry and could help to control new populations. It is a common method of management in Australia and New Zealand. Their consumption is indiscriminate and could result in the loss of other desirable species. This is particularly true in riparian areas.
	Burning is only effective if the root sprouts are controlled by other methods, such as chemicals, when they resprout after the burn.
Biological	Blackberry leaf rust fungus (*Phragmidium violaceum*) was discovered in 2005 on the coast of Oregon and has since spread through most of the counties. It appears to have been accidentally introduced. It partially to fully defoliates Himalaya blackberry and evergreen blackberry (*Rubus laciniatus*) and also reduces tip rooting.

> The fungus is native to Europe, the Middle East and Africa and has been used for years to control native blackberry plants in Australia and New Zealand. It is not an approved biocontrol agent yet and has not shown sustained control of Himalaya blackberry over a wide region.

CHEMICAL CONTROL

The following specific use information is based on reports by researchers and land managers. Other trade names may be available, and other compounds also are labeled for this weed. Directions for use may vary between brands; see label before use. Herbicides are listed by mode of action and then alphabetically. The order of herbicide listing is not reflective of the order of efficacy or preference. Excellent control information, both chemical and non-chemical, can be obtained at http://www.ipm.ucdavis.edu/PMG/PESTNOTES/pn7434.html#MANAGEMENT and http://extension.oregonstate.edu/catalog/pdf/em/em8894.pdf.

GROWTH REGULATORS	
Dicamba *Banvel, Clarity*	**Rate:** 1 to 2 pt product/acre (0.5 to 1 lb a.e./acre) **Timing:** Postemergence, to weed regrowth in the late summer or fall following a mowing or tillage treatment. **Remarks:** Dicamba provides only suppression of growth. It is a broadleaf-selective herbicide often combined with other active ingredients, particularly 2,4-D. Tank mix combinations with glyphosate are also more effective. It may injure grasses at higher rates. Do not apply when outside temperatures exceed 80°F. Do not exceed 64 oz product/acre per year.
Fluroxypyr *Vista XRT*	**Rate:** 22 oz product/acre (7.7 oz a.e./acre) **Timing:** Postemergence when target plants are growing rapidly. **Remarks:** Reduced control occurs if plants are under stressed growth conditions.
Triclopyr *Garlon 3A, Garlon 4 Ultra, Pathfinder II* Aminopyralid + triclopyr (*Capstone*)	**Rate:** Broadcast foliar treatment: 4 pt product (*Garlon 4 Ultra*)/acre (2 lb a.e./acre). Spot treatment: 0.75 to 1% *Garlon 4 Ultra* or 1% *Garlon 3A*; thoroughly cover the foliage. Basal bark treatment: 20% *Garlon 4 Ultra* mixed with basal oil or seed oil; *Pathfinder II* is a ready-to-use triclopyr/oil mix. Dormant stem and leaf treatment: 1% v/v solution of *Garlon 4 Ultra* with 2 to 3% v/v crop oil concentrate or seed oil. For *Capstone* use 8 to 9 pt product/acre. **Timing:** Postemergence in mid-summer or early fall after flowering and start of fruit set. Basal bark applications can be made almost any time of the year, even after leaves have senesced (aged, dried, and fallen from plant). In areas where people frequently harvest the fruit of wild blackberries, a mid-fall basal bark treatment might be desirable to avoid human contact with the chemical. For dormant stem and leaf treatment apply to dormant leaves and stems in late fall and winter in a 3% crop oil concentrate mixture. Spray the plant until it is thoroughly wet but not to the point of runoff. Like basal bark treatments, the timing of this technique prevents human contact with the herbicide during berry-picking season. **Remarks:** Foliage or stems (dormant stem application) must be thoroughly wet. Triclopyr is broadleaf-selective and safe on most grasses. It is most effective on smaller plants and has little or no residual activity. For basal bark treatment, thoroughly cover a 12 to 15-in basal section of the stem with spray but not to the point of runoff. *Garlon 3A* and other amine formulations are registered for aquatic use. Ester formulations (e.g., *Garlon 4 Ultra*) may volatilize if applied in warm temperatures. Application in some counties and grape-growing areas may be restricted. Sometimes aminopyralid + triclopyr (*Capstone*) or glyphosate and triclopyr (1% solution each) are used to achieve better control.
AROMATIC AMINO ACID INHIBITORS	
Glyphosate *Roundup, Accord XRT II*, and others	**Rate:** Broadcast foliar treatment: 2 to 3 qt product (*Roundup ProMax*)/acre (2.25 to 3.4 lb a.e./acre). Spot treatment: 0.5 to 1.5% v/v solution. **Timing:** Postemergence in late summer to early fall when canes are growing rapidly, have reached full leaf maturity, and after berries are formed. Fall treatments must be made before a killing frost. **Remarks:** Fall treatment symptoms may not show before frost. Retreatment may be necessary for complete control. Trailing blackberry is more difficult to control. Glyphosate controls grasses in the treated area as well as other vegetation. To obtain good control complete foliage coverage (spray-to-wet) is essential. Burning or mowing 40 to 60 days after spraying with glyphosate increases the level of control and also contributes to pasture establishment by removing stem debris. Shoots recovering from sublethal glyphosate treatment tend to die more quickly when subjected to heavy grazing. Sometimes glyphosate

and triclopyr (1% each in solution) are used in combination to achieve better control.

BRANCHED-CHAIN AMINO ACID INHIBITORS

Metsulfuron
Escort

Rate: 0.5 to 1 oz product/acre (0.3 to 0.6 oz a.i./acre)

Timing: Postemergence, to fully leafed-out vegetation before fall leaf coloration.

Remarks: Metsulfuron is primarily active on broadleaf species. Apply only to pasture, rangeland, and non-crop sites. Do not apply when plants are under stressed growing conditions. Metsulfuron can be used in a premix with aminopyralid (*Opensight*) or a tank mix with triclopyr for better control. Metsulfuron and its formulations are not registered for use in California.

Sulfometuron
Oust and others

Rate: 3 to 4 oz product/acre (2.25 to 3 oz a.i./acre)

Timing: Early postemergence when target plants are germinating or actively growing. Will only be effective on very small plants and not on fully mature plants.

Remarks: Add a surfactant at 0.25% v/v for improved control.

PHOTOSYNTHETIC INHIBITORS

Hexazinone
Velpar L

Rate: 3 to 4 gal product/acre (6 to 8 lb a.i./acre)

Timing: Preemergence or postemergence when plants are germinating or actively growing.

Remarks: Hexazinone is used as a nonselective herbicide in non-cropland areas and as a selective herbicide in reforestation practices. It only suppresses the growth of Himalaya blackberry. Use higher rates on fine textured soils and soils with high organic matter. Do not apply to frozen ground. Non-target plants may be adversely affected from drift and run-off. Apply when there is adequate moisture for activation. Hexazinone can be mixed with triclopyr for better control. High rates of hexazinone can create bare ground, so only use high rates in spot treatments.

Tebuthiuron
Spike

Rate: 20 lb product (*Spike 20P*)/acre (4 lb a.i./acre)

Timing: Preemergence before the start of spring growth or before expected seasonal rainfall.

Remarks: Do not apply tebuthiuron at more than 20 lb/acre. Do not apply more than 10 lb/acre in areas that receive 20 inches or less of annual rainfall. May injure non-target species. Follow restrictions on the label for use around desirable plants.

Rumex acetosella L.
Red sorrel (sheep sorrel)

Family: Polygonaceae
Range: Throughout the United States, including all western states.
Habitat: Agronomic and vegetable crops, pastures, fields, roadsides, gardens, landscaped areas, nursery crops, orchards, vineyards, turf, grasslands, open disturbed sites in forest and forest plantations, coastal dunes, disturbed riparian areas, and other disturbed sites. Red sorrel generally thrives on sites that are infrequently disturbed, and reportedly requires open soil for germination. It tolerates serpentine and moderately acidic soils.
Origin: Native to Europe.
Impact: Red sorrel occurs nearly worldwide and is a weed of pastures, natural areas, and various crops. In natural areas, it can displace native grasses and forbs. The foliage contains variable amounts of oxalates and under certain conditions can be toxic to livestock, particularly horses and sheep, when large quantities are ingested within a short period. However, most animals avoid consuming large amounts of the sour-flavored foliage if more palatable forage is available. The oxalic acid in red sorrel can also cause dermatitis in some animals.
California Invasive Plant Council (Cal-IPC) Inventory: Moderate Invasiveness

Red sorrel is an erect perennial with slender stems to 1.5 ft tall. Plants typically grow in clonal patches of a single sex that can persist for several years. It has creeping roots, arrowhead-shaped leaves, and male and female flowers on separate plants (dioecious). Red sorrel can be identified by its reddish flowering stems and distinctly arrowhead-shaped leaves. The first and subsequent few leaves are ovate, with a membranous sheathing stipule at each node (ochrea). Later leaves are generally arrowhead-shaped with outward-spreading basal lobes. The lower leaves are mostly arrowhead-shaped with narrow outward-spreading basal lobes. The foliage consists of glabrous alternate leaves, mostly near the base of the plant. The stipules are long and membranous with a silver-white sheathing.

The male flowers of red sorrel are initially yellowish-green and the female flowers are reddish and drooping. Plants reproduce vegetatively from creeping roots and by seed. Under suitable conditions, root fragments 0.5 inch long or more can regenerate into new plants. The seeds are dispersed widely by animals, water, and agricultural and landscape operations. Some seeds survive ingestion by livestock and birds. Buried seeds can survive for more than 25 years.

NON-CHEMICAL CONTROL

Mechanical (pulling, cutting, disking)	Control of red sorrel can be difficult because of its creeping rhizomes and long-lived seeds, but is most effective when infestations are caught early. Careful hand pulling (removing all plant parts) can be effective when conducted before the root and rhizome systems are established. On a small scale this can be done effectively with a pick or shovel.
	Plants are too short to be affected by mowing or grazing.
	Where possible, repeated cultivation during the dry season may reduce patches by weakening the root system. Cultivation must be at short intervals to deplete root reserves, but plants must be allowed time to produce 2 or 3 inches of new green tissue between cultivations for maximum depletion of reserves. Occasional cultivation can disperse root fragments and may result in an increase in the population, especially under moist conditions.
	Liming and the addition of nitrogen fertilization can increase soil pH and enhance the growth of other plants. This can reduce red sorrel infestations.
Cultural	There is no evidence of grazing as an effective tool to control red sorrel. Although red sorrel is potentially poisonous to livestock because of the presence of soluble oxalates, it is grazed by sheep, cattle, and mule deer. In Idaho red sorrel increases under heavy grazing regimes and decreases under light grazing regimes. However, in Oregon the frequency of red sorrel was not affected by late season cattle grazing in a riparian mountain meadow.

	No studies were found that used prescribed fire as a tool to control red sorrel. Studies suggest that controlled burning generally has little effect and may actually increase red sorrel abundance.
	Reduction of disturbance will leave less open area for seed germination. Seeding a competitive crop after cultivation may further reduce the abundance of red sorrel.
Biological	There are no biological control efforts for the management of red sorrel.

CHEMICAL CONTROL

The following specific use information is based on published papers and reports by researchers and land managers. Other trade names may be available, and other compounds also are labeled for this weed. Directions for use may vary between brands; see label before use. Herbicides are listed by mode of action and then alphabetically. The order of herbicide listing is not reflective of the order of efficacy or preference.

GROWTH REGULATORS	
Clopyralid *Transline*	**Rate:** 0.3 to 1.3 pt product/acre (2 to 8 oz a.e./acre) **Timing:** Postemergence, to small, rapidly growing weeds. **Remarks:** Clopyralid is a broadleaf-selective herbicide with a fairly short soil residual activity. It is very safe on grasses.
Dicamba *Banvel, Clarity*	**Rate:** 0.5 to 1 pt product/acre (0.25 to 0.5 lb a.e./acre) **Timing:** Postemergence when red sorrel has new foliage, usually from mid-fall to mid-spring. Spring applications control spring-germinating seedlings better than earlier treatment timings. **Remarks:** Dicamba kills red sorrel seedlings and most of the old plants. It prevents surviving plants from setting seed.
Other growth regulator tank mixes and premixes	There are a number of tank mixes and premixes that have been shown to give effective control of red sorrel, including aminopyralid + metsulfuron (*Opensight*) at 2 to 2.5 oz product/acre, 2,4-D + dicamba + metsulfuron (*Cimarron Max*), triclopyr + clopyralid (*Redeem R&P*), picloram + fluroxypyr (*Surmount*), and picloram + 2,4-D (*Grazon P+D*). All formulations that include metsulfuron or picloram are not registered for use in California.
AROMATIC AMINO ACID INHIBITORS	
Glyphosate *Roundup, Accord XRT II, and others*	**Rate:** 50% v/v solution of glyphosate (*Roundup ProMax*) in water using a wiper application **Timing:** Postemergence treatment when plants are growing rapidly but before seed production. **Remarks:** A wiper application can prevent damage to other non-target species. Apply the wiper only to the foliage of red sorrel.
PHOTOSYNTHETIC INHIBITORS	
Hexazinone *Velpar L*	**Rate:** 2 to 6 pt product/acre (0.5 to 1.5 lb a.i./acre) **Timing:** Preemergence or postemergence. **Remarks:** Hexazinone is nonselective at high rates but selective at low rates (e.g., 1 to 2 pt product/acre). It is not considered the most effective product for red sorrel control. In some studies, control with hexazinone appeared promising at first but red sorrel recovered and produced a large number of seeds. High rates of hexazinone can create bare ground, so only use high rates in spot treatments.

Rumex crispus L.; curly dock
Rumex obtusifolius L.; broadleaf dock

Curly and broadleaf dock

Family: Polygonaceae
Range: Curly dock is found throughout the U.S., including every western state. Broadleaf dock is found in most of the western states, except Nevada, Wyoming and North Dakota.
Habitat: Ditches, roadsides, wetlands, meadows, riparian areas, alfalfa and pasture fields (especially with poor drainage), orchards and other disturbed moist areas. Plants prefer wet to moist soils but tolerate periods of dryness, and can grow in most climates and soil types.
Origin: *R. crispus*, Eurasia; *R. obtusifolia*, western Europe.
Impact: Both species can be very competitive and outcompete more desirable vegetation for water, nutrients and light. Under certain conditions, they can accumulate soluble oxalates making them toxic to livestock.
California Invasive Plant Council (Cal-IPC) Inventory: *R. crispus*, Limited Invasiveness

Both species are erect perennials from 1.5 to 3 ft tall, occasionally to 5 ft tall. Curly dock leaves are lanceolate and up to 20 inches long, whereas broadleaf dock has lanceolate-ovate leaves that are up to 30 inches long. Leaf width is the most diagnostic feature to separate the two species. Curly dock has dark green, hairless leaves that are relatively narrow, with curly or wavy margins, while broadleaf dock has wider leaves with smooth to finely ruffled margins. As members of the buckwheat family, these plants have a characteristic membranous sheath at the leaf base and usually swollen nodes. Both docks have a taproot that extends deep into the soil, enabling them to survive drought periods and outcompete other vegetation. The plant bolts from the rosette in late spring.

The flower stalk is round in cross-section, hairless, and ribbed in both species. Stems are usually unbranched below the flower head. The flowers are not showy; they are small, greenish and appear in whorled clusters at the end of the upright stems. Flower clusters of broadleaf dock are not as dense and closely spaced as those of curly dock. After the flowers senesce, the fruit takes on a characteristic rusty-brown color and can remain on the plant over winter. Plants reproduce primarily by seed which falls close to the plant. The fruit can also disperse long distances by water, as they have a corky appendage called a callosity that allows them to float. Both species are prolific seed producers. Seed remains viable for 20 years; some sources suggest it can survive for over 50 years.

NON-CHEMICAL CONTROL

Mechanical (pulling, cutting, disking)	Curly and broadleaf dock are difficult to control by hand-pulling because of their deep taproot. The root usually breaks off and plants can regenerate from the portion left in the soil. Cutting them off at least 2 inches below the soil surface with a shovel or other implement is more effective.
	Continual mowing before seeding can be effective in reducing seed production, but most habitats are not amenable to mowing.
	Both docks can be controlled mechanically through tillage but the habitats where they are found are not usually conducive to cultivation.
Cultural	As the docks are not readily eaten by livestock, intensive grazing will likely result in an increase in populations. In some cases, the high levels of soluble oxalates are toxic to grazing species.
	Prescribed burning is not effective for the control of the perennial dock species.
	Improving drainage, when feasible, can help to control dock species.
Biological	No insect or disease biocontrol agents are available for use on any *Rumex* species in North America.

CHEMICAL CONTROL
The following specific use information is based on published papers and reports by researchers and land managers. Other trade names may be available, and other compounds also are labeled for this weed. Directions for use may vary between brands; see label before use. Herbicides are listed by mode of action and then alphabetically. The order of herbicide listing is not reflective of the order of efficacy or preference.

GROWTH REGULATORS	
2,4-D Several names	**Rate:** 0.5 to 2 pt product/acre (0.24 to 0.95 lb a.e./acre) **Timing:** Postemergence, to rapidly growing plants. Smaller plants are more easily controlled. **Remarks:** 2,4-D is broadleaf-selective and has no soil activity. It is often combined with other active ingredients, e.g., dicamba. Do not apply ester formulations when outside temperatures exceed 80°F. Aquatic registered formulations are available for use close to water.
Aminocyclopyrachlor + chlorsulfuron *Perspective*	**Rate:** 4.75 to 8 oz product/acre **Timing:** Postemergence similar to aminopyralid. **Remarks:** *Perspective* provides broad-spectrum control of many broadleaf species. Although generally safe for grasses, it may suppress or injure certain annual and perennial grass species. Do not treat in the root zone of desirable trees and shrubs. Do not apply more than 11 oz product/acre per year. At this high rate, cool-season grasses will be damaged, including bluebunch wheatgrass. Not yet labeled for grazing lands. Add an adjuvant to the spray solution. This product is not approved for use in California and some counties of Colorado (San Luis Valley).
Aminopyralid *Milestone*	**Rate:** 4 to 7 oz product/acre (1 to 1.75 oz a.e./acre) **Timing:** Postemergence in spring to rapidly growing plants, or treat soil shortly before plants emerge. **Remarks:** Aminopyralid is a broadleaf herbicide similar to picloram, but more selective. It is safe on most grasses but will kill most legumes and members of the Asteraceae. It has some soil residual activity and is registered for application up to water's edge.
Aminopyralid + 2,4-D, *Forefront HL* Aminopyralid + metsulfuron, *Opensight* Aminopyralid + triclopyr, *Capstone*	**Rate:** 1.5 to 2.1 pt *Forefront HL*/acre; 1.5 to 2 oz *Opensight*/acre; 4 to 6 pt *Capstone*/acre **Timing:** Postemergence, to rapidly growing plants. **Remarks:** The tank mixes are all broadleaf-selective. *Opensight* is not registered for use in California.
Clopyralid *Transline*	**Rate:** 0.67 to 1.33 pt product/acre (4 to 8 oz a.e./acre) **Timing:** Postemergence, to rapidly growing plants. **Remarks:** Clopyralid is a broadleaf herbicide similar to picloram, but more selective on the docks. It is very safe on grasses but will kill most legumes and members of the Asteraceae.
Clopyralid + 2,4-D *Curtail*	**Rate:** 2 to 4 qt product/acre **Timing:** Postemergence, to rapidly growing plants. **Remarks:** The mixture is broadleaf-selective with a wide range of susceptible species.
Dicamba *Banvel, Clarity*	**Rate:** 1 to 2 qt product/acre (1 to 2 lb a.e./acre) **Timing:** Postemergence, to rapidly growing plants. Smaller plants are more easily controlled. Use higher rates for larger plants. **Remarks:** Dicamba is a broadleaf-selective herbicide often combined with other active ingredients. It may injure grasses at higher rates. Do not apply when outside temperatures exceed 80°F. *Overdrive,* a premix of dicamba with diflufenzopyr, has been reported to be effective on both curly and broadleaf docks. Apply postemergence to rapidly growing plants, 4 to 8 oz product/acre. Diflufenzopyr is an auxin transport inhibitor which causes dicamba to accumulate in shoot and root meristems, increasing its activity. Higher rates should be used on biennials or when treating perennial weeds. Add a non-ionic surfactant to the treatment solution at 0.25% v/v or a methylated seed oil at 1% v/v solution.
Fluroxypyr *Vista XRT*	**Rate:** 6 oz product/acre (2.1 oz a.e./acre) **Timing:** Postemergence when the target plants are growing rapidly. For optimum control add 0.25 to 0.5% seed oil surfactant. **Remarks:** Reduced control occurs if plants are under stressed growth conditions.
Picloram *Tordon 22K*	**Rate:** 1 to 2 qt product/acre (0.5 to 1 lb a.e./acre) **Timing:** Postemergence, to rapidly growing plants before the bloom stage.

	Remarks: Most broadleaf plants are susceptible, but relatively safe on established grasses. Use non-ionic surfactant at 0.25%. It has a relatively long residual activity. May injure young or germinating grasses. Do not apply near trees, or where soil is highly permeable and where water table is high. Also available as a premix with 2,4-D (*Grazon P+D*) or fluroxypyr (*Surmount*). Picloram is a restricted use herbicide. Picloram and its formulations are not registered for use in California.
Triclopyr *Garlon 3A, Garlon 4 Ultra*	**Rate:** 0.33 to 1.5 gal *Garlon 3A*/acre (1 to 4.5 lb a.e./acre), 0.25 to 1 gal *Garlon 4 Ultra*/acre (1 to 4 lb a.e./acre) **Timing:** Postemergence, to rapidly growing plants. **Remarks:** Broadleaf-selective, safe on most grasses. Most effective on smaller plants and has little or no residual activity. *Garlon 3A* and other amine formulations are registered for aquatic use. *Garlon 4 Ultra* is formulated as a low volatile ester. However, in warm temperatures, spraying onto hard surfaces such as rocks or pavement can increase the risk of volatilization and off-target damage. Can be used in a tank mix with clopyralid or 2,4-D and is available in premixed formulations with these chemicals.
AROMATIC AMINO ACID INHIBITORS	
Glyphosate *Roundup, Rodeo, Aquamaster*, and others	**Rate:** 2 to 3.3 qt product (*Roundup ProMax*)/acre (2.25 to 3.7 lb a.e./acre) **Timing:** Postemergence, to rapidly growing plants. **Remarks:** Glyphosate is a nonselective herbicide. It has no soil activity and its effectiveness is increased by addition of ammonium sulfate. Aquatic registered formulations, e.g., *Rodeo* and *Aquamaster*, are available for use close to water.
BRANCHED-CHAIN AMINO ACID INHIBITORS	
Chlorsulfuron *Telar*	**Rate:** 1 to 2.6 oz product/acre (0.75 to 1.95 oz a.i./acre) **Timing:** Postemergence from bud to bloom stage or to fall rosettes. **Remarks:** Chlorsulfuron provides residual control up to one year after treatment. It has mixed selectivity, but is generally safe on grasses. Always use a surfactant.
Imazapic *Plateau*	**Rate:** 8 to 12 oz product/acre (2 to 3 oz a.e./acre) **Timing:** Preemergence or postemergence. **Remarks:** Mixed selectivity, tends to select for members of the Asteraceae and some grasses. It has some soil residual activity. In postemergence applications, use a methylated seed oil surfactant at 0.25%. Imazapic is not registered for use in California.
Imazapyr *Arsenal, Habitat, Stalker, Chopper, Polaris*	**Rate:** 3 to 4 pt product/acre (0.75 to 1 lb a.e./acre) **Timing:** Shortly after emergence. **Remarks:** Nonselective herbicide; long soil residual activity, leaves more bare ground than other treatments, even a year after application. *Habitat* is an aquatic registered formulation available for use close to water.
Metsulfuron *Escort*	**Rate:** 0.5 to 1 oz product/acre (0.3 to 0.6 oz a.i./acre) **Timing:** Postemergence from bud to bloom stage or to fall rosettes. **Remarks:** Do not apply when plants are under stressed growing conditions. In addition to a premix with aminopyralid, metsulfuron can also be used in a premix with 2,4-D + dicamba (*Cimarron Max*). Metsulfuron and its formulations are not registered for use in California.
Sulfometuron *Oust* and others	**Rate:** 1.33 to 2 oz product/acre (1 to 1.5 oz a.i./acre) **Timing:** Preemergence or early postemergence before or during the rainy season, when the target plants are germinating or actively growing. **Remarks:** Add a surfactant at 0.25% v/v for improved control.
PHOTOSYNTHETIC INHIBITORS	
Hexazinone *Velpar DF*	**Rate:** 2 to 5.33 lb product/acre (1.5 to 4 lb a.i./acre) **Timing:** Postemergence, to rapidly growing plants. **Remarks:** Apply when there is adequate moisture for activation. A broad-spectrum herbicide which kills or injures many desirable grasses or forbs. High rates of hexazinone can create bare ground, so only use high rates in spot treatments.

Ruppia maritima L.

Widgeongrass

Family: Potamogetonaceae or Ruppiaceae
Range: Coastal states, including California, Oregon and Washington.
Habitat: Marshes, ponds, sloughs, tidal estuaries, ditches, canals. Typically inhabits brackish, alkaline, or saline waters to several meters deep.
Origin: A widespread native with a nearly worldwide distribution.
Impacts: Widgeongrass is a valuable food and habitat plant for wildlife and is not considered a weed in most natural areas. However, it can be weedy in ditches, irrigation channels, and other controlled aquatic systems.

Widgeongrass is a submersed aquatic perennial, often in brackish water, with linear leaves and rhizomes. Like most submerged plants, the foliage is glabrous. The stems are branched, up to 3 ft long, and rooting at the nodes. The leaves are narrow and strap-like (0.5 mm wide) and sessile, bright green with most leaves alternate but some opposite. Leaves can be 4 inches long. Rhizomes are slender.

Inflorescences are submerged, consisting of two minute flowers on a straight stalk. The flowers lack sepals and petals. The carpel stalks elongate as fruits develop. Flowers are self-pollinated underwater; the pollen is transported on the surface of clinging bubbles. The fruiting heads are umbel-like, with each fruit on a long stalk (to nearly 2 inches long). Plants reproduce vegetatively from rhizomes and stem fragments and by seed. Seed production can be high. Seeds are dispersed with water, mud, and by clinging to the feet, fur, or feathers of animals. Seeds lack a hard coat and germinate in spring. It is unknown how long seeds can survive in the soil substrate.

NON-CHEMICAL CONTROL

Mechanical (pulling, cutting, dredging)	Repeated mechanical harvesting can help reduce stem densities, but escaped stem fragments can drift elsewhere and develop into new plants. Removing and destroying stem fragments from recreational equipment, such as boat propellers, docking lines, and fishing gear can help prevent the spread of widgeongrass. Several types of "bottom barriers" are available and are used to cover and smother specific infested areas. Materials used include polyvinyl chloride sheets, small-mesh screens and natural fibers such as jute. Bottom barriers are best installed in spring before plants produce large biomass and exceed 10 inches tall.
Cultural	Dewatering infested areas during periods of high temperature in summer can suppress regrowth. Suppression of flowering and seed production by dewatering may reduce subsequent spring growth.
Biological	The sterile triploid grass carp (white amur) is a relatively nonselective herbivorous fish that will consume widgeongrass if it is growing in fresh water. However, grass carp do not tolerate saline conditions. Usually grass carp first consumes its preferred submersed plants such as native pondweeds.

CHEMICAL CONTROL

Little information has been published on the control of widgeongrass. The following specific use information is based on reports by researchers and land managers. Other trade names may be available, and other compounds also are labeled for this weed. Directions for use may vary between brands; see label before use. Herbicides are listed by mode of action and then alphabetically. The order of herbicide listing is not reflective of the order of efficacy or preference.

BRANCHED-CHAIN AMINO ACID INHIBITORS	
Imazamox *Clearcast*	**Rate:** In-water applications: 200 ppb. Dewatered (drawdown) applications: 64 oz product/acre (8 oz a.i./acre); first flush of water in canals must NOT be used for irrigation. **Timing:** Apply to water in early spring to early summer (rapid growth). Dewatered applications should be made in late winter at least 14 days before water will be reintroduced.

	Remarks: Use an approved surfactant.
CONTACT PHOTOSYNTHETIC INHIBITORS	
Diquat *Reward*	**Rate:** In-water applications: 0.1 to 0.25 ppm **Timing:** Apply to water in late spring to early summer. Diquat is a fast-acting contact herbicide that can be effective in mid to late summer, but if biomass is large, only a portion of the infested sites should be treated to minimize effects of decreasing dissolved oxygen. **Remarks:** Diquat is quickly bound to, and becomes inactivated on, suspended clay particles and it should not be used in moderately or highly turbid water.
NON-HERBICIDAL CHEMICALS	
Dyes or colorants *Aquashade*	Although technically not herbicides, dyes and colorants control submerged aquatic plants by absorbing light in the water column and reducing photosynthesis. Applications should be made in early spring and repeated to maintain the concentration recommended on the label. Colorants are not as effective on well-established plants in mid- to late summer.

Salix spp.
Willows

Salix laevigata (female)

Family: Salicaceae
Range: There are many native species of willow throughout the United States.
Habitat: Willows occur in places with moist soils such as floodplains, ditch and stream banks, the borders of lakes and ponds, meadows, swamps, and fens.
Origin: Species from the genus *Salix* are native worldwide, including North America, Europe, and Asia.
Impact: While there are occasionally problematic non-native willows, in most cases native willows can become weedy when they occur in irrigation canals and drainage ditches, waterways, and controlled aquatic systems. Due to their shallow root system they can also cause flooding and bank erosion by displacing water from the stream bed. Willows also cause a seasonal influx of organic matter into the stream system when the leaves fall, resulting in reduced water quality.

Salix exigua (male)

Willow species are usually winter deciduous shrubs or small trees that grow up to 20 ft in height. The bark is bitter tasting and has antipyretic (fever reducer) properties. The smaller branches are often flexible and can be hairy or glabrous. The leaves are bright green or glaucous, alternate, elliptically shaped, with entire to toothed margins. The woody roots are spreading and can develop new shoots from basal rootstock.

Willow flowers from March to May. The flowers are dioecious (male and female flowers on separate plants). The inflorescences are catkins with many flowers; flowers have a single bract but no sepals or petals. The female flowers have simple pistils while the male flowers consist only of stamens. The flowers are insect- or wind-pollinated. Seed capsules open by two valves and contain numerous elliptic seeds 1 mm long with a tuft of long hair at the apex. Seeds are wind- or water-dispersed. Like other small-seeded, wind-dispersed riparian species, the seeds likely survive only a few months under field conditions.

NON-CHEMICAL CONTROL

Mechanical (pulling, cutting, disking)	Hand pulling of seedlings less than 1.5 ft tall is a very effective means of controlling young stands. Small root fragments left from pulling small willows should not result in resprouting, but stem fragments will resprout.
	Cutting is only effective if paired with herbicide treatment, including cut stump or foliar treatment of new growth.
	Heavy equipment including bulldozing is impractical in wet areas due to disturbance to sensitive habitat. Willows will also sprout from small stem pieces left behind or pushed into the ground by the equipment.
Cultural	Neither grazing nor burning is an effective control method for willows.
Biological	Biocontrol agents are being studied for some of the willow species. Some promising agents include scale insects and some walkingstick-like insects in the genus *Astroma*. An agent that targets a specific species will be difficult to approve due to the mixture of native and non-native willow species in many areas and the frequency of hybridization between willows.

CHEMICAL CONTROL
The following specific use information is based on published papers and reports by researchers and land managers. Other trade names may be available, and other compounds also are labeled for this weed. Directions for use may vary between brands; see label before use. Herbicides are listed by mode of action and then alphabetically. The order of herbicide listing is not reflective of the order of efficacy or preference.

Foliar or Preemergence: Use only for plants which are less than 6 ft tall and are not immediately next to sensitive native species and waterways. Apply before leaves start to fall.

Cut Stump: Useful for treating large individual trees, or where the presence of desirable species and water preclude foliar application. Is an effective method year-round.

Hack-and-squirt: Using a hand axe or drill, make cuts 6 to 18 inches above the ground at the same level about 2 to 4 inches apart and 1 inch deep slanting downward to prevent herbicide leakage. Stem injection can be slow and labor-intensive, but it is effective year round and important for protecting sensitive native flora as well as preventing chemical runoff into waterways. Leave the tree undisturbed for 12 months before removal to insure that the treatment was successful.

GROWTH REGULATORS	
2,4-D *Weedar 64* and several other names	**Rate:** Foliar broadcast treatment: 2 to 3 qt product/acre in 100 gal water (1.9 to 2.85 lb a.e./acre) **Timing:** Postemergence when the target plants are growing rapidly and the leaves are fully developed. **Remarks:** Thoroughly wet the stem and foliage of target plants. If the target plants are above 6 to 8 ft and in a dense population then a higher rate may be necessary. Use the aquatic formulations of 2,4-D on willows close to a water source.
Triclopyr *Garlon 3A*, *Garlon 4 Ultra*	**Rate:** 4 to 8 qt product (*Garlon 3A*)/acre or 3 to 6 qt *Garlon 4 Ultra*/acre (3 to 6 lb a.e./acre) **Timing:** Apply when willows are growing rapidly. **Remarks:** Foliage, stems and root collars must be thoroughly covered with spray. Triclopyr is broadleaf-selective and safe on most grasses. It is most effective on smaller plants and has little or no residual activity. *Garlon 3A* and other amine formulations are registered for aquatic use. *Garlon 4 Ultra* is formulated as a low volatile ester. However, in warm temperatures, spraying onto hard surfaces such as rocks or pavement can increase the risk of volatilization and off-target damage.
AROMATIC AMINO ACID INHIBITORS	
Glyphosate *Rodeo*, *Aquamaster*	**Rate:** 3 to 4 qt product (*Roundup ProMax*)/acre (3.375 to 4.5 lb a.e./acre) **Timing:** Postemergence after full leaf expansion during late summer or early fall before the first frost. **Remarks:** Use a higher rate on larger plants or in dense areas of growth. Glyphosate is a nonselective herbicide. It has no soil activity and its effectiveness is increased by addition of ammonium sulfate. Aquatic registered formulations, e.g., *Rodeo* and *Aquamaster*, are available for use close to water.
BRANCHED-CHAIN AMINO ACID INHIBITORS	
Imazapyr *Habitat*	**Rate:** 4 to 6 pt product/acre (1 to 1.5 lb a.e./acre) **Timing:** Postemergence, to fully leafed-out brush between spring and fall. **Remarks:** Postemergence applications require the addition of a non-ionic surfactant or a crop oil concentrate. Imazapyr is a nonselective herbicide. It has long soil residual activity and leaves more bare ground than other treatments, even a year after application. *Habitat* is an aquatic registered formulation available for use close to water.
Metsulfuron *Escort*	**Rate:** 1 to 3 oz product/acre (0.6 to 1.8 oz a.i./acre) **Timing:** Postemergence, to fully leafed-out brush between spring and fall. **Remarks:** Full spray coverage of the foliage and terminal growing points is required. Use a surfactant. Apply only to rangeland and non-crop sites. Not registered in aquatic areas or in California.
PHOTOSYNTHETIC INHIBITORS	
Hexazinone *Velpar L*	**Rate:** Broadcast treatment: 2 to 4 gal product/acre (4 to 8 lb a.i./acre). Spot treatment at base of plant (basal-soil): undiluted product at a rate of 2 to 4 ml/inch of stem diameter. **Timing:** From pre-bud break in late winter until new growth hardens off at the end of summer. **Remarks:** Hexazinone is used as a nonselective herbicide in non-cropland areas and as a selective herbicide in reforestation practices. Do not apply to frozen ground. Apply when there is adequate moisture for activation. Not registered for use in aquatic areas. High rates of hexazinone can create bare ground, so only use high rates in spot treatments.

Salsola tragus L.; Russian-thistle
Salsola paulsenii Litv.; barbwire Russian-thistle

Russian-thistle (tumbleweed) and barbwire Russian-thistle

S. tragus

S. paulsenii

Family: Chenopodiaceae
Range: Russian-thistle is found in all contiguous states except Florida. Barbwire Russian-thistle is found in Arizona, California, Colorado, Nevada, New Mexico, Oregon, and Utah.
Habitat: Disturbed sites, waste places, roadsides, fields, disturbed natural and semi-natural plant communities. Grow best on loose, sandy soils. Can tolerate arid and alkaline soils; competitive where moisture limits growth of other species.
Origin: Native to Eurasia.
Impact: Both plants are invasive in arid natural areas. They often grow along roads, creating a visual barrier. The blowing skeletons interfere with traffic and can lodge against fences and structures, creating a fire hazard. Russian-thistle is an alternate host for the beet leafhopper (*Circulifer tenellus*) that can carry beet curly-top virus. Some populations in the Pacific Northwest are resistant to sulfonylurea herbicides (e.g., chlorsulfuron, sulfometuron); resistance to triazines (atrazine, simazine) also is suspected.
Western states listed as Noxious Weed: Both species are listed in California
California Invasive Plant Council (Cal-IPC) Inventory: Limited Invasiveness

    Russian-thistle and barbwire Russian-thistle are bushy, spiny summer annuals. Russian-thistle grows to 3 ft tall or more, often more or less spherical in shape. Barbwire Russian-thistle grows to 20 inches tall and is usually wider than tall. They are not true thistles in the Asteraceae. The two species are similar and can hybridize. They have rigid, upward-curving branches and reduced, stiff, prickly upper stem leaves (bracts) at maturity. Russian-thistle usually appears bluish-green while barbwire Russian-thistle is more yellow-green. These plants have a deep taproot and spreading lateral roots, and are able to extract deep soil moisture. Under certain conditions, both can accumulate oxalates to levels that are toxic to livestock, especially sheep, but immature plants in moderation can provide an extra source of forage for livestock on arid rangeland.
    Both species flower in summer to early fall. Flowers are small and inconspicuous, without petals, mostly solitary in leaf axils. The sepals are wing-like, often brownish to pinkish to deep red. At senescence, plants become gray to brown. The main stems of Russian-thistle break off at ground level, allowing plants to disperse seeds as they tumble with the wind. Skeletons can persist for a year or more and are often found along fences and other structures, where they can pose a fire hazard. Barbwire Russian-thistle is less likely to become a tumbleweed and most seeds fall near the parent plant. Most seeds germinate the following spring and survive only 1 year in the field. A few seeds may survive up to 3 years. Seeds require very little moisture to germinate. Seedlings require loose soil for successful establishment.

NON-CHEMICAL CONTROL

Mechanical (pulling, cutting, disking)	Russian-thistle plants can be pulled out or hoed just below ground level before seed set. Hand-pulling or hoeing is effective in small infestations.
	Mowing older plants tends to cause the plants to grow low, but repeated mowing may provide control. Mowing just before flower maturation has worked in some cases. Mowing after seed set will disperse the seed.
	Tillage will control both seedling and larger plants, but to control an infestation cultivation must be repeated until the short-lived soil seed bank (< 2 years) becomes depleted. However, tillage increases disturbance, which favors Russian-thistle germination and establishment.

Cultural	Livestock can graze and suppress young plants, but this should not be their only forage.
Both species do poorly in areas with other vegetation. Establish desirable plants such as competitive perennial grasses in disturbed or open areas or after controlling Russian-thistle. Russian-thistle competes poorly in situations with firm, regularly irrigated soil and is rarely a problem in managed landscapes.
Monitor areas downwind of existing infestations, particularly at obstructions like fences and washes. To prevent opening up new invasion sites, avoid disking in abandoned areas.
Biological	The leaf mining moth *Coleophora klimeschiella* and the stem boring moth *Coleophora parthenica* have been released as biocontrol agents, but control of Russian-thistle with these insects has been poor.

CHEMICAL CONTROL

The following specific use information is based on published papers and reports by researchers and land managers. Other trade names may be available, and other compounds also are labeled for this weed. Directions for use may vary between brands; see label before use. *Salsola* plants are difficult to control with postemergence herbicides late in the season, when plants are spiny and hardened off. Herbicides are listed by mode of action and then alphabetically. The order of herbicide listing is not reflective of the order of efficacy or preference.

GROWTH REGULATORS	
2,4-D
Several names	**Rate:** 1 to 2 qt product/acre (0.95 to 1.9 lb a.e./acre)
Timing: Postemergence, to young plants. Plants should be rapidly growing and unstressed.
Remarks: 2,4-D is broadleaf-selective with no soil activity. It is often combined with other active ingredients, e.g., chlorsulfuron, dicamba. Do not apply when outside temperatures exceed 80°F.
Aminocyclopyrachlor + chlorsulfuron
Perspective	**Rate:** 4.75 to 8 oz product (*Perspective*)/acre
Timing: Postemergence or preemergence. Postemergence applications are most effective when applied to plants from the seedling to the mid-rosette stage.
Remarks: Provides broad-spectrum control of many broadleaf species. Although generally safe for grasses, it may suppress or injure certain annual and perennial grass species. Do not treat in the root zone of desirable trees and shrubs. Do not apply more than 11 oz product/acre per year. At this high rate, cool-season grasses will be damaged, including bluebunch wheatgrass. Not yet labeled for grazing lands. Add an adjuvant to the spray solution. This product is not approved for use in California and some counties of Colorado (San Luis Valley).
Aminopyralid
Milestone	**Rate:** 7 oz product/acre (1.75 oz a.e./acre)
Timing: Preemergence only; does not control Russian-thistle as a postemergence application.
Remarks: Can be used in combination with other herbicides such as *Accord XRT II* or *Landmark XP* for total vegetation control.
Dicamba
Banvel, Clarity	**Rate:** 1 to 4 pt product/acre (0.5 to 2 lb a.e./acre)
Timing: Postemergence, to rapidly growing plants. More effective on smaller plants. Use higher rates for larger plants.
Remarks: Broadleaf-selective herbicide often combined with other active ingredients. May injure grasses at higher rates. Do not apply when outside temperatures exceed 80°F.
Overdrive, a premix of dicamba with diflufenzopyr, has been reported to be effective on Russian-thistle. Diflufenzopyr is an auxin transport inhibitor which causes dicamba to accumulate in shoot and root meristems, increasing its activity. *Overdrive* is applied postemergence at 4 to 8 oz product/acre rapidly growing plants. Higher rates should be used on large annuals. Add a non-ionic surfactant to the treatment solution at 0.25% v/v or a methylated seed oil at 1% v/v solution.
Picloram
Tordon 22K	**Rate:** 8 to 12 oz product/acre (2 to 3 oz a.e./acre)
Timing: Postemergence, to young, rapidly growing plants.
Remarks: Broadleaf selective. Relatively safe on established grasses, but may injure young or germinating grasses. Long residual activity, may leach. Use non-ionic surfactant at 0.25%. Also effective mixed with dicamba or 2,4-D. Do not apply near trees, or where soil is highly permeable and where water table is high. Restricted use herbicide. Not registered for use in California.
Triclopyr
Garlon 3A, Garlon 4	**Rate:** 1.33 to 2.67 qt *Garlon 3A*/acre, 1 to 2 qt *Garlon 4 Ultra*/acre (1 to 2 lb a.e./acre)
Timing: Postemergence to rapidly growing plants. Most effective on smaller plants. Use higher rates

Ultra	for larger plants. **Remarks:** Broadleaf-selective, safe on most grasses. Little or no residual activity. *Garlon 4 Ultra* is formulated as a low volatile ester. However, in warm temperatures, spraying onto hard surfaces such as rocks or pavement can increase the risk of volatilization and off-target damage.
AROMATIC AMINO ACID INHIBITORS	
Glyphosate *Roundup, Accord XRT II,* and others	**Rate:** 1 to 1.5 qt product (*Roundup ProMax*)/acre (1.1 to 1.7 lb a.e./acre) for seedlings; 1.5 to 4 qt product (*Roundup ProMax*)/acre (1.7 to 4.5 lb a.e./acre) for larger plants and plants under stress; 0.5% to 1% v/v solution for spot treatments. **Timing:** Postemergence to rapidly growing plants before seed set. **Remarks:** Glyphosate is a nonselective herbicide. It has no soil activity. Drought stress will reduce glyphosate activity. Its effectiveness is increased by addition of ammonium sulfate.
BRANCHED-CHAIN AMINO ACID INHIBITORS	
Chlorsulfuron *Telar*	**Rate:** 1 to 2 oz product/acre (0.75 to 1.5 oz a.i./acre) **Timing:** Preemergence to early postemergence. **Remarks:** Mixed selectivity; generally safe on grasses, but fall application may injure bromes. Most effective preemergence. Use a surfactant for postemergence applications. It has long soil residual activity. Some populations have developed resistance to related herbicides; where resistance is suspected, use other herbicides or combinations.
Imazapic *Plateau*	**Rate:** 8 to 12 oz product/acre (2 to 3 oz a.e./acre) **Timing:** Fall or spring, from preemergence until plants are 3 inches tall. **Remarks:** Mixed selectivity; tends to favor species in the Asteraceae, as well as some grasses. It has some soil residual activity. In postemergence applications, use a methylated seed oil surfactant at 0.25%. Some populations have developed resistance to related herbicides; where resistance is suspected, use other herbicides or combinations. Not registered for use in California.
Imazapyr *Arsenal, Habitat, Stalker, Chopper, Polaris*	**Rate:** 2 to 3 pt product/acre (0.5 to 0.75 lb a.e./acre) **Timing:** Preemergence or postemergence. **Remarks:** Nonselective herbicide with long soil residual activity. Leaves more bare ground than other treatments, even a year after application. Some populations have developed resistance to related herbicides; where resistance is suspected, use other herbicides or combinations.
Propoxycarbazone-sodium *Canter R+P*	**Rate:** 0.9 to 1.2 oz product/acre (0.63 to 0.84 oz a.i./acre) **Timing:** Postemergence to small, rapidly growing plants. **Remarks:** A broad-spectrum herbicide that will control many species. Provides only partial control of Russian-thistle. Perennial grass species vary in tolerance. A non-ionic surfactant should be added at 0.25 to 0.5% v/v solution. Some populations have developed resistance to related herbicides; where resistance is suspected, use other herbicides or combinations.
Sulfometuron *Oust* and others	**Rate:** 2 to 8 oz product/acre (1.5 to 6 oz a.i./acre) **Timing:** Preemergence to early postemergence. **Remarks:** Mixed selectivity; fairly safe on native perennial grasses, especially wheatgrass. Other desirable grasses may be stunted, stressed, or injured. Good for revegetation use. Fairly long soil residual activity. Some populations have developed resistance to related herbicides; where resistance is suspected, use other herbicides or combinations.
PHOTOSYNTHETIC INHIBITORS	
Hexazinone *Velpar L*	**Rate:** 4 to 6 pt product/acre (1 to 1.5 lb a.i./acre) **Timing:** Preemergence to early postemergence. **Remarks:** Mixed selectivity, fairly long soil residual activity. Active both foliar and soil applied. In soil applications, rates will vary with soil texture and soil organic matter. Best results are obtained if applied when soil is moist. Hardwood trees near application site can absorb this chemical through the roots and may be injured or killed. Do not spray near the root zone of desirable hardwood trees or shrubs. High rates of hexazinone can create bare ground, so only use high rates in spot treatments.

Salvia aethiopis L.

Mediterranean sage

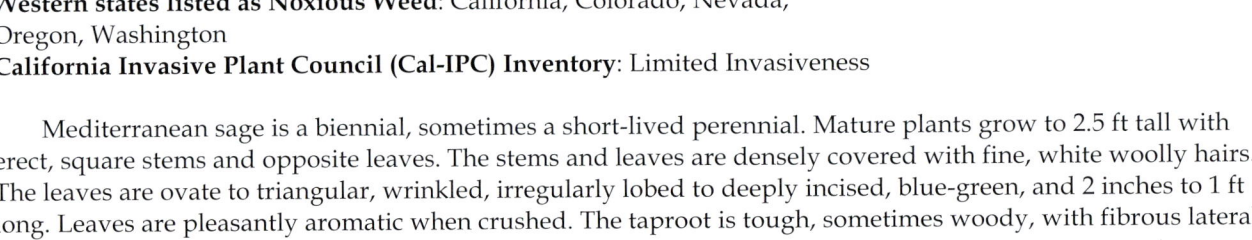

Family: Lamiaceae
Range: Northeastern California, Arizona, Idaho, Utah, Colorado, Oregon, and Washington
Habitat: Degraded big sagebrush communities, rangeland, disturbed sites, pastures, roadsides, and occasionally dryland crops. Often associated with sites dominated by annual grasses such as medusahead.
Origin: Native to Europe.
Impacts: Mediterranean sage has spread over 1.3 million acres in the western United States with new infestations occurring each year. It is unpalatable to livestock, but is not considered toxic. It can spread rapidly in degraded big sagebrush communities. Wind-blown plants can lodge in large masses along fencerows.
Western states listed as Noxious Weed: California, Colorado, Nevada, Oregon, Washington
California Invasive Plant Council (Cal-IPC) Inventory: Limited Invasiveness

Mediterranean sage is a biennial, sometimes a short-lived perennial. Mature plants grow to 2.5 ft tall with erect, square stems and opposite leaves. The stems and leaves are densely covered with fine, white woolly hairs. The leaves are ovate to triangular, wrinkled, irregularly lobed to deeply incised, blue-green, and 2 inches to 1 ft long. Leaves are pleasantly aromatic when crushed. The taproot is tough, sometimes woody, with fibrous lateral roots.

Flowering stems are highly branched near the top forming broad panicles with whorls of 5 to 10 yellow to white flowers. Each is encircled by silvery-haired bracts with pointed tips. Plants reproduce only by seed and large plants may produce 50,000 to 100,000 seeds. Seed dispersal occurs when mature plants break near the soil surface and tumble in the wind, spreading seed for long distances. When seeds become moist, a mucilaginous cover imbibes water to protect them from desiccation. Little is known about seed longevity in the soil, but it is expected that the seeds survive for several years.

NON-CHEMICAL CONTROL

Mechanical (pulling, cutting, disking)	With small infestations, hand-pulling or digging is feasible and effective. When digging, sever the root below the soil surface.
	Plants regrow after mowing. Repeated mowing in the bolting to early flowering stage can reduce seed production.
	Tillage is effective at controlling existing plants, but it is impractical on most sites due to rocky, uneven terrain.
Cultural	Most livestock avoid grazing Mediterranean sage. Overgrazing promotes spread.
	Fire is not an effective control.
	Promoting competitive vegetation can slow spread and help prevent establishment.
Biological	The root-feeding biocontrol weevil, *Phrydiuchus tau*, is a promising long-term management strategy for Mediterranean sage. Successful establishment of *P. tau* has been documented in Oregon, Idaho, and northern California. Larval feeding damages flower shoot buds and root crowns. Adults can cause minor defoliation of rosette leaves. No non-target effects have been reported.

CHEMICAL CONTROL
The following specific use information is based on published papers and reports by researchers and land managers. Other trade names may be available, and other compounds also are labeled for this weed. Directions for use may vary between brands; see label before use. Herbicides are listed by mode of action and then alphabetically. The order of herbicide listing is not reflective of the order of efficacy or preference.

GROWTH REGULATORS	
2,4-D Several names	**Rate:** 1 to 2 qt product/acre (0.95 to 1.9 lb a.e./acre) **Timing:** Postemergence from rosette to beginning of bolting stage. Most effective on small rosettes. **Remarks:** 2,4-D is often tank-mixed with chlorsulfuron or dicamba. It does not control large bolting plants. 2,4-D is broadleaf-selective and safe on most grasses, but has minimal soil activity. Do not apply ester formulation when outside temperatures exceed 80°F. Add a surfactant.
Aminocyclopyrachlor + chlorsulfuron *Perspective*	**Rate:** 4.75 to 8 oz product (*Perspective*)/acre **Timing:** Postemergence or preemergence. Postemergence applications are most effective when applied to plants from the seedling to the mid-rosette stage. **Remarks:** *Perspective* provides broad-spectrum control of many broadleaf species. Although generally safe for grasses, it may suppress or injure certain annual and perennial grass species. Do not treat in the root zone of desirable trees and shrubs. Do not apply more than 11 oz product/acre per year. At this high rate, cool-season grasses will be damaged, including bluebunch wheatgrass. Not yet labeled for grazing lands. Add an adjuvant to the spray solution. This product is not approved for use in California and some counties of Colorado (San Luis Valley).
Aminopyralid *Milestone*	**Rate:** 5 to 7 oz product/acre (1.25 to 1.75 oz a.e./acre) **Timing:** Postemergence from the rosette to young bolting stage. **Remarks:** Aminopyralid has a longer soil residual activity than clopyralid. It is safe on most grasses, although preemergence application at high rates can greatly suppress some annual grasses, such as medusahead. Applications can decrease seed production in some annual and perennial grass species. Other premix formulations of aminopyralid can also be used. These include *Opensight* (aminopyralid + metsulfuron; 1.5 to 2.5 oz product/acre) and *Forefront HL* (aminopyralid + 2,4-D; 1.5 to 2 pt product/acre); apply at the rosette to bolting stage.
Clopyralid *Transline*	**Rate:** 0.67 to 1.33 pt product/acre (4 to 8 oz a.e./acre). Use higher rate for older plants or dense stands. **Timing:** Postemergence from the rosette to young bolting stage. Most effective for young plants. **Remarks:** Clopyralid is very safe on grasses, but will injure many members of the Asteraceae, particularly thistles, and legumes, including clovers. Most other broadleaf species and all grasses are not injured.
Clopyralid + 2,4-D *Curtail*	**Rate:** 2 to 4 qt *Curtail*/acre **Timing:** Same as for clopyralid. **Remarks:** Add a non-ionic surfactant.
Dicamba *Banvel, Clarity*	**Rate:** 0.5 to 2 pt product/acre (0.25 to 1 lb a.e./acre). Use higher rates for large rosettes and bolting plants. **Timing:** Postemergence from rosette to beginning of bolting stage. **Remarks:** Dicamba is a broadleaf-selective herbicide often combined with other active ingredients. Add a surfactant.
Picloram *Tordon 22K*	**Rate:** 1 to 2 pt product/acre (4 to 8 oz a.e./acre). **Timing:** Preemergence and postemergence. With postemergence application, treat at rosette to early bolting stage, when plants are growing rapidly. **Remarks:** Picloram controls a wide range of broadleaf species and has relatively long soil residual activity. Although well-developed grasses are not usually injured by labeled use rates, some applicators have noted that young grass seedlings with fewer than four leaves may be killed. Do not apply near trees. *Tordon 22K* is a federally restricted use pesticide. Picloram is not registered for use in California.
AROMATIC AMINO ACID INHIBITORS	
Glyphosate *Roundup, Accord XRT II*, and others	**Rate:** 1 to 2 qt product (*Roundup ProMax*)/acre (1.1 to 2.25 lb a.e./acre). Spot treatment, 1.5% v/v solution. **Timing:** Postemergence, to rapidly growing plants from the rosette to early bolting stage. **Remarks:** Glyphosate will not kill seeds or inhibit germination the following season. Glyphosate has no soil activity. In addition, it is nonselective and may result in bare ground conditions that are

	susceptible to weed recruitment. In areas with desirable vegetation, use spot treatment. Glyphosate is a good control option if reseeding is planned shortly after application, as it will not injure seedlings emerging after application. Add a surfactant when using a formulation where it is not already included (e.g., *Rodeo*, *Aquamaster*).
BRANCHED-CHAIN AMINO ACID INHIBITORS	
Chlorsulfuron *Telar*	**Rate:** 1 to 2.6 oz product/acre (0.75 to 1.95 oz a.i./acre) **Timing:** Postemergence from the rosette to bolting stage. **Remarks:** Always use a surfactant. Included with aminocyclopyrachlor in *Perspective*.
Metsulfuron *Escort*	**Rate:** 1 oz product/acre (0.6 oz a.i./acre) **Timing:** Postemergence from the rosette to bolting stage. **Remarks:** Metsulfuron has similar activity to chlorsulfuron. Metsulfuron has some soil residual activity. Always use a surfactant. Metsulfuron can be tank-mixed with 2,4-D at 1 to 2 pt product/acre. Other premix formulations of metsulfuron can be used at similar application timing. These include *Cimarron Max* (metsulfuron + dicamba + 2,4-D) and *Cimarron X-tra* (metsulfuron + chlorsulfuron). Metsulfuron is not registered for use in California.

Salvinia molesta Mitch.
Giant salvinia

Photo courtesy of D.W. Kratville

Family: Salviniaceae
Range: Many of the southern states, particularly Texas. In the western United States, it is found in California and Arizona.
Habitat: Streams, lakes, ponds, ditches, and even rice fields. Thrives in slightly acidic, high nutrient, warm, slow-moving freshwater. Resistant to periods of low temperature, dewatering, and elevated pH levels. Low tolerance to salinity.
Origin: Native to South America and brought to North America through the aquarium trade.
Impacts: Giant salvinia grows rapidly to cover the surface of lakes and streams, spreading aggressively by vegetative fragments. It forms floating mats that shade and crowd out native plants. Thick mats reduce oxygen content and degrade water quality for fish and other aquatic organisms. Mats impede boating, fishing, and swimming, and clog water intakes for irrigation and electrical generation.
Western states listed as Noxious Weed: Arizona, California, Colorado, Nevada, Oregon
California Invasive Plant Council (Cal-IPC) Inventory: High Invasiveness (Alert)

Giant salvinia is a free floating aquatic fern. Depending on the climate, it can either be a perennial or an annual (in non-tropical regions). Its fronds occur in whorls of three, two floating and one submerged. The submerged frond is finely dissected and functions as roots. The floating fronds are positioned opposite each other and are round to oblong in shape. The upper surface has rows of cylindrical papillae. Each papilla has four hairs at its distal end that are joined together at their tips to form what looks like an inverted egg-beater. This cage-like structure is an effective air trap giving the plant buoyancy in the water. The fronds are light to medium green, often with brownish edges in mature plants, and with a distinctive fold in the center. The plant exhibits great morphological variation depending on the conditions of habitat (such as space and nutrient availability), and ranges from a slender floating specimen with fronds less than 0.5 inch wide to fronds up to 2.5 inches wide.

Reproduction is by vegetative growth and fragmentation. As many as five lateral buds can be found at one node. Populations can double every 2 days in the wild and small ponds have been completely covered with giant salvinia in as little as 6 weeks from the point of invasion. Although giant salvinia produces an egg-shaped spore sac along the submersed frond, all spores are infertile. Plants can also be carried overland on anything entering infested waters.

NON-CHEMICAL CONTROL

Mechanical (pulling, cutting, disking)	Manual removal of the weed is only carried out to alleviate water blockages in the short term, and not as a permanent solution. This is because reinfestation is certain and the cost of manual control is high. Manual removal is only practical in the early stages of invasion.
	Typically, floating booms and wire nets have some value in containing salvinia infestations. However, such equipment is subject to breakage under the weight of large windblown mats.
	Habitat modification is a useful strategy in modified areas or areas in which the use of herbicides is unacceptable. Dams, human-made lakes, canals and other water bodies can be drained (or the water level reduced) to strand and dry out giant salvinia.
Cultural	Boaters and anglers can help prevent spread by removing all aquatic plants from propellers, intakes, trailers and gear before leaving a launch area. Always blow out jet ski intakes and wash boats and equipment land-side before traveling to a new waterway.
Biological	A weevil, *Cyrtobagous salviniae*, has been used successfully in in at least 10 countries and has been released in Florida, Texas and Louisiana. It is native to south-eastern Brazil. Feeding and damage by the salvinia weevil is dependent on levels of nitrogen in the plant. In South Africa, Botswana, and India, where the weevils have been introduced, salvinia has been reduced to 1% of its former area. However, in the Northern Territory, Australia, high water temperatures have been associated with the failure of the weevil to control the plant. This may account for its inability to establish in California or Arizona.

CHEMICAL CONTROL

The following specific use information is based on published papers and reports by researchers and land managers. Other trade names may be available, and other compounds also are labeled for this weed. Directions for use may vary between brands; see label before use. Herbicides are listed by mode of action and then alphabetically. The order of herbicide listing is not reflective of the order of efficacy or preference.

AROMATIC AMINO ACID INHIBITORS	
Glyphosate *Aquamaster, Rodeo*	**Rate:** Broadcast treatment: 3 to 3.75 qt product (*Rodeo* or *Aquamaster*)/acre (3 to 4 lb a.e./acre) in water plus an aquatic approved surfactant containing 0.1% v/v non-ionic organosilicone and 0.25% v/v non-ionic spreader sticker. Spot treatment: 2% solution v/v *Aquamaster* and water plus 0.5 to 2% v/v of a non-ionic surfactant to thoroughly wet all leaves **Timing:** Postemergence when plants are growing rapidly. **Remarks:** Apply spray solution to completely wet the target weed. Do not spray to runoff. Glyphosate is a nonselective systemic herbicide. Repeated applications may be necessary for complete control.
BRANCHED-CHAIN AMINO ACID INHIBITORS	
Penoxsulam *Galleon*	**Rate:** 2 to 5.6 oz product/acre (0.5 to 1.4 oz a.i./acre) in water plus an aquatic approved surfactant **Timing:** Postemergence when plants are growing rapidly. **Remarks:** Apply spray solution to completely wet the target weed. Do not spray to runoff. Penoxsulam is a selective systemic herbicide and should not be applied in areas where it will be diluted rapidly. *Galleon* will take 60 to 120 days or longer to completely kill the target plants.
PIGMENT SYNTHESIS INHIBITORS	
Fluridone *Sonar*	**Rate:** 1.3 qt product/acre (1.3 lb a.i./acre) plus an approved aquatic surfactant **Timing:** Postemergence when plants are growing rapidly. **Remarks:** Apply spray solution to completely wet the target weed. Do not spray to runoff. *Sonar* is a selective systemic herbicide that is slow-acting and requires an extended contact period. Rapid water movement or any condition that results in rapid dilution of fluridone will reduce its effectiveness.
Contact photosynthetic inhibitors	
Diquat *Reward*	**Rate:** Spot treatment: 0.5% solution v/v *Reward* and water plus 0.25 to 1% v/v of an approved aquatic wetting agent. Broadcast treatment: 0.5 to 2 gallons product/acre (1 to 4 lb a.i./acre) in water plus 1 to 2 pt/acre of an approved wetting agent. **Timing:** Postemergence when plants are growing rapidly. **Remarks:** Apply spray solution to completely wet the target weed. Do not spray to runoff. Make additional applications if treating densely-packed weeds or mats. Weed escapes are best controlled if repeat applications are made within 2 weeks of the first treatment.
Flumioxazin *Clipper*	**Rate:** 6 to 12 oz product/acre (3 to 6 oz a.i./acre) in water plus an approved aquatic surfactant **Timing:** Early postemergence when plants are young and rapidly growing. **Remarks:** Apply spray solution to completely wet the target weed. Do not spray to runoff. Flumioxazin is a broad-spectrum contact herbicide. Water pH needs to be below 8.5 or flumioxazin will rapidly degrade and lose effectiveness.

Sapium sebiferum L.
   (= *Triadica sebifera* (L.) Small)

Chinese tallowtree

Family: Euphorbiaceae
Range: Primarily found in the southeastern United States, from North Carolina to Texas. More recently becoming an issue in California.
Habitat: Disturbed and undisturbed bottomlands, fields, coastal prairies, riparian areas, and wetlands. Tolerates shade, drought, saline, flooded conditions, and temperatures as low as 5°F. Grows best on well-drained clay-peat soil, but can tolerate a wide range of soil conditions.
Origin: Native to China and Japan where it has been cultivated for seed oil production for more than a thousand years. It is thought that Benjamin Franklin introduced Chinese tallowtree into the United States in 1776. Significant planting occurred throughout the Gulf Coast in the early 1900s both as a seed crop and as an ornamental. Since then it has escaped from cultivation, is extremely invasive in much of the lower southeast, and is currently expanding its range.
Impacts: Chinese tallowtree can aggressively invade disturbed and undisturbed terrestrial, wetland, and riparian plant communities. The tree is most problematic in the southeastern U.S., where large tracts of coastal prairie have been replaced by stands of Chinese tallowtree. Stands replace native vegetation and can significantly alter the soil nutrient status. The milky sap and unripe fruits are mildly toxic to humans and livestock when ingested.
California Invasive Plant Council (Cal-IPC) Inventory: Moderate Invasiveness (Alert)

Chinese tallowtree is a fast-growing deciduous tree that often grows to about 20 ft, but can reach 40 to 50 ft tall. It is freely branching, with an open, airy appearance. The leaves are pendant, diamond shaped, abruptly pointed at the tip, and 2 to 3 inches long. In fall the leaves turn brilliant shades of scarlet, orange, yellow and maroon. Like most members of the spurge family, broken twigs and leaf stems exude a milky latex sap.

The flowers are produced in yellowish-green catkins on the branch tips. The fruits are 3-lobed brown capsules that open to reveal three white, waxy seeds that resemble popcorn. Chinese tallowtree has a tremendous reproductive potential. They reach reproductive age in as little as 3 years and can remain productive for 100 years. A mature tree may annually produce an average of 100,000 seeds that are spread mainly by birds and water. Most seeds survive over a year under field conditions, but little is known of the seedbank longevity. In addition to prolific seed production, Chinese tallowtree resprouts from stumps, and roots readily develop shoots.

NON-CHEMICAL CONTROL

Mechanical (pulling, cutting, disking)	Hand pulling can remove seedlings and small saplings, but this technique is generally not effective or practical for established trees.
	Cutting is most effective before flowering to prevent seed production. Because tallowtree spreads by suckering, resprouts are common after treatment. Cutting should be combined with an herbicide treatment or with multiple cuttings over a period of years. Cut trees at ground level with power or manual saws. Trees standing in water may be successfully killed by cutting them below the water line.
	Heavy equipment can be effectively used to control tallow trees on canal banks and in areas where soil disturbance and nonselective species removal are not important considerations. Stumps remaining following such treatment will require herbicide application to prevent regrowth.
	Girdling can be an effective treatment where the use of herbicides is impractical. Using a hatchet, make a cut through the bark encircling the base of the tree, approximately 6 inches above the ground. Be sure that the cut goes well into or below the cambium layer. This method will kill the top of the tree but resprouts are common. Follow-up treatments for many years may be required until roots are exhausted, so this method is not recommended for large populations.
Cultural	Grazing (foliage is toxic to cattle), burning or flooding are not effective management options for controlling Chinese tallowtree.

| Biological | No biological control agents have been released for the control of Chinese tallowtree. |

CHEMICAL CONTROL

The following specific use information is based on published papers and reports by researchers and land managers. Other trade names may be available, and other compounds also are labeled for this weed. Directions for use may vary between brands; see label before use. Herbicides are listed by mode of action and then alphabetically. The order of herbicide listing is not reflective of the order of efficacy or preference.

GROWTH REGULATORS	
Picloram + 2,4-D *Tordon 22K* + 2,4-D amine Picloram + triclopyr *Tordon 22K* + *Remedy Ultra*	**Rate:** Foliar spot treatment: 1 qt *Tordon 22K*, plus 4 pt 2,4-D amine (1.9 lb a.e./acre) or 1 pt *Remedy Ultra*, plus 0.5% v/v surfactant. Apply to thoroughly wet all leaves. **Timing:** Postemergence in spring or fall when conditions are favorable for plant growth. **Remarks:** High rates of picloram can give long-term soil residual control for broadleaves. *Tordon 22K* is a federally restricted use pesticide. It is not registered for use in California.
Triclopyr *Garlon 3A, Garlon 4 Ultra, Pathfinder II*	**Rate:** Low volume foliar treatment: 2% v/v solution of triclopyr and water plus 0.5% surfactant; apply to thoroughly wet all leaves. Cut stump treatment: 20% *Garlon 4 Ultra* in 80% oil carrier, or 50% *Garlon 3A* in water. Basal bark treatment: 20 to 30% *Garlon 4 Ultra* in 70 to 80% oil carrier, or *Pathfinder II* as a ready to use formulation. Stem injection treatment: one cut per every 3 inches of stem diameter, and 1 ml of undiluted *Garlon 3A* added to each cut. For clumps, one hack per every 6 inches of total stem diameter. Treat the largest stems. Basal cut stump treatment: 20% *Garlon 4 Ultra* in 80% oil carrier. **Timing:** Foliar treatments are best when leaves are fully expanded. Cut stump, basal cut stump, basal bark, and stem injection treatments can be applied as long as the ground is not frozen, but are best when used in late summer or early fall, before leaf drop. **Remarks:** Triclopyr is a selective herbicide for broadleaf species and will not damage desirable grasses growing nearby. Cut stump treatment: cut stems horizontally at or near ground level, and immediately apply herbicide solution to cover the outer 20% of the stump face. Suckering from the roots typically occurs after cutting, but the treatment should control most resprouts. Basal bark treatment: spray the lower trunk, including the root collar, to a height of 12 to 15 inches from the ground; the spray should thoroughly wet the lower stem but not to the point of runoff. Stem injection treatment: be sure that each cut goes well into or below the cambium layer. Trees should not be cut for at least one month following basal bark and stem injection treatments. Triclopyr can be used as a premix with aminopyralid (*Capstone*) at 5 to 8% v/v solution for spot treatments.
AROMATIC AMINO ACID INHIBITORS	
Glyphosate *Roundup, Accord XRT II*, and others	**Rate:** Foliar treatment: 2% v/v solution of *Roundup ProMax* (or other trade name with a similar concentration of glyphosate) and water plus 0.5% surfactant to thoroughly wet all leaves. Cut stump treatment: undiluted *Roundup* (or other trade name) or 50% of herbicide concentrate in water. Stem injection treatment: one cut per every 3 inches of stem diameter, and 1 ml of undiluted herbicide added to each cut. For clumps, one hack per every 6 inches 6f total stem diameter. Treat the largest stems. **Timing:** Postemergence foliar treatments are best when leaves are fully expanded. Cut stump and stem injection treatments can be applied as long as the ground is not frozen, but is best when used in late summer or early fall, before leaf drop. **Remarks:** Glyphosate is a nonselective systemic herbicide. Cut stump and stem injection applications are as described for triclopyr. Trees should not be cut for at least 4 months after foliar and stem injection treatments.
BRANCHED-CHAIN AMINO ACID INHIBITORS	
Imazapyr *Arsenal, Habitat, Stalker, Chopper, Polaris*	**Rate:** Low volume foliar treatment: 1% v/v solution of *Stalker* and water plus 0.5% surfactant to thoroughly wet all leaves. Cut stump treatment: 10% *Stalker* in 90% oil carrier. Basal bark treatment: 20% v/v *Stalker* or *Chopper* formulation in 80% oil carrier. Stem injection treatment: one cut per every 3 inches of stem diameter, and 1 ml of undiluted herbicide (*Arsenal* or *Habitat*) added to each cut. For clumps, one hack per every 6 inches of total stem diameter. Treat the largest stems. **Timing:** Best in late summer to early fall, but before leaf drop. Avoid stem injection between March and April. **Remarks:** Imazapyr is a soil residual herbicide and may result in bare ground around trees for some time

	after treatment. Cut stump, basal bark and stem injection applications are as described for triclopyr. Trees should not be cut for at least 4 months after foliar, basal bark and stem injection treatments.
PHOTOSYNTHETIC INHIBITORS	
Hexazinone *Velpar L*	**Rate:** Broadcast soil treatment: 2 to 4 gal/acre (4 to 8 lb a.i./acre). Basal (soil) single stem treatment: undiluted product at a rate of 2 to 4 ml per inch of stem diameter, applied to the soil surface within 3 ft of the stem. **Timing:** Preemergence from late winter through summer. **Remarks:** Hexazinone is a residual herbicide applied as a broadcast or basal-soil treatment for brush control. Basal (soil) single stem treatment: one squirt of spot gun per 1 inch stem diameter. High rates of hexazinone can create bare ground, so only use high rates in spot treatments.
Tebuthiuron *Spike*	**Rate:** Basal (soil) single stem treatment: up to 7.5 lb product (*Spike 80DF*)/acre (6 lb a.i./acre) **Timing:** Soil treatments can be applied anytime except when the soil is frozen or saturated with moisture. Applications should be made before the start of spring growth or before expected seasonal rainfall. **Remarks:** Tebuthiuron is a surface applied, soil-active product intended for total vegetation control in non-cropland. For best control, do not disturb plants for 2 years after application.

Schinus molle L.; Peruvian peppertree
Schinus terebinthifolius Raddi; Brazilian peppertree

Peppertrees

S. molle

S. terebinthifolius

Family: Anacardiaceae
Range: In the western U.S., both species are only found in California. They are also problematic in many tropical areas, including Hawaii.
Habitat: Canyons, washes, slopes, riparian areas, fields, and along roadsides. Plants grow best where some soil moisture is available during the warm season.
Origin: Peruvian peppertree is native to the riparian habitats of Peru. Brazilian peppertree is native to the dry grasslands of southern Brazil. The peppertrees are common landscape ornamentals that were introduced to the U.S. 100 to 200 years ago. Both species have escaped cultivation and become invasive in some areas.
Impacts: Peruvian peppertree is more widespread than Brazilian peppertree in California, but appears to be less problematic. Peruvian peppertree is susceptible to black scale (*Saissetia oleae*), a pest of citrus. Plants can cause dermatitis in sensitive individuals. Fruits are used to make a drink in South America. Brazilian peppertree is locally invasive in certain riparian areas of southern California and has aggressively colonized hundreds of thousands of acres in Florida. Brazilian peppertree foliage can be toxic to horses and cattle when ingested, and direct contact with the sap can cause contact dermatitis in sensitive individuals. Fruits are readily consumed and dispersed by wildlife, particularly birds. Brazilian peppertree fruits are sometimes sold as "pink peppercorns" in the U.S. However, ingestion of fruits in large quantities can cause severe digestive tract irritation in animals and humans.
California Invasive Plant Council (Cal-IPC) Inventory: Both species are Limited Invasiveness

Peppertrees are evergreen shrubs to trees with alternate, glabrous, aromatic odd-pinnate-compound leaves. Peruvian peppertree can grow to 60 ft tall, whereas Brazilian peppertree is generally shorter, to a maximum of 35 ft tall. The leaves of Peruvian peppertree are 4 to 12 inches long with 15 to 59 slender leaflets per leaf. Brazilian peppertree leaves are about 4 to 8 inches long with only seven leaflets per leaf.

In both species the male and female flowers develop on separate trees (dioecious). Inflorescences are panicles of numerous greenish-white flowers mostly 1 to 3 mm long. Flowers are insect-pollinated and the fruits are spherical, berry-like, 4 to 8 mm diameter and pink to red with one seed. Plants reproduce by seed and sometimes vegetatively from root sprouts. Most seeds remain viable for less than 1 year after dispersal.

NON-CHEMICAL CONTROL

Mechanical (pulling, cutting, disking)	Small seedling plants can be removed manually and this has been a successful technique in many natural areas, provided it does not cause significant disturbance that can favor peppertree establishment. Entire saplings, including root systems, can be pulled up by hand, but by the time the plant is several feet tall, hand pulling may no longer be possible. In this case, larger saplings can be removed with a weed wrench.
	For larger plants, control is more difficult as they can resprout from the base. Heavy equipment such as bulldozers, front end loaders, root rakes and other specialized equipment can be used, but the entire root system must be removed to prevent resprouting, because root pieces as small as 0.25 inch diameter can resprout. This is often not suitable in natural areas.
	Mechanical operations that disrupt the vegetation while placing large volumes of peppertree seed on the soil will likely lead to reinvasion, perhaps at greater density than the original population.
Cultural	The seeds of *Schinus* cannot tolerate heat, thus will not germinate following a fire. However, basal trunk and root sprouting can be aggressive after burning. Once *Schinus* saplings attain a height of 3 ft most plants

	are able to survive fire by coppicing and will grow more rapidly than competing native hardwoods, thus increasing dominance of the stand. Burning at 5-year intervals, however, has been shown to exclude *Schinus* from pine forests of Florida. Another important consideration for prescribed burning of *Schinus* species is that they belong to the Anacardiaceae, which is the same family as poison oak and ivy. Thus, the sap and smoke from the burning may irritate or cause an allergic response in sensitive individuals.
Biological	As of 2010, biological control agents have not been released for peppertrees, although release has been recommended in Florida for a sap-sucking thrip species (*Pseudophilotrips ichini*) and a defoliating sawfly (*Heteroperreyia hubrichi*). An application has also been submitted for release of a leaflet roller (*Episimus unguiculus*). Brazilian pepper trees that were defoliated 5 times in 3 years had significantly fewer berries than Brazilian pepper trees that were not defoliated or were defoliated less frequently. It is felt that targeting biological control efforts on juveniles may be more effective than more mature plants.

CHEMICAL CONTROL

The following specific use information is based on reports by researchers and land managers. Other trade names may be available, and other compounds also are labeled for this weed. Directions for use may vary between brands; see label before use. Herbicides are listed by mode of action and then alphabetically. The order of herbicide listing is not reflective of the order of efficacy or preference. Excellent recommendations for controlling Brazilian peppertree can be found at http://www.fleppc.org/Manage_Plans/schinus.pdf.

GROWTH REGULATORS	
Dicamba *Banvel, Clarity*	**Rate:** 1.8 to 5% v/v solution for spot applications **Timing:** Postemergence, generally in late summer or autumn when plants are growing actively and translocating sugars to the below-ground tissues. **Remarks:** Results are inconsistent and often good control is not achieved with dicamba. It does not appear to be the best option for control of peppertrees.
Fluroxypyr *Vista XRT*	**Rate:** 1% v/v solution for spot applications **Timing:** Postemergence, generally in late summer or autumn when plants are growing actively and translocating sugars to the below-ground tissues. **Remarks:** Fluroxypyr can be used for both foliar applications and basal bark treatments, but no data are available on either treatment for the control of peppertrees.
Triclopyr *Garlon 3A, Garlon 4 Ultra*	**Rate:** Foliar treatment: 1.5 to 3% v/v solution with *Garlon 4 Ultra* and 15% with *Garlon 3A*. Cut stump treatment: 20% v/v *Garlon 4 Ultra* in water, or undiluted *Garlon 3A* or 50% *Garlon 3A* in water. Basal bark treatment: 20 to 30% v/v *Garlon 4 Ultra* in basal oil, applied to the bark around the bottom 1 ft of the tree. Lower rates (14%) have been shown to be successful by others. Stem injection: for large trees, undiluted *Garlon 3A* or *Garlon 4 Ultra* can be effective and will prevent regrowth. When using a drill technique, it is important to drill all the folds and scars to ensure that the entire cambium has been accessed. **Timing:** Postemergence, generally in late summer or autumn when plants are growing actively and translocating sugars to the below-ground tissues. **Remarks:** Triclopyr has little to no soil residual activity, but may damage other broadleaf species, especially with aerial foliar applications to larger trees. Peppertrees can be effectively controlled by cutting and treating the stumps with herbicide. The trunk should be cut as close to the ground as possible, and triclopyr should be applied to the cambium within 5 minutes. Treat before plants begin to fruit.
AROMATIC AMINO ACID INHIBITORS	
Glyphosate *Roundup, Accord XRT II*, and others	**Rate:** Broadcast foliar treatment: 1.3 to 3.3 qt product (*Roundup ProMax*)/acre (1.5 to 3.7 lb a.e./acre). Spot treatment to seedlings and young plants: 1% v/v solution. Foliar application to older plants: > 1.5% v/v solution. Cut stump treatment and stem injection: 40% of concentrated product. **Timing:** Postemergence, generally in late summer or autumn when plants are growing actively and translocating sugars to the below-ground tissues. **Remarks:** Glyphosate is nonselective and has no soil residual activity. Peppertrees can be effectively controlled by cutting and treating the stumps with herbicide. The trunk should be cut as close to the ground as possible. Apply glyphosate to the cambium within 5 minutes. Treat before plants begin to fruit.
BRANCHED-CHAIN AMINO ACID INHIBITORS	
Imazapyr	**Rate:** Foliar treatment broadcast: 2 to 4 pt product/acre (0.5 to 1 lb a.e./acre). Foliage spot treatment: 0.5 to 1% v/v solution. Cut stump treatment: 20% of concentrate. Stem injection treatment: undiluted

Arsenal, Habitat, Chopper, Stalker, Polaris	concentrate. **Timing:** Late summer or autumn when plants are growing actively and translocating sugars to the below-ground tissues. **Remarks:** All treatment methods can be effective. Imazapyr can have long soil residual activity.
PHOTOSYNTHETIC INHIBITORS	
Hexazinone *Velpar L*	**Rate:** 2 gal product/acre (4 lb a.i./acre) **Timing:** Preemergence at the beginning of the rainy season to allow for movement of herbicide into soil and uptake by roots. **Remarks:** Hexazinone is a root-absorbed herbicide that can be very effective for the control of Brazilian peppertree. However, it is slow-acting and has residual activity. This residual activity is beneficial for long-term control but can pose problems to non-target species. For widely scattered plants, where access to the main stem is difficult, basal spot treatments are easily applied. High rates of hexazinone can create bare ground, so only use high rates in spot treatments.
Tebuthiuron *Spike 20P*	**Rate:** 20 lb product/acre (4 lb a.i./acre), or 0.25 to 1 oz product/6 inches of basal stem diameter (0.1 to 0.25 oz a.i./6 inches of basal stem diameter) **Timing:** Preemergence at the beginning of the rainy season to allow for movement of herbicide into soil and uptake by roots. **Remarks:** Tebuthiuron has very long soil residual activity. Selective control around target plant can be achieved by placing the pelleted form of the herbicide at the base of the plant.

Schismus arabicus Nees; Arabian mediterraneangrass
Schismus barbatus (Loefl. ex L.) Thell.; common mediterraneangrass

Mediterraneangrasses

Schismus arabicus
Schismus barbatus

Family: Poaceae
Range: Southwestern U.S., including California, Arizona, Nevada, New Mexico and Utah. Common mediterraneangrass is also found in Texas.
Habitat: Open, disturbed and undisturbed areas, roadsides, desert and semi-desert shrubland, dry river beds, mud flats, waste places. Common mediterraneangrass is also found in coastal scrub.
Origin: Both species are native to southern Europe, but some authorities believe they are also native to southwestern Asia and North Africa.
Impacts: Mediterraneangrasses can displace native annual vegetation and help to increase the fire frequency in desert ecosystems. Fires fueled only by dried *Schismus* species are generally patchy and seldom burn hot enough to ignite small shrubs. However, in areas where larger exotics also grow around the bases of shrubs, *Schismus* species can carry fire across the open areas to ignite the larger exotics, which often burn with intensity great enough to kill the shrubs. In some areas, an abundance of *Schismus* species may contribute to the type conversion of desert shrubland into annual grassland.
California Invasive Plant Council (Cal-IPC) Inventory: Both species are Limited Invasiveness

Both species are generally tufted cool-season annuals in the desert and semi-desert regions of California. They are low growing, reaching 8 inches tall, and have fine foliage and small, dense panicles to about 2 inches long. The ligules consist of a ring of hairs about 0.5 to 1 mm long. Mediterraneangrasses also have a distinctive hair collar margin and membranous auricles 2 to 4 mm long on the lower leaves.

Flowers are present from March through May and are self-compatible. Spikelets are small (4 to 6.5 mm long), lack awns, and have 3 to 8 florets per spikelet. Plants reproduce only by seed. Seedlings mature rapidly as the temperature warms. Arabian mediterraneangrass seedling roots can tolerate complete dryness for about a month and continue to grow normally upon rehydration. Seeds generally fall near the parent plant, but some plants with intact panicles can break away at ground level and tumble in the wind, dispersing florets and seeds. There is no information on seed longevity in the soil seedbank, but only a small percentage of the seedbank appear to germinate in a given year. However, with such small seeds, it is not likely that the seedbank remains for a very long time.

NON-CHEMICAL CONTROL

Mechanical (pulling, cutting, disking)	Because of their small size, hand removal or string trimming of mediterraneangrasses is impractical and can cause significant disruption of the soil surface, which may promote further weed establishment. Plowing, disking, or scraping may initially reduce surface biomass, but soil disturbance and reduced shading select for the reinvasion of these grasses.
Cultural	Mediterraneangrasses are good forage species and livestock grazing can remove their biomass. However, in the long term, grazing will increase the population of the annual grasses due to reduced cover and increased disturbance.
	Prescribed fire generally promotes the growth of mediterraneangrasses. The small seeds that fall to the ground are not damaged by the high temperatures of the fire. In addition, fire removes the litter layer and can increase soil nutrients, thus leading to increases in annual grasses. Thus, prescribed burning is not recommended for the control of mediterraneangrasses.
Biological	No biological control agents are available for the control of mediterraneangrasses. A black smut, *Ustilago aegyptica*, can form on the spikelets. Natural infestations of smut do not seem to be widespread or severe enough to significantly affect populations.

CHEMICAL CONTROL

The following specific use information is based on reports by researchers and land managers. Other trade names may be available, and other compounds also are labeled for this weed. Directions for use may vary between brands; see label before use. Herbicides are listed by mode of action and then alphabetically. The order of herbicide listing is not reflective of the order of efficacy or preference.

LIPID SYNTHESIS INHIBITORS	
Clethodim *Select, Envoy*	**Rate:** 1 to 2 pt product (*Envoy*)/acre (2 to 4 oz a.i./acre) **Timing:** Postemergence before plants produce viable seeds. **Remarks:** Clethodim is grass-selective and will not damage broadleaf species. It also has no soil activity. Include crop oil concentrate surfactant or non-ionic surfactant. Clethodim has not been tested on mediterraneangrasses, but is expected to provide control. Note that *Envoy* formulation is 1 lb a.i./gallon, *Select* is 2 lb a.i./gallon.
Fluazifop *Fusilade*	**Rate:** Broadcast foliar treatment: 1 to 1.5 pt product/acre (4 to 6 oz a.i./acre). Spot treatment: 0.5% v/v product. **Timing:** Early postemergence; best before boot stage. **Remarks:** Fluazifop is grass-selective and will not damage broadleaf species. It also has no soil activity. Include crop oil concentrate surfactant or non-ionic surfactant. Fluazifop has not been tested on mediterraneangrasses, but is expected to provide control.
Sethoxydim *Poast*	**Rate:** 1.5 to 2 pt product/acre (4.5 to 6 oz a.i./acre) **Timing:** Early postemergence; best before boot stage. **Remarks:** Sethoxydim is grass-selective and has no effect on broadleaf species. It also has no soil activity. Include crop oil concentrate surfactant. Sethoxydim has not been tested on mediterraneangrasses, but is expected to provide control.
AROMATIC AMINO ACID INHIBITORS	
Glyphosate *Roundup, Accord XRT II*, and others	**Rate:** Broadcast foliar treatment: 1 to 2 pt product (*Roundup ProMax*)/acre (0.56 to 1.1 lb a.e./acre). Spot treatment: 1% v/v solution. **Timing:** Postemergence in the beginning of the season when plants are growing rapidly. **Remarks:** Low volume applications are most effective. The small surface area of the plants makes good coverage difficult. Retreatment may be necessary. Glyphosate is a nonselective herbicide. Spot applications are most often recommended.
BRANCHED-CHAIN AMINO ACID INHIBITORS	
Imazapic *Plateau*	**Rate:** 4 to 12 oz product/acre (1 to 3 oz a.e./acre) **Timing:** Preemergence in fall or postemergence in early spring. In colder climates, spring applications after snow melt are better than fall treatments. **Remarks:** Imazapic has mixed selectivity and tends to favor members of the Asteraceae. It has some soil residual activity. Imazapic can tie up in litter and its efficacy may be very much reduced under situations where there is heavy thatch on the soil surface. It has not been tested on mediterraneangrasses, but is expected to provide control. Imazapic is not registered for use in California.
Sulfometuron *Oust* and others	**Rate:** 2 to 6.67 oz product/acre (1.5 to 5 oz a.i./acre) **Timing:** Preemergence or postemergence. Fall and spring applications can both be effective, but fall applications may give full season control. **Remarks:** Sulfometuron has mixed selectivity. It can cause minor damage to some native perennial grasses and has a fairly long soil residual. Higher rates may increase control but will also give more bare ground. Sulfometuron has not been tested on mediterraneangrasses, but is expected to provide control.

Sesbania punicea (Cav.) Benth.

Red sesbania

Family: Fabaceae
Range: Not widespread in the western United States. Occurs throughout the southeast from Virginia to eastern Texas and a few localities in California.
Habitat: Riparian corridors, coastal plains and disturbed sites such as roadsides, ditches, canals and areas adjacent to ornamental plantings. In regions with long, dry summers, red sesbania invades moist areas. Survives occasional freezes but not harsh winters. It is most likely to spread to wildlands adjacent to or downstream from ornamental plantings.
Origin: Native to South America (Argentina, Brazil, Paraguay, and Uruguay). Introduced to the United States as an ornamental.
Impacts: Red sesbania grows rapidly and forms dense stands so thick that access to riparian areas becomes difficult to impossible. It displaces native vegetation used by wildlife and contributes to bank erosion and flooding. Large infestations can decrease water flow and reduce recreational uses. Sesbania can fix nitrogen, which enables the plant to colonize and dominate areas with poor soil. Increased soil nitrogen fertility gives a competitive advantage to other non-native weeds that thrive on high nitrogen levels. Foliage, flowers and seeds contain sesbanimides and saponins that are toxic to humans and animals when ingested. A dose of less than 0.1% of body weight in seeds ingested over a period of days can be lethal.
California Invasive Plant Council (Cal-IPC) Inventory: High Invasiveness (Alert)

Red sesbania is a deciduous shrub or small tree, up to 12 ft tall. The leaves are 3 to 8 inches long, alternate, pinnate-compound and often drooping. Leaflets are mostly even, 10 to 40 per leaf, oblong, 0.5 to 1 inch long and end in a tiny pointed tip. The bark is gray to reddish brown in color and covered with lenticels (airy aggregation of cells that function as a pore, providing a medium for gas exchange between the internal tissues and atmosphere).

The fruit and flowers of *sesbania* are characteristic of the legume family. Showy coral or red flowers are 0.5 to 1 inch long and hang in clusters up to 10 inches long. The distinctive seed pods are longitudinally 4-winged, oblong, 2 to 4 inches long and are often dispersed by water where the pods float and the wings act like sails. Pods are sharply pointed, contain 4 to 10 seeds separated by partitions and make a characteristic rattling sound when shaken. Reproduction is solely by seed production. Plants generally begin to fruit at 2 to 3 years of age and individual trees can survive for up to 15 years. The seed bank is often limited, with seed longevity less than 3 years.

Non-chemical control

Mechanical (pulling, cutting, disking)	Hand pulling can remove seedlings and young plants. The root system is not very large, especially in waterlogged situations, so pulling is relatively easy.
	Cutting sesbania to ground level in spring before it flowers will reduce the number of seeds produced and will deplete the plant's energy reserves. The effectiveness of mechanical methods is increased as *Sesbania punicea* does not produce root sprouts when the shoot is damaged. Stump sprouting can occur, however, and cutting should be combined with an herbicide treatment or with multiple cuttings over a period of years to maximize efficacy. Cut shrubs at ground level with power or manual saws.
	Heavy equipment can be effective but is often not practical, as populations are frequently in waterlogged soils or near riparian areas where access can be limited or difficult. Stumps remaining following such treatment will require herbicide application to prevent regrowth.
Cultural (burning, grazing)	Grazing is not recommended as sesbania is toxic when ingested.
	A technique known as "flaming" or "blanching" which consists of passing a flame over the plant is very effective at controlling emerging seedlings, but not resprouting stumps, especially those in areas with a high water table. Prescribed burning can be used to regulate *S. punicea* in grasslands when conditions are dry, but

	only seedlings and small plants are killed outright. Large plants are mostly only scorched and usually resprout.
Biological	No USDA-approved biocontrol agents exist for this species; however three biocontrol agents are used against *Sesbania punicea* in South Africa. A very successful biological control program has kept red sesbania under control in many parts of South Africa since the 1980s.
	The program includes three introduced agent species; *Trichapion lativentre*, a bud-feeding weevil that feeds on the leaflets as adults and develops within the flower buds as larvae; *Rhyssomatus marginatus*, a weevil whose larvae destroy the ripening seeds within the pods and whose adults feed on the leaves, flowers and meristems of the plants; and *Neodiplogrammus quadrivittatus*, a large stem-boring weevil whose larvae tunnel in the stems and branches causing structural damage, especially to vascular tissues, which eventually kills the plants. All three beetles act in combination to improve control. The bud feeder destroys almost all (> 98%) of the flowers and reduces seed production dramatically. The seed feeder destroys 84% of the seeds that are produced in spite of damage caused by the bud feeder. Together these two species reduce seed production by > 99.8%, rendering the plants almost sterile.

CHEMICAL CONTROL

The following specific use information is based on published papers and reports by researchers and land managers. Other trade names may be available, and other compounds also are labeled for this weed. Directions for use may vary between brands; see label before use. Herbicides are listed by mode of action and then alphabetically. The order of herbicide listing is not reflective of the order of efficacy or preference.

GROWTH REGULATORS	
Triclopyr *Garlon 3A, Garlon 4 Ultra, Pathfinder II*	**Rate:** Foliar treatment: 0.5% v/v solution of *Garlon 4 Ultra* to thoroughly wet all leaves. Cut stump treatment: 0.5 to 1.5% *Garlon 4 Ultra* v/v in water, or 3% *Garlon 3A* v/v in water. *Pathfinder II* is a ready to use formulation. **Timing:** Apply when plants are growing rapidly. **Remarks:** Triclopyr is a selective herbicide to control broadleaf species. For cut stump treatments, cut stems horizontally at or near ground level. Apply herbicide solution immediately after cutting.
Other growth regulator tank mixes and premixes	Tank mixes that include 2,4-D and dicamba, 2,4-D and aminopyralid (*Forefront HL*), and triclopyr and fluroxypyr (*PastureGard*) have also been shown to effectively control red sesbania.
AROMATIC AMINO ACID INHIBITORS	
Glyphosate *Rodeo, Aquamaster*	**Rate:** Foliar treatment: 1 to 1.5% v/v solution of *Rodeo* or *Aquamaster* in water, applied to thoroughly wet all leaves. Cut stump treatment: 10% *Roundup* (or other trade name) v/v in water. **Timing:** Apply when plants are growing rapidly. Foliar treatments should be made in late summer or early fall. For cut stump treatment, application in late summer, early fall or dormant season provides best control. Treatment should occur immediately after cutting. **Remarks:** Glyphosate is a nonselective systemic herbicide with no soil activity. Plants should not be cut for at least 4 months after foliar treatments.
BRANCHED-CHAIN AMINO ACID INHIBITORS	
Imazapyr *Habitat*	**Rate:** Cut stump treatment: 2% v/v solution of *Habitat*. **Timing:** Best when used in late summer to early fall, but before leaf drop. **Remarks:** Imazapyr is a soil residual herbicide and may result in bare ground around trees for some time after treatment. Cut stump applications are as described for triclopyr. *Habitat* is the aquatic formulation of imazapyr and is recommended in most situations.

Silybum marianum (L.) Gaertn.
Blessed milkthistle

Family: Asteraceae
Range: Found in much of the western United States, especially southwestern Oregon and western Washington, California, Nevada, Arizona, New Mexico, and Colorado.
Habitat: Grows in disturbed sites, roadsides, pastures, fields, agronomic crops, waste places, orchards, and trail margins in chaparral and woodlands. Grows best on fertile soils. Blessed milkthistle often occurs in dense, competitive stands. Plant size is very dependent on moisture.
Origin: Native to the Mediterranean region.
Impact: Plants develop large (up to 3 ft in diameter) rosettes that block light to nearby vegetation and suppress germination and growth. Plants can reach 6 to 9 ft in height, and skeletons continue to stand for several months, keeping an area bare of other vegetation. Infestations can be dense and dominant in pastures. The spiny nature of the plant can cause physical injury to livestock.
Western states listed as Noxious Weed: Oregon, Washington
California Invasive Plant Council (Cal-IPC) Inventory: Limited Invasiveness

Blessed milkthistle is an erect winter or summer annual (rarely a biennial) that generally grows to 6 ft tall. The seedlings (first true leaves) and mature plants have prickles and nearly glabrous (smooth), shiny green leaves. The distinguishing characteristic is the upper leaf surfaces, which are conspicuously variegated in white.

The flowerheads consist of numerous large pink to purple disk flowers. Plants reproduce only by seed. Most seeds germinate with the first fall rains, but germination can last through winter until early spring. Seeds disperse only a short distance by wind, but are dispersed longer distances by human movement in crop seed and feed contaminates. Seeds can survive at least 9 years under field conditions.

NON-CHEMICAL CONTROL

Mechanical (pulling, cutting, disking)	Cultivation can control seedlings. Mowing mature plants before flowers open can help control stands. Tillage can be an effective control option for younger plants.
Cultural	Grazing is typically not an option for control, as plants are generally too spiny for animals to use as forage. Because plants develop early in the season, burning is not an effective control option and can encourage seed germination and establishment.
Biological	The seedhead weevil (*Rhinocyllus conicus*) was released in southern California as a biocontrol agent for several thistle species. Although it is fairly common on plants throughout the western United States and can reduce seed production, it has not provided effective control of blessed milkthistle. In addition, the weevil attacks several native thistle species.

CHEMICAL CONTROL
The following specific use information is based on published papers and reports by researchers and land managers. Other trade names may be available, and other compounds also are labeled for this weed. Directions for use may vary between brands; see label before use. Herbicides are listed by mode of action and then alphabetically. The order of herbicide listing is not reflective of the order of efficacy or preference.

GROWTH REGULATORS

2,4-D

Several names

Rate: 3 to 4 pt product/acre (1.43 to 1.9 lb a.e./acre)

Timing: Postemergence in spring or fall to young rapidly growing plants.

Remarks: 2,4-D is a broadleaf-selective herbicide with no soil residual activity. Use fall treatments to control rosettes. Use spring treatments before flower stalk elongates. Annual treatments are needed to control seedlings. Pasture legumes are injured or eliminated at these rates.

Aminocyclopyrachlor + chlorsulfuron

Perspective

Rate: 4.75 to 8 oz product/acre

Timing: Postemergence when target plants are growing rapidly.

Remarks: *Perspective* provides broad-spectrum control of many broadleaf species. Although generally safe for grasses, it may suppress or injure certain annual and perennial grass species. Do not treat in the root zone of desirable trees and shrubs. Do not apply more than 11 oz product/acre per year. At this high rate, cool-season grasses will be damaged, including bluebunch wheatgrass. Not yet labeled for grazing lands. Add an adjuvant to the spray solution. This product is not approved for use in California and some counties of Colorado (San Luis Valley).

Aminopyralid

Milestone

Rate: 3 to 5 oz product/acre (0.75 to 1.25 oz a.e./acre)

Timing: Postemergence in spring or early summer to rosettes or bolting plants or in fall to seedlings and rosettes.

Remarks: Aminopyralid is a broadleaf-selective herbicide. It has moderate soil residual activity. A non-ionic surfactant at 1 to 2 qt per 100 gal of spray solution can enhance control under adverse environmental conditions.

Clopyralid

Transline

Rate: 0.25 to 1 pt product/acre (1.5 to 6 oz a.e./acre)

Timing: Postemergence from the seedling to the bud stage. Best if applied to rapidly growing weeds.

Remarks: Clopyralid is very safe on grasses, but will injure many members of the Asteraceae, particularly thistles, and can also injure legumes, including clovers. Most other broadleaf species and all grasses are not injured.

Clopyralid + 2,4-D amine

Curtail

Rate: 1 to 5 qt product/acre

Timing: Postemergence, to rapidly growing plants after most basal leaves emerge but before bud stage.

Remarks: With CRP applications, it is used only in established grass. For best results, wait at least 20 days after application before disturbing treated areas (cultivation, mowing, fertilization with shank-type applicators) to allow thorough translocation. Apply in enough total spray volume to ensure good coverage.

Dicamba

Banvel, Clarity

Rate: 1 to 2 pt product/acre (0.5 to 1 lb a.e./acre)

Timing: Postemergence before flower stalk lengthens on established plants and for seedling control. Make fall applications to control rosettes.

Remarks: Dicamba is a broadleaf-selective herbicide with little soil residual activity. It is safe on grasses. It may be necessary to repeat applications for several years to control new emerging seedlings.

Dicamba is available premixed with diflufenzopyr in a formulation called *Overdrive*. Diflufenzopyr is an auxin transport inhibitor which causes dicamba to accumulate in shoot and root meristems, increasing its activity. *Overdrive* is applied postemergence at 4 to 8 oz product/acre to plants in the rosette stage. Add a non-ionic surfactant to the treatment solution at 0.25% v/v or a methylated seed oil at 1% v/v solution. Use higher rates on thistles that have bolted.

Picloram

Tordon 22K

Rate: 1 pt product/acre (4 oz a.e./acre)

Timing: Postemergence in fall or spring before plants bolt.

Remarks: Picloram is selective on broadleaf species and has long soil residual activity. Soil residual may last over 1 year after a 0.25 lb a.e./acre application. Follow-up applications may be necessary to control escaped plants. Picloram is a restricted use herbicide. It is not registered for use in California.

AROMATIC AMINO ACID INHIBITORS

Glyphosate + 2,4-D

Campaign

Rate: Broadcast foliar treatment: 1 to 2 pt product/acre. Spot treatment: 1 to 2% v/v solution.

Timing: Postemergence, to plants in the rosette stage in spring or before freeze-up in fall.

	Remarks: This combination is a nonselective product and neither compound has soil activity. There are no grazing restrictions if *Campaign* is used for spot treatments in less than 10% of the total grazed area. Do not cut forage for hay within 30 days of application.
BRANCHED-CHAIN AMINO ACID INHIBITORS	
Chlorsulfuron *Telar*	**Rate:** 1 oz product/acre (0.75 oz a.i./acre) **Timing:** Postemergence, to young rapidly growing weeds. **Remarks:** Chlorsulfuron is a broad-spectrum herbicide, but is considered to be relatively safe on most grasses. It has fairly long soil residual activity. Do not apply to frozen ground. Maintain constant agitation while mixing product with water. Add 0.25% v/v of a non-ionic surfactant to spray mixture. Avoid contact with sensitive crops. Do not treat powdery, dry soils and light, sandy soils if rain is not likely after treatment.
Metsulfuron *Escort*	**Rate:** 1 oz product/acre (0.6 oz a.i./acre) **Timing:** Postemergence, to rapidly growing plants. **Remarks:** Metsulfuron is a broadleaf-selective herbicide and is safe on most grasses. It has some soil activity. Using a non-ionic or silicone-based surfactant can increase its effectiveness. Metsulfuron is not registered for use in California.
Rimsulfuron *Matrix*	**Rate:** 4 oz product/acre (1 oz a.i./acre) **Timing:** Preemergence or early postemergence. **Remarks:** Rimsulfuron controls several annual grasses and broadleaves. Perennial grasses are tolerant to fall applications when established and grown under dryland conditions. Application to rapidly growing or irrigated perennial grasses may result in their injury or death. Rimsulfuron provides soil residual control in cool climates but degrades rapidly under warm conditions. Rimsulfuron will not control summer annual weeds when applied in fall or spring. Add a surfactant when applying postemergence. Rimsulfuron must be activated by rainfall or irrigation of at least half an inch. For the best results, rainfall should occur within 2 to 3 weeks of application and under cooler temperatures. Do not apply more than 4 oz product/acre per year.

Sinapis arvensis L.
Wild mustard

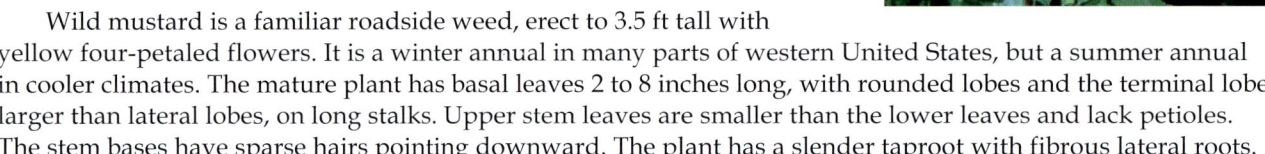

Family: Brassicaceae
Range: Throughout the western United States.
Habitat: Disturbed places, roadsides, fields, pastures, agronomic crops, orchards, vineyards, ditch banks, dry washes.
Origin: Native to Europe.
Impacts: Wild mustard can form dense patches that outcompete desirable plants. It is most common in croplands, roadsides and waste areas, as well as urban environments, but can also be a problem in some natural areas. Wild mustard seeds contain alkaloids that can be toxic to livestock if ingested in large quantities.
California Invasive Plant Council (Cal-IPC) Inventory: Limited Invasiveness

Wild mustard is a familiar roadside weed, erect to 3.5 ft tall with yellow four-petaled flowers. It is a winter annual in many parts of western United States, but a summer annual in cooler climates. The mature plant has basal leaves 2 to 8 inches long, with rounded lobes and the terminal lobe larger than lateral lobes, on long stalks. Upper stem leaves are smaller than the lower leaves and lack petioles. The stem bases have sparse hairs pointing downward. The plant has a slender taproot with fibrous lateral roots.

Wild mustard flowers in late winter to early spring, occasionally in early fall. The flowers appear in racemes. The four petals are pale to bright yellow, 8 to 12 mm long. Flowers are followed by pods (siliques) which are linear, with a persistent beak at the tip, on a stalk that is shorter than the pod. The pods are ascending, not laying close to the stem, 1 to 1.5 inches long, and often slightly constricted between seeds. After senescence, brown stems with pod remnants can persist for a few months. Seeds generally fall near the parent plants. Seeds can survive at least 11 years under field conditions.

NON-CHEMICAL CONTROL

Mechanical (pulling, cutting, disking)	Manual removal or cultivation before seeds develop can control populations. Physical removal is easiest during the seedling stage. Control methods implemented over a period of years will eventually exhaust the seedbank.
	Mowing during the bud to bloom stage can help suppress seed production.
Cultural	Wild mustard is palatable to livestock, but if eaten in excess can cause gastric distress. Grazing is not an effective control method.
	Burning is not an effective control method, as the plants remain succulent into the seeding stage and they produce too early in the season to use broad scale fire. Flaming can be used to control young plants.
Biological	There are no biological control agents available for this weed, primarily because of its close relationship with other economically important mustard crops.

CHEMICAL CONTROL

The following specific use information is based on published papers and reports by researchers and land managers. Other trade names may be available, and other compounds also are labeled for this weed. Directions for use may vary between brands; see label before use. Herbicides are listed by mode of action and then alphabetically. The order of herbicide listing is not reflective of the order of efficacy or preference.

GROWTH REGULATORS	
2,4-D Several names	**Rate:** 1 to 2 qt product/acre (0.95 to 1.9 lb a.e./acre)
	Timing: Postemergence, to rapidly growing plants. Most effective on smaller plants.
	Remarks: 2,4-D is broadleaf-selective. Effective control may require repeat applications. It has no soil activity. Use a surfactant. Do not apply ester formulations when outside temperatures exceed 80°F. 2,4-D can be mixed with various other compounds (e.g., dicamba, triclopyr, carfentrazone), either in

	tank mixes or in commercial combinations.
Aminocyclopyrachlor + chlorsulfuron *Perspective*	**Rate:** 1.75 to 2.75 oz product/acre **Timing:** Postemergence in spring up to flowering. **Remarks:** *Perspective* provides broad-spectrum control of many broadleaf species. Although generally safe for grasses, it may suppress or injure certain annual and perennial grass species. Do not treat in the root zone of desirable trees and shrubs. Do not apply more than 11 oz product/acre per year. At this high rate, cool-season grasses will be damaged, including bluebunch wheatgrass. Not yet labeled for grazing lands. Add an adjuvant to the spray solution. This product is not approved for use in California and some counties of Colorado (San Luis Valley).
Aminopyralid + metsulfuron *Opensight*	**Rate:** 1.5 to 2 oz product/acre **Timing:** Preemergence in fall or postemergence when target plants are in the seedling to rosette stage. **Remarks:** Not registered for use in California.
Dicamba *Banvel, Clarity*	**Rate:** 1 to 2 pt product/acre (0.5 to 1 lb a.e./acre) **Timing:** Postemergence, to rapidly growing plants. Most effective on smaller plants. **Remarks:** Dicamba is a broadleaf-selective herbicide often combined with other active ingredients. It is effective earlier in the season than 2,4-D. It can be tank-mixed with 2,4-D. Has very limited soil residual. Do not apply when outside temperatures exceed 80°F. *Overdrive,* a premix of dicamba with diflufenzopyr, has been reported to be effective on some mustard species. Diflufenzopyr is an auxin transport inhibitor which causes dicamba to accumulate in shoot and root meristems, increasing its activity. *Overdrive* is applied postemergence at 4 to 8 oz product/acre rapidly growing plants. Higher rates should be used on large annuals. Add a non-ionic surfactant to the treatment solution at 0.25% v/v or a methylated seed oil at 1% v/v solution.
Picloram *Tordon 22K*	**Rate:** 1 qt product/acre (8 oz a.e./acre) **Timing:** Preemergence in winter, or early postemergence in late fall or spring. **Remarks:** Picloram has a very long residual activity and should provide 2 years of control. It is a broadleaf-selective herbicide and will not generally injure grasses. Wild mustard is listed on the *Tordon 22K* label, though few other recommendations indicate it is the desired option for control. However, picloram can be used in a premix with fluroxypyr (*Surmount*) for the control of mustards. Picloram is a restricted use herbicide. Picloram formulations are not registered for use in California.
Triclopyr *Garlon 3A, Garlon 4 Ultra*	**Rate:** 0.33 to 1.33 gal *Garlon 3A* /acre, or 0.25 to 1 gal *Garlon 4 Ultra*/acre of (1 to 4 lb a.e./acre) **Timing:** Postemergence, to rapidly growing weeds, up to bud stage. **Remarks:** Triclopyr is broadleaf-selective and safe on most grasses. It is most effective on smaller plants. *Garlon 4 Ultra* is formulated as a low volatile ester. However, in warm temperatures, spraying onto hard surfaces such as rocks or pavement can increase the risk of volatilization and off-target damage.
Triclopyr + 2,4-D *Crossbow*	**Rate:** 1 qt product/acre **Timing:** Postemergence, to small, rapidly growing weeds. **Remarks:** Include non-ionic surfactant.
AROMATIC AMINO ACID INHIBITORS	
Glyphosate *Roundup, Accord XRT II,* and others	**Rate:** 1 to 2 pt product (*Roundup ProMax*)/acre (0.56 to 1.1 lb a.e./acre) **Timing:** Postemergence, to rapidly growing plants from rosette to bud stage. **Remarks:** Glyphosate has no soil activity and is nonselective. Its effectiveness is increased by addition of ammonium sulfate.
BRANCHED-CHAIN AMINO ACID INHIBITORS	
Chlorsulfuron *Telar*	**Rate:** 0.25 to 0.5 oz product/acre (0.19 to 0.375 oz a.i./acre) **Timing:** Preemergence to early postemergence. **Remarks:** Chlorsulfuron has mixed selectivity and is generally safe on grasses. It is most effective preemergence. Use a surfactant for postemergence applications. It has fairly long soil residual activity. Some populations have developed resistance to related herbicides; where resistance is

	suspected, use other herbicides or combinations.
Chlorsulfuron + metsulfuron or sulfometuron *Cimarron X-tra* or *Landmark XP*	**Rate:** 0.5 oz *Cimarron X-tra*/acre; 0.9 oz *Landmark XP*/acre **Timing:** Postemergence, to rapidly growing plants. Most effective on smaller plants. **Remarks:** Mixed selectivity. *Cimarron X-tra* is not registered for use in California.
Imazapic *Plateau*	**Rate:** 4 to 6 oz product/acre (1 to 1.5 oz a.e./acre) **Timing:** Preemergence in fall to postemergence in spring. **Remarks:** Imazapic has mixed selectivity and tends to favor species in the Asteraceae, as well as some grasses. In postemergence applications, use a methylated seed oil surfactant at 1.5 to 2 pt product/acre. It has some soil residual activity. Not registered for use in California.
Metsulfuron *Escort*	**Rate:** 0.33 to 0.5 oz product/acre (0.2 to 0.3 oz a.i./acre) **Timing:** Postemergence, to young, rapidly growing weeds in spring before flowering, or in fall to new rosettes. **Remarks:** Mixed selectivity, generally safe on grasses. Use a surfactant. Can be tank-mixed with 2,4-D and/or dicamba, or with chlorsulfuron. Not registered for use in California.
Propoxycarbazone-sodium *Canter R+P*	**Rate:** 0.9 to 1.2 oz product/acre (0.63 to 0.84 oz a.i./acre) **Timing:** Postemergence to small, rapidly growing plants. **Remarks:** Propoxycarbazone is a broad-spectrum herbicide that will control many species, including wild mustard. Perennial grass species vary in tolerance. A non-ionic surfactant should be added at 0.25 to 0.5% v/v solution.
Rimsulfuron *Matrix*	**Rate:** 2 to 4 oz product/acre (0.5 to 1 oz a.i./acre) **Timing:** From preemergence in fall to early postemergence in spring. **Remarks:** Rimsulfuron controls several annual grasses and broadleaves. Perennial grasses are tolerant to fall applications when established and grown under dryland conditions. Application to rapidly growing or irrigated perennial grasses may result in their injury or death. It provides soil residual control in cool climates but degrades rapidly under warm conditions. Rimsulfuron will not control summer annual weeds when applied in fall or spring. Add a surfactant when applying postemergence.
Sulfosulfuron *Outrider*	**Rate:** 0.75 to 2 oz product/acre (0.56 to 1.5 oz a.i./acre) **Timing:** Early postemergence, winter to early spring, when desirable perennials are dormant. **Remarks:** Sulfosulfuron has mixed selectivity but is fairly safe on native perennial grasses, especially wheatgrasses. To be most effective it may be necessary to add a non-ionic surfactant. Sulfosulfuron has fairly long soil residual activity.
PHOTOSYNTHETIC INHIBITORS	
Hexazinone *Velpar L*	**Rate:** 2 to 4 pt product/acre (0.5 to 1 lb a.i./acre) **Timing:** Preemergence to early postemergence. **Remarks:** Hexazinone has both foliar and soil activity. Its selectivity is mixed. Use higher rates on fine soils or high organic matter soils, or when weeds are under stress. It also has fairly long soil residual activity. Hardwood trees near application site can be damaged when they absorb this chemical through the roots. High rates of hexazinone can create bare ground, so only use high rates in spot treatments.

Sisymbrium altissimum L.; tumble mustard
Sisymbrium irio L.; London rocket

Tumble mustard and London rocket

S. altissimum

Family: Brassicaceae
Range: London rocket is found primarily in the southwestern United States, including California, Nevada, Arizona, Utah, Colorado, and New Mexico. Tumble mustard is found in every contiguous state, including all western states.
Habitat: Both species are found in disturbed areas such as abandoned fields, roadsides, orchards. They are also found in deserts and riparian areas.
Origin: Still unclear but thought to have originated in Europe and Eurasia.
Impact: Both species mature earlier in the year than native species, giving them a competitive advantage. They are widespread in cultivated agricultural fields within their ranges. They can replace native annuals in wildland settings. Like other mustards, London rocket can harbor diseases and pests that attack closely related crops in the mustard family.
California Invasive Plant Council (Cal-IPC) Inventory: *S. irio*, Moderate Invasiveness

S. irio

London rocket is an erect annual or winter annual forb that grows up to 2 to 3 ft tall, more commonly to around 20 inches. Its stems are glabrous or slightly pubescent and have several branches near the base of the plant. They are green and sometimes have a purple tinge. The basal leaves are about 6 inches long with a pronounced midvein. They are deeply lobed, ovate to lanceolate, and alternately positioned along the stem. Leaves along the upper stem are smaller and are narrow or oblong in shape.

Tumble mustard is a winter or summer annual, or even sometimes a biennial, to 4.5 ft tall, with ascending linear pods. Its foliage is sparsely long-hairy and the lower leaves can be 6 inches long, with a triangular shaped terminal lobe and lanceolate lateral lobes. The upper leaves are dissected into long linear lobes.

The inflorescences are terminal racemes. The flowers are bright to pale yellow with four petals. The petals of London rocket are 2.5 to 4 mm long and those of tumble mustard are twice as long. The seed pods grow in an ascending pattern on the upper stem and are linear and cylindrical. The seed pods of London rocket are 1.25 to 1.5 inches long and those of tumble mustard are 2 to 4 inches long. Both species reproduce only by seed and are capable of producing several thousand seeds per plant. The seeds of London rocket disperse when the fruit split and drop the seed to the soil beneath the parent plant. The seeds of tumble mustard disperse when senesced stems break off at ground level and tumble with the wind. Buried seeds of both species survive up to about 10 years.

NON-CHEMICAL CONTROL

Mechanical (pulling, cutting, disking)	Hand pulling is a viable control method if the population is small and isolated. However, it is time-consuming and difficult to use when managing widespread infestations. For smaller plants in the early rosette stage, a hula hoe can be effective.
	Mowing can reduce populations provided it is done before the seeds become viable to eventually exhaust the seedbank.
	Cultivation is very effective when done before seed production.
	Solarization has been shown to be an effective method of killing London rocket plants and seeds in areas with hot summers.

Cultural	Grazing is effective if timed to prevent seed production. Grazing with sheep is preferred over cattle because they graze lower on the plant. Additionally, meat and milk can become tainted when cows consume large quantities of these mustards. Burning is effective when done before seed production, provided there is sufficient fuel to carry a fire. Seed on the soil surface can potentially be killed by the heat from the fire; however, research in Australia showed that burning standing wheat stubble does not achieve a temperature high enough to kill mustard seeds. Temperatures did reach high enough levels to kill seeds when crop residue was concentrated in windrows. Burning is not usually recommended because the fuel needed to carry a fire would likely be after seed production was completed.
Biological	No microbial pathogens or insect biocontrol agents are available for the control of either *Sisymbrium* species.

CHEMICAL CONTROL

The following specific use information is based on published papers and reports by researchers and land managers. Other trade names may be available, and other compounds also are labeled for this weed. Directions for use may vary between brands; see label before use. Herbicides are listed by mode of action and then alphabetically. The order of herbicide listing is not reflective of the order of efficacy or preference.

GROWTH REGULATORS	
2,4-D Several names	**Rate:** 1 to 2 qt product/acre (0.95 to 1.9 lb a.e./acre) **Timing:** Postemergence, to rapidly growing plants. Most effective on smaller plants. **Remarks:** 2,4-D is broadleaf-selective. Effective control may require repeat applications. It has no soil activity. Use a surfactant. Do not apply ester formulations when outside temperatures exceed 80°F. It is heavily restricted in grape-growing areas. 2,4-D can be mixed with various other compounds (e.g., dicamba, triclopyr, carfentrazone), either in tank mixes or in commercial combinations.
Aminocyclopyrachlor + chlorsulfuron *Perspective*	**Rate:** 1.75 to 2.75 oz product/acre **Timing:** Postemergence in spring up to flowering. **Remarks:** Provides broad-spectrum control of many broadleaf species. Generally safe for grasses, but may suppress or injure certain annual and perennial grass species. Do not treat in the root zone of desirable trees and shrubs. Do not apply more than 11 oz product/acre per year. At this high rate, cool-season grasses will be damaged, including bluebunch wheatgrass. Not yet labeled for grazing lands. Add an adjuvant to the spray solution. This product is not approved for use in California and some counties of Colorado (San Luis Valley).
Aminopyralid + metsulfuron *Opensight*	**Rate:** 1.5 to 2 oz product/acre **Timing:** Preemergence in fall or postemergence when target plants are in the seedling to rosette stage. **Remarks:** Not registered for use in California.
Dicamba *Banvel, Clarity*	**Rate:** 1 to 2 pt product/acre (0.5 to 1 lb a.e./acre) **Timing:** Postemergence, to rapidly growing plants. Most effective on smaller plants. **Remarks:** Dicamba is a broadleaf-selective herbicide often combined with other active ingredients. It is effective earlier in the season than 2,4-D. It can be tank-mixed with 2,4-D. Has very limited soil residual. Avoid drift to sensitive crops. Do not apply when outside temperatures exceed 80°F. *Overdrive*, a premix of dicamba with diflufenzopyr, has been reported to be effective on some mustard species. Diflufenzopyr is an auxin transport inhibitor which causes dicamba to accumulate in shoot and root meristems, increasing its activity. *Overdrive* is applied postemergence at 4 to 8 oz product/acre to rapidly growing plants. Higher rates should be used on large annuals. Add a non-ionic surfactant to the treatment solution at 0.25% v/v or a methylated seed oil at 1% v/v solution.
Triclopyr + 2,4-D *Crossbow*	**Rate:** 1 qt product/acre **Timing:** Postemergence, to small, rapidly growing weeds. **Remarks:** Both herbicides are active on only broadleaf species. Include non-ionic surfactant.
AROMATIC AMINO ACID INHIBITORS	
Glyphosate *Roundup, Accord XRT II*, and others	**Rate:** 1 to 2 pt product (*Roundup ProMax*)/acre (0.56 to 1.1 lb a.e./acre) **Timing:** Postemergence, to rapidly growing plants from rosette to bud stage. **Remarks:** Glyphosate has no soil activity and is nonselective. Its effectiveness is increased by addition

of ammonium sulfate.

BRANCHED-CHAIN AMINO ACID INHIBITORS

Chlorsulfuron

Telar

Rate: 0.25 to 0.5 oz product/acre (0.19 to 0.375 oz a.i./acre)

Timing: Preemergence to early postemergence.

Remarks: Chlorsulfuron has mixed selectivity and is generally safe on grasses. It is most effective preemergence. Use a surfactant for postemergence applications. It has fairly long soil residual activity.

Chlorsulfuron + metsulfuron or sulfometuron

Cimarron X-tra or Landmark XP

Rate: 0.5 oz *Cimarron X-tra*/acre; 0.9 oz *Landmark XP*/acre

Timing: Postemergence, to rapidly growing plants. Most effective on smaller plants.

Remarks: Mixed selectivity. *Cimarron X-tra* is not registered for use in California.

Imazapic

Plateau

Rate: 4 to 6 oz product/acre (1 to 1.5 oz a.e./acre)

Timing: Preemergence in fall to postemergence in spring.

Remarks: Mixed selectivity, tends to favor species in the Asteraceae, as well as some grasses. In postemergence applications, use a methylated seed oil surfactant at 0.25% v/v. Some soil residual activity. Not registered for use in California.

Imazapyr

Arsenal, Habitat, Stalker, Chopper, Polaris

Rate: Spot treatment: 1 to 5% v/v solution (*Arsenal*), depending on volume of treatment solution

Timing: Preemergence in fall to postemergence in spring.

Remarks: Imazapyr is a broad-spectrum herbicide with long soil residual activity. In postemergence applications, use a methylated seed oil surfactant at 0.25% v/v solution.

Metsulfuron

Escort

Rate: 0.5 to 2 oz product/acre (0.3 to 1.2 oz a.i./acre)

Timing: Preemergence or postemergence to young, rapidly growing weeds in spring before flowering, or in fall to new rosettes.

Remarks: Mixed selectivity, generally safe on grasses. Some soil residual activity. Use a surfactant. Can be tank-mixed with 2,4-D and/or dicamba, or with chlorsulfuron. Not registered for use in California.

Propoxycarbazone-sodium

Canter R+P

Rate: 0.9 to 1.2 oz product/acre (0.63 to 0.84 oz a.i./acre)

Timing: Postemergence, to small, rapidly growing plants.

Remarks: Broad-spectrum herbicide that will control many species, including both London rocket and tumble mustard. Perennial grass species vary in tolerance. A non-ionic surfactant should be added at 0.25 to 0.5% v/v solution.

Rimsulfuron

Matrix

Rate: 2 to 4 oz product/acre (0.5 to 1 oz a.i./acre)

Timing: Preemergence in fall to early postemergence in spring.

Remarks: Controls several annual grasses and broadleaves. Perennial grasses are tolerant to fall applications when established and grown under dryland conditions. Application to rapidly growing or irrigated perennial grasses may result in their injury or death. Provides soil residual control in cool climates but degrades rapidly under warm conditions. Rimsulfuron will not control summer annual weeds when applied in fall or spring. Add a surfactant when applying postemergence.

Sulfosulfuron

Outrider

Rate: 0.75 to 2 oz product/acre (0.56 to 1.5 oz a.i./acre)

Timing: Early postemergence, winter to early spring, when desirable perennials are dormant.

Remarks: Mixed selectivity, but fairly safe on native perennial grasses, especially wheatgrasses. To be most effective it may be necessary to add a non-ionic surfactant. Fairly long soil residual activity.

PHOTOSYNTHETIC INHIBITORS

Hexazinone

Velpar L

Rate: 2 to 4 pt product/acre (0.5 to 1 lb a.i./acre)

Timing: Preemergence to early postemergence.

Remarks: Both foliar and soil activity. Its selectivity is mixed. Use higher rates on fine soils or high organic matter soils, or when weeds are under stress. Fairly long soil residual activity. Hardwood trees near application site can be damaged when they absorb this chemical through the roots. High rates of hexazinone can create bare ground, so only use high rates in spot treatments.

Sorghum halepense (L.) Pers.

Johnsongrass

Family: Poaceae
Range: Nearly all contiguous states, including all western states.
Habitat: Disturbed sites, roadsides, fields, agronomic and vegetable crops. Grow best on fertile, moist, well-drained soils in warm temperate to subtropical regions where some warm-season moisture is available. Also found in orchards, vineyards, cotton fields, and ditchbanks. Occasionally in undisturbed wildlands, particularly riparian areas.
Origin: Native to the Mediterranean region.
Impact: Johnsongrass invades riverbank communities and disturbed sites, particularly fallow fields and forest edges, where it crowds out native species and slows succession. It quickly dominates the herbaceous flora and reduces plant diversity. Johnsongrass grows rapidly, is highly competitive with other plants, and can be difficult to control. In addition, it can hybridize with commercial sorghum cultivars. Healthy plants can provide good forage for livestock, but the foliage can produce toxic amounts of hydrocyanic acid when exposed to frost, stressed by drought, or damaged by trampling or herbicides. Under these conditions, the foliage can be poisonous to livestock when ingested. Young shoots and second growth are typically more dangerous than uncut mature plants because cyanide levels are higher. Dried plant material does not lose its toxicity, but well-cured hay from healthy mature plants is usually safe. Under certain conditions, plants may accumulate toxic levels of nitrates. Weedy sorghums are subject to various bacterial, fungal, and nematode infections. They also serve as alternate hosts for the sorghum midge (*Contarinia sorghicola*) and the viruses that cause sugarcane mosaic virus, maize chlorotic dwarf virus, and corn stunt disease.
Western states listed as Noxious Weed: California, Colorado, Idaho, Nevada, Oregon, South Dakota, Utah, Washington

Johnsongrass is a coarse, typically tufted perennial grass to 6 ft tall with tillers from the crown. The stems are erect and unbranched and the leaves are rolled in the bud, flat, glabrous to sparsely hairy, especially near the ligules. The leaf margins are scabrous with a conspicuously whitish midvein. The ligules are membranous, 3 to 6 mm long, with a fringe of hairs at the top. Johnsongrass has vigorous, coarse rhizomes up to 0.5 inch thick and 6 ft long. The rhizomes are whitish with large brown or purplish brown scales at the nodes, often rooting at the nodes. The seedlings resemble young corn seedlings, but can be distinguished by carefully removing young seedlings from the soil and examining the attached seed.

Inflorescences are open pyramid-shaped panicles 4 to 20 inches long. They are initially pale green or greenish-violet, but often mature to a dark reddish- or purplish-brown. Some panicles shed spikelets (shatter) at maturity. The lemma has awns that are bent, twisted, early deciduous. Johnsongrass reproduces both by seed and vegetatively from rhizomes. The seeds primarily fall near the parent plant. Some seeds survive ingestion by birds and mammals. Dormant seed can survive for at least 6 years under field conditions, and it has been estimated that some seed may remain viable for up to 15 years.

NON-CHEMICAL CONTROL

Mechanical (pulling, cutting, disking)	Hand pulling is considered too slow and not an effective control method unless all rhizomes are removed or new sprouts are controlled. Large mature plants are almost impossible to pull by hand. Rhizomes break easily and are often left in the soil where they will resprout. Best results are obtained in early spring when soil is moist and rhizomes are least likely to break.
	Johnsongrass does not tolerate repeated, close mowing. Such a mowing regime can kill johnsongrass seedlings, prevent seed production, and reduce rhizome growth and regrowth of shoots. In most cases, however, mowing does not kill or eliminate established plants.
	Repeated tillage can be an effective control strategy. Smaller rhizome fragments are brought to the soil surface, where they are susceptible to desiccation and freezing winter temperatures. Shoots developing from

	these rhizome fragments are less vigorous. If cultivation is not repeated, the infestation can spread, since broken rhizome segments can produce roots and shoots. Repeated tillage (e.g. six times at 2-week intervals during the growing season) prevents rhizome development and reduces johnsongrass populations. Tilling can be used on some sites such as bottomlands and old fields. Shallow plowing helps control johnsongrass by breaking up rhizome systems, exposing rhizomes to the sun or killing frosts, and depleting carbohydrate reserves. First plowing is in spring (May), followed by similar plowings every 3 to 6 weeks.
Cultural	Grazing and burning are not effective for the control of johnsongrass.
	Prevent johnsongrass from becoming established in new areas. This can be accomplished by preventing the production of seed, the spread of rhizomes from infested to uninfested areas, and by controlling seedlings originating from shattered seed.
Biological	No biological control agents are currently available for the management of johnsongrass.

CHEMICAL CONTROL

The following specific use information is based on published papers and reports by researchers and land managers. Other trade names may be available, and other compounds also are labeled for this weed. Directions for use may vary between brands; see label before use. Herbicides are listed by mode of action and then alphabetically. The order of herbicide listing is not reflective of the order of efficacy or preference.

LIPID SYNTHESIS INHIBITORS	
Clethodim *Select*, *Envoy*	**Rate:** 9 to 32 oz product (*Envoy*)/acre (2.25 to 8 oz a.i./acre) **Timing:** Postemergence, to rapidly growing johnsongrass; make the first application when rhizome johnsongrass is 12 to 24 inches tall and the second, if necessary, when it is 6 to 18 inches tall. **Remarks:** May require several applications with proper timing to insure effectiveness. Note that *Envoy* formulation is 1 lb a.i./gallon, *Select* is 2 lb a.i./gallon.
Fluazifop *Fusilade*	**Rate:** 1 to 1.5 pt product/acre (4 to 6 oz a.e./acre) **Timing:** Postemergence, to rapidly growing johnsongrass 8 to 18 inches tall but before boot stage. **Remarks:** Apply with 1% v/v crop oil concentrate or 0.25% v/v non-ionic surfactant. Fluazifop acts very slowly, taking 2 to 4 weeks to show effectiveness. Do not apply to stressed grasses. If weed regrows, repeat application at 4 to 6 oz a.e./acre. Maximum use rate per year varies by state.
Sethoxydim *Poast*	**Rate:** 1.5 to 2.5 pt product/acre (4.5 to 7.5 oz a.e./acre) **Timing:** Postemergence, to rapidly growing johnsongrass; make the first application at 1.5 to 2.5 pt product/acre when johnsongrass is 10 inches tall and the second application, if necessary, at 1 to 1.5 pt product/acre when plants are 8 inches tall. **Remarks:** Apply with 2 pt/acre crop oil concentrate. Sethoxydim acts very slowly, taking 2 to 4 weeks to show effectiveness. Do not apply to stressed grasses.
AROMATIC AMINO ACID INHIBITORS	
Glyphosate *Roundup*, *Accord XRT II*, and others	**Rate:** 2 to 3 qt product (*Roundup ProMax*)/acre (2.25 to 3.375 lb a.e./acre) **Timing:** Postemergence, to plants that are growing rapidly and about 18 inches tall to the early flowering stage. Fall applications are also effective before seed production. **Remarks:** Glyphosate is a nonselective herbicide with no soil activity. Multiple applications will be required. After mowing, glyphosate can be applied at 2 lb product/acre within 2 to 3 weeks. Spot control is not effective unless surrounding seed sources are also eliminated. Control is not as effective when applied to moisture-stressed plants.
BRANCHED-CHAIN AMINO ACID INHIBITORS	
Imazapic *Plateau*	**Rate:** 12 oz product/acre (3 oz a.e./acre) **Timing:** Postemergence at the late-boot or early bloom stage. **Remarks:** Use a methylated seed oil at 1 qt/acre. Do not exceed 25 gal/acre spray volume. Imazapic is not registered for use in California.
Imazapic + glyphosate *Journey*	**Rate:** 21 to 32 oz product/acre **Timing:** Postemergence when grass is 18 to 24 inches tall. **Remarks:** Add a suitable surfactant to the spray mix. Not registered for use in California.

Imazapyr *Arsenal, Habitat, Stalker, Chopper, Polaris*	**Rate:** 2 to 3 pt product (*Habitat*)/acre (8 to 12 oz a.e./acre) **Timing:** Postemergence, to perennial johnsongrass. Residual activity may control emerging seedlings. **Remarks:** Untreated plants can occasionally be affected by the uptake of imazapyr through movement into the topsoil.
Propoxycarbazone-sodium *Canter R+P*	**Rate:** 0.9 to 1.2 oz product/acre (0.63 to 0.84 oz a.i./acre) **Timing:** Postemergence from the 2-leaf to 2-tiller stage when plants are growing rapidly. **Remarks:** Propoxycarbazone is a broad-spectrum herbicide that will control many species, and will give partial control of johnsongrass. Perennial grass species vary in tolerance. A non-ionic surfactant should be added at 0.25 to 0.5% v/v solution.
Rimsulfuron *Matrix*	**Rate:** 3 to 4 oz product/acre (0.75 to 1 oz a.i./acre) **Timing:** Postemergence, to seedlings. **Remarks:** Rimsulfuron gives good to excellent control, depending on the size of the plant. It controls several annual grasses and broadleaves. Perennial grasses are tolerant to fall applications when established and grown under dryland conditions. Application to rapidly growing or irrigated perennial grasses may result in injury or death of the crop. It provides soil residual control in cool climates but degrades rapidly under warm conditions. Rimsulfuron will not control summer annual weeds when applied in fall or spring. Add a surfactant when applying postemergence.
Sulfosulfuron *Outrider*	**Rate:** 0.75 to 2 oz product/acre (0.56 to 1.5 oz a.i./acre) **Timing:** Postemergence in spring or fall. **Remarks:** Sulfosulfuron is a broad-spectrum herbicide that may damage other non-target plants. For best weed control do not mow or graze 2 weeks before or after treatment. The use of this chemical in areas where soils are permeable, particularly where the water table is shallow, may result in groundwater contamination.

Spartina alterniflora Loisel. and hybrids; smooth cordgrass
Spartina anglica C.E. Hubbard; common cordgrass
Spartina densiflora Brongn.; dense-flowered cordgrass
Spartina patens (Aiton) Muhlenb.; salt-meadow cordgrass

Cordgrasses

Family: Poaceae
Range: Coastal estuaries of California, Oregon and Washington. *Spartina anglica* is not reported from Oregon and *S. densiflora* is not reported from Washington.
Habitat: Restricted to marine salt marsh and mud flat habitats.
Origin: *S. alterniflora* was unintentionally introduced from the Atlantic coast. *S. anglica* is native to the United Kingdom. *S. densiflora* is native to southern South America and *S. patens* is native to the southeastern U.S.
Impacts: One plant can develop into a large, dense, circular clonal patch up to about 80 ft in diameter. Numerous individuals on barren mud flats can spread until no open space remains. This can significantly impact shorebird feeding habitat. Because smooth cordgrass can grow in deeper water, it can colonize open mud flats that are normally devoid of vegetation. Mud flats densely populated with smooth cordgrass do not provide suitable habitat for foraging shorebirds. Smooth cordgrass also displaces native vegetation higher on the shore in salt marshes.
Western states listed as Noxious Weed: All species are listed in Oregon and Washington
California Invasive Plant Council (Cal-IPC) Inventory: *S. alterniflora* and hybrids, *S. densiflora*, High Invasiveness (Alert); *S. anglica*, Moderate Invasiveness (Alert); *S. patens*, Limited Invasiveness

Cordgrasses are perennial grasses that can grow to 8 ft tall. Smooth cordgrass has extensive creeping rhizomes and spreads in a circular patch from the point of establishment. In contrast, the other cordgrasses are bunchgrasses. In the southern San Francisco Bay, smooth cordgrass readily hybridizes with California cordgrass (*S. foliosa*), and the hybrid is more fecund and invasive than the non-native parent plant. Cordgrass leaves are 8 to 20 inches long and generally lack hairs. The ligules consist of a fringe of hairs 0.5-2 mm long.

Inflorescences are panicles 4 to 16 inches long, consisting of 5 to 30 spike-like branches. Cordgrasses are primarily outcrossers. Plants reproduce by seed and vegetatively from creeping rhizomes and rhizome fragments. Both the seeds and rhizome fragments can disperse with water. The seeds typically do not survive for more than one year. Seedheads are generally susceptible to infection by an ergot fungus, which can limit viable seed production.

NON-CHEMICAL CONTROL

Mechanical (pulling, cutting, disking)	Bunchgrass type spartina species are easier to control using mechanical methods compared to the creeping forms (i.e., *S. alterniflora* and hybrids).
	Hand pulling of small patches can be effective. However, seedlings begin to tiller in the first season. For dense-flowered cordgrass one tool used was a metal-bladed brush cutter that cut into the shallow rhizomes in the top 4 inches of the marsh.
	Cutting alone can reduce populations and stem density, but will not give effective control. Combining cutting and smothering are reported to be highly effective against *S. anglica*, achieving around 98% control.
	Mechanical removal with heavy equipment has been used in various estuaries, although it is labor intensive and expensive. As with hand pulling, all rhizome fragments must be removed to prevent resprouting. Many other mechanical tools have been tried, including covering, digging, crushing, disking, tilling. Tillage and disking are best applied in winter and one study showed that it took 3 to 6 consecutive years of treatment to reach 99% control. Summer was the best time to apply crushing but this treatment required 9 to 10 years of

	application.
Cultural	Few cultural controls are available, although covering small patches with black plastic for a long period suppressed regrowth. Grazing and burning are not appropriate tools in the habitats where spartina invades.
Biological	*Prokelisia marginata* is a planthopper native to the Atlantic and Gulf coasts of North America. It has been shown to oviposit on spartina leaves and cause scars near the base of the plant. However, there does not appear to be any significant effect on the control of spartina by the insect.

CHEMICAL CONTROL

The following specific use information is based on published information or reports by researchers and land managers. Other trade names may be available, and other compounds also are labeled for this weed. Directions for use may vary between brands; see label before use. Herbicides are listed by mode of action and then alphabetically. The order of herbicide listing is not reflective of the order of efficacy or preference.

AROMATIC AMINO ACID INHIBITORS	
Glyphosate *Rodeo,* *Aquamaster*	**Rate:** Broadcast foliar treatment: 2 to 4 gal product *Rodeo* or *Aquamaster*/acre (8 to16 lb a.e./acre). Spot treatment: 2 to 5% v/v solution through hand-held spray equipment, or 33% solution with wiper applicators. **Timing:** Postemergence, to rapidly growing cordgrass any time from late June until first killing frost. **Remarks:** High rates of glyphosate are needed. Treat at least 6 hours before tidewater will cover plants. Glyphosate precipitates in water high in divalent and trivalent salts. Thus, applications to marine species can be inconsistent and not as effective as imazapyr. Debris and silt on cordgrass also reduce performance. Glyphosate can be used in combination with imazapyr.
BRANCHED-CHAIN AMINO ACID INHIBITORS	
Imazapyr *Habitat*	**Rate:** Broadcast foliar treatment: 4 to 6 pt product/acre (1 to 1.5 lb a.e./acre). Spot treatment: 2.5 to 7.5% v/v solution **Timing:** Postemergence mid-season from mid-June to mid-September was considered the best timing in a study in Washington. **Remarks:** Imazapyr is the preferred treatment and has shown the most consistent and effective results. Aerial applications are most often used when large infestations are being treated, but backpack sprayers can be used for smaller patches. Add suitable adjuvant to spray solution. A list of adjuvants can be found at http://www.spartina.org/rfq/RFQ-Att4_Products.pdf.

Spartium junceum L.
Spanish broom

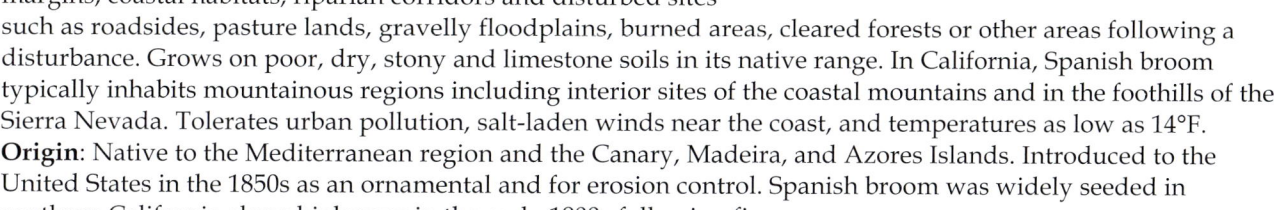

Family: Fabaceae
Range: Along the Pacific coast from Washington to southern California and Hawaii. Also found in other states as a cultivated ornamental (Utah, the Intermountain west, and Texas).
Habitat: Grasslands, shrublands, oak woodlands, forest margins, coastal habitats, riparian corridors and disturbed sites such as roadsides, pasture lands, gravelly floodplains, burned areas, cleared forests or other areas following a disturbance. Grows on poor, dry, stony and limestone soils in its native range. In California, Spanish broom typically inhabits mountainous regions including interior sites of the coastal mountains and in the foothills of the Sierra Nevada. Tolerates urban pollution, salt-laden winds near the coast, and temperatures as low as 14°F.
Origin: Native to the Mediterranean region and the Canary, Madeira, and Azores Islands. Introduced to the United States in the 1850s as an ornamental and for erosion control. Spanish broom was widely seeded in southern California along highways in the early 1900s following fires.
Impacts: Spanish broom grows rapidly and forms dense stands that most wildlife find impenetrable and unpalatable. Dense stems make regeneration of most other plant species difficult or impossible, and the accumulation of woody biomass creates a dangerous fire hazard. Broom can fix nitrogen, which increases soil fertility and gives a competitive advantage to other non-native weeds that thrive on high nitrogen levels.
Western states listed as Noxious Weed: California, Oregon, Washington
California Invasive Plant Council (Cal-IPC) Inventory: High Invasiveness

Spanish broom is a rapidly growing, deciduous shrub, 10 to 15 ft tall. The stems are long, smooth, slender, cylindrical and erect with few branches, appearing rush-like. Spanish broom leaves are small, 0.5 to 1 inch long, oval, and smooth-margined. Leaves are ephemeral, remaining on the plant for 4 months or less, giving the plant a leafless appearance. Because the leaves are small and deciduous, the stems are green and are the primary photosynthetic tissue of the plant.

The yellow, pea-like flowers are produced on current-year shoots in terminal racemes with several flowers. The flowers are large, up to 1 inch long, and grow on short stalks on both sides of the main stem. Reproduction is by seed and plants begin flowering from 18 months to 3 years of age. Seeds are produced in slightly flattened pods 1.5 to 4 inches long. When mature, pods are dark brown, covered with long, silky, silvery hairs, and contain 10 to 18 seeds. Seed dispersal occurs when pods eject the seeds several feet from the plant. Seeds can remain viable in the soil for up to 30 years. Large soil seedbanks often accumulate making long term control difficult. Shrubs may live for up to 30 years.

NON-CHEMICAL CONTROL

Mechanical (pulling, cutting, disking)	Hand pulling can remove seedlings and small shrubs, but once established this technique is generally not effective. For larger established shrubs, a weed wrench or other woody weed extractor can be used. Care must be taken to extract the entire root or stump sprouting will occur. Best results are achieved when soil is moist. Disturbing the soil can stimulate the seedbank.
	Cutting broom off before it flowers will reduce seed production and will deplete the plant's energy reserves. Resprouts are common after treatment. Cutting broom at the end of the dry season can help reduce resprouting from the root crown. Cutting should be combined with an herbicide treatment or with multiple cuttings over a period of years. Cut shrubs at ground level with power or manual saws.
	Heavy equipment can be effectively used to control broom in areas where soil disturbance and nonselective species removal are not important considerations. Stumps remaining following such treatment will require herbicide application to prevent regrowth.
Cultural	Grazing is not considered an effective control option. Flowers and seeds of brooms contain quinolizidine alkaloids and can be toxic to humans and livestock when ingested. Foliage may be mildly toxic and is unpalatable to most livestock, except goats. Goats confined to a small area can help control resprouting stands after a cutting or burn treatment.

	Burning alone is not an effective method for controlling broom. Although burning can remove large amounts of debris, it can increase the population as it removes competitive vegetation, releases nutrients into the soil, and stimulates the germination of broom seeds left in the soil. Burning is more effective if followed with an herbicide application, subsequent burnings, and/or revegetation using desirable species. It is important to employ a control strategy following a burn, otherwise the broom population in subsequent years may become worse than before.
Biological	Three insects have been introduced as biological control agents on broom. These include the Scotch broom seed beetle (*Bruchidius villosus*), the Scotch broom seed weevil (*Apion fuscirostre*), and the Scotch broom twig miner moth (*Leucoptera spartifoliella*). The latter two species are specific to Scotch broom, while the seed beetle also attacks Portuguese broom, Spanish broom, and French broom.

CHEMICAL CONTROL

The following specific use information is based on published papers and reports by researchers and land managers. Other trade names may be available, and other compounds also are labeled for this weed. Directions for use may vary between brands; see label before use. Herbicides are listed by mode of action and then alphabetically. The order of herbicide listing is not reflective of the order of efficacy or preference.

GROWTH REGULATORS	
Picloram *Tordon 22K*	**Rate:** Broadcast foliar treatment: 2 qt product/acre (non-cropland) or 1 qt product per acre (rangeland), plus 0.25 to 0.5% v/v surfactant. **Timing:** Postemergence foliar treatments are best when plants are growing rapidly at or beyond early to full bloom stage. **Remarks:** High levels of picloram can give long-term soil activity for broadleaves. Picloram is a restricted use herbicide. It is not registered for use in California.
Triclopyr *Garlon 3A, Garlon 4 Ultra, Pathfinder II*	**Rate:** Spot treatment: 0.75 to 1.5% v/v solution of *Garlon 4 Ultra*, or 1 to 1.5% *Garlon 3A* and water plus 0.25 to 0.5% v/v surfactant; apply to thoroughly wet all leaves. Low volume/thinline treatment: 10% v/v solution of *Garlon 4 Ultra* plus 20% v/v ethylated crop oil in water. Cut stump treatment: 20% v/v *Garlon 4 Ultra* in water, or undiluted *Garlon 3A* or 50% *Garlon 3A* in water. Basal bark or basal cut stump treatment: 20% v/v *Garlon 4 Ultra* in 20% v/v ethylated crop oil and water, or *Pathfinder II* as a ready-to-use formulation. **Timing:** Postemergence when plants are growing rapidly. Cut stump and basal bark treatments can be applied any time as long as the ground is not frozen. **Remarks:** Triclopyr is a selective herbicide for broadleaf species and will not damage desirable grasses growing nearby. For cut stump treatments, cut stems horizontally at or near ground level. For *Garlon 3A*, apply herbicide solution immediately after the stump is cut. Suckering from the roots typically occurs after cutting, but the treatment should control most resprouts. For basal bark treatment, treat to a height of 12 to 18 inches from the ground. Thorough wetting is necessary for good control. Plants should not be cut for at least one month after basal bark treatments.
Triclopyr + 2,4-D *Crossbow*	**Rate:** Spot treatment: 0.5 to 1.5% v/v solution of *Crossbow* and water to thoroughly wet all leaves. **Timing:** Postemergence when plants are growing rapidly. **Remarks:** *Crossbow* in water forms an emulsion (not a solution), and separation may occur unless the spray mixture is agitated continuously.
AROMATIC AMINO ACID INHIBITORS	
Glyphosate *Roundup, Accord XRT II*, and others	**Rate:** Spot treatment: 1.5 to 2% v/v solution of *Roundup ProMax* (or other trade name with similar concentration of glyphosate) in water to thoroughly wet all leaves. Low volume/thinline treatment: 10% v/v solution of *Roundup* (or other trade name) in water. Cut stump treatment: 25 to 50% v/v *Roundup* (or other trade name) in water; higher rate can reduce resprouting but may exceed label rate if stands are dense. **Timing:** Postemergence when plants are growing rapidly. Foliar treatments should be made in late summer or early fall. For cut stump treatment, application in late summer, early fall or dormant season provides best control. Treatment should occur immediately after cutting. **Remarks:** Glyphosate is a nonselective systemic herbicide. It gives good control with some resprouts. Plants should not be cut for at least 4 months after foliar treatments. Cut stump applications are as described for triclopyr.

BRANCHED-CHAIN AMINO ACID INHIBITORS	
Imazapyr *Arsenal, Habitat, Stalker, Chopper, Polaris*	**Rate:** Spot treatment: 1 to 2% v/v solution of *Stalker* plus 0.25 to 0.5% surfactant v/v in water; apply to thoroughly wet all leaves. Low volume/thinline treatment: 10% v/v solution of *Stalker* plus 20% v/v ethylated crop oil in water. Cut stump treatment: 20% v/v solution of *Stalker* plus 20% v/v ethylated crop oil in water or 20% *Habitat* v/v in 80% water carrier. Basal bark treatment: 20% v/v solution of *Stalker* plus 20% v/v ethylated crop oil in water. **Timing:** Postemergence when plants are growing rapidly. Best when used in late summer to early fall. **Remarks:** Imazapyr is a soil residual herbicide and may result in bare ground around plants for some time after treatment. Cut stump and basal bark applications are as described for triclopyr. Plants should not be cut for at least 4 months after basal bark treatment.

Sphaerophysa salsula (Pall.) DC.

Swainsonpea

Family: Fabaceae
Range: Much of the western United States, with the exception of North and South Dakota. Considered eradicated in California.
Habitat: Disturbed sites, roadsides, irrigation ditches, cultivated crops. It has a high potential for establishment along streams, irrigation canals, waste ways, pastures, and meadows with a high water table. Has become a problem in some poorly drained, marshy, or saline areas of the western U.S. Most often found as a weed where alfalfa is grown for seed, because the seeds of the two species closely resemble each other.
Origin: Native to Asia. It may have been introduced to the United States for forage or soil stabilization.
Impacts: Mostly a problem for alfalfa seed producers. Can somewhat impact native plant diversity along riparian areas. Can impact mineral cycling or nutrient dynamics of an area. It is also unpalatable to livestock and wildlife.
Western states listed as Noxious Weed: California, Nevada, Oregon, Washington

Swainsonpea is an herbaceous perennial to 5 ft tall, with pinnate-compound leaves. Plants often develop an extensive system of vigorous, woody, creeping, horizontal roots that develop new shoots. Stems are erect to ascending and covered with short white hairs. The leaves are odd pinnate-compound with 15 to 23 leaflets that are oblong to ovate and up to 1 inch long. The upper surface of leaflets is mostly glabrous, while the lower surface is covered with short, white hairs. Roots are associated with nitrogen-fixing bacteria. Plants are often long-lived.

The pea-like flowers are about 0.5 inch long and brick-red to orange-red. They are arranged in axillary racemes near the stem tips. The pods inflate at maturity and become ovoid to spherical, 0.75 to 1.5 inches long with a short stalk-like base. The pods are indehiscent with numerous seeds. Plants spread by seeds and lateral creeping roots. Entire pods disperse as units with seeds enclosed. A large proportion of the seed is hard-coated and requires scarification to germinate. Like other members of the Fabaceae, the seeds likely survive for many years in the seedbank.

NON-CHEMICAL CONTROL

Mechanical (pulling, cutting, disking)	Small populations can be controlled by manually removing individual plants, including as much of the root system as possible, followed by frequent removal of root sprouts and seedlings.
	Mowing may reduce seed production, but will not completely control this species.
	Tillage may be ineffective due to an extensive creeping root system that sends up numerous shoots, and may spread severed rootstocks to new areas.
Cultural	Grazing may reduce seed production, but will not completely control this species. Cattle likely prefer the seed pods of swainsonpea, and seed viability probably remains high after passing through animals. Therefore, cattle should be removed from areas after seed production. Like many other legumes, the seeds are extremely hard and may be viable in the soil for many years.
	Burning is not an effective control method, as fire stimulates root sprouting.
Biological	Currently, no registered biocontrol agent for swainsonpea is available in the United States.

CHEMICAL CONTROL

The following specific use information is based on published papers and reports by researchers and land managers. Other trade names may be available, and other compounds also are labeled for this weed. Directions for use may vary between brands; see label before use. Herbicides are listed by mode of action and then alphabetically. The order of herbicide listing is not reflective of the order of efficacy or preference.

GROWTH REGULATORS	
2,4-D Several names	**Rate:** 1 qt product/acre (0.95 lb a.e./acre) **Timing:** Postemergence, to rapidly growing plants, particularly at the budding stage. **Remarks:** 2,4-D is a broadleaf herbicide with no soil residual activity. Repeated applications at the label rate for 2 or more years may be effective. Do not apply ester formulations when outside temperatures exceed 80°F. 2,4-D appears to be the herbicide most commonly used to control swainsonpea.
Aminopyralid *Milestone*	**Rate:** 5 to 7 oz product/acre (1.25 to 1.75 oz a.e./acre) **Timing:** Postemergence in early bloom or during fall when translocation of carbohydrates to the roots is maximized. **Remarks:** Aminopyralid is a broadleaf herbicide similar to picloram, but more selective and with shorter soil residual activity. It is generally very safe on grasses. Broadcast applications will also provide preemergence control of germinating seeds.
Clopyralid *Transline*	**Rate:** 0.25 to 1.33 pt/acre (1.5 to 8 oz a.e./acre) **Timing:** Postemergence in early bloom or during fall when translocation of carbohydrates to the roots is maximized. **Remarks:** Clopyralid selectively controls certain broadleaf species, particularly members of the Asteraceae and Fabaceae. It is safe on grasses. Clopyralid has some soil residual activity, but not as long as aminopyralid. Repeat treatments will be necessary until seedbank is depleted, possibly several years.
Dicamba *Banvel*, *Clarity*	**Rate:** 2 to 3 qt/acre (2 to 3 lb a.e./acre) **Timing:** Postemergence in early bloom or during fall when translocation of carbohydrates to the roots is maximized. **Remarks:** Dicamba is broadleaf-selective and has short soil activity. Do not apply when outside temperatures exceed 80°F.
Picloram *Tordon 22K*	**Rate:** 1 to 2 qt product/acre (0.5 to 1 lb a.e./acre) **Timing:** Postemergence in spring or fall. **Remarks:** Picloram is one of the most effective chemical control options. It has long soil residual, so broadcast applications will also control germinating seed. *Tordon 22K* is a federally restricted use pesticide. Picloram is not registered for use in California.
Triclopyr *Garlon 3A*, *Garlon 4 Ultra*	**Rate:** 1 to 1.5 pt *Garlon 4 Ultra* /acre (0.5 to 0.75 lb a.e./acre) or 1 qt *Garlon 3A* (0.75 lb ae/acre) **Timing:** Postemergence in early bloom or during fall when translocation of carbohydrates to the roots is maximized. **Remarks:** Triclopyr is a growth regulator herbicide with little or no soil residual activity. It is broadleaf-selective and typically does not harm grasses. *Garlon 4 Ultra* is formulated as a low volatile ester. However, in warm temperatures, spraying onto hard surfaces such as rocks or pavement can increase the risk of volatilization and off-target damage.
AROMATIC AMINO ACID INHIBITORS	
Glyphosate *Roundup*, *Accord XRT II*, and others	**Rate:** 2 qt product (*Roundup ProMax*)/acre (2.25 lb a.e./acre) **Timing:** Postemergence in early bloom or during fall when translocation of carbohydrates to the roots is maximized. **Remarks:** Glyphosate has no soil activity and is nonselective. Repeated applications will probably be necessary. This species is a good candidate for wiper applications at 33% to 50% v/v solution. Regrowth is likely and reapplication may be necessary.

Sporobolus indicus (L.) R. Br.

Smutgrass

Family: Poaceae
Range: California, Oregon, Hawaii. Other regions of the United States.
Habitat: Roadsides, turf, ditches, pastures, especially irrigated pastures, turf, occasionally agronomic crop fields and other disturbed open places.
Origin: Native to tropical America.
Impacts: Coarse and less palatable than other pasture species, so livestock avoid grazing it. As a result, smutgrass may invade and take over irrigated pastures.

Smutgrass is a warm-season perennial bunchgrass up to 3.5 ft tall, with dense, slender spikelike panicles. Especially in humid areas, the panicles and upper leaves are sometimes infected with a black fungus commonly referred to as smut. The mature plant is mostly hairless, except for a few tiny hairs on the collar margins. The stems are spreading to erect, round in cross-section, wiry, branched, tough and fibrous at the base. The root system is fibrous. The plant forms dense tufts that enlarge by developing new stems around the perimeter.

This plant usually flowers in summer to fall, but can go into flower as early as spring in mild winter areas. The panicles are long and thin, generally 4 to 20 inches long, sometimes as long as 32 inches, and 5 to 10 mm wide at the base. They are often gray-green or purplish in color.

Smutgrass spreads by producing large numbers of tiny seeds. When damp, the seeds are sticky and can disperse by clinging to the fur, feathers, and feet of animals, the shoes and clothing of people, vehicle tires, tools, and agricultural equipment. Seeds can also spread with water, mud, or as a seed contaminant. Seeds can survive in the soil seedbank for more than 2 years.

In California, smutgrass often invades irrigated pastures, especially those that are heavily grazed and/or poorly drained. Livestock generally avoided consuming smutgrass unless more palatable forage is unavailable. Smutgrass is most problematic in southern California and in the southern United States.

NON-CHEMICAL CONTROL

Mechanical (pulling, cutting, disking)	Small plants can be hand pulled from wet soil. If individuals are noticed in a new area, they should be pulled as soon as possible.
	Mowing is not effective alone, and mowing after flowering can spread seeds. However, mowing early in the growing season can remove old top growth, making it easier to treat regrowth with herbicides.
	Cultivation when the soil is dry can kill mature plants, but also disturbs the soil, allowing the seedbank to reinfest the site.
Cultural	Smutgrass is palatable, but not a preferred forage. Intensive grazing (e.g., 42 cows/acre for 21 days) can overcome grazing preferences and can help to suppress smutgrass. Since livestock preferentially graze other species, a moderate grazing will reduce the height of desirable species, leaving tall smutgrass, which can be selectively controlled by herbicide wiping.
	Burning alone is not an effective control. Burning can remove old biomass, making smutgrass more palatable for grazing the following spring. Burning may also facilitate subsequent wiper treatments.
	Stopping irrigation for summer and allowing a pasture to dry out has been shown to kill adult smutgrass, but the infestation reestablishes from the seedbank.
Biological	There have been no biological control efforts for smutgrass in the United States.

CHEMICAL CONTROL

The following specific use information is based on published papers and reports by researchers and land managers. Other trade names may be available, and other compounds also are labeled for this weed. Directions for use may vary between brands; see label before use. Herbicides are listed by mode of action and then alphabetically. The order of herbicide listing is not reflective of the order of efficacy or preference.

AROMATIC AMINO ACID INHIBITORS

Glyphosate *Roundup, Accord XRT II*, and others	**Rate:** Broadcast foliar treatment: 2 to 3 qt product (*Roundup ProMax*)/acre (2.25 to 3.375 lb a.e./acre). Spot treatment: 3% product v/v solution. Wiper treatment: 33% to 50% of concentrated product in water. **Timing:** Postemergence, to rapidly growing plants. Wiper treatments are most effective when flowering panicles are tall but still green, before seed set (e.g., July in California's Central Valley). **Remarks:** Glyphosate is nonselective, so may kill desirable competitors. It has no soil activity. Effectiveness is increased by addition of ammonium sulfate. Some formulations (*Rodeo* and *Aquamaster*) are registered for use in or near water.

BRANCHED-CHAIN AMINO ACID INHIBITORS

Imazapic *Plateau*	**Rate:** 10 to 12 oz product/acre (2.5 to 3 oz a.e./acre) **Timing:** Postemergence in spring after plants have greened up. **Remarks:** Imazapic has mixed selectivity and tends to favor Asteraceae and some grasses. Safe for most native grasses, but higher rates may suppress seed of some cool-season grasses. Use methylated seed oil surfactant at 0.25%. Imazapic has some soil residual activity. Depending on conditions, it may provide only suppression. Not registered for use in California.

PHOTOSYNTHETIC INHIBITORS

Hexazinone *Velpar L*	**Rate:** 2.75 to 4.5 pt product/acre (11 to 18 oz a.i./acre) for pasture and rangeland; 3 to 4 gal product/acre (6 to 8 lb a.i./acre) for non-crop use **Timing:** Preemergence, or postemergence to rapidly growing plants. **Remarks:** Hexazinone is less effective on heavy soils. It generally provides only suppression at pasture/rangeland rates. Since it can move in soil, it should be kept away from roots of desired trees. High rates of hexazinone can create bare ground, so only use high rates in spot treatments.

Taeniatherum caput-medusae (L.) Nevski
(= *Elymus caput-medusae* L. [Jepson Manual 2012])

Medusahead

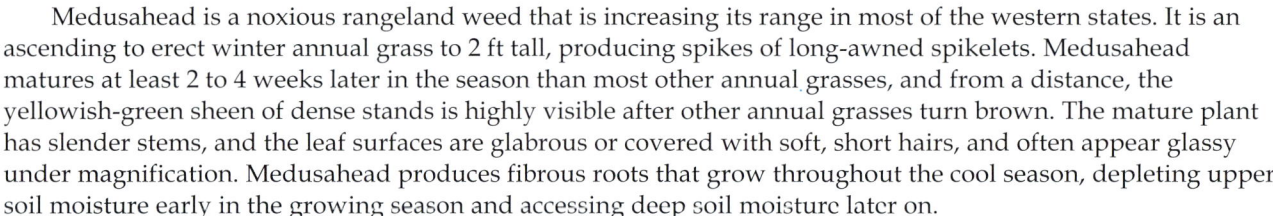

Family: Poaceae
Range: Arizona, California, Idaho, Nebraska, Nevada, Oregon, Utah, Washington; a few locations in the northeastern states.
Habitat: Disturbed sites, grassland, rangeland, openings in chaparral, oak woodlands, and rarely in agronomic fields. Generally in areas that receive at least 9 inches of rain per year, so not common in the low desert. Grows best on clay soils or where deep soil moisture is available late in the growing season.
Origin: Native to the Mediterranean region.
Impact: Dense stands displace desirable vegetation and reduce livestock and wildlife carrying capacity. Unpalatable to livestock except during the early growth stages. The stiff awns and hard florets can injure eyes, nostrils, and mouths of grazing animals. Birds and rodents usually avoid feeding on the seeds. Senesced plants form a dense layer of thatch that takes a couple of years to decompose. The thatch layer changes the temperature and moisture dynamics of the soil, reduces seed germination of other species, and creates fuel for wildfires.
Western states listed as Noxious Weed: California, Colorado, Nevada, Oregon, Utah
California Invasive Plant Council (Cal-IPC) Inventory: High Invasiveness

Medusahead is a noxious rangeland weed that is increasing its range in most of the western states. It is an ascending to erect winter annual grass to 2 ft tall, producing spikes of long-awned spikelets. Medusahead matures at least 2 to 4 weeks later in the season than most other annual grasses, and from a distance, the yellowish-green sheen of dense stands is highly visible after other annual grasses turn brown. The mature plant has slender stems, and the leaf surfaces are glabrous or covered with soft, short hairs, and often appear glassy under magnification. Medusahead produces fibrous roots that grow throughout the cool season, depleting upper soil moisture early in the growing season and accessing deep soil moisture later on.

Medusahead flowers in early summer, often after other annual plants have senesced. Its spikes are 0.5 to 2 inches long excluding awns, and the main spike axis does not break apart at maturity. The fertile seeds have long, often twisted, awns 1 to 3 inches long; the seeds and awns are covered with minute, upward-pointing barbs and are rough to the touch. Seed production is usually prolific. Some florets (seeds) can remain attached to spikes long after plants turn brown. This characteristic allows seeds to disperse by clinging to the feet, fur, and feathers of animals and the shoes and clothing of people.

Most seeds germinate in fall after the first rain, but some seeds remain dormant or germinate in winter or spring. Seeds can germinate in dense litter under low moisture conditions. Seedlings remain attached to the long-awned floret. Seedlings can survive desiccation of the primary root and develop adventitious roots when moisture becomes available. Most seed appears to germinate or lose viability within 2 years in the field.

NON-CHEMICAL CONTROL

Mechanical (pulling, cutting, disking)	There are mixed reports on the effectiveness of mowing. Early-season mowing is likely to be ineffective and may harm other species. Late-season mowing, at the boot to early flowering stage, may help to suppress medusahead. However, mowing after seed set will disperse the seeds.
	In areas where medusahead has built up a heavy thatch, removal of the thatch by raking, tillage, or burning can reduce dominance by medusahead and can help other plant species to get established.
	Tillage (disking and plowing) will control existing medusahead plants, as well as burying seed and breaking up deep thatch layers. Tillage should be accomplished before seed set. In rangeland or wildland areas, the increased potential for soil erosion, loss of soil moisture, loss of organic matter, and loss of macrobiotic crusts may outweigh the weed control benefits of tillage. These factors should be considered before applying tillage over a large area.
Cultural	The use of fire has given mixed results for medusahead control. Burning in low-elevation, warm-winter sites

	(e.g., California's Central Valley and foothills) can be extremely effective. Burns are conducted when medusahead is beginning to head out but before seed drop, when most desirable plants have already dropped seed. Two years of burning can nearly eliminate an infestation. In contrast, burning in high-elevation, cool-winter sites usually fails to control medusahead. It is thought that because of reduced ecosystem productivity and a shorter growing season at these sites, there isn't adequate fuel to carry a fire hot enough to kill medusahead seeds. Because of its high silica content, livestock generally avoid grazing medusahead as it approaches maturity. However, heavy grazing in spring, when medusahead is still palatable, can reduce, but not eliminate, an infestation. To limit seed dispersal, animals should be removed before the plants mature. Spring grazing is especially effective in areas where dried medusahead litter has been previously burned or grazed. Best control is achieved if plants are grazed in the stage of boot to flowerhead emergence. In experimental trials, medusahead populations have been reduced using confined grazing with sheep. In recent trials, fertilizing pastures with nitrogen improved the palatability of medusahead, making it attractive to cattle and resulting in medusahead suppression. As discussed above, thatch removal by raking, tillage, or burning can favor establishment of desirable plants. This can also improve efficacy of subsequent applications of some soil-applied herbicides, particularly imazapic.
Biological	Crown rot fungus (*Fusarium culmorum*), an endemic fungus of dry soils in the western states, is a potential biocontrol agent. However, there are no studies to show its effectiveness.

CHEMICAL CONTROL

The following specific use information is based on published papers and reports by researchers and land managers. Other trade names may be available, and other compounds also are labeled for this weed. Directions for use may vary between brands; see label before use. Herbicides are listed by mode of action and then alphabetically. The order of herbicide listing is not reflective of the order of efficacy or preference.

GROWTH REGULATORS	
Aminopyralid *Milestone*	**Rate:** 7 to 14 oz product/acre (1.75 to 3.5 oz a.e./acre) **Timing:** Preemergence in fall. **Remarks:** A broadleaf-selective herbicide that is safe on most grasses. There is a 2(ee) Supplemental label for this use in Arizona, California, Colorado, Idaho, Oregon, Washington, Wyoming, and Utah. Research in California's Central Valley showed that 14 oz of *Milestone* (spot treatment rate)/acre gave ~90% control of medusahead, and 7 oz/acre gave ~60% control.
AROMATIC AMINO ACID INHIBITORS	
Glyphosate *Roundup, Accord XRT II*, and others	**Rate:** 0.75 to 1 pt product (*Roundup ProMax*)/acre (0.42 to 0.56 lb a.e./acre) for early-season selective control in shrubland or other perennial systems; 1 to 2 qt product (*Roundup ProMax*)/acre (1.1 to 2.25 lb a.e./acre) for late-season, nonselective control. **Timing:** For selective control, apply postemergence in spring after all seedlings are up and before heading; the tillering stage is ideal. For late-season, nonselective control, apply to rapidly growing plants before seeds are produced. **Remarks:** Glyphosate is a nonselective herbicide with no soil activity.
BRANCHED-CHAIN AMINO ACID INHIBITORS	
Imazapic *Plateau*	**Rate:** 4 to 12 oz product/acre (1 to 3 oz a.e./acre) **Timing:** Fall or spring. In warm-winter areas, fall applications may be most effective. In colder climates, spring applications after snow melt are better. **Remarks:** Imazapic has some soil residual activity and mixed selectivity. It tends to favor members of the Asteraceae and some grasses. Use a spray adjuvant for postemergence applications. Effects vary depending on soil texture and soil organic matter. Heavy soils and high organic matter may require higher rates. Imazapic also can tie up in litter, and its efficacy is reduced under situations where there is lots of thatch on the soil surface. Not registered for use in California.
Rimsulfuron *Matrix*	**Rate:** 4 oz product/acre (1 oz a.i./acre) **Timing:** Preemergence (fall) to early postemergence (early spring). **Remarks:** Rimsulfuron controls several annual grasses and broadleaves. Perennial grasses are tolerant to fall

	applications when established and grown under dryland conditions. Application to rapidly growing or irrigated perennial grasses may result in their injury or death. It provides soil residual control in cool climates but degrades rapidly under warm conditions. Rimsulfuron will not control summer annual weeds when applied in fall or spring. Add a surfactant when applying postemergence.
Sulfometuron *Oust* and others	**Rate:** 0.75 to 1.5 oz product/acre (0.56 to 1.13 oz a.i./acre) **Timing:** Preemergence to early postemergence. Preemergence (fall) applications are generally more effective. **Remarks:** Sulfometuron is a broad-spectrum herbicide that is fairly safe on native perennial grasses. This can be an advantage in revegetation use. Use lower rates in arid environments, higher rates in wetter areas (> 20 inches rainfall) and on high organic matter soils. It has fairly long soil residual activity.
Sulfometuron + chlorsulfuron *Landmark XP*	**Rate:** 1.5 to 2.25 oz product/acre **Timing:** Preemergence, in fall or after soil thaws in spring. **Remarks:** See sulfometuron.

Tamarix aphylla (L.) Karst.; athel tamarisk
Tamarix chinensis Lour.; Chinese tamarisk
Tamarix gallica L.; French tamarisk
Tamarix parviflora DC.; smallflower tamarisk
Tamarix ramosissima Ledeb.; saltcedar, and hybrids of *T. ramosissima*, *T. gallica*, and *T. chinensis*

Saltcedar and tamarisk

Family: Tamaricaceae
Range: All western and southwestern states, some species widely planted.
Habitat: River, lake and pond margins, washes, roadsides, ditches, flats, sand dunes, desert springs. Grows best in alkaline soil, but tolerates salinity and acidity. Mature plants survive desert heat, below-freezing temperatures, periodic flooding, drought, burning.
Origin: Native to eastern Asia, northern Africa, the Middle East, India, and southern Europe. *Tamarix* species were introduced as landscape ornamentals and have escaped cultivation in many states, especially in the southwest. Athel tamarisk is still widely sold as an ornamental, but is not as invasive as the other species.
Impacts: All species are facultative phreatophytes that can use both surface and groundwater. The presence of numerous trees along riparian corridors or around desert springs can seriously reduce underground water tables and surface water availability, drying up wetlands, and reducing flows. Roots extract salts from deep soil layers and excrete it from the leaves. Salt is deposited on the soil surface with the leaf litter. The increased salinity of the upper soil profile inhibits the growth, survival, and recruitment of desirable native vegetation. Although some animals will seek cover or nest in *Tamarix* thickets, most wildlife does not consume *Tamarix* foliage, fruits, or seeds. *Tamarix* species can increase flooding in riparian areas by narrowing channel width. In addition, the plants are flammable and can introduce fire into wetland and riparian communities that are not adapted to periodic burning.
Western states listed as Noxious Weed: *T. parviflora*, California, Colorado, Montana, Nevada, New Mexico, North Dakota, South Dakota, Wyoming. *T. ramosissima* and hybrids, California, Colorado, Montana, Nevada, New Mexico, North Dakota, Oregon, South Dakota, Utah, Washington, Wyoming. *T. chinensis*, California, Colorado, Montana, New Mexico, North Dakota, South Dakota, Wyoming. *T. gallica*, California, Montana, New Mexico, South Dakota, Wyoming
California Invasive Plant Council (Cal-IPC) Inventory: *T. parviflora* and *T. ramosissima*, High Invasiveness; *T. aphylla*, Limited Invasiveness

Tamarix species are small trees or shrubs with tiny scale- or awl-like leaves. Smallflower tamarisk is the shortest of the species and grows to about 15 ft tall. Saltcedar, Chinese tamarisk, and French tamarisk grow to about 20 ft tall. All four of these species are deciduous and have awl-like twig leaves that strongly overlap each other and have acute tips. The foliage of saltcedar, Chinese tamarisk, and French tamarisk is usually more bluish-green than smallflower tamarisk. Athel tamarisk grows to 40 ft tall and is evergreen or semi-evergreen. Its twig leaves are scale-like, not or barely overlapping, and appear like small segments along a stem. *Tamarix* species typically develop an efficient, deep root system (> 15 ft deep) and have a high evapotranspiration rates in arid climates during the warm season. This occurs because once the roots are within the water table they branch profusely into numerous lateral roots to several feet long.

The inflorescences of *Tamarix* species consist of racemes, mostly 3 to 5 mm wide, typically simple, but occasionally compound and panicle-like. The flowers have a nectar disc at the base which is useful for species identification. Flowers of saltcedar, Chinese tamarisk, and French tamarisk are white, pale or dark pink, mostly with 5 sepals, petals, and stamens. Flowers of smallflower tamarisk are typically pale to dark pink, mostly with 4 sepals, petals, and stamens. Athel tamarisk flowers are white to pale pink, mostly with 5 sepals, petals, and stamens. Plants all reproduce by seed, although seed production is not common in athel tamarisk, and sometimes vegetatively from root sprouts and stem fragments. Stem fragments can take root when buried in a moist substrate, such as might occur with a flooding event. The fruit is a small capsule, often less than 5 mm long, with numerous minute seeds. One mature saltcedar plant can produce about 500,000 seeds per year. Seeds disperse primarily with wind and water, and germination occurs shortly after seed dispersal in spring through summer. Seeds lack a dormancy period, and most germinate within 24 hours after contacting water. Seeds typically survive for only 5 weeks. Individual trees can live 75 to 100 years or more.

Although saltcedar (*T. ramosissima*) and Chinese tamarisk (*T. chinensis*) do not hybridize in their native range, they appear to hybridize extensively in the United States. This hybrid is the most common invasive *Tamarix* species from Oklahoma to Washington to California. Seed viability is high in the hybrids. Less extensively, hybrids between saltcedar and Chinese tamarisk with smallflower tamarisk (*T. parviflora*), Canary Island tamarisk (*T. canariensis*) and French tamarisk (*T. gallica*) also occur. The abundance of invasive hybrids may explain the confusion associated with the identification of *Tamarix* species in the western states.

NON-CHEMICAL CONTROL

Mechanical (pulling, cutting, disking)	Mechanical control methods include mowing, burning, chopping, chaining, and disking. However, these methods usually only suppress saltcedar temporarily and will not eradicate infestations. Saltcedar is also able to resprout vigorously from the root crown following mechanical control methods. These methods can be labor intensive and expensive and may be more effective on small infestations.
	Hand pulling can be an effective way to control tamarisk in situations where plants are small, where access is difficult, or where herbicides cannot be used. Hand pulling is generally used to control new *Tamarix* seedlings and small plants around isolated desert springs in national parks after the larger plants have been killed with other techniques.
	Mowing is occasionally useful to reduce the volume of tamarisk before treatment with herbicide, especially in relatively level sites where prescribed burning is not feasible. However, a single cutting of tamarisk is ineffective, because tamarisks resprout vigorously. By comparison, cutting combined with herbicide treatment can be a very effective integrated approach. In addition, cutting tamarisk can reduce consumption of ground water, through removal of transpiring leaves.
	Heavy equipment can be used to remove entire plants. However, this is expensive, and any fragments that move into the water column may resprout and form new populations. This technique also causes considerable soil disturbance and ecosystem disruption. A root plow pulled by a bulldozer has become a standard method for saltcedar control, providing good to excellent control. Root plowing is most effective when the soil is relatively dry and when combined with follow-up treatments such as hand grubbing resprouts or applying herbicides. Root plowing may affect desirable vegetation and could lead to wind erosion. Another technique uses site preparation tractors or skid steers equipped with forestry mulching attachments, such as a hydro-ax. These are adapted to forestry brush and tree clearing. The hydro-ax can mow or chip either living or dead saltcedar at about 1 acre/hr on level terrain.
Cultural	Cattle, goats, and sheep will graze saltcedar plants if desirable vegetation is lacking. Saltcedar has little nutritional value and cattle will only graze young seedlings early in the year. Goats might be able to control dense stands of tamarisk where little native vegetation is present, particularly if the stands are cut or burned first, with goats eating the regrowth.
	As a stand-alone strategy, burning has not been successful. Saltcedar is generally top-killed by burning, but plants readily resprout from the remaining root crown and adventitious buds on the lateral roots. Repeated yearly burns can suppress saltcedar and kill some of the plants after 3 to 4 years. Furthermore, burning may suppress saltcedar infestations by eliminating the closed canopy, slowing the rate of invasion, and allowing desirable vegetation to respond, thereby increasing biodiversity. Prescribed burns can be followed up with herbicide treatments to control resprouting plants. One strategy is to cut 20 to 25% of the largest tamarisk plants in stands several months before burning to create enough dry ground fuel to carry a fire. For greatest efficacy, burning should be conducted during the hottest part of summer, when plants experience the greatest water stress.
	If floods occur during the period of poplar/cottonwood (*Populus* spp.) or willow (*Salix* spp.) seeding (which precedes saltcedar seeding), then natural revegetation will consist of almost pure stands of those, with very little saltcedar. Young seedlings of saltcedar can be controlled by flooding for 1 month.
Biological	The release of the saltcedar leaf beetle (*Diorhabda carinulata*) from China has made significant impacts on many populations of saltcedar. This insect feeds on the leaves of saltcedar and slowly reduces plant vigor.

> Tamarisk does not usually die from a single defoliation from tamarisk beetles, and it can resprout within several weeks of defoliation. Repeated defoliation of individual tamarisk trees can lead to severe dieback the next season and death of the tree within several years. Data indicate that 4 years of defoliation can result in about 60% mortality. Biological control will not eradicate tamarisk but it has the potential to suppress tamarisk populations by 75 to 85%.
>
> Since its release, the beetle has defoliated tens of thousands of acres of tamarisk in Nevada, Utah, Colorado and Wyoming. The insect spreads rapidly but is poorly adapted to more southern regions of the U.S. This led to further exploration to find better adapted biotypes or species of *Diorhabda*. In this process, five sibling species were found to comprise a new *Diorhabda* species group. Each species has unique biogeographical traits suiting them to different regions invaded by tamarisk in North America. Four species were previously tested and verified as specific feeders of tamarisk, and released in North America under the name *D. elongata*. *D. carinulata* from China and Kazakhstan (old name *D. e. deserticola*), is successfully suppressing tamarisk over large areas of the Great Basin desert. The true *D. elongata* is establishing well in California and parts of west Texas and is best suited for Mediterranean habitats. *D. sublineata* has potential for the Sonoran and southeastern Chihuahuan deserts. *D. carinata* should be best suited for the Great Plains grasslands and the Mojave and northern Chihuahuan deserts. A fifth species, *D. meridionalis*, may be suited to subtropical maritime deserts but has yet to be safety tested.
>
> One controversial aspect of the release of the tamarisk beetle is that defoliation can locally reduce nesting habitat for riparian woodland birds until native woodland flora recover. This has primarily centered on the federally endangered southwestern willow flycatcher, *Empidonax traillii* ssp. *extimus*. The beetles have defoliated some tamarisk nest trees of the southwestern willow flycatcher on the Virgin River in southern Utah, and actions to protect the flycatcher are under consideration. In some areas, tamarisk may be replaced by grasslands or shrublands, resulting in losses of riparian forest habitat for birds.

CHEMICAL CONTROL

The following specific use information is based on publications and reports by researchers and land managers. These are the products that provide effective control. Those that do not provide sufficient control have been omitted from the table. Other trade names may be available, and other compounds also are labeled for this weed. Directions for use may vary between brands; see label before use. Herbicides are listed by mode of action and then alphabetically. The order of herbicide listing is not reflective of the order of efficacy or preference. The best publication for the management of saltcedar species is published through Colorado State University, entitled *Tamarisk: Best Management Practices in Colorado Watersheds*.

GROWTH REGULATORS	
Triclopyr *Garlon 3A, Garlon 4 Ultra, Pathfinder II*	**Rate:** Cut stump treatment: 50% to undiluted *Garlon 3A* (in water) or 25 to 100% *Garlon 4 Ultra* (in oil). Basal bark treatment: 20 to 30% *Garlon 4 Ultra* in oil on young trees without well-developed bark. **Timing:** While many sources indicate that applications can be applied year round, it is best to apply in summer or fall when plants are still growing and not water stressed. At this time, the greatest amount of herbicide will translocate to the below-ground tissues. **Remarks:** Cut stump treatments can be very effective. Cut stems horizontally at or near ground level, and immediately apply herbicide solution to cover the outer 20% of the stump face. Can be mixed with a color dye to determine which trees have been treated. Basal bark treatments should be made to smaller trees with thin bark. Spray the lower trunk, including the root collar, to a height of 12 to 15 inches from the ground; the spray should thoroughly wet the lower stem but not to the point of runoff. For smaller trees 3 to 4 ft tall, a foliar treatment of 7 oz aminopyralid (*Milestone*) + 3 qt *Garlon 4 Ultra*/acre with 0.25% non-ionic surfactant or 1 qt/acre seed oil surfactant gives good control. Follow-up treatment of resprouts with this mixture will be necessary. This mixture is selective and will not injure desirable grasses.
AROMATIC AMINO ACID INHIBITORS	
Glyphosate *Rodeo, Aquamaster*	**Rate:** Broadcast foliar treatment: 1.5 to 3.3 qt product (*Roundup ProMax*)/acre (1.7 to 3.7 lb a.e./acre). Cut stump treatment: 100% of concentrated product to wet the cambial area of the stump face. **Timing:** Broadcast treatments should be made in late summer or early fall when plants are translocating carbohydrates to the below-ground tissues. Cut stump treatments can be made year-round but avoid treatment under drought conditions. **Remarks:** Glyphosate provides only partial control of *Tamarix* species. Because the herbicide precipitates out when in contact with divalent and trivalent salts, the salty excretions on the foliar glands will reduce the effectiveness of glyphosate. Foliar treatment with glyphosate will probably be most effective if applied

shortly after a rainfall event.

BRANCHED-CHAIN AMINO ACID INHIBITORS

Imazapyr *Arsenal AC, Habitat, Stalker, Chopper, Polaris*	**Rate:** Broadcast foliar treatment: 1 qt product (*Arsenal AC*)/acre or 2 qt product (*Habitat*)/acre (1 lb a.i./acre). Spot treatment: 1% v/v *Habitat* spray-to-wet or 3 to 5% for low volume treatments in 10 gal water per acre (GPA). Cut stump treatment: 10% v/v of concentrate on outer cambial region. **Timing:** Late summer or early fall when plants are fully expanded and are translocating carbohydrates to the below-ground tissues. **Remarks:** Imazapyr is the most widely used herbicide to control saltcedar. Both conventional and low-volume applications can give good control. For foliar broadcast or spot treatments add 1% v/v of a non-ionic surfactant or methylated seed oil. Imazapyr is relatively nonselective and can damage or kill other desirable non-target species through direct contact or drift. Low volume treatments are applied to the upper portions of the plant, with coverage not exceeding 30%. Low volume treatments can be very effective on previously burned, recovering plants 6 to 10 ft tall. Spot treatments can be made using a drizzle gun. In addition to ground equipment, treatments can be made by aerial equipment. Helicopter treatments are considered better than fixed-wing as they allow for more consistent results. Herbicide activity may be reduced as saltcedar height and stem number increase. Plants should not be removed for at least 2 years to ensure good control. ALS-resistant kochia has invaded some locations after control of *Tamarix* with imazapyr.
Imazapyr + glyphosate	**Rate:** Broadcast foliar treatment: 1.5 to 2 pt *Habitat*/acre plus 1 to 2 pt glyphosate product/acre. Spot foliar treatment: 0.5% v/v of each product for most consistent results. Apply with 0.25% v/v non-ionic surfactant. **Timing:** Postemergence treatments should be made in late summer or early fall when plants are translocating carbohydrates to the below-ground tissues. **Remarks:** This combination is nonselective. Spot treatments can be made using a drizzle gun. Plants should not be removed for at least 2 years to ensure good control.

Tanacetum vulgare L.
Common tansy

Family: Asteraceae
Range: Most contiguous states, including every western state.
Habitat: Disturbed places, gardens, yards, roadsides, fence rows, pastures, waste places, forest clearings, and meadows.
Origin: Native to temperate Eurasia. Brought to the U.S. for horticultural and medicinal reasons.
Impacts: Heavy consumption could be dangerous to humans and livestock because the plant's volatile oil contains thujone, a substance that can cause convulsions and miscarriages, as well as skin irritation. Common tansy has low palatability and livestock poisoning is rare. It frequently spreads following intensive grazing and soil disturbance. Dense stands reduce production of forage for livestock and wildlife. Infestations are problematic when trying to restore disturbed sites to desirable vegetation.
Western states listed as Noxious Weed: Colorado, Montana, Washington, Wyoming
California Invasive Plant Council (Cal-IPC) Inventory: Moderate Invasiveness

Common tansy is an erect, aromatic perennial to 5 ft tall. The foliage is glandular, glabrous to sparsely hairy. Dense clumps of stems are sometimes weakly woody. The leaves are alternate, deeply pinnate-lobed, sessile or short-stalked, evenly dotted with flat or sunken glands. Primary divisions of the leaves are narrow, mostly 4 to 10 paired. Secondary divisions are narrow and toothed. Creeping roots are thick, extensive, with numerous lateral roots. New shoots from creeping roots emerge close to the parent plant forming dense clumps.

Common tansy is easily recognized by the flat-topped clusters of small, button-like yellow flowers. Plants can produce more than 2,000 seeds. Plants reproduce primarily by seed, but also can spread by creeping roots. Seeds mainly disperse short distances by falling to the ground beneath the parent plant. There is no information on seed viability and longevity; however, in one study 75% of seeds collected in fall germinated after 1 month of cold stratification, indicating high viability.

NON-CHEMICAL CONTROL

Mechanical (pulling, cutting, disking)	With small infestations hand-pulling is feasible, especially when soils are moist. Wear gloves when hand-pulling to avoid contact dermatitis from oily foliage. Revisit the site to pull any resprouts. Hoeing and shallow cultivation are not effective at killing existing plants since plants resprout from root fragments.
	Mowing will not kill established plants, but reports suggest mowing shortly before bloom can reduce seed production. Mowing can be used to reduce litter cover before herbicide application.
	Tillage can spread root fragments with regenerative buds. Common tansy is not normally a problem in cultivated crop fields, thus frequent tillage over an extended time will likely reduce competitiveness.
Cultural	Common tansy is toxic in large quantities and abortions in cattle have been reported in mid-west states. In Montana, most classes of livestock and some wildlife have been observed grazing common tansy with no known adverse effects. Sheep and goats can eat significant amounts of common tansy without toxic effects. Sheep have been used to manage common tansy in Montana. Clipping data suggest sheep may consume up to 90% of common tansy aboveground biomass while consuming similar quantities of perennial grasses. Overgrazing promotes common tansy in pastures.
	Prescribed burning is not effective. Following fire, common tansy can quickly resprout from root buds not affected by heat. Prescribed burning can be used to reduce litter cover before herbicide application.
	Promoting competitive vegetation can slow the spread of common tansy.
Biological	A biological control effort was launched in 2006, but no biological control agents are available to date.

CHEMICAL CONTROL

The following specific use information is based on published papers and reports by researchers and land managers. Other trade names may be available, and other compounds also are labeled for this weed. Directions for use may vary between brands; see label before use. Herbicides are listed by mode of action and then alphabetically. The order of herbicide listing is not reflective of the order of efficacy or preference.

GROWTH REGULATORS	
2,4-D Several names	**Rate:** 1 to 2 qt product/acre (0.95 to 1.9 lb a.e./acre) **Timing:** Postemergence, to rapidly growing plants before flowering. **Remarks:** 2,4-D gives best results with wiper applications. It provides only partial control in most university trials. 2,4-D is broadleaf-selective and is safe on most grasses. It has minimal soil activity. Do not apply ester formulation when outside temperatures exceed 80°F.
Aminocyclopyrachlor + chlorsulfuron *Perspective*	**Rate:** 4.75 to 8 oz product (*Perspective*)/acre **Timing:** Postemergence. Most effective when applied to plants in the flower bud stage. **Remarks:** *Perspective* provides broad-spectrum control of many broadleaf species. Although generally safe for grasses, it may suppress or injure certain annual and perennial grass species. Do not treat in the root zone of desirable trees and shrubs. Do not apply more than 11 oz product/acre per year. At this high rate, cool-season grasses will be damaged, including bluebunch wheatgrass. Not yet labeled for grazing lands. Add an adjuvant to the spray solution. This product is not approved for use in California and some counties of Colorado (San Luis Valley).
Aminopyralid + metsulfuron *Opensight*	**Rate:** 2.5 to 3.3 oz product/acre **Timing:** Postemergence, when plants are at bud or later. **Remarks:** Not registered for use in California.
Aromatic amino acid inhibitors	
Glyphosate *Roundup, Accord XRT II,* and others	**Rate:** 1 to 2 qt product (*Roundup ProMax*)/acre (1.1 to 2.25 lb a.e./acre). Spot treatment, 1.5% v/v solution. Wiper treatment: 33 to 50% of concentrated product. **Timing:** Postemergence, to rapidly growing plants before flowering. **Remarks:** Best results occur with wiper applications. Glyphosate will not kill seeds or inhibit germination the following season. Glyphosate has no soil activity and is nonselective. It can create bare ground conditions that make the area susceptible to weed recruitment. In areas with desirable vegetation, use spot treatment. Glyphosate is a good control option if reseeding is planned shortly after application, as it will not injure seedlings emerging after application. Add a surfactant or methylated seed oil when using a formulation where it is not already included (e.g., *Rodeo, Aquamaster*).
BRANCHED-CHAIN AMINO ACID INHIBITORS	
Chlorsulfuron *Telar*	**Rate:** 1 to 2.6 oz product/acre (0.75 to 1.95 oz a.i./acre) **Timing:** Postemergence at flower bud stage. **Remarks:** Most established perennial grasses are tolerant. Always use a surfactant.
Metsulfuron *Escort*	**Rate:** 1 to 2 oz product/acre (0.6 to 1.2 oz a.i./acre) **Timing:** Postemergence at flower bud stage. **Remarks:** Metsulfuron is an effective treatment for common tansy control. Always use a surfactant. Other premix formulations of metsulfuron can be used at similar application timing. These include *Cimarron Max* (metsulfuron + dicamba + 2,4-D), *Opensight* (metsulfuron + aminopyralid) and *Cimarron X-tra* (metsulfuron + chlorsulfuron). Metsulfuron is not registered for use in California.

Torilis arvensis (Huds.) Link

Hedgeparsley

Family: Apiaceae
Range: Washington, Oregon, California, Idaho, and Utah.
Habitat: Woodlands, pastures, fields, forest margins, and disturbed sites such as roadsides and ornamental landscapes. Can tolerate full sun to dense shade and grows on most soil types.
Origin: Native to central and southern Europe.
Impacts: Hedgeparsley has bristly fruiting structures that can be a nuisance to livestock, pets, and humans. The burs stick to the fur and hair of animals and can cause mechanical injury by lodging in the nose, eyes, and ears of pets and livestock.
Western states listed as Noxious Weed: Washington
California Invasive Plant Council (Cal-IPC) Inventory: Moderate Invasiveness

Hedgeparsley is an upright annual weed that grows 6 to 24 inches tall. It germinates with the first fall rains and its lacy green foliage often makes the plant inconspicuous amongst grasses and forbs. The leaves are alternate, mostly pinnate-dissected 2 to 3 times, on stalks up to 3 inches long. Leaves are sparsely covered with short flattened hairs.

Plants produce white flowers, 2 to 3 mm wide in compound umbels 2 to 3 inches across. Each flower produces an oblong fruit 2 to 4 mm long, covered with minutely barbed, hook-tipped bristles. The fruit is initially rosy to whitish green in appearance, but later turns brown. Reproduction is entirely by seed. Although there is no data on the longevity of the seed in the soil, it is expected that they would survive a few years.

NON-CHEMICAL CONTROL

Mechanical (pulling, cutting, disking)	Hand pulling is effective on small incipient populations. Pulling is most effective before flowering in late spring when plants are elongated and soil is still moist.
	Mowing or disking at flowering stage can provide good control. Resprouts may occur after mowing and a secondary treatment may be required.
Cultural	Grazing can provide some control, if grazed at a high stocking density before flowering.
	There is no information on the effectiveness of control with prescribed burning.
Biological	There are no biological control programs for the management of *Torilis arvensis*.

CHEMICAL CONTROL

The following specific use information is based on reports by researchers and land managers. Other trade names may be available, and other compounds also are labeled for this weed. Directions for use may vary between brands; see label before use. Herbicides are listed by mode of action and then alphabetically. The order of herbicide listing is not reflective of the order of efficacy or preference.

GROWTH REGULATORS	
2,4-D Several names	**Rate:** Broadcast treatment: 2 to 4 pt product/acre (0.95 to 1.9 lb a.e./acre).
	Timing: Postemergence when plants are growing rapidly. Applications in spring provide best control.
	Remarks: 2,4-D is a selective herbicide for broadleaf species and will not damage desirable grasses growing nearby. This rate has been shown to give good control of other broadleaf weeds in rangeland. Good coverage is necessary.
Triclopyr *Garlon 4 Ultra*	**Rate:** Broadcast treatment: 0.5 to 1 qt product/acre (0.5 to 1 lb a.e./acre). Spot treatment: 1 % v/v solution *Garlon 4 Ultra* and water applied to thoroughly wet all leaves.

	Timing: Postemergence when plants are growing rapidly. Applications in spring provide best control.
	Remarks: Triclopyr is a selective herbicide for broadleaf species and will not damage desirable grasses growing nearby.
AROMATIC AMINO ACID INHIBITORS	
Glyphosate *Roundup, Accord XRT II*, and others	**Rate:** Broadcast treatment: 1 to 2 qt product (*Roundup ProMax*)/acre (1.1 to 2.25 lb a.e./acre). Spot treatment: 1.5 to 2% v/v solution *Roundup* (or other trade name) and water applied to thoroughly wet all leaves. **Timing:** Postemergence when plants are growing rapidly. Applications in early spring provide best control. **Remarks:** Glyphosate is a nonselective systemic herbicide and has no soil activity.
BRANCHED-CHAIN AMINO ACID INHIBITORS	
Chlorsulfuron *Telar*	**Rate:** Broadcast treatment: 1 to 1.5 oz product/acre (0.75 to 1.13 oz a.i./acre). **Timing:** Preemergence or early postemergence. **Remarks:** Chlorsulfuron is a selective broadleaf herbicide used preemergence or postemergence in non-cropland areas. It has fairly long soil residual activity.
Imazapic *Plateau*	**Rate:** Broadcast treatment: 4 to 6 oz product/acre (1 to 1.5 oz a.e./acre). **Timing:** Preemergence or early postemergence. **Remarks:** Imazapic is a selective postemergence herbicide effective for controlling broadleaf weeds and some grasses. It has some soil residual activity. Imazapic is not registered for use in California.

Toxicodendron diversilobum (Torr. & A. Gray) E. Greene

Pacific poison-oak

Family: Anacardiaceae
Range: Baja California to British Columbia. West of the Cascade Range in Washington and Oregon; ubiquitous in California west of the Sierra Nevada. Also along the western side of Nevada.
Habitat: Mixed evergreen forests, woodlands, chaparral, coastal sage scrub, and riparian zones. It is one of the most widespread shrubs in California. It generally occurs on acid soils, but is not limited to any particular soil type, texture or drainage pattern. Pacific poison-oak is typically found at less than 5,000 ft elevation and grows on all aspects. It can tolerate drought, fire, and low temperatures.
Origin: Native to the Pacific Coast of the western United States from British Columbia to Baja California.
Impacts: One of the most hazardous native plants in the western states. It can be problematic wherever people are likely to contact the plant such as along trails or during brush removal around homes, along rights-of-way, fire breaks, construction sites, etc. All plant parts, except the pollen, have resin canals that contain the phenolic compound urushiol. Direct contact with bruised, broken, or insect-damaged plant parts, including dormant leafless stems, or contact with items such as tools, clothing, gloves, and pets that have had direct contact with plants can cause allergenic contact dermatitis in sensitive individuals. Smoke from burning material can cause severe respiratory irritation if inhaled. Sensitivity to Pacific poison-oak often increases with repeated exposure.

Pacific poison-oak is a native deciduous shrub to 12 ft tall, with compound leaves that typically consist of three (sometimes five) leaflets. Plants are sometimes vine-like with stems to 75 ft long, and may climb trees or other support structures by adventitious roots and/or wedging stems within grooves or crevices of the support. The bright green leaves are round to ovate, diversely lobed or toothed and resemble oak leaves.

Small white flowers occur in leaf axils, with male and female flowers on separate plants. In fall the leaves turn brilliant shades of scarlet, orange, yellow and maroon. Pacific poison-oak reproduces vegetatively from root sprouts and by seed. Reproduction by root suckers is most prevalent after disturbance such as fire or browsing by animals. Suckering can also be extensive after being cut or damaged. Seeds are produced in smooth, berrylike fruits and are most commonly spread by birds.

NON-CHEMICAL CONTROL

Mechanical (pulling, cutting, disking)	Protective clothing, including washable cotton gloves over plastic gloves, can help prevent exposure to the resin that causes contact dermatitis. Even with these protective measures, hand pulling is not generally recommended, as exposure can cause severe contact dermatitis in susceptible individuals. Mechanical removal of plants, including the root systems, is most effective when the soil is moist. Any stumps or rhizomes left behind will resprout. Hand pulling can remove seedlings, but once underground rhizomes have developed, this technique is generally not effective.
Cultural	Repeated grazing by sheep and/or goats will eventually kill the plant by exhausting the root carbohydrate reserves. Burning is not recommended as it does not kill the root system and the smoke is hazardous to human health.
Biological	As a native plant of the western United States, there are no efforts to develop biological control programs for Pacific poison-oak.

CHEMICAL CONTROL

The following specific use information is based on published papers and reports by researchers and land managers. Other trade names may be available, and other compounds also are labeled for this weed. Directions for use may vary between brands; see label before use. Herbicides are listed by mode of action and then alphabetically. The order of herbicide listing is not reflective of the order of efficacy or preference.

GROWTH REGULATORS

Dicamba	**Rate:** Broadcast treatment: 2 qt product/acre (2 lb a.i./acre). Cut stump treatment: 25 to 50% dicamba in 50 to 75% water.

Banvel, Clarity		**Timing:** Postemergence foliar treatments are best when plants are growing rapidly. Best when used in late summer to early fall, but before leaf drop. For cut stump treatment, application in late summer, early fall or dormant season provides the best control. Treat immediately after cutting.
		Remarks: Selective herbicide for broadleaf species; will not damage desirable grasses growing nearby. Basal bark applications are as described for triclopyr. Plants should not be cut for at least 4 months after basal bark treatments. Roots may sucker after cutting, but the treatment should control most resprouts.
Picloram *Tordon 22K*		**Rate:** Foliar treatment: 1 qt product/acre. Efficacy is improved with the addition of 3 to 5 qt/acre of *Garlon 4 Ultra*, or 4 to 8 qt of *Garlon 3A*/acre.
		Timing: End of summer to beginning of fall when plants are growing rapidly, but before leaf drop.
		Remarks: High rates can give long-term soil residual activity for broadleaves. Also available as a premix with 2,4-D (*Grazon P+D*). Restricted use pesticide. Picloram and its formulations are not registered for use in California.
Triclopyr *Garlon 3A,* *Garlon 4 Ultra,* *Pathfinder II*		**Rate:** Low volume foliar treatment: 1 to 5% v/v solution of *Garlon 4 Ultra*, or 1 to 3% *Garlon 3A* in water with a 0.5% v/v surfactant to thoroughly wet all leaves. Cut stump treatment: 25% *Garlon 4 Ultra* in 75% oil carrier, or undiluted *Garlon 3A* or 50% *Garlon 3A* in water. Basal bark treatment: 20% *Garlon 4 Ultra* in 80% oil carrier, or *Pathfinder II* as a ready-to-use formulation. For basal cut stump treatment: 25% *Garlon 4 Ultra* in 75% oil carrier. For cut stump treatment: undiluted *Garlon 3A*, or 50% *Garlon 3A* in water.
		Timing: Postemergence when plants are growing rapidly. Cut stump, basal cut stump, and basal bark treatments can be applied anytime as long as the ground is not frozen, but are best when used in late summer or early fall, before leaf drop.
		Remarks: Selective herbicide for broadleaf species; will not damage desirable grasses growing nearby. For cut stump treatments, cut stems horizontally at or near ground level and immediately apply herbicide solution. Suckering from the roots typically occurs after cutting, but the treatment should control most resprouts. For basal bark treatment, spray the lower trunk, including the root collar, to a height of 12 to 15 inches from the ground; the spray should thoroughly wet the lower stem but not to the point of runoff. Plants should not be cut for at least 1 month following basal bark treatment.
Triclopyr +2,4-D *Crossbow*		**Rate:** Foliar treatment: 1 to 1.5% solution of *Crossbow* and water. Apply to thoroughly wet all leaves.
		Timing: Postemergence when plants are growing rapidly.
		Remarks: *Crossbow* in water forms an emulsion (not a solution), and separation may occur unless the spray mixture is agitated continuously.
AROMATIC AMINO ACID INHIBITORS		
Glyphosate *Roundup, Accord XRT II*, and others		**Rate:** Foliar treatment: 2% v/v solution of *Roundup ProMax* (or other trade name with similar concentration of glyphosate) in water to thoroughly wet all leaves. Low volume spot treatment: 5 to 10% v/v solution of *Roundup* (or other trade name) in water. Spray coverage should be uniform over at least 50% of the foliage. Cut stump treatment: undiluted *Roundup* or 50% v/v *Roundup* (or other trade name) in water.
		Timing: Postemergence foliar treatments are best when plants are growing rapidly at or beyond bloom stage, particularly after fruits form. Treat in late summer or early fall before leaves lose green color. Suckering from the roots might occur the following year. For cut stump treatment, application in late summer, early fall or dormant season provides the best control. Treat immediately after cutting.
		Remarks: Nonselective systemic herbicide with no soil activity. Gives good control with some resprouts. Plants should not be cut for at least 4 months after foliar treatments. Cut stump applications are as described for triclopyr.
BRANCHED-CHAIN AMINO ACID INHIBITORS		
Imazapyr *Arsenal, Habitat,* *Stalker, Chopper,* *Polaris*		**Rate:** Cut stump treatment: 20% *Stalker* or *Chopper* formulation in 80% oil carrier or 20% *Arsenal* or *Habitat* in 80% water carrier. Basal bark treatment: 20% *Stalker* or *Chopper* formulation in 80% oil carrier.
		Timing: Best when used for stem treatments in late summer to early fall, but before leaf drop.
		Remarks: Imazapyr is a soil residual herbicide and may result in bare ground around plants for some time after treatment. Cut stump and basal bark applications are as described for triclopyr. Plants should not be cut for at least 4 months after basal bark treatments. Roots may sucker after cutting, but the treatment should control most resprouts.

Trapa natans L.

Water chestnut

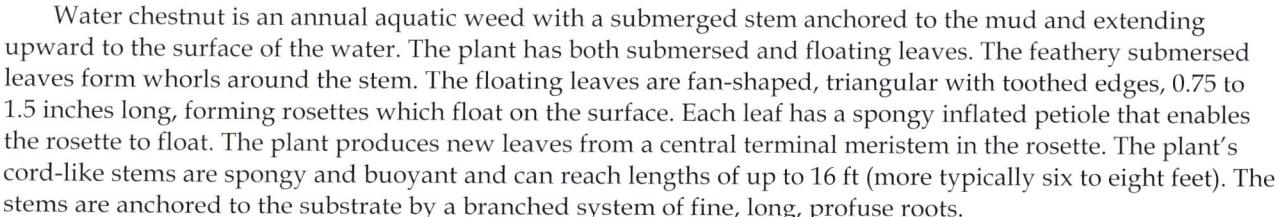

Family: Trapaceae
Range: Not yet present in the western United States. Currently invasive in the northeastern United States, but is expected to eventually be introduced into other areas of the country.
Habitat: Lakes, ponds, canals, and slow water; grows best in shallow, nutrient-rich lakes and rivers.
Origin: Originally thought to be native to warm temperate parts of Eurasia and Africa; more recent work considers it of Asian origin. *Trapa natans* produces a nut-like fruit that can be cooked or eaten raw and has been cultivated in China and India for at least 3,000 years. Water chestnut was first introduced to North America in the 1870s, where it is known to have been grown in a botanical garden at Harvard University in 1877. The plant escaped cultivation and was found growing in the Charles River by 1879.
Impacts: Water chestnut is very competitive with desirable aquatic species in shallow waters with soft, muddy bottoms. It can develop nearly impenetrable mats across wide areas of water, which can block light penetration, reduce oxygen levels leading to fish kills, and impact fishing and recreational activities. The nuts drift to shore where their sharp spines may hurt bare feet.
Western states listed as Noxious Weed: Arizona, Oregon, Washington

Water chestnut is an annual aquatic weed with a submerged stem anchored to the mud and extending upward to the surface of the water. The plant has both submersed and floating leaves. The feathery submersed leaves form whorls around the stem. The floating leaves are fan-shaped, triangular with toothed edges, 0.75 to 1.5 inches long, forming rosettes which float on the surface. Each leaf has a spongy inflated petiole that enables the rosette to float. The plant produces new leaves from a central terminal meristem in the rosette. The plant's cord-like stems are spongy and buoyant and can reach lengths of up to 16 ft (more typically six to eight feet). The stems are anchored to the substrate by a branched system of fine, long, profuse roots.

Inconspicuous flowers are borne in the axils of younger leaves above the water, in the center of the rosette. The flowers have four white petals 0.3 inches long. Once the flowers are pollinated by insects, the flower stalks curve downward and the fruit develops under water. Each flower produces a large, woody nut with four barbed spines derived from the flower sepals. When mature, the fruits fall from the plant and sink to the bed where they have a dormancy period of 4 months. The horns may act as anchors to limit the movement of the seeds, keeping them in suitable depths of water. The seeds overwinter at the bottom of the water body and germinate during the warm season. The newly developed shoot grows to the water surface, where the rosette is formed. A single seed may give rise to 10 to 15 rosettes, each producing up to 15 to 20 seeds. Reports indicate that seeds can remain viable for 5 to 12 years. However, most seeds probably germinate in the first 2 years. Water chestnut can disperse both vegetatively and by seed. The rosettes can detach from their stems and float to another area, or more often the nuts are dispersed by currents or waves.

NON-CHEMICAL CONTROL

Mechanical (pulling, cutting, disking)	Hand removal is an effective means for eradication of smaller populations; *Trapa natans* roots are easily pulled. It is important to remove the whole plant because floating plants can spread seeds downstream. The potential of *Trapa natans* seeds to lay dormant for up to 12 years makes total eradication difficult. Hand harvesting from canoes and raking have been effective and are a means to promote community involvement.
Cultural	The aggressive and competitive characteristics of *Trapa natans* make cultural practices problematic. Dewatering alone has not been found to destroy the seed bank. Prevention of new infestations and early detection are the best approaches. Since the *Trapa* genus contains plants that are in the food trade, public education about the risk and impacts of releases can help prevent introductions.
Biological	A leaf beetle (*Galerucella birmanica*) and moth (*Nymphuline pyralid*) found feeding on water chestnut in its native range were thought to have biocontrol potential but were found to be generalist feeders. Other insects observed feeding on the plant have not been shown to give effective control. The grass carp *Ctenopharyngodon idella* can be used to control water chestnut but is a nonselective herbivore that will

almost certainly harm native species.

CHEMICAL CONTROL
The following specific use information is based on reports by researchers and land managers. Other trade names may be available, and other compounds also are labeled for this weed. Directions for use may vary between brands; see label before use.

GROWTH REGULATORS	
2,4-D *Weedar 64,* *Aqua Kleen*	**Rate:** Broadleaf foliar treatment: 2 to 4 pt product/acre (0.95 to 1.9 lb a.e./acre) in a 100 to 400 gal solution/acre. Granular (*Aqua Kleen*) treatment: 100 to 200 lb product/acre **Timing:** Postemergence, to foliage or as granular treatment from spring to summer. **Remarks:** Check label for restrictions and limits of 2,4-D in water used for various purposes. 2,4-D is typically used only by state and federal action agencies or state and federal programs. Check for changes in labeling for other herbicides such as ALS inhibitors. Triclopyr has also been used for controlling water chestnut, but the efficacy is similar or worse than 2,4-D.

Tribulus terrestris L.
Puncturevine

Family: Zygophyllaceae
Range: Most contiguous states, including all western states. Also found nearly worldwide.
Habitat: Disturbed places, roadsides, railways, cultivated crops, orchards, vineyards, waste places, and walkways. Prevalent in areas with a hot summer. Grows best on dry, sandy soils, but tolerates most soil types. Killed by freezing temperatures.
Origin: Native to the Mediterranean region.
Impacts: Typically found along roadsides or in waste areas. Impacts in wildlands and rangelands are probably minor. More of a nuisance to humans or a threat to livestock due to its toxicity. During the first half of the 1900s, puncturevine was one of the most problematic roadside weeds in the U.S., as its burs easily punctured the tires of early cars. The foliage contains a number of saponin compounds that can be toxic to livestock, especially sheep, when ingested in quantity. It is also a noxious weed in many regions of Australia.
Western states listed as Noxious Weed: Arizona, California, Colorado, Idaho, Nevada, Oregon, Washington

Puncturevine is a prostrate summer annual with green to reddish-brown stems, spreading radially from a central crown. It has even-pinnate-compound leaves, with 3 to 7 pairs of leaflets per leaf.

Plants produce small, solitary yellow flowers, which develop into burs with stout spines that can injure people and animals, as well as puncture bicycle tires. Puncturevine reproduces only by seed. It produces 5-sided woody burs, about 5 to 12 mm in diameter, which separate into wedge-shaped nutlets, each with 2 stout spines and several short prickles. These spiny nutlets–with seeds inside–disperse on vehicle tires, shoes and clothing, and the fur, feathers, and feet of animals. Puncturevine seed germination requires warm temperatures. Buried seed usually remain viable for 3 to 6 years. Seedlings emerge from early spring through summer, often in flushes following increased soil moisture. Seedlings develop a deep taproot within a few weeks. Flowers may be produced within 3 weeks and burs within 6 weeks of germination.

NON-CHEMICAL CONTROL

Mechanical (pulling, cutting, disking)	Hand-pulling is feasible when the population is low, especially when soils are moist and the vines are sufficiently long to allow pulling. However, plants should be pulled before the spiny fruit develop.
	Mowing is ineffective because of the low growth form of the plant.
	Hoeing and shallow cultivation are effective at killing existing plants, and should be initiated before flowering and seed production. Shallow tilling to 1 inch or less is sufficient, particularly when plants are small. Deep tillage is not recommended, because it may bury seed deep in the soil profile, where they will survive longer.
Cultural	Neither grazing nor burning are recommended practices as plants are poisonous and the low growing habit precludes fire as a management option.
	Planting competitive vegetation or the application of 4 to 6 inches of mulch have been used to suppress puncturevine infestations. Areas with bare ground are idea sites for puncturevine growth, even with only limited amounts of rainfall.
	In areas with an existing seed bank, seed can be harvested by placing carpet or other "sticky" material on boards or a roller that are then applied to or rolled over the soil. With several passes, most of the seed on the soil can be removed.
Biological	In 1961, the stem weevil (*Microlarinus lypriformis*) and seed weevil (*Microlarinus lareynii*) were introduced from Italy as biocontrol agents. They have proved to be a successful management tool for puncturevine. A resurgence of the weed occurred again in central California in the mid-1990s, prompting the rearing and reintroduction of the seed weevil. Today, the insects help to maintain puncturevine populations at low levels. However, the insects are sensitive to prolonged periods of frost. Thus, populations of puncturevine may increase after years with cold winters. This is more prevalent in northern states.

CHEMICAL CONTROL

The following specific use information is based on published papers and reports by researchers and land managers. Other trade names may be available, and other compounds also are labeled for this weed. Directions for use may vary between brands; see label before use. Herbicides are listed by mode of action and then alphabetically. The order of herbicide listing is not reflective of the order of efficacy or preference.

GROWTH REGULATORS	
2,4-D Several names	**Rate:** 2 to 4 pt product/acre (0.95 to 1.9 lb a.e./acre) **Timing:** Postemergence every 3 weeks during the growing season or when new seedlings appear. **Remarks:** 2,4-D is a broadleaf herbicide with no soil activity. Seeds will germinate throughout summer when moisture is available. Retreatment may be necessary if new germination occurs. Avoid drift to sensitive crops.
Aminocyclopyrachlor + chlorsulfuron *Perspective*	**Rate:** 4.75 to 8 oz product/acre **Timing:** Postemergence, to rapidly growing young plants. **Remarks:** *Perspective* provides broad-spectrum control of many broadleaf species. Although generally safe for grasses, it may suppress or injure certain annual and perennial grass species. Do not treat in the root zone of desirable trees and shrubs. Do not apply more than 11 oz product/acre per year. At this high rate, cool-season grasses will be damaged, including bluebunch wheatgrass. Not yet labeled for grazing lands. Add an adjuvant to the spray solution. This product is not approved for use in California and some counties of Colorado (San Luis Valley).
Dicamba *Banvel, Clarity*	**Rate:** 1 to 2 pt product/acre (0.5 to 1 lb a.e./acre) **Timing:** Postemergence, to rapidly growing young plants. Older plants are more difficult to control. **Remarks:** Dicamba is a broadleaf-selective herbicide often combined with other active ingredients. Retreat as needed, but do not exceed 2 qt product/acre during growing season. Dicamba may leach in sandy soils. Do not apply when outside temperatures exceed 80°F. *Overdrive,* a premix of dicamba with diflufenzopyr, has been reported to be effective on puncturevine. Diflufenzopyr is an auxin transport inhibitor which causes dicamba to accumulate in shoot and root meristems, increasing its activity. *Overdrive* is applied postemergence at 4 to 8 oz product/acre rapidly growing plants. Higher rates should be used on large annuals. Add a non-ionic surfactant to the treatment solution at 0.25% v/v or a methylated seed oil at 1% v/v solution.
Fluroxypyr *Vista XRT*	**Rate:** 22 oz product/acre (7.7 oz a.e/acre) **Timing:** Postemergence, before budding when plants are still small and rapidly growing. **Remarks:** Do not apply more than 22 oz product/acre per year.
Picloram + 2,4-D *Tordon 101M*	**Rate:** 2 to 4 pt product/acre **Timing:** Postemergence, before flowering when plants are growing rapidly. **Remarks:** Picloram has long soil residual activity. *Tordon 101M* is a federally restricted use pesticide. Picloram is not registered for use in California.
AROMATIC AMINO ACID INHIBITORS	
Glyphosate *Roundup, Accord XRT II,* and others	**Rate:** Broadcast treatment: 2 to 4 pt product (*Roundup ProMax*)/acre (1.1 to 2.25 lb a.e./acre). Spot treatment: 0.5% to 1% v/v solution. **Timing:** Postemergence, to rapidly growing young plants. Older plants are more difficult to control. **Remarks:** Glyphosate is a nonselective herbicide. It has no soil activity and its effectiveness is increased by the addition of ammonium sulfate to prevent the active ingredient from precipitating with cationic salts. Add non-ionic surfactant.
BRANCHED-CHAIN AMINO ACID INHIBITORS	
Chlorsulfuron *Telar*	**Rate:** 1 to 2.6 oz product/acre (0.75 to 1.95 oz a.i./acre) **Timing:** Preemergence or early postemergence. **Remarks:** Chlorsulfuron has mixed selectivity, but is generally safe on grasses. Use a surfactant for postemergence applications. It has long soil residual activity.
Imazapic *Plateau*	**Rate:** 8 to 12 oz product/acre (2 to 3 oz a.e./acre) **Timing:** Early postemergence when plants are growing rapidly.

	Remarks: Imazapic is primarily active on annual grasses. Add 1 qt/acre methylated seed oil. Do not exceed 25 gal/acre spray volume. Imazapic is not registered for use in California.
Imazapyr *Arsenal, Habitat, Stalker, Chopper, Polaris*	**Rate:** 2 to 3 pt product/acre (8 to 12 oz a.e./acre) **Timing:** Preemergence or postemergence. **Remarks:** Imazapyr is a nonselective herbicide and has long soil residual activity. Treatments often result in more bare ground than other treatments, even 1 year after application.
Rimsulfuron *Matrix*	**Rate:** 4 oz product/acre (1 oz a.i./acre) **Timing:** Preemergence to postemergence, when target plants are 1 to 3 inches wide. **Remarks:** Rimsulfuron controls several annual grasses and broadleaves. Perennial grasses are tolerant to fall applications when established and grown under dryland conditions. Application to rapidly growing or irrigated perennial grasses may result in their injury or death. It provides soil residual control in cool climates but degrades rapidly under warm conditions. Rimsulfuron will not control summer annual weeds when applied in fall or spring. Add a surfactant when applying postemergence. It must be activated by rainfall or irrigation of at least half an inch. For the best results, rainfall should occur within 2 to 3 weeks of application and under cooler temperatures. Do not apply more than 4 oz product/acre per year.
CONTACT PHOTOSYNTHETIC INHIBITORS	
Paraquat *Gramoxone*	**Rate:** Broadcast treatment: 1.33 to 2.67 pt product/acre (5.3 to 10.7 oz a.i./acre). Spot treatment: 0.25% to 0.5% v/v solution. **Timing:** Postemergence, to rapidly growing young plants. Older plants are more poorly controlled. Activity appears to be better when applied under cloudy conditions and worst under full sunlight. **Remarks:** Paraquat is a contact herbicide that is nonselective on annuals. Good coverage is necessary for effective control. It has no soil activity and is a restricted-use herbicide.

Trichostema lanceolatum Benth.

Vinegarweed

Family: Lamiaceae
Range: Washington, Oregon, and throughout California, except the Great Basin and desert regions.
Habitat: Grasslands, oak woodlands, and disturbed sites such as roadsides and newly planted vineyards and orchards.
Origin: Native to California, Oregon and Washington.
Impacts: Vinegarweed is often considered a rangeland weed because it is unpalatable to livestock and populations may increase with grazing pressure. It contains aromatic volatile oils that make it unpalatable to grazing and foraging animals. It is, however, an important component of native grassland communities, particularly as a pollen source for bees and other insects.

Vinegarweed is a highly aromatic native summer annual that typically inhabits dry open disturbed places and overgrazed grasslands. Vinegarweed has soft-hairy foliage with a strong vinegar-like scent and lanceolate leaves, 0.8 to 3 inches long, with 3 veins from the base.

The bilateral purple flowers are in racemes and do not appear whorled. Flowers have a corolla tube that bends upward, 4 stamens, and a style that is much longer than the corolla and curved downward. Plants reproduce only by seed that are primarily dispersed by falling to the ground below the parent plant. Seeds probably survive for several years in the soil.

NON-CHEMICAL CONTROL

Mechanical (pulling, cutting, disking)	Hand pulling is very effective on small populations. Pulling is most effective before flowering in late spring when plants are elongated and soil is still moist. Gloves should be worn, as the plants contain a resin with a strong odor.
Cultural	Grazing is not considered an effective control option. The leaves contain volatile oils that make the plant unpalatable to grazing and foraging animals.
	Vinegarweed is a late season plant and does not form dense enough stands to allow for burning as a management tool.
Biological	No biological control programs have been developed for vinegarweed, as it is a desirable native species in most situations.

CHEMICAL CONTROL

The following specific use information is based on reports by researchers and land managers. Other trade names may be available, and other compounds also are labeled for this weed. Directions for use may vary between brands; see label before use. Herbicides are listed by mode of action and then alphabetically. The order of herbicide listing is not reflective of the order of efficacy or preference.

GROWTH REGULATORS	
2,4-D Several names	**Rate:** Broadcast treatment: 2 to 4 pt product/acre (0.95 to 1.9 lb a.e./acre)
	Timing: Postemergence, when plants are growing rapidly. Applications in spring provide best control.
	Remarks: This rate has been shown to give good control of other broadleaf weeds in rangeland. Good coverage is necessary. The ester formulations of 2,4-D generally are more effective than the amines for vinegarweed control.
Picloram *Tordon 22K*	**Rate:** Broadcast treatment: 1 to 2 pt product/acre (4 to 8 oz a.e./acre) plus 0.25 to 0.5% v/v surfactant.
	Timing: Postemergence, when plants are growing rapidly. Applications in spring provide best control.
	Remarks: High levels of picloram can give long-term soil activity for broadleaves. *Tordon 22K* is a federally restricted use pesticide. It is not registered for use in California.
Triclopyr	**Rate:** Broadcast treatment: 1 to 2 qt product/acre (1 to 2 lb a.e./acre). Spot treatment: 1 to 1.5% v/v

Garlon 3A, Garlon 4 Ultra, Remedy Ultra	solution *Garlon 4 Ultra* or *Remedy Ultra* and water, applied to thoroughly wet all leaves.
	Timing: Postemergence when plants are growing rapidly. Applications in spring provide best control.
	Remarks: Triclopyr is a selective herbicide for broadleaf species and will not damage desirable grasses growing nearby.
AROMATIC AMINO ACID INHIBITORS	
Glyphosate *Roundup, Accord XRT II*, and others	**Rate:** Broadcast treatment: 1 to 2 qt product (*Roundup ProMax*)/acre (1.1 to 2.25 lb a.e./acre). Spot treatment: 1.5 to 2% v/v solution *Roundup* (or other trade name) and water, applied to thoroughly wet all leaves.
	Timing: Postemergence when plants are growing rapidly. Applications in early spring provide best control.
	Remarks: Glyphosate is a nonselective systemic herbicide with no soil activity.

Tripidium ravennae (L.) H. Scholz
   (= *Saccharum ravennae* (L.) L., *Erianthus ravennae* (L.) Beauv.)

Ravennagrass

Family: Poaceae
Range: Southwestern states, including California, Arizona, New Mexico, Utah and Colorado.
Habitat: Wetlands and river channels, as well as ditches and other riparian areas. It has become a problem in the Grand Canyon.
Origin: Native to Eurasia. Introduced as a large ornamental perennial grass.
Impacts: Ravennagrass can form impenetrable stands that originate from under other vegetation. It reduces native plant diversity, including rare and endangered species in these sites.
California Invasive Plant Council (Cal-IPC) Inventory: Moderate Invasiveness

Ravennagrass is a large, tufted perennial with flowering stems to 12 ft tall and leaves that are distributed on the stem, up to the base of the inflorescence. It is similar in appearance to pampasgrass and jubatagrass. Ravennagrass is distinguished by having blade bases that are very densely covered with long, fuzzy, tawny hairs that typically hide the ligule and upper blade base surface.

The feather-like plumes or inflorescences are in a panicle up to 2 ft long and often purplish-bronze turning to silver-gray in fall. The tiny seeds are dispersed long distances by both wind and water. It is not known how long the seeds can survive in the soil, but because of their small size it is expected that they don't survive much more than 1 year.

NON-CHEMICAL CONTROL

Mechanical (pulling, cutting, disking)	Manual removal of individual plants is possible provided all root and rhizome fragments are also removed. The National Park Service staff and volunteers report removal of over 25,000 plants in the Grand Canyon. It is now considered rare throughout the canyon. Control techniques are very similar to those published for *Cortaderia* species.
Cultural	Because of the proximity to riparian areas, there are no cultural control options that have been reported.
Biological	No biological control agents are available for ravennagrass control.

CHEMICAL CONTROL

The following specific use information is based on reports by researchers and land managers. Other trade names may be available, and other compounds also are labeled for this weed. Directions for use may vary between brands; see label before use.

AROMATIC AMINO ACID INHIBITOR	
Glyphosate *Rodeo*, *Aquamaster*	**Rate:** Most control efforts use spot treatments of a 5% glyphosate v/v solution of *Rodeo* or *Aquamaster* (2.5% a.e.). **Timing:** Postemergence, to plants that are fully mature, but before flowers produce viable seeds. This is likely during the early summer months. **Remarks:** Glyphosate is a nonselective herbicide with no soil activity. It has proven to be the most effective chemical control option for ravennagrass. There have also been reports that a 5% glyphosate solution mixed with 1% imazapyr (*Habitat*) will give very effective control.

Typha spp.
Cattails

Typha latifolia

Family: Typhaceae
Range: All western states.
Habitat: Freshwater wetlands throughout North America including lakeshores, river backwaters, ditches, bogs, fresh or brackish marshes, lakes, and ponds. Cattails tolerate nutrient rich, acidic, alkaline, and slightly saline conditions; generally not shade tolerant.
Origin: Most species are native to North America. Some populations are hybrids.
Impacts: Cattail control is largely dependent on land management goals. Cattails can behave like aggressive introduced weeds, but they are a native element in a variety of natural communities and can provide valuable wildlife habitat. Solid stands can limit biodiversity in wetlands, decrease recreation opportunities, and impede water movement. With sedimentation or changes in hydrology, shallow wetlands, ponds, and slow-moving streams may become vulnerable to cattail monocultures that eliminate open water.

Cattails are emergent perennials that can grow to 10 ft tall. The stems are erect, unbranched, rigid and solid. The leaves are alternate, most near stem bases, linear, thick, and spongy. Rhizomes are tough, creeping, branched, white with fibrous scale leaves. Roots from rhizomes are fibrous and shallow.

Inflorescences are cigar-shaped flower spikes, usually taller than the leaves and densely covered with numerous tiny male and female unisexual flowers. The male flowers are produced above the female flowers on the spike, wind pollinated, and self-compatible. One plant can produce approximately 250,000 soft downy seeds in fall. Seeds are said to remain viable in the seed-bank for up to 100 years, but it more likely that they only survive a couple of years. Seeds germinate primarily in late spring. Most seedlings emerge from the substrate in water to 14 inches deep, but some may emerge from water up to 2.5 ft deep. Local reproduction is primarily vegetative from rhizomes. Most rhizomes survive for less than 3 years. Rhizome fragments frequently disperse with water or substrate movement.

NON-CHEMICAL CONTROL

Mechanical (pulling, cutting, disking)	Cattails can be successfully controlled by physical removal. Hand-pulling or cutting cattails at the end of flowering followed by submergence of all cattail stems gave good control in several studies. Cutting plants below the water-line two to three times before flowering was also effective. It is important to remove all dead and live cattail stems.
	Crushing, shearing, or disking is effective for severing the aerenchyma (air filled cells) that link rhizomes with the leaves. To reduce plant survival, however, these techniques must be combined with flooding to induce stress from anaerobic starch conversion.
Cultural	If possible, removing sedimentation and increasing water depth will usually discourage cattail monocultures. Maintaining water levels 1.4 to 2 ft over the tops of existing shoots can kill established plants within a couple of years. Water depths over 2 ft can discourage cattail recruitment and seedling survival. Narrow-leaved cattail (*Typha angustifolia* L.) grows in deeper water and water levels need to be 4 ft or deeper to prevent seedling survival.
	Fire provides little or no cattail control. However, burning followed by reflooding to about 1 ft in spring controlled cattail in one study. Fire can be helpful for litter cleanup.
Biological	There are no available biological control agents for cattail control, primarily since the species are native to North America.

CHEMICAL CONTROL

The following specific use information is based on published papers and reports by researchers and land managers. Other trade names may be available, and other compounds also are labeled for this weed. Directions

for use may vary between brands; see label before use. Herbicides are listed by mode of action and then alphabetically. The order of herbicide listing is not reflective of the order of efficacy or preference.

AROMATIC AMINO ACID INHIBITORS	
Glyphosate *Aquamaster, Rodeo* and others	**Rate:** 3 to 4 qt product *Rodeo* or *Aquamaster*/acre (3 to 4 lb a.e./acre). Spot treatment: 2 % v/v. **Timing:** Postemergence, at flowering after heads are formed and before frost. **Remarks:** Glyphosate will not kill seeds or inhibit germination the following season. Glyphosate has no soil activity. Allow 7 days or longer before clipping or tillage. Flooding after herbicide application improved control in several studies. Adding a surfactant or emulsifier is recommended as cattails have a thick waxy coating on the leaf. Retreatment with herbicides is often necessary for complete control.
BRANCHED-CHAIN AMINO ACID INHIBITORS	
Imazamox *Clearcast*	**Rate:** 2 to 4 pt product/acre (0.25 to 0.5 lb a.i./acre) **Timing:** Postemergence, from new growth through killing frost. **Remarks:** Imazamox has mixed selectivity and controls several broadleaf and annual grass species. It is registered for control of vegetation in and around aquatic and non-cropland sites. It has irrigation and water use restrictions. Add a spray adjuvant, such as a methylated seed oil, to improve control.
Imazapyr *Habitat*	**Rate:** 0.5 to 2 qt product/acre (0.25 to 1 lb a.e./acre) **Timing:** Postemergence from boot to flowering. **Remarks:** *Habitat* is registered for aquatic use. Also effective following early season mowing and/or disking. It is a nonselective herbicide. Imazapyr has long soil residual activity and may leave more bare ground than other treatments, even a year after application. Add a spray adjuvant. Do not apply in the root zone of desirable trees.

Ulex europaeus L.
Gorse

Family: Fabaceae
Range: Although gorse occurs along the Atlantic coast from Virginia to Massachusetts, it is most prevalent along the Pacific coast from California to British Columbia and is also found at high elevation on two Hawaiian Islands.
Habitat: Grasslands, shrublands, forest margins, coastal habitats and disturbed sites such as roadsides, pasture lands, gravelly floodplains, burned areas, and cleared forests. Grows well on shady slopes with high soil moisture and good drainage. Frost-damaged plants can resprout from the crown. Does not survive severely cold winters or arid climates. It grows best on acidic soil and tolerates many soil types, including serpentine, but seldom grows in high calcium soils.
Origin: Native to western Europe and introduced as an ornamental or hedge shrub.
Impacts: Gorse often forms dense, impenetrable thickets that exclude desirable vegetation and increase fire risk. Mature plants contain about 2 to 4% flammable oils. Gorse can fix nitrogen, enabling the plant to colonize and dominate areas with poor soil. The plant also produces abundant leaf litter that can acidify the upper soil layers. Soil is often bare between individual plants, increasing erosion on steep slopes where gorse has replaced grasses or forbs. Plants are spiny and mostly unpalatable when mature, thus reducing pasture quality.
Western states listed as Noxious Weed: California, Oregon, Washington
California Invasive Plant Council (Cal-IPC) Inventory: High Invasiveness

Gorse is a dense, spiny, evergreen shrub to 5 ft tall, with yellow, pea-like flowers. The stems are highly branched, alternate and terminate with a green spine 1.5 to 2.5 inches long. Juvenile plants have soft, grey-green stems with trifoliate leaves. Mature shrubs appear leafless with simple leaves modified into stiff, curved, awl-like spines 0.25 to 1 inch long. Spines and leaves have a waxy coating and end in a sharp yellow point. Gorse has an extensive, lateral root system that contains nitrogen-fixing root nodules.

Plants begin flowering from 18 months to 3 years of age. Reproduction is by seed which are produced in small, hairy pods 0.5 to 0.75 inches long. Pods become black when mature, and upon drying, disperse seeds by ejecting them several feet from the plant. Seeds can remain viable in the soil for up to 30 years. Large soil seedbanks often accumulate, making long term control difficult. Shrubs may live for up to 30 years.

NON-CHEMICAL CONTROL

Mechanical (pulling, cutting, disking)	Hand pulling can remove seedlings and small shrubs, but once established this technique is generally not effective.
	Cutting gorse off before it flowers will reduce seed production and deplete the plant's energy reserves. Resprouts are common after treatment. Cutting should be combined with an herbicide treatment or with multiple cuttings over a period of years. Cut shrubs at ground level with power or manual saws.
	Heavy equipment can be effectively used to control gorse in areas where soil disturbance and nonselective species removal are not important considerations. Stumps that remain following such treatment will require herbicide application to prevent regrowth.
Cultural	Repeated grazing by goats and/or sheep can greatly reduce seedling establishment and crown regrowth. In a long-term study, the best control of gorse was achieved by first burning gorse stands, followed by grazing goats or a 2:1 mix of goats and sheep at 10 or more animals/acre. On areas of unburned gorse, sustained goat stocking for 4 to 5 years provided good control in some situations. Once animals are removed, new seedlings must be controlled.
	Burning alone does not kill the root system and resprouts are common after treatment. Burning often stimulates a flush of seedling germination after the first rains. Following a burn with herbicide application provides good control.
Biological	The gorse seed weevil (*Exapion ulicis*) and spider mite (*Tetranychus lintearius*) are biocontrol agents that have become established in California. The seed weevil reduces seed production but cannot kill established stands. The seed weevil was first introduced to New Zealand in 1931, and was widely

> established by 1935. It has destroyed approximately 35% of the seed crop annually since then. Heavy mite *(Tetranychus lintearius)* infestations can kill branches and reduce overall plant vigor and are apparent by the dense webbing that covers the foliage.

CHEMICAL CONTROL

The following specific use information is based on published papers and reports by researchers and land managers. Other trade names may be available, and other compounds also are labeled for this weed. Directions for use may vary between brands; see label before use. Herbicides are listed by mode of action and then alphabetically. The order of herbicide listing is not reflective of the order of efficacy or preference.

GROWTH REGULATORS	
Picloram *Tordon 22K*	**Rate:** Broadcast treatment: 1 to 2 qt product/acre (0.5 to 1 lb a.e./acre). Spot treatment: 0.5% v/v solution and water plus 0.25 to 0.5% v/v surfactant to thoroughly wet all leaves. **Timing:** Postemergence foliar treatments are best when plants are growing rapidly at or beyond early to full bloom stage. **Remarks:** High levels of picloram can give long-term soil activity for broadleaves. Picloram is a restricted use herbicide. It is not registered for use in California.
Triclopyr *Garlon 3A, Garlon 4 Ultra, Pathfinder II*	**Rate:** Low volume spot treatment: 0.5 to 2% v/v solution of *Garlon 4 Ultra*, or 0.5 to 2% *Garlon 3A* and water plus 0.25 to 0.5% v/v surfactant to thoroughly wet all leaves. Cut stump treatment: 25% *Garlon 4 Ultra* in 75% oil carrier, or undiluted *Garlon 3A* or 50% *Garlon 3A* in water. Basal bark treatment: 20% *Garlon 4 Ultra* in 80% oil carrier, or undiluted *Pathfinder II* as a ready-to-use formulation. Basal cut stump treatment: 25% *Garlon 4 Ultra* in 75% oil carrier. **Timing:** Postemergence when plants are growing rapidly. Cut stump, basal cut stump, and basal bark treatments can be applied anytime as long as the ground is not frozen, but are best used in late summer or early fall. **Remarks:** Triclopyr is a selective herbicide for broadleaf species and will not damage desirable grasses growing nearby. For cut stump treatments, cut stems horizontally at or near ground level, and immediately apply herbicide solution. Suckering from the roots typically occurs after cutting, but the treatment should control most resprouts. Basal bark treatment: spray the lower trunk, including the root collar, to a height of 12 to 15 inches from the ground; the spray should thoroughly wet the lower stem but not to the point of runoff. Plants should not be cut for at least 1 month following basal bark treatment.
Triclopyr + 2,4-D *Crossbow*	**Rate:** Spot treatment: 1 to 1.5% v/v solution of *Crossbow* and water to thoroughly wet all leaves. **Timing:** Postemergence when plants are growing rapidly. **Remarks:** *Crossbow* in water forms an emulsion (not a solution), and separation may occur unless the spray mixture is agitated continuously.
AROMATIC AMINO ACID INHIBITORS	
Glyphosate *Roundup, Accord XRT II,* and others	**Rate:** Spray-to-wet spot treatment: 1.5 to 2% v/v solution of *Roundup ProMax* (or other trade name with similar concentration of glyphosate) in water to thoroughly wet all leaves. Low volume spot treatment: 5 to 10% v/v solution of *Roundup* (or other trade name) in water. Spray coverage should be uniform with at least 50% of the foliage contacted. Cut stump treatment: undiluted product or 50% v/v in water. **Timing:** Postemergence when plants are growing rapidly. Foliar treatments should be made in late summer or early fall. For cut stump treatment, application in late summer, early fall or dormant season provides best control. Treatment should occur immediately after cutting. **Remarks:** Nonselective systemic herbicide; gives good control with some resprouts. Plants should not be cut for at least 4 months after foliar treatments. Cut stump applications are as described for triclopyr.
BRANCHED-CHAIN AMINO ACID INHIBITORS	
Metsulfuron *Escort*	**Rate:** Broadcast foliar treatment: 1 oz product/acre (0.6 oz a.i./acre) plus 0.25% v/v surfactant. **Timing:** Postemergence foliar treatments are best when plants are growing rapidly at or beyond early to full bloom stage. **Remarks:** Although metsulfuron has some preemergent activity, best results are generally obtained when applied to the foliage during active growth. Metsulfuron is not registered for use in California.

Ventenata dubia (Leers) Coss. & Dur.

North African wiregrass

Family: Poaceae
Range: Many western states, particularly the northwestern states including Washington, Oregon, California, Idaho, Montana, Wyoming, and Utah.
Habitat: Bunchgrass, sagebrush, and meadow communities. Also found in Conservation Reserve Program lands, pastures and grass hayfields.
Origin: Native to Eurasia and North Africa.
Impacts: *Ventenata dubia* can outcompete perennial bunchgrasses but the mechanism for this ability is not clear. It is high in silica (2.7%)–making it poorly palatable to grazing animals–but not as high as medusahead (6.4%). Litter can build up on the soil surface, likely because of the higher silica content. Plants dry early in the season, similar to downy brome, and should pose similar risks with respect to fire.

North African wiregrass is a short-statured winter annual grass, 6 to 28 inches tall, with narrow leaves 1 to 2 mm wide. It is shallowly rooted, typically within the top 1 to 2 inches of soil. Ligules are membranous and 1 to 4 mm long.

The spikelets resemble a small wild oat. Spikelets usually have 3 florets with at least 1 floret with a twisted awn, 3/4 to 1 inch long that is attached to the lemma at 1/2 the lemma length. Seeds disperse primarily by falling to the ground soon after maturity. Seeds readily germinate in fall, approximately 2 weeks after downy brome emerges. Seed burial studies suggest longevity in the field of less than 2 years.

NON-CHEMICAL CONTROL

Mechanical (pulling, cutting, disking)	Hand pulling when the soil is moist can uproot this shallowly rooted species. Small infestations of a few square yards potentially could be pulled, but this is too labor intensive with large infestations.
Cultural	Grazing for weed control is unlikely to be effective given the low palatability of North African wiregrass. Using grazing to maintain competitive perennial grasses should indirectly provide some suppression.
	Prescribed burning in summer would be expected to provide some control of *Ventenata dubia*, provided it was conducted while the seeds remained on the plant. There is no direct evidence to support this, but results with similar species suggest that it would likely be effective.
Biological	There are no biological control efforts for the management of *Ventenata dubia*.

CHEMICAL CONTROL

The following specific use information is based on reports by researchers and land managers. Other trade names may be available, and other compounds also are labeled for this weed. Directions for use may vary between brands; see label before use. Herbicides are listed by mode of action and then alphabetically. The order of herbicide listing is not reflective of the order of efficacy or preference.

AROMATIC AMINO ACID INHIBITORS	
Glyphosate	**Rate:** 0.75 pt product (*Roundup ProMax*)/acre (0.42 lb a.e./acre)
Roundup, Accord XRT II, and others	**Timing:** Postemergence. Fall application is the best timing to kill plants and prevent injury to many natives. Spring application will reduce seed production.
	Remarks: Glyphosate at this low rate can be effective in fall and prevent long-lasting injury to perennial bunchgrasses. Multiple years of treatment are necessary. The herbicide is nonselective and has no soil activity.

BRANCHED-CHAIN AMINO ACID INHIBITORS	
Imazapic	**Rate:** 5 to 6 oz product/acre (1.25 to 1.5 oz a.e./acre)
Plateau	**Timing:** Postemergence to seedlings in early spring or fall.
	Remarks: If warm-season perennial grasses are present, apply in fall when they are dormant and annual

	grasses have germinated. This timing will damage cool-season perennial grasses. If applied in pasture and desirable grasses are injured, a half rate of a balanced fertilizer applied the following spring may facilitate recovery. Imazapic has some soil residual activity. Imazapic is not registered for use in California.
Rimsulfuron *Matrix*	**Rate:** 4 oz product/acre (1 oz a.i./acre) **Timing:** Preemergence or postemergence in fall. **Remarks:** Rimsulfuron controls several annual grasses and broadleaves. Perennial grasses are tolerant to fall applications when established and grown under dryland conditions. Application to rapidly growing or irrigated perennial grasses may result in their injury or death. It provides soil residual control in cool climates but degrades rapidly under warm conditions. Rimsulfuron will not control summer annual weeds when applied in fall or spring. Add a surfactant when applying postemergence.
Sulfometuron *Oust* and others	**Rate:** 0.75 to 2.25 oz product/acre (0.56 to 1.7 oz a.i./acre) **Timing:** Either preemergence or postemergence in fall or early spring. **Remarks:** Sulfometuron is a broad-spectrum herbicide and will control many species. A combination of sulfometuron and chlorsulfuron (*Landmark XP*) at 0.75 oz product/acre has also been shown to be very effective.
Sulfosulfuron *Outrider*	**Rate:** 0.75 oz product/acre (0.56 oz a.i./acre) **Timing:** Either preemergence or postemergence in fall. **Remarks:** If meadow foxtail is present, it will be removed with this application. Planting of some wheatgrasses is possible in the following spring but other species should not be planted until the following fall or spring.
PHOTOSYNTHETIC INHIBITORS	
Hexazinone *Velpar L*	**Rate:** 2 to 6 qt product/acre (1 to 3 lb a.i./acre), check hexazinone formulation to make sure active ingredient rate is correct **Timing:** Preemergence in fall or very early spring. **Remarks:** There is no direct evidence that hexazinone will control *Ventenata dubia*, but it is generally effective against annual grasses and would be expected to be effective. High rates of hexazinone can create bare ground, so only use high rates in spot treatments.

Verbascum thapsus L.; common mullein
Verbascum blattaria L.; moth mullein

Common and moth mullein

Family: Scrophulariaceae
Range: All western states.
Habitat: Open areas including rangeland, forest clearings, meadows, and roadsides. Prefers, but is not limited to, disturbed sites with well-drained soils.
Origin: Native to Eurasia.
Impacts: Populations can spread rapidly and form dense stands in disturbed areas. Unpalatable to livestock due to their woolly leaves. Established stands are extremely difficult to control due to their abundant, long-lived seed bank.

Common mullein is a biennial, short-lived perennial or, rarely, an annual to 7 ft tall. Moth mullein is a biennial to 4 ft tall. Both species exist as basal rosettes until they develop a single tall flowering stem at maturity. Common mullein rosettes grow up to 2 ft in diameter with oblong to oblanceolate, gray-green woolly leaves. Moth mullein rosettes grow to 16 inches in diameter with oblanceolate, bright green leaves that are irregularly-toothed and sparsely hairy to glabrous. Both mullein species form a deep tap root.

The common mullein inflorescence is a spike-like raceme with yellow flowers 1 to 1.5 inches wide. The moth mullein inflorescence is a spike-like raceme with white or pale yellow flowers with fine purple hairs covering stamen filaments. Each plant can produce over 100,000 seeds. The seeds can survive over 100 years in the soil. Most seed falls near the parent plant. Seedling establishment is dependent on periodic disturbance. These are ephemeral plants that are often displaced by other plants in undisturbed, densely vegetated sites. Studies have shown common mullein establishment is greatly enhanced in bare ground areas.

NON-CHEMICAL CONTROL

Mechanical (pulling, cutting, disking)	Hand-pulling before seed set is an effective control method for mullein plants growing on loose soils. When digging, sever the root below the soil surface. Soil disturbance stimulates recruitment.
	Repeated mowing in the bolting to early flowering stage can reduce seed production. Mowing low-growing rosettes is not effective as rosettes quickly regrow if mowing is discontinued.
	Tillage is effective for controlling existing plants, but soil disturbance stimulates recruitment. Common mullein is not competitive in areas that experience annual cropping.
Cultural	Common mullein has low palatability at all growth stages. Livestock typically avoid grazing mullein if palatable forage is available. A California study reported the abundance of common mullein in a cow's summer diet ranged from 0% to 3.5%. Common mullein is common on over-grazed sites.
	Fire is not an effective control and often can dramatically stimulate recruitment from the seedbank. Thus, fire can be used to manage the seedbank if the new seedlings are controlled after they germinate.
	After a disturbance event, common mullein populations are typically ephemeral, and in time the abundance normally decreases. Promoting competitive vegetation and minimizing soil disturbance are likely the best control options. Competitive vegetation that shades and competes for early season moisture reduces mullein seedling establishment.
Biological	A curculionid weevil (*Gymnaetron tetrum*), specific to *Verbascum thapsus*, was accidentally introduced to North America from Europe before 1937. The larvae mature in the seed capsules and destroy up to 50% of the seeds.

CHEMICAL CONTROL

The following specific use information is based on research papers and reports by land managers. Other trade names may be available, and other compounds also are labeled for this weed. Directions for use may vary between brands; see label before use. Herbicides are listed by mode of action and then alphabetically. The order of herbicide listing is not reflective of the order of efficacy or preference.

GROWTH REGULATORS	
Aminocyclopyrachlor + chlorsulfuron *Perspective*	**Rate:** 4.75 to 8 oz *Perspective*/acre **Timing:** Postemergence or preemergence. Postemergence applications are most effective when applied from the rosette to early bolting stage. **Remarks:** *Perspective* provides broad-spectrum control of many broadleaf species. Although generally safe for grasses, it may suppress or injure certain annual and perennial grass species. Do not treat in the root zone of desirable trees and shrubs. Do not apply more than 11 oz product/acre per year. At this high rate, cool-season grasses will be damaged, including bluebunch wheatgrass. Not yet labeled for grazing lands. Add an adjuvant to the spray solution. This product is not approved for use in California and some counties of Colorado (San Luis Valley).
Aminopyralid *Milestone*	**Rate:** 7 oz product/acre (1.75 oz a.e./acre) **Timing:** Postemergence from the rosette to young bolting stage. **Remarks:** Safe on most grasses, although preemergence application at high rates can greatly suppress some annual grasses, such as medusahead. Applications can decrease seed production in some annual and perennial grass species. For postemergence applications, add a non-ionic surfactant at 0.25 to 0.5% v/v to aid in absorption into the woolly leaves. Other premix formulations of aminopyralid can be used. These include *Opensight* (aminopyralid + metsulfuron) (see below) and *Forefront HL* (aminopyralid + 2,4-D). *Forefront HL* is applied at 1.5 to 2.1 pt product/acre during the rosette stage.
Aminopyralid + metsulfuron *Opensight*	**Rate:** 2 to 3.3 oz product/acre **Timing:** Spring or fall. Apply 2 oz product/acre to plants in the rosette stage, or 2.5 to 3.3 oz product/acre for bolting plants less than 12 inches tall. **Remarks:** Not registered for use in California.
Fluroxypyr *Vista XRT*	**Rate:** 22 oz product/acre (7.7 oz a.e./acre) **Timing:** Postemergence, from seedling state to young bolting stage. **Remarks:** Fluroxypyr provides only suppression of the *Verbascum* species. It is a broadleaf-selective herbicide with no soil residual activity. Add a surfactant or seed oil surfactant.
Picloram *Tordon 22K*	**Rate:** 1 qt product/acre (8 oz a.e./acre). **Timing:** Preemergence and postemergence. With postemergence applications, treat at rosette to early bolting stage, when plants are growing rapidly. Add a surfactant to aid absorption into the woolly leaves. **Remarks:** Picloram controls a wide range of broadleaf species and has relatively long soil residual activity. Although well-developed grasses are not usually injured by labeled use rates, some applicators have noted that young grass seedlings with fewer than four leaves may be killed. Do not apply near trees. Picloram is a restricted use herbicide. It is not registered for use in California. On the label, *Tordon 22K* at 1 to 1.5 pt + 2,4-D at 1 lb a.e./acre is listed as controlling common mullein.
AROMATIC AMINO ACID INHIBITORS	
Glyphosate *Roundup, Accord XRT II*, and others	**Rate:** 2 qt product (*Roundup ProMax*)/acre (2.25 lb a.e./acre). Spot treatment: 2 % v/v solution **Timing:** Postemergence, from seedling to late bolting stage. **Remarks:** Glyphosate will provide control only during the year of application; it has no soil activity and will not kill seeds or inhibit germination the following season. Glyphosate is nonselective. It can create bare ground conditions that are susceptible to weed recruitment. In areas with desirable vegetation, use spot treatment. Glyphosate is a good control option if reseeding is planned shortly after application, as it will not affect seedlings emerging after application. Add a surfactant when using a formulation where it is not already included (e.g., *Rodeo, Aquamaster*).

BRANCHED-CHAIN AMINO ACID INHIBITORS	
Chlorsulfuron *Telar*	**Rate:** 1 to 2.6 oz product/acre (0.75 to 1.95 oz a.i./acre) **Timing:** Postemergence from seedling to bolting stage. **Remarks:** Chlorsulfuron has mixed selectivity and is generally safe on grasses. Always use a surfactant. Also included with aminocyclopyrachlor in *Perspective*. *Telar* can be used near water, but cannot be applied to water.
Imazapyr *Arsenal, Habitat, Stalker, Chopper, Polaris*	**Rate:** 1 to 2 qt product/acre (0.5 to 1 lb a.e./acre) **Timing:** Preemergence and postemergence. **Remarks:** Imazapyr has soil residual activity and may impact restoration efforts. Add a spray adjuvant.
Metsulfuron *Escort*	**Rate:** 1 to 2 oz product/acre (0.6 to 1.2 oz a.i./acre) **Timing:** Postemergence from seedling to bolting stage. **Remarks:** Metsulfuron has similar activity to chlorsulfuron. Metsulfuron has some soil residual activity. Always use a surfactant. Other premix formulations of metsulfuron can be used at similar application timing. These include *Cimarron Max* (metsulfuron + dicamba + 2,4-D) and *Cimarron X-tra* (metsulfuron + chlorsulfuron). Metsulfuron is not registered for use in California.

Veronica anagallis-aquatica L.

Water speedwell

Family: Scrophulariaceae
Range: Throughout the United States.
Habitat: Near aquatic environments, including ditches, slow streams, pond and stream margins, wet meadows, and marshy places.
Origin: Native to Europe.
Impacts: Can form dense stands in aquatic systems and impede the flow of water. Also weedy in Canada and South America.

Water speedwell is an emergent biennial or perennial to 3 ft tall, with glabrous opposite leaves. Plants typically develop a system of rhizomes and/or stolons. The stems are erect to spreading, sometimes creeping and often rooting at lower nodes. The leaves are usually sessile and 1 to 4 inches long with a smooth to serrate margin.

Inflorescences are axillary racemes opposite each other. Flowers are deeply 4-lobed, 5 to 10 mm in diameter, and pale lavender-blue with darker lines. The fruit are flattened, nearly round capsules with a distinctive notch at the apex. Plants reproduce by seed and vegetatively from rhizomes and/or stolons. Seeds disperse with water and soil and can germinate year round under favorable conditions. Seeds are small, and while there is no information on their longevity in the soil seedbank, it is expected that they would not survive more than a couple of years.

NON-CHEMICAL CONTROL

Mechanical (pulling, cutting, disking)	Although plants are very brittle, care must be taken when using hand removal techniques, as fragmented plants can be the source of new infestations in aquatic settings. Because plants are nearly always found in aquatic setting, tillage is not typically appropriate.
Cultural	Because water speedwell is most often found in ditches, canals and other aquatic areas, it is not appropriate to use grazing or prescribed burning as a control option.
Biological	There are no biological control agents available for water speedwell.

CHEMICAL CONTROL

Little information has been published on the control of water speedwell. Information on control was obtained though literature on herbicides used to control other speedwell species. It is expected that they will also have activity on water speedwell.

The following specific use information is based on reports by researchers and land managers. Other trade names may be available, and other compounds also are labeled for this weed. Directions for use may vary between brands; see label before use.

AROMATIC AMINO ACID INHIBITORS	
Glyphosate *Rodeo,* *Aquamaster*	**Rate:** 1 to 3 lb product (*Rodeo* or *Aquamaster*)/acre (0.5 to 1.5 lb a.e./acre). Wiper treatment: 33 to 50% of concentrated product. **Timing:** Postemergence, to rapidly growing plants. **Remarks:** Glyphosate is a nonselective product with no soil activity. A wiper applicator can give more selective control.

Vinca major L.
Big periwinkle

Family: Apocynaceae
Range: Primarily California, but also Oregon, Washington, Idaho, Utah, Arizona, New Mexico and much of the southern and eastern United States.
Habitat: Riparian corridors, moist woodlands, forest margins, coastal habitats, and disturbed sites such as roadsides and old homesteads. Grows best under moist, shady conditions on sandy to medium loam soil, with acidic to neutral pH. Can also tolerate drought, full sun, heavy clay and slightly alkaline soils. Foliage is susceptible to frost damage.
Origin: Native to central Europe and the Mediterranean region. Introduced to the United States in the 1700s as an ornamental and for medicinal uses.
Impacts: Under favorable conditions, plants spread invasively and can develop a dense ground cover that outcompetes other vegetation in natural areas. Big periwinkle is becoming a dominant woodland understory in many areas of California. Infestations around old homesteads have been present for many years and serve as nurseries for further spread. Some plants in the dogbane (Apocynaceae) family are extremely toxic, although poisoning due to the ingestion of big periwinkle is poorly documented.
California Invasive Plant Council (Cal-IPC) Inventory: Moderate Invasiveness

Big periwinkle is an herbaceous perennial groundcover with milky sap, trailing stems to 3 ft long, and ascending to erect flower-bearing stems to 1.5 ft tall. The leaves are dark glossy green, 2 to 3 inches long, oval in shape and slightly pointed at the tip.

Big periwinkle produces showy lavender-blue funnel-shaped solitary flowers. However, the species is considered sterile with only a few documented seedlings encountered. Thus, reproduction occurs vegetatively from trailing stems that root at the tips and from stem fragments. Plants and stem fragments disperse through human activities, such as landscape plantings and careless disposal of yard waste. Under favorable conditions, stem cuttings left on the ground can take root. In riparian areas, water currents can fragment stems and carry them downstream where they may root if lodged in a suitable location.

NON-CHEMICAL CONTROL

Mechanical (pulling, cutting, disking)	Hand pulling is very effective if careful attention is paid to removing all stems, root nodes and stolons. Repeated removal efforts over multiple years may allow desirable vegetation to colonize the area.
	Because big periwinkle can resprout and establish from stem fragments, mowing or cutting is not recommended.
Cultural	Grazing is not considered an effective control option. The stems contain milky latex that makes the plant unpalatable to grazing and foraging animals.
	Burning is also not practical, nor is it an effective control tool for big periwinkle control.
Biological	Big periwinkle is still a very common urban ground cover. Thus, there are no biological control agents available for its control.

CHEMICAL CONTROL

The following specific use information is based on published papers and reports by researchers and land managers. Other trade names may be available, and other compounds also are labeled for this weed. Directions for use may vary between brands; see label before use. Herbicides are listed by mode of action and then alphabetically. The order of herbicide listing is not reflective of the order of efficacy or preference.

GROWTH REGULATORS

Picloram *Tordon 22K*	**Rate:** Broadcast foliar treatment: 2 qt product/acre (1 lb a.e./acre) plus 0.25 to 0.5% v/v surfactant applied to thoroughly wet all leaves.
	Timing: Postemergence foliar treatments are best when plants are growing rapidly.

	Remarks: High levels of picloram can give long-term soil activity for broadleaves. Picloram is a restricted use herbicide. It is not registered for use in California.
Triclopyr *Garlon 4 Ultra*	**Rate:** Foliar treatment: 1.12 to 2.25 pt product/acre of *Garlon 4 Ultra* (0.56 to 1.12 lb a.e./ac) and water plus 1% v/v surfactant, applied to thoroughly wet all leaves. Low volume/thinline treatment: 25% v/v solution of *Garlon 4 Ultra* plus 20% v/v crop seed oil in water. Wiper treatment: 25% v/v solution of *Garlon 4 Ultra* plus 20% v/v ethylated crop oil in water. **Timing:** Postemergence when plants are growing rapidly. Applications in spring provide the best control. **Remarks:** Triclopyr is selective for broadleaf species and will not damage desirable grasses growing nearby. *Garlon 4 Ultra* is formulated as a low volatile ester. However, in warm temperatures, spraying onto hard surfaces such as rocks or pavement can increase the risk of volatilization and off-target damage.
AROMATIC AMINO ACID INHIBITORS	
Glyphosate *Roundup, Accord XRT II*, and others	**Rate:** Foliar treatment: 2.5 to 5 qt product (*Roundup ProMax*)/acre (2.8 to 5.6 lb a.e./acre) in water, applied to thoroughly wet all leaves. Low volume/thinline treatment: 25% v/v solution of *Roundup* (or other trade name) in water. **Timing:** Postemergence when plants are growing rapidly. Applications in late summer or early fall provide the best control. **Remarks:** Glyphosate is a nonselective systemic herbicide with no soil activity. It gives good control with some resprouts.
BRANCHED-CHAIN AMINO ACID INHIBITORS	
Imazapyr *Arsenal, Habitat, Stalker, Chopper, Polaris*	**Rate:** Foliar treatment: 2.25 to 4.5 pt product/acre of *Stalker* (0.56 to 1.12 lb a.e./acre) and water plus 1% v/v surfactant, applied to thoroughly wet all leaves. Low volume/thinline treatment: 25% v/v solution of *Stalker* plus 20% v/v ethylated crop oil in water. Wiper treatment: 25% v/v solution of *Stalker* plus 20% v/v ethylated crop oil in water. **Timing:** Postemergence when plants are growing rapidly. Best when used in late summer to early fall. **Remarks:** Imazapyr is a soil residual herbicide and may result in bare ground around plants for some time after treatment.

Vulpia myuros (L.) C. Gmel.
(= *Festuca myuros* L. [Jepson Manual 2012])

Rattail fescue

Family: Poaceae
Range: Much of the western United States, except North and South Dakota, Wyoming and Colorado. Widespread in the coastal states.
Habitat: Disturbed and undisturbed open areas, including dry and seasonally wet sites, roadsides, rangeland, pastures, fields, occasionally agronomic fields, grassland, slopes, washes, open areas in many plant communities, including chaparral and open woodland. It tolerates drought, some shade, very poor sandy soil, shallow soil, and acidic soil.
Origin: Uncertain; thought to have originated from Europe.
Impacts: Rattail fescue is considered to be a poor forage grass for livestock. At maturity, the sharp florets can injure the mouths, nostrils, and eyes of grazing animals. In mixed-conifer forests of the Sierra Nevada, rattail fescue can outcompete conifer seedlings for root space, nutrients, and water. Rattail fescue has also been shown to interfere with the establishment and/or growth of rare native California herbs in vernal pools or other sensitive sites [e.g., San Diego mesamint (*Pogogyne abramsii*) and large-flowered fiddleneck (*Amsinckia grandiflora*)]. It is also a major problem in pastures and agronomic crops in southern Australia. A commercial cultivar is sometimes used as a cover crop, for erosion control, or to reestablish vegetation on highly degraded sites, such as mined areas.
California Invasive Plant Council (Cal- IPC) Inventory: Moderate Invasiveness

Rattail fescue is a winter annual to 2 ft tall. Rattail fescue is one of the widespread, non-native annual grasses that have naturalized throughout much of the western United States. The leaves are rolled in bud, 1 to 3 mm wide, usually less than 8 inches long, with the upper surface glabrous, scabrous, or covered with short stiff hairs. The ligules are membranous and short, 0.2 to 0.6 mm long, often longer on the sides. The root system is fibrous and typically shallow.

The inflorescence of rattail fescue is a narrow, spike-like panicle, mostly 2 inches to 1 ft long, and less than 1 inch wide. It is fairly dense, and sometimes one-sided. Plants reproduce only by seed, which disperse short distances by falling to the ground beneath the parent plant. Longer dispersal can occur when seeds cling to the fur of animals. In Australia, the seedbank is reported to survive for about 3 years.

The genus *Vulpia* is occasionally included within the *Festuca* genus, but unlike most *Festuca*, *Vulpia* species are annual. There are other problematic non-native *Vulpia* species, including squirreltail fescue (*Vulpia bromoides* (L.) Gray) and six-weeks fescue (*Vulpia octoflora* (Walter) Rydb.), but they are not nearly as common as rattail fescue. In addition, there is a native species, small fescue (*Vulpia microstachys* (Nutt.) Munro), which is considered a desirable species and often included in restoration seed mixes.

NON-CHEMICAL CONTROL

Mechanical (pulling, cutting, disking)	Hand pulling rattail fescue may be effective early in spring before the seed sets, but this is very labor intensive and is only used on small infestations. Minimize soil disturbance when hand pulling to avoid stimulating new germination. Mowing or string trimming can also be used to reduce rattail fescue if done before the seed sets in early summer.
	Because rattail fescue is shallowly rooted, it tends to be intolerant of tillage. Deep cultivation to bury the seeds in fall before the first rain, followed by manual removal of seedlings, can help control this species in crop fields. Most seeds do not survive burial. Shallow cultivation to stimulate germination in fall or spring, followed by a subsequent cultivation to kill seedlings, can also be effective.
Cultural	Rattail fescue is favored by most disturbances, including fire. Fire has limited use in controlling rattail fescue. Spring prescribed burning conducted in the boot or dough stage, before seed release, may temporarily reduce rattail fescue by destroying the current-year seed crop. However, early spring fire may also kill the current-year seeds of native plants.

	Rattail fescue is not particularly palatable and often persists in areas that are heavily grazed. On California annual rangelands, rattail fescue may increase under moderate to heavy grazing at the expense of more palatable grasses. In an Australian study, however, rattail fescue was significantly reduced by strategically timed heavy grazing for a few weeks in spring to reduce seed production and for a period in fall to limit seedling survival.
Biological	No known biological agents are available for control of any species of *Vulpia*.

CHEMICAL CONTROL

The following specific use information is based on published papers or reports by researchers and land managers. Other trade names may be available, and other compounds also are labeled for this weed. Directions for use may vary between brands; see label before use. Herbicides are listed by mode of action and then alphabetically. The order of herbicide listing is not reflective of the order of efficacy or preference.

LIPID SYNTHESIS INHIBITORS	
Sethoxydim, fluazifop, and clethodim *Poast, Fusilade,* and *Select* or *Envoy*	**Rate:** 1.5 to 2 pt product/acre **Timing:** Postemergence, to young plants. **Remarks:** These are all grass-selective herbicides that have no effect on broadleaf species and no soil activity. Apply with a crop oil concentrate surfactant. Partial control was achieved with all these products on rattail fescue, but they are not typically effective on members of the genus *Vulpia* or *Festuca*.
AROMATIC AMINO ACID INHIBITORS	
Glyphosate *Roundup, Accord XRT II,* and others	**Rate:** 1.5 to 2.5 pt product (*Roundup ProMax*)/acre (0.84 to 1.4 lb a.e./acre) for early postemergence treatment. For sequential treatments, apply at 1 to 1.5 pt product/acre early and late postemergence. **Timing:** Best applied postemergence to plants in the early to mid-tiller stages of growth (2 to 5 tillers per plant). Control with glyphosate is less satisfactory at the seedling stage, or after the plants develop 5 to 10 tillers. In other studies, a late fall application of glyphosate, timed after annual grass seedling emergence but before emergence of native perennial grass seedlings, gave good control of rattail fescue and promoted native forbs. **Remarks:** Glyphosate is only marginally effective in controlling rattail fescue and often requires the use of a higher application rate for adequate control. Glyphosate is a nonselective herbicide. It has no soil activity. Add ammonium sulfate to spray solution to ensure good activity by lowering the pH and preventing glyphosate from complexing with divalent cations and reducing its concentration in the spray solution.
BRANCHED-CHAIN AMINO ACID INHIBITORS	
Imazapic *Plateau*	**Rate:** 8 to 12 oz product/acre (2 to 3 oz a.e./acre) **Timing:** Preemergence in late summer to fall, or postemergence in fall to very early winter. In colder climates, spring applications after snow melt are better than fall treatments. The stage of growth of annual grasses can be critical to the effectiveness of imazapic. Late winter applications when annual grasses have 1 to 4 leaves and have not tillered can be effective, but control is not as effective on larger grasses. **Remarks:** Imazapic has been shown to be very effective on rattail fescue with a preemergence application. Postemergence application can also provide control, but control is only partial. Imazapic has mixed selectivity and tends to favor members of the Asteraceae. It has long soil residual activity. It can tie up in litter, so its efficacy may be reduced under situations where there is heavy thatch on the soil surface. Imazapic is not registered for use in California.
Propoxycarbazone-sodium *Canter R+P*	**Rate:** 0.9 to 1.2 oz product/acre (0.63 to 0.84 oz a.i./acre) **Timing:** Postemergence, from the 2-leaf to 2-tiller stage when plants are growing rapidly. **Remarks:** Propoxycarbazone is a broad-spectrum herbicide that will control many species. It will provide only partial control of rattail fescue. Perennial grass species vary in tolerance. A non-ionic surfactant should be added at 0.25 to 0.5% v/v solution.
Sulfometuron *Oust* and others	**Rate:** 1.33 to 5 oz product/acre (1 to 3.75 oz a.i./acre). For comparison, rates for control of annual bromes are 0.75 to 2 oz product/acre in locations with less than 20 inches of annual rainfall, and 3 to 5 oz product/acre in areas with more than 20 inches of annual rainfall.

	Timing: Preemergence or postemergence. Fall and spring applications can both be effective.
	Remarks: Sulfometuron has been used to control rattail fescue in Australian pastures. Sulfometuron has mixed selectivity. Minor damage can occur to some native perennial grasses. It has a fairly long soil residual. Higher rates may increase control but will also give more bare ground.
Sulfosulfuron *Outrider*	**Rate:** 0.55 oz product/acre (0.4 oz a.i./acre)
	Timing: Postemergence. Early stages of growth appear to be more susceptible.
	Remarks: Results are variable, depending on site and year. Control with a single treatment can range from 30 to 95%.
CONTACT PHOTOSYNTHETIC INHIBITORS	
Paraquat *Gramoxone*	**Rate:** 0.75 to 2 pt product/acre (3 to 8 oz a.i./acre)
	Timing: Postemergence, to rapidly growing plants. Most effective on plants at the beginning of the flowering stage.
	Remarks: Paraquat gives fair control of rattail fescue (about 80%) when applied at anthesis. Paraquat is nonselective on annual species. It is a non-systemic herbicide that only kills contacted foliage and has no soil activity. Paraquat is a restricted use herbicide that is highly toxic to animals. The formulation has a stenching agent.

Xanthium strumarium L.; common cocklebur
Xanthium spinosum L.; spiny cocklebur

Common and spiny cocklebur

Xanthium strumarium

Xanthium spinosum

Family: Asteraceae
Range: Common cocklebur occurs throughout the west and in all contiguous states. Spiny cocklebur is found throughout much of the western United States, except Wyoming, North and South Dakota.
Habitat: Woodlands, pastures, fields, forest margins, coastal habitats and disturbed sites such as roadsides, ornamental landscapes, agricultural fields, and urban waste areas; also common along riparian areas. The cockleburs can grow in most environments and can tolerate many soil types.
Origin: Both species are native to North America.
Impacts: Cocklebur is a widespread weed in temperate to subtropical regions nearly worldwide. Ingestion of seedlings and seeds at 1% of body weight or more can be fatally toxic to livestock, especially pigs and calves. Seed and sprouts contain a high concentration of a diterpene glycoside that causes an acute metabolic disorder characterized by a sudden drop in blood glucose and an increase of certain liver enzymes. In humans, handling cocklebur can cause contact dermatitis in sensitive individuals. Spiny cocklebur has sharp, stout spines which can cause injury, and the burs of both species may stick to fur or clothing.
Western states listed as Noxious Weed: *Xanthium spinosum*, Oregon, Washington

Common and spiny cocklebur are native plants that produce large burs covered with hook-tipped prickles. Both species occur as weeds in many areas throughout the world.

Common cocklebur is a summer annual to 4.5 ft tall, with green leaves and stems without spines. The stems are thick, branched, often slightly fleshy, reddish or black-spotted, or tinged dull red. The leaves have a distinctive scent, are broadly triangular, 1 to 6 inches long and wide, often weakly three-lobed, with three main veins from the base. Upper and lower surfaces are green and rough to the touch. Plants are highly variable within and between populations.

Spiny cocklebur is a summer annual to 3 ft tall, with leaves up to 3 inches long divided into three irregular lobes. The upper surface of the leaf is shiny dark green with short appressed hairs that are denser on the veins. The lower surface is pale green and downy. Leaves are arranged alternately on the stem. The stems are erect and slightly curved with many branches. The stems have yellow/green 3-pronged spines at the base of each leaf or branch.

Both species reproduce by seed. Male flowers are produced in terminal spikes at the ends of branches or in the upper leaf axils. They are small, green to rusty red and inconspicuous. Female flowers form lower on the stem at the leaf axils and at the nodes; they become hard prickly burs at maturity. Spiny cocklebur has yellowish-green, nearly cylindrical burs, mostly 0.5 inch long and 0.2 to 0.4 inches wide. Common cocklebur has ellipsoid burs, 0.5 to 1.3 inches long, green to yellowish brown, with two conspicuous thick straight or curved beaks at the apex. The burs contain two seeds that can survive up to 3 years under field conditions. Burs disperse primarily with water – they can float for up to 1 month – and by clinging to animals and other objects.

NON-CHEMICAL CONTROL

Mechanical (pulling, cutting, disking)	Hand pulling is effective on small incipient populations. Pulling is most effective before bur development and seed dispersal. Cocklebur can cause dermatitis in sensitive individuals. Individuals handling cocklebur should wear protective clothing to prevent contact.
	Mowing or disking at flowering stage will control cockleburs. Resprouts may occur after mowing and a secondary treatment may be required. Cut-off plants with immature burs can still develop viable seed.

Cultural	Neither grazing nor burning is considered an effective control option. Seeds and foliage contain a glycoside that can be fatally toxic to livestock.
Biological	In the western U.S., there are no efforts to use biological control agents, as both species are native and considered desirable in most natural communities. Worldwide, 60 different species have been identified that attack cocklebur. Several species of insects have been introduced to Australia to control *Xanthium strumarium*, but results have generally been disappointing. The most promising control species appears to be *Nupserha antennata*, a beetle native to India and Pakistan. Another potential biocontrol agent for *Xanthium strumarium* is *Oedopa*, which feeds exclusively on *Xanthium*.
	Fourteen species of fungi infect *Xanthium* in the U.S. and Canada. The rust *Puccinia xanthii*, which occurs throughout the U.S., southern Canada, parts of Europe, and India, is an obligate parasite on species of *Xanthium* and *Ambrosia*. It attacks all aerial parts of the plant except the flowers. Fungal and bacterial pathogens have had some success in controlling *Xanthium strumarium* in India.
	Cuscuta pentagona (dodder), a higher plant parasite, has been found on cocklebur. *Orobanche ramosa* (broomrape) is another parasitic plant found on a variety of cultivated and weedy plants, including *Xanthium*.

CHEMICAL CONTROL

The following specific use information is based on published papers and reports by researchers and land managers. Other trade names may be available, and other compounds also are labeled for this weed. Directions for use may vary between brands; see label before use. Herbicides are listed by mode of action and then alphabetically. The order of herbicide listing is not reflective of the order of efficacy or preference.

GROWTH REGULATORS	
2,4-D Several names	**Rate:** Broadcast treatment: 2 to 4 pt product/acre (0.95 to 1.9 lb a.e./acre). **Timing:** Postemergence when plants are growing rapidly. Applications in spring provide best control. **Remarks:** 2,4-D is selective for broadleaf species and will not damage desirable grasses growing nearby. Good coverage is necessary. 2,4-D can be tank-mixed with dicamba, and is available in a premix with triclopyr (*Crossbow*).
Aminopyralid *Milestone*	**Rate:** Broadcast treatment: 3 to 5 oz product/acre (0.75 to 1.25 oz a.e./acre) plus 0.25 to 0.5% v/v surfactant. **Timing:** Early postemergence when plants are small and rapidly growing. **Remarks:** Aminopyralid is a selective herbicide for broadleaf species and will not damage desirable grasses growing nearby. Other options include aminopyralid in a premix with 2,4-D (*Forefront HL*, 1.5 to 2.1 pt product/acre) or metsulfuron (*Opensight*, 1.5 to 2 oz product/acre). The formulation with metsulfuron is not registered for use in California.
Clopyralid *Transline*	**Rate:** Broadcast treatment: 4 to 10 oz product/acre (1.5 to 3.75 oz a.e./acre) plus 0.25 to 0.5% v/v surfactant, applied to thoroughly wet all leaves. **Timing:** Early postemergence when plants are small and rapidly growing. **Remarks:** Clopyralid is a selective herbicide for broadleaf species and will not damage desirable grasses growing nearby. Clopyralid can be tank mixed with triclopyr (*Garlon 4 Ultra*) for control of cockleburs.
Dicamba *Banvel, Clarity*	**Rate:** Broadcast treatment: 0.5 to 1.5 pt product/acre (0.25 to 0.75 lb a.e./acre) plus 0.25 to 0.5% v/v surfactant. **Timing:** Early postemergence when plants are small and rapidly growing. **Remarks:** Dicamba is a selective herbicide for broadleaf species and will not damage desirable grasses growing nearby. Dicamba can be tank mixed with 2,4-D. *Overdrive,* a premix of dicamba with diflufenzopyr, has been reported to be effective on common cocklebur. Diflufenzopyr is an auxin transport inhibitor which causes dicamba to accumulate in shoot and root meristems, increasing its activity. *Overdrive* is applied postemergence at 4 to 8 oz product/acre. Higher rates should be used on large annuals. Add non-ionic surfactant to the treatment solution at 0.25% v/v, or methylated seed oil at 1% v/v solution.
Fluroxypyr *Vista XRT*	**Rate:** Broadcast treatment: 11 oz product/acre (3.4 oz a.e./acre) plus 0.25 to 0.5% v/v surfactant. To optimize control use 0.25 to 0.5% v/v seed oil surfactant. **Timing:** Postemergence, when plants are growing rapidly.

	Remarks: Fluroxypyr is a selective herbicide for broadleaf species and will not damage desirable grasses growing nearby.
Picloram *Tordon 22K*	**Rate:** Broadcast treatment: 1 to 2 pt product/acre (4 to 8 oz a.e./acre) plus 0.25 to 0.5% v/v surfactant. **Timing:** Postemergence, when plants are growing rapidly. **Remarks:** High levels of picloram can give long-term soil activity for broadleaves. Also available in premixes with 2,4-D (*Grazon P+D*) or fluroxypyr (*Surmount*). Picloram is a restricted use herbicide. Picloram and all formulations including picloram are not registered for use in California.
Triclopyr *Garlon 4 Ultra, Remedy Ultra*	**Rate:** 2 pt product/acre (1 lb a.e./acre) **Timing:** Early postemergence, when plants are small and rapidly growing. **Remarks:** Triclopyr is a selective herbicide for broadleaf species and will not damage desirable grasses growing nearby. *Remedy Ultra* is the formulation used in rangelands.
AROMATIC AMINO ACID INHIBITORS	
Glyphosate *Roundup, Accord XRT II,* and others	**Rate:** Broadcast treatment: 1 to 2 qt product (*Roundup ProMax*)/acre (1.1 to 2.25 lb a.e./acre). Spot treatment: 1.5 to 2% v/v solution *Roundup* (or other trade name) and water. **Timing:** Postemergence when plants are growing rapidly. Applications in early spring provide best control. **Remarks:** Glyphosate is a nonselective systemic herbicide.
BRANCHED-CHAIN AMINO ACID INHIBITORS	
Imazapic *Plateau*	**Rate:** Broadcast treatment: 4 to 6 oz product/acre (1 to 1.5 oz a.i./acre) plus 0.25 to 0.5% v/v surfactant. **Timing:** Preemergence or early postemergence. **Remarks:** Imazapic is a selective postemergence herbicide effective for controlling broadleaf weeds and some grasses. It is not registered for use in California.
Imazapyr *Arsenal, Habitat, Stalker, Chopper, Polaris*	**Rate:** Broadcast treatment: 3 to 4 pt product/acre (0.75 to 1 lb a.e./acre) plus 0.25 to 0.5% v/v surfactant. **Timing:** Preemergence or postemergence. **Remarks:** Imazapyr is effective for controlling several broadleaf and grass weeds. It is not very selective and may produce bare ground at rates above 1 pt product/acre.
Metsulfuron *Escort*	**Rate:** Broadcast treatment: 0.33 to 0.5 oz product/acre (0.2 to 0.3 oz a.i./acre) plus 0.25 to 0.5% v/v surfactant. **Timing:** Early postemergence. **Remarks:** Metsulfuron is a selective herbicide for broadleaf species and will not damage desirable grasses growing nearby. Metsulfuron is also available in a premix with 2,4-D + dicamba (*Cimarron Max*). Metsulfuron and its formulations are not registered for use in California.
Propoxycarbazone-sodium *Canter R+P*	**Rate:** 0.9 to 1.2 oz product/acre (0.63 to 0.84 oz a.i./acre) **Timing:** Postemergence, to small, rapidly growing plants. **Remarks:** Propoxycarbazone is a broad-spectrum herbicide that will control many species. It will provide only partial control of common cocklebur. Perennial grass species vary in tolerance. A non-ionic surfactant should be added at 0.25 to 0.5% v/v solution.
Sulfosulfuron *Outrider*	**Rate:** Broadcast treatment: 0.75 to 1.33 oz product/acre (0.56 to 1 oz a.i./acre) plus 0.25 to 0.5% v/v surfactant. **Timing:** Preemergence or postemergence. **Remarks:** Sulfosulfuron is a selective, systemic herbicide for many annual and perennial weeds.

Zigadenus spp.
(= *Toxicoscordion* spp.)

Deathcamas

Zigadenus venosus

Family: Liliaceae
Range: Members of the genus *Zigadenus* are found throughout the United States, including the western states.
Habitat: The specific habitat varies with the species. Meadow deathcamas (*Z. venenosus* S. Wats.) inhabits moist or dry places in grassland, forest, coastal scrub, and rocky slopes. Foothill deathcamas (*Z. paniculatus* (Nutt.) S. Wats.) inhabits dry places in sagebrush scrub, grassland, and coniferous forest. Other species inhabit chaparral, open woodlands, desert, seasonally moist sites, and marshes.
Origin: All species are native to North America.
Impacts: The foliage and bulbs contain an array of steroidal alkaloids and are toxic to livestock. The quantity of alkaloids in plant materials varies between species. Meadow and foothill deathcamas are the most toxic species and are poisonous to all livestock. Ingestion of meadow deathcamas at about 0.4% of body weight or more can produce toxicity symptoms, usually within 24 hours of ingestion. Sheep are generally the most susceptible to poisoning. Cattle can be poisoned, especially in early spring when deathcamas is one of the first plants to green up. Humans have been poisoned when they mistake deathcamas for wild onion. Deathcamas species are generally not considered weeds in natural areas.

All *Zigadenus* species are perennials with simple stems to nearly 3 ft tall. Plants exist as a rosette of linear leaves before flowering stems develop in spring. The leaf blades of deathcamas are basal, smooth, and V-shaped in cross-section. The leaves are more succulent than grasses. Roots arise from a dark brown to black coated, onion-like bulb 0.5 to 2.5 inches in diameter. The bulbs of deathcamas lack the distinctive odor of onions. Above-ground growth commences in fall or spring, depending on the region. The bulbs can persist in a dormant state for decades with little growth or flowering.

Flowers of deathcamas are mostly erect, bisexual and/or staminate, in panicles or racemes. Each fruit produces numerous seeds, and the dead stems with capsule remnants can persist into winter. This can aid in identifying infested areas. Most seeds fall near the parent plant. Seed viability has been reported above 80%. Mean germination was 48% under field conditions in Central Cascades Ponderosa Pine forest. Little information is available on seed longevity in the soil.

NON-CHEMICAL CONTROL

Mechanical (pulling, cutting, disking)	Hand-pulling and digging are effective controls but often impractical if infestations span a large area. Remove the bulb to prevent regrowth.
	Plants quickly regrow after mowing. Mowing can be used to remove green foliage in early spring but is not an effective control.
	Repeated tillage can suppress deathcamas, but plants often resprout from bulbs left in the soil.
Cultural	After a fire deathcamas often regrows from protected bulbs in the soil, although summer burning has been shown to suppress deathcamas the following year.
	Deathcamas is toxic to livestock. Avoid grazing when adequate forage is not available or when deathcamas populations are dense. Deathcamas poisoning normally occurs in early spring as deathcamas greens up before most grasses. Animals usually will not freely choose deathcamas provided other forage is available.
	Improving range conditions and reseeding perennial grasses can help suppress deathcamas.
Biological	No biological control agents are available for the control of any deathcamas species, primarily because they are all native to North America.

CHEMICAL CONTROL

The following specific use information is based on published papers and reports by researchers and land managers. Other trade names may be available, and other compounds also are labeled for this weed. Directions for use may vary between brands; see label before use.

GROWTH REGULATORS	
2,4-D Several names	**Rate:** 1.5 to 2 qt product/acre (1.43 to 1.9 lb a.e./acre)
	Timing: Postemergence when plants have 3 to 6 leaves in spring.
	Remarks: Broadleaf-selective and safe on most grasses. 2,4-D has minimal soil activity. Do not apply ester formulation when outside temperatures exceed 80°F. Add a surfactant.

Susceptibility charts for additional species

These susceptibility charts suggest control methods for western invasive plants that are less widespread than species in the full Weed Reports. These charts are separated into grasses and sedges, herbaceous forbs, submerged and floating aquatic species, and woody plants. While we don't include full weed reports for these species, the susceptibility charts do have qualitative information on mechanical, cultural, and chemical control options and the best timing or growth stage for each treatment. As with the plants in the full weed reports, we have included what we consider to be likely control options, based on results or observation with related species. We also include whether or not the herbicides are considered foliar or soil active (or both) and their half-life range (or average). These susceptibility charts are a quick way to determine potential control options for about 100 species.

TABLE		PAGE
2	**Grasses and sedges** not included in the weed reports – **non-chemical** control	434
3	**Grasses and sedges** not included in the weed reports – **chemical** control	436
4	**Herbaceous forbs** not included in the weed reports – **non-chemical** control	438
5	**Herbaceous forbs** not included in the weed reports – **chemical** control	446
6	**Woody plants** not included in the weed reports – **non-chemical** control	454
7	**Woody plants** not included in the weed reports – **chemical** control	456
8	**Aquatic plants** not included in the weed reports – **non-chemical** control	458
9	**Aquatic plants** not included in the weed reports – **chemical** control	460

Table 2. **Non-chemical** control of **grasses and sedges** not included in the weed reports.

FAMILY / Species	CULTURAL	
	Grazing	Prescribed burning
CYPERACEAE (sedge)		
Scirpus acutus (= *Schoenoplectus acutus*) Hardstem bulrush, tule	P	P
POACEAE (grass)		
Agrostis avenacea Pacific bentgrass	F sheep may be effective in western US	P will kill existing stand but will spread seed in newly disturbed sites
Agrostis stolonifera Creeping bentgrass	P	P
Anthoxanthum odoratum Sweet vernalgrass	P	P*
Briza maxima Big quakinggrass	G* before seed drop	G* before seed drop
Cenchrus spp. Sandburs	P spiny seeds	P
Dactylis glomerata Orchardgrass	P	P
Eragrostis spp. Lovegrasses, stinkgrasses	P	NIA
Festuca arundinacea Tall fescue	P	P
Paspalum dilatatum Dallisgrass	P	P*
Poa pratensis Kentucky bluegrass	P	G in spring, timing is critical; multiple years may be necessary
Polypogon monspeliensis Rabbitfoot polypogon	F	G
Secale cereale Common rye	G likely requires multiple grazing to prevent seed production	G* before seed drop
Setaria spp. Foxtails	P	NIA
Stipa capensis Mediterranean steppegrass	G* before flower heads appear	NIA

E = Excellent control, generally better than 95%
G = Good control, 80-95%
F = Fair control, 50-80%
P = Poor control, below 50%
* = Likely based on results of observations of related species

Control includes effects within the season of treatment. Control is followed by best timing, if known, when efficacy is **E** or **G**.

MECHANICAL		
Mowing and cutting	**Tillage**	**Grubbing, digging** or **hand pulling**
G* FLW with all stubble submerged for >2 wks	**P**	**G** when soil is moist
P	**G**	**G**
P	**P** may spread plant as it reproduces from stolons	**F** may spread plant as it reproduces from stolons
F* best conducted when soil is dry and before seed production	**F** for badly infested fields that require replanting	**G** for small patches; remove rhizomes
G* easy to pull, best just before viable seed production	**E** just before viable seed production	**E** just before viable seed production
P can reduce seed production, but plants are often low growing	**E** before seed production	**E** before seed production
P can stimulate tillering, repeated mowing will suppress	**P** may spread plant	**G** non-spreading bunchgrass can be pulled
P can reduce seed production over time	**E** before seed production	**E** before seed production
P	**F** for badly infested fields that require replanting	**P**
P	**G** for seedlings	**F**
P	**G** if repeated several times; **P** with one treatment	**P**
F	**G**	**G**
G best conducted when soil is dry and before seed production	**E** before seed production	**E** just before viable seed production
P	**E** before seed production	**E** before seed production
G* before viable seed are produced	**E*** expected to be effective on annual species	**E*** good for small patches, easy to pull

FLW = flowering
NIA = No information available

Table 3. **Chemical control** of **grasses and sedges** not included in the weed reports.

FAMILY / Species	HERBICIDES (Telar) Chlorsulfuron	(Select, Envoy) Clethodim	(Fusilade) Fluazifop	(Roundup and others) Glyphosate	(Velpar) Hexazinone
CYPERACEAE (sedge)					
Scirpus acutus; = *Schoenoplectus acutus*; Hardstem bulrush, tule	NIA	NIA	NIA	E Su, Fa; 2,4-D is also effective	NIA
POACEAE (grass)					
Agrostis avenacea Pacific bentgrass	NIA	G*	G*	E	G
Agrostis stolonifera Creeping bentgrass	P	G	G	E	NIA
Anthoxanthum odoratum Sweet vernalgrass	NIA	E* Su	E* Su	E* Su	E*
Briza maxima Big quakinggrass	NIA	E* active young growth	E* active young growth	E active growth	E*
Cenchrus spp. Sandburs	P	E	G	E	G
Dactylis glomerata Orchardgrass	P	E	E	E	F / E on seedlings
Eragrostis spp. Lovegrasses, stinkgrasses	P	E	E	E	NIA
Festuca arundinacea Tall fescue	F–G	G	F	E	F–G
Paspalum dilatatum Dallisgrass	P	E	G	E	G–E
Poa pratensis Kentucky bluegrass	P	E Sp	F	E Sp	G Sp suppression
Polypogon monspeliensis Rabbitfoot polypogon	F	E	G	E	NIA
Secale cereale Common rye	P	E	E	G–E Sp	E
Setaria spp. Foxtails	F	E	G	E	F–G
Stipa capensis Mediterranean steppegrass	E*	E*	E*	E*	NIA
Foliar activity	+	+	+	+	+
Soil activity	O	▼	▼	▼	O
Range/average half-life of herbicide in the soil (days)	28-42	3	15	none	90

E = Excellent control, generally better than 95%
G = Good control, 80-95%
F = Fair control, 50-80%
P = Poor control, below 50%
* = Likely based on results of observations of related species

Control includes effects within the season of treatment. Control is followed by best timing, if known, when efficacy is **E** or **G**.

(Plateau) Imazapic	(Habitat, Arsenal, others) Imazapyr	(Matrix) Rimsulfuron	(Poast) Sethoxydim	(Oust) Sulfometuron	(Outrider) Sulfosulfuron	Organic products
NIA	E Su, Fa	NIA	NIA	NIA	NIA	P
NIA	NIA	NIA	G*	G*	E	P
NIA	NIA	NIA	G	NIA	E	P
NIA	E* Su	NIA	E* Su	E*	NIA	G plant oils (Bioganic®) gave >86% control
E*	E*	NIA	E* active young growth	E*	NIA	NIA
G applied post	E	NIA	E	NIA	NIA	NIA
F only suppression	E	NIA	E	NIA	NIA	P*
G	E	G	E	NIA	NIA	NIA
G–E	E*	F post-emergence	G–E seedings	E Sp, Fa	NIA	P
F–G suppression	G–E	P	F	P	NIA	P
P	E	NIA	P	E	NIA	P
P	NIA	NIA	E	NIA	NIA	NIA
F better tank mixed with glyphosate	G–E*	G early Fa or Sp	E	E	NIA	NIA
G	E	E	E	G except green foxtail	NIA	NIA
P	NIA	NIA	E*	NIA	NIA	NIA
+	+	+	+	+	+	+
O	O	O	▼	O	O	▼
120	25-142	20	4-5	20-28	12	NIA

Fa = Fall
Sp = Spring
Su = Summer

NIA = No information available

+ indicates herbicide can have postemergence foliar activity
O indicates that the herbicide is considered to have soil residual activity
▼ indicates that it does not have a long enough residual activity in the soil to give control

Table 4. **Non-chemical** control of **herbaceous forbs** not included in the weed reports.

FAMILY / Species	CULTURAL	
	Grazing	Prescribed burning
AMARANTHACEAE (amaranth)		
Amaranthus spp. Pigweeds, amaranths	**P** increases with grazing, known to cause nitrate accumulation in livestock	NIA
APIACEAE (carrot)		
Carum carvi Common caraway	**P** not a good forage, will increase with grazing	NIA
ARACEAE (arum)		
Zantedeschia aethiopica Calla or arum lily	**P** can be toxic to animals	**P**
ASTERACEAE (sunflower)		
Ambrosia acanthicarpa Annual bursage	**F** often regrows after grazing	**P** often increases after fire; usually green during fire season
Ambrosia artemisiifolia Common ragweed	**F** often regrows after grazing	**P** often increases after fire; green most of the summer, other fuel needed to carry a fire in infested areas
Arctotheca calendula Capeweed	**P**	**P** will likely resprout quickly
Artemisia absinthium Absinth wormwood	**P** increases with overgrazing	**F** may increase infestations, but four consecutive year spring burns reduced population
Chrysanthemum coronarium Crowndaisy	**G** can be grazed early in the season	NIA
Chrysothamnus nauseosus Rabbitbrush	**F** often regrows after grazing and may be considered desirable forage	**P** considered a fire adapted species that can be enhanced by fire
Cichorium intybus Chicory	**F** often regrows after grazing and may be considered desirable forage	NIA
Conyza bonariensis Hairy fleabane	**F**	**G**
Conyza canadensis Horseweed	**F**	**G**
Cotula australis, C. coronopifolia Brassbuttons	**P** avoided by grazers	**P** not typically in areas conducive to burning
Erechtites spp. Burnweeds	NIA	**P** not typically in areas conducive to burning
Helianthus annuus Common sunflower	**F** will graze new shoots	**P** fire creates conditions favorable to sunflower establishment
Iva axillaris Povertyweed	**P** heavy grazing often promotes invasion	**P** plants resprout from roots after fire
Lactuca serriola Prickly lettuce	**G**	**G**

E = Excellent control, generally better than 95%
G = Good control, 80-95%
F = Fair control, 50-80%
P = Poor control, below 50%
* = Likely based on results of observations of related species

Control includes effects within the season of treatment. Control is followed by best timing, if known, when efficacy is **E** or **G**.

MECHANICAL		
Mowing and cutting	**Tillage**	**Grubbing, digging or hand pulling**
F single mowing can suppress seed production but plants often regrow; best control if cut at ground level or multiple times	**E** seedling can be controlled with a single cultivation when soil is dry	**E** pull or dig plants before FLW
P plants will often rebloom below cutting height	**G** repeated cultivation effective	**G** hand pulling can control small populations
G must be repeated	**P** will spread rhizomes	**G** must remove all tubers and rhizomes
P plants will often rebloom after mowing	**E** expect new flush of plants if soil moisture is available after tillage	**E** remove as much of the plant and root as possible to prevent regrowth
P plants will often rebloom after mowing, try to cut at ground level	**E** expect new flush of plants if soil moisture is available after tillage	**E** remove as much of the plant and root as possible to prevent regrowth
P may facilitate further spread	**P**	**G** all parts must be removed or plant can spread
F can prevent seed production	**G** needs to be repeated several times during the season, if conducted one Fa is better than Sp	**G** hand pull or dig up roots when soil is moist, must remove all roots
G annual species, would be expected to be significantly damaged by mowing or cutting	**E** tillage should control annual if conducted when soil is dry	**G** annual species should be easily controlled by hand pulling
P can regrow following mowing	**F** must be fairly deep and may need to be repeated	**E** remove as much of the plant and root as possible to prevent regrowth
F can regrow after mowing but can prevent seed production	**G** deep tillage is effective	**E** remove as much of the plant and root as possible to prevent regrowth
G	**F**	**G**
G	**F**	**G**
F sprawls and may root at stem nodes	**G** not typically in areas where tillage can be used, but would expect to give control	**F** sprawls and may root at stem nodes
G as an annual or weak perennial, cutting should provide good control	**G** not typically in areas where tillage can be used, but would expect to give control	**G** fairly easy to remove by hand pulling
P can reduce seed production	**E** control plants before FLW, germination after tillage is common	**E** pull or dig plants before FLW
P perennial that resprouts from creeping roots	**P** root fragments resprout	**F** must remove roots to be effective
P	**G**	**G**

FLW = flowering
NIA = No information available
Fa = Fall
Sp = Spring

Table 4 continued: Non-chemical control of herbaceous forbs

FAMILY	CULTURAL	
Species	Grazing	Prescribed burning
Senecio jacobaea Tansy ragwort	**P–F** only sheep show tolerance to toxins in plant	**P–G** sufficient fuel will provide kill of seedlings, but established plants will resprout
Taraxacum officinale Dandelion	**P–F** only sheep are effective; plants tend to be favored by grazing because the basal rosette of leaves escape damage	**P** plants resprout and fire creates conditions favorable to invasion
BRASSICACEAE (mustard)		
Alyssum alyssoides Yellowtuft	NIA	NIA
Berteroa incana Hoary alyssum	**P** toxic to horses	NIA
Brassica rapa Birdsrape mustard	**G** livestock health problems from grazing brassicas are relatively rare and can largely be avoided by good agronomic and grazing management	**P** fire creates conditions favorable to invasion
Cakile maritima Sea rocket	**F** somewhat palatable	NIA
Capsella bursa-pastoris Shepherd's-purse	**F**	**P**
Chorispora tenella Blue mustard	**F**	**P**
Descurainia sophia Flixweed	**F** moderately palatable to livestock	**P** fire creates conditions favorable to flixweed invasion but can be used to reduce seed production
Hesperis matronalis Damesrocket	NIA	**G** with sufficient fuel, burning can kill seedlings; established plants will resprout
Lepidium perfoliatum Clasping pepperweed	**F**	**P** fire creates conditions favorable to invasion
Lobularia maritima Sweet alyssum	**P**	**P**
Malcolmia africana African mustard	**F***	**P***
Thlaspi arvense Field pennycress	**P**	**P** fire creates conditions favorable to invasion
CARYOPHYLLACEAE (pink)		
Saponaria officinalis Bouncingbet	**P** can be toxic to livestock	**P**
CHENOPODIACEAE (goosefoot)		
Chenopodium album and *C. berlandieri* Common lambsquarter and netseed (giant) lambsquarter	**P**	**F**

E = Excellent control, generally better than 95%
G = Good control, 80-95%
F = Fair control, 50-80%
P = Poor control, below 50%
* = Likely based on results of observations of related species

Control includes effects within the season of treatment. Control is followed by best timing, if known, when efficacy is **E** or **G**.

	MECHANICAL	
Mowing and cutting	**Tillage**	**Grubbing, digging** or **hand pulling**
P–F side shoots may emerge, must be conducted before FLW	**P** root fragments resprout	**G** roots need to be removed, needs to be repeated 2-3 times a year before FLW
P basal leaves escape cutting in most situations and plants regrow	**P** survival is high because of its ability to regenerate from root sections, repeat treatment is necessary	**G** the upper portion of the root must be entirely removed
F just before FLW	**G** repeat treatment	**E** small patches only
F just before FLW	**G** repeat treatment	**E** small patches only
E cutting at soil surface should be effective	**E** target seedlings and rosettes, germination after tillage is common when soil moisture is present	**E** remove before seed production
E cutting at soil surface should be effective	**G** not generally practical in dune settings	**E** easy to remove with hand pulling, must be repeated
P	**G**	**G**
P mow close to the soil surface	**E**	**E** remove before seed production
P can regrow following mowing, mow shortly before FLW and cut near ground level	**E** target seedlings and rosettes, germination after tillage is common when soil moisture is present	**E** hand-pulling or digging are effective
F can prevent seed production, but plants often survive, 2-5 years mowing needed to	**F** biennial to perennial likely to resprout after tillage	**E** hand pulling or use of dandelion puller effective when soil is moist
P mow shortly before FLW and cut near ground level	**E** effective control, germination after tillage is common when soil moisture is available	**E** remove before seed production
P low growing plants will escape injury	**E**	**G**
P* mow close to the soil surface	**E***	**E*** remove before seed production
P mow before FLW, matures early in the season	**E** till before FLW	**E** remove before FLW
P	**F** resprouts from rhizomes	**F** resprouts from rhizomes
F	**F**	**G**

FLW = flowering
NIA = No information available

Table 4 continued: Non-chemical control of herbaceous forbs

FAMILY	CULTURAL			
Species	Grazing		Prescribed burning	
Sarcobatus vermiculatus Greasewood	F	increases with grazing, known to cause oxalate poisoning in sheep and cattle	P	resprouts after fire
FABACEAE (pea or bean)				
Lathyrus latifolius Everlasting peavine	P	can be toxic to livestock	P	as a perennial it will regrow after burning
Medicago polymorpha California burclover	F		F	
Melilotus albus White sweetclover	P	palatable and nutritious to livestock	P	fire creates conditions favorable to invasion
Trifolium hirtum Rose clover	P		P	fire will flush seedbank
GERANIACEAE (geranium)				
Geranium robertianum Herb-robert	NIA		NIA	
IRIDACEAE (iris)				
Crocosmia x crocosmiflora Crocosmia or Montbretia	P		P	
Watsonia meriana Watsonia	G	grazing reported effective in Australia, but may be toxic to horses	P	
LAMIACEAE (mint)				
Lamium amplexicaule Henbit	F		F	
LILIACEAE (lily)				
Asparagus asparagoides Bridal creeper	F	in Australia some control with sheep grazing	P	readily resprouts, may be partially killed with very intense fires
LYTHRACEAE (loosestrife)				
Lythrum hyssopifolium Hyssop loosestrife	P		P	resprouts, usually found in wet areas
Lythrum virgatum Wand loosestrife	P*		P*	resprouts, usually found in wet areas
MALVACEAE (mallow)				
Malva neglecta Common mallow	P		P	
PLANTAGINACEAE (plantain)				
Plantago lanceolata Buckhorn plantain	P	can reduce growth only slightly	P	
POLYGONACEAE (buckwheat)				
Emex spinosa Spiny emex	NIA		NIA	
Polygonum arenastrum Prostrate knotweed	P		P	

E = Excellent control, generally better than 95%
G = Good control, 80-95%
F = Fair control, 50-80%
P = Poor control, below 50%

* = Likely based on results of observations of related species
Control includes effects within the season of treatment. Control is followed by best timing, if known, when efficacy is **E** or **G**.

MECHANICAL		
Mowing and cutting	**Tillage**	**Grubbing, digging** or **hand pulling**
G chaining can damage population and reduce density	**P** not practical in most areas	NIA
P mow before FLW, matures early in the season, mowing difficult do to long, viney stems	**G** repeated cultivation can control plants	**G** pull or dig young plants early before FLW
F	**G**	**G**
P regrows rapidly if mowed early in the growing season	**E** germination after tillage common when soil moisture is available	**E** pull or dig young plants in moist soils; difficult to pull when mature
P	**E**	**F** may root from nodes on sprawling stems
F sprawling nature makes it hard to cut all parts above ground	**E** most areas not conducive to tillage, but expected to be effective on this annual	**E** weak roots allows it to be pulled easily
P resprouts quickly; must be conducted repeatedly to weaken corm	**P** will spread corms	**G** small patches can be dug up, but corms must be removed
P resprouts quickly; must be conducted repeatedly to weaken corm	**P** will spread corms	**G** small patches can be dug up, but corms must be removed
F	**G**	**E**
P resprouts readily	**P** will disperse rhizomes and spread plant	**F** must remove all rhizome fragments
P	**P** will disperse resprouting plant parts	**F** must remove all roots and underground stems, and remove material from site
P*	**P*** will disperse resprouting plant parts	**F*** must remove all roots and underground stems, and remove material from site
P grows prostrate	**G** soil must be dry to prevent resprouting at nodes	**F** may resprout from crown
P may give some reduction of growth, but low growing habit will reduce control	**G**	**G** difficult but much more effective than cutting
G as an annual it would be expected to be controlled by cutting at right timing	**G** repeated cultivation can control plants	**E** appears easy to hand pull small infestations
P prostate growth form makes it impossible to mow	**E** germination after tillage common	**E**

FLW = flowering
NIA = No information available

Table 4 continued: Non-chemical control of herbaceous forbs

FAMILY	CULTURAL	
Species	Grazing	Prescribed burning
Polygonum convolvulus Wild buckwheat	**P**	**P** fire creates conditions favorable to establishment
RANUNCULACEAE (buttercup)		
Clematis vitalba Old-man's-beard	**G** sheep will feed on seedlings and young vines at ground level	NIA
Ranunculus acris Tall buttercup	**P** toxic to animals	**P** usually found in moist meadows and pastures not conducive to burning, if burned plants will resprout
Ranunculus repens Creeping buttercup	**P**	**P**
Ranunculus testiculatus (=*Ceratocephala testiculata*) Bur buttercup	**P** toxic to animals	NIA
SCROPHULARIACEAE (figwort)		
Digitalis purpurea Foxglove	**P**	**P** smoke from plant is toxic, areas of infestation not conducive to burning
SOLANACEAE (nightshade)		
Solanum elaeagnifolium Silverleaf nightshade	**P** toxic to livestock, goats can handle 25% in forage consumed, may reduce fruit production	**P** below ground structures not affected
URTICACEAE (nettle)		
Urtica dioica Stinging nettle	**P** not readily grazed by livestock	**P** below ground rhizomes not affected
VERBENACEAE (vervain)		
Verbena bonariensis Tall vervain	NIA	NIA
Verbena litoralis Seashore vervain	NIA	NIA
ZYGOPHYLLACEAE (caltrop)		
Peganum harmala Harmal, African rue	**P** not palatable to livestock	**P** recovery will occur rapidly
Zygophyllum fabago Syrian beancaper	**P** not palatable to livestock	**P** below ground structures not affected

E = Excellent control, generally better than 95%
G = Good control, 80-95%
F = Fair control, 50-80%
P = Poor control, below 50%
* = Likely based on results of observations of related species

Control includes effects within the season of treatment. Control is followed by best timing, if known, when efficacy is **E** or **G**.

	MECHANICAL		
Mowing and cutting		**Tillage**	**Grubbing, digging or hand pulling**
P mow before seed production		**G** germination after tillage is very common; tillage favors invasion	**G** remove plants before FLW
P except seedlings		**F**	**G** roots much be removed from site, will regenerate from stem fragments
F mowing can reduce seed production, frequent mowing reduces infestation		**P–F** need to severely damage taproot, some recovery or spread may occur	**G** must remove all roots and rhizomes, best in moist soils, best for small patches
P low growing plants will escape injury and quickly recover		**F–G** must be conducted before roots become well established	**F** creeping roots, only effective on small patches, remove all stem fragments
G early maturing so must be cut early with very low cutting height		**E** must be done early before seed production	**E**
P would have to be repeated several times		**P**	**G** hand pulling can be effective in spring when soils are moist
F can reduce infestation and seed production, plants will develop rosettes below mower blades		**P** deep root system can recover, must be repeated several times to be effective	**P** will resprouting from root system
P regrows rapidly if mowed early in the growing season		**P–F** likely requires repeated tillage for several years	**P–F** extensive rhizomes and stinging hairs make hand pulling difficult
NIA		NIA	NIA
NIA		NIA	NIA
P recover will occur from underground reproductive structures		**P** roots will fragment and recover	**G** roots must be removed or new plants will form
NIA		NIA	**G** young plants can be pulled in moist soil. **P** on older plants due to resprouting

FLW = flowering
NIA = No information available

Table 5. **Chemical control** of **herbaceous forbs** not included in the weed reports.

FAMILY / Species	(several names) 2,4-D	(Perspective) Aminocyclopyrachlor + chlorsulfuron	(Milestone) Aminopyralid	(Telar) Chlorsulfuron	(Transline) Clopyralid	(Banvel, Clarity) Dicamba	(Roundup, others) Glyphosate
AMARANTHACEAE (amaranth)							
Amaranthus spp. Pigweeds, amaranths	G	P	P–G	E	P	E	E
APIACEAE (carrot)							
Carum carvi Common caraway	G–E	E	G–E[1]	NIA	F	P–F	P
ARACEAE (arum)							
Zantedeschia aethiopica Calla lily	G	NIA	NIA	G[4]	NIA	NIA	NIA
ASTERACEAE (sunflower)							
Ambrosia acanthicarpa Annual bursage	E	E	E	P	G*	E	E*
Ambrosia artemisiifolia Common ragweed	E	E	E	P	G	E	E
Arctotheca calendula Capeweed	G	NIA	NIA	NIA	NIA	NIA	G
Artemisia absinthium Absinth wormwood	G–E	E	E	NIA	E	G–E	E
Chrysanthemum coronarium Crowndaisy	NIA	E*	E*	E	E*	NIA	E*
Chrysothamnus nauseosus Rabbitbrush	G[8]	G	F	P	P	P	G
Cichorium intybus Chicory	E	E	E	P	E	E	P
Conyza bonariensis Hairy fleabane	E	E	E	E*	E	E	E[9]
Conyza canadensis Horseweed	E	E	E	E	G–E	E	E[9]
Cotula australis, *C. coronopifolia* Brassbuttons	E*	E*	E*	E*	E*	E*	E*
Erechtites spp. Burnweeds	NIA	NIA	NIA	NIA	NIA	NIA	E*
Helianthus annuus Common sunflower	E	E	E	E	E	E	G
Iva axillaris Povertyweed	G	NIA	G	F	NIA	G	NIA
Lactuca serriola Prickly lettuce	E	E	E	E	E	E	E
Senecio jacobaea Tansy ragwort	E	E	E	G	E	E	E
Taraxacum officinale Dandelion	G	E	E	F	G	E	E

1. Better with 2,4-D or metsulfuron
2. Better with 2,4-D
3. Often with 2,4-D or dicamba
4. Apply at FLW stage
5. Apply before FLW stage
6. Seedlings only
7. Top kill only
8. Ester formulation
9. Except when resistant
10. With chlorsulfuron

Propoxycarbazone-sodium is not included in the table but will provide control of blue mustard (*Chorispora tenella*), bur buttercup (*Ranunculus testiculatus*), field pennycress (*Thlaspi arvense*), flixweed (*Descurainia sophia*), and silverleaf nightshade (*Solanum elaeagnifolium*), as well as partial control of common ragweed (*Ambrosia artemisiifolia*), buckhorn plantain (*Plantago lanceolata*), and wild buckwheat (*Polygonum convolvulus*).

(Velpar) Hexazinone	(Plateau) Imazapic	(Habitat, Arsenal, others) Imazapyr	(Escort) Metsulfuron	(Gramoxone) Paraquat	(Tordon 22K) Picloram	(Matrix) Rimsulfuron	(Oust) Sulfometuron	(Outrider) Sulfosulfuron	(Garlon) Triclopyr
E	E	E	E	E	E	E	E	NIA	E
NIA	NIA	NIA	G–E[2]	NIA	E[3]	NIA	NIA	NIA	F
NIA	NIA	NIA	G[4]	F[5]	NIA	NIA	NIA	NIA	NIA
NIA	G*	E*	P	P[6]	E	NIA	G*	NIA	G*
NIA	G	E	P	NIA	E	G	E	NIA	G
NIA	NIA	NIA	NIA	F[7]	NIA	NIA	NIA	NIA	G
NIA	NIA	NIA	P	P	E	NIA	NIA	NIA	NIA
NIA	NIA	E	E	NIA	E*	NIA	E	NIA	NIA
NIA	P	G	P	P	P–G	P	NIA	NIA	P–F
NIA	P	P	P	P[6]	E	P	NIA	P	F
E	P	E	E	G	E	E	E*	E*	E
E	P	E	E	G	E	E	E	E	E
E*	P*	E*	E*	G*	E*	E*	E*	E*	E*
NIA	NIA	NIA	NIA	NIA	NIA	NIA	NIA	NIA	NIA
G	F	E	G*	P[6]	E	E	E	E	E
NIA	NIA	NIA	F	NIA	NIA	NIA	NIA	NIA	NIA
E	P	E	E	G	E	E	NIA	NIA	E
F	P	E	E	P	E	NIA	E	E[10]	E
E	F	E	E	P	E	F	E	NIA	E

E = Excellent control, generally better than 95%
G = Good control, 80-95%
F = Fair control, 50-80%
P = Poor control, below 50%

* = Likely based on results of observations of related species
Control includes effects within the season of treatment.
Control is followed by best timing, if known, when efficacy is **E** or **G**.

NIA = No information available

Table 5 continued: Chemical control of herbaceous forbs

FAMILY / Species	(several names) 2,4-D	(Perspective) Aminocyclopyrachlor + chlorsulfuron	(Milestone) Aminopyralid	(Telar) Chlorsulfuron	(Transline) Clopyralid	(Banvel, Clarity) Dicamba	(Roundup, others) Glyphosate
BRASSICACEAE (mustard)							
Alyssum alyssoides Yellowtuft	NIA	NIA	P	E	P	NIA	E
Berteroa incana Hoary alyssum	NIA	NIA	P	E	P	NIA	NIA
Brassica rapa Birdsrape mustard	E	P	P	E	P	E	E
Cakile maritima Sea rocket	NIA	P	P	E	P	NIA	E FOL, WCK, FLW
Capsella bursa-pastoris Shepherd's-purse	E	P	P	E	P	E	E
Chorispora tenella Blue mustard	F	P	P	E	P	G	E
Descurainia sophia Flixweed	E	P	P	E	P	E	E
Hesperis matronalis Damesrocket	F–G	P	P	E	P	F–G	E
Lepidium perfoliatum Clasping pepperweed	E	P	P	E	P	E	E
Lobularia maritima Sweet alyssum	E*	NIA	NIA	G*	NIA	G*	E*
Malcolmia africana African mustard	F*	P*	P*	E*	P*	G*	E*
Thlaspi arvense Field pennycress	E	NIA	NIA	E	NIA	G	E
CARYOPHYLLACEAE (pink)							
Saponaria officinalis Bouncingbet	NIA	NIA	NIA	E	NIA	G	G
CHENOPODIACEAE (goosefoot)							
Chenopodium album Common lambsquarter	E	NIA	E	E	P	E	E
Chenopodium berlandieri Netseed (giant) lambsquarter	E*	NIA	E	E*	P	E*	E*
Sarcobatus vermiculatus Greasewood	G–E FOL, Su	NIA	NIA	G FOL	P	G FOL	NIA
FABACEAE (pea or bean)							
Lathyrus latifolius Everlasting peavine	NIA	E	E	E	E	NIA	E
Medicago polymorpha California burclover	F	E	E	E	E	E	F
Melilotus albus White sweetclover	E	E	E	E	E	E	F
Trifolium hirtum Rose clover	G	E	E	F	E	E	G

2. Better with 2,4-D
6. Seedlings only

E = Excellent control, generally better than 95%
G = Good control, 80-95%
F = Fair control, 50-80%
P = Poor control, below 50%

(Velpar) Hexazinone	(Plateau) Imazapic	(Habitat, Arsenal, others) Imazapyr	(Escort) Metsulfuron	(Gramoxone) Paraquat	(Tordon 22K) Picloram	(Matrix) Rimsulfuron	(Oust) Sulfometuron	(Outrider) Sulfosulfuron	(Garlon) Triclopyr
NIA	E	NIA	E	NIA	NIA	NIA	NIA	NIA	NIA
NIA	NIA	NIA	E	NIA	NIA	NIA	E	NIA	NIA
E	E	E	E	G	E	E	E	E*	NIA
NIA	NIA	NIA	NIA	NIA	NIA	E*	E*	E*	NIA
E	E	E	E	G	E	E	E	E	E
E	E	E	E	P[6]	NIA	NIA	G*	E*	NIA
G	E	E	E	G	NIA	E	E	G	NIA
NIA	E	NIA	E	NIA	NIA	G*	E*	E*	E
E	E	E	E	P[6]	P	NIA	E	G*	NIA
NIA	NIA	NIA	NIA	G*	NIA	G*	E*	E*	G*
E*	E*	E*	E*	P*[6]	NIA	NIA	G*	E*	NIA
E	NIA	NIA	E	P[6]	E	NIA	NIA	G	NIA
F	NIA	NIA	NIA	NIA	E	NIA	E	NIA	NIA
E	E	E	E	E	E	G	NIA	E	E
E*	E*	E*	E*	E*	E*	G*	NIA	NIA	E*
NIA	NIA	NIA	G–E FOL	P	G[2] FOL, Su	NIA	NIA	NIA	F FOL
NIA	P	E	E	NIA	E	NIA	NIA	NIA	E
G	G	G	G	F	G	G	E	NIA	E
NIA	P	E	E	P[6]	E	NIA	G–E	NIA	E
E	P	F	E	NIA	E	E	E	NIA	E

  * = Likely based on results of observations of related species. Control includes effects within the season of treatment. Control is followed by best timing, if known, when efficacy is **E** or **G**.

NIA = No information available
Su = Summer
FLW = flowering
FOL = foliar treatment
WCK = wick

Table 5 continued: Chemical control of herbaceous forbs

FAMILY / Species	(several names) 2,4-D	(Perspective) Aminocyclopyrachlor + chlorsulfuron	(Milestone) Aminopyralid	(Telar) Chlorsulfuron	(Transline) Clopyralid	(Banvel, Clarity) Dicamba	(Roundup, others) Glyphosate
GERANIACEAE (geranium)							
Geranium robertianum — Herb-robert	E	NIA	P	E	P	E	E
IRIDACEAE (iris)							
Crocosmia x crocosmiflora — Crocosmia or Montbretia	NIA	P	P	NIA	P	NIA	E FOL, WCK
Watsonia meriana — Watsonia	NIA	P	P	P	P	NIA	E
LAMIACEAE (mint)							
Lamium amplexicaule — Henbit	F	NIA	E	NIA	P	G	E
LILIACEAE (lily)							
Asparagus asparagoides — Bridal creeper	NIA	NIA	NIA	P	P	NIA	G[12] FOL, WCK
LYTHRACEAE (loosestrife)							
Lythrum hyssopifolium — Hyssop loosestrife	G	NIA	NIA	NIA	NIA	G[14]	E
Lythrum virgatum — Wand loosestrife	G*	NIA	NIA	NIA	NIA	G[14]*	E*
MALVACEAE (mallow)							
Malva neglecta — Common mallow	G	P	P	E	P	G–E	G
PLANTAGINACEAE (plantain)							
Plantago lanceolata — Buckhorn plantain	G	P	P	NIA	P	E	G
POLYGONACEAE (buckwheat)							
Emex spinosa — Spiny emex	G	NIA	NIA	E	NIA	E	G
Polygonum arenastrum — Prostrate knotweed	G	NIA	G*	NIA	F	E	G
Polygonum convolvulus — Wild buckwheat	G	NIA	G*	G	E	E	F
RANUNCULACEAE (buttercup)							
Clematis vitalba — Old-man's-beard	E*	NIA	E*	NIA	G	NIA	E
Ranunculus acris — Tall buttercup	E	E*	E	E	NIA	E	E
Ranunculus repens — Creeping buttercup	E	E	G–E	E	NIA	E	E
Ranunculus testiculatus (=*Ceratocephala testiculata*) — Bur buttercup	G	E*	G	E	NIA	E	E

6. Seedlings only
11. With glyphosate
12. High rate
13. Max label rate
14. Usually mixed with 2,4-D

E = Excellent control, generally better than 95%
G = Good control, 80-95%
F = Fair control, 50-80%
P = Poor control, below 50%
* = Likely based on results of observations of related species

(Velpar) Hexazinone	(Plateau) Imazapic	(Habitat, Arsenal, others) Imazapyr	(Escort) Metsulfuron	(Gramoxone) Paraquat	(Tordon 22K) Picloram	(Matrix) Rimsulfuron	(Oust) Sulfometuron	(Outrider) Sulfosulfuron	(Garlon) Triclopyr
NIA	G	NIA	NIA	NIA	NIA	NIA	E*	NIA	E
NIA	NIA	NIA	E[11] WCK	NIA	NIA	NIA	NIA	NIA	NIA
NIA	NIA	NIA	P	NIA	NIA	NIA	NIA	NIA	NIA
F	G	E	E	E	NIA	E	NIA	G	NIA
NIA	NIA	NIA	G[13] FOL	P	P	NIA	P	NIA	P
P	NIA	E	E*	P	NIA	NIA	NIA	NIA	E*
P*	NIA	E*	E*	P*	NIA	NIA	NIA	NIA	E*
G	NIA	G	NIA	P[6]	NIA	G–E	E	NIA	P
G	G	NIA	E	P	E	G	E	NIA	G
NIA	NIA	NIA	E	E	E	NIA	NIA	NIA	NIA
E	E	E	E	P[6]	NIA	F	NIA	NIA	NIA
NIA	NIA	E	G	P[6]	E	NIA	NIA	NIA	NIA
NIA	E*	E	G	NIA	E*	NIA	NIA	NIA	G
NIA	NIA	NIA	E	NIA	NIA	NIA	NIA	E*	NIA
NIA	NIA	NIA	E	NIA	E	NIA	NIA	E*	G
E	E	NIA	E	NIA	E	NIA	E	NIA	E

Control includes effects within the season of treatment. Control is followed by best timing, if known, when efficacy is **E** or **G**.

NIA = No information available

FOL = foliar treatment
WCK = wick

Table 5 continued: Chemical control of herbaceous forbs

FAMILY Species	(several names) 2,4-D	(Perspective) Aminocyclopyrachlor + chlorsulfuron	(Milestone) Aminopyralid	(Telar) Chlorsulfuron	(Transline) Clopyralid	(Banvel, Clarity) Dicamba	(Roundup, others) Glyphosate
SCROPHULARIACEAE (figwort)							
Digitalis purpurea Foxglove	P–F	P	P	NIA	P	NIA	F–G
SOLANACEAE (nightshade)							
Solanum elaeagnifolium Silverleaf nightshade	E	NIA	E	NIA	P	E	F–E
URTICACEAE (nettle)							
Urtica dioica Stinging nettle	E	NIA	E	NIA	P	E	E
VERBENACEAE (vervain)							
Verbena bonariensis Tall vervain	E	NIA	NIA	NIA	NIA	NIA	P–F*
Verbena litoralis Seashore vervain	E	NIA	NIA	NIA	NIA	NIA	P–F[16]
ZYGOPHYLLACEAE (caltrop)							
Peganum harmala Harmal, African rue	NIA	NIA	NIA	NIA	NIA	NIA	F–G[17]
Zygophyllum fabago Syrian beancaper	P	NIA	NIA	NIA	NIA	F	G
Foliar activity	+	+	+	+	+	+	+
Soil activity	▼	O	O	O	O	▼	▼
Range/average half-life of herbicide in the soil (days)	10	80-164	35	28-42	12-70	<14	-

15. Postemergence
16. Better on young plants
17. Often mixed with 2,4-D
18. Also tebuthiuron

E = Excellent control, generally better than 95%
G = Good control, 80-95%
F = Fair control, 50-80%
P = Poor control, below 50%
* = Likely based on results of observations of related species

(Velpar) Hexazinone	(Plateau) Imazapic	(Habitat, Arsenal, others) Imazapyr	(Escort) Metsulfuron	(Gramoxone) Paraquat	(Tordon 22K) Picloram	(Matrix) Rimsulfuron	(Oust) Sulfometuron	(Outrider) Sulfosulfuron	(Garlon) Triclopyr
NIA	NIA	NIA	G	NIA	G	NIA	NIA	NIA	P
NIA	NIA	G-E	NIA	NIA	E	NIA	NIA	NIA	P-G
NIA	NIA	G	NIA	P	E	NIA	NIA	NIA	E
P*	F*[15]	E*	NIA	NIA	NIA	NIA	P*	NIA	NIA
P*	F*[15]	E*	NIA	NIA	NIA	NIA	P*	NIA	NIA
E[18]	NIA	E	E	NIA	NIA	NIA	NIA	NIA	NIA
NIA	NIA	NIA	G	NIA	G	NIA	NIA	NIA	NIA
▲	+	+	+	+	+	▲	▲	+	+
O	O	O	O	▼	O	O	O	O	▼
90	120	25-142	7-42	-	20-300	20	20-28	12	10-46

Control includes effects within the season of treatment. Control is followed by best timing, if known, when efficacy is E or G.

NIA = No information available

+ indicates herbicide can have postemergence foliar activity

▲ indicates that it is only considered a preemergence herbicide

O indicates that the herbicide is considered to have soil residual activity

▼ indicates that it does not have a long enough residual activity in the soil to give control

Table 6. **Non-chemical control** of **woody plants** not included in the weed reports.

FAMILY	Species	CULTURAL	
		Grazing	
AQUIFOLIACEAE (holly)	*Ilex aquifolium* English holly	P	
ASTERACEAE (sunflower)	*Artemisia tridentata* Sagebrush	P	overgrazing increases populations
	Euryops multifidus Sweet resinbush	P	not considered palatable to livestock
BIGNONIACEAE (bignonia)	*Catalpa bignonioides* Catalpa	P	
BUDDLEJACEAE (buddleja)	*Buddleja davidii* Butterflybush	F	goats have been reported to eat the plant and suppress populations
CLUSIACEAE or HYPERICACEAE (St. Johnswort)	*Hypericum canariensis* Canary Island hypericum	NIA	
FABACEAE (pea or bean)	*Acacia melanoxylon* Black acacia	P	
MYOPORACEAE (myoporum)	*Myoporum laetum* Myoporum	P	plants are toxic to animals
OLEACEAE (olive)	*Olea europaea* Olive	P	often leads to spread
PITTOSPORACEAE (pittosporum)	*Pittosporum undulatum* Victorian box	NIA	
ROSACEAE (rose)	*Crataegus monogyna* English hawthorn	P	very thorny
	Prunus cerasifera Cherry plum	NIA	
	Pyracantha spp. Pyracanthas or firethorns	P	very thorny
ULMACEAE (elm)	*Ulmus pumila* Siberian elm	NIA	

E = Excellent control, generally better than 95%
G = Good control, 80-95%
F = Fair control, 50-80%
P = Poor control, below 50%
* = Likely based on results of observations of related species

Control includes effects within the season of treatment. Control is followed by best timing, if known, when efficacy is **E** or **G**.

	MECHANICAL	
Prescribed burning	**Hand removal, weed wrench, cutting**	**Heavy equipment removal**
F one burn, need to repeat to be effective	**P** plant will resprout	**G** small plant can be dug in moist soil, weed wrench on smaller
G repeated burning will reduce populations	**E** does not resprout well	**F** chaining operations can reduce density
G used in integrated approach with herbicide treatment to recovered plants	**G** need to follow-up control newly germinating seedlings or treat regrowth following mowing	**E** remove most of root system
Typically found in riparian areas not conducive to burning	**G** weed wrench can be used to remove young plants, soil should be moist, remove entire root	**P** generally growing in sensitive areas
NIA	**G** seedlings can be hand pulled, young shrubs can be dug up, older plants	NIA
NIA	**F** large root system, soil needs to be very moist, small seedlings pulled by hand	NIA
P fire stimulates recruitment	**P** resprouts from base and has spreading root system	NIA
P resprouts after fire	**F** readily resprouts when roots left in soil	NIA
P older plant resprout, but may kill younger plants	**G** hand pull seedlings, weed wrench can be used to remove young plants	NIA
G will kill most adult plants and destroy the soil seed bank	**G** Sp; seedlings can be hand pulled, young plants dug out or removed with weed wrench	**G** bulldozer has been used in Australia to fell trees. This was followed by burning thickets
P often increases after fire	**G** early Su; young plants can be hand pulled or removed by a weed wrench	NIA
P will resprout	**G** small plants can be hand pulled, young plants pulled with a weed wrench,	**G** bulldozer can uproot plants
P will resprout	**G** weed wrench can be used to remove young plants	NIA
F* adults will resprout, but fire can kill saplings	**E** girdling stem kills trees, Su, seedlings hand pulled, small trees removed with weed wrench	NIA

Sp = Spring
Su = Summer
NIA = No information available

Table 7. **Chemical control** of **woody plants** not included in the weed reports.

FAMILY	Species	(several names) 2,4-D	(Roundup, others) Glyphosate
AQUIFOLIACEAE (holly)	*Ilex aquifolium* English holly	**E*** INJ	**E*** CS, INJ, Su
ASTERACEAE (sunflower)	*Artemisia tridentata* Sagebrush	**E** Sp, ester form	**G*** FOL
	Euryops multifidus Sweet resinbush	NIA	NIA
BIGNONIACEAE (bignonia)	*Catalpa bignonioides* Catalpa	NIA	**E** CS, FOL, INJ, Su, Fa
BUDDLEJACEAE (buddleja)	*Buddleja davidii* Butterflybush	NIA	**E** CS, Su **G** FOL, Su, Fa
CLUSIACEAE or **HYPERICACEAE** (St. Johnswort)	*Hypericum canariensis* Canary Island hypericum	NIA	**E** CS
FABACEAE (pea or bean)	*Acacia melanoxylon* Black acacia	NIA	**G*** FOL **E** CS
MYOPORACEAE (myoporum)	*Myoporum laetum* Myoporum	NIA	**E** CS
OLEACEAE (olive)	*Olea europaea* Olive	**F** CS	**G–E** CS, INJ, Su, Fa
PITTOSPORACEAE (pittosporum)	*Pittosporum undulatum* Victorian box	NIA	**E** CS, INJ Su, Fa wick on seedlings
ROSACEAE (rose)	*Crataegus monogyna* English hawthorn	NIA	**G–E** CS, FOL
	Prunus cerasifera Cherry plum	**G–E*** FOL mixed with picloram; BB with ester, Su; CS or INJ, Su	**E*** FOL, Fa
	Pyracantha spp. Pyracanthas or firethorns	**F*** FLO	**E** CS, INJ, Su, Fa
ULMACEAE (elm)	*Ulmus pumila* Siberian elm	**E** FOL, Su; BB	**G** FOL **E** CS, Su, Fa
Foliar activity		+	+
Soil activity		▼	▼
Range/average half-life of herbicide in the soil (days)		10	-

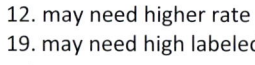

12. may need higher rate
19. may need high labeled rate

E = Excellent control, generally better than 95%
G = Good control, 80-95%
F = Fair control, 50-80%
P = Poor control, below 50%
* = Likely based on results of observations of related species

Control includes effects within the season of treatment.
Control is followed by best timing, if known, when efficacy is **E** or **G**.

(Velpar) **Hexazinone**	(Habitat, Arsenal, others) **Imazapyr**	(Tordon 22K) **Picloram**	(Spike) **Tebuthiuron**	(Garlon) **Triclopyr**
NIA	E CS, FOL, INJ, Su	E CS, INJ, Su	E*[12]	E* BB
NIA	NIA	P FOL	E Fa	G FOL E CS
NIA	NIA	G	NIA	NIA
NIA	E CS, INJ, Su, Fa	NIA	NIA	E* FOL, CS, BB, INJ, Su, Fa
G used in New Zealand	E FOL, INJ, Sp, Su; CS, Su	NIA	NIA	E CS, Su; BB, Fa G FOL, Su
NIA	NIA	NIA	NIA	E FOL, BB
E*	P* typically poor on legume species	E* FOL	E*[12]	E FOL, CS, BB
NIA	NIA	NIA	NIA	NIA
P	E CS, INJ, Su, Fa	P FOL E CS	P	E FOL, CS, BB
NIA	NIA	E CS, INJ, Su, Fa	NIA	E INJ
E	E FOL	NIA	E*[19]	E FOL, CS
E*	E FOL	G–E* FOL, mixed with 2,4-D	E*[19]	E FOL, CS, INJ, BB, Fa
E*	E CS, INJ, Su, Fa	E Sp or Fa mixed with 2,4-D (FOL) or triclopyr (BB, CS, FOL)	E*[19]	F–G BB G FOL, CS
E Sp	E CS, BB	E FOL mixed with 2,4-D	E*[19] Sp, Fa	E FOL, CS, INJ, BB, Su, Fa
▲	+	+	▲	+
O	O	O	O	O
90	25-142	20-300	360-450	10-46

NIA= No information available

Possible application methods
BB = basal bark
CS = cut stump
FOL = foliar
INJ = stem injection

Sp = Spring
Su = Summer
Fa = Fall

FLO = full leafed out

+ indicates herbicide can have postemergence foliar activity
▲ indicates that it is only considered a preemergence herbicide
O indicates that the herbicide is considered to have soil residual activity
▼ indicates that it does not have

Table 8. **Non-chemical control** of **aquatic plants** not included in the reports.

FAMILY / Species	MECHANICAL	
	Cutting	Hand pulling or vacuuming
BUTOMACEAE (flowering rush)		
Butomus umbellatus Flowering rush	**P** plants produce no viable seed but resprout with underground rhizomes	**P** plants produce no viable seed but underground rhizomes difficult to remove
CABOMBACEAE (watershield)		
Brasenia schreberi Watershield	**F** but plants will regrow	**G** if roots and rhizomes are removed
HYDROCHARITACEAE (waterweed)		
Elodea canadensis Common elodea	**F** but will stimulate regrowth and fragments will spread the plant	**G**
NYMPHAEACEAE (waterlily)		
Nuphar lutea (= *Nuphar polysepala*) Yellow pondlily	**F** but plants will regrow	**G** if roots and rhizomes are removed
Nymphaea odorata Fragrant waterlily	**F** but plants will regrow	**G** if roots and rhizomes are removed
ZANNICHELLIACEAE (Horned pondweed)		
Zannichellia palustris Horned pondweed	**G** but plants will regrow quickly	**G** if roots and rhizomes are removed; plants will regrow from seed

E = Excellent control, generally better than 95%.
G = Good control, 80-95%.
F = Fair control, 50-80%.
P = Poor control, below 50%.
* = Likely based on results of observations of related species.

Control includes effects within the season of treatment. Control is followed by best timing, if known, when efficacy is **E** or **G**.

	CULTURAL			BIOLOGICAL
Benthic barrier	**Drawdown**		**Shading**	**Grass carp**
P	P		P	NIA
F but only if barriers are in place by early spring	**P** plants will recover from rhizomes		**P** can reduce size, but only with extreme shade	**P**
G if barriers are in place by early spring	**F** unless drawdown allows complete drying of sediments for several weeks. Seed may survive.		**P** plants are well adapted to low light. (Some dyes can be useful if applied early and often)	**E**
F but only if barriers are in place by early spring	**P** plants will recover from rhizomes		**P** can reduce size, but only with extreme shade	**P**
F but only if barriers are in place by early spring	**P** plants will recover from rhizomes		**P** can reduce size, but only with extreme shade	**P**
G if barriers are in place by early spring	**F** unless drawdown allows complete drying of sediments for several weeks. Seed may survive		**P** plants are well adapted to low light (some dyes can be useful if applied early and often)	**E***

NIA = No information available

Table 9. **Chemical control** of **aquatic plants** not included in reports.

FAMILY / Species	HERBICIDE (several names) 2,4-D	(Magnicide) Acrolein	(Tradewind) Bispyribac-sodium	(several names) Copper formulations	(Reward, others) Diquat	(Aquathol, Hydrothol) Endothall
BUTOMACEAE (flowering rush)						
Butomus umbellatus Flowering rush	F–G with triclopyr; variable results	P	NIA	P	P	F–G variable results
CABOMBACEAE (watershield)						
Brasenia schreberi Watershield	E	NIA	NIA	P	G	P
HYDROCHARITACEAE (waterweed)						
Elodea canadensis Common elodea	P	E	NIA	E	E	E
NYMPHAEACEAE (waterlily)						
Nuphar luteum (= *Nuphar polysepala*) Yellow pondlily	E	NIA	NIA	P	G	P
Nymphaea odorata Fragrant waterlily	E	NIA	F	P	G	P
ZANNICHELLIACEAE (horned pondweed)						
Zannichellia palustris Horned pondweed	NIA	E	NIA	P	G	G
Avg. water half-life (days)	7	0.2-7.5	9	Indefinitely	1-7[20]	4-7

20 = Binds tightly to soil particles and is inactivated quickly in turbid water.

E = Excellent control, generally better than 95%
G = Good control, 80-95%
F = Fair control, 50-80%
P = Poor control, below 50%
* = Likely based on results of observations of related species

HERBICIDE							DYE
(Clipper) Flumioxazin	(Sonar) Fluridone	(Rodeo, Aquamaster) Glyphosate	(Clearcast) Imazamox	(Habitat) Imazapyr	(Galleon) Penoxsulam	(Garlon 3A, Renovate) Triclopyr	Aquashade
F–G variable results	NIA	NIA	F–G variable results	F–G variable results	NIA	F–G variable results	P
G	F* partial control	G	E	E	G*	G*	P
E	E	NIA	NIA	NIA	F*	P	G
F	F*	E can be tank mixed with imazapyr	G	NIA	G	G	P
F*	F*	G* can be tank mixed with imazapyr	E	E	G	G	P
NIA	G*	NIA	NIA	NIA	NIA	NIA	G
0.7 at pH 7	20-270	12-70[20]	5-15	2-3	3	0.1-0.5	-

Control includes effects within the season of treatment.
Control is followed by best timing, if known, when efficacy is E or G.

NIA = No information available

Control Options

MECHANICAL CONTROL OR PHYSICAL REMOVAL

Mechanical control techniques physically damage shoots, roots, or root crowns of plants to the point where they can no longer survive. Alternatively, mechanical control options can remove the entire plant. These methods include hand-pulling, hoeing, string trimming (weed whipping or weed whacking), tilling, mowing, cutting, lopping, chainsawing, grubbing, digging, chaining, shredding, roller chopping, vacuuming and removing plants with heavy equipment such as bulldozers or backhoes. These techniques can be expensive and disrupt the soil, creating disturbed sites prone to reinvasion or invasion of other species. While most are used in terrestrial settings, some, such as hand-pulling, vacuuming, shredding, and cutting, can also be used in aquatic systems. In terrestrial systems, most mechanical treatments work best on relatively level terrain.

MANUAL REMOVAL

Hand-pulling, string trimming, shoveling, and hoeing are valuable tools for preventing initial infestations or controlling small populations of potentially important weed species. These are also effective for the control of small infestations at the fringe of a major infestation. String trimming is most effective when used as late as possible, but before plants produce viable seed. Hand-pulling, shovels, hoes, and other hand tools (e.g., pick-ax or mattock) are more effective in loose or moist soil where it is easier to completely remove the crown or root buds. They also are commonly used in a follow-up control program where only a few plants remain after a couple of years of more intensive management. Hand weeding techniques are typically more effective on annual species, and less effective on perennials. However, for small infestations of perennials, repeated manual removal can eventually eradicate the population. This may be most effective in loose soils, after rains have made the substrate workable, or in muddy soils. All vegetative reproductive fragments must be removed from the soil to prevent resprouting. On harder soils, control with hand pulling is more difficult, as it is nearly impossible to remove all rhizome or underground reproductive fragments.

Under some circumstances, manual methods of control can be successful with woody species. However, they are very labor-intensive and require continual follow-up control efforts. Small seedlings of all woody species can be removed by hand. Once plants get larger, hand removal is not usually successful. Such a technique is suitable only for small infestations and around trees and shrubs where other methods are not practical.

MOWING OR CUTTING

Mowing is often used to control annuals. It is not always successful for controlling biennials and perennials, but on occasion can reduce seed production and provide suppression if used repeatedly. Mowing can promote a more even balance of forage species by increasing light for the lower canopy, which can include desirable legumes. In addition, it can remove the flowering stems of late-season undesirable invasives, thus preventing or reducing new seed deposits into the soil seedbank. When desirable perennial grasses are present, mowing can maintain their vigor and remove unpalatable older growth.

Mowing can also be used to remove thatch or litter buildup (e.g., medusahead, yellow starthistle or perennial pepperweed). Removal of the thatch layer can increase the competitiveness of other desirable species and can also dramatically improve the efficacy of a subsequent herbicide treatment.

If timed correctly with biennial and perennial plants, mowing can prevent flowering and seed set, reduce plant vigor, and deplete carbohydrate reserves. Repeated mowing can be especially effective and may give advantages to desirable perennial grasses. Mowing should be conducted when plants are still green but when seasonal soil moisture is almost exhausted. On the negative side, it can result in dissemination of seeds and can stimulate production of new stems from vegetative buds below the cut surface.

Timing is critical to the success of mowing. Mowing too early increases light penetration without removing a significant proportion of weed biomass. This benefits weedy species by stimulating rapid recovery of growth. Mowing too late will not prevent the production of viable seed, resulting in a similar or greater weed problem in the following year. The optimum time for mowing most annual species is in the flowering stage before seed development. Soil moisture can also influence the effectiveness of mowing. Generally, mowing when soil moisture is low will reduce the potential for regrowth and new seed production.

Mowing options are limited in areas with steep terrain. In some cases, mowing is not feasible due to the risk of physical damage of equipment and the fire hazard caused by sparks generated from contact of the equipment with rocks. Even when mowing is employed as a control technique, it is not always successful and can decrease the reproductive efforts of insect biocontrol agents, injure late growing native forb species, and reduce fall and winter forage for wildlife and livestock.

In addition to mowing, there are other cutting techniques to control weeds, particularly perennial or woody species. Girdling trees involves manually cutting away bark and cambial tissues around the trunks of undesirable trees. This is a relatively inexpensive method and is done with an ordinary ax in the spring when the trees are rapidly growing. This method is often unsuccessful as trees generally resprout below the girdle unless the cut is treated with an herbicide. In some cases, trees have been able to recover their cambial connections and survive a girdling treatment. Removing stems by brush cutters, power saws, axes, machetes, loppers, and clippers is also insufficient to prevent resprouting of sucker-sprouting invasive woody species. As a consequence, mechanical removal will be ineffective unless all stems are cut at least several times per year to exhaust their reserve food supply. The best time to cut is when plants begin to flower.

HEAVY EQUIPMENT

Mechanical removal using heavy equipment can be effective under some circumstances, but can severely damage sensitive environments. For woody invaders, the use of a bulldozer or backhoe can remove the entire root crown (root plowing). Eradication of perennial or woody species using these techniques is extremely difficult, even with use of a backhoe. Deep rhizomes or root fragments buried under the alluvium readily resprout.

For larger infestations of some perennial species (e.g., giant reed or saltcedar) on accessible terrain, heavy cutting tools (rotary brush-cutter, chainsaw or tractor-mounted mower) can reduce biomass and provide better access to sites for ground application of herbicides. Thus, the combined use of heavy equipment with follow-up herbicide treatment can provide more successful results.

The disadvantages to the use of heavy equipment include its high cost and significant disturbance to the surrounding area. In addition, heavy equipment is of limited use on complex or sensitive terrain, or on severe slopes, and may interfere with reestablishment of native plants and animals. In irrigation ditches or ponds where heavy equipment is available, dredging can be used effectively to remove emerged aquatic species, although this too often leads to resprouting from fragments left behind.

TILLAGE

One of the most common mechanical weed control techniques used in agricultural systems is cultivation or tillage. Tillage equipment can include plows or discs, which control annual weeds by burying plant parts, including seeds. In contrast, the use of harrows, knives, and sweeps will damage root systems or separate shoots from roots. Tillage must be conducted when the surface soil is dry, otherwise plant fragments will regrow and can exacerbate the problem. Despite its effectiveness in the control of annual weeds, tillage can increase atmospheric dust levels and soil erosion. Tillage is not generally practiced on rangelands or wildlands and can spread perennial invasive plants that disperse vegetatively by rhizomes or roots when cut.

CULTURAL CONTROL

Cultural control practices in rangelands most often include fire, grazing, or revegetation efforts. All these strategies involve manipulation of disturbance regimes to suppress invasive plants and enhance desirable vegetation.

GRAZING

While overgrazing without periodic rest can selectively reduce desirable grass competitiveness and increase invasion of less palatable or poisonous species, appropriate grazing can shift a plant community toward more desired species. The effectiveness of grazing can depend on the class of livestock used as well as stocking rates, grazing intensity, and frequency.

Livestock Class. Successful weed management by grazing can depend on the type of grazer and grazing timing. Because most animals have preferences for certain plants, grazing systems can be designed with different classes of livestock or different timing of grazing to increase utilization of weedy plants. For example, the foraging behavior of both cattle and goats is conducive to the management of yellow starthistle when plants are grazed at the bolting stage, whereas only goats are effective when plants are in the spiny stage of growth. The ideal time to graze is when the noxious species are most susceptible to defoliation or when the impact on desirable vegetation is minimal.

Stocking Rates. Stocking rates of livestock can be adjusted to maximize weed management, avoid overgrazing, and minimize soil compaction. In some cases, low stocking rates allow cattle to graze preferred plants repeatedly and avoid less palatable weed species. This may result in higher weed infestations. In contrast, if the weedy species is preferred, then lower stocking rates can reduce weed populations. Thus, depending upon the rangeland and the weed composition, areas with low cattle stocking rates may have either higher or lower weed cover than areas that are overgrazed or rotationally grazed. Rangelands, pastures, or grasslands with high stock densities can have the advantage of forcing cattle to graze less preferable species that they typically would avoid. This can lead to more uniform composition of plant species and more competitive desirable species. In contrast, overgrazing without periodic rest can stress more desirable species and reduce their competitiveness with weeds.

Grazing Intensity. High intensive and short duration grazing, practiced on a rotational basis (called mob stocking) is a management system widely adopted in other countries. In this system, pastures or rangelands are intensively grazed for a few days, often with the use of electric fencing. The area is subsequently allowed to recover for about a month before grazing is repeated. This system typically results in more uniform and complete forage use, as well as greater weed utilization. With intensive grazing, forage is not completely grazed and recovery occurs rapidly. As an added benefit, the remaining forage reduces light penetration to the soil surface and provides additional weed suppression. In many cases, intensive livestock grazing can provide much better control of specific weeds than season-long cattle grazing.

Intensive grazing can also have some negative impacts. For example, trampling and removing overstory vegetation can disturb soil and enhance weed seed germination by moving seed to the soil surface. In addition, it can break up thatch, remove competition, and provide increased light penetration to weeds. High grazing intensity can also lead to increased soil compaction, which can reduce the vigor of desired pasture species and lead to rapid weed establishment.

PRESCRIBED BURNING

Fire plays an important role in the formation and maintenance of rangeland and grassland ecosystems. As with any disturbance, the effects of fire are influenced by its frequency, intensity, seasonal timing, and interactions with other disturbances. In general, prescribed fire is most successful for the control of late season annuals and will typically stimulate perennials that resprout from the base.

In using prescribed burns to control invasive plants, the main objective is to deplete the soil seedbank or destroy their reproductive structures. To control annual species with fire, it is critical to either kill plants before their seeds become viable or destroy the seeds before they disperse. Thus burns should be conducted when the target plant seeds are still held in the canopy, where they will be exposed to direct flames, but the seeds of desirable species have dispersed to the ground. Seeds on the soil surface are not generally exposed to lethal temperatures of grassland fires. For perennial or herbaceous plants with protected meristems, e.g., rosettes or rhizomes, the burn must be hot enough to damage the vegetative reproductive tissues and prevent resprouting. For this reason, prescribed burning is rarely effective for the management of perennials and in most cases can stimulate their growth.

In addition to invasive weed control, prescribed burning can also be used in fall to remove thatch, recycle nutrients trapped in the dried vegetation, stimulate early growth of desirable species in the spring, and obtain more uniform grazing within grasslands. However, burning designed to control invasive species in rangelands generally needs to be conducted in late spring or summer. Unfortunately, this is when the risks of escaped fires are greatest. Moreover, air quality problems and liability issues can also present a problem when burns are conducted near populated areas. In areas where biological control agents are present, burning may cause damage to these insect populations. In some areas, burning can lead to rapid invasion by other undesirable species with wind dispersed seeds, particularly members of the sunflower family.

Since multiple burns are not usually practical or permitted and fuel loads may not be sufficient to allow multiple year burns, integrated approaches are often more appropriate than using burning as a sole control option. Thus, prescribed burning often requires a follow-up program to prevent escaped or isolated plants from completing their life cycle. Where the seedbank is short-lived, a follow-up program may take only a couple of years; in other cases it may take longer. For woody species, burning will not provide a sole method for management; most species, such as saltcedar, tree-of-heaven, and Himalaya blackberry, will readily resprout after fire. However, fire can be a valuable part of an integrated approach for thinning heavy infestations before follow-up application of herbicide to control resprouts or new seedlings.

COMPETITION

Reestablishment of desirable, competitive plant species is the best long-term sustainable method to suppress weeds or inhibit plant invasions. This can also improve a site's forage value and enhance wildlife habitat.

Revegetation is often most successful when used as part of an Integrated Pest Management (IPM) program. It is essential to control invasive plants to create an environment more conducive to establishment of desired vegetation that will fill the niche vacated by the controlled invader. When rangeland deterioration is severe and few desirable species are present, it may be necessary to revegetate the site to reclaim the productive potential. In most cases, revegetation can be expensive. Selecting the right species is critical to success of rangeland revegetation programs. Seeded species must be capable of establishing and should be adapted to soil conditions, elevation, and climate on the site. Planted species selected should be competitive with invasive plants, and contribute to improved ecosystem function. Land managers are often faced with the decision of whether to use native or introduced plant materials that differ in cost, establishment success, and resistance to reinvasion.

WATER MANAGEMENT

In aquatic systems, it may be possible to manipulate water level at a particular time of the year to achieve control of aquatic species. This is most practical in irrigation canals or some lakes, reservoirs and ponds where water levels can be controlled. A drawdown (dewater) involves lowering the water to expose sediments in the areas near the shore. The degree of plant control from a drawdown is dependent on plant species and temperature. A drawdown accompanied by freezing temperatures provides the greatest aquatic plant control. Raising the water level can also provide some level of control for a few emergent aquatic species. For example, flooding of purple loosestrife or perennial pepperweed is effective if plants can be inundated throughout the growing season. The drawback to drawdown, however, is that it is also detrimental to most other emergent species and can promote the spread of the weeds to shallow areas.

BENTHIC BARRIER

Benthic barriers can be economical to control small weedy paths in aquatic areas. Although they are usually made of impermeable plastics, other materials such as jute can be used. The barriers primarily control aquatic weeds by blocking light penetration to the sediment surface. While plastic barriers need to be removed after a period of time, jute and other biodegradable natural barrier materials can be left in place to decompose. Benthic barriers do not require registration with US EPA or state agencies and have minimal impact on the water column and non-target organisms. The disadvantages to benthic barriers are that sediment accumulation may occur on top of the barriers where weeds may then germinate and grow. In addition, the barriers may be lifted up by gas bubbles that are produced from the sediment. Although jute is biodegradable, it often does not provide adequate control of weed propagules.

BIOLOGICAL CONTROL

Classical biological control involves relocation of natural enemies, typically insects, of the invasive plant from their native habitats onto plants in their invaded habitats. The long-term goal of these programs is to exert sufficient stress on the target plant to reduce plant competitive ability and dominance. With insect agents this can be achieved by boring into roots, shoots and stems, defoliation, seed predation, or extracting plant fluids. Synchrony in the life cycles of host plant and agent, adaptation of the agent to a new climate and habitats, ability of the agent to find the host, capacity of the agent to reproduce rapidly, and the nature, extent, and timing of the damage caused by the biocontrol agent are among the factors that determine biocontrol agent effectiveness.

Biological control agents (including insects, pathogens or fish) are mobile and are expected to move from the release area and spread throughout the region. As a result, this control method is not specific to an invaded site or weed infestation. The goal of biological control is to establish self-sustaining populations of beneficial organisms that build up high numbers on the target weed. It is hoped that attack by the biological control agents will reduce the invasiveness of the host weed and result in substantial reduction in its abundance. A key aspect of the control agents is their high level of specificity to the target weed. For this, many years of research are necessary to find the appropriate natural enemies and to perform the necessary host specificity tests. Once completed, the results are submitted for review by USDA-APHIS, the agency that approves permits for the introduction of living organisms.

While biological control efforts have targeted a number of habitats, the majority of released agents have been directed towards invasive plants of rangelands and aquatic systems. Of the nearly 100 examples worldwide where weed biocontrol programs have been underway for a sufficient period to assess control, only 28% have resulted in control that could be rated as sometimes complete, while 35% have shown no control of the target plant. Thus, despite its great benefits, biological control efforts are often unsuccessful. However, the long-term benefits of an effective biological control agent can far exceed the development costs. The results from a successful biological control agent last longer than other management techniques and reduce the need for subsequent herbicide treatments or other mechanical or cultural control methods. When considering both the successful and unsuccessful expenditures on the development of biological control efforts in Australia, the economic benefits become apparent with a 23:1 cost benefit ratio.

The most common biological control agents in the western United States are insects, but a few plant pathogens have also been released and others are being investigated. In addition to these host specific organisms, a non-selective fish, grass carp (*Ctenopharyngodon idella*), has been used successfully to control some aquatic species, including hydrilla, in canals and ponds. The grass carp is an herbivorous fresh water species that can give relatively inexpensive and long-term (decades) control of aquatic plants. It has the disadvantage of being difficult to contain and it will feed on most plants in the system, including desirable native species. Thus, it is typically used in systems where water is used by humans.

CHEMICAL CONTROL

Herbicides are generally classified by their mode of action, selectivity, and location of application (foliar or soil). Soil-applied herbicides generally target emerging seedlings, whereas foliar-applied herbicides control plants ranging from seedlings to fully mature. Most herbicides used in terrestrial systems are foliarly active, while a few are only soil active. Many important herbicides used to control invasive and weedy species in terrestrial areas have both soil and foliar activity.

AQUATIC HERBICIDES

Herbicides labeled for aquatic use can be classified as either contact or systemic (see Table 13, p. 512) for characteristics of aquatic herbicides). Contact herbicides act rapidly on the tissues contacted, typically causing extensive cellular damage and membrane leakage. These compounds do not translocated to the below sediment vegetative reproductive structures such as root crowns, roots, or rhizomes of perennial plants. As such, they are considered more effective on annual weeds. Systemic herbicides are translocated throughout the above and below sediment portions of the plant. They require a longer time to exert their effect but are much more effective on perennial plants compared to contact herbicides.

Complexed copper compounds include a variety of chelated formulations of copper that keep the active ingredient in solution much longer than unchelated copper sulfate. Because copper rapidly precipitates, especially in harder water, it is not as effective for control of algae and some macrophytes (particularly hydrilla) as the chelated formulations.

For control of submerged aquatic species, applications of herbicides are made directly to the water column. This can be surface or subsurface applications as a solution, crystal, or granule formulation. The herbicide is rapidly and directly taken up by the plant tissues within the water column, as these plants do not contain a waxy cuticle barrier on their epidermis. To be efficient, applicators must have a thorough knowledge of the exchange rate of the water, as well as the herbicide concentration and proper exposure time required to obtain effective control of the target species. The response of different plant species to different herbicides becomes a function of these factors. With this knowledge, an effective herbicide concentration can be used that will also minimize potential environmental problems and reduce costs.

TERRESTRIAL AND RIPARIAN HERBICIDES

In natural areas and rangelands, herbicides are the most frequently used tool for the control of invasive and weedy plants. Unlike cropland environments where all plants, except the crop, are considered to be weeds, wildlands and rangelands often have only one or a few invasive plants that are the target of control measures.

These plants are generally growing in association with several desirable species. Thus, selectively is critical when developing an herbicide control program and most herbicides are selective only within certain rates, environmental conditions, and methods of application.

There are several application methods used to treat invasive and weedy plants of natural areas and rangelands, including fixed-wing aircraft, helicopter, ground applicators, backpack sprayers, and wick applicators. Broadcast applications are typically conducted using aircraft, tractors, ATVs, or other specialized equipment. These techniques can cover more ground quickly and are more economical, but can damage sensitive non-target species. For this reason, they are best used in areas with large infestations and few, if any, sensitive desirable species present. Both preemergence and postemergence herbicides are applied broadcast, and in forested areas granule formulations of preemergence herbicides can be used to prevent injury to desirable trees. In some cases, residual thatch can influence the effectiveness of herbicides with preemergence activity. For example, imazapic can bind to standing thatch or other dried debris on the soil surface, thus reducing the effectiveness of the application.

When selectivity is required, but cannot be achieved with broadcast applications, directed or spot treatments are the most appropriate method of herbicide application. In addition, spot treatments can be used to control early weed invasions or to prevent the spread of small infestations. To achieve this, herbicides are generally applied with backpack sprayers or wick applicators. This almost always requires the use of postemergence herbicides. As an exception, the pellet formulation of tebuthiuron can be applied preemergence at the base of a weedy woody species. Directed applications often require more labor and are more expensive compared to broadcast treatments. Applications can be made at higher rates but at lower volumes, or using a spray-to-wet method which uses higher volumes, but lower rates. It is also possible to achieve selective control of a particular weed with otherwise non-selective or relatively non-selective postemergence herbicides by using a wiper applicator (ropewick or weed wick). These can be either hand held or vehicle mounted boom wipers. As a benefit, this application method reduces the potential for herbicide drift and injury to adjacent sensitive agricultural crops and can be used to selectively control invasive species around vernal pools, streams and other bodies of water, or in areas with rare and endangered species or other desirable plants.

The most commonly used herbicides on rangelands and wildlands are 1) auxin-like growth regulators (2,4-D, aminocyclopyrachlor, aminopyralid, clopyralid, dicamba, fluroxypyr, picloram, and triclopyr) that selectively control broadleaf species; 2) glyphosate, a non-selective foliar-applied aromatic amino acid inhibiting herbicide that has no activity in the soil; and 3) imidazolinone and sulfonylurea herbicides (chlorsulfuron, imazapic, imazapyr, metsulfuron, rimsulfuron, and sulfometuron) that disrupt the synthesis of branched-chain amino acids essential for plant growth. While other herbicides occasionally used in terrestrial non-crop areas can inhibit photosynthesis (hexazinone, paraquat, tebuthiuron) or lipid synthesis (clethodim, fluazifop-P-butyl, sethoxydim), they are not as widely used.

HERBICIDES FOR WOODY PLANT CONTROL

Although young woody plants can be treated with foliar or, occasionally, wick-applied herbicides, larger and taller infestations are often difficult to broadcast treat. Exceptions to this would include aerial applications to near or complete monotypic stands of an invasive species such as saltcedar, in areas difficult to access with ground equipment. For the most part, however, taller woody vegetation is often spot or directly treated with glyphosate, imazapyr, or triclopyr. These herbicides are often applied directly to stem tissues by cut stump, basal cut stump, stem injection (hack-and-squirt) or low volume basal bark treatments. Cut stump treatments are often used on very large trees capable of resprouting. Applications of typically undiluted herbicide (glyphosate, triclopyr, or imazapyr) are applied to the cambium region around the edge of a freshly cut tree. Basal cut stump applications are made using basal bark mixtures and applied to both the cut surface of a stem and to the bark of the stem. Stem injection application has been shown to be most effective with the herbicide imazapyr. In most cases, undiluted herbicide is added to one or several cuts, depending on the diameter of the trunk. Low volume basal bark treatments are generally made with diluted ester forms of either imazapyr or triclopyr (20 to 30% in 70 to 80$ basal oil). They are usually applied to trees with stems < 6 in in diameter and a thin bark. It is critical to use the ester formulation with oil to penetrate the bark. For this reason, glyphosate is never used in basal bark treatments.

INTEGRATED PEST MANAGEMENT (IPM)

There are several mechanical, cultural, biological and chemical control options that have been developed for the management of invasive and weedy plants on non-crop areas. To be successful, these management strategies require careful planning that incorporates a long-term approach consisting of prevention programs, education materials and activities, economical, and in most cases, multi-year integrated management approaches that not only control the invasive plant, but also improve the degraded plant community, enhance the utility of the ecosystem, and provide resistance to reinvasion or establishment of other invasive plants.

In wildland and rangeland systems, the use of a single control method for invasive plant management is often unsustainable. Removing invasive plant species with mechanical, cultural, or chemical methods may only open niches for other undesirable species to establish unless desirable species are present to fill the vacated niches. In many cases, a successful long-term management program should be designed to include combinations of the various approaches provided in this book. It is advantageous to begin an invasive plant management program when infestations are just beginning and small. Under this situation, eliminating them might be the only management necessary because the native or desirable plant community can then recover. When infestations are large and well-established, IPM strategy, and perhaps even revegetation program will need to be incorporated into a long-term management plan.

Integrated invasive plant management emphasizes the recovery of ecosystem function that can include energy flow, nutrient cycling, soil water retention, and other functions. Thus, this sustainable approach provides a context for managing invasive plants at an ecosystem-centered level, rather than focusing on the control of a specific invasive plant or the use of a pest control technology. For this reason, all available tools should be considered during development of integrated weed management programs and those selected should optimize attainment of specific management objectives by the most economical means.

Control Equipment and Techniques

Manual and mechanical techniques can include a wide variety of tools and methods. They are generally used to control relatively small populations and can be time consuming, labor intensive and expensive, if the labor force in not made up of volunteers. The specific type of tool used can be very specific to the target species and the terrain. Often, these techniques must be repeated to insure that reestablishment or recovery does not occur. As such, it is not uncommon for areas that are controlled by mechanical or manual means to be heavily trampled and disturbed. To be most effective, this technique will require either another follow-up technique (e.g., herbicide treatment) or should receive several visits to control new seedlings of the same or different weedy species. It is also important to note that clothing and equipment should be thoroughly cleaned and inspected to prevent movement of seeds and other weedy propagules to non-infested sites.

MANUAL CONTROL

HAND PULLING OR REMOVAL

Annuals and some tap-rooted perennials are effectively controlled by hand pulling. Larger plants are more difficult to remove and other tools are more appropriate. Free-floating aquatic plants, such as water hyacinth, duckweeds, giant salvinia, or *Azolla* (mosquitofern) are also controlled by manual removal, but it is very difficult to collect all of these plants and, due to their perennial life cycle which allows them to recover fairly quickly. For perennial species, as much of the root needs to be removed as possible to prevent resprouting.

The advantages of this technique are that it can reduce damage to non-target species and the cost of equipment is typically low. However, it is generally used on small areas, is labor intensive, and can cause physical injury and exhaustion to individuals laboring for long periods. To minimize disturbance to the environment, soil should be replaced when possible and the number of people in any site should be limited. For safety, workers should wear gloves, as many plants can cause skin irritation or cuts, a long-sleeve shirt, and long pants.

TOOLS FOR HAND REMOVAL

There are many tools that can be used to remove invasive or weedy plants. Hoes and other simple hand-held tools are good for annuals, but more substantial tools are necessary for perennials. In wildland areas, many of the tools for removing perennials are designed to grip the weed stem and provide leverage necessary to pull the roots out. These tools can vary in size and strength, which must match the size and difficulty in removing the specific weed species. One problem with the use of these tools is that they can be heavy and difficult to carry in remote areas. In addition, they will fragment plants and not remove the roots when used in dry firm soils. They are best used when soils are sandy or moist.

Pulaski. This tool is a combination of an axe on one end and an adze (similar to a sharp sturdy hoe) on the other end. The handle can be of wood, plastic or fiberglass. The Pulaski is often the tool of choice for constructing firebreaks. It is also commonly used for constructing trails. In invasive plant work, it can be very effective for the removal of large bunchgrasses, such as pampasgrass and jubatagrass. The axe end can quarter the plant, while the adze end can be used to remove the root quarters from the soil.

Root Talon. The Root Talon is somewhat similar to the Pulaski, but is more lightweight. On one end is a pick-ax and on the other end is a claw-like gripping device that can grab the base of the stem of a plant. The handle can then provide leverage to pull out the root. The tool is lighter and less expensive than the weed wrench and is best used for removing smaller plants, including shallow-rooted tree saplings or herbaceous plants.

Weed Wrench. A Weed Wrench™ (commonly referred to throughout this handout as just weed wrench) is a manually operated tool that clasps the base of the stem with opposable jaws mounted on a metal base plate, and a long handle that is pulled to extract the plant from the ground. There are four sizes and the largest model can extract plants with trunks up to 4 in (10 cm) in diameter. The weed wrench is often used on smaller woody plants, rather than perennial herbaceous species, which have deeper roots that can easily break off and resprout. It is much heavier and more powerful than a Pulaski or Root Talon. Weed wrenches are more effective when used on moist soil, as it is easy to remove the root compared to dry firm soils.

CUTTING TOOLS

There are many different cutting tools that are used to control non-crop weeds. Although they do not always kill the plant, they can reduce seed production, reduce above ground biomass, and slow the growth of plants. Many perennial and woody plants resprout quickly following cutting.

Loppers, Pruners, Machete, Saws, Spades, etc. Most cutting tools are lighter, less expensive, and smaller to carry compared to the previously described hand removal tools. For this reason, they are restricted to stem diameters that are an inch or less. When used at the right time of year, control can be fairly effective. Some studies have shown that the best time to cut plants is from the beginning to the middle of the dry season when resprouting is limited due to lack of soil moisture. Such cutting techniques can be used to clear dense brush or cut away individual plants. Anvil-type pruners are better than scissor-type or bypass pruners, as they can tolerate more twisting and physical abuse. Folding saws can also be useful and are very light weight. The triple-bevel tooth design on the blade is very efficient for cutting wood. Good saws will have impulse-hardened teeth that last a long time. Some plants can be cut below the ground surface with a flat-nosed spade. This works well for taprooted weeds, especially when all the shoot buds are removed. The best way to use a spade is to push it as far below the ground as possible before cutting and sever the plant below the root crown.

Girdling. Girdling is often used to control trees or shrubs that have a single trunk. It is most effective on woody species that do not resprout from the base. Unfortunately, most of the problematic woody plants do resprout from the base and are not controlled by a girdling treatment. Girdling requires that the bark be cut as a strip along the entire diameter of the trunk. The cut must be deep enough to sever the vascular cambium, which is the living tissue of the tree. Sometimes parallel cuts can be made about 3 in apart along the circumference of the trunk. The cut can be made using a knife, axe, machete, or saw.

Shredding. Shredding can be used with heavy equipment, such as masticators, in terrestrial areas, or with equipment such as Amphibious Terminator, the AquaPlant Terminator, and the Cookie Cutter in aquatic environments. Shredders chop up plants, such as water hyacinth, into fragments too small to reproduce. In contrast to harvesting, where plants are removed from the water mostly intact, shredding keeps the chopped fragments in the water to decompose. This costs less than harvesting but can maintain a high nutrient load in the water that can result in algae blooms or rapid recovery of missed plants. Shredding in the spring causes fewer effects on water quality because the plants are small early in the growing season. Another downside of shredding is that the many fragments of hyacinth left behind may regrow.

Tillage. Tillage kills annual plants in one of two ways. Shallow tillage using knives, sweeps, harrows, shallow disking, etc. detach roots from shoots and causes the plants to desiccate. This method should always be used in dry soils so that the fragmented plants do not produce adventitious roots from the nodes. The best timing is when the soil is dry, but new seeds have not yet been produced. The second tillage method is deep plowing. This kills plants by burying them deep in the soil profile. Plowing is used in agricultural systems but is rarely used in natural areas. The use of shallow tillage practices is also not common in natural areas, but is occasionally used in restoration sites to give newly seeded desirable species better conditions to establish. However, tillage causes severe disturbance which can facilitate invasion of other weedy species. Thus, it is best used in already highly degraded systems. Tillage is not generally used to control perennials, particularly those with rhizomes or creeping roots. With these species, tillage will spread the plant fragments around and exacerbate the problem. Many types of tillage equipment are available. Equipment ranges from small hand-pushed models, to tractor mounted power-driven tillers. The appropriate model depends on the size and type of the habitat.

CULTURAL CONTROL

SOIL SOLARIZATION

Solarizing the soil requires the use of clear plastic placed over moist soil under high sunlight and warm conditions in the late spring or summer. It is a common practice in very high cash crops, such as strawberries, but is not often used in natural areas. The clear plastic (typically polyethylene films) allows the sun's radiation to penetrate and heat the soil. However, the heat becomes trapped and cannot penetrate the plastic. This leads to high temperatures in the soil which can kill plant seeds and other microbial organisms or insects, but only to a couple of inches deep. Since most weed seed germinate near the soil surface, it can greatly reduce the soil seedbank. Because of its effect on many organisms, soil solarization can have a multi-year impact on biological, physical, and chemical properties of the soil and this can impact the establishment and growth of native species.

The technique is most effective against cooler season annual species, and much less effective on perennials or even warm season annuals.

PLASTIC MULCHES

As opposed to clear plastic, black plastics do not heat the soil to any degree, but they do block light which can also suppress the establishment or germination of weeds. Like solarization, the technique is only used on small areas primarily to control annuals, though it has proven successful on some herbaceous perennial species.

GRAZING

Overgrazing or improper grazing has been well recognized to increase weed problems in rangelands. When used properly, however, grazing can be an effective way of suppressing invasive or weedy plants. Grazing is often a desirable technique in areas that are excessively large and too expensive to apply an herbicide or where herbicides cannot be applied due to label restrictions or various policy limitations. Although there are some drawbacks to grazing, including the problems associated with overgrazing, spreading weed seed, and creating soil disturbance, there are also many benefits. For example, light timely grazing can reduce annual invasive grasses around vernal pools and heavy short duration grazing can break up soil and litter layers and expose the soil for germination of more desirable native plant species.

Animal Choice. The importance of choosing the right livestock type for managing a particular weedy species was already discussed. Cattle, horses, goats, sheep, and even geese may be used to control weeds. For example, horses can graze some invasive species, but tend to be more selective feeders compared to cattle or sheep. In addition, horses are often more susceptible to some poisonous species, such as yellow starthistle and Russian knapweed, compared to other animals. Cattle tend to prefer grasses, but will readily feed on broadleaf species, provided they can wrap their tongues around the plant and pull them out. Sheep and goats prefer broadleaf herbs and are better adapted to grazing in steep and rocky terrain. Sheep, in particular, do not graze an area as uniformly as cattle. Thus, various types of herding or fencing techniques are used to concentrate the animals for more even foraging. Goats eat a greater variety of plants than other livestock and are browsers that often stand on their hind legs to feed. This allows them to more effectively control woody species. It is important to note that some seeds can pass through the digestive system of animals intact and viable. This can lead to dispersal of seeds. To prevent this, animals should not be moved to weed-free areas for about 5 to 9 days after feeding in an infested area so that any viable seeds can pass through the system. As a preventative measure, it is best to graze weedy populations before they produce viable seeds. Predation can impact grazers. Sheep, goats and geese are more susceptible to predators compared to horses or cattle.

Timing. As previously discussed, animals should be grazed for control of a specific weed at the time when the target plant is most susceptible to damage. This is generally before seed production and when soil moisture is low and the plant is least likely to recover. For example, medusahead can be suppressed when grazed just before they exude their flowering head, but no control is achieved with grazing too early in the season. While this gives the greatest opportunity for management, it may not be the time of the season when the nutrient level is at its peak or when the plant is the most palatable.

Duration and Intensity. The duration, stocking rates, and intensity of grazing can be critical to the success of a weed management system. For example, when grazing intensity is too low, animals are likely to feed on preferred species, which may not be the target plant. This is particularly true for species that are less palatable or a somewhat spiny. Grazing too heavily can damage all the desirable species that might otherwise recover following lighter grazing practices. To achieve the proper intensity and duration may require a rotational grazing strategy, where the animals can be grazing at short duration and high intensity from one infestation to another. Rotational grazing often requires special temporary fencing techniques, particularly portable electric fences. While this is much easier for goats and sheep, larger animals, such as cattle and horses are difficult to contain with this type of fencing, unless it is built taller, sturdier, and with more wires. Of course, this can also be more expensive. Salt licks have been used successfully to concentrate animal impact in a particular area. For goats, it is possible to tie the lead animal to a tether to prevent the other animals from leaving a specific site.

FLAMING

Flaming for control of non-crop weeds is very different from prescribed burning (described in more detail below). Flaming is conducted in the cooler moister time of the year and never during the hot and dry season. In fact, flaming can even be conducted while it is raining. The principle of flaming is to quickly heat the plant and destroy cell integrity. This causes cellular leakage and death of the tissues. Flaming does not burn the plant or cause ground disturbance by catching the vegetation on fire. The only visible evidence of change is that the leaves deepen slightly in color, then appear waterlogged, and finally wilt. Because flaming works as a contact treatment, it is only effective on small annuals or seedlings of perennials and even woody plants. More specifically, it can kill plants that have all of their growing points exposed to the heat of the flame. Plants, such as grasses, are harder to kill because their growing points are at the soil surface or perhaps even protected beneath the soil a short distance. The best timing for the control of weeds using flaming is in the winter or early spring after they have germinated (cotyledon to 5 leaf stage), yet conditions are still moist.

While flaming requires only a single fairly quick pass, it is a slower process compared to herbicide treatments. In addition, the user must take great care in handling the equipment, as the flame is extremely hot and can cause severe personal injury if the user makes direct contact with the heat source.

Although more expensive, flaming can be used repeatedly to control perennial species by eventually exhausting the plants' reserves. The principle is similar to repeated mowing or cutting and requires that the application timing coincides with low below ground carbohydrate reserves. To achieve this, flaming should be repeated about every 2 to 3 weeks. This may not always be possible in some areas where conditions do not remain wet enough for an extended time period.

There are two types of flaming equipment, based on the way fuel is delivered to the torch. In Vapor Withdrawal Systems the fuel, liquid propane gas (LPG), is converted to a gas in the tank, and travels through the hose to the torch as a gas. When using the larger size vapor torches, the expansion of the liquid fuel to gas causes the tank and the torch handle to get very cold. Gloves are recommended. When flaming in wet or cold weather, the torch handle and the cylinder may frost up over time. Since wet and cold conditions are conducive for flaming in wildlands, vapor systems are best suited for intermittent use. Icing can be reduced by using a smaller size torch, or by using a larger fuel cylinder (although this is heavier to carry).

In Liquid Withdrawal Systems the fuel is delivered as a liquid all the way to the tip of the torch, and is vaporized there. Liquid systems are not as susceptible to icing up, which makes them better suited for continuous use. Both the cylinder and the torch are made specifically for liquid withdrawal. It is important to note that a liquid cylinder cannot be used with a vapor torch, or a vapor cylinder with a liquid torch.

Small torches with capacities of 50,000 to 100,000 BTU are suitable for spot flaming or small jobs. They are usually used with a 10 or 20 lb cylinder which can be easily carried in one hand. A set-up is also available which incorporates a 10 lb cylinder into a backpack frame, which is especially useful in difficult terrain. Torches are available with capacities all the way up to 2,000,000 BTU, but torches with capacities over 750,000 BTU are intended primarily for burning off areas of dense vegetation. For flaming large areas on fairly level ground, a liquid system with a 40 lb cylinder mounted on a hand dolly is a good choice. Larger cylinders are available and can be mounted on trailers, tractors, or ATV's. Large torches allow the user to work faster, but are more difficult to control around non-target species. In addition, they use more fuel. A 100,000 BTU torch uses 2 to 3 lb/hr, and a 500,000 BTU torch uses about 10 lb/hr.

PRESCRIBED BURNING
Fire can be a very valuable tool for stimulating desirable native species, particularly perennial grasses and legumes. It can also be used to remove the thatch layer, recycle the nutrients, increase light exposure and heat the soil earlier in the season. All these factors can stimulate germination of more desirable plants. The use of fire as a control tool should depend on its effectiveness on the target species, the timing of application, and its effect on the desired vegetation. In other words, will a prescribed burn achieve the land management objectives? Thus, it is critical to know the effects of a prescribed burn to ensure that it does not increase the population of the weedy species, lead to the establishment of new invasives, reduce a desirable species, impact sensitive animal or insect populations important to ecosystem function, or cause other environmental problems such as loss of topsoil through increased wind or water erosion. These can be influenced by weather, topography, available fuel, and timing of the burn.

The timing of a prescribed burn can be critical to its success. For example, invasive annual grasses (e.g., medusahead, barb goatgrass, ripgut brome) must be burned before they initiate seed drop and yellow starthistle should be burned before its seeds become viable. Other burns designed to control seedlings of woody plants should be conducted early in the season when sufficient fuel is available, but the target plants are not too large to survive the burn.

When possible, restoring the historic fire regime is likely to be most conducive to the competitiveness of native vegetation. This may not always be possible due to many factors associated with landscape changes, land use requirement, proximity to urban areas, and other fire policies within an area. In some cases, for example with medusahead, repeated burning is the most effective control option, but it difficult to accomplish because of fire restrictions.

Most prescribed fires can be conducted with the help of local or state fire agencies. These can be part of a training program or a larger more developed fire program. As such, the cost to conduct prescribed burns can vary dramatically from one area to another.

Some weedy or invasive species are not controlled by fire and, in fact, their germination is stimulated with a prescribed burn. In these cases, it may be necessary to integrate a burn with other control options, such as an herbicide treatment, to ensure that the plant does not become a major problem in subsequent years. This is particularly true when fuel loads in the second year are too low to carry a fire. The same principle is true when an unplanned catastrophic fire occurs in an area. Land managers may have to implement a subsequent treatment to either prevent new invasions or suppress weedy species that are adapted to fire. In some cases, an accidental burn can give control of an invasive plant in that area. Land managers should take advantage of these situations in the following years to further suppress the weed by employing other control techniques. It is also important to note that staging areas for equipment in large catastrophic fires often have vehicles from all over the western United States. It is not uncommon for these areas to develop new invasive weed problems the following season. All equipment, trucks, engines, tools, clothes and should be cleaned of weed seeds before entering or leaving the site.

BIOLOGICAL CONTROL

Biological control agents almost never eradicate weedy or invasive plants. It is hoped that biocontrol programs can reduce the impact of the weed to a level that is below the threshold of economic damage or harm to the ecosystem. Although there are some risks associated with biological control (as previously discussed), it is typically viewed as an environmentally friendly and economically sustainable way to control pest organisms. Worldwide, there have been well over 1000 intentional releases of over 400 species of biocontrol agents targeting about 150 weeds in at least 75 countries. Most of these are targeted against invasive plants of rangeland, forests, wildlands, and aquatic areas. This is primarily because these systems require more selective control of a specific plant compared to agricultural systems that may have many different weedy species. In the western United States, about 109 insects and three rust pathogen biocontrol agents have been released to control 46 plant species [see Table 14 (p. 513), Table 15 (p. 518), and Table 16 (p. 519) on biocontrol agents and target species in the western United States)]. Of the target species, seven are aquatic and 39 are terrestrial. The grass carp is also considered a biocontrol agent, but is much less host specific compared to both insects and pathogens. It is used to target several species of aquatic plants.

Nearly all the biological control programs in the western United States are considered classical biological control, where one or more species of biocontrol agents is introduced from the pest's native range. Augmentative biocontrol approach uses mass releases of predators, herbivores, or pathogens that are already present but whose effects on the target are normally limited by their ability to reproduce and spread. Because the bioagent either fails to establish or does not remain abundant enough to control the pest, they must be reared and released again each time the pest population erupts. Thus, new releases are made to increase the populations. While this approach is generally used to control insect pests, it can be used on rare occasions for plant pests. An example of an augmentative biocontrol program with weeds would be the grass carp program. Grass carp are first sterilized before release to ensure that they do not become problem species themselves.

Classical biocontrol programs often fail, primarily because the bioagents either never establish in the field, or they do not cause enough damage to the target pest to reduce the impact below a threshold level.

INTEGRATION OF BIOCONTROL WITH OTHER CONTROL METHODS

In addition to biological control, many other management techniques, including mechanical, cultural and chemical control options can be used to control invasive plants. Integration of these strategies with biological control can provide more effective long-term management of a particular weed. Herbicides do not generally impact insect populations or pathogens and, therefore, can be used without compromising the effectiveness of biocontrol agents. In some cases, particularly with aquatic weeds, the combination of sublethal herbicide concentrations and biological control agents can act synergistically. Other studies have used integrated management systems that combine biocontrol agents and prescribed burning. Although burning in spring or summer can kill exposed biological control agents, both insects and pathogens are mobile organisms and have the opportunity to readily reoccupy the treated site. In other situations, timely burning can increase available nitrogen or remove the soil litter layer that benefits the population growth of biocontrol agents. Likewise, mowing and other mechanical treatments can be timed or adjusted to enhance biocontrol. Even adding nutrients to an infested site may enhance the biocontrol effort. For example, adding nitrogen to nutrient poor waters infested by giant salvinia increased the nutritional quality of the weed and provided better nutritional value for the biocontrol agents. This resulted in increased insect fecundity and higher densities of the bioagent. Other examples show that biological control in concert with competitive desirable non-target plants can also improve the control of some invasive plant species and prevent subsequent seed set.

CHEMICAL CONTROL

Before using any control method, including herbicides, it is important to evaluate whether the technique will do more economic or conservation good than harm and not endanger the health and safety of the applicators or other key organisms in the environment. When applying herbicides, there are several factors that need to be considered. This includes the distance from the application perimeter to water bodies, depth to groundwater, susceptibility of non-target species to the herbicide, particularly rare and endangered species. Appropriate and realistic buffer zones should be established around sensitive sites. In many cases these are dictated by the specific policies of the area.

HERBICIDE MOVEMENT AND TIMING OF APPLICATION

Timing of herbicide applications can determine the effectiveness of the treatment. The best time to apply a specific herbicide will depend upon the pathway in which it is translocated and the target site where it acts. These same characteristics will, to a large degree, determine whether the herbicide is applied to the foliage, stem, or soil.

The rapid activity of contact herbicide precludes their ability to translocate much in plants. In contrast, systemic herbicides are slower acting, thus giving them the opportunity to translocate in the xylem (apoplast) or phloem (symplast) to sites not directly in contact with the spray solution. A contact herbicide moves very little in plants and kills only plant parts in close proximity to the point of chemical contact. Parts that are not contacted (e.g., roots) may, however, die because they are deprived of essential compounds obtained from contacted parts (e.g., leaves). For contact herbicides to be effective, adequate distribution of the herbicide spray solution over the foliage is essential. Contact herbicides are most useful for control of annuals, or perennials in the seedling stage, thus the timing of application has to be in early growth stages. In more mature plants and perennials, contact herbicides cause only temporary suppression and plants resprout quickly from protected stem buds or underground reproductive structures. Nearly all organic herbicides have contact activity. In aquatic systems, many herbicides have contact activity because they can come into close proximity with much of the plant surface without the need to translocate.

The apoplast, or non-living portions of the plant, is the primary pathway that water moves from the soil to the foliage. This includes the cell walls and the xylem. The driving force for this upward movement is the removal of water from the leaves by transpiration. Transpiration of water from the leaves acts as a wick, bringing more water and herbicides up from the roots. Although an herbicide may be considered xylem mobile, it must at some time enter the living tissues or symplast to exert its phytotoxic effect. Xylem mobile herbicides are usually soil-applied, accumulate in the leaves, and are photosynthetic inhibitors. Examples include hexazinone and tebuthiuron. Because these herbicides are primarily absorbed by the roots, application timing can depend on the soil mobility of the herbicide. For example, hexazinone is soil mobile and can be applied before the rainy season, where it will then leach into the root zone providing shrub control in reforestation efforts. Other xylem mobile herbicides are most effective on annuals when applied before germination. These compounds will be in the root zone as seedlings develop.

The symplast consists of the living tissues of the plant. This encompasses the network of connecting cytoplasms throughout the plant, including the phloem. It is primarily within the plant cell symplast that sugars move from photosynthetically active tissues, principally the leaves, to growing tissues including the apical meristems, expanding young leaves, rapidly elongating stems, developing flowers and fruits, and root tips. Sugars can also accumulate in storage tissues, particularly underground root crowns, taproots, tubers, corms, rhizomes and bulbs. Phloem mobile herbicides primarily move along the same pathway as sugars, although they also move to some degree in the xylem. Because sugars are primarily produced in the leaves, most phloem mobile herbicides are applied to the foliage and are active on both annuals and perennials, including woody species.

When a perennial or woody plant with underground reproductive organs is treated with a postemergence, phloem mobile herbicide, translocation of sugars and the herbicide to belowground parts is most rapid when large amounts of carbohydrates are moving toward the roots. This usually occurs after full leaf development. When perennial plants are beginning to emerge in the spring or woody plants are leafing-out, postemergence application of a phloem mobile herbicide will not provide effective control, as very little herbicide will translocate to the below ground vegetative reproductive structures. Thus, timing of foliar herbicide application is important for the control of perennial weeds, and generally, the most effective treatment time is in summer to early fall. Because these herbicides accumulate at the growing points and storage organs, they are used to control annuals and perennials, including woody plants. For this reason they are, by far, the most common group of herbicides used for the control of invasive plants.

ADJUVANTS

An adjuvant is any material added to an herbicide spray solution to enhance or modify the performance of the solution. Adjuvants include surfactants, stickers, extenders, activators, compatibility agents, buffers and acidifiers, deposition aids, de-foaming agents, thickeners, and dyes.

Adjuvants can be divided into two general categories: spray modifiers and activators. Spray modifiers are those adjuvants that change or modify the surface tension of the spray solution. Activator adjuvants primarily influence the absorption of the herbicide by a direct interaction with the plant cuticle, a thick, waxy layer present on leaves and stems of most plants. The most common types of adjuvants are surfactants (the word was derived from "surface active agents"). Another commonly used adjuvant type is oil. In addition to oils being used as adjuvants they can also be used as contact herbicides (particularly organic materials) and as carriers for synthetic herbicides. A third adjuvant type is ammonium sulfate, which is a salt that is commonly used as a fertilizer. Ammonium sulfate acts like a buffer to bring the pH of the water solution down. This will allow the sulfate to precipitate out the divalent and trivalent cations, such as magnesium, calcium, and aluminum. These cations can react with glyphosate to cause the herbicide to precipitate out of the solution. When the ammonium sulfate solution is added before glyphosate, it will remove these cations and increase the activity of glyphosate.

Surfactants. Surfactants are the most important adjuvants. They facilitate or enhance the emulsifying, dispersing, spreading, sticking or wetting properties of liquids (spray modifiers). In addition, they can directly influence the absorption of herbicides by changing the cuticle characteristics of plants (activators). Simply put, surfactants are capable of binding to two types of surfaces, such as oil and water. Some herbicide formulations come with a surfactant already added, whereas in others formulations, surfactants can be added before application. Whether a surfactant should be added will be determined by the type of herbicide being applied and the target plant. Herbicide labels often give recommendations for appropriate surfactants. For example, the aquatic formulations of glyphosate do not contain a surfactant, though one is needed to get maximum effect of the herbicide. The label will recommend surfactants that can be used with these aquatic formulations.

An important aspect of surfactants is to reduce the surface tension of a spray solution to allow better contact between the spray droplet and the plant surface. This will enhance the ability of the herbicide active ingredient to absorb into the plant tissues. Surfactants achieve this by:

- causing more uniform spreading of the spray solution and uniform wetting of the plant;
- helping spray droplets stick to the plant, resulting in less runoff;
- allowing droplets to penetrate hairs, scales, or other surface projections;
- partially solubilizing the waxy plant cuticle substances;
- preventing crystallization of the herbicide on the leaf surface by keeping it in solution.

Organosilicone surfactants cause substantially greater spreading of the spray drop than typical surfactants. This can be good with plant surfaces that are hard to cover, but can also reduce activity if the spray tension is reduced so much that the spray solution drips off the plant surface.

Oils. Oils used in agriculture are of two primary types: refined mineral oil (petroleum-based) and seed oils (sometimes called vegetable oils). Mineral oils are specific "cuts" from the distillation of petroleum. These products, such as diesel, are not recommended today as they are far more toxic and cause more environmental damage compared to methylated or ethylated seed oils. Seed oils are commonly mixed with emulsifiable concentrates when applying herbicides as a basal bark treatment to woody plants. Seed oils can be more viscous in colder temperatures and may not be as useful as petroleum oils for basal applications during cold weather.

Ammonium sulfate. Ammonium sulfate used as an adjuvant can help overcome decreased herbicidal activity due to antagonism with another component in the spray solution. Calcium, sodium, potassium, and magnesium salts in the water used as a spray carrier can reduce the efficacy of many herbicides, most notably glyphosate. When glyphosate is added to high pH (alkaline) water with high levels of salts, particularly di- or tri-valent cations such as calcium, magnesium, and aluminum, the herbicide readily precipitates out and is no longer active. Adding ammonium sulfate to the spray solution before the herbicide is added reduces the pH of the solution, causing the sulfate to bind with the cations and to precipitate. When glyphosate is subsequently added, more of the herbicide stays in solution, giving increased activity.

Dyes. Dyes are often recommended as an adjuvant in spray solutions. Though they can be somewhat messy, they let the applicator know what areas or individual plants they have already treated. This can greatly prevent double application to an area or skips in treatment. Furthermore, the addition of a dye has a tendency to increase the safety of the applicator, as the dye is easily noticed on equipment or clothing when care is not taken. Some premixed herbicides include a dye (e.g., *Pathfinder II*® includes the active ingredient triclopyr, a surfactant, and a dye). Ester-based herbicides like *Garlon 4 Ultra* require oil-soluble dyes. Dyes are inexpensive and available at most garden stores.

APPLICATION METHODS

Herbicides can be applied to the soil, directly into the water column, or on the foliage or stems of plants. Once applied, herbicides must enter the plant and move to the tissues where they exert their biochemical activity. Entry of an herbicide into a plant is generally through the roots or leaves, depending on where the herbicide is applied. The tissues where the herbicide is most easily absorbed, the pathway of translocation, and the site of action determine whether a compound is applied to the soil, stem, foliage, or directly into the water column (some herbicides).

As previously discussed, application of herbicides can be as broadcast treatments over large areas, directed or spot applications to individual plants or small patches, or by specialized wicking techniques. Treatment can be made through aerial, boat or ground equipment, or by individuals carrying backpack sprayers or wicks.

Postemergence foliar treatment is one of the most common herbicide application techniques. This application is made to the leaves of newly emerged annual seedlings or to larger established herbaceous perennials and even some woody species. Absorption through the waxy cuticle is the most important means of herbicide entry into the leaf. Although its primary function is to protect leaf surfaces from water and gas loss, it also acts as a significant barrier to the penetration of water-soluble herbicides. Most herbicides are polar (amine or other salt formulations) and, thus, are less likely to penetrate the waxy cuticle. In these cases, a surfactant is often added to, or formulated with, the herbicide to increase uptake. Some postemergence herbicides are developed as non-polar ester formulations that can easily penetrate the cuticular waxes. Herbicides such as 2,4-D, triclopyr, and imazapyr are available in both salt and ester formulations, each being used in different habitats, or with different timings and application techniques.

Postemergence herbicides can be applied as broadcast treatments or directed (spot) applications to control early weed invasions or to prevent the spread of small infestations. Broadcast treatments are best in areas with large infestations and few, if any, sensitive/susceptible non-target species present. In many cases, an herbicide may impact a non-target species, but that species will recover over time. They can be made aerially with fixed-winged aircraft or helicopters, ground applicators using a variety of vehicles, or boats in aquatic settings. Directed or spot treatments have the advantage of selectively removing individuals or patches of targeted invasive species and although they can be more labor intensive and expensive, they tend to use less total herbicide per area compared to broadcast applications.

Although aquatic herbicides are generally applied by broadcast methods, including aerially or by boat, granular formulations can be used in a similar manner as a spot treatment, as the pellet is slowly dissolved in the target zone where the weed is present. In terrestrial systems, directed treatments are generally applied with backpack sprayers. These foliar treatments are directed at the species of interest.

Directed applications can be applied using different techniques. Spray-to-wet treatments use high spray volumes, but relatively low herbicide rates. They require better coverage compared to low volume treatments, which use higher concentrations of herbicide. Low-volume treatments require more precision, but can save labor because they require fewer refills of the backpack sprayer.

Another recently developed application technique is the drizzle method, in which a fine spray stream is applied with a spray gun. Unlike spray-to-wet applications with small evenly spaced droplets, a drizzle application deposits a few large, sparsely distributed droplets on the plant leaves. It is similar to the low volume method, using high herbicide concentrations but reducing labor costs with fewer reloadings of the sprayer, again allowing workers to cover more ground in less time. As another advantage, the spray stream reduces drift and allows the applicator to treat invasive plants on steep banks, or areas with limited access.

It is also possible to achieve selective control of a particular invasive plant with otherwise non-selective or relatively non-selective postemergence herbicides by employing a wick applicator. These can be either hand-held or vehicle-mounted boom wipers, usable either for spot treatments or broadcast applications. As a benefit, this application method reduces the potential for herbicide drift and injury to adjacent sensitive areas. With this technology, herbicides are applied at high concentrations directly to the foliage. Hand held wicks can target individual plants, while selectivity using boom wipers relies on a height difference between the taller weedy plant and the lower growing desirable vegetation.

Preemergence treatment is the application of an herbicide to the soil before the emergence of the seedling or, in some instances, adult perennials. Although a preemergence herbicide can be absorbed by emerging shoot tissues, the primary route of entry into the plant is through the roots. Preemergence treatments are typically broadcast applied using liquid formulations. On occasion, granular herbicide formulations can be applied aerially, particularly in forested or woody areas or in areas where herbicide drift can be a problem. Preemergence herbicides can also be used to make spot treatments to smaller incipient infestations. Preemergence herbicides require incorporation by rainfall to reach the root zone of the target plant.

Although many herbicides are only effective when applied postemergence or preemergence, there are a number that can be both active in the soil or with foliar application. This is particularly true for herbicides that are used to control many invasive herbaceous plants. These compounds generally give postemergence control of seedlings and rosettes, as well as soil residual activity for at least a couple of months until spring rainfall is completed. Such compounds include aminocyclopyrachlor, aminopyralid, clopyralid, picloram, chlorsulfuron, metsulfuron, imazapic, and imazapyr.

FOLIAR TREATMENT TECHNIQUES

These methods apply herbicide directly to the leaves and stems of a plant. An adjuvant or surfactant is often needed to enable the herbicide to penetrate the plant cuticle. There are several types of foliar application tools available.

Boom Applicators. A boom applicator can be mounted on an airplane, helicopter, or vehicle, but can also be hand carried. When mounted on an ATV, tractor or other vehicle they can broadcast apply herbicides over a large area quickly. Boom applications are rarely made as spot treatments, but can be used to treat patches of weeds. Hand held boom equipment has a spray bar in the shape of a "T" and typically has between two and five nozzles. They can spray widths of 4 to 10 ft. Boom applications almost always need to be carried above the height of the weed population for proper coverage. This application method is more susceptible to vapor or spray drift compared to spot applications. In addition, unless a very selective herbicide is used, such applications are more likely to cause damage to susceptible, non-target plants.

Specialized boom applicators, such as the Boominator or Boom Buster, use a single nozzle but put out a wide and uniform spray pattern. They can cover large areas with very even and large droplet size over uneven terrain. The advantage is that they have a lower risk of drift. They can be hand held or attached to an ATV. Since they are broadcast applications, they are more likely to cause non-target plant damage unless a selective herbicide is used.

Single Nozzle Applicators. Most single nozzle applications are used for spot or directed treatment. They generally include backpack sprayers, but on occasion can employ a spray bottle. Because the spray solution is directed at the target species, this technique can use a non-selective herbicide selectively. Crook-necked squirt bottles and similar equipment can be ordered from laboratory supply companies and are easy to carry over distances and through dense vegetation.

Drizzle Applicator. The drizzle application method has been widely used in Hawaii for selective control of invasive plants in tropical plant communities where there are many species intertwined and it is difficult to move far off the pathway. The method employs a backpack sprayer and a spray gun fitted with an orifice disk (0.5 mm). At 20 psi, the spray gun shoots a fine stream 20 to 30 ft that breaks into large droplets when it contacts the plant. Unlike foliar applications, where the entire plant is sprayed-to-wet with small evenly spaced droplets, the drizzle technique is comprised of a few large, sparsely distributed droplets on the plant leaf surface. The drizzle technique applies herbicide at considerably lower volumes than foliar applications, ranging from 2 to 5 gal/acre. Because of the low volume, the spray solution should contain a high concentration (10 to 20% herbicide product in the spray solution) of the herbicide. Typically, glyphosate, triclopyr or imazapyr are used for this treatment and the herbicide is mixed with about 20% seed oil in water to aid in penetration. As the gun is moved from side to side, the spray stream breaks up into large droplets, which "drizzle" on the plants.

This application technique greatly reduces labor costs, including the number of required spray tank refills. In addition, the long spray reach decreases walking time to target plants allowing the applicator to cover more ground. Furthermore, it also increases applicator ability to treat plants on steep banks or in areas with limited access. While it does not provide complete safety from drift or overspray, the risk is considerably lower than conventional foliar treatment with smaller droplets.

Wick Applicators. Wick applicators can be vehicle-mounted or hand-held. Typically, glyphosate is used at between 33 and 75% of concentrate with wick applications. Vehicle-mounted wicks can be adjusted in their height above the ground to obtain selective control of taller undesirable vegetation. When there is no height differential among species, selectivity can be lost. Hand-held wicks have a reservoir in the long handle that can be filled with the herbicide solution. The base is made of a material similar to a paint roller. The roller-like material is wiped onto the weedy annuals or perennials. For large bunchgrass species, the base of all tillers should be contacted by the herbicide to get effective control. Use of a wick eliminates the possibility of spray drift or droplets falling on non-target plants. However, herbicide can drip or dribble from some wicks. Hand-held wicks can easily be obtained on the internet or in many garden shops.

Some techniques that are variations of the wick application include dipping a heavy cotton glove, worn over a thick rubber/latex (or nitrile) glove, directly into an herbicide solution and then applying the herbicide by grabbing the parts of the target plant were the treatment is intended. This allows for a high level of precision with little to no runoff or drift.

STEM TREATMENT FOR WOODY PLANTS

The three most common methods for treating large woody invasive plants include cut stump, stem injection and basal bark. Each of these techniques has its advantages and disadvantages. Not all woody species respond the same to either these treatment techniques or the herbicides used in their application. All three methods are used only on woody plants that resprout when cutting is used alone. The best treatment timing for all three methods is generally in mid-summer to early fall when photosynthates are translocating at their maximum rate to the below ground reproductive structures.

Cut Stump. Cut stump treatments can be used for controlling shrubs with multiple stems to very large trees. The stems are cut down with either loppers (shrubs) or a chainsaw and a concentrated herbicide solution is applied to the outside cambial region of the cut stem immediately after cutting before wound healing begins. It is not necessary to treat the center of the trunk in large trees, as this is deadwood. For smaller trees of multiple stemmed woody species, it may not be possible to only treat the cambium region. For vines, small stemmed shrubs, or small trees, the entire stump can be treated. Stump treatments are most effective during periods of active growth with the goal of eliminating resprouts from the base of the stem. In areas where aesthetics are important, cut stump treatments are desirable because stems can be completely removed with no dead material left on the ground after treatment. The other advantage to this method is that no follow-up work is typically needed.

The herbicides most typically used for cut stump treatment include glyphosate, imazapyr and triclopyr. They do not work equally on all species, and without a specific study on individual species, it is hard to predict which of these herbicides, if any, will be most effective. The method of herbicide application can vary. Some use a single nozzle sprayer, others a squirt bottle, or even a paint brush. The herbicide should not be allowed to run off the stump to the ground. In many cases, it is not possible to get 100% control, and some resprouts are expected the following year. Depending on the vigor of the resprouts, these can be treated by cutting, basal bark applications, or foliar applications.

One of the advantages of the cut stump treatment is that very large trees can be controlled and the technique is very selective. In addition, relatively little herbicide is required. The disadvantage is that all the operations much be conducted at the same time, including cutting, treatment, and generally removing the cut material to prevent a fire hazard.

Basal Cut Stump. This application technique combines cut surface and low volume basal bark methods. Low volume basal bark mixtures are used by applying the spray to both the cut surface and to the bark or collar region of cut stumps. See the low volume basal bark section for details on the mixtures.

Stem Injection. In stem injection (or "hack-and- squirt"), a hatchet, axe or machete is used to make a series of downward cuts in the bark around the circumference of the tree trunk. For most species, it takes one cut for every 3 in of trunk diameter. About one milliliter of undiluted herbicide is added per incision immediately after each cut. The herbicide can be applied using a squirt bottle, syringe, or adapting veterinary equipment designed to inject antibiotics into livestock. Like the cut stump treatment, glyphosate or the water soluble formulations of imazapyr and triclopyr are used. Application timing must be in summer or early fall when sap flow is downwards towards the roots. This technique does not work on every species, but has been shown to be effective on some important woody invasives. A similar frilling technique can also be employed. This requires that the trunk of the tree be completely girdled so the cambium is exposed under the bark. Undiluted herbicide can be applied into the frills marks.

Some land managers use a cordless drill rather than cutting tools, particularly for trees with dbh (diameter at breast height) greater than 3 in. Drills and bits need to be sturdy and have enough battery power to last for prolonged time period. One hole should be drilled for each inch of stem diameter. Like a hack treatment, the drill angle should be downward and through the cambium to catch and hold the herbicide solution in the cambium for maximum uptake.

Other variations on the stem injection technique include the use of the EZ-Ject Lance. Herbicide pellets can be injected into the trunk of a tree using this 5 ft long metal tube. Though it is a convenient way of applying an herbicide, the technique is far more expensive than other stem injection methods.

The advantage of the stem injection technique is that the operation can be split into treatment in summer or fall, and cutting and removal in the following spring. In addition, little herbicide is used and it is possible to leave some dead trees as snags for wildlife use.

It may also be helpful to use spray paint to mark trees that either need to be treated or that were treated to save time. Alternatively, if personnel are separated into teams that either cut or drill the stem or a team that treats the cuts or holes, it may be helpful to spray circles around the cuts of holes so that the applicators can easily find them.

Basal Bark. In shrubs with multiple small stems or in smaller trees (stem diameter generally < 6 in), concentrated chemical can be applied directly to the basal 12 to 18 in of green stems using a basal bark application method. The basal bark treatment is not effective on older plants with thick corky bark. The herbicide can be applied with a backpack sprayer, hand-held bottle, or a wick, although a backpack sprayer is the fastest method. The most common basal bark application method is low volume for example using 20 to 30% herbicide triclopyr ester in a 70 to 80% basal oil and sprayed on the stem or trunk to a height of 12 to 15 in from the ground in a manner that thoroughly wets the lower stems, including the root collar area, but not to the point of runoff.

As with the other stem techniques, application timing must be later in the season when carbohydrates are moving to the below ground reproductive structures and can be used much of the year as long as the bark is not saturated with water or frozen. Because the herbicide has to penetrate through the thin bark layer to reach the cambium, herbicides in a concentrated form must be applied in an oil carrier, and only oil soluble formulations of herbicides can be used. The most common herbicide used in basal bark treatments is the emulsifiable concentrate formulation of imazapyr or triclopyr. Such formulations can be volatile; however, so basal bark treatments should be performed only on calm, cool days. Because glyphosate is not soluble in oil, it is not effective using this technique.

Basal bark treatments leave a dead standing plant. For wildland areas where snags are encouraged for wildlife habitat, this or the stem injection method can be a preferred control technique. In other areas where snags could result in a fire or health hazard, removal of the tree should be about one year after treatment.

Shrubs with an upright growth habit generally have more accessible stems at the base and are easier to treat. More dense shrubs or thickets are difficult to treat and the labor costs can be considerably higher. One advantage of a basal bark treatment is the very low risk of off-site movement in spray drift or overspray. This can be an important factor in sensitive areas where susceptible desirable species are in close proximity to the target plants.

FACTORS THAT INFLUENCE HERBICIDE SELECTIVITY

Most herbicides can also kill desirable vegetation under a particular set of conditions. What determines selectivity is the total amount of herbicide that reaches a sensitive metabolic site. An herbicide is selective to a particular vegetation type and/or plant species only within certain limits, typically determined by complex interactions between the plant, the herbicide, and the environment. Thus, control of weeds by selective herbicides is usually relative, not absolute. Because selectivity is typically rate dependent, it is important to use herbicides at the recommended rate. Lower rates may not control the weeds, and higher rates may result in loss of selective control and injure non-target plants.

Role of the Plant in Selectivity. Plant susceptibility to herbicides is influenced by the growth stage and rate, morphology, and genetics of each species or particular ecotype of a species. As a general rule, younger annual plants are easier to kill than older ones. For both annuals and perennials, rapidly growing plants are more susceptible to herbicides than slower-growing ones. For this reason, many herbicides tend to be less effective when applied under environmental stress conditions, where growth rates are reduced. In addition, height difference between an invasive species and desirable vegetation can be the basis for selectivity using wick applicators or directed sprays.

Plant morphology is also a common factor that can determine the susceptibility of a species to a particular herbicide. The location of the growing points can play a major role in the activity of different herbicides. The growing points of grasses and broadleaf rosettes are located at the base of the plant and are protected from contact herbicides by the surrounding leaves. These same plants may be more susceptible to soil-applied herbicides. Perennial plants have protected buds below the soil surface that can escape contact herbicide injury, whereas many seedling dicot species have exposed susceptible growing points at the top of the young plant and in the leaf axils.

Leaf features, such as the shape, angle, arrangement, surface area, hairs and wax thickness can influence the amount of herbicide that will contact the plant and be available for absorption. Spray solution tends to bounce or run off when applied to a plant with narrow vertical leaves. A broadleaf plant with horizontal, flat, wide leaves can more easily retain the spray solution. Plants with fewer overlapping leaves will contact less herbicide than plants with numerous non-overlapping leaves. A dense layer of leaf hairs can hold the herbicide droplets away from the leaf surface, whereas a thin layer of leaf hairs causes the chemical to stay on the leaf surface longer than normal. Waxy surfaces adhere and absorb less of the herbicide solution.

Genetic differences among species can lead to differences in the rate of metabolic degradation of an herbicide, potential for a species to develop herbicide resistance, binding characteristics at the enzymatic site of action, and the ability of a plant to absorb and transport the herbicide to the site of action.

Role of the Herbicide in Selectivity. Among the factors influencing selectivity, the amount of herbicide applied to an area is perhaps most critical to obtaining proper selectivity. At high enough concentrations, most herbicides will kill all plants. At sufficiently low doses, herbicides will not kill any plants. Slight variations in the molecular configuration of an herbicide will change its properties, which in turn modify its effect on plants, soil longevity, and rate of metabolic detoxification. Similarly, altering the formulation of the product (e.g., granular, ester, salt) or adding a particular surfactant can narrow or broaden the spectrum of species an herbicide will control.

By placing the herbicide in contact with the weedy plant, it is possible to achieve selectivity with an otherwise non-selective compound. This can be accomplished with both preemergence and postemergence herbicides. With preemergence herbicides, for example, some compounds are not inherently selective but may be made to function selectively because of their location in the soil. Such selectivity depends upon a difference in seed germination depth, seedling rooting habit between desirable and invasive plants, and herbicide mobility in the soil. With foliar-applied herbicides, this can be accomplished by using shielded sprays, directed sprays, or wick applicators. Treatment timing can also achieve the effect of placement selectivity. Non-selective foliar herbicides are commonly used selectively to control emerged annual plants before emergence of more desirable species or bud break in woody species.

Role of the Environment in Selectivity. Environmental factors often account for much of the inconsistency in the performance of herbicides. There are many environmental conditions that influence herbicide selectivity and efficacy, including temperature, humidity, and precipitation. High temperatures can reduce the amount of time that an herbicide will remain in solution on the leaf surface, increase vaporization of some compounds, contribute to co-distillation where the herbicide evaporates with the water, and increase the rate of soil degradation through enhanced microbial activity. Cold temperatures can have the opposite effect.

Too much soil moisture can increase herbicide leaching, while too little can slow the movement of the herbicide into the root zone of germinating plants. With foliar active postemergence herbicides, rain during or soon after applications may wash herbicides off the leaves and reduce uptake and effectiveness.

Any environmental condition that directly affects the cuticle has some effect on herbicide absorption of postemergence compounds. For the most part, maximum weed control occurs when a foliar-treated herbicide is applied under warm, humid conditions with adequate soil moisture. In contrast, minimum control generally occurs when plants are water-stressed at cool temperatures with low humidity. When plants are under drought stress, photosynthesis slows down and growth is reduced. This will lower translocation rates and reduce herbicide accumulation at the site of action. Other forms of stress, including nutrient, light, acidity, or temperature stress can also reduce the efficacy of herbicides on plants.

LITERATURE

Bossard, C.C., J.M. Randall and M.C. Hoshovsky (eds). 2000. Invasive Plants of California's Wildlands. California Exotic Pest Plant Council. University of California Press, Berkeley. 360 pp.

Brooks, M.L., C.M. D'Antonio, D.M. Richardson, J.B. Grace, J.E. Keeley, J.M. DiTomaso, R.J. Hobbs, M. Pellant, and D. Pyke. 2004. Effects of invasive alien plants on fire regimes. BioScience 54(7):677-688.

California Weed Science Society. 2002. Principles of Weed Control. 3rd Ed. Thomson Publications, Fresno, California. 630 pp.

DiTomaso, J.M. 2000. Invasive weeds in rangelands: species impacts, and management. Weed Science 48:255-265.

DiTomaso, J.M. 2006. Using prescribed burning in integrated strategies. Pages 19-27, In The Use of Fire as a Tool for Controlling Invasive Plants. DiTomaso, J.M. and D.W. Johnson, eds. California Invasive Plant Council, Publ. #2006-01, Berkeley, CA.

DiTomaso, J.M. 2006. Control of invasive plants with prescribed fire. Pages 7-18, In The Use of Fire as a Tool for Controlling Invasive Plants. DiTomaso, J.M. and D.W. Johnson, eds. California Invasive Plant Council, Publ. #2006-01, Berkeley, CA.

Duncan, C.L. and J.K. Clark (eds.) 2005. Invasive Plnats of Range and Wildlands and Their Environmental, Economic, and Societal Impacts. Weed Science Society of America, Lawrence, KS. 222 pp.

Madsen, J.D. 2000. Advantages and disadvantages of aquatic plant management techniques. US Army Corps of Engineers. Engineer Res. and Dev. Center. Publ. ERDC/EL MP-00-1

Moore, K. 2010. The weed worker's toolbelt. Cal-IPC News. 18(1):6-7.

Pacific Northwest Weed Management Handbook. 2010. Oregon State University. http://pnwhandbooks.org/weed/

Radosevich, S.R., J.S. Holt and C.M. Ghersa. 2007. Weed and Invasive Plant Management Approaches, Methods, and Tools. Page 259-306. In, Ecology of Weeds and Invasive Plants. 3rd Ed. John Wiley & Sons, Inc. New York.

Ross, M.A. and C.A. Lembi. 2009. Applied Weed Science: including the ecology and management of invasive plants. 3rd Ed. Pearson Prentice Hall, New Jersey. 561 pp.

Sheley, R. and J. Petroff (eds). 1999. Biology and Management of Noxious Rangeland Weeds. Oregon State University Press. Corvallis, OR. 438 pp.

Tu, Mandy; Hurd, Callie; Randall, John M.; and The Nature Conservancy, "Weed Control Methods Handbook: Tools & Techniques for Use in Natural Areas" (2001). All U.S. Government Documents (Utah Regional Depository). Paper 533. http://digitalcommons.usu.edu/govdocs/533

University of Nebraska. 2011 Guide for Weed Management in Nebraska. University of Nebraska Cooperative Extension. EC 130. http://www.ianrpubs.unl.edu/epublic/live/ec130/build/ec130%20table%20of%20contents.pdf

Vencill, W.K. (ed.). 2007. Herbicide Handbook. 9th edition. Weed Science Society of America. Lawrence, KS.

Weeds CRC. 2004. Introductory Weed Management Manual. CRC for Australian Weed Management, Adelaide, Australia.

Weed Management Handbook. 2006-2007. Montana State University Extension Service. http://www.invasive.org/weedcd/pdfs/wmh.pdf

Environmental Concerns and Safety

Nearly all weed control methods influence both the abiotic and biotic components of the ecosystem. These include various mechanical control techniques, cultural practices such as grazing, burning, flooding or revegetation, the introduction of biological weed control agents, and the use of chemicals. While it is possible to minimize these risks, it is important to recognize that all strategies pose some level of risk to the environment, as well as animal and human health.

MECHANICAL CONTROL STRATEGIES

HAND PULLING AND HOEING

Hand pulling and hoeing are labor intensive and relatively ineffective for the control of perennial weeds (with exception of the weed wrench). In addition, repeated access for follow up treatments can result in establishing trails, severe soil disturbance, and wildlife harassment. Soil disturbance around the removed plants can create an ideal site for reestablishment of new seedlings or rapid invasion of another undesirable species. In addition, trampling of habitat by large numbers of people, particularly volunteers, in these sites can damage sensitive native species and further disturb the soil. The potential also exists for serious physical injury when removing plants with spines, prickles, or razor-like or spiny leaf margins, particularly noxious thistles. In other cases, handling plants that contain toxins or skin allergens can expose an individual to their poisonous effects. These risks can be minimized by wearing appropriate protective clothing and gloves. Additional risks of using these techniques can include contact with venomous snakes, insects, and other wildlife. Repeated treatments are generally necessary when using hand pulling or hoeing techniques, and this can exacerbate the potential problems described above.

BULLDOZING AND OTHER HEAVY EQUIPMENT

Bulldozers and backhoes can be used to remove larger shrubs or trees, such as saltcedar (*Tamarix* spp.), from infested habitats. These types of heavy equipment dramatically alter soil structure, create large disturbed sites susceptible to re- or new invasions, and can have a negative impact on the associated animals and insects in the treated areas. Furthermore, heavy equipment increases soil compaction in the traveled sites and produces a considerable amount of exhaust from fuel consumption. Other motorized equipment, such as chainsaws, can present a significant human health risk. The use of heavy equipment is often very expensive, particularly when repeated treatments are necessary.

TILLAGE OR CULTIVATION

Tillage is more common to agricultural areas than to non-crop areas. On occasion, tillage can be used in rangelands in revegetation programs, and also along roadsides and utility rights-of-way. Tillage can utilize plows or discs that can control annual weeds by burying plant parts. This is more effective on annuals than perennials. In contrast, harrows, knives, and sweeps can be used to damage root systems or to separate shoots from roots. This technique must be applied when the surface soil is dry, or fragmented plant segments will regrow and possibly amplify the problem. While a single till will often control annuals, repeated tillage is necessary for the control of many perennials. Despite its effectiveness in the control of annual weeds, tillage can expose soil to both wind erosion, under dry conditions, and water erosion, under extremely wet conditions. It addition, it can alter soil structure, prolong the longevity of noxious weeds by burying seeds deep in the soil profile, and in many cases, enhance a perennial weed problem by spreading rhizome fragments or stimulating emergence of new shoots from underground stems. Heavy equipment produces fuel exhausts and can increase the atmospheric discharge of soil particles, commonly referred to as PM10 (particulate matter < 10 microns in diameter).

MOWING

Mowing is a popular control technique along highways and in recreational areas. It has less impact on the environment than tillage. However, like tillage, it can also lead to fuel exhaust and increase in PM10. In this case, the particles are very small plant fragments, often detached hairs. On rocky ground, mowing can present a significant fire hazard when the metal blades create sparks from contact with rocks or other hard objects.

Perhaps the greatest risk with mowing is the impact on plant composition. Proper timing can minimize these risks, whereas mowing at the wrong time can increase noxious weed populations. Mowing can also be detrimental to desirable insect populations, including biological control agents that also target the same plant.

CULTURAL CONTROL METHODS

GRAZING

Grazing with cattle, sheep or goats can be an effective method of controlling noxious weeds. The success of this method often depends upon proper timing. Intensive overgrazing has also led to the invasion of many of our rangeland weeds. In some cases, grazing can select for a particular weed or group of weeds, particularly poisonous plants or plants armed with spines or prickles. In contrast, grazing can be very non-selective and may endanger sensitive non-target species. Goats are typically browsers and can effectively control certain noxious species. However, when confined they can intensively forage both desirable and undesirable species and may even strip the bark off trees. Livestock can also trample desirable sensitive species and can spread noxious weeds over a wide range when seeds become attached to hair or when they remain intact after passing through the digestive system.

COMPETITIVE RESEEDING PROGRAMS

Reintroducing competitive species into infested non-crop areas as part of a control program is essential to the sustainable management of noxious weeds. In the most desirable cases, competitive, endemic, native species should be reestablished. However, in most cases, particularly rangeland environments, endemic native species do not appear to be capable of outcompeting noxious weeds. Even in areas where natives are reintroduced, they may not be genotypically endemic to these habitats. In addition, once established, many of these species, especially the perennial grasses, develop into near monocultures. This can have a dramatic impact on total plant and animal diversity within these sites. Finally, successful reseeding programs often utilize drills attached to tractors. Like tillage and mowing, this process produces exhaust fumes and can lead to increased soil compaction.

BURNING

Four major risks are associated with prescribed burning as a method of controlling non-crop or natural areas weeds. First, air quality issues, including PM10 emissions, can be a significant problem when burns are conducted adjacent to urban areas. This potential problem can be avoided or by conducting burns only in more isolated regions not adjacent to populated sites. Public relations problems can be minimized by educating residents of the intended goals of the project before the burn. Second, a major risk of prescribed burning is the potential of fire escapes. This is particularly true when burns are conducted during the summer months. This can be minimized by proper preparation and thorough involvement of local or state fire departments. Third, perhaps the most overlooked risk of burning is the impact fire may have on small animals and insects unable to escape the burn. Finally, burning may increase soil erosion and impact the plant composition within a site. Species that complete their life cycle before the burn will be selected for, while those with later flowering times will be selected against.

BIOLOGICAL CONTROL

Biological control is typically considered to be environmentally safe, energy self-sufficient, cost-effective, and often self-sustaining. Unfortunately, far less than half of the biocontrol efforts in the United States have demonstrated some level of success. Despite the overwhelmingly positive aspects of biocontrol, some risks do exist. Most biocontrol agents introduced to the United States are native to other continents. Although we often study the host specificity of these organisms under quarantine conditions, little is known of their impact on the ecosystem as a whole, including other insect populations. In some cases, accidental introductions of pathogens or insects can occur when biocontrol agents are released.

In addition to the risk of damaging desirable non-target species, weed biological control agents pose other potential risks. Grass carp (*Ctenopharyngodon idella*), native to the Amur River of China, is an herbaceous fish that provides excellent control of aquatic weeds, particularly hydrilla (*Hydrilla verticillata*). Each individual fish can eat enough aquatic plants to grow 3 to 5 lb per day and adults may weigh as much as 100 lb at maturity. However, the presence of this bottom feeding species in aquatic environments can reduce water quality. In addition, their rapid growth rates often lead to crowding out of desirable game fish and elimination of protective cover for young fish. As another example, the endangered willow flycatcher, *Empidonax trailii extimus*, uses saltcedar to nest in areas where willows have been displaced. The release of *Diorhabda* spp., leaf feeding beetles specific for saltcedar control, is still of concern because of the threat it has on flycatcher populations.

CHEMICAL CONTROL

Herbicides are the most widely used method for controlling weeds, both in agricultural and non-crop environments, and are generally considered the most economic and effective. The potential risks associated with herbicide use have been widely publicized both in the scientific literature and the public press. Although these risks are often greatly exaggerated, improper use of herbicides can lead to several potential problems, including spray or vapor drift, water contamination, animal or human toxicity, selection for herbicide resistance in weeds, and reduction in plant diversity.

SPRAY AND VAPOR DRIFT

Herbicide drift may injure susceptible crops, ornamentals, or non-target native species. Drift can also cause non-uniform application in a field and/or reduce efficacy of the herbicide in controlling weeds.

Several factors influence drift, including spray droplet size, wind and air stability, humidity and temperature, physical properties of herbicides and their formulations, and method of application. For example, the amount of herbicide lost from the target area and the distance it moves both increase as wind velocity increases. Under inversion conditions, when cool air is near the surface under a layer of warm air, little vertical mixing of air occurs. Spray drift is most severe under these conditions since small spray droplets will fall slowly and move to adjoining areas even with very little wind. Low relative humidity and high temperature cause more rapid evaporation of spray droplets between sprayer and target. This reduces droplet size, resulting in increased potential for spray drift.

Vapor drift can occur when an herbicide volatilizes from a liquid or solid to a gas. The formulation and volatility of the compound will determine its vapor drift potential. Potential of vapor drift is greatest under high temperatures and with ester formulations. Ester formulations of 2,4-D are very susceptible to vapor drift and should not be applied at temperatures above 80°F.

Herbicides are applied by airplane, helicopter, ground sprayer, or roller and rope-wick applicators. Nozzle height controls the distance a droplet must fall before reaching the weeds or soil. Less distance means less travel time and less drift. Wind velocity is often greater as height above ground increases, so droplets from nozzles close to the ground would be exposed to lower wind speed. Applications are more likely to be above the inversion layer when herbicides are aerially applied. This will not allow herbicides to mix with lower air layers and increase long distance drift.

A number of measures can be taken to minimize the potential for herbicide drift. Chemical treatments should be made under calm or low wind conditions, preferably when humidity is high and temperatures are relatively low. Ground equipment reduces the risk of drift, and wick applicators nearly eliminate it. Use of the correct formulation under a particular set of conditions is important. For example, applying an ester formulation of postemergence herbicides during the hotter periods of the summer is not recommended.

GROUNDWATER AND SURFACE WATER CONTAMINATION

Most herbicide groundwater contamination results from "point sources." Point source contaminations include spills or leaks at storage and handling facilities, improperly discarding containers, and rinsing equipment in loading and handling areas, often times into adjacent drainage ditches. Point sources are characterized by discrete, identifiable locations discharging relatively high local concentrations. These contaminations can be avoided through proper calibration, mixing, and cleaning of equipment.

Non-point source groundwater contaminations of herbicides are relatively uncommon. They can occur, however, when a mobile herbicide is applied in areas with a shallow water table. In this situation, the choice of an appropriate herbicide or alternative control strategy can prevent contamination of the water source.

Surface water contamination with herbicides can occur when herbicides are applied intentionally or accidentally into ditches, irrigation channels or other bodies of water, or when soil-applied herbicides are carried away in runoff to surface waters. Direct application into water sources is generally used for control of aquatic species. In these cases, there is a restriction period before the use of this water for human activities. Accidental contamination of surface waters can occur when irrigation ditches are sprayed with herbicides or when buffer zones around water sources are not wide enough. In many situations, alternative methods of herbicide treatment, including wick application, will greatly reduce the risk of surface water contamination.

Loss of a preemergence herbicide through erosion may occur when a heavy rain follows a chemical treatment. It is possible to minimize herbicide runoff to surface waters by carefully monitoring weather forecasts before applying herbicides. Applications of preemergence herbicides should not be made when forecasts call for heavy rainfall. Predictions of precipitation between 0.5 and 1 in should allow a preemergence herbicide to percolate into the soil profile, thus minimizing the subsequent risk of surface runoff.

HERBICIDE RESISTANCE

Selection for herbicide-resistant weed biotypes is greatly accelerated with the continuous use of the same herbicide or a group of herbicides with a similar site of action at a site. For example, widespread resistance to the sulfonylurea herbicides (sulfometuron and chlorsulfuron) has been reported for Russian thistle (*Salsola tragus*) along California highways. In most natural areas the occurrence of herbicide resistant weeds is rare because repeated applications year after year are not typical.

EFFECTS OF HERBICIDES ON PLANT DIVERSITY

Continuous broadcast use of one herbicide or a combination will often select for plant species demonstrating greatest tolerance. Since even selective herbicides tend to injure several species, repeated use will eventually have a negative impact on plant diversity. This can be minimized or avoided by employing an integrated weed management approach.

FATE OF HERBICIDES IN THE ENVIRONMENT

Herbicides can be taken up by plants and either kill the plant or become metabolized by the plant. In addition, there are several other potential outcomes of an herbicide in the environment. These include three categories; degradation, immobilization, and movement in the soil, water or atmosphere.

DEGRADATION

Herbicides can degrade to small non-phytotoxic molecules in the soil by three common mechanisms, photodecomposition, chemical decomposition, or microbial degradation. These mechanisms determine the half-life of an herbicide (see Table 10, p. 487). The half-life is the time it takes half of the active molecule to break down and become inactive. Since these processes depend on rainfall patterns, soil moisture, organic matter content, soil texture, pH, sunlight, and temperature the ability of an herbicide to degrade can vary dramatically from one region to another and from one season to another. For this reason, half-life values are often presented as a range, with an average value.

Photodegradation. Photodegradation refers to decomposition by sunlight. Herbicides can photodegrade in the soil, but generally photodecompose much faster in clear water where they are exposed to more light. Compounds that absorb light in the UV and color spectrum tend to photodegrade faster than clear compounds that only absorb light in the UV spectrum. For this reason, art paintings should not be exposed to full sunlight and camera flashes are not usually allowed in art galleries. Herbicides such as trifluralin or oryzalin, which are not typically used in natural areas, are yellow to orange and photodegrade very rapidly. Photodegradation, however, is not the most common method of herbicide breakdown for terrestrial herbicides used to control natural area weeds.

Chemical Decomposition. Chemical decomposition is usually a slow process and is driven by chemical reactions, including hydrolyzation (reaction with hydrogen, usually in the form of water), oxidation (reaction with oxygen), and dissociation (loss of an ammonium or other chemical group from the parent molecule).

Microbial Degradation. Microbial degradation is the most common method by which herbicides are degraded in the soil. Bacteria, fungi and actinomycetes are the primary organisms that degrade herbicides. Herbicides can be degrades by specific organisms or by a general category of microbes and the rate at which an herbicide degrades depends on the microbial community, as well as the soil temperature, moisture content, and organic matter content. With some herbicides, repeated application leads to the development of enhanced degradation. In this case, subsequent applications of the herbicide are degraded faster than when the herbicide was applied the first time. This is because the specific microbe populations that degrade the herbicide build up in the soil and break the compound down faster with subsequent applications.

Table 10. Mechanisms of herbicide degradation in soil, potential for photochemical decomposition, and average half-life of selected herbicides registered for use in terrestrial natural areas.

Chemical family	Herbicide	Soil degradation[a] Microbial	Soil degradation[a] Chemical	Potential for photo-decomposition[b]	Avg. half-life (days)	Vapor pressure at 20°C or higher (Pa)
Aryloxyphenoxy propionic acid	Fluazifop-P-butyl	••	••	L	15	5.4×10^{-5}
Benzoic acid	Dicamba	•••	•	L	<14	4.5×10^{-3}
Bipyridylium	Paraquat	•	•	M	1000[c]	1×10^{-4}
Cyclohexanedione	Clethodim	•••	•	L	3	1.33×10^{-5}
Cyclohexanedione	Sethoxydim	•••	•	L	5	2.13×10^{-5}
Glycine	Glyphosate	•••	•	L	47[c]	3.99×10^{-5}
Imidazolinone	Imazapyr	•••	•	L	90	1.3×10^{-5}
Imidazolinone	Imazapic	•••	•	L	120	1.3×10^{-5}
Phenoxy acetic acid	2,4-D	•••	•	L	10	1.9×10^{-5}
Pyridine carboxylic acid	Aminopyralid	•••	•	L	35	2.56×10^{-8}
Pyridine carboxylic acid	Clopyralid	•••	•	L	40	1.36×10^{-3}
Pyridine carboxylic acid	Fluroxypyr	•••	•	L	36	2×10^{-5} ester
Pyridine carboxylic acid	Picloram	•	•••	M	90	8.2×10^{-5}
Pyridine carboxylic acid	Triclopyr	•••	•	M in water L otherwise	30	1.6×10^{-4}
Pyrimidine	Aminocyclopyrachlor	•••	•	L	80 to 164	4.89×10^{-6}
Sulfonylurea	Chlorsulfuron	•	•••	L	28 to 42	3.9×10^{-9}
Sulfonylurea	Metsulfuron	•	•••	L	30	3.3×10^{-10}
Sulfonylurea	Rimsulfuron	•	•••	L	20	1.5×10^{-6}
Sulfonylurea	Sulfometuron	•	•••	L	24	7.2×10^{-12}
Triazine	Hexazinone	•••	•	L	90	2.7×10^{-5}
Ureas	Tebuthiuron	•••	•	L	360	2.7×10^{-4}

a: ••• = primary method, •• = secondary method, • = minor method
b: H = high potential, M = medium potential, L = low potential
c: Binds tightly to soil and is not available for breakdown or uptake by plants

IMMOBILIZATION THROUGH SORPTION TO SOIL

Sorption to Soil. The sorptive capacity of a soil is related to its soil texture (i.e., sand, silt and clay content) and amount of organic matter because these components determine the amount of surface area in a soil. Soils with greater surface area typically have greater amounts of herbicide sorption. Sand has far less surface area compared to silt and silt has far less surface area compared to clay and organic matter. In addition, clay and organic matter particles have a net negative charge (measured as cation exchange capacity), which facilitates the binding of many herbicides.

Adsorption. Adsorption refers to the binding of herbicides by soil particles, which prevents their subsequent movement in the soil. Immobilization of an herbicide in soils can be weak or strong. For herbicides like paraquat and glyphosate, binding to the charged particles in the soil can be so strong that the herbicide not only is immobile, but it cannot be taken up by plants or degraded by microbes. Although the half-life of these two compounds seems long, it is irrelevant since the compound is not biologically available once absorbed. Most herbicides, however, bind much more weakly to soil particles. In essence they bind and release over and over again in a reversible process. Usually this is due to a weaker attachment, known as hydrogen bonding. This binding or adsorption process prevents the herbicide from leaching through the soil profile. Like paraquat and glyphosate, herbicides with higher adsorption rates are not as accessible to microbial degradation and thus, often have longer soil residual activity.

Water-soluble herbicides generally have low adsorption capacities, and are usually more mobile in the environment and more available for plant uptake, microbial metabolism and other degradation processes. Ester formulations, in general, are relatively insoluble in water and are more strongly adsorbed to soil particles. The strength of the adsorption process determines the equilibrium between herbicide molecules bound to the soil and herbicide molecules dissolved in the soil water. Thus, one of the most important factors determining the fate of an herbicide in the environment is its solubility in water. For many preemergence compounds, it is important for an herbicide to be retained near the soil surface, as this is the regions where most weed seeds are germinating. In other cases, where a preemergence herbicide is applied to control older more mature plants or woody species, it is important for the compound to be more water soluble and leach to greater soil depths where the roots are located. The pH of the soil can also greatly affect adsorption. At higher pH values, herbicides such as the imidazolinones are adsorbed less and are much more mobile.

Soil Sorption Coefficients. Soil sorption coefficients, when available, are better indicators of soil sorption and herbicide mobility in soil water than water solubility. Values used to measure the tendency of an herbicide to move with soil water are K_D and K_{OC} sorption coefficients. Low values indicate that most of the herbicide is in the water phase, thus increasing the potential for leaching. In contrast, high values indicate most of the herbicide is tightly adsorbed by the soil and not mobile. For example, paraquat and glyphosate are very soluble herbicides yet both have high soil sorption coefficients and do not move once they reach the soil. Paraquat has a positive charge and tightly binds to the negatively charged soil particles and glyphosate is rapidly inactivated by precipitation with iron and aluminum complexes in the soil, as well as its ability to bind to soil particles. Soil sorption coefficients for herbicides (along with herbicide water solubility) are examples of the useful information that can be obtained from the Herbicide Handbook, published by the Weed Science Society of America (www.wssa.net.)

MOVEMENT

Movement through the environment occurs when herbicides occur in surface or subsurface water runoff, are volatilized from spray particles, soil or plant surfaces during or after application, or leach down into the soil. These processes can occur simultaneously and continuously in the environment. Soil water movement can incorporate and distribute herbicides in the surface layers of the soil improving weed control or can move herbicides deep into the soil profile where they can injure desirable plants, reduce weed control, or move into groundwater.

Leaching. Leaching is a physical process involving the movement of herbicides downward through the soil profile in water. The amount of movement is directly influenced by the adsorptive ability of the herbicide (i.e., the amount of herbicide in the soil solution) and the amount of water movement in the soil. As previously discussed, soil texture, organic matter and environmental conditions influence the sorptive capacity and permeability of soils and the potential for water movement and herbicide leaching. For example, sandy soils are wetted to a greater depth by a given amount of water compared to clay soils. Macropores can also facilitate rapid movement of water, and potentially herbicide, to deeper soil depths and perhaps ground water. It is important to note that large amounts of water applied slowly in single applications cause greater movement than similar amounts scattered over an extended time period. Furthermore, the first irrigation or rainfall event after herbicide application is more important to movement than subsequent irrigation or rainfall.

Volatilization. Off-site movement of an herbicide can occur through spray drift. This is when the herbicide particle is moved in the water or carrier that is applied to the target plants. Spray drift is very different from volatilization. Volatility is the physical change of a solid or liquid to a gas. Some herbicides can volatilize in the soil to more evenly distribute the compound and increase efficacy. However, most discussion of volatilization refers to loss of the herbicide from the treatment area, thus reducing control and increasing potential off-site problems.

Herbicides can be classified according to their vapor pressure, which is a measure of their volatility. Higher vapor pressures lead to rapid volatilization, by which a compound transitions into the gas phase and moves off-site in the wind. The units of vapor pressure are in pascal (Pa). Ethanol, for example, is extremely volatile and has a vapor pressure of 5830 Pa. Water has a vapor pressure of 3162 Pa. In contrast, most herbicides have very low vapor pressures (Table 10, p. 487), although the ester formulations of herbicides are higher than the amine or salt formulations. Volatility generally increases with increasing temperature and soil moisture.

Water solubility and vapor pressure are related in their effect on volatilization. Highly soluble herbicides are less subject to loss from volatility than herbicides of low solubility with the same vapor pressure. Sorption also affects losses from volatility because herbicides that are tightly bound to soil (have high soil sorption coefficients) are less likely to be lost through volatilization.

Volatilization most often occurs during application, but also can occur after the herbicide has been deposited on plants or the soil surface. Volatility of an herbicide can cause serious economic losses when 1) poor weed control results due to the loss of herbicides from the target site, 2) herbicides move to non-target sites and cause injury, or 3) herbicides cause unintended pollution of the environment. Of the herbicides used widely in natural area situations, only the ester forms of 2,4-D and triclopyr (Note: *Garlon 4 Ultra* is a low volatile ester formulation) can volatilize and potentially cause problems. To decrease the potential for volatilization avoid spray contact with impervious surfaces, such as roads and rocks.

WHAT TO CONSIDER WHEN CHOOSING AN HERBICIDE

For natural areas, the choice of an herbicide will depend upon several factors, related to the properties of the herbicide, target weed, environmental conditions, and other considerations. Clearly the herbicide needs to be able to control the target species, but should also provide safety to non-target organisms, including plants and animals. Furthermore, it should not be highly susceptible to drift or volatilization, and should not have a high risk of moving into groundwater or surface water. Its persistence in the soil should be appropriate to the time necessary to provide good control of the target weed. In general, it would be the goal of a land manager to be successful in managing an invasive species with the lowest possible negative impact on the environment. In many, although not all, cases, herbicides can provide this over other techniques, but proper selection and use can help ensure a positive outcome.

REGISTRATION OF HERBICIDES

All herbicides must be federally registered by the US Environmental Protection Agency (EPA) under Section 3 of the Federal Insecticide, Fungicide and Rodenticide Act (FIFRA). Herbicides must also be registered by individual states for use within that state. There are two provisions within the federal law that allow states to obtain specific pesticide uses that are not available under federal registration. These include:

Section 18; Emergency exemptions from registration

Section 24(c); Special local needs registrations, often called a SLN label

SECTION 18

EPA has applied this section to exempt states from the provisions of FIFRA that regulate the manner in which a pesticide is made available for use or how it is used. An emergency exemption authorizes a state to permit use of a pesticide to control a pest problem, when the needed pesticide either is not federally registered or, if registered, it does not have residue tolerances established for the food/feed crops to be treated. EPA regulations for Section 18 provide four types of emergency exemptions: specific, crisis, public health, and quarantine. Generally, only specific and crisis apply to weedy plants.

Specific Exemptions. EPA can authorize specific exemptions for uses of pesticides when the emergency condition involves the need to control a pest that can cause severe economic losses in crops or to reduce risks to the environment or to beneficial organisms, including endangered or threatened species. The process requires that the state department of agriculture submit a formal request for a specific exemption. However, the request must originate from or on behalf of the entities affected by the emergency. EPA considers biological, economic, environmental fate and human health effects before they approve of the exemption. The submittal and review process may take up to 4 months and the duration of a specific exemption is for 1 year. However, the EPA can authorize the exemption for several years provided that progress is being made to provide full registration of the chemical for that use. A supplemental special needs label is required from the pesticide manufacturer. This label includes directions for use and restrictions and conditions of the approved specific exemption.

Crisis Exemptions. Under the crisis provisions of Section 18, a state can authorize use of a pesticide in emergency conditions, if the emergency requires that applications must be made before a specific exemption could be granted. Before this use, the state must notify EPA of its intent 24 to 36 hours in advance. If a specific exemption request for the emergency use is not already under review at EPA, the state must submit such a request within 15 days of enacting the crisis exemption.

SECTION 24(C)

The state department of agriculture is the lead agency responsible for registering pesticides for special local needs (SLN) under Section 24(c) of FIFRA. SLN is defined as, "an existing or imminent pest problem within a State for which the State lead agency, based upon satisfactory supporting information, has determined that an appropriate federally registered pesticide is not sufficiently available." Under Section 24(c), states are authorized to allow a new use for a federally registered pesticide product. When a state grants an SLN registration, the EPA is informed and provided with a letter of notification and a copy of the accepted label. EPA has 90 days to contact the state about the registration. SLN registrations are generally used for growers of minor crops, who often have limited access to pest management options, including pesticides.

HERBICIDE LABELS

It is critical to be familiar with the label before applying an herbicide. Labels are approved by EPA and contain information on application rates and requirements, pesticide handling and environmental safety information, protective clothing and equipment, cleaning instructions, and many other aspects critical to making a proper herbicide application. Labels are provided when the products are purchased, but are also available on the internet.

RECORD KEEPING

While some states require that detailed records be taken for each herbicide application, it is good common sense to do this whether or not it is a state requirement. At minimum, written record should include all plants/areas treated, amounts and types of herbicide used, and dates of application. Such information will be very useful in evaluating the efficacy of control, determining the impact on non-target species, monitoring the project's success, improving methodology, identifying mistakes, or determining potential environmental issues that may develop.

MINIMIZING PESTICIDE EXPOSURE

It is important to minimize exposure to pesticides during and after application. Personal Protective Equipment (PPE) is listed on the label. Applicators are required to wear this equipment when applying that particular pesticide. When using volunteers, the minimum PPE, even when not required on the label, should include rubber boots, protective aprons, sturdy overalls, or suits (e.g., disposable Tyvek™ suits), rubber gloves, and safety glasses or goggles. With some herbicides, respirators are required.

Following an application, the label may specify a Restricted-Entry Interval (REI). This requires that all workers or people be kept out of the pesticide treated area for a specific length of time following the application. The REI for a given product may differ depending on the several factors, including the toxicity of the formulation or compound, climate and application method. When more than one pesticide product is applied, use the longest REI. REIs for herbicides generally are between 12 and 48 hours.

CLEANING AND DISPOSING OF EQUIPMENT AND CONTAINERS

After use, application equipment and empty herbicide containers should be triple rinsed with clean water in the treatment area. For containers, about 10% of the container volume should be used for each rinse. An alternative method of rinsing is to use a pressure system that can be easier and more effective than a triple wash system. The rinsate from the wash can be added to the spray mix and applied according to the label. Left over spray mix or rinsate that is not applied to the treatment area will have to be disposed of as hazardous waste, so it is much easier to plan accordingly and use the right amount of herbicide, as well as spray out the rinsate in the treated areas. Do not mix different rinsates or dump rinsate on the ground or into storm drains.

To dispose of containers it is often possible to use a state herbicide container recycling program. In parts of California and some other states one may be required to get written permission from your County Agriculture Commissioner to dispose of containers. Call your local Commissioner for details. If no specific agri-chemical container recycling program is available, puncture the empty container to prevent its further use before disposing of it. In most areas, small numbers of empty, triple-rinsed containers can be disposed as trash for pick-up or taken to the local dump, unless the label states otherwise. If the herbicide label states that the container may not be disposed of in regular sanitary landfills, call the county or municipal waste department for information on Hazardous Material Collection dates.

Application equipment should be thoroughly cleaned before storing. Other protective gear, including gloves, boots, etc. should also be thoroughly rinsed and washed in mild soap. When personal protection gear is cleaned the applicator or handler should wash their hands and any herbicide-exposed areas of their bodies with soap and water, before taking a shower. Contaminated clothes should be removed as soon as possible and be cleaned and dried separately from other uncontaminated clothes.

PESTICIDE SPILLS AND CLEANUP

Maximum care should be taken to prevent spills. Most spills occur when pesticides are being transported or when containers leak in storage. When a spill does occur, immediate action is required, including reporting the spill. Spill cleanup kits are required at pesticide storage areas or where pesticides are handled or transported. Response to an herbicide spill may vary with size and location of the spill, but the sooner the spill can be contained, absorbed, and disposed of the less chance it will cause health or environmental problems. The primary steps to dealing with spills are to 1) control the spill by protecting yourself and others and stop the leak, 2) contain and confine the spill, protect water sources, absorb the liquid or cover the dry material, and 3) clean up the spill, neutralize and decontaminate the site, and clean the contaminated equipment and PPE.

TOXICITY

When used improperly, herbicides can pose a health risk. This can be minimized with proper safety techniques. When proper safety precautions are taken, human exposure to herbicides used in natural areas should be minimal. Although animals can also be at some risk from herbicide exposure, most herbicides registered for use in natural areas are relatively non-toxic to wildlife. To prevent injury to wildlife, care should be taken to apply these compounds at labeled rates.

The greatest risk of pesticide exposure is when mixing and applying herbicides or when workers unintentionally reenter a treated area too soon following treatment. The route of herbicides entry into the body is most commonly through the skin, referred to as dermal, or by inhalation. It is also possible to accidentally splash or expose the eyes to herbicides during mixing or application. Dermal exposure often occurs when contaminated gloves contact other parts of the body.

An individual's reaction to a particular herbicide can depend on several factors, including the herbicide itself, the exposure dose, the part of the body exposed, and the rate at which the compound is metabolized and excreted from the body. Not all individuals react similarly to a particular herbicide or dose. In general, small individuals, particularly children, are more sensitive to a given dose than larger people.

The mechanism by which herbicides cause toxicity is generally by blocking biochemical processes or disrupting cell membranes. Sublethal exposure generally causes no response. The effects are often localized and can include irritation to the eyes, nose, or throat, or can be more general when the compound enters the blood stream. Symptoms can occur immediately after exposure or develop gradually. Common symptoms of low-level exposure to some herbicides can include skin and eye irritation, headache, and nausea, whereas higher doses can cause blurred vision, dizziness, heavy sweating, weakness, stomach pain, vomiting, diarrhea, extreme thirst, and blistered skin, as well as behavioral alterations such as apprehension, restlessness, and anxiety.

NON-CROP HERBICIDES AND THEIR CHARACTERISTICS

Many important characteristic of 27 herbicides used in non-crop areas, including aquatic systems, are listed in Table 11, next page, 492. This list should be referred to routinely when considering the control options for individual weedy or invasive plant species. Table 12, page 510, lists rainfastness (how long herbicides should be on plant foliage before rainfall) and grazing / haying restrictions for foliar-applied herbicides.

The concept of active ingredient (a.i.) and acid equivalent (a.e.) is critical to proper calibration of herbicides. This is because not all formulations of the same herbicide contain the same concentration of that active compound. Thus, a 1% solution of more concentrated formulation would contain more herbicide in the spray tank compared to a 1% solution of a less concentrated formulation. For this reason, it is important to mix an herbicide based on the proper rate of a.i. or a.e. This is not always the case, as most labels will also tell you how much product to add to the spray tank per acre treated. Although this book does not provide detailed instructions on how to properly calibrate spray equipment, such information can be found in a number of other reference books or on-line resources, including the Weed Research and Information online video website (wric.ucdavis.edu).

In brief, the term active ingredient (a.i.) refers to the proportion of an herbicide formulation that is made up of the actual herbicidal compound. The rest of the formulation is made up of other ingredients, primarily inert compounds or carrier material. The use of the term active ingredient is generally used for herbicides that are not formulated as acids. For example, the active ingredient for all solid formulations (pellets, crystals, granules) is in a.i. and this is usually represented as the percentage a.i. per dry weight. In other cases, the a.i. of a liquid formulation is represented by the pounds of the active material per gallon of product (lb a.i./gal). Since a gallon is around 8 lb it is easy to estimate that a 4 lb a.i./gal herbicide has about 50% of the material as active herbicide. However, the label will always say exactly what percentage of the formulation in made up of the herbicide. The percentage of active ingredient or acid equivalent in the formulation is usually listed below these percentages in a separate table or paragraph.

Acid equivalent (a.e.) is always used for herbicides that are weak acids. Most foliar applied herbicides are weak acids, including all the growth regulator (auxinic) compounds. If these herbicides (negative charge) were formulated as the acid form with a proton (H^+) as the counter ion they would not be very soluble in water or in oil and would not stay in solution. For this reason it is necessary to formulate weak acid herbicides as salts, or esters to keep the herbicide in solution. The salt will dissociate from the active herbicide molecule readily in solution, so that the herbicide has a negative charge. The ester formulation, however, stays as an ester in solution. This makes it better able to move through the waxy cuticle and into the plant. Once the ester form of the herbicide enters the plant, the ester group is cleaved off the parent molecule, allowing the acid to move into the cell and affect the plant.

Acid equivalent (a.e.) is used instead of active ingredient (a.i.) for some herbicidesbecause the molecular weight of the specific salt or ester group can vary dramatically in size. When attached to the active herbicide molecule, the salt or ester group plus the active herbicide is the a.i. However, that value is misleading because only the active herbicide molecule has any activity in the plant. Thus, a.e. does not consider the molecular weight of the salt or ester, but only considers the active herbicide molecule itself. When a salt formulation of a weak acid herbicide is added to water, the salt group (called the counter ion) dissociates from the active herbicide molecule. Thus, it is really more important to know the acid equivalent (a.e.) when you are trying to accurately calibrate your herbicide solution. With the a.e. value, calibration can be standardized regardless of the herbicide formulation used.

Table 11. **Characteristics of herbicides** used for control of natural-area weeds.

2,4-D

Trade names	*Navigate* (10% acid equivalent; aquatic formulation) *Weedar*, *Weedone* DMA 4 IVM, and many other trade names, often tank mixed with other herbicides (various concentrations but generally about 4 lb acid equivalent/gal)
Chemical family	Phenoxy carboxylic acid
Manufacturer	Dow AgroSciences, Bayer CropScience, Nufarm, and many others
Formulation	Water soluble or emulsifiable concentrate. Formulated as acids, salts, and esters and sold as liquids, water-soluble powders, granules (aquatic formulation), and pellets.
Use areas	Rangeland, wildland, pasture, forest, CRP, certain aquatic situations
Application method	Postemergence
Use rates	0.25 to 2.0 lb acid equivalent/acre, see specific label for concentration. 100 to 200 lb product/acre in aquatic systems.
Soil persistence (range of half-life)	Average half-life is 10 days. In natural, clear water, half-life is 7 to 14 days.
Potential for leaching into surface and ground water	Potentially mobile, but rapid soil degradation keeps herbicide close to soil surface.
Volatilization	Acid formulations non-volatile, ester formulations highly volatile
Selectivity	Annual and perennial broadleaf species and some woody species
Plant growth stage	Best results when treating rapidly growing plants, particularly seedlings in annuals. In perennials, treat plants later in season when sugars are translocating to underground reproductive structures.
Translocation	Very phloem mobile
Mechanism of action	Growth regulator herbicide, mimics naturally occurring auxins
Important considerations	Do not use the ester formulations when ambient temperatures are expected to reach 80°F or higher.

Acrolein

Trade names	*Magnacide H* (92% active ingredient)
Chemical family	None
Manufacturer	Baker Hughes
Formulation	Water soluble
Use areas	Aquatic systems only, irrigation canals
Application method	Directly to water in irrigation canals
Use rates	Up to 15 ppm
Soil persistence (range of half-life)	Half-life in water depends on temperature and pH, but ranges between 5 and 180 hours. In canals from several western states, the half-life ranged between 5.5 and 10.2 hours.
Potential for leaching into surface and ground water	Applied directly to herbicide. Not considered susceptible to leaching into groundwater as it adsorbs quickly to soil particles.
Volatilization	Highly volatile, but not considered volatile after application directly to water.
Selectivity	Submerged aquatic plants, smaller floating aquatic plants, and algae.
Plant growth stage	Plants must be near or fully grown as herbicide will have to contact tissues to show effects.
Translocation	Absorbed readily by aquatic plants due to lack of cuticle. Does not translocate well to below ground structures, thus acts as a contact herbicide.
Mechanism of action	A general cell toxicant that reacts with sulfhydryl groups, destroying enzymes and disrupting plant metabolic systems. Causes rapid cell destruction and death.
Important considerations	Acrolein is also very toxic to all animals. Fish are very sensitive to acrolein and the herbicide has a Restricted Danger label. Being phased out because of environmental and health issues.

Aminocyclopyrachlor

Trade names	*Plainview* (tank mix with sulfometuron) *Perspective* (tank mix with chlorsulfuron) *Streamline* (tank mix with metsulfuron) *Viewpoint* (tank mix with metsulfuron and imazapyr)
Chemical family	Pyrimidine
Manufacturer	DuPont
Formulation	Water soluble
Use areas	Utility, rights-of-way, non-crop areas, industrial sites and natural areas
Application method	Both pre- and postemergence to weeds
Use rates	See labels as aminocyclopyrachlor is only registered for use in tank mixes
Soil persistence (range of half-life)	Average soil half-life ranged from 80 to 164 days.
Potential for leaching into surface and ground water	Limited, may leach into ground water if used in areas where soils are permeable, particularly where the water table is shallow
Volatilization	Not volatile
Selectivity	Broadleaf species, but particularly member of the Asteraceae (sunflower family) and Fabaceae (pea family). Can also control certain species in other families including the Apiaceae (carrot family) and Polygonaceae (knotweed family). Some activity preemergence on medusahead and downy brome, as well as woody plants. Broader spectrum than clopyralid and aminopyralid.
Plant growth stage	For annuals, best to apply in seedling stage as this will provide both postemergence and preemergence activity. For perennials, plants should be fully expanded to ensure movement to underground vegetative parts.
Translocation	Readily translocated in the phloem and accumulates at below and above ground growing points or storage organs.
Mechanism of action	Growth regulator herbicide, mimics naturally occurring auxins.
Important considerations	Herbicide is not yet registered in all states. In addition, it is not sold alone, though most tests for efficacy have been conducted for only aminocyclopyrachlor. Do not use treated plant residue as mulch for vegetable crops. The current labels are all for vegetation management use only and cannot be used where grazing or feeding will take place. DuPont has submitted rangeland/pasture labels and expect first registrations in 2013. The labels will be different than the current VM labels, including different rates, additional weeds listed, and more information on grass tolerance. Aminocyclopyrachlor degrades slowly in compost and can injure sensitive plants when used as a mulch or fertilizer source.

Aminopyralid

Trade names	*Milestone* and *Milestone VM* (2 lb acid equivalent/gal) *Capstone* (tank mix with triclopyr) *Forefront HL* and *PasturAll HL* (tank mixes with 2,4-D) *Opensight* and *Chaparral* (tank mix with metsulfuron)
Chemical family	Pyridine carboxylic acid
Manufacturer	Dow AgroSciences
Formulation	Water soluble
Use areas	Conservation Reserve Program, rangelands, non-irrigation ditch banks, natural areas and wildlands, non-crop and rights-of-ways, grazed areas, and more
Application method	Both pre- and postemergence to weeds
Use rates	3 to 7 oz product/acre broadcast, up to 14 oz product/acre spot treatment
Soil persistence (range of half-life)	Average half-life of about 35 days in eight study sites in North America.
Potential for leaching into surface and ground water	Limited, may leach into ground water if used in areas where soils are permeable, particularly where the water table is shallow

Table 11 continued: Characteristics of herbicides

Volatilization	Not volatile
Selectivity	Broadleaf species, but particularly member of the Asteraceae (sunflower family) and Fabaceae (legume family). Can also control certain species in other families including the Apiaceae (carrot family), Solanaceae (nightshade family), and Polygonaceae (knotweed family). Some activity preemergence on medusahead. Typically better than clopyralid on tough to control perennials in the Asteraceae.
Plant growth stage	For annuals, best to apply in seedling stage as this will provide both postemergence and preemergence activity. For perennials, plants should be fully expanded to ensure movement to underground vegetative parts. In some perennials, fall applications to dried material give good control of new bud growth.
Translocation	Readily translocated in the phloem and accumulates at below and above ground growing points or storage organs.
Mechanism of action	Growth regulator herbicide, mimics naturally occurring auxins.
Important considerations	Can use to water's edge. No grazing restrictions. Do not use treated plant residue as mulch for vegetable crops. Clippings from treated areas should not be used as compost; this herbicide degrades slowly in compost and can be a problem when treated plants are used as a mulch or fertilizer source in sensitive crops or landscapes.

Bispyribac-sodium

Trade names	*Tradewind* (80% active ingredient)
Chemical family	Pyrimidinyloxybenzoic acid
Manufacturer	Valent
Formulation	Soluble powder
Use areas	Aquatic areas, lakes, ponds, reservoirs, marshes, ditches and canals
Application method	Applied as a subsurface application for submerged aquatic weeds or as a surface application targeting floating and emergent aquatic weeds.
Use rates	Apply between 20 and 45 ppb for direct application to water and up to 8 oz product/acre for surface broadcast applications.
Soil persistence (range of half-life)	Half-life between 46 and 109 days in soil and 9 days in an aquatic system.
Potential for leaching into surface and ground water	Low leaching potential as the herbicide is readily adsorbed to the soil.
Volatilization	Negligible
Selectivity	Broad spectrum control of submersed and floating vegetation in aquatic systems.
Plant growth stage	Applications should be made to plants after they begin active growth. For mature perennial emergent species, application should be made later in the season when translocation of sugars is to underground storage tissues.
Translocation	Absorbed directly into plant tissues through water or foliarly to emerged vegetation. Translocated in the xylem, but primarily in the phloem and accumulates at growing points.
Mechanism of action	Inhibits the synthesis of branched chain amino acids (leucine, valine, and isoleucine) by blocking the activity of the enzyme acetolactate synthase (ALS).
Important considerations	For aquatic use only. Not yet registered in much of United States. Expected registration in 2012.

Chlorsulfuron

Trade names	*Telar* (75% active ingredient) *Landmark* (tank mix with sulfometuron) (25% active ingredient) *Perspective* (tank mix with aminocyclopyrachlor) (15.8% active ingredient) *Cimarron X-tra* and *Cimarron Plus* (tank mix with metsulfuron) (37.5% or 15% active ingredient)
Chemical family	Sulfonylurea
Manufacturer	DuPont

Formulation	Dry flowable
Use areas	Pasture, range, CRP, non-crop industrial sites
Application method	Pre- and postemergence
Use rates	0.25 to 3 oz product/acre; 1.33 oz product/acre maximum grazed areas
Soil persistence (range of half-life)	28 to 42 days
Potential for leaching into surface and ground water	Chlorsulfuron is moderately persistent and highly mobile and has potential to enter surface waters from runoff. The very low application rate and microbial breakdown suggest that it has little potential to enter ground water.
Volatilization	Non-volatile
Selectivity	Broad spectrum, many broadleaf and grass species, both annual and perennial, but is best on broadleaf species.
Plant growth stage	Best results when weeds are treated in the bud to bloom or fall rosette stage.
Translocation	Translocates in the xylem, as well as the phloem.
Mechanism of action	Inhibits the synthesis of branched chain amino acids (leucine, valine, and isoleucine) by blocking the activity of the enzyme acetolactate synthase (ALS).
Important considerations	*Telar* can be used near water, but cannot be applied to water.

Clethodim

Trade names	*Select* (2 lb active ingredient/gal) *Envoy* (0.97 lb active ingredient/gal)
Chemical family	Cyclohexanedione
Manufacturer	Valent
Formulation	Emulsifiable concentrate
Use areas	Uncultivated non-ag areas, fallow areas
Application method	Postemergence
Use rates	9 to 32 oz product/acre
Soil persistence (range of half-life)	Average half-life in soil is 3 days
Potential for leaching into surface and ground water	Very low risk
Volatilization	Negligible
Selectivity	Specific to annual and perennial grasses.
Plant growth stage	For annual species, best applied when plants are young and rapidly growing. For perennials, best applied when plants are fully expanded and photosynthates are moving to below ground vegetative structures. Do not apply herbicide when plants are under stress, particularly drought stress.
Translocation	Translocation is slow, but movement is primarily in the phloem where it accumulates in the growing points, typically the shoots.
Mechanism of action	Inhibits fatty acid synthesis by blocking the activity of the enzyme acetyl CoA carboxylase (ACCase).
Important considerations	Clethodim is not registered for use in many non-crop areas. Do not apply near water.

Clopyralid

Trade names	*Transline* and *Reclaim* (3 lb acid equivalent/gal) *Curtail* (tank mix with 2,4-D) *Confront* (tank mixes with triclopyr) *WideMatch* (tank mix with fluroxypyr)

Table 11 continued: Characteristics of herbicides

Chemical family	Pyridine carboxylic acid
Manufacturer	Dow AgroSciences
Formulation	Water soluble
Use areas	CRP, non-crop areas, industrial manufacturing and storage sites, rights-of-way, wildlands, rangelands, tree plantations, and grass pastures
Application method	Both pre- and postemergence to weeds
Use rates	4 to 21 oz product/ acre (0.33 to 1.33 pt product/acre)
Soil persistence (range of half-life)	Half-life in soil ranges between 12 and 70 days, with average of 40 days
Potential for leaching into surface and ground water	Moderate leaching potential, particularly in areas with shallow water table.
Volatilization	Not volatile
Selectivity	Broadleaf species, but particularly member of the Asteraceae (sunflower family) and Fabaceae (legume family). Can also control certain species in other families including the Apiaceae (carrot family), Solanaceae (nightshade family), and Polygonaceae (knotweed family). Some activity preemergence on medusahead. Not as effective as aminopyralid on tough to control perennials in the Asteraceae, but safer on oak seedlings and saplings.
Plant growth stage	For annuals, best to apply in seedling stage as this will provide both postemergence and preemergence activity. For perennials, plants should be fully expanded to ensure movement to underground vegetative parts.
Translocation	Readily translocated in the phloem and accumulates at below and above ground growing points or storage organs.
Mechanism of action	Growth regulator herbicide, mimics naturally occurring auxins.
Important considerations	Cannot apply near water. No grazing restrictions. Do not use treated plant residue as mulch for vegetable crops. Clippings from treated areas should not be used as compost; this herbicide degrades slowly in compost and can be a problem when treated plants are used as a mulch or fertilizer source in sensitive crops or landscapes.

Copper and Copper Chelate

Trade names	*Cutrine* (9% active ingredient) *Komeen* (8% active ingredient) *Harpoon* (8% active ingredient) Many others (e.g., *Aquatrine, Nautique, K-Tea, Captain, Clearigate*)
Chemical family	Inorganic compound, no recognized chemical family
Manufacturer	SePRO, Applied Biochemists and others
Formulation	Crystal or granule that is water soluble
Use areas	Aquatic environments
Application method	Direct application to water
Use rates	0.2 to 1 ppm
Soil persistence (range of half-life)	Elemental copper cannot be degraded.
Potential for leaching into surface and ground water	Copper has low leaching ability due to strong adsorption to soil.
Volatilization	None
Selectivity	Primarily used to control algae, provides control of many submerged and floating-leaf weeds.
Plant growth stage	Rapidly growing vegetation or algae.
Translocation	Once absorbed by plant, copper can be translocated to some degree.

Mechanism of action	Although low levels of copper are needed for growth, high concentrations cause an imbalance with other enzyme metal cofactors, resulting in enzyme blockage. Cell destruction occurs quickly following exposure.
Important considerations	Copper is a heavy metal and can accumulate in the environment with continuous use. Chelated forms of copper are used in hard waters to prevent precipitation.

Dicamba

Trade names	*Banvel* (4 lb acid equivalent/gal) *Clarity* (4 lb acid equivalent/gal) *Vanquish* (4 lb acid equivalent/gal)
Chemical family	Benzoic acid
Manufacturer	DuPont, Riverdale, BASF
Formulation	Water soluble
Use areas	Rangelands, pastures, non-ag areas, fallow areas, rights-of-way, utility sites
Application method	Postemergence
Use rates	0.5 to 4 pt product/acre
Soil persistence (range of half-life)	Typically with a half-life <14 days, depending on conditions.
Potential for leaching into surface and ground water	Fairly mobile in the soil but degrades rapidly. Risk of leaching low.
Volatilization	Volatilization can vary by formulation, but not generally volatile unless applied at very high ambient temperatures.
Selectivity	Annual and perennial broadleaf species.
Plant growth stage	Best results when treating rapidly growing plants, particularly seedlings in annuals. In perennials, treat plants later in season when sugars are translocating to underground reproductive structures.
Translocation	Very phloem mobile and accumulates at below and above ground growing points or storage organs.
Mechanism of action	Growth regulator herbicide, mimics naturally occurring auxins.
Important considerations	Not typically used alone. Can be tank mixed. Do not apply near water.

Diquat

Trade names	*Reward, Diquat, Knockout* (37.3% active ingredient; 2 lb active ingredient/gal) *Weedtrine* (8.53% active ingredient; 0.5 lb active ingredient/gal)
Chemical family	Bipyridylium
Manufacturer	Syngenta, Alligare, Applied Biochemists, Nufarm Americas
Formulation	Water soluble
Use areas	Aquatic (lakes, ponds, drainage ditches and other slow moving water)
Application method	Directly applied to water for postemergence activity on free floating or submerged aquatic weeds.
Use rates	No more than 2 gal product/surface acre of water.
Soil persistence (range of half-life)	1000 days, but binds so strongly to clay that it is not biologically available for breakdown or uptake by plants.
Potential for leaching into surface and ground water	Registered for aquatic use, considered immobile in soil.
Volatilization	Not volatile
Selectivity	Effective only on aquatic annuals, will knock back perennials but they will recover over time. Used for control of submerged and free floating aquatic weeds.

Table 11 continued: Characteristics of herbicides

Plant growth stage	Best results when plants are actively photosynthesizing and growing and when water temperatures have reached or exceed about 50 F.
Translocation	Very slow translocation, considered a contact herbicide.
Mechanism of action	Inhibits photosynthesis at the photosystem I (PS I) reaction site.
Important considerations	Application to muddy water will result in reduced control because the herbicide will be bound to soil particles.

Endothall	
Trade names	*Aquathol* (40.3 active ingredient, 28.6% acid equivalent; for granular 63% active ingredient, 44.7% acid equivalent) *Cascade* (40.3% active ingredient, 28.6% acid equivalent) *Hydrothol* (53% active ingredient, 23.36% acid equivalent; for granular 11.2% active ingredient, 5% acid equivalent)
Chemical family	No recognized family
Manufacturer	United Phosphorus
Formulation	Water soluble and granular
Use areas	Aquatic (ponds, lakes and canals)
Application method	Primarily postemergence for control of algae and submerged aquatic vegetation.
Use rates	*Aquathol K* (0.5 to 5 ppm); *Hydrothol* 0.05 to 2.5 ppm
Soil persistence (range of half-life)	Half-life in water ranges from 3 to 7 days.
Potential for leaching into surface and ground water	Registered for aquatic use. Does not persist long in soil or water.
Volatilization	Non-volatile
Selectivity	Most effective on aquatic annuals. Will knock back submerged perennials but they will recover over time.
Plant growth stage	Applications should be made soon after emergence of new vegetation
Translocation	Little translocation in xylem, considered a contact herbicide.
Mechanism of action	Not well understood. Inhibits lipid and protein biosynthesis, and causes membrane leakage.
Important considerations	Seven day restriction on use of water for drinking, washing food or irrigation.

Fluazifop-P-Butyl	
Trade names	*Fusilade DX* (2 lb acid equivalent/gal)
Chemical family	Aryloxyphenoxy propionic acid
Manufacturer	Syngenta
Formulation	Emulsifiable concentrate
Use areas	Fallow land and other non-crop areas, wastelands, uncultivated non-ag areas
Application method	Postemergence
Use rates	16 to 24 oz product/acre
Soil persistence (range of half-life)	Average half-life in the soil is 15 days.
Potential for leaching into surface and ground water	Very low risk
Volatilization	Negligible
Selectivity	Specific to annual and perennial grasses.

Plant growth stage	For annual species, best applied when plants are young and rapidly growing. For perennials, best applied when plants are fully expanded and photosynthates are moving to below ground vegetative structures. Do not apply herbicide when plants are under stress, particularly drought stress.
Translocation	Translocation is slow, but movement is primarily in the phloem where it accumulates in the growing points, typically the shoots.
Mechanism of action	Inhibits fatty acid synthesis by blocking the activity of the enzyme acetyl CoA carboxylase (ACCase).
Important considerations	Not registered for many non-crop areas. Check label before use.

Flumioxazin

Trade names	*Clipper* (51% active ingredient)
Chemical family	Dicarboximide
Manufacturer	Valent
Formulation	Water dispersible granule
Use areas	Aquatic areas (canals, drainage ditches, fresh water ponds, lakes, marshes and reservoirs)
Application method	Primarily postemergence for control of submerged aquatic vegetation. Foliar applications on free floating aquatic species.
Use rates	Use rate of 100 to 400 ppb of active ingredient for submerged weeds and 6 to 12 oz product/surface acre for floating or emergent weeds.
Soil persistence (range of half-life)	It is degraded by hydrolysis and has a half-life of 17.5 min in water at pH 9.0 compared to a half-life of 16.1 h and 4.1 d at pH 7.0 and 5.0, respectively.
Potential for leaching into surface and ground water	Registered for aquatic use. Rapid breakdown reduces accumulation in sediment and risk of leaching.
Volatilization	Negligible
Selectivity	Most effective aquatic herbicide on submerged perennials, some free floating aquatic species, and some algae.
Plant growth stage	Applications should be made to rapidly growing plants at an early growth stage.
Translocation	Considered a fast-acting contact herbicide.
Mechanism of action	Inhibits the enzyme protoporphyrinogen oxidase (Protox), which is a precursor to the synthesis of chlorophyll.
Important considerations	Aquatic use only. Loses herbicidal activity in water with pH greater than 8.5. Should not be applied to flowing water. Not registered in many areas, but is expected to be registered for use in 2012.

Fluridone

Trade names	*Sonar* (5% or 41.7% active ingredient; 4 lb active ingredient /gal) *Avast!* (41.7% active ingredient) *Sonarone* (5% active ingredient) *Whitecap* (41.7% active ingredient)
Chemical family	No recognized family
Manufacturer	SePRO
Formulation	Water soluble and granular
Use areas	Aquatic (fresh water ponds, lakes, reservoirs, drainage canals and irrigation canals)
Application method	Primarily postemergence for control of submerged aquatic vegetation.
Use rates	Maximum use rate of 150 ppb
Soil persistence (range of half-life)	Average half-life is about 20 days in aerobic water, 9 mo in anaerobic water and about 90 days in the hydrosoil.

Table 11 continued: Characteristics of herbicides

Potential for leaching into surface and ground water	Registered for aquatic use
Volatilization	Non-volatile
Selectivity	Most effective aquatic herbicide on submerged perennials. Can also give control of free floating aquatics, and some emersed plants.
Plant growth stage	Applications should be made to mature plants that are rapidly growing.
Translocation	Phloem mobile
Mechanism of action	Bleaching agent that inhibits the enzyme phytoene desaturase in the carotenoid biosynthesis pathway.
Important considerations	To achieve effective control a minimum of 45 days (up to 90 days) of herbicide contact is required.

Fluroxypyr

Trade names	*Vista XRT* (2.8 lb acid equivalent/gal) Often tank mixed with many other active ingredients and with numerous trade names
Chemical family	Pyridinyloxy alkanoic acid
Manufacturer	Dow AgroSciences
Formulation	Emulsifiable concentrate
Use areas	Industrial sites, fallow areas, rangelands within rights-of-way, and pine plantations
Application method	Postemergence
Use rates	6 to 22 oz product/acre
Soil persistence (range of half-life)	Half-life in North America ranges between 11 and 38 days.
Potential for leaching into surface and ground water	Generally not considered to have a high potential for ground or surface water contamination at it is only moderately persistent and applied postemergence.
Volatilization	Negligible
Selectivity	Controls annual and perennial broadleaf weeds and woody brush.
Plant growth stage	Best results when treating rapidly growing plants, particularly seedlings in annuals. In perennials, treat plants later in season when sugars are translocating to underground reproductive structures. For woody species late summer or fall applications are best when using cut stump, stem injection, and basal bark treatments.
Translocation	Readily translocated in the phloem and accumulates at below and above ground growing points or storage organs.
Mechanism of action	Growth regulator herbicide, mimics naturally occurring auxins. Also interferes with the ability of plants to metabolize nitrogen and produce enzymes.
Important considerations	Typically used as a tank mix with other active ingredients.

Glyphosate

Trade names	*Roundup ProMax* (4.5 lb acid equivalent/gal) Other *Roundup* formulations can differ in acid equivalent *Rodeo* (4 lb acid equivalent/gal; lacks a surfactant) *Accord XT II* (5.4 lb acid equivalent/gal; has a surfactant package included) *Aquamaster* (4 lb acid equivalent/gal; lacks a surfactant) Many other names
Chemical family	Glycine or organophosphorus
Manufacturer	Monsanto and others
Formulation	Water soluble
Use areas	Rights-of-way, non-crop areas, riparian areas, emerged aquatic vegetation, forests, rangelands, and other wildland areas.

Application method	Postemergence, wick, cut stump, stem injection
Use rates	1 to 7 qt product/acre for broadcast application, up to undiluted for cut stump treatments, 25 to 70% of concentrate in wick applications.
Soil persistence (range of half-life)	Although glyphosate has an average half-life of 47 days, it is not biologically available to microbes or plants due to its high adsorptive capacity on most soils.
Potential for leaching into surface and ground water	Very low risk of movement due to high adsorption to soil particles.
Volatilization	Negligible
Selectivity	Broad spectrum control of annual and perennial grasses and broadleaf species, as well as many woody species.
Plant growth stage	Best results when treating rapidly growing plants, particularly seedlings in annuals. In perennials, treat plants later in season when sugars are translocating to underground reproductive structures. For woody species late summer or fall applications are best when using cut stump or stem injection treatments.
Translocation	Translocated in the phloem and accumulates at above and below ground growing points and storage areas.
Mechanism of action	Inhibits the activity of enolpyruvyl shikimate 3-phosphate (EPSP) synthase which catalyzes the synthesis of three essential aromatic amino acids in plants; tyrosine, phenylalanine, and tryptophan.
Important considerations	Glyphosate has no soil activity. Foliar applications can be made using many different techniques. Can damage non-target species due to its non-selective nature. Both aquatic and terrestrial formulations available.
Hexazinone	
Trade names	*Velpar L* (2 lb active ingredient/gal) *Velpar DF* (dispersible granules) (75% active ingredient by weight) *Velpar ULW* (75% active ingredient by weight) *Pronone* 10G (10% active ingredient by weight)
Chemical family	Triazine
Manufacturer	DuPont
Formulation	Water Dispersible Liquid or granule
Use areas	Used in rangeland, pastures and non-crop areas, Christmas trees plantations, forestry site preparation and release areas, and industrial areas.
Application method	Typically applied preemergence
Use rates	2 to 10 qt product/acre; 0.3 to 2.3 lb product (*Velpar DF*)/acre
Soil persistence (range of half-life)	Fairly long soil persistence with half-life of about 90 days.
Potential for leaching into surface and ground water	Fairly mobile in the soil due to high water solubility. Known to leach through soil into ground water under certain conditions.
Volatilization	Not volatile
Selectivity	Most often used for brush control, but is considered a broad spectrum herbicide controlling many annual, biennial and perennial broadleaf and grass weeds, and woody plants.
Plant growth stage	Apply preemergence to germination or emergence or to soil where plants are rapidly growing.
Translocation	Rapidly transported in the xylem (water conducting tissues) after root absorption.
Mechanism of action	Inhibits photosynthesis at photosystem II (PSII). Blocks the flow of electons through PSII causing accumulation of toxic free radicals.
Important considerations	Hexazinone can be used near water, but not in water. It is permissible to treat intermittent drainage, intermittently flooded low lying sites, seasonally dry flood plains and transitional areas between upland and lowland sites when no water is present. It is also permissible to treat marshes, swamps and bogs after water has receded as well as seasonally dry flood deltas.

Table 11 continued: Characteristics of herbicides

Imazamox

Trade names	*Clearcast* (1 lb acid equivalent/gal)
Chemical family	Imidazolinone
Manufacturer	SePRO and BASF
Formulation	Water soluble
Use areas	Aquatic areas, lakes, ponds, reservoirs, marshes, ditches and canals
Application method	Applied directly into water or onto emergent foliage of aquatic plants or exposed sediment after drawdown.
Use rates	Not to exceed 500 ppb for direct application to water and 0.5 lb acid equivalent/acre (2 qt product/acre) for foliar treatments.
Soil persistence (range of half-life)	Half-life in water ranges from 5 to 15 days, depending on water clarity, depth, and available sunlight.
Potential for leaching into surface and ground water	Low leaching potential as the herbicide is readily adsorbed to the soil.
Volatilization	Not volatile
Selectivity	Broad spectrum control of submersed, floating, and emergent vegetation in aquatic systems.
Plant growth stage	Applications should be made to plants after they begin active growth. For mature perennial emergent species, application should be made later in the season when translocation of sugars is to underground storage tissues.
Translocation	Absorbed directly into plant tissues through water or foliarly to emerged vegetation. Translocated in the xylem, but primarily in the phloem and accumulates at growing points.
Mechanism of action	Inhibits the synthesis of branched chain amino acids (leucine, valine, and isoleucine) by blocking the activity of the enzyme acetolactate synthase (ALS).
Important considerations	For aquatic use only. Not yet registered in much of United States. Expected registration in 2012.

Imazapic

Trade names	*Plateau* (2 lb acid equivalent/gal)
Chemical family	Imidazolinone
Manufacturer	BASF
Formulation	Water soluble
Use areas	Pastures, rangeland, non-crop areas, and conifer plantation site preparation
Application method	Active both pre- and postemergence.
Use rates	2 to 12 oz product/acre
Soil persistence (range of half-life)	Average half-life reported to be 120 days. However, field results on annual grass control suggest that the half-life based on herbicidal activity is much less than this value in western states.
Potential for leaching into surface and ground water	Studies show the herbicide remains in the top 1 to 1.5 ft of the soil and has low potential for leaching into groundwater, but is susceptible to surface runoff.
Volatilization	Negligible
Selectivity	Broad spectrum herbicide on both broadleaf and grass weeds, but is particularly effective on invasive annual grasses such as downy brome and medusahead.
Plant growth stage	Apply before weeds have emerged or to small rapidly growing plants.
Translocation	Translocates in the xylem, as well as the phloem.
Mechanism of action	Inhibits the synthesis of branched chain amino acids (leucine, valine, and isoleucine) by blocking the activity of the enzyme acetolactate synthase (ALS).
Important considerations	When heavy thatch or litter is present, the herbicide can bind to these residues and effectiveness will be reduced. Do not use around water. Not registered for use in California.

Imazapyr

Trade names	*Arsenal* (4 lb acid equivalent/gal) *Habitat* (2 lb acid equivalent/gal) *Chopper* (2 lb acid equivalent/gal) *Stalker* (2 lb acid equivalent/gal) *Polaris* (2 and 4 lb acid equivalent/gal)
Chemical family	Imidazolinone
Manufacturer	BASF
Formulation	Water soluble and emulsifiable concentrate. Formulated as salts and as emulsifiable concentrates.
Use areas	Riparian areas, emerged aquatic vegetation, forests, and other wildland areas
Application method	Preemergence, postemergence, basal bark (ester formulations), cut stump, and stem injection
Use rates	4 to 40 oz product/acre, undiluted in stem injection treatments
Soil persistence (range of half-life)	Field studies give a range of half-lives from 25 to 142 days, depending on soil characteristics and environmental conditions. Half-life in shallow ponds is between 2 and 3 days.
Potential for leaching into surface and ground water	Studies show the herbicide remains in the top 1 to 1.5 ft of the soil and has low potential for leaching into groundwater, but is susceptible to surface runoff.
Volatilization	Negligible
Selectivity	Broad spectrum herbicide that controls many grasses and broadleaf species, including annuals, biennials, perennials and woody species.
Plant growth stage	Best results when treating rapidly growing plants, particularly seedlings in annuals. In perennials, treat plants later in season when sugars are translocating to underground reproductive structures. For woody species late summer or fall applications are best when using cut stump, stem injection, and basal bark treatments.
Translocation	Translocates in the xylem, as well as the phloem.
Mechanism of action	Inhibits the synthesis of branched chain amino acids (leucine, valine, and isoleucine) by blocking the activity of the enzyme acetolactate synthase (ALS).
Important considerations	Active ingredient may leach out of dead roots of treated plants and be picked up by adjacent non-target species. Use only water soluble formulations around aquatic areas.

Metsulfuron

Trade names	*Escort* (60% active ingredient) *Streamline* (tank mix with aminocyclopyrachlor) *Cimarron* (tank mix with either dicamba and 2,4-D or chlorsulfuron) *Viewpoint* (tank mix with imazapyr and aminocyclopyrachlor) *Oust Extra* (tank mix with sulfometuron) *Opensight* and *Chaparral* (tank mix with aminopyralid)
Chemical family	Sulfonylurea
Manufacturer	DuPont
Formulation	Dispersible granule
Use areas	Pasture, range, non-crop sites
Application method	Pre- and postemergence
Use rates	0.1 to 3 oz product/acre
Soil persistence (range of half-life)	7 to 42 days, with typical half-life of 30 days
Potential for leaching into surface and ground water	Very little leaching below the surface horizon was detected in several soil types.
Volatilization	Non-volatile

Table 11 continued: Characteristics of herbicides

Selectivity	Primarily controls annual and perennial broadleaf species.
Plant growth stage	Best results when weeds are treated in the bud to bloom or fall rosette stage.
Translocation	Translocates in the xylem, as well as the phloem.
Mechanism of action	Inhibits the synthesis of branched chain amino acids (leucine, valine, and isoleucine) by blocking the activity of the enzyme acetolactate synthase (ALS).
Important considerations	Do not apply near water. Not registered for use in California.

Paraquat	
Trade names	*Gramoxone* (30.1% active ingredient, 2 lb active ingredient/gal) *Firestorm* (43.8% active ingredient, 4.14 lb active ingredient/gal)
Chemical family	Bipyridylium
Manufacturer	Syngenta, Chemtura
Formulation	Water soluble
Use areas	Pasture, forests
Application method	Postemergence
Use rates	1 to 4 pt product/acre
Soil persistence (range of half-life)	1000 days, but binds so strongly to clay that it is not biologically available for breakdown or uptake by plants.
Potential for leaching into surface and ground water	Considered immobile in soil.
Volatilization	Non-volatile
Selectivity	Annual broadleaf and grass weeds, but more active on broadleaf species.
Plant growth stage	Best results when treating seedlings that are actively photosynthesizing and growing.
Translocation	Very slow translocation, considered a contact herbicide.
Mechanism of action	Inhibits photosynthesis at the photosystem I (PS I) reaction site.
Important considerations	Complete coverage of target plants is required to obtain effective control.

Penoxsulam	
Trade names	*Galleon* (2 lb active ingredient/gal)
Chemical family	Sulfonamide or triazolopyrimidine
Manufacturer	SePRO
Formulation	Water soluble
Use areas	Aquatic areas, lakes, ponds, reservoirs, marshes, ditches and canals
Application method	Applied directly into water or onto emergent foliage of aquatic plants or exposed sediment after drawdown.
Use rates	Not to exceed 150 ppb
Soil persistence (range of half-life)	Aquatic herbicide that does not persist in soil as it is readily adsorbed. Susceptible to photodegradation and half-life in natural water was 3 days.
Potential for leaching into surface and ground water	Low leaching potential as the herbicide is readily adsorbed to the soil.
Volatilization	Not volatile
Selectivity	Broad spectrum control of broadleaf, grass and sedges in aquatic systems.
Plant growth stage	Applications should be made to plants after they begin active growth. Application to mature plants will require higher application rates and longer exposure periods.
Translocation	Absorbed directly into plant tissues through water, and translocated in the xylem, as well as the phloem.
Mechanism of action	Inhibits the synthesis of branched chain amino acids (leucine, valine, and isoleucine) by blocking

	the activity of the enzyme acetolactate synthase (ALS).
Important considerations	For aquatic use only. Concentration must drop below 1 ppb before treated water can be used for other purposes.

Picloram

Trade names	*Tordon* 22K (2 lb acid equivalent/gal) *Grazon P+D*, *Tordon 101 Mixture* (=*Tordon 101M*), *Tordon RTU*, and *Pathway* (tank mixes with 2,4-D) *Surmount* (tank mix with fluroxypyr)
Chemical family	Pyridine carboxylic acid
Manufacturer	Dow AgroSciences
Formulation	Water soluble
Use areas	Conservation Reserve Program areas, non-crop sites, forests, rangelands, utility areas, rights-of-way, wildlands
Application method	Both pre- and postemergence to weeds, cut stump treatment on woody species
Use rates	Not to exceed 1 lb acid equivalent per acre (2 qt of *Tordon 22K*/acre)
Soil persistence (range of half-life)	Average half-life in soil of 90 days, with a range of 20 to 300 days, depending on climate and soils.
Potential for leaching into surface and ground water	Can leach in soils and has the potential to contaminate groundwater. Leach more prevalent in sandy soils with low organic matter content.
Volatilization	Negligible
Selectivity	Many broadleaf herbaceous and woody plants.
Plant growth stage	For annuals, best to apply in seedling stage as this will provide both postemergence and preemergence activity. For perennials, plants should be fully expanded to ensure movement to underground vegetative parts. For woody species late summer or fall applications are best when using cut stump treatments.
Translocation	Readily translocated in the phloem and accumulates at below and above ground growing points or storage organs.
Mechanism of action	Growth regulator herbicide, mimics naturally occurring auxins.
Important considerations	Long residual herbicide that can leach into groundwater in some areas and may restrict revegetation of sensitive desirable plants. Products with >1 lb a.e./gal are federally restricted use products and applicators need special licensing to apply them. Not registered for use in California.

Propoxycarbazone-sodium

Trade names	*Canter R+P* (70% active ingredient)
Chemical family	Sulfonylaminocarbonyltriazolilnone
Manufacturer	Wilbur-Ellis
Formulation	Water dispersable granule
Use areas	Rangeland, grass pastures and Conservatrion Reserve Program, as well as restoration sites
Application method	Postemergence
Use rates	Not to exceed 1.2 oz product/acre
Soil persistence (range of half-life)	Typical half-life of 9 days, persistence is short
Potential for leaching into surface and ground water	Low threat for leaching into ground water
Volatilization	Non-volatile
Selectivity	Broad spectrum, primarily controls grasses, but can also control many broadleaf, both annual and perennial.

Table 11 continued: Characteristics of herbicides

Plant growth stage	Late fall or spring to rapidly growing plants. Best control occurs when applications are made before grass weeds tiller and broadleaf weeds are smaller than 2 in in diameter.
Translocation	Translocates in the xylem, as well as the phloem.
Mechanism of action	Inhibits the synthesis of branched chain amino acids (leucine, valine, and isoleucine) by blocking the activity of the enzyme acetolactate synthase (ALS).
Important considerations	Do not apply near water.

Rimsulfuron

Trade names	*Matrix* (25% active ingredient)
Chemical family	Sulfonylurea
Manufacturer	DuPont
Formulation	Dry flowable
Use areas	Rangeland, industrial non-crop sites and for restoration sites
Application method	Preemergence
Use rates	1 to 4 oz product/acre
Soil persistence (range of half-life)	Typical half-life of 20 days, persistence is relatively short
Potential for leaching into surface and ground water	Low threat for leaching into ground water
Volatilization	Non-volatile
Selectivity	Broad spectrum, many broadleaf and grass species, both annual and perennial.
Plant growth stage	Late fall to early winter when soil temperatures are cool and rainfall is available to activate the product.
Translocation	Translocates in the xylem, as well as the phloem.
Mechanism of action	Inhibits the synthesis of branched chain amino acids (leucine, valine, and isoleucine) by blocking the activity of the enzyme acetolactate synthase (ALS).
Important considerations	Do not apply near water.

Sethoxydim

Trade names	*Poast* (1.5 lb acid equivalent/gal)
Chemical family	Cyclohexanedione
Manufacturer	BASF
Formulation	Emulsifiable concentrate
Use areas	Uncultivated non-agricultural areas, rights-of-way, fallow lands
Application method	Postemergence
Use rates	1 to 2.5 pt product/acre
Soil persistence (range of half-life)	Average half-life in soils is 4 to 5 days, but can range from a few hours to 25 days.
Potential for leaching into surface and ground water	Very low risk
Volatilization	Negligible
Selectivity	Specific to annual and perennial grasses
Plant growth stage	For annual species, best applied when plants are young and rapidly growing. For perennials, best applied when plants are fully expanded and photosynthates are moving to below ground vegetative structures. Do not apply herbicide when plants are under stress, particularly drought stress.

Translocation	Translocation is slow, but movement is primarily in the phloem where it accumulates in the growing points, typically the shoots.
Mechanism of action	Inhibits fatty acid synthesis by blocking the activity of the enzyme acetyl CoA carboxylase (ACCase).
Important considerations	Sethoxydim is not registered for use in many non-crop areas. Do not apply near water.

Sulfometuron

Trade names	*Oust* (75% active ingredient) *Landmark* (tank mix with chlorsulfuron) (50% active ingredient) *Oust Extra* (tank mix with metsulfuron) (56.25% active ingredient)
Chemical family	Sulfonylurea
Manufacturer	DuPont
Formulation	Dispersible granule
Use areas	Rangeland, forest, industrial non-crop sites
Application method	Preemergence
Use rates	0.75 to 2.25 oz product/acre
Soil persistence (range of half-life)	Typical half-life of 20 to 28 days, but half-life is extended under cold temperatures.
Potential for leaching into surface and ground water	Mobility in soil is greater at higher soil pH and lower organic matter content.
Volatilization	Non-volatile
Selectivity	Broad spectrum to non-selective, controls many broadleaf and grass species, both annual and perennial.
Plant growth stage	Fall or spring, before moisture expectation and plant growth, but not when soil is frozen.
Translocation	Translocates in the xylem, as well as the phloem.
Mechanism of action	Inhibits the synthesis of branched chain amino acids (leucine, valine, and isoleucine) by blocking the activity of the enzyme acetolactate synthase (ALS).
Important considerations	Do not apply near water. Herbicide can move off-site on light wind-blown arid soil.

Sulfosulfuron

Trade names	*Outrider* (75% active ingredient)
Chemical family	Sulfonylurea
Manufacturer	Monsanto
Formulation	Dispersible granule
Use areas	Rangeland, pasture, non-crop sites
Application method	Both pre- and postemergence
Use rates	0.75 to 2 oz product/acre
Soil persistence (range of half-life)	Average half-life of 12 days, but ranges between 2 to 6 days under aerobic and 7 to 28 days under anaerobic conditions.
Potential for leaching into surface and ground water	Mobile in soil but short half life reduces risk of movement to groundwater, except in areas with shallow water tables or very permeable soils.
Volatilization	Negligible
Selectivity	Broad spectrum to non-selective, controls many annual and perennial grasses, and primarily annual broadleaf species.
Plant growth stage	For annuals, treat young plants when they are rapidly growing and not under stress.
Translocation	Translocates in the xylem, as well as the phloem.

Table 11 continued: Characteristics of herbicides

Mechanism of action	Inhibits the synthesis of branched chain amino acids (leucine, valine, and isoleucine) by blocking the activity of the enzyme acetolactate synthase (ALS).
Important considerations	Do not apply near water. Not registered for use in California.

Tebuthiuron

Trade names	*Spike 20P* (20% active ingredient) *Spike 80DF* (80% active ingredient)
Chemical family	Substituted urea
Manufacturer	Dow AgroSciences
Formulation	Pellet (P) or dry flowable (DF)
Use areas	Utility areas, pastures, rights-of-way, rangeland, wildland areas
Application method	Preemergence application as granular or pellet. Can be applied postemergence, but is not as effective.
Use rates	Up to 4 lb active ingredient/acre, or 6 lb active ingredient/acre for spot treatments
Soil persistence (range of half-life)	Long soil persistence with half-life generally between 12 and 15 months in areas receiving adequate rainfall and much longer in low rainfall areas.
Potential for leaching into surface and ground water	Only a potential leaching problem where soils are permeable and water tables are shallow. Not generally found to move laterally into surface water (< 5 ft).
Volatilization	Negligible
Selectivity	Used for woody plant control.
Plant growth stage	Dormant season application timing is often recommended. Must have sufficient rainfall after treatment to move herbicide into root zone.
Translocation	Readily absorbed by the roots and translocated to the foliage through the xylem.
Mechanism of action	Inhibits photosynthesis at photosystem II (PSII). Blocks the flow of electons through PSII causing accumulation of toxic free radicals.
Important considerations	Very persistent herbicide that can damage or kill most trees. Generally only used for spot treatments to selectively control target species. Broadcast applications can be used for sagebrush control to improve sage grouse habitat. High use rates can give total vegetation control. Do not apply near water.

Triclopyr

Trade names	*Garlon 3A* (3 lb acid equivalent/gal) *Garlon 4 Ultra* (4 lb acid equivalent/gal) *RemedyUltra* (4 lb acid equivalent/gal) *Renovate* (3 lb acid equivalent/gal) *Pathfinder II* (0.75 lb acid equivalent/gal) *Crossbow* (tank mix with 2,4-D) *Capstone* (tank mix with aminopyralid) *PastureGard HL* (tank mix with fluroxypyr)
Chemical family	Pyridinyloxy alkanoic acid
Manufacturer	Dow AgroSciences
Formulation	Water soluble or emulsifiable concentrate. Formulated as acids, salts, and esters and sold as liquids
Use areas	Utility areas, rights-of-way, rangelands, forests, natural areas, aquatic and riparian areas (aquatic registrations only)
Application method	Postemergence, cut stump, stem injection, and basal bark (ester formulations only). Aquatic formulations can be applied to emergent vegetation or directly to water.
Use rates	For terrestrial and riparian areas 1 to 8 qt *Garlon 4 Ultra*/acre or 1 to 12 qt *Garlon 3A*/acre. In aquatic systems, 1.5 to 6 lb acid equivalent/acre (2 to 8 qt product/acre). Maximum use rate in grazed areas is 2 lb a.e./acre)

Soil persistence (range of half-life)	Average soil half-life of 30 days, but ranges between 10 and 46 days, depending soil characteristics, moisture, and temperature. In clear water, half-life is about 10 hours at 25°C (77°F).
Potential for leaching into surface and ground water	Generally not considered to have a high potential for ground or surface water contamination at it is only moderately persistent and applied postemergence.
Volatilization	Negligible for amine formulations, but ester formulations are low volatile.
Selectivity	Controls woody and herbaceous broadleaf species.
Plant growth stage	Best results when treating rapidly growing plants, particularly seedlings in annuals. In perennials, treat plants later in season when sugars are translocating to underground reproductive structures. For woody species late summer or fall applications are best when using cut stump, stem injection, and basal bark treatments.
Translocation	Readily translocated in the phloem and accumulates at below and above ground growing points or storage organs.
Mechanism of action	Growth regulator herbicide, mimics naturally occurring auxins.
Important considerations	To decrease the potential for volatility avoid spray contact with impervious surfaces, such as roads and rocks, with increasing ambient air temperatures.

Table 12. Rainfastness and grazing and hay harvesting restrictions for herbicides registered for terrestrial use.

Common name	Rainfastness	Grazing and hay harvesting restrictions
2,4-D	6 to 8 hour period suggested for amine formulations, while 1 hour is suggested for the ester formulations	Do not graze lactating dairy animals on treated areas within 7 days after treatment. Do not cut forage for hay within 7 days of application.
Aminocyclopyrachlor	At least a 48 hour period is recommended	Because the herbicide is tank mixed and not used alone, do not graze or feed forage, hay, or straw from treated areas to livestock. DuPont expects to have grazing tolerance and registrations sometime next year.
Aminopyralid	No indication on label, but likely 2 to 6 hours period is best when applying for postemergence control	No grazing or hay harvest restrictions. Do not use treated cut foliage as mulch. See label for further information on use restrictions for hay, manure or other materials.
Chlorsulfuron	Not important when used preemergence, but when used postemergence likely a 48 hour period is recommended, but as low as 4 hours has been noted	No grazing or hay harvest restrictions for any livestock, including lactating animals, with application rates up to 1.33 oz/acre, which is the maximum use rate in grazed areas.
Clethodim	Do not apply if rain is expected within 1 hour	15 days minimum following application is required before grazing, feeding or harvesting for forage or hay.
Clopyralid	A 6 hour rain-free period is suggested for best control when using postemergence	No grazing or hay harvest restrictions. Do not use treated cut foliage as mulch.
Dicamba	Rainfall or irrigation occurring within 4 hours after postemergence applications may reduce the effectiveness	For lactating dairy animals there is a 7-day grazing and 37 day hay harvest restriction at rates up to 1 pt (0.5 lb a.e./ acre). For rates up to 1 qt (1 lb a.e./acre) it is 21 days before grazing and 51 days restriction before hay harvest. Up to 2 qt (2 lb a.e./acre) it is 40 days before grazing and 70 days restriction before hay harvest. There is no waiting period between treatment and grazing for non-lactating dairy animals.
Fluazifop	Rainfast period is only 1 hour	In pastures the withholding period for grazing is 21 days.
Fluroxypyr	No information on label.	No grazing restrictions for livestock, including lactating or non-lactating dairy animals.
Glyphosate	Label notes that rainfall or irrigation after application may reduce effectiveness, but 0.5 to 4 hours is noted in other reports, depending on formulation	No waiting period required between treatment and feeding or grazing for 2 qt/acre or less of *Roundup ProMax*. At rates greater than 2 qt/acre, remove livestock before application and wait 8 weeks after application before grazing or harvesting.
Hexazinone	Preemergence, no rainfastness period necessary	Recently changed from 30-day grazing restriction to no grazing restrictions for beef and dairy cattle. Haying restrictions are 38 days after application up to 1.125 lb a.i./acre, but 60 days for rates above 1.125 lb a.i./acre for both grazing and haying.

Common name	Rainfastness	Grazing and hay harvesting restrictions
Imazapic	Rainfast period is only 1 hour when used for postemergence control	No grazing restrictions for lactating dairy animals or beef and non-lactating dairy animals. There is a 7 day restriction before hay harvest.
Imazapyr	Only 1 hour for postemergence use, often used for preemergence treatment	No grazing restrictions. Do not cut for hay for 7 days after application. Spot applications to grass pasture and rangeland may not exceed more than 10% of the area to be grazed or cut for hay.
Metsulfuron	Label indicates that weed and brush control may be reduced if rainfall occurs soon after application	No grazing or haying restriction for rates of 1.67 oz/acre or less. At rates from 1.67 to 3.33 oz/acre, forage grasses may be cut for hay, fodder or green forage and fed to livestock, including lactating animals, 3 days after treatment.
Paraquat	Rain occurring 15 to 30 minutes or more after application will have no effect on the activity	Grazing or pre-harvest interval is 60 days in alfalfa, clover, and other legumes. Do not graze livestock after application or before burning in pastures.
Picloram	Often used preemergence, for postemergence treatment 2 to 6 hours is recommended	Do not graze lactating dairy animals on treated areas within 14 days after treatment. When applying more than 1 qt/acre (0.5 lb a.e./acre) do not cut grass for feed within 14 days of treatment. There are no restrictions for rates below 1 qt/acre.
Propoxycarbozone-sodium	Rainfast 4 hours after application	Do not cut area for hay within 7 days of treatment.
Rimsulfuron	Typically applied preemergence, no rainfastness period necessary	Do not graze or feed forage, grain or fodder from treated areas to livestock for one year.
Sethoxydim	Rainfast period is only 1hour	For alfalfa cover crops, do not apply within 7 days of grazing, feeding, or cutting for forage, or within 14 days of cutting for alfalfa for hay.
Sulfometuron	Preemergence, no rainfastness period necessary	Grazing and cut forage restrictions of 12 months post-application apply.
Sulfosulfuron	Rainfast period is 2 hours for postemergence use	There are no grazing restrictions. Do not mow or harvest for 2 weeks before or 2 weeks after treatment.
Tebuthiuron	Preemergence, no rainfastness period necessary	There are no grazing restrictions. Do not cut hay for livestock feed for 12 months after treatment.
Triclopyr	No rainfastness period indicated in label, but likely to be similar to acid and ester formulations of 2,4-D, which indicate 6 to 8 hour period for amine formulation and 1 hour for ester formulation	Except for lactating dairy animals, there are no grazing restrictions following application. This is currently in the process of being changed. Do not allow lactating dairy animals to graze treated areas until the next growing season following application. Do not harvest hay for 14 days after application. Grazed areas of non-cropland and forestry sites may be spot treated if they comprise no more than 10% of the total grazable area. Maximum use rate of 2 lb a.e./acre for grazed areas except for basal and cut stump applications.

Table 13. Characteristics of **aquatic herbicides** registered, or soon to be registered, for use in the western United States.

Herbicide	Contact or systemic	(h or d) Half-life in water	Use or maximum rate allowed in water	Aquatic use	Target plant groups
2,4-D	Systemic	7 to 14 d	20 to 40 lb a.e./acre, to 2 ppm	Floating leaf and submerged	Dicots
Acrolein	Contact	5.5 to 10.2 h	Up to 15 ppm	Submerged	Non-selective, best on annuals
Bispyribac-sodium	Systemic	9 d	20 to 45 ppb	Floating leaf and submerged	Broad spectrum
Copper/copper chelate	Contact	Copper not degraded, 3 d for complexed copper	0.2 to 1 ppm	Submerged	Algae, some submerged and floating leaf
Diquat	Contact	1 to 7 d, binds with suspended soil particles in water	2 gal product/acre water, to 2 ppm	Floating leaf and submerged	Broad spectrum, best on annuals
Endothall	Contact	3 to 7 d	0.05 to 5 ppm	Submerged	Broad spectrum, best on annuals
Flumioxazin	Contact	16.1 h at pH 7.0; 4.1 d at pH 5.0	100 to 400 ppb a.i.	Floating leaf and submerged	Broad spectrum on submerged, floating leaf and some algae
Fluridone	Systemic	20 d	Up to 150 ppb	Submerged	Broad spectrum
Glyphosate	Systemic	14 d	Up to 7.5 pt product/acre, to 0.2 ppm	Emerged and floating leaf	Non-selective
Imazamox	Systemic	5 to 15 d	Up to 500 ppb	Submersed, floating leaf, and emergent	Broad spectrum
Imazapyr	Systemic	2 to 3 d	4 to 40 oz product/acre	Emergent	Broad spectrum
Penoxsulam	Systemic	3 d	Up to 150 ppb	Submersed, floating leaf, and emergent	Broad spectrum
Triclopyr	Systemic	10 h	1.5 to 6 lb acid equivalent/acre (2 to 8 qt/acre), to 2.5 ppm	Submerged, emerged and floating leaf	Dicots

Table 14. **Biological control** programs or established agents for **non-crop weeds** in the western United States and a qualitative evaluation of their effectiveness. Effectiveness can vary dramatically from one location to another.

Family/Species Common name	(pathogen noted) Agent	Type of organism	Part of plant attacked	(control ability or potential) Effectiveness
APIACEAE				
Conium maculatum Poison-hemlock	*Agonopterix alstroemeriana*	Moth	Feeds on foliage, buds, and flowers	Poor in CA; Good to Excellent in other western states
ASTERACEAE				
Acroptilon repens Russian knapweed	*Aulacidea acroptilonica*	Wasp	Develops galls on stem	Unknown
	Jaapiella ivannikova	Midge	Develops galls on growing points	Unknown
	Subanguina picridis	Nematode	Gall former	Poor
	Urophora kasachstanica	Gall fly	Larvae feed on seed	Unknown
	Urophora xanthippe	Gall fly	Larvae feed on seed	Unknown
Carduus acanthoides Plumeless thistle	*Rhinocyllus conicus*	Weevil	Larvae feed on seed	Good
	Trichosirocalus horridus	Weevil	Larvae feeds on roots and root crowns	Good
	Urophora solstitialis	Gall fly	Larvae feed on seed	Unknown
Carduus nutans Musk thistle	*Cheilosia corydon*	Fly	Larvae mine into stem	Unknown
	Psylliodes chalcomera	Beetle	Defoliator	Unknown
	Puccinia carduorum (pathogen)	Rust fungus	Damage foliage	Good
	Rhinocyllus conicus	Weevil	Larvae feed on seed	Excellent
	Trichosirocalus horridus	Weevil	Larvae feeds on roots and root crowns	Unknown
	Urophora solstitialis	Gall fly	Larvae feed on seed	Unknown
Carduus pynocephalus Italian thistle	*Cheilosia corydon*	Fly	Larvae mine into stem	Unknown
	Rhinocyllus conicus	Weevil	Larvae feed on seed	Fair
	Trichosirocalus horridus	Weevil	Larvae feeds on roots and root crowns	Fair
Carduus tenuiflorus Slenderflower thistle	*Cheilosia corydon*	Fly	Larvae mine into stem	Fair
	Puccinia carduorum (pathogen)	Rust fungus	Damage foliage	Unknown
	Rhinocyllus conicus	Weevil	Larvae feed on seed	Good
	Trichosirocalus horridus	Weevil	Larvae feeds on roots and root crowns	Unknown
Centaurea calcitrapa Purple starthistle	*Bangasternus fausti*	Weevil	Larvae feed on seed	Poor
	Larinus minutus	Weevil	Larvae feed on seed	Poor
	Terellia virens	Fly	Larvae feed on seed	Poor
Centaurea cyanus Bachelor's button	*Chaetorellia australis*	Fly	Larvae feed on seed	Fair
	Urophora quadrifasciata	Gall fly	Larvae feed on seed	Unknown
Centaurea debeauxii spp. *thuillieri* Meadow knapweed	*Bangasternus fausti*	Weevil	Larvae feed on seed	Poor
	Cyphocleonus achates	Weevil	Root borer, gall former	Poor
	Larinus minutus	Weevil	Larvae feed on seed	Fair
	Larinus obtusus	Weevil	Larvae feed on seed	Good
	Metzneria paucipunctella	Moth	Larvae feed on seed	Unknown
	Sphenoptera jugoslavica	Beetle	Root borer and gall former	Poor
	Urophora quadrifasciata	Gall fly	Larvae feed on seed	Fair

Table 14 continued: Biological control for non-crop weeds

Family/Species Common name	(pathogen noted) Agent	Type of organism	Part of plant attacked	(control ability or potential) Effectiveness
Centaurea diffusa Diffuse knapweed	*Agapeta zoegana*	Moth	Larvae bore into root	Unknown
	Bangasternus fausti	Weevil	Larvae feed on seed	Good
	Chaetorellia acrolophi	Fly	Larvae feed on seed	Unknown
	Cyphocleonus achates	Weevil	Root borer, gall former	Good
	Larinus minutus	Weevil	Larvae feed on seed	Excellent
	Larinus obtusus	Weevil	Larvae feed on seed	Poor
	Metzneria paucipunctella	Moth	Larvae feed on seed	Unknown
	Pterolonche inspersa	Moth	Root borer	Poor
	Sphenoptera jugoslavica	Beetle	Root borer and gall former	Good
	Terellia virens	Fly	Larvae feed on seed	Unknown
	Urophora affinis	Gall fly	Larvae feed on seed	Good
	Urophora quadrifasciata	Gall fly	Larvae feed on seed	Good
Centaurea iberica Iberian starthistle	*Bangasternus fausti*	Weevil	Larvae feed on seed	Unknown
Centaurea jacea Brown knapweed	*Larinus obtusus*	Weevil	Larvae feed on seed	Fair
	Urophora quadrifasciata	Gall fly	Larvae feed on seed	Unknown
Centaurea nigra Black knapweed	*Urophora quadrifasciata*	Gall fly	Larvae feed on seed	Unknown
Centaurea solstitialis Yellow starthistle	*Bangasternus orientalis*	Weevil	Larvae feed on seed	Poor
	Ceratapion basicorne	Weevil	Rosette feeder and root borer	Unknown
	Chaetorellia australis	Fly	Larvae feed on seed	Poor
	Chaetorellia succinea	Fly	Larvae feed on seed	Good
	Eustenopus villosus	Weevil	Larvae feed on seed	Good
	Larinus curtus	Weevil	Larvae feed on seed	Poor
	Puccinia jaceae var. *solstitialis* (pathogen)	Rust fungus	Damage foliage	Poor
	Urophora jaculata	Gall fly	Larvae feed on seed	Poor
	Urophora quadrifasciata	Gall fly	Larvae feed on seed	Poor
	Urophora sirunaseva	Gall fly	Larvae feed on seed	Poor
Centaurea stoebe Spotted knapweed	*Agapeta zoegana*	Moth	Root borer	Fair
	Bangasternus fausti	Weevil	Larvae feed on seed	Fair
	Chaetorellia acrolophi	Fly	Larvae feed on seed	Poor
	Cyphocleonus achates	Weevil	Root borer, gall former	Fair
	Larinus minutus	Weevil	Larvae feed on seed	Excellent
	Larinus obtusus	Weevil	Larvae feed on seed	Good
	Metzneria paucipunctella	Moth	Larvae feed on seed	Good
	Pelochrista medullana	Moth	Root borer	Poor
	Pterolonche inspersa	Moth	Root borer	Fair
	Sphenoptera jugoslavica	Beetle	Root borer and gall former	Fair
	Terellia virens	Fly	Larvae feed on seed	Fair

Table 14 continued: Biological control for non-crop weeds

Family/Species Common name	(pathogen noted) Agent	Type of organism	Part of plant attacked	(control ability or potential) Effectiveness
Centaurea stoebe	*Urophora affinis*	Gall fly	Larvae feed on seed	Good
Spotted knapweed	*Urophora quadrifasciata*	Gall fly	Larvae feed on seed	Good
Centaurea virgata ssp. *squarrosa*	*Agapeta zoegana*	Moth	Root borer	Unknown
	Bangasternus fausti	Weevil	Larvae feed on seed	Excellent
Squarrose knapweed	*Cyphocleonus achates*	Weevil	Root borer, gall former	Poor
	Larinus minutus	Weevil	Larvae feed on seed	Excellent
	Pterolonche inspersa	Moth	Root borer	Poor
	Sphenoptera jugoslavica	Beetle	Root borer and gall former	Fair
	Terellia virens	Fly	Larvae feed on seed	Poor
	Urophora affinis	Gall fly	Larvae feed on seed	Poor
	Urophora quadrifasciata	Gall fly	Larvae feed on seed	Poor
Chondrilla juncea	*Bradyrrhoa gilveolella*	Moth	Root borer	Poor
Rush skeletonweed	*Cystiphora schmidti*	Midge	Root borer and gall former	Excellent
	Eriophyes chondrillae	Gall mite	Destroys young developing tissues	Excellent
	Puccinia chondrillina (pathogen)	Rust fungus	Damage foliage	Good
Cirsium arvense	*Altica carduorum*	Flea beetle	Larvae defoliate plants	Poor
Canada thistle	*Ceutorhynchus litura*	Weevil	Larvae feeds on roots and root crowns	Fair
	Rhinocyllus conicus	Weevil	Larvae feed on seed	Fair
	Urophora cardui	Gall fly	Larvae bore into stems	Fair
Cirsium vulgare	*Cheilosia corydon*	Fly	Larvae mine into stem	Unknown
Bull thistle	*Rhinocyllus conicus*	Weevil	Larvae feed on seed	Poor
	Trichosirocalus horridus	Weevil	Larvae feeds on roots and root crowns	Unknown
	Urophora stylata	Gall fly	Larvae feed on seed	Good
Onopordum acanthium	*Rhinocyllus conicus*	Weevil	Larvae feed on seed	Poor
Scotch thistle	*Trichosirocalus horridus*	Weevil	Larvae feeds on roots and root crowns	Unknown
Senecio jacobaea	*Botanophila seneciella*	Fly	Larvae feed on seed	Fair
Tansy ragwort	*Longitarsus jacobaeae*	Flea beetle	Defoliator and root borer	Excellent
	Pegohylemyia seneciella	Fly	Larvae feed on seed	Poor
	Tyria jacobaeae	Moth	Defoliator	Excellent
Silybum marianum	*Rhinocyllus conicus*	Weevil	Larvae feed on seed	Fair
Blessed milkthistle				
CHENOPODIACEAE				
Halogeton glomeratus	*Coleophora klimeschiella*	Moth	Leaf miner	Poor
Halogeton	*Coleophora parthenica*	Moth	Stem miner	Poor
Salsola tragus	*Aceria salsolae*	Blister mite	Feeds on young growing tips	Unknown
Russian thistle	*Coleophora klimeschiella*	Moth	Leaf miner	Poor
	Coleophora parthenica	Moth	Stem miner	Poor

Table 14 continued: Biological control for non-crop weeds

Family/Species Common name	(pathogen noted) Agent	Type of organism	Part of plant attacked	(control ability or potential) Effectiveness
CLUSIACEAE				
Hypericum perforatum Common St. Johnswort	Agrilus hyperici	Beetle	Root and stem borer	Excellent
	Aplocera plagiata	Moth	Defoliator	Fair
	Chrysolina hyperici	Beetle	Defoliator	Excellent
	Chrysolina quadrigemina	Beetle	Defoliator	Excellent
	Chrysolina varians	Beetle	Defoliator	Poor
	Zeuxidiplosis giardi	Midge	Forms galls at leaf buds	Poor
CONVOLVULACEAE				
Convolvulus arvensis Field bindweed	Aceria malherbae	Eriophyid mite	Forms galls on buds and young leaf tips	Fair
	Tyta luctuosa	Moth	Defoliator	Poor
EUPHORBIACEAE				
Euphorbia esula Leafy spurge	Aphtona abdominalis	Flea beetle	Defoliator and root feeder	Unknown
	Aphthona cyparissiae	Flea beetle	Defoliator and root feeder	Good
	Aphthona czwalinae	Flea beetle	Defoliator and root feeder	Good
	Aphthona flava	Flea beetle	Defoliator and root feeder	Fair
	Aphthona lacertosa	Flea beetle	Defoliator and root feeder	Excellent
	Aphthona nigriscutis	Flea beetle	Defoliator and root feeder	Good
	Chamaesphecia crassicornis	Moth	Root borer	Unknown
	Chamaesphecia empiformis	Moth	Root borer	Unknown
	Chamaesphecia hungarica	Moth	Root borer	Unknown
	Chamaesphecia tenthrediniformis	Moth	Root borer	Unknown
	Hyles euphorbiae	Moth	Defoliator	Good
	Oberea erythrocephala	Beetle	Stem and root borer	Good
	Spurgia esula	Midge	Forms galls on shoot tips	Unknown
FABACEAE				
Cytisus scoparius Scotch broom	Aceria genistae	Eriophyid gall mite	Feeds on tissue sap	Unknown
	Apion fuscirostre	Weevil	Larvae feed on seed	Poor
	Bruchidius villosus	Beetle	Larvae feed on seed	Fair in some areas
	Exapion fuscirostre	Weevil	Larvae feed on seed	Fair in some areas
	Leucoptera spartifoliella	Moth	Stem and twig miner	Poor
Genista monspessulana French broom	Bruchidius villosus	Beetle	Larvae feed on seed	Unknown
Ulex europaeus Gorse	Exapion ulicis	Weevil	Larvae feed on seed	Fair
	Tetranychus lintearius	Spider mite	Feeds on tissue sap	Poor
LAMIACEAE				
Salvia aethiopis Mediterranean sage	Phrydiuchus spilmani	Weevil	Larvae feeds on roots and root crowns	Unknown
	Phrydiuchus tau	Weevil	Larvae feeds on roots and root crowns	Good

Table 14 continued: Biological control for non-crop weeds

Family/Species Common name	(pathogen noted) **Agent**	**Type of organism**	**Part of plant attacked**	(control ability or potential) **Effectiveness**
LYTHRACEAE				
Lythrum salicaria Purple loosestrife	Galerucella calmariensis	Beetle	Defoliator	Excellent
	Galerucella pusilla	Beetle	Defoliator	Excellent
	Hylobius transversovittatus	Weevil	Root borer	Fair
	Nanophyes marmoratus	Weevil	Adults and larvae feel on buds and seed	Fair
POACEAE				
Arundo donax Giant reed	Tetramesa romana	Galling wasp	Larvae feed on stem tissue	Unknown
Spartina alterniflora Smooth cordgrass	Prokelisia marginata	Plant hopper	Sap sucker	Low
SCROPHULARIACEAE				
Linaria dalmatica Dalmatian toadflax	Brachypterolus pulicarius	Flower beetle	Adult feed on shoots, larvae feed on flowers	Poor
	Calophasia lunula	Moth	Defoliator	Fair
	Eteobalea intermediella	Moth	Root borer	Poor
	Eteobalea serratella	Moth	Root borer	Poor
	Gymnetron antirrhini	Weevil	seed head	Unknown
	Gymnetron linariae	Weevil	root-galling	Unknown
	Mecinus janthiniformis	Weevil	Stem borer	Excellent
Linaria vulgare Yellow toadflax	Brachypterolus pulicarius	Flower beetle	Adult feed on shoots, larvae feed on flowers	Fair
	Calophasia lunula	Moth	Defoliator	Fair
	Gymnetron antirrhini	Weevil	Larvae feed on seed	Fair
	Gymnetron linariae	Weevil	root-galling	Unknown
	Mecinus janthinus	Weevil	Stem borer	Unknown
TAMARICACEAE				
Tamarix parviflora Smallflower tamarisk	Diorhabda elongata	Beetle	Defoliator	Excellent
Tamarix ramosissima and hybrids Saltcedar	Diorhabda carinulata	Beetle	Defoliator	Excellent
ZYGOPHYLLACEAE				
Tribulus terrestris Puncturevine	Microlarinus lareynii	Weevil	Seed feeder	Excellent
	Microlarinus lypriformis	Weevil	Stem borer	Excellent

Table 15. **Biological control** programs or established agents for **aquatic weeds** in the western United States and a qualitative evaluation of their effectiveness. Effectiveness can vary dramatically from one location to another.

Family/Species Common name	Common name	(pathogen noted) Agent	Type of organism	Part of plant attacked	(control ability or potential) Effectiveness
AMARANTHACEAE					
Alternanthera philoxeroides Alligatorweed		Arcola malloi	Moth	Stem borer	Poor
		Agasicles hygrophila	Flea beetle	Defoliator	Excellent in some areas
		Amynothrips andersoni	Thrip	Defoliator	Poor
		Vogtia malloi	Moth	Stem borer	Fair
ARACEAE					
Pistia stratiotes Water lettuce		Neohydronomus affinis	Weevil	Adults and larvae feed on foliage	Unknown
		Spodoptera pectinicornis	Moth	Larvae feed on foliage	Unknown
AZOLLACEAE					
Azolla filiculoides Mosquitofern		Stenopelmus rufinasus	Weevil	Frond feeder	Excellent
HALORAGACEAE					
Myriophyllum spicatum Eurasian watermilfoil		Euhrychiopsis lecontei	Weevil	Stem feeder	Excellent in some areas
HYDROCHARITACEAE					
Hydrilla verticillata Hydrilla		Bagous affinis	Weevil	Tuber feeder	Poor
		Bagous hydrillae	Weevil	Stem borer	Unknown
		Hydrellia balcuinasi	Fly	Leaf miner	Poor
		Hydrellia pakistanae	Fly	Leaf miner	Poor
PONTEDERIACEAE					
Eichhornia crassipes Water hyacinth		Neochetina bruchi	Weevil	Leaf petiole feeders	Excellent in some areas
		Neochetina eichhorniae	Weevil	Leaf petiole feeders	Excellent in some areas
		Sameodes alligutalis	Moth	Larvae feed on foliage	Poor
		Niphograpta affinis	Weevil	Defoliator	Poor
SALVINIACEAE					
Salvinia molesta complex Giant salvinia		Cyrtobagous salviniae	Weevil	Foliage feeder	Excellent

Table 16. **Grass carp** feeding preferences and effect on control.

Preferred species easily controlled	Moderately preferred species occasionally controlled	Least preferred species not generally controlled
Cabomba caroliniana (fanwort; submerged)	Aquatic grasses (emergent)	*Alternanthera philoxeroides* (alligatorweed; emergent)
Ceratophyllum demersum (coontail; submerged)	*Azolla* spp. (mosquitofern; free floating)	*Brasenia schreberi* (watershield; floating leaves)
Egeria densa (Brazilian egeria; submerged)	*Hydrocotyle* spp. (pennywort; emergent or with floating leaves)	Bulrushes, cattails and large grasses (emergent)
Elodea canadensis (Elodea; submerged)	*Lemna* spp. (duckweeds; free floating)	*Eichhornia crassipes* (water hyacinth; free floating)
Hydrilla verticillata (hydrilla; submerged)	*Myriophyllum spicatum* (Eurasian watermilfoil; submerged)	*Eleocharis* spp. (spikerushes, submerged or emergent)
Myriophyllum aquaticum (parrotfeather; partly emerged, mostly submerged)	*Salvinia molesta* complex (giant salvinia; free floating)	Filamentous algae (submerged)
Najas spp. (naiads; submerged)	*Wolffia* spp. (watermeal)	*Nitella* spp. (stonewort; submerged algae)
Potamogeton spp. (pondweeds; many submerged, others with floating leaves)		*Nuphar polysepala* (yellow pondlily; floating leaves)
Ruppia spp. (widgeongrass; submerged)		*Nymphaea* spp. (water lilies; floating leaves)
		Nymphoides peltata (yellow floatingheart; floating leaves)
		Polygonum spp. (smartweeds; emergent)

Table 17. Important **internet sites** for herbicides.

Information	Website
Crop Data Management Systems (CDMS) a searchable database of print-on-demand pesticide labels including many SLN 24(c)	http://www.cdms.net/manuf/default.asp
Greenbook a searchable database of print-on-demand pesticide labels, MSDS sheets, and supplemental lables	http://www.greenbook.net/
National Pesticide Information Center (NPIC) a source of scientific, unbiased information	http://npic.orst.edu
Pesticide toxicology information at *EXTOXNET*	http://extoxnet.orst.edu
University of California Statewide Integrated Pest Management Program (UC IPM) educational materials: pesticide safety publications	http://www.ipm.ucdavis.edu/IPMPROJECT/pesttrain.html

Index

Species with full reports in the Weed Reports section are indicated in bold.

absinth wormwood (*Artemisia absinthium*): Herbaceous Forbs, 438, 446
Acacia melanoxylon (black acacia): Woody Plants, 454, 456
acacia
    black or blackwood acacia (*Acacia melanoxylon*): Woody Plants, 454, 456
    false acacia (**black locust**): ***Robinia pseudoacacia***, 335
Acetosa oblongifolia (*Rumex obtusifolius*): ***Rumex crispus & obtusifolius***, 346
Acetosella spp.: ***Rumex acetosella***, 344
Achnatherum capense (*Stipa capensis*): Grasses & Sedges, 434, 436
Acroptilon repens (**Russian knapweed**), 16, 513
Aegilops spp.
    cylindrica (**jointed goatgrass**): ***Aegilops*** **spp.**, 19
        [= *caudata, tauschii*]
    triuncialis (**barb goatgrass**): ***Aegilops*** **spp.**, 19
        [= *persica, squarrosa, triaristata*]
African
    asparagus fern (*Asparagus asparagoides*): Herbaceous Forbs, 442, 450
    mustard (**Saharan mustard**): ***Brassica tournefortii***, 64
    mustard (*Malcolmia africana*): Herbaceous Forbs, 440, 448
    rue (*Peganum harmala*): Herbaceous Forbs, 444, 452
    sage (**Mediterranean sage**): ***Salvia aethiopis***, 356, 516
    veldtgrass (**purple veldtgrass**): ***Ehrharta*** **spp.**, 165
Ageratina adenophora (**croftonweed**), 21
agrimony, sticky (**croftonweed**): ***Ageratina adenophora***, 21
Agropyron repens: ***Elytrigia repens***, 172
Agrostis spp.
    alopecuroides (*Polypogon monspeliensis*): Grasses & Sedges, 434, 436
    avenacea (Pacific bentgrass): Grasses & Sedges, 434, 436
        [= *retrofracta*]
    indica: ***Sporobolus indicus***, 390
    miliacea: ***Piptatherum miliaceum***, 315
    stolonifera (creeping bentgrass): Grasses & Sedges, 434, 436
        [= *alba, maritima, palustris*]
Ailanthus spp.
    altissima (**tree-of-heaven**), 23
    glandulosa: ***Ailanthus altissima***, 23
Aira spp.
    capensis (*Ehrharta calycina*): ***Ehrharta*** **spp.**, 165
    holcus-lanata: ***Holcus lanatus***, 212
Aleppo grass (**johnsongrass**): ***Sorghum halepense***, 380
alfalfa, poor man's (**kochia**): ***Bassia*** **spp.**, 54
alfilaree, alfilaria (**redstem filaree**): ***Erodium cicutarium***, 177
Algerian ivy (*Hedera canariensis*): ***Hedera*** **spp.**, 202
Alhagi spp.
    pseudalhagi (**camelthorn**), 25
        [= *camelorum, maurorum*]
alkali swainsonpea (**swainsonpea**): ***Sphaerophysa salsula***, 388
Alliaria petiolata (**garlic mustard**), 27
alligatorweed: ***Alternanthera philoxeroides***, 29, 518
Alopecurus spp.
    aristatus (*Polypogon monspeliensis*): Grasses & Sedges, 434, 436
    monspeliensis (*Polypogon monspeliensis*): Grasses & Sedges, 434, 436
alta fescue (*Festuca arundinacea*): Grasses & Sedges, 434, 436
Alternanthera spp.
    philoxeroides (**alligatorweed**), 29, 518
        [= *paludosa, philoxerina*]
alyssum
    Alyssum alyssoides (yellowtuft): Herbaceous Forbs, 440, 448

hoary alyssum (*Berteroa incana*): Herbaceous Forbs, 440, 448
sweet (yellowtuft) alyssum (*Alyssum alyssoides*): Herbaceous Forbs, 440, 448
sweet alyssum (*Lobularia maritima*): Herbaceous Forbs, 440, 448
yellow alyssum (*Alyssum alyssoides*): Herbaceous Forbs, 440, 448
amaranth, amaranth pigweed (*Amaranthus* spp.): Herbaceous Forbs, 438, 446
Amaranthus spp. (pigweeds, amaranths): Herbaceous Forbs, 438, 446
Ambrosia spp.
    acanthicarpa (annual bursage): Herbaceous Forbs, 438, 446
    artemisiifolia (common ragweed): Herbaceous Forbs, 438, 446
American
    cancer (**common pokeweed**): ***Phytolacca americana***, 311
    elodea (*Elodea canadensis*): Aquatic Plants, 458, 460
    pepper (**Peruvian peppertree**): ***Schinus*** **spp.**, 364
    pokeweed (**common pokeweed**): ***Phytolacca americana***, 311
    pondweed (*Potamogeton nodosus*): ***Potamogeton natans & nodosus***, 326
    stinging nettle (*Urtica dioica*): Herbaceous Forbs, 444, 452
Ammophila arenaria (= *A. arundinacea*) (**European beachgrass**), 31
Amorpha fruticosa (**indigobush**), 33
Amsinckia **spp. (fiddleneck)**, 35
    menziesii var. *intermedia* (**coast fiddleneck**): ***Amsinckia*** **spp.**, 35
        [= *arizonica, arvensis, attenuate, californica, campestris, demissa, echinata, hanseni, intactilis, intermedia, longituba, obovallata, valens*]
    menziesii var. *menziesii* (**Menzies fiddleneck**): ***Amsinckia*** **spp.**, 35
        [= *copelandii, eatonii, helleri, hirticaulis, micrantha, parviflora, retrorsa*]
anacharis (**Brazilian egeria**): ***Egeria densa***, 163
Anacharis spp.
    canadensis (*Elodea canadensis*): Aquatic Plants, 458, 460
    densa: ***Egeria densa***, 163
Anchusa arvensis (small bugloss): see ***Echium plantagineum***, 159
Andean pampasgrass (**jubatagrass**): ***Cortaderia*** **spp.**, 130
Andropogon halepensis: ***Sorghum halepense***, 380
anise, sweet anise, aniseed (**fennel**): ***Foeniculum vulgare***, 187
annual
    beardgrass (*Polypogon monspeliensis*): Grasses & Sedges, 434, 436
    bursage, burweed (*Ambrosia acanthicarpa*): Herbaceous Forbs, 438, 446
    dogtail (**hedgehog dogtailgrass**): ***Cynosurus echinatus***, 145
    false-brome: ***Brachypodium distachyon***, 57
    fescue, zorro (**rattail fescue**): ***Vulpia myuros***, 425
    ryegrass (**Italian ryegrass**): ***Lolium*** **spp.**, 260
    sunflower (*Helianthus annuus*): Herbaceous Forbs, 438, 446
Anthoxanthum odoratum (sweet vernalgrass): Grasses & Sedges, 434, 436
Anthriscus **spp.**
    caucalis (**bur chervil**): ***Anthriscus*** **spp.**, 39
        [= *neglecta, scandicina, vulgaris*]
    sylvestris (**wild chervil**): ***Anthriscus*** **spp.**, 39
apple, silverleaf bitter (*Solanum elaeagnifolium*): Herbaceous Forbs, 444, 452
Arabian mediterraneangrass (*Schismus arabicus*): ***Schismus*** **spp.**, 367
Arabian schismus (**Arabian mediterraneangrass**): ***Schismus*** **spp.**, 367
arabiangrass (**Arabian mediterraneangrass**): ***Schismus*** **spp.**, 367
Arctotheca spp.
    calendula (capeweed): Herbaceous Forbs, 438, 446

calendulacea: A. calendula, Herbaceous Forbs, 438, 446
Arctotis calendula (Arctotheca calendula): Herbaceous Forbs, 438, 446
Argentine vervain (*Verbena bonariensis*): Herbaceous Forbs, 444, 452
Artemisia spp.
    absinthium (absinth wormwood): Herbaceous Forbs, 438, 446
    tridentata (sagebrush): Woody Plants, 454, 456
artichoke
    thistle: ***Cynara cardunculus***, 138
    desert or wild artichoke (**artichoke thistle**): ***Cynara cardunculus***, 138
arum lily (*Zantedeschia aethiopica*): Herbaceous Forbs, 438, 446
***Arundo* spp.**
    dioeca (**Cortaderia selloana**): ***Cortaderia* spp.**, 130
    donax (**giant reed**), 40
    glauca, latifolia, sativa: ***Arundo donax***, 40, 517
    littoralis: ***Ammophila arenaria***, 31
    selloana (**Cortaderia selloana**): ***Cortaderia* spp.**, 130
***Asclepias* spp.**
    fascicularis (**Mexican whorled milkweed**): ***Asclepias* spp.**, 43
    [= *macrophylla, mexicana*]
    speciosa (**showy milkweed**): ***Asclepias* spp.**, 43
Asian mustard (**Saharan mustard**): ***Brassica tournefortii***, 64
Asparagus asparagoides (bridal creeper): Herbaceous Forbs, 442, 450
asparagus fern, African (*Asparagus asparagoides*): Herbaceous Forbs, 442, 450
aspen, Florida (**Chinese tallowtree**): ***Sapium sebiferum***, 361
asphodel, pink (**onionweed**): ***Asphodelus fistulosus***, 45
***Asphodelus* spp.**
    fistulosus (**onionweed**), 45
    tenuifolius: ***Asphodelus fistulosus***, 45
asses' thistle (**Scotch thistle**): ***Onopordum acanthium***, 292, 515
asthmaweed (*Conyza bonariensis*): Herbaceous Forbs, 438, 446
athel tamarisk (= athel pine, athel tree) (*Tamarix aphylla*): ***Tamarix* spp.**, 395
Atlantic
    cordgrass (**smooth cordgrass**): ***Spartina* spp.**, 383, 517
    ivy (***Hedera hibernica***): ***Hedera* spp.**, 202
***Atriplex* spp.**
    denticulata, flagellaris: ***Atriplex semibaccata***, 47
    semibaccata (**Australian saltbush**), 47
Australian
    cheesewood (*Pittosporum undulatum*): Woody Plants, 454, 456
    fireweed (*Erechtites* spp.): Herbaceous Forbs, 438, 446
    ryegrass (**Italian ryegrass**): ***Lolium* spp.**, 260
    saltbush: ***Atriplex semibaccata***, 47
    spinach (*Chenopodium berlandieri*): Herbaceous Forbs, 440, 448
    waterbuttons (*Cotula australis*): Herbaceous Forbs, 438, 446
Austrian peaweed (**swainsonpea**): ***Sphaerophysa salsula***, 388
***Avena* spp.**
    barbata (**slender oat**): ***Avena* spp.**, 49
    [= *alba, hirsuta, sativa, strigosa*]
    fatua (**wild oat**): ***Avena* spp.**, 49
    [= *hybrida*]
awn grass, common (*Stipa capensis*): Grasses & Sedges, 434, 436
azolla (**mosquitofern**): ***Azolla* spp.**, 52, 518
***Azolla* spp.** (**mosquitofern**), 52, 518
azzarola (*Crataegus monogyna*): Woody Plants, 454, 456
babies' slippers (**birdsfoot trefoil**): ***Lotus corniculatus***, 263
baby rose (**multiflora rose**): ***Rosa* spp.**, 338
baby's breath (= babysbreath gypsophila): ***Gypsophila paniculata***, 198
bachelor buttons (*Centaurea cyanus*): 513
bachelor's button (**baby's breath**): ***Gypsophila paniculata***, 198
bacon-weed (*Chenopodium album*): Herbaceous Forbs, 440, 448
ball cress (**hairy whitetop**): ***Cardaria* spp.**, 76
bamboo
    bamboo reed (**giant reed**): ***Arundo donax***, 40, 517

    Japanese bamboo (**Japanese knotweed**): ***Polygonum cuspidatum* et al.**, 318
    Mexican bamboo (**Japanese knotweed**): ***Polygonum cuspidatum* et al.**, 318
bank thistle (**bull thistle**): ***Cirsium vulgare***, 122, 515
barb goatgrass (barbed goatgrass) (*Aegilops triuncialis*): ***Aegilops* spp.**, 19
barbwire Russian-thistle (*Salsola paulsenii*): ***Salsola* spp.**, 353
barilla (**halogeton**): ***Halogeton glomeratus***, 200, 515
barley
    foxtail: ***Hordeum jubatum***, 216
    hare (*Hordeum murinum* ssp. *leporinum*): ***Hordeum marinum & murinum***, 218
    Mediterranean (*Hordeum marinum* ssp. *gussonianum*): ***Hordeum marinum & murinum***, 218
    mouse (**hare barley**): ***Hordeum marinum & murinum***, 218
    seaside (**Mediterranean barley**): ***Hordeum marinum & murinum***, 218
    squirreltail (**foxtail barley**): ***Hordeum jubatum***, 216
    wild (**hare barley**): ***Hordeum marinum & murinum***, 218
barleygrass (**hare barley**): ***Hordeum marinum & murinum***, 218
barrel medic (*Medicago polymorpha*): Herbaceous Forbs, 442, 448
basin big sagebrush (*Artemisia tridentata*): Woody Plants, 454, 456
bassia, fivehook (= bassia) (*Bassia hyssopifolia*): ***Bassia* spp.**, 54
***Bassia* spp.**
    hyssopifolia (**fivehook bassia**): ***Bassia* spp.**, 54
    scoparia (**kochia**): ***Bassia* spp.**, 54
    sieversiana (***Bassia scoparia***): ***Bassia* spp.**, 54
beachgrass, European: ***Ammophila arenaria***, 31
bead-podded mustard (*Chorispora tenella*): Herbaceous Forbs, 440, 448
bean
    castor bean (**castorbean**): ***Ricinus communis***, 333
    hog's or jupiter's bean (**black henbane**): ***Hyoscyamus niger***, 226
beancaper, Syrian (*Zygophyllum fabago*): Herbaceous Forbs, 444, 452
bean tree, Indian (*Catalpa bignonioides*): Woody Plants, 454, 456
bear-bind (*Polygonum convolvulus*): Herbaceous Forbs, 444, 450
bearded creeper (**common crupina**): ***Crupina vulgaris***, 134
beardgrass
    annual or tawny beardgrass (*Polypogon monspeliensis*): Grasses & Sedges, 434, 436
bee nettle (*Lamium amplexicaule*): Herbaceous Forbs, 442, 450
beggar's-lice (**houndstongue**): ***Cynoglossum officinale***, 143
bell-shaped knotweed (**Himalayan knotweed**): ***Polygonum cuspidatum* et al.**, 318
belvedere, red belvedere, belvedere-cypress (**kochia**): ***Bassia* spp.**, 54
bent
    carpet bent, creeping bent, redtop bent (*Agrostis stolonifera*): Grasses & Sedges, 434, 436
bentgrass
    creeping bentgrass (*Agrostis stolonifera*): Grasses & Sedges, 434, 436
    Pacific bentgrass (*Agrostis avenacea*): Grasses & Sedges, 434, 436
    seaside bentgrass (*Agrostis stolonifera*): Grasses & Sedges, 434, 436
Bermuda buttercup (**buttercup oxalis**): ***Oxalis pes-caprae***, 295
bermudagrass: ***Cynodon dactylon***, 141
berry
    prairie berry (*Solanum elaeagnifolium*): Herbaceous Forbs, 444, 452
    saltbush (**Australian saltbush**): ***Atriplex semibaccata***, 47
    scrambling berry saltbush (**Australian saltbush**): ***Atriplex semibaccata***, 47
    white wax berry (**Chinese tallowtree**): ***Sapium sebiferum***, 361
Berteroa incana (hoary alyssum): Herbaceous Forbs, 440, 448
betty, sweet (*Saponaria officinalis*): Herbaceous Forbs, 440, 448
big
    periwinkle: ***Vinca major***, 423
    quakinggrass (*Briza maxima*): Grasses & Sedges, 434, 436
    sagebrush (*Artemisia tridentata*): Woody Plants, 454, 456
    taper (**common mullein**): ***Verbascum* spp.**, 419

INDEX

bigleaf periwinkle (**big periwinkle**): ***Vinca major***, 423
bindweed
    black or climbing (*Polygonum convolvulus*): Herbaceous Forbs, 444, 450
    field: *Convolvulus arvensis*, 127, 516
    ivy (*Polygonum convolvulus*): Herbaceous Forbs, 444, 450
    knot (*Polygonum convolvulus*): Herbaceous Forbs, 444, 450
    small (**field bindweed**): ***Convolvulus arvensis***, 127, 516
bird thistle (**bull thistle**): ***Cirsium vulgare***, 122, 515
birdfoot deervetch (**birdsfoot trefoil**): ***Lotus corniculatus***, 263
birdgrass (*Polygonum arenastrum*): Herbaceous Forbs, 442, 450
birdsfoot trefoil (= broadleaf birdsfoot trefoil): ***Lotus corniculatus***, 263
bird's-nest (**wild carrot**): ***Daucus carota***, 149
birdsrape mustard (*Brassica rapa*): Herbaceous Forbs, 440, 448
bitter
    buttons (**common tansy**): ***Tanacetum vulgare***, 399
    dock (**broadleaf dock**): ***Rumex crispus & obtusifolius***, 346
    silverleaf bitter apple (*Solanum elaeagnifolium*): Herbaceous Forbs, 444, 452
black
    acacia (*Acacia melanoxylon*): Woody Plants, 454, 456
    bindweed (*Polygonum convolvulus*): Herbaceous Forbs, 444, 450
    henbane: *Hyoscyamus niger*, 226
    knapweed: *Centaurea nigra*, 514
    locust: *Robinia pseudoacacia*, 335
    mustard: *Brassica nigra*, 61
    thistle (**bull thistle**): ***Cirsium vulgare***, 122, 515
blackberry, Himalaya or **Himalayan: *Rubus armeniacus***, 341
blackwood acacia (*Acacia melanoxylon*): Woody Plants, 454, 456
bladderpod (**red sesbania**): ***Sesbania punicea***, 369
blessed milkthistle: *Silybum marianum*, 371, 515
blind nettle (*Lamium amplexicaule*): Herbaceous Forbs, 442, 450
bloomfell (**birdsfoot trefoil**): ***Lotus corniculatus***, 263
blowball (*Taraxacum officinale*): Herbaceous Forbs, 440, 446
blue
    daisy (*Cichorium intybus*): Herbaceous Forbs, 438, 446
    gum (**Tasmanian blue gum**): ***Eucalyptus globulus***, 180
    heliotrope (*Heliotropium amplexicaule*): see ***Myosotis latifolia***, 279
    mustard (*Chorispora tenella*): Herbaceous Forbs, 440, 448
    periwinkle (**big periwinkle**): ***Vinca major***, 423
    sailors (*Cichorium intybus*): Herbaceous Forbs, 438, 446
    Tasmanian blue gum: *Eucalyptus globulus*, 180
    thistle (**bull thistle**): ***Cirsium vulgare***, 122, 515
bluebell, Riverina (**vipers bugloss**): ***Echium plantagineum***, 159
bluegrass
    bulbous bluegrass: *Poa bulbosa*, 316
    Kentucky bluegrass (*Poa pratensis*): Grasses & Sedges, 434, 436
    winter bluegrass (**bulbous bluegrass**): ***Poa bulbosa***, 316
blueweed (chicory) (*Cichorium intybus*): Herbaceous Forbs, 438, 446
blueweed (echium): *Echium vulgare*, 161
bog rush (*Juncus effusus*): *Juncus* spp., 238
Bohemian knotweed (*Polygonum x bohemicum*): ***Polygonum cuspidatum* et al.**, 318
bottlebrush (**horsetail, scouringrush**): ***Equisetum* spp.**, 175
bottlegrass (*Setaria* spp.): Grasses & Sedges, 434, 436
bouncingbet (*Saponaria officinalis*): Herbaceous Forbs, 440, 448
bouquet-violet (**purple loosestrife**): ***Lythrum salicaria***, 271, 517
box, Victorian (*Pittosporum undulatum*): Woody Plants, 454, 456
bozzleweed (*Iva axillaris*): Herbaceous Forbs, 438, 446
***Brachypodium* spp.**
    distachyon (**annual false-brome**), 57
    sylvaticum (**slender false-brome**), 59
Brasenia schreberi (watershield): Aquatic Plants, 458, 460
brassbuttons
    common (*Cotula coronopifolia*): Herbaceous Forbs, 438, 446
    small (*Cotula australis*): Herbaceous Forbs, 438, 446
    southern (*Cotula australis*): Herbaceous Forbs, 438, 446
***Brassica* spp.**
    arvensis: ***Sinapis arvensis***, 374
    campestris (*Brassica rapa*): Herbaceous Forbs, 440, 448
    geniculata: ***Hirschfeldia incana***, 211
    kaber: ***Sinapis arvensis***, 374
    nigra (**black mustard**), 61
    rapa (birdsrape mustard): Herbaceous Forbs, 440, 448
    tournefortii (**Saharan mustard**), 64
brassica, hairy (**shortpod mustard**): ***Hirschfeldia incana***, 211
Brazilian
    egeria: *Egeria densa*, 163
    peppertree (*Schinus terebinthifolius*): ***Schinus* spp.**, 364
    watermilfoil (**parrotfeather**): ***Myriophyllum aquaticum***, 281
    waterweed (**Brazilian egeria**): ***Egeria densa***, 163
bridal creeper (*Asparagus asparagoides*): Herbaceous Forbs, 442, 450
brier
    hip (**dog rose**): ***Rosa* spp.**, 338
    dog brier (**dog rose**): ***Rosa* spp.**, 338
    wild brier (**dog rose**): ***Rosa* spp.**, 338
    witch's brier (**dog rose**): ***Rosa* spp.**, 338
bristlegrass (foxtails): *Setaria* spp., Grasses & Sedges, 434, 436
bristly
    dogtail grass (**hedgehog dogtailgrass**): ***Cynosurus echinatus***, 145
    foxtail (*Setaria* spp.): Grasses & Sedges, 434, 436
    oxtongue: *Picris echioides*, 313
    thistle (**plumeless thistle**): ***Carduus* spp.**, 79, 513
brittlewort (*Nitella* spp.): ***Chara, Nitella* spp.**, 114
Briza spp.
    eragrostis (*Eragrostis* spp.): Grasses & Sedges, 434, 436
    maxima (big quakinggrass): Grasses & Sedges, 434, 436
    megastachya (*Eragrostis* spp.): Grasses & Sedges, 434, 436
broad waterweed (*Elodea canadensis*): Aquatic Plants, 458, 460
broadleaf (**broad-leaved**)
    birdsfoot trefoil (**birdsfoot trefoil**): ***Lotus corniculatus***, 263
    cattail (**cattail**): ***Typha* spp.**, 413
    dock (*Rumex obtusifolius*): ***Rumex crispus & obtusifolius***, 346
    forget-me-not: *Myosotis latifolia*, 279
    pepperweed (**perennial pepperweed**): ***Lepidium latifolium***, 244
    pondweed (**floatingleaf pondweed**): ***Potamogeton natans & nodosus***, 326
    toadflax (**Dalmatian toadflax**): ***Linaria dalmatica***, 254, 517
brome
    downy (*Bromus tectorum*): ***Bromus diandrus* et al.**, 66
    drooping (**downy brome**): ***Bromus diandrus* et al.**, 66
    false (**annual false-brome**): ***Brachypodium distachyon***, 57
    foxtail (**red brome**): ***Bromus diandrus* et al.**, 66
    great (**ripgut brome**): ***Bromus diandrus* et al.**, 66
    Japanese (*Bromus japonicus*): ***Bromus hordeaceus, japonicus***, 71
    purple false (**annual false-brome**): ***Brachypodium distachyon***, 57
    red (*Bromus madritensis* ssp. *rubens*): ***Bromus diandrus* et al.**, 66
    ripgut (*Bromus diandrus*): ***Bromus diandrus* et al.**, 66
    smooth: *Bromus inermis*, 69
    soft (*Bromus hordeaceus*): ***Bromus hordeaceus, japonicus***, 71
    stiff brome (**annual false-brome**): ***Brachypodium distachyon***, 57
***Bromus* spp.**
    arundinaceus (*Festuca arundinacea*): Grasses & Sedges, 434, 436
    diandrus (**ripgut brome**): ***Bromus diandrus* et al.**, 66
    [= *gussonei, rigidus*]
    distachya (**annual false-brome**): ***Brachypodium distachyon***, 57
    glomeratus (*Dactylis glomerata*): Grasses & Sedges, 434, 436
    hordeaceus (**soft brome**): ***Bromus hordeaceus, japonicus***, 71
    [= *confertus*]
    inermis (**smooth brome**), 69
    japonicus (**Japanese brome**): ***Bromus hordeaceus, japonicus***, 71

[= *abolini, chiapporianus, commutatus, cyri, gedrosianus, multiflorus, patulus, pendulus, unilateralis, vestitus*]
    ***madritensis* ssp. *rubens* (red brome): *Bromus diandrus* et al.**, 66
    [= *rubens*]
    ***tectorum* (downy brome): *Bromus diandrus* et al.**, 66
broncograss (**downy brome**): ***Bromus diandrus* et al.**, 66
broom
    Cape broom (**French broom**): ***Genista monspessulana***, 191, 516
    English broom (**Scotch broom**): ***Cytisus scoparius***, 147, 516
    French broom: *Genista monspessulana*, 191, 516
    Montpellier broom (**French broom**): ***Genista monspessulana***, 191, 516
    prickly broom (**gorse**): ***Ulex europaeus***, 415, 516
    Scotch broom: *Cytisus scoparius*, 147, 516
    Spanish broom: *Spartium junceum*, 385
    thorn broom (**gorse**): ***Ulex europaeus***, 415, 516
    weavers broom (**Spanish broom**): ***Spartium junceum***, 385
brown knapweed: *Centaurea jacea*, 514
***Bryonia alba* (white bryony)**, 73
bryony, white: *Bryonia alba*, 73
buck plantain (*Plantago lanceolata*): Herbaceous Forbs, 442, 450
buckhorn plantain (*Plantago lanceolata*): Herbaceous Forbs, 442, 450
buckthorn, coast (**coast fiddleneck**): ***Amsinckia* spp.**, 35
buckwheat, wild (*Polygonum convolvulus*): Herbaceous Forbs, 444, 450
***Buddleja davidii* (butterflybush)**: Woody Plants, 454, 456
buffaloweed (*Ambrosia artemisiifolia*): Herbaceous Forbs, 438, 446
buffelgrass: *Pennisetum ciliare*, 297
bugle-lily (*Watsonia meriana*): Herbaceous Forbs, 442, 450
bugloss
    bugloss (**bristly oxtongue**): ***Picris echioides***, 313
    bugloss-picris (**bristly oxtongue**): ***Picris echioides***, 313
    purple bugloss (**vipers bugloss**): ***Echium plantagineum***, 159
    purple vipers bugloss (**vipers bugloss**): ***Echium plantagineum***, 159
    small bugloss (*Anchusa arvensis*): see *Echium plantagineum*, 159
    vipers bugloss: *Echium plantagineum*, 159
    vipers bugloss (**blueweed**): ***Echium vulgare***, 161
bulbil (*Watsonia meriana*): Herbaceous Forbs, 442, 450
bulbous bluegrass: *Poa bulbosa*, 316
bulbous canarygrass (**hardinggrass**): ***Phalaris aquatica***, 303
bull thistle: *Cirsium vulgare*, 122, 515
bullhead (**puncturevine**): ***Tribulus terrestris***, 407, 517
bulrush, hardstem (*Scirpus acutus*): Grasses & Sedges, 434, 436
bur
    buttercup (*Ranunculus testiculatus*): Herbaceous Forbs, 444, 450
    chervil (*Anthriscus caucalis*): *Anthriscus* spp., 39
    parsley (**hedgeparsley**): ***Torilis arvensis***, 399
    thistle (**bull thistle**): ***Cirsium vulgare***, 122, 515
    dog bur (**houndstongue**): ***Cynoglossum officinale***, 143
    noogoora bur (**common cocklebur**): ***Xanthium* spp.**, 428
    sheep bur (**common cocklebur**): ***Xanthium* spp.**, 428
burclover or California burclover (*Medicago polymorpha*): Herbaceous Forbs, 442, 448
burdock, sea (**common cocklebur**): ***Xanthium* spp.**, 428
burgrass (*Cenchrus* spp.): Grasses & Sedges, 434, 436
burningbush, Mexican (**kochia**): ***Bassia* spp.**, 54
burnut, ground (**puncturevine**): ***Tribulus terrestris***, 407, 517
burnweed (*Erechtites* spp.): Herbaceous Forbs, 438, 446
burr medic (*Medicago polymorpha*): Herbaceous Forbs, 442, 448
bursage, annual (*Ambrosia acanthicarpa*): Herbaceous Forbs, 438, 446
burweed (bur-weed)
    annual or flatspine burweed (*Ambrosia acanthicarpa*): Herbaceous Forbs, 438, 446
    hedgehog burweed (**common cocklebur**): ***Xanthium* spp.**, 428
    yellow burweed (**coast fiddleneck**): ***Amsinckia* spp.**, 35
bush lupine, yellow (*Lupinus arboreus*): *Lupinus* spp., 267
bushman's burnweed (*Erechtites* spp.): Herbaceous Forbs, 438, 446

butter and eggs (**yellow toadflax**): ***Linaria vulgaris***, 257, 517
butter daisy (**oxeye daisy**): ***Leucanthemum vulgare***, 247
buttercup
    Bermuda: *Oxalis pes-caprae*, 295
    bur (*Ranunculus testiculatus*): Herbaceous Forbs, 444, 450
    buttercup oxalis: *Oxalis pes-caprae*, 295
    common (*Ranunculus acris*): Herbaceous Forbs, 444, 450
    creeping (*Ranunculus repens*): Herbaceous Forbs, 444, 450
    hornseed (*Ranunculus testiculatus*): Herbaceous Forbs, 444, 450
    meadow (*Ranunculus acris*): Herbaceous Forbs, 444, 450
    showy (*Ranunculus acris*): Herbaceous Forbs, 444, 450
    swamp (*Ranunculus repens*): Herbaceous Forbs, 444, 450
    tall (*Ranunculus acris*): Herbaceous Forbs, 444, 450
    testiculate (*Ranunculus testiculatus*): Herbaceous Forbs, 444, 450
butterflybush (*Buddleja davidii*): Woody Plants, 454, 456
butterweed (*Conyza canadensis*): Herbaceous Forbs, 438, 446
butterwort, curveseed (*Ranunculus testiculatus*): Herbaceous Forbs, 444, 450
button thistle (**bull thistle**): ***Cirsium vulgare***, 122, 515
button, bachelor's (**baby's breath**): ***Gypsophila paniculata***, 198
button, soldier's (*Cotula australis*): Herbaceous Forbs, 438, 446
buttonbur (**common cocklebur**): ***Xanthium* spp.**, 428
buttons, bachelor (*Centaurea cyanus*): 513
buttons, bitter (**common tansy**): ***Tanacetum vulgare***, 399
buttonweed (*Malva neglecta*): Herbaceous Forbs, 442, 450
cabbage thistle (**blessed milkthistle**): ***Silybum marianum***, 371, 515
cabomba (**fanwort**): ***Cabomba caroliniana***, 74
***Cabomba caroliniana* (fanwort)**, 74
Cakile maritima (sea rocket): Herbaceous Forbs, 440, 448
California
    burclover (*Medicago polymorpha*): Herbaceous Forbs, 442, 448
    fern (**poison hemlock**): ***Conium maculatum***, 125, 513
    nettle (*Urtica dioica*): Herbaceous Forbs, 444, 452
    peppertree (**Peruvian peppertree**): ***Schinus* spp.**, 364
    slender nettle (*Urtica dioica*): Herbaceous Forbs, 444, 452
    waterprimrose (**creeping waterprimrose**): ***Ludwigia* spp.**, 265
calla lily (*Zantedeschia aethiopica*): Herbaceous Forbs, 438, 446
caltrop (**puncturevine**): ***Tribulus terrestris***, 407, 517
camelthorn: *Alhagi pseudalhagi*, 25
Canada or Canadian
    fleabane (*Conyza canadensis*): Herbaceous Forbs, 438, 446
    thistle: *Cirsium arvense*, 119, 515
    horsetail (*Conyza canadensis*): Herbaceous Forbs, 438, 446
    pondweed (*Elodea canadensis*): Aquatic Plants, 458, 460
    waterweed (*Elodea canadensis*): Aquatic Plants, 458, 460
Canary Island
    date palm (*Phoenix canariensis*): *Phoenix & Washingtonia*, 307
    hypericum (*Hypericum canariensis*): Woody Plants, 454, 456
canarygrass
    bulbous canarygrass (**hardinggrass**): ***Phalaris aquatica***, 303
    reed canarygrass: *Phalaris arundinacea*, 305
    tuberous canarygrass (**hardinggrass**): ***Phalaris aquatica***, 303
cancer jalap (**common pokeweed**): ***Phytolacca americana***, 311
cancer, American (**common pokeweed**): ***Phytolacca americana***, 311
candle-wick (**common mullein**): ***Verbascum* spp.**, 419
candy grass (*Eragrostis* spp.): Grasses & Sedges, 434, 436
cane
    giant cane (**giant reed**): ***Arundo donax***, 40, 517
    Indian cane (**red sorrel**): ***Rumex acetosella***, 344
    maiden cane (**johnsongrass**): ***Sorghum halepense***, 380
    reed cane (**giant reed**): ***Arundo donax***, 40, 517
    wild cane (**giant reed**): ***Arundo donax***, 40, 517
canola (**wild mustard**): ***Sinapis arvensis***, 374
cape or Cape
    broom (**French broom**): ***Genista monspessulana***, 191, 516
    cowslip (**buttercup oxalis**): ***Oxalis pes-caprae***, 295

dandelion or gold (*Arctotheca calendula*): Herbaceous Forbs, 438, 446
   -ivy: ***Delairea odorata***, 152
     sorrel (**buttercup oxalis**): ***Oxalis pes-caprae***, 295
       -weed (*Arctotheca calendula*): Herbaceous Forbs, 438, 446
Capsella bursa-pastoris (shepherd's-purse): Herbaceous Forbs, 440, 448
caraway, common (*Carum carvi*): Herbaceous Forbs, 438, 446
card teasel, card thistle (*Dipsacus fullonum* or *D. sativus*): ***Dipsacus* spp.**, 154
***Cardaria* spp.**
   chalepensis (lens-podded whitetop): ***Cardaria* spp.**, 76
   draba (hoary cress): ***Cardaria* spp.**, 76
   latifolia: ***Lepidium latifolium***, 244
   pubescens (hairy whitetop): ***Cardaria* spp.**, 76
cardoon (**artichoke thistle**): ***Cynara cardunculus***, 138
***Carduus* spp.**
   acanthoides (plumeless thistle): ***Carduus* spp.**, 79, 513
     [= *fortior*]
   arvensis: ***Cirsium arvense***, 119, 515
   lanceolatus, vulgaris (**bull thistle**): ***Cirsium vulgare***, 122, 515
   leiophyllus (***Carduus nutans***): ***Carduus* spp.**, 79, 513
   mariae, marianus: ***Silybum marianum***, 371, 515
   nutans (musk thistle): ***Carduus* spp.**, 79, 513
     [= *thoermeri*]
   pycnocephalus (Italian thistle): ***Carduus* spp.**, 79, 513
   tenuiflorus (slenderflower thistle): ***Carduus* spp.**, 79, 513
carelessweed (*Amaranthus* spp.): Herbaceous Forbs, 438, 446
carnation spurge (*Euphorbia terracina*): ***Euphorbia* spp.**, 182
carnationweed, Geraldton (**carnation spurge**): ***Euphorbia* spp.**, 182
Carolina geranium (*Geranium carolinianum*): see ***Geranium* spp.**, 193
Carota sativa: ***Daucus carota***, 149
carpet bent (*Agrostis stolonifera*): Grasses & Sedges, 434, 436
Carpobrotus edulis (= *edule*) (**iceplant**), 83
carrot
   fern (**poison hemlock**): ***Conium maculatum***, 125, 513
   weed (*Cotula australis*): Herbaceous Forbs, 438, 446
   wild carrot: ***Daucus carota***, 149
***Carthamus* spp.**
   lanatus (**woolly distaff thistle**), 85
   maculatum: ***Silybum marianum***, 371, 515
Carum carvi (common caraway): Herbaceous Forbs, 438, 446
case plant, case weed (*Capsella bursa-pastoris*): Herbaceous Forbs, 440, 448
Caspian manna (**camelthorn**): ***Alhagi pseudalhagi***, 25
cassilata (**black henbane**): ***Hyoscyamus niger***, 226
castor oil plant (**castorbean**): ***Ricinus communis***, 333
castorbean: ***Ricinus communis***, 333
catalpa (*Catalpa bignonioides*): Woody Plants, 454, 456
Catalpa bignonioides (catalpa): Woody Plants, 454, 456
cat's clover (**birdsfoot trefoil**): ***Lotus corniculatus***, 263
catsear
   common catsear (*Hypochaeris radicata*): ***Hypochaeris* spp.** 230
   glabrous (**smooth catsear**): ***Hypochaeris* spp.**, 230
   hairy (**common catsear**): ***Hypochaeris* spp.**, 230
   long-rooted (**common catsear**): ***Hypochaeris* spp.**, 230
   rough (**common catsear**): ***Hypochaeris* spp.**, 230
   smooth catsear (*Hypochaeris glabra*): ***Hypochaeris* spp.**, 230
   spotted (**common catsear**): ***Hypochaeris* spp.**, 230
cattail: ***Typha* spp.**, 413
catweed (**croftonweed**): ***Ageratina adenophora***, 21
Caucalis spp.
   arvensis, capensis: ***Torilis arvensis***, 401
   carota, daucus: ***Daucus carota***, 149
   scandicina (*Anthriscus caucalis*): ***Anthriscus* spp.**, 39
Cenchrus ciliaris: ***Pennisetum ciliare***, 297
Cenchrus spp. (sandburs): Grasses & Sedges, 434, 436
***Centaurea* spp.**
   calcitrapa (purple starthistle): ***Centaurea calcitrapa* & *iberica***, 87, 513
   crupina: ***Crupina vulgaris***, 134
   cyanus (bachelor buttons): 513
   debeauxii (**meadow knapweed**), 90, 513
     [= *jacea x nigra, jacea* ssp. *pratensis, x moncktonii, x pratensis*]
   diffusa (**diffuse knapweed**), 93, 514
   iberica (**Iberian starthistle**): ***Centaurea calcitrapa* & *iberica***, 87, 514
   jacea (brown knapweed): 514
   melitensis (**Malta starthistle**), 97, 514
   nigra black (knapweed): 514
   picris, repens: ***Acroptilon repens***, 16, 513
   solstitialis (**yellow starthistle**), 100, 514
   stoebe (**spotted knapweed**), 105, 514
     [= *biebersteinii, maculosa*]
   virgata ssp. *squarrosa* (**squarrose knapweed**), 109
     [= *squarrosa, triumfetti*]
Centromadia spp.: see ***Hemizonia* spp.**, 205
Ceratophyllum demersum (**coontail**), 112
cereal rye (*Secale cereale*): Grasses & Sedges, 434, 436
Chaetochloa spp. (*Setaria* spp.): Grasses & Sedges, 434, 436
Chamaeraphis viridis (*Setaria* spp.): Grasses & Sedges, 434, 436
chara (*Chara* spp.): ***Chara, Nitella* spp.**, 114
***Chara* spp.** (chara): ***Chara, Nitella* spp.**, 114
charlie, creeping (**field bindweed**): ***Convolvulus arvensis***, 127, 516
cheatgrass, cheat, downy cheat (*Bromus tectorum*): ***Bromus diandrus, madritensis, tectorum***, 66
cheeseplant, cheeseweed, cheeses (*Malva neglecta*): Herbaceous Forbs, 442, 450
cheesewood, Australian (*Pittosporum undulatum*): Woody Plants, 454, 456
Chenopodium spp.
   album (common lambsquarter): Herbaceous Forbs, 440, 448
     [= *amaranticolor, browneanum, centrorubrum, giganteum, lanceolatum, suecicum*]
   berlandieri (netseed lambsquarter): Herbaceous Forbs, 440, 448
     [= *biforme, congestum, lucidum, triangulare*]
   scoparia (*Bassia scoparia*): ***Bassia* spp.**, 54
cherry plum (*Prunus cerasifera*): Woody Plants, 454, 456
chervil, bur (*Anthriscus caucalis*): ***Anthriscus* spp.**, 39
chervil, wild (*Anthriscus sylvestris*): ***Anthriscus* spp.**, 39
chess
   downy chess or early chess (**downy brome**): ***Bromus diandrus, madritensis, tectorum***, 66
   Japanese chess (**Japanese brome**): ***Bromus hordeaceus, japonicus***, 71
   slender chess (**downy brome**): ***Bromus diandrus, madritensis, tectorum***, 66
   soft chess (*Bromus hordeaceus*): ***Bromus hordeaceus, japonicus***, 71
chestnut, water: ***Trapa natans***, 405
chicken tree (**Chinese tallowtree**): ***Sapium sebiferum***, 361
chicory (*Cichorium intybus*): Herbaceous Forbs, 438, 446
China lettuce (*Lactuca serriola*): Herbaceous Forbs, 438, 446
Chinaman's greens (*Amaranthus* spp.): Herbaceous Forbs, 438, 446
Chinese
   elm (**tree-of-heaven**): ***Ailanthus altissima***, 23
   tallowberry (**Chinese tallowtree**): ***Sapium sebiferum***, 361
   tallowtree: ***Sapium sebiferum***, 361
   tamarisk (*Tamarix chinensis*): ***Tamarix* spp.**, 395
Chondrilla juncea (**rush skeletonweed**), 116, 515
Chorispora tenella (blue mustard): Herbaceous Forbs, 440, 448
Christmas holly (*Ilex aquifolium*): Woody Plants, 454, 456
Christmasberry (**Brazilian peppertree**): ***Schinus* spp.**, 364
Chrysanthemum spp.
   coronarium (crowndaisy): Herbaceous Forbs, 438, 446
   leucanthemum: ***Leucanthemum vulgare***, 247
   uliginosum: ***Tanacetum vulgare***, 399

INDEX

vulgare: ***Tanacetum vulgare***, 399
Chrysothamnus nauseosus (rabbitbrush): Herbaceous Forbs, 438, 446
Cichorium intybus (chicory): Herbaceous Forbs, 438, 446
cigar tree (*Catalpa bignonioides*): Woody Plants, 454, 456
cinquefoil
    roughfruit (**sulfur cinquefoil**): ***Potentilla recta***, 328
    sulfur or sulphur: ***Potentilla recta***, 328
***Cirsium* spp.**
    arvense (**Canada thistle**), 119, 515
        [= *incanum, ochrolepidium, setosum*]
    maculatum: ***Silybum marianum***, 371, 515
    vulgare (**bull thistle**), 122, 515
        [= *abyssinicum, lanceolatum*]
clasping cress, pepperweed, or peppercress (*Lepidium perfoliatum*): Herbaceous Forbs, 440, 448
Clematis vitalba (old-man's-beard): Herbaceous Forbs, 444, 450
clematis, evergreen (*Clematis vitalba*): Herbaceous Forbs, 444, 450
climbing bindweed (*Polygonum convolvulus*): Herbaceous Forbs, 444, 450
closed-leaved pondweed (**leafy pondweed**): ***Potamogeton* & *Stuckenia***, 323
clotbur (**common cocklebur**): ***Xanthium* spp.**, 428
clotbur, spiny (**spiny cocklebur**): ***Xanthium* spp.**, 428
clotweed (**spiny cocklebur**): ***Xanthium* spp.**, 428
clover
    cat's clover (**birdsfoot trefoil**): ***Lotus corniculatus***, 263
    hairy clover (*Trifolium hirtum*): Herbaceous Forbs, 442, 448
    honey clover (*Melilotus albus*): Herbaceous Forbs, 442, 448
    pin clover (**redstem filaree**): ***Erodium cicutarium***, 177
    rose clover (*Trifolium hirtum*): Herbaceous Forbs, 442, 448
    tree clover (*Melilotus albus*): Herbaceous Forbs, 442, 448
clusterberry, red (*Cotoneaster lacteus*): *Cotoneaster* spp., 132
cluster-flowered verbena (*Verbena bonariensis*): Herbaceous Forbs, 444, 452
coakum (**common pokeweed**): ***Phytolacca americana***, 311
coarse fescue (*Festuca arundinacea*): Grasses & Sedges, 434, 436
coast
    buckthorn (**coast fiddleneck**): ***Amsinckia* spp.**, 35
    dandelion (**common catsear**): ***Hypochaeris* spp.**, 230
    fiddleneck (*Amsinckia menziesii* var. *intermedia*): ***Amsinckia* spp.**, 35
    fireweed (*Erechtites* spp.): Herbaceous Forbs, 438, 446
    nettle (*Urtica dioica*): Herbaceous Forbs, 444, 452
    sandbur (*Cenchrus* spp.): Grasses & Sedges, 434, 436
cocklebur
    cocklebur (**spiny cocklebur**): ***Xanthium* spp.**, 428
    common cocklebur (*Xanthium strumarium*): ***Xanthium* spp.**, 428
    dagger cocklebur (**spiny cocklebur**): ***Xanthium* spp.**, 428
    rough cocklebur (**common cocklebur**): ***Xanthium* spp.**, 428
    spiny cocklebur (*Xanthium spinosum*): ***Xanthium* spp.**, 428
cocksfoot (*Dactylis glomerata*): Grasses & Sedges, 434, 436
cockspur, yellow (**yellow starthistle**): ***Centaurea solstitialis***, 100, 514
coffeeweed (chicory) (*Cichorium intybus*): Herbaceous Forbs, 438, 446
coffeeweed (red sesbania) (**red sesbania**): ***Sesbania punicea***, 369
coltstail (*Conyza canadensis*): Herbaceous Forbs, 438, 446
common
    awn grass (*Stipa capensis*): Grasses & Sedges, 434, 436
    brassbuttons (*Cotula coronopifolia*): Herbaceous Forbs, 438, 446
    caraway (*Carum carvi*): Herbaceous Forbs, 438, 446
    catsear (*Hypochaeris radicata*): ***Hypochaeris* spp.**, 230
    cattail (**cattail**): ***Typha* spp.**, 413
    cocklebur (*Xanthium strumarium*): ***Xanthium* spp.**, 428
    cordgrass (*Spartina anglica*): ***Spartina* spp.**, 383
    crupina: *Crupina vulgaris*, 134
    dandelion (*Taraxacum officinale*): Herbaceous Forbs, 440, 446
    duckweed (*Lemna* spp.): ***Lemna, Spirodela,* & *Wolffia***, 242
    elodea (*Elodea canadensis*): Aquatic Plants, 458, 460

eucalyptus (**Tasmanian blue gum**): ***Eucalyptus globulus***, 180
fennel (**fennel**): ***Foeniculum vulgare***, 187
fiddleneck (**coast fiddleneck**): ***Amsinckia* spp.**, 35
fig (**fig**): ***Ficus carica***, 185
floating pondweed (**floatingleaf pondweed**): ***Potamogeton natans* & *nodosus***, 326
forget-me-not (**broadleaf forget-me-not**): ***Myosotis latifolia***, 279
foxglove (*Digitalis purpurea*): Herbaceous Forbs, 444, 452
foxtail (**hare barley**): ***Hordeum marinum* & *murinum***, 218
gorse (**gorse**): ***Ulex europaeus***, 415, 516
henbane (**black henbane**): ***Hyoscyamus niger***, 226
henbit (*Lamium amplexicaule*): Herbaceous Forbs, 442, 450
horehound (**white horehound**): ***Marrubium vulgare***, 273
horsetail (**horsetail, scouringrush**): ***Equisetum* spp.**, 175
houndstongue (**houndstongue**): ***Cynoglossum officinale***, 143
knotweed (*Polygonum arenastrum*): Herbaceous Forbs, 442, 450
knotweed (**pale smartweed**): ***Polygonum persicaria* & *lapathifolium***, 320
lambsquarter (*Chenopodium album*): Herbaceous Forbs, 440, 448
linaria (**yellow toadflax**): ***Linaria vulgaris***, 257, 517
mallow (*Malva neglecta*): Herbaceous Forbs, 442, 450
mediterraneangrass (*Schismus barbatus*): ***Schismus* spp.**, 367
mullein (*Verbascum thapsus*): ***Verbascum* spp.**, 419
mustard (*Brassica rapa*): Herbaceous Forbs, 440, 448
mustard (**wild mustard**): ***Sinapis arvensis***, 374
pampasgrass (**pampasgrass**): ***Cortaderia* spp.**, 130
pokeweed: *Phytolacca americana*, 311
quickgrass (**bermudagrass**): ***Cynodon dactylon***, 141
ragweed (*Ambrosia artemisiifolia*): Herbaceous Forbs, 438, 446
reed: *Phragmites australis*, 309
rush (*Juncus effusus*): *Juncus* spp., 238
rye (*Secale cereale*): Grasses & Sedges, 434, 436
saltwort (**Russian-thistle**): ***Salsola* spp.**, 353, 515
sandbur (*Cenchrus* spp.): Grasses & Sedges, 434, 436
scouringrush (**horsetail, scouringrush**): ***Equisetum* spp.**, 175
sorrel (**red sorrel**): ***Rumex acetosella***, 344
spikeweed (**spikeweed**): ***Hemizonia* spp.**, 205
St. Johnswort: *Hypericum perforatum*, 228, 516
storksbill (**redstem filaree**): ***Erodium cicutarium***, 177
sunflower (*Helianthus annuus*): Herbaceous Forbs, 438, 446
tansy: ***Tanacetum vulgare***, 399
tarweed (**spikeweed**): ***Hemizonia* spp.**, 205
teasel (*Dipsacus fullonum*): ***Dipsacus* spp.**, 154
thistle (**bull thistle**): ***Cirsium vulgare***, 122, 515
toad rush: *Juncus* spp., 238
toadflax (**yellow toadflax**): ***Linaria vulgaris***, 257, 517
velvetgrass: *Holcus lanatus*, 212
waterweed (**Brazilian egeria**): ***Egeria densa***, 163
wild lettuce (*Lactuca serriola*): Herbaceous Forbs, 438, 446
buttercup (*Ranunculus acris*): Herbaceous Forbs, 444, 450
compact-headed thistle (**Italian thistle**): ***Carduus* spp.**, 79, 513
compass plant (*Lactuca serriola*): Herbaceous Forbs, 438, 446
Conium maculatum (**poison-hemlock**), 125, 513
***Convolvulus* spp.**
    arvensis (**field bindweed**), 127, 516
        [= *ambiguens, incanus*]
Conyza spp.
    bonariensis (hairy fleabane): Herbaceous Forbs, 438, 446
        [= *ambigua, linearis, linifolia*]
    canadensis (horseweed): Herbaceous Forbs, 438, 446
    South American conyza (*C. bonariensis*): Herbaceous Forbs, 438, 446
coontail: *Ceratophyllum demersum*, 112
copal tree (**tree-of-heaven**): ***Ailanthus altissima***, 23
cordgrasses: ***Spartina* spp.**, 383
corn thistle (**Canada thistle**): ***Cirsium arvense***, 119, 515
cornbind (**field bindweed**): ***Convolvulus arvensis***, 127, 516

cornbind (black bindweed) (*Polygonum convolvulus*): Herbaceous Forbs, 444, 450
cornbind, dullseed (*Polygonum convolvulus*): Herbaceous Forbs, 444, 450
Cortaderia spp.
    jubata (**jubatagrass**): *Cortaderia* spp., 130
        [= *atacamensis*]
    selloana (**pampasgrass**): *Cortaderia* spp., 130
        [= *argentea, dioica*]
Cotoneaster spp. (cotoneasters), 132
    franchetii (**orange cotoneaster**): *Cotoneaster* spp., 132
    lacteus (**Parney's cotoneaster**): *Cotoneaster* spp., 132
        [= *lactea, parneyi*]
    pannosus (**silverleaf cotoneaster**): *Cotoneaster* spp., 132
        [= *pannosa*]
cotoneasters: *Cotoneaster* spp., 132
    milkflower cotoneaster (**Parney's cotoneaster**): *Cotoneaster* spp., 132
    orange cotoneaster (*Cotoneaster franchetii*): *Cotoneaster* spp., 132
    Parney's cotoneaster (*Cotoneaster lacteus*): *Cotoneaster* spp., 132
    rockspray cotoneaster (**Parney's cotoneaster**): *Cotoneaster* spp., 132
    silverleaf cotoneaster (*Cotoneaster pannosus*): *Cotoneaster* spp., 132
cotton
    palm (**Mexican fan palm**): *Phoenix* & *Washingtonia*, 307
    thistle (**Scotch thistle**): *Onopordum acanthium*, 292, 515
    Scotch cottonthistle (**Scotch thistle**): *Onopordum acanthium*, 292, 515
Cotula spp.
    australis (southern brassbuttons): Herbaceous Forbs, 438, 446
    coronopifolia (brassbuttons): Herbaceous Forbs, 438, 446
couch or couchgrass
    bermudagrass: *Cynodon dactylon*, 141
    quackgrass: *Elytrigia repens*, 172
cow lily (*Nuphar lutea*): Aquatic Plants, 458, 460
cow sorrel (**red sorrel**): *Rumex acetosella*, 344
cowslip, cape (**buttercup oxalis**): *Oxalis pes-caprae*, 295
cranesbill, cut-leaved (**cutleaf geranium**): *Geranium* spp., 193
cranesbill, dove's-foot (*Geranium molle*): see *Geranium* spp., 193
Crataegus spp.
    monogyna (English hawthorn): Woody Plants, 454, 456
        [= *apiifolia, curvisepala, dissecta, oxyacantha*]
creek nettle (*Urtica dioica*): Herbaceous Forbs, 444, 452
creeper, bearded (**common crupina**): *Crupina vulgaris*, 134
creeper, bridal (*Asparagus asparagoides*): Herbaceous Forbs, 442, 450
creeping
    amaranth (*Amaranthus* spp.): Herbaceous Forbs, 438, 446
    bent (*Agrostis stolonifera*): Grasses & Sedges, 434, 436
    bentgrass (*Agrostis stolonifera*): Grasses & Sedges, 434, 436
    buttercup (*Ranunculus repens*): Herbaceous Forbs, 444, 450
    charlie (**field bindweed**): *Convolvulus arvensis*, 127, 516
    crowfoot (*Ranunculus repens*): Herbaceous Forbs, 444, 450
    jenny (**field bindweed**): *Convolvulus arvensis*, 127, 516
    jenny (**garden loosestrife**): *Lysimachia vulgaris*, 269
    primrose-willow (**creeping waterprimrose**): *Ludwigia* spp., 265
    saltbush (**Australian saltbush**): *Atriplex semibaccata*, 47
    thistle (**Canada thistle**): *Cirsium arvense*, 119, 515
    waterprimrose (**creeping waterprimrose**): *Ludwigia* spp., 265
cress
    ball cress (**hairy whitetop**): *Cardaria* spp., 76
    clasping cress (*Lepidium perfoliatum*): Herbaceous Forbs, 440, 448
    globe-podded hoary cress (*Cardaria draba* or *C. pubescens*): *Cardaria* spp., 76
    heart-podded hoary cress (**hoary cress**): *Cardaria* spp., 76
    hoary cress (*Cardaria draba, C. chalepensis,* or *C. pubescens*): *Cardaria* spp., 76
    lens-podded hoary cress (**lens-podded whitetop**): *Cardaria* spp., 76
crimson fountaingrass: *Pennisetum setaceum*, 301

crispate-leaved or crisped pondweed (**curlyleaf pondweed**): ***Potamogeton* & *Stuckenia*,** 323
crocosmia (*Crocosmia* x *crocosmiflora*): Herbaceous Forbs, 442, 450
Crocosmia x *crocosmiflora* (crocosmia): Herbaceous Forbs, 442, 450
croftonweed: *Ageratina adenophora*, 21
crossflower (*Chorispora tenella*): Herbaceous Forbs, 440, 448
crowfoot
    creeping (*Ranunculus repens*): Herbaceous Forbs, 444, 450
    tall (*Ranunculus acris*): Herbaceous Forbs, 444, 450
    tubercled (*Ranunculus testiculatus*): Herbaceous Forbs, 444, 450
crowndaisy (*Chrysanthemum coronarium*): Herbaceous Forbs, 438, 446
crowtoes (**birdsfoot trefoil**): *Lotus corniculatus*, 263
crunch-weed (**wild mustard**): *Sinapis arvensis*, 374
Crupina vulgaris (**common crupina**), 134
crupina, common: *Crupina vulgaris*, 134
Cuba grass (**johnsongrass**): *Sorghum halepense*, 380
cultivated radish (**radish**): *Raphanus* spp., 331
cultivated teasel (**Fuller's teasel**): *Dipsacus* spp., 154
culver-foot (*Geranium molle*): see *Geranium* spp., 193
curled-leaved pondweed (**curlyleaf pondweed**): ***Potamogeton* & *Stuckenia*,** 323
curltop ladysthumb (**pale smartweed**): ***Polygonum persicaria* & *lapathifolium*,** 320
curly dock, curled or curly-leaved dock (*Rumex crispus*): *Rumex crispus* & *obtusifolius*, 346
curlyleaf pondweed (*Potamogeton crispus*): *Potamogeton* & *Stuckenia*, 323
curlytop knotweed (**pale smartweed**): ***Polygonum persicaria* & *lapathifolium*,** 320
curse, Dutch (**oxeye daisy**): *Leucanthemum vulgare*, 247
curse, Paterson's (**vipers bugloss**): *Echium plantagineum*, 159
curveseed butterwort (*Ranunculus testiculatus*): Herbaceous Forbs, 444, 450
Cuscuta japonica (**Japanese dodder**), 136
cutleaf or cut-leaved
    burnweed (*Erechtites* spp.): Herbaceous Forbs, 438, 446
    coast fireweed (*Erechtites* spp.): Herbaceous Forbs, 438, 446
    cranesbill (**cutleaf geranium**): *Geranium* spp., 193
    fireweed (*Erechtites* spp.): Herbaceous Forbs, 438, 446
    geranium (*Geranium dissectum*): *Geranium* spp., 193
    teasel (*Dipsacus laciniatus*): *Dipsacus* spp., 154
Cynara cardunculus (**artichoke thistle**), 138
Cynodon dactylon (**bermudagrass**), 141
Cynoglossum officinale (**houndstongue**), 143
Cynosurus echinatus (**hedgehog dogtailgrass**), 145
cypress
    flowering cypress (**athel tamarisk**): *Tamarix* spp., 395
    mock cypress (**kochia**): *Bassia* spp., 54
    summer cypress (**kochia**): *Bassia* spp., 54
***Cytisus* spp.**
    monspessulanus: *Genista monspessulana*, 191, 516
    scoparius (**Scotch broom**), 147, 516
Dactylis glomerata (orchardgrass): Grasses & Sedges, 434, 436
dagger cocklebur (**spiny cocklebur**): *Xanthium* spp., 428
daisy
    blue (*Cichorium intybus*): Herbaceous Forbs, 438, 446
    butter, dog, dun, field, horse, maudlin, moon, or white (**oxeye daisy**): *Leucanthemum vulgare*, 247
    crown or garland (*Chrysanthemum coronarium*): Herbaceous Forbs, 438, 446
    oxeye: *Leucanthemum vulgare*, 247
    red (**orange hawkweed**): *Hieracium* spp., 209
dallisgrass (*Paspalum dilatatum*): Grasses & Sedges, 434, 436
Dalmatian toadflax: *Linaria dalmatica*, 254, 517
damesrocket (*Hesperis matronalis*): Herbaceous Forbs, 440, 448
dandelion

coast (**common catsear**): *Hypochaeris* spp., 230
common (*Taraxacum officinale*): Herbaceous Forbs, 440, 446
dandelion (*Taraxacum officinale*): Herbaceous Forbs, 440, 446
false (*Hypochaeris glabra* or *H. radicata*): *Hypochaeris* spp., 230
date palm, Canary Island (*Phoenix canariensis*): *Phoenix & Washingtonia*, 307
Daubentonia punicea: *Sesbania punicea*, 369
Daucus carota (**wild carrot**), 149
deadly hemlock (**poison hemlock**): *Conium maculatum*, 125, 513
deadnettle (*Lamium amplexicaule*): Herbaceous Forbs, 442, 450
deadnettle, henbit (*Lamium amplexicaule*): Herbaceous Forbs, 442, 450
deathcamas: *Zigadenus* spp., 431
death-weed (*Iva axillaris*): Herbaceous Forbs, 438, 446
deervetch, birdfoot (**birdsfoot trefoil**): *Lotus corniculatus*, 263
Delairea odorata (**cape-ivy**), 152
dense waterweed (**Brazilian egeria**): *Egeria densa*, 163
dense-flowered cordgrass (*Spartina densiflora*): *Spartina* spp., 383
Descurainia sophia (flixweed): Herbaceous Forbs, 440, 448
desert
   artichoke (**artichoke thistle**): *Cynara cardunculus*, 138
   lovegrass (*Eragrostis* spp.): Grasses & Sedges, 434, 436
   mustard (**London rocket**): *Sisymbrium* spp., 377
   nightshade (*Solanum elaeagnifolium*): Herbaceous Forbs, 444, 452
devil, Mexican (**croftonweed**): *Ageratina adenophora*, 21
devil's
   grass (**bermudagrass**): *Cynodon dactylon*, 141
   paintbrush (**orange hawkweed**): *Hieracium* spp., 209
   plague (**wild carrot**): *Daucus carota*, 149
   thorn (*Emex spinosa*): Herbaceous Forbs, 442, 450
devil's-grass (**rush skeletonweed**): *Chondrilla juncea*, 116, 515
devil's-grass (**quackgrass**): *Elytrigia repens*, 172
devil's-weed (*Iva axillaris*): Herbaceous Forbs, 438, 446
diffuse knapweed: *Centaurea diffusa*, 93, 514
diffuse lovegrass (*Eragrostis* spp.): Grasses & Sedges, 434, 436
Digitalis purpurea (foxglove): Herbaceous Forbs, 444, 452
Digitaria dactylon: *Cynodon dactylon*, 141
Digitaria dilatata (*Paspalum dilatatum*): Grasses & Sedges, 434, 436
Dipsacus spp.
   fullonum (**common teasel**): *Dipsacus* spp., 154
   laciniatus (**cutleaf teasel**): *Dipsacus* spp., 154
   sativus (**Fuller's teasel**): *Dipsacus* spp., 154
   sylvestris (*Dipsacus fullonum*): *Dipsacus* spp., 154
distaff thistle (**woolly distaff thistle**): *Carthamus lanatus*, 85
ditchbur (**common cocklebur**): *Xanthium* spp., 428
Dittrichia graveolens (stinkwort), 157
dock
   bitter (**broadleaf dock**): *Rumex crispus & obtusifolius*, 346
   broadleaf (*Rumex obtusifolius*): *Rumex crispus & obtusifolius*, 346
   curly, curled, or curly-leaved (*Rumex crispus*): *Rumex crispus & obtusifolius*, 346
   narrowleaf (**curly dock**): *Rumex crispus & obtusifolius*, 346
   round (*Malva neglecta*): Herbaceous Forbs, 442, 450
   sour (**red sorrel**): *Rumex acetosella*, 344
   sour (**curly dock**): *Rumex crispus & obtusifolius*, 346
   velvet (**common mullein**): *Verbascum* spp., 419
   yellow (**curly dock**): *Rumex crispus & obtusifolius*, 346
dockleaf smartweed (**pale smartweed**): *Polygonum persicaria & lapathifolium*, 320
dodder, giant (**Japanese dodder**): *Cuscuta japonica*, 136
dodder, Japanese: *Cuscuta japonica*, 136
dog
   -berry (**dog rose**): *Rosa* spp., 338
   brier (**dog rose**): *Rosa* spp., 338
   bur (**houndstongue**): *Cynoglossum officinale*, 143
   daisy (**oxeye daisy**): *Leucanthemum vulgare*, 247
   -eared sorrel (**red sorrel**): *Rumex acetosella*, 344
   grass (**quackgrass**): *Elytrigia repens*, 172
   rose (*Rosa canina*): *Rosa* spp., 338
dog's tongue (**houndstongue**): *Cynoglossum officinale*, 143
dog's tooth grass (**bermudagrass**): *Cynodon dactylon*, 141
dogtail grass, bristly (**hedgehog dogtailgrass**): *Cynosurus echinatus*, 145
dogtail, annual (**hedgehog dogtailgrass**): *Cynosurus echinatus*, 145
dogtail, hedgehog (**hedgehog dogtailgrass**): *Cynosurus echinatus*, 145
dogtailgrass, hedgehog: *Cynosurus echinatus*, 145
donax reed (**giant reed**): *Arundo donax*, 40, 517
doorweed (*Polygonum arenastrum*): Herbaceous Forbs, 442, 450
dovefoot geranium (*Geranium molle*): see *Geranium* spp., 193
dove's-foot cranesbill (*Geranium molle*): see *Geranium* spp., 193
dove's-foot geranium (*Geranium molle*): see *Geranium* spp., 193
downy brome, cheat, or chess (*Bromus tectorum*): *Bromus diandrus, madritensis, tectorum*, 66
downy thistle (**Scotch thistle**): *Onopordum acanthium*, 292, 515
drooping brome (**downy brome**): *Bromus diandrus, madritensis, tectorum*, 66
duckmeat (*Spirodela* spp.): *Lemna, Spirodela, & Wolffia*, 242
duckweed (*Lemna, Spirodela* spp.): *Lemna, Spirodela, & Wolffia*, 242
duckweed, giant (*Spirodela* spp.): *Lemna, Spirodela, & Wolffia*, 242
dun daisy (**oxeye daisy**): *Leucanthemum vulgare*, 247
Dutch curse (**oxeye daisy**): *Leucanthemum vulgare*, 247
dwarf elm (*Ulmus pumila*): Woody Plants, 454, 456
dwarf mallow (*Malva neglecta*): Herbaceous Forbs, 442, 450
dyer's woad: *Isatis tinctoria*, 236
early chess (**downy brome**): *Bromus diandrus, madritensis, tectorum*, 66
early tamarisk (**smallflower tamarisk**): *Tamarix* spp., 395, 517
Echium spp.
   lycopsis: *Echium plantagineum*, 159
   menziesii (*Amsinckia menziesii* var. *menziesii*): *Amsinckia* spp., 35
   plantagineum (**vipers bugloss**), 159
   vulgare (**blueweed**), 161
echium, purple (**vipers bugloss**): *Echium plantagineum*, 159
edible fig (**fig**): *Ficus carica*, 185
egeria (**Brazilian egeria**): *Egeria densa*, 163
Egeria densa (**Brazilian egeria**), 163
egeria, Brazilian: *Egeria densa*, 163
eggleaf spurge (**oblong spurge**): *Euphorbia* spp., 182
eglantine (**sweetbriar rose**): *Rosa* spp., 338
Egyptian millet (**johnsongrass**): *Sorghum halepense*, 380
Ehrharta spp. (**veldtgrasses**), 165
Eichhornia crassipes (**water hyacinth**), 167, 518
Eichhornia speciosa: *Eichhornia crassipes*, 167, 518
Elaeagnus angustifolia (**Russian-olive**), 169
elephant grass (**giant reed**): *Arundo donax*, 40, 517
elm
   Chinese (**tree-of-heaven**): *Ailanthus altissima*, 23
   dwarf (*Ulmus pumila*): Woody Plants, 454, 456
   Siberian (*Ulmus pumila*): Woody Plants, 454, 456
Elodea spp.
   canadensis (common elodea): Aquatic Plants, 458, 460
      [= *brandegae, ioensis, linearis, planchonii*]
   densa: *Egeria densa*, 163
   verticillata: *Hydrilla verticillata*, 221, 518
elodea
   American elodea (*Elodea canadensis*): Aquatic Plants, 458, 460
   common elodea (*Elodea canadensis*): Aquatic Plants, 458, 460
   Florida elodea (**hydrilla**): *Hydrilla verticillata*, 221, 518
   leafy elodea (**Brazilian egeria**): *Egeria densa*, 163
Elymus caput-medusae: *Taeniatherum caput-medusae*, 392
Elymus repens: *Elytrigia repens*, 172
Elytrigia repens (**quackgrass**), 172
Emex spinosa (spiny emex): Herbaceous Forbs, 442, 450
emex, spiny (*Emex spinosa*): Herbaceous Forbs, 442, 450

English
- broom (**Scotch broom**): *Cytisus scoparius*, 147, 516
- hawthorn (*Crataegus monogyna*): Woody Plants, 454, 456
- holly (*Ilex aquifolium*): Woody Plants, 454, 456
- ivy (***Hedera helix***): *Hedera* spp., 202
- plantain (*Plantago lanceolata*): Herbaceous Forbs, 442, 450
- ryegrass (**perennial ryegrass**): *Lolium* spp., 260
- thistle (*Lactuca serriola*): Herbaceous Forbs, 438, 446

equisetum (**horsetail, scouringrush**): *Equisetum* spp., 175
***Equisetum* spp.** (**horsetail, scouringrush**), 175
eragrostis (*Eragrostis* spp.): Grasses & Sedges, 434, 436
eragrostis, low (*Eragrostis* spp.): Grasses & Sedges, 434, 436
Erechtites spp. (burnweeds): Herbaceous Forbs, 438, 446
erechtites (*Erechtites* spp.): Herbaceous Forbs, 438, 446
erect veldtgrass (*Ehrharta erecta*): *Ehrharta* spp., 165
Erianthus ravennae: ***Tripidium ravennae***, 412
Ericameria nauseosa (*Chrysothamnus nauseosus*): Herbaceous Forbs, 438, 446
Erigeron spp.
- *bonariensis, linifolium, undulatus* (*Conyza bonariensis*): Herbaceous Forbs, 438, 446
- *canadensis, pusillus* (*Conyza canadensis*): Herbaceous Forbs, 438, 446

***Erodium cicutarium* (redstem filaree)**, 177
escobilla (**Peruvian peppertree**): *Schinus* spp., 364
***Eucalyptus globulus* (Tasmanian blue gum)**, 180
eucalyptus, common (**Tasmanian blue gum**): *Eucalyptus globulus*, 180
Eupatorium spp.: see ***Ageratina adenophora***, 21
eupatory (**croftonweed**): *Ageratina adenophora*, 21
***Euphorbia* spp.**
- *esula* (**leafy spurge**): *Euphorbia* spp., 182, 516
  [= *discolor, gmelinii, virgata*]
- *oblongata* (**oblong spurge**): *Euphorbia* spp., 182
- *terracina* (**carnation spurge**): *Euphorbia* spp., 182

Eurasian watermilfoil: *Myriophyllum spicatum*, 283, 518
European
- beachgrass: *Ammophila arenaria*, 31
- morningglory (**field bindweed**): *Convolvulus arvensis*, 127, 516
- pennyroyal (**pennyroyal**): *Mentha pulegium*, 275
- searocket (*Cakile maritima*): Herbaceous Forbs, 440, 448

Euryops multifidus (sweet resinbush): Woody Plants, 454, 456
Euryops subcarnosus (*Euryops multifidus*): Woody Plants, 454, 456
evergreen clematis (*Clematis vitalba*): Herbaceous Forbs, 444, 450
evergreen millet (**johnsongrass**): *Sorghum halepense*, 380
everlasting peavine (*Lathyrus latifolius*): Herbaceous Forbs, 442, 448
everlasting thorn (*Pyracantha* spp.): Woody Plants, 454, 456
face-clock (*Taraxacum officinale*): Herbaceous Forbs, 440, 446
fairy moss (**mosquitofern**): *Azolla* spp., 52, 518
faitours-grass (**leafy spurge**): *Euphorbia* spp., 182, 516
false
- acacia (**black locust**): *Robinia pseudoacacia*, 335
- brome (**annual false-brome**): *Brachypodium distachyon*, 57
- brome, purple (**annual false-brome**): *Brachypodium distachyon*, 57
- dandelion (**smooth** or **common catsear**): *Hypochaeris* spp., 230
- Guinea grass (**johnsongrass**): *Sorghum halepense*, 380
- pepper (**Peruvian peppertree**): *Schinus* spp., 364
- sandalwood (*Myoporum laetum*): Woody Plants, 454, 456
- starthistle (**woolly distaff thistle**): *Carthamus lanatus*, 85

false-brome, annual: *Brachypodium distachyon*, 57
false-brome, slender: *Brachypodium sylvaticum*, 59
fan palm, Mexican (*Washingtonia robusta*): *Phoenix* & *Washingtonia*, 307
fan palm, Washington (**Mexican fan palm**): *Phoenix* & *Washingtonia*, 307
fanweed (*Thlaspi arvense*): Herbaceous Forbs, 440, 448
fanwort: *Cabomba caroliniana*, 74
farmer's foxtail (**hare barley**): *Hordeum marinum* & *murinum*, 218

fat-hen (*Chenopodium album*): Herbaceous Forbs, 440, 448
faux poivrier (Brazilian peppertree): *Schinus* spp., 364
fennel, common or sweet: *Foeniculum vulgare*, 187
fennel-leaf pondweed (sago pondweed): *Potamogeton* & *Stuckenia*, 323
fern
- California fern (**poison hemlock**): *Conium maculatum*, 125, 513
- carrot fern (**poison hemlock**): *Conium maculatum*, 125, 513
- horsetail fern (**horsetail, scouringrush**): *Equisetum* spp., 175
- Nebraska fern (**poison hemlock**): *Conium maculatum*, 125, 513
- parsley fern (**common tansy**): *Tanacetum vulgare*, 399
- water fern (**mosquitofern**): *Azolla* spp., 52, 518

fescue
- alta, coarse, reed, or tall fescue (*Festuca arundinacea*): Grasses & Sedges, 434, 436
- fescue (*Festuca arundinacea*): Grasses & Sedges, 434, 436
- **foxtail fescue (rattail fescue)**: *Vulpia myuros*, 425
- **rattail fescue** (rat's-tail, rat-tailed, red-tailed): *Vulpia myuros*, 425
- sixweeks fescue (**rattail fescue**): *Vulpia myuros*, 425

Festuca spp.
- *arundinacea* (tall fescue): Grasses & Sedges, 434, 436
- *barbata* (***Schismus barbatus***): *Schismus* spp., 367
- *elatior* (*Festuca arundinacea*): Grasses & Sedges, 434, 436
- *glomerata* (*Dactylis glomerata*): Grasses & Sedges, 434, 436
- *megalura*: *Vulpia myuros*, 425
- *myuros*: *Vulpia myuros*, 425
- *perennis* (*Lolium perenne*): *Lolium* spp., 260

fetid nightshade (**black henbane**): *Hyoscyamus niger*, 226
***Ficus carica* (fig)**, 185
fiddleneck: *Amsinckia* spp., 35
- **coast fiddleneck (coast fiddleneck)**: *Amsinckia* spp., 35
- common fiddleneck (**coast fiddleneck**): *Amsinckia* spp., 35
- fireweed fiddleneck (**coast fiddleneck**): *Amsinckia* spp., 35
- **Menzies fiddleneck (*Amsinckia menziesii* var. *menziesii*)**: *Amsinckia* spp., 35
- small-flowered fiddleneck (**Menzies fiddleneck**): *Amsinckia* spp., 35

field
- **bindweed**: *Convolvulus arvensis*, 127, 516
- daisy (**oxeye daisy**): *Leucanthemum vulgare*, 247
- kale (**wild mustard**): *Sinapis arvensis*, 374
- morningglory (**field bindweed**): *Convolvulus arvensis*, 127, 516
- mustard (*Brassica rapa*): Herbaceous Forbs, 440, 448
- mustard (**wild mustard**): *Sinapis arvensis*, 374
- pennycress (*Thlaspi arvense*): Herbaceous Forbs, 440, 448
- sandbur (*Cenchrus* spp.): Grasses & Sedges, 434, 436
- sorrel (**red sorrel**): *Rumex acetosella*, 344

fiery thorn (*Pyracantha* spp.): Woody Plants, 454, 456
fig, common or edible: *Ficus carica*, 185
fig, Hottentot: *Carpobrotus edulis*, 83
filaree, redstem: *Erodium cicutarium*, 177
fingerweed (**coast fiddleneck**): *Amsinckia* spp., 35
fireball (**kochia**): *Bassia* spp., 54
firethorn (*Pyracantha* spp.): Woody Plants, 454, 456
fireweed
- Australian fireweed (*Erechtites* spp.): Herbaceous Forbs, 438, 446
- cutleaf fireweed (*Erechtites* spp.): Herbaceous Forbs, 438, 446
- fireweed (**kochia**): *Bassia* spp., 54
- fireweed fiddleneck (**coast fiddleneck**): *Amsinckia* spp., 35
- little fireweed (*Erechtites* spp.): Herbaceous Forbs, 438, 446
- New Zealand fireweed (*Erechtites* spp.): Herbaceous Forbs, 438, 446
- rancher's fireweed (**coast fiddleneck**): *Amsinckia* spp., 35
- toothed coast fireweed (*Erechtites* spp.): Herbaceous Forbs, 438, 446

Fitch's tarweed (*Hemizonia fitchii*): *Hemizonia* spp., 205
five-fingers, sulphur (**sulfur cinquefoil**): *Potentilla recta*, 328
fivehook bassia (*Bassia hyssopifolia*): *Bassia* spp., 54
fivehorn smotherweed (**fivehook bassia**): *Bassia* spp., 54

five-stamen tamarisk (**saltcedar**): ***Tamarix* spp.**, 395, 517
flag, yellow (**yellowflag iris**): ***Iris pseudacorus***, 234
flannel plant (**common mullein**): ***Verbascum* spp.**, 419
flannel-leaf (**common mullein**): ***Verbascum* spp.**, 419
flatspine burweed (*Ambrosia acanthicarpa*): Herbaceous Forbs, 438, 446
flatweed (*Hypochaeris glabra* or *H. radicata*): ***Hypochaeris* spp.**, 230
flaxgrass (**wild oat**): ***Avena* spp.**, 49
fleabane
    Canada fleabane (*Conyza canadensis*): Herbaceous Forbs, 438, 446
    flaxleaved fleabane (*Conyza bonariensis*): Herbaceous Forbs, 438, 446
    hairy fleabane (*Conyza bonariensis*): Herbaceous Forbs, 438, 446
fleeceflower, Japanese (**Japanese knotweed**): ***Polygonum cuspidatum* et al.**, 318
flickertail (**foxtail barley**): ***Hordeum jubatum***, 216
flixweed (*Descurainia sophia*): Herbaceous Forbs, 440, 448
floating
    marshpennywort (**floating pennywort**): ***Hydrocotyle ranunculoides***, 224
    pennywort: ***Hydrocotyle ranunculoides***, 224
    pondweed, common (**floatingleaf pondweed**): ***Potamogeton natans & nodosus***, 326
    primrose-willow (**creeping waterprimrose**): ***Ludwigia* spp.**, 265
floatingheart, yellow: ***Nymphoides peltata***, 291
floatingleaf pondweed (*Potamogeton natans*): ***Potamogeton natans & nodosus***, 326
Florida
    aspen (**Chinese tallowtree**): ***Sapium sebiferum***, 361
    elodea (**hydrilla**): ***Hydrilla verticillata***, 221, 518
    holly (**Brazilian peppertree**): ***Schinus* spp.**, 364
flowering cypress (**athel tamarisk**): ***Tamarix* spp.**, 395
flowering rush (*Butomus umbellatus*): Aquatic Plants, 458, 460
Foeniculum vulgare (**fennel**), 187
    [= *F. foeniculum*, *F. officinale*]
fog grass (**common velvetgrass**): ***Holcus lanatus***, 212
fog, Yorkshire (**common velvetgrass**): ***Holcus lanatus***, 212
foothill deathcamas (*Zigadenus paniculatus*): ***Zigadenus* spp.**, 431
forget-me-not
    broadleaf forget-me-not: ***Myosotis latifolia***, 279
    common or wood forget-me-not (**broadleaf forget-me-not**): ***Myosotis latifolia***, 279
    yellow forget-me-not (**coast fiddleneck**): ***Amsinckia* spp.**, 35
fountaingrass, crimson: ***Pennisetum setaceum***, 301
fountaingrass, tender (**crimson fountaingrass**): ***Pennisetum setaceum***, 301
four o'clock, heartleaf (**wild four-o'clock**): ***Mirabilis nyctaginea***, 277
four-o'clock, wild: ***Mirabilis nyctaginea***, 277
foxglove, common or purple (*Digitalis purpurea*): Herbaceous Forbs, 444, 452
foxtail
    bristly, giant, golden, green, nodding, tall, yellow foxtail (*Setaria* spp.): Grasses & Sedges, 434, 436
    common or farmer's foxtail (**hare barley**): ***Hordeum marinum & murinum***, 218
    foxtail barley: ***Hordeum jubatum***, 216
    foxtail brome (**red brome**): ***Bromus diandrus, madritensis, tectorum***, 66
    foxtail fescue (**rattail fescue**): ***Vulpia myuros***, 425
    foxtail millet (*Setaria* spp.): Grasses & Sedges, 434, 436
    foxtails (pigeongrass) (*Setaria* spp.): Grasses & Sedges, 434, 436
    (*Setaria* spp.): Grasses & Sedges, 434, 436
fragrant waterlily (*Nymphaea odorata*): Aquatic Plants, 458, 460
freeway iceplant: ***Carpobrotus edulis***, 83
French broom: ***Genista monspessulana***, 191, 516
French tamarisk (*Tamarix gallica*): ***Tamarix* spp.**, 395
Frenchweed (*Thlaspi arvense*): Herbaceous Forbs, 440, 448
fringe, water (**yellow floatingheart**): ***Nymphoides peltata***, 291

frogbit (**common catsear**): ***Hypochaeris* spp.**, 230
frogbit, smooth: ***Limnobium laevigatum***, 252
frost-blite (*Chenopodium album*): Herbaceous Forbs, 440, 448
Fuller's
    teasel (*Dipsacus sativus*): ***Dipsacus* spp.**, 154
    herb (*Saponaria officinalis*): Herbaceous Forbs, 440, 448
    thistle (**bull thistle**): ***Cirsium vulgare***, 122, 515
furze (**gorse**): ***Ulex europaeus***, 415, 516
Galega officinalis (**goatsrue**), 189
garden loosestrife: ***Lysimachia vulgaris***, 269
garden tansy (**common tansy**): ***Tanacetum vulgare***, 399
garget (**common pokeweed**): ***Phytolacca americana***, 311
garland daisy (*Chrysanthemum coronarium*): Herbaceous Forbs, 438, 446
garlic mustard: ***Alliaria petiolata***, 27
Genista juncea: ***Spartium junceum***, 385
Genista monspessulana (**French broom**), 191, 516
gentle annie (*Cenchrus* spp.): Grasses & Sedges, 434, 436
Geraldton carnationweed (**carnation spurge**): ***Euphorbia* spp.**, 182
***Geranium* spp.**
    carolinianum (Carolina geranium): see ***Geranium* spp.**, 193
    dissectum (**cutleaf geranium**): ***Geranium* spp.**, 193
    laxum (*Geranium dissectum*): ***Geranium* spp.**, 193
    molle (dovefoot geranium): see ***Geranium* spp.**, 193
    purpureum (**little robin**): ***Geranium* spp.**, 193
    robertianum (herb-robert): Herbaceous Forbs, 442, 450; also see ***Geranium* spp.**, 193
geranium, Carolina geranium (*Geranium carolinianum*): see ***Geranium* spp.**, 193
    cutleaf geranium (*Geranium dissectum*): ***Geranium* spp.**, 193
    dovefoot geranium (*Geranium molle*): see ***Geranium* spp.**, 193
    robert or Robert's geranium (*Geranium robertianum*): Herbaceous Forbs, 442, 450; also see ***Geranium* spp.**, 193
German-ivy (**Cape ivy**): ***Delairea odorata***, 152
giant
    bristlegrass (*Setaria* spp.): Grasses & Sedges, 434, 436
    cane (**giant reed**): ***Arundo donax***, 40, 517
    creek nettle (*Urtica dioica*): Herbaceous Forbs, 444, 452
    dodder (**Japanese dodder**): ***Cuscuta japonica***, 136
    duckweed (*Spirodela* spp.): ***Lemna, Spirodela, & Wolffia***, 242
    foxtail (*Setaria* spp.): Grasses & Sedges, 434, 436
    hogweed: ***Heracleum mantegazzianum***, 207
    knotweed (**Sakhalin knotweed**): ***Polygonum cuspidatum* et al.**, 318
    plumeless thistle (**plumeless thistle**): ***Carduus* spp.**, 79, 513
    plumeless thistle (**musk thistle**): ***Carduus* spp.**, 79, 513
    reed (**giant reed**): ***Arundo donax***, 40, 517
    salvinia: ***Salvinia molesta***, 359
    whiteweed (**perennial pepperweed**): ***Lepidium latifolium***, 244
giraffe head (*Lamium amplexicaule*): Herbaceous Forbs, 442, 450
glabrous catsear (**smooth catsear**): ***Hypochaeris* spp.**, 230
Glebionis coronaria (*Chrysanthemum coronarium*): Herbaceous Forbs, 438, 446
globe-podded hoary cress (*Cardaria draba* or *C. pubescens*): ***Cardaria* spp.**, 76
glossy privet: ***Ligustrum lucidum***, 250
glovewort (**houndstongue**): ***Cynoglossum officinale***, 143
Glyceria declinata (**waxy mannagrass**), 196
goatgrass, barb or **barbed** (*Aegilops triuncialis*): ***Aegilops* spp.**, 19
goatgrass, jointed (*Aegilops cylindrica*): ***Aegilops* spp.**, 19
goathead (**puncturevine**): ***Tribulus terrestris***, 407, 517
goatsrue: ***Galega officinalis***, 189
goatweed (**common St. Johnswort**): ***Hypericum perforatum***, 228, 516
golden
    flower (**oxeye daisy**): ***Leucanthemum vulgare***, 247
    foxtail (*Setaria* spp.): Grasses & Sedges, 434, 436
    starthistle (**yellow starthistle**): ***Centaurea solstitialis***, 100, 514
goldens (**oxeye daisy**): ***Leucanthemum vulgare***, 247

goosefoot, wall (*Chenopodium berlandieri*): Herbaceous Forbs, 440, 448
goosefoot, white (*Chenopodium album*): Herbaceous Forbs, 440, 448
gorse: *Ulex europaeus*, 415, 516
gorse, common (**gorse**): *Ulex europaeus*, 415, 516
gosmore (**common catsear**): *Hypochaeris* spp., 230
grass poly (*Lythrum hyssopifolium*): Herbaceous Forbs, 442, 450
grass rush: *Juncus* spp., 238
grass, pin (**redstem filaree**): *Erodium cicutarium*, 177
greasewood (*Sarcobatus vermiculatus*): Herbaceous Forbs, 442, 448
great
    brome (**ripgut brome**): *Bromus diandrus, madritensis, tectorum*, 66
    oxeye (**oxeye daisy**): *Leucanthemum vulgare*, 247
    ragweed (*Ambrosia artemisiifolia*): Herbaceous Forbs, 438, 446
    water speedwell (**water speedwell**): *Veronica anagallis-aquatica*, 422
greater periwinkle (**big periwinkle**): *Vinca major*, 423
green
    amaranth (*Amaranthus* spp.): Herbaceous Forbs, 438, 446
    bristlegrass (*Setaria* spp.): Grasses & Sedges, 434, 436
    foxtail (*Setaria* spp.): Grasses & Sedges, 434, 436
greens, Chinaman's (*Amaranthus* spp.): Herbaceous Forbs, 438, 446
greenvine (**field bindweed**): *Convolvulus arvensis*, 127, 516
grim-the-collier (**orange hawkweed**): *Hieracium* spp., 209
ground burnut (**puncturevine**): *Tribulus terrestris*, 407, 517
ground honeysuckle (**birdsfoot trefoil**): *Lotus corniculatus*, 263
groundsel, ivy (**Cape ivy**): *Delairea odorata*, 152
Guinea grass, false (**johnsongrass**): *Sorghum halepense*, 380
gum succory (**rush skeletonweed**): *Chondrilla juncea*, 116, 515
gum, blue (**Tasmanian blue gum**): *Eucalyptus globulus*, 180
gum, Tasmanian blue: *Eucalyptus globulus*, 180
Gypsophila paniculata (**baby's-breath**), 198
gypsophila, babysbreath (**baby's breath**): *Gypsophila paniculata*, 198
gypsophyll, tall (**baby's breath**): *Gypsophila paniculata*, 198
gypsy-combs (*Dipsacus fullonum* or *D. sativus*): *Dipsacus* spp., 154
gypsyflower (**houndstongue**): *Cynoglossum officinale*, 143
hairgrass (**North African wiregrass**): *Ventenata dubia*, 417
hairy
    brassica (**shortpod mustard**): *Hirschfeldia incana*, 211
    catsear (**common catsear**): *Hypochaeris* spp., 230
    clover (*Trifolium hirtum*): Herbaceous Forbs, 442, 448
    fleabane (*Conyza bonariensis*): Herbaceous Forbs, 438, 446
    mustard (**shortpod mustard**): *Hirschfeldia incana*, 211
    whitetop (**Cardaria pubescens**): *Cardaria* spp., 76
halogeton: *Halogeton glomeratus*, 200, 515
Halogeton glomeratus (**halogeton**), 200, 515
hardheads: *Acroptilon repens*, 16, 513
hardinggrass: *Phalaris aquatica*, 303
hardstem bulrush or tule (*Scirpus acutus*): Grasses & Sedges, 434, 436
hardy pampasgrass (**ravennagrass**): *Tripidium ravennae*, 412
hare barley (*Hordeum murinum* ssp. *leporinum*): *Hordeum marinum & murinum*, 218
harmal or harmel (*Peganum harmala*): Herbaceous Forbs, 444, 452
hawkweeds: *Hieracium* spp., 209
    kingdevil hawkweed (*Hieracium piloselloides*): *Hieracium* spp., 209
    meadow hawkweed (*Hieracium caespitosum*): *Hieracium* spp., 209
    mouse-ear hawkweed (*Hieracium pilosella*): *Hieracium* spp., 209
    orange hawkweed (*Hieracium aurantiacum*): *Hieracium* spp., 209
    queendevil hawkweed (*Hieracium glomeratum*): *Hieracium* spp., 209
    yellow hawkweed (**meadow hawkweed**): *Hieracium* spp., 209
hawthorn
    English, oneseed, or singleseed (*Crataegus monogyna*): Woody Plants, 454, 456
heartleaf four o'clock (**wild four-o'clock**): *Mirabilis nyctaginea*, 277
heart-podded hoary cress (**hoary cress**): *Cardaria* spp., 76
heart-weed (**ladysthumb**): *Polygonum persicaria & lapathifolium*, 320
Hedera spp.
    algeriensis (*Hedera canariensis*): *Hedera* spp., 202

    canariensis (**Algerian ivy**): *Hedera* spp., 202
    helix (**English ivy**): *Hedera* spp., 202
    hibernica (**Atlantic ivy**): *Hedera* spp., 202
hedge nettle (*Urtica dioica*): Herbaceous Forbs, 444, 452
hedge pink (*Saponaria officinalis*): Herbaceous Forbs, 440, 448
hedgehog
    bur-weed (**common cocklebur**): *Xanthium* spp., 428
    dogtail (**hedgehog dogtailgrass**): *Cynosurus echinatus*, 145
    dogtailgrass: *Cynosurus echinatus*, 145
    grass (*Cenchrus* spp.): Grasses & Sedges, 434, 436
hedge-hoggy (**hedgehog dogtailgrass**): *Cynosurus echinatus*, 145
hedgeparsley: *Torilis arvensis*, 401
    spreading hedgeparsley (**hedgeparsley**): *Torilis arvensis*, 401
    upright hedgeparsley (**hedgeparsley**): *Torilis arvensis*, 401
Helianthus spp.
    annuus (common sunflower): Herbaceous Forbs, 438, 446
    [= *aridus, lenticularis*]
heliotrope, blue (*Heliotropium amplexicaule*): see *Myosotis latifolia*, 279
Heliotropium amplexicaule (blue heliotrope): see *Myosotis latifolia*, 279
Hemizonia spp.
    fitchii (**Fitch's tarweed**): *Hemizonia* spp., 205
    parryi (**Parry's tarweed**): *Hemizonia* spp., 205
    pungens (**spikeweed**): *Hemizonia* spp., 205
    virgata: *Holocarpha virgata*, 214
hemlock
    deadly hemlock (**poison hemlock**): *Conium maculatum*, 125, 513
    poison hemlock: *Conium maculatum*, 125, 513
    spotted hemlock (**poison hemlock**): *Conium maculatum*, 125, 513
henbane, black: *Hyoscyamus niger*, 226
henbane, common (**black henbane**): *Hyoscyamus niger*, 226
henbell (**black henbane**): *Hyoscyamus niger*, 226
henbit, henbit deadnettle, common henbit (*Lamium amplexicaule*): Herbaceous Forbs, 442, 450
Heracleum mantegazzianum (**giant hogweed**), 207
heraldic thistle (**Scotch thistle**): *Onopordum acanthium*, 292, 515
herb sophia (*Descurainia sophia*): Herbaceous Forbs, 440, 448
herbe de Cuba (**johnsongrass**): *Sorghum halepense*, 380
herb-robert (*Geranium robertianum*): Herbaceous Forbs, 442, 450; also see *Geranium* spp., 193
heronsbill (**redstem filaree**): *Erodium cicutarium*, 177
Hesperis matronalis (**damesrocket**): Herbaceous Forbs, 440, 448
Heteranthera formosa: *Eichhornia crassipes*, 167, 518
Hieracium spp. (**hawkweeds**)*Hieracium* spp., 209
    aurantiacum (**orange hawkweed**): *Hieracium* spp., 209
    caespitosum (**meadow hawkweed**): *Hieracium* spp., 209
    glomeratum (**queendevil hawkweed**): *Hieracium* spp., 209
    pilosella (**mouse-ear hawkweed**): *Hieracium* spp., 209
    piloselloides (**kingdevil hawkweed**): *Hieracium* spp., 209
    pratense (*Hieracium caespitosum*): *Hieracium* spp., 209
Himalaya or Himalayan
    blackberry: *Rubus armeniacus*, 341
    knotweed (*Polygonum polystachyum*): *Polygonum cuspidatum* et al., 318
hindhead (**common tansy**): *Tanacetum vulgare*, 399
Hippochaete hyemalis (**horsetail, scouringrush**): *Equisetum* spp., 175
Hirschfeldia incana (**shortpod mustard**), 211
hoary
    alyssum (*Berteroa incana*): Herbaceous Forbs, 440, 448
    cress (**lens-podded whitetop**): *Cardaria* spp., 76
    cress (*Cardaria draba*): *Cardaria* spp., 76
    cress (**hairy whitetop**): *Cardaria* spp., 76
    cress, globe-podded (*Cardaria draba* or *C. pubescens*): *Cardaria* spp., 76
    cress, heart-podded (**hoary cress**): *Cardaria* spp., 76
    cress, lens-podded (**lens-podded whitetop**): *Cardaria* spp., 76
    mustard (**shortpod mustard**): *Hirschfeldia incana*, 211

nettle (*Urtica dioica*): Herbaceous Forbs, 444, 452
hogbite (**rush skeletonweed**): ***Chondrilla juncea***, 116, 515
hog's-bean (**black henbane**): ***Hyoscyamus niger***, 226
hogweed, giant: ***Heracleum mantegazzianum***, 207
Holcus lanatus (**common velvetgrass**), 212
holly
    Christmas holly (*Ilex aquifolium*): Woody Plants, 454, 456
    English holly (*Ilex aquifolium*): Woody Plants, 454, 456
    Florida holly (**Brazilian peppertree**): ***Schinus*** spp., 364
hollyleaf naiad (*Najas marina*): ***Najas*** spp., 286
Holocarpha virgata (**virgate tarweed**), 214
holy thistle (**blessed milkthistle**): ***Silybum marianum***, 371, 515
honey clover (*Melilotus albus*): Herbaceous Forbs, 442, 448
honeysuckle, ground (**birdsfoot trefoil**): ***Lotus corniculatus***, 263
hop o' my thumb (**birdsfoot trefoil**): ***Lotus corniculatus***, 263
Hordeum spp.
    caput-medusae: ***Taeniatherum caput-medusae***, 392
    jubatum (**foxtail barley**), 216
    marinum ssp. *gussonianum* (**Mediterranean barley**): ***H. marinum & murinum***, 218
        [= *geniculatum, gussonianum, hystrix*]
    murinum ssp. *leporinum* (**hare barley**): ***H. marinum & murinum***, 218
        [= *leporinum*]
horehound, white (or common): ***Marrubium vulgare***, 273
horned pondweed (*Zannichellia palustris*): Aquatic Plants, 458, 460
hornseed buttercup (*Ranunculus testiculatus*): Herbaceous Forbs, 444, 450
hornwort: ***Ceratophyllum demersum***, 112
horse
    daisy (**oxeye daisy**): ***Leucanthemum vulgare***, 247
    sorrel (**red sorrel**): ***Rumex acetosella***, 344
    thistle (*Lactuca serriola*): Herbaceous Forbs, 438, 446
horsenettle, white (*Solanum elaeagnifolium*): Herbaceous Forbs, 444, 452
horsepipes (**horsetail, scouringrush**): ***Equisetum*** spp., 175
horsetail
    Canadian horsetail (*Conyza canadensis*): Herbaceous Forbs, 438, 446
    common horsetail (**horsetail, scouringrush**): ***Equisetum*** spp., 175
    horsetail: ***Equisetum*** spp., 175
    horsetail fern (**horsetail, scouringrush**): ***Equisetum*** spp., 175
horseweed (*Conyza canadensis*): Herbaceous Forbs, 438, 446
Hottentot fig: ***Carpobrotus edulis***, 83
houndsbane (**white horehound**): ***Marrubium vulgare***, 273
houndstongue: ***Cynoglossum officinale***, 143
hurnal (*Peganum harmala*): Herbaceous Forbs, 444, 452
hyacinth, water: ***Eichhornia crassipes***, 167, 518
hydrilla: ***Hydrilla verticillata***, 221, 518
Hydrilla verticillata (= *H. lithuanica*) (**hydrilla**), 221, 518
Hydrocotyle ranunculoides (**floating pennywort**), 224
        [= *batrachiodes, natans*]
hyoscyamus (**black henbane**): ***Hyoscyamus niger***, 226
Hyoscyamus niger (**black henbane**), 226
Hypericum spp.
    canariensis (Canary Island hypericum): Woody Plants, 454, 456
    Canary Island Hypericum (*H. canariensis*): Woody Plants, 454, 456
    perforatum (**common St. Johnswort**), 228, 516
Hypochaeris glabra (**smooth catsear**): ***Hypochaeris*** spp., 230
Hypochaeris radicata (**common catsear**): ***Hypochaeris*** spp., 230
hyssop loosestrife or lythrum (*Lythrum hyssopifolium*): Herbaceous Forbs, 442, 450
hyssop-leaved echinopsilon (**fivehook bassia**): ***Bassia*** spp., 54
Iberian starthistle (*Centaurea iberica*): ***Centaurea calcitrapa & iberica***, 87, 514
iceplant or freeway iceplant: ***Carpobrotus edulis***, 83
Ilex aquifolium (English holly): Woody Plants, 454, 456

Illinois pondweed (*Potamogeton illinoensis*): ***Potamogeton & Stuckenia***, 323
Indian
    bean tree (*Catalpa bignonioides*): Woody Plants, 454, 456
    cane (**red sorrel**): ***Rumex acetosella***, 344
    pond lily (*Nuphar lutea*): Aquatic Plants, 458, 460
    teasel (**Fuller's teasel**): ***Dipsacus*** spp., 154
indigobush: ***Amorpha fruticosa***, 33
inkberry (**common pokeweed**): ***Phytolacca americana***, 311
insane root (**black henbane**): ***Hyoscyamus niger***, 226
Iris pseudacorus (**yellowflag iris**), 234
iris, yellowflag or paleyellow: ***Iris pseudacorus***, 234
Irish ivy (*Hedera hibernica*): ***Hedera*** spp., 202
iron weed (**perennial pepperweed**): ***Lepidium latifolium***, 244
Isatis tinctoria (**dyer's woad**), 236
isband (*Peganum harmala*): Herbaceous Forbs, 444, 452
Italian
    ivy (**Cape ivy**): ***Delairea odorata***, 152
    plumeless thistle (**Italian thistle**): ***Carduus*** spp., 79, 513
    ryegrass (*Lolium multiflorum*): ***Lolium*** spp., 260
    thistle (**slenderflower thistle**): ***Carduus*** spp., 79, 513
    thistle (*Carduus pycnocephalus*): ***Carduus*** spp., 79, 513
Iva axillaris (**povertyweed**): Herbaceous Forbs, 438, 446
ivy
    Algerian ivy (*Hedera canariensis*): ***Hedera*** spp., 202
    Atlantic ivy (*Hedera hibernica*): ***Hedera*** spp., 202
    ivy bindweed (*Polygonum convolvulus*): Herbaceous Forbs, 444, 450
    Cape ivy: ***Delairea odorata***, 152
    English ivy (*Hedera helix*): ***Hedera*** spp., 202
    German ivy (**Cape ivy**): ***Delairea odorata***, 152
    ivy groundsel (**Cape ivy**): ***Delairea odorata***, 152
    Irish ivy (*Hedera hibernica*): ***Hedera*** spp., 202
    Italian ivy (**Cape ivy**): ***Delairea odorata***, 152
    parlor ivy (**Cape ivy**): ***Delairea odorata***, 152
    water ivy (**Cape ivy**): ***Delairea odorata***, 152
jack, spiny three-cornered (*Emex spinosa*): Herbaceous Forbs, 442, 450
jalap, cancer (**common pokeweed**): ***Phytolacca americana***, 311
jane, salvation (**vipers bugloss**): ***Echium plantagineum***, 159
Japanese
    bamboo (**Japanese knotweed**): ***Polygonum cuspidatum*** et al., 318
    bristlegrass (*Setaria* spp.): Grasses & Sedges, 434, 436
    brome (*Bromus japonicus*): ***Bromus hordeaceus, japonicus***, 71
    chess (**Japanese brome**): ***Bromus hordeaceus, japonicus***, 71
    dodder: ***Cuscuta japonica***, 136
    fleeceflower (**Japanese knotweed**): ***Polygonum cuspidatum*** et al., 318
    knotweed (*Polygonum cuspidatum*): ***Polygonum cuspidatum*** et al., 318
    rose (**multiflora rose**): ***Rosa*** spp., 338
jenny, creeping (**field bindweed**): ***Convolvulus arvensis***, 127, 516
jenny, creeping (**garden loosestrife**): ***Lysimachia vulgaris***, 269
Jim Hill mustard (**tumble mustard**): ***Sisymbrium*** spp., 377
johnsongrass: ***Sorghum halepense***, 380
jointed
    goatgrass (*Aegilops cylindrica*): ***Aegilops*** spp., 19
    wild radish (**wild radish**): ***Raphanus*** spp., 331
jubatagrass (*Cortaderia jubata*): ***Cortaderia*** spp., 130
Juncus effusus (**soft rush**): ***Juncus*** spp., 238
Juncus patens (**spreading rush**): ***Juncus*** spp., 238
Juncus spp. (**rushes**), 238
junegrass (**downy brome**): ***Bromus diandrus, madritensis, tectorum***, 66
jupiter's-bean (**black henbane**): ***Hyoscyamus niger***, 226
kale, field (**wild mustard**): ***Sinapis arvensis***, 374
kale, wild (**wild radish**): ***Raphanus*** spp., 331
kedlock (**wild mustard**): ***Sinapis arvensis***, 374
kellup-weed (**oxeye daisy**): ***Leucanthemum vulgare***, 247

INDEX

Kentucky bluegrass (*Poa pratensis*): Grasses & Sedges, 434, 436
kikuyugrass: ***Pennisetum clandestinum***, 299
kingdevil hawkweed (*Hieracium piloselloides*): *Hieracium* spp., 209
kinghead (*Ambrosia artemisiifolia*): Herbaceous Forbs, 438, 446
klamathweed (**common St. Johnswort**): *Hypericum perforatum*, 228, 516
knapweed
    black (*Centaurea nigra*): 514
    brown (*Centaurea jacea*): 514
    diffuse knapweed: ***Centaurea diffusa***, 93, 514
    meadow knapweed: ***Centaurea debeauxii***, 90, 513
    Russian knapweed: ***Acroptilon repens***, 16, 513
    spotted knapweed: ***Centaurea stoebe***, 105, 514
    squarrose knapweed: ***Centaurea virgata* ssp. *squarrosa***, 109, 515
    white knapweed (**diffuse knapweed**): *Centaurea diffusa*, 93, 514
knot bindweed (*Polygonum convolvulus*): Herbaceous Forbs, 444, 450
knotgrass (**quackgrass**): *Elytrigia repens*, 172
knotweed
    bell-shaped (**Himalayan knotweed**): *Polygonum cuspidatum* et al., 318
    Bohemian (*Polygonum x bohemicum*): *Polygonum cuspidatum* et al., 318
    common (prostrate knotweed) (*Polygonum arenastrum*): Herbaceous Forbs, 442, 450
    common (**pale smartweed**): *Polygonum persicaria* & *lapathifolium*, 320
    curlytop (**pale smartweed**): *Polygonum persicaria* & *lapathifolium*, 320
    giant (*Polygonum sachalinense*): *Polygonum cuspidatum* et al., 318
    Himalayan (*Polygonum polystachyum*): *Polygonum cuspidatum* et al., 318
    Japanese (*Polygonum cuspidatum*): *Polygonum cuspidatum* et al., 318
    oval-leaf (*Polygonum arenastrum*): Herbaceous Forbs, 442, 450
    prostrate (*Polygonum arenastrum*): Herbaceous Forbs, 442, 450
    Sakhalin (*Polygonum sachalinense*): *Polygonum cuspidatum* et al., 318
    spotted (**ladysthumb**): *Polygonum persicaria* & *lapathifolium*, 320
kochia (*Bassia scoparia*): *Bassia* spp., 54
Kochia spp.
    alata, parodii, scoparia, sieversiana, trichophila (*Bassia scoparia*): *Bassia* spp., 54
    hyssopifolia (***Bassia hyssopifolia***): *Bassia* spp., 54
konpeito-gusa (*Cenchrus* spp.): Grasses & Sedges, 434, 436
lace, Queen Anne's (**wild carrot**): *Daucus carota*, 149
Lactuca scariola (*Lactuca serriola*): Herbaceous Forbs, 438, 446
Lactuca serriola (prickly lettuce): Herbaceous Forbs, 438, 446
Lady Campbell weed (**vipers bugloss**): *Echium plantagineum*, 159
lady's purse (*Capsella bursa-pastoris*): Herbaceous Forbs, 440, 448
lady's thistle (**blessed milkthistle**): *Silybum marianum*, 371, 515
ladysthumb (*Polygonum persicaria*): *Polygonum persicaria* & *lapathifolium*, 320
ladysthumb, curltop (**pale smartweed**): *Polygonum persicaria* & *lapathifolium*, 320
Lagarosiphon major (**oxygenweed**), 240
lambsquarter
    common (*Chenopodium album*): Herbaceous Forbs, 440, 448
    netseed (*Chenopodium berlandieri*): Herbaceous Forbs, 440, 448
Lamium amplexicaule (henbit): Herbaceous Forbs, 442, 450
lanceleaf plantain (*Plantago lanceolata*): Herbaceous Forbs, 442, 450
large
    periwinkle (**big periwinkle**): *Vinca major*, 423
    quakinggrass (*Briza maxima*): Grasses & Sedges, 434, 436
    watergrass (*Paspalum dilatatum*): Grasses & Sedges, 434, 436
    white waterlily (*Nymphaea odorata*): Aquatic Plants, 458, 460
largefruit amaranth (*Amaranthus* spp.): Herbaceous Forbs, 438, 446

late tamarisk (**saltcedar**): ***Tamarix*** spp., 395, 517
Lathyrus latifolius (everlasting peavine): Herbaceous Forbs, 442, 448
leafy
    elodea (**Brazilian egeria**): *Egeria densa*, 163
    pondweed (*Potamogeton foliosus*): *Potamogeton* & *Stuckenia*, 323
    spurge (*Euphorbia esula*): *Euphorbia* spp., 182, 516
Lehmann lovegrass (*Eragrostis* spp.): Grasses & Sedges, 434, 436
Lemna spp. (**duckweed**): ***Lemna, Spirodela,* & *Wolffia***, 242
lens-podded whitetop or hoary cress (*Cardaria chalepensis*): *Cardaria* spp., 76
Lenticula minor (*Lemna* spp.): *Lemna, Spirodela,* & *Wolffia*, 242
lentil, water (*Lemna* spp.): *Lemna, Spirodela,* & *Wolffia*, 242
Leontodon taraxacum (*Taraxacum officinale*): Herbaceous Forbs, 440, 446
Lepidium spp.
    appelianum (*Cardaria pubescens*): *Cardaria* spp., 76
    chalepense (*Cardaria chalepensis*): *Cardaria* spp., 76
    draba (*Cardaria draba*): *Cardaria* spp., 76
    latifolium (**perennial pepperweed**), 244
    perfoliatum (clasping pepperweed): Herbaceous Forbs, 440, 448
    repens (*Cardaria chalepensis*): *Cardaria* spp., 76
leporinum barley (**hare barley**): ***Hordeum marinum* & *murinum***, 218
lettuce
    China, prickly, or wild (*Lactuca serriola*): Herbaceous Forbs, 438, 446
    water (*Pistia stratiotes*): 518
Leucanthemum leucanthemum: ***Leucanthemum vulgare***, 247
Ligusticum foeniculum: ***Foeniculum vulgare***, 187
Ligustrum lucidum (**glossy privet**), 250
lilac, summer (*Buddleja davidii*): Woody Plants, 454, 456
lily
    arum lily (*Zantedeschia aethiopica*): Herbaceous Forbs, 438, 446
    calla lily (*Zantedeschia aethiopica*): Herbaceous Forbs, 438, 446
    cow lily (*Nuphar lutea*): Aquatic Plants, 458, 460
    fragrant waterlily (*Nymphaea odorata*): Aquatic Plants, 458, 460
    yellow pond lily (*Nuphar lutea*): Aquatic Plants, 458, 460
Limnobium laevigatum (**smooth frogbit**), 252
Limnobium spongia: ***Limnobium laevigatum***, 252
Linaria spp.
    dalmatica (**Dalmatian toadflax**), 254, 517
    genistifolia ssp. *dalmatica*: *Linaria dalmatica*, 254, 517
    vulgaris (**yellow toadflax**), 257, 517
linaria, common (**yellow toadflax**): *Linaria vulgaris*, 257, 517
lions-tooth (*Taraxacum officinale*): Herbaceous Forbs, 440, 446
little
    fireweed (*Erechtites* spp.): Herbaceous Forbs, 438, 446
    lovegrass (*Eragrostis* spp.): Grasses & Sedges, 434, 436
    robin (*Geranium purpureum*): *Geranium* spp., 193
Lobularia maritima (sweet alyssum): Herbaceous Forbs, 440, 448
lobularia, seaside (*Lobularia maritima*): Herbaceous Forbs, 440, 448
locust, black: *Robinia pseudoacacia*, 335
locust, yellow (**black locust**): *Robinia pseudoacacia*, 335
Lolium spp.
    arundinaceum (*Festuca arundinacea*): Grasses & Sedges, 434, 436
    multiflorum (**Italian ryegrass**): ***Lolium*** spp., 260
        [= *perenne* ssp. *multiflorum, temulentum*]
    perenne (**perennial ryegrass**): ***Lolium*** spp., 260
        [= *brasilianum, canadense*]
London rocket (*Sisymbrium irio*): *Sisymbrium* spp., 377
long-flowered veldtgrass (*Ehrharta longiflora*): *Ehrharta* spp., 165
long-fruited wild turnip (**Saharan mustard**): *Brassica tournefortii*, 64
longleaf or long-leaved pondweed (**American pondweed**): *Potamogeton natans* & *nodosus*, 326
long-rooted catsear (**common catsear**): *Hypochaeris* spp., 230
loosestrife
    garden: *Lysimachia vulgaris*, 269
    garden (*Lysimachia punctata*): see *Lysimachia vulgaris*, 269

hyssop (*Lythrum hyssopifolium*): Herbaceous Forbs, 442, 450
    purple: *Lythrum salicaria*, 271, 517
    wand (*Lythrum virgatum*): Herbaceous Forbs, 442, 450
    yellow (*Lysimachia vulgaris* or *L. punctata*): ***Lysimachia vulgaris***, 269
lopgrass (**soft brome**): ***Bromus hordeaceus, japonicus***, 71
lotus (**birdsfoot trefoil**): ***Lotus corniculatus***, 263
Lotus corniculatus (**birdsfoot trefoil**), 263
lovegrass (*Eragrostis* spp.): Grasses & Sedges, 434, 436
lover's pride (**ladysthumb**): ***Polygonum persicaria* & *lapathifolium***, 320
lovevine (**field bindweed**): ***Convolvulus arvensis***, 127, 516
low
    amaranth (*Amaranthus* spp.): Herbaceous Forbs, 438, 446
    eragrostis (*Eragrostis* spp.): Grasses & Sedges, 434, 436
    mallow (*Malva neglecta*): Herbaceous Forbs, 442, 450
ludwigia: *Ludwigia* spp., 265
Ludwigia spp. (**waterprimroses**), 265
lupine: *Lupinus* spp., 267
lupine, yellow bush (*Lupinus arboreus*): ***Lupinus*** spp., 267
Lupinus arboreus (**yellow bush lupine**): ***Lupinus*** spp., 267
Lupinus spp. (**lupine**), 267
Lyall nettle (*Urtica dioica*): Herbaceous Forbs, 444, 452
Lysimachia punctata (garden or yellow loosestrife): see ***Lysimachia vulgaris***, 269
Lysimachia vulgaris (**garden loosestrife**), 269
Lythrum spp.
    adsurgens (*Lythrum hyssopifolium*): Herbaceous Forbs, 442, 450
    hyssopifolium (hyssop loosestrife): Herbaceous Forbs, 442, 450
    salicaria (**purple loosestrife**), 271
    virgatum (wand loosestrife): Herbaceous Forbs, 438, 446
lythrum, hyssop (*Lythrum hyssopifolium*): Herbaceous Forbs, 442, 450
lythrum, purple (**purple loosestrife**): ***Lythrum salicaria***, 271, 517
malcolm stock (*Malcolmia africana*): Herbaceous Forbs, 440, 448
Malcolmia africana (African mustard): Herbaceous Forbs, 440, 448
mallow
    common, dwarf, low, roundleaf, or running mallow (*Malva neglecta*): Herbaceous Forbs, 442, 450
Malta starthistle: ***Centaurea melitensis***, 97
Malva neglecta (common mallow): Herbaceous Forbs, 442, 450
manna, Caspian (**camelthorn**): ***Alhagi pseudalhagi***, 25
manna, Persian (**camelthorn**): ***Alhagi pseudalhagi***, 25
mannagrass, waxy: ***Glyceria declinata***, 196
mare's tail (*Conyza canadensis*): Herbaceous Forbs, 438, 446
marguerite (**oxeye daisy**): ***Leucanthemum vulgare***, 247
marram grass (**European beachgrass**): ***Ammophila arenaria***, 31
marrube (**white horehound**): ***Marrubium vulgare***, 273
Marrubium vulgare (**white horehound**), 273
marsh elder, small-flowered (*Iva axillaris*): Herbaceous Forbs, 438, 446
marshpennywort, floating (**floating pennywort**): ***Hydrocotyle ranunculoides***, 224
marvel (**white horehound**): ***Marrubium vulgare***, 273
mat amaranth (*Amaranthus* spp.): Herbaceous Forbs, 438, 446
mat sandbur (*Cenchrus* spp.): Grasses & Sedges, 434, 436
matweed (*Polygonum arenastrum*): Herbaceous Forbs, 442, 450
maudlin daisy (**oxeye daisy**): ***Leucanthemum vulgare***, 247
maudlinwort (**oxeye daisy**): ***Leucanthemum vulgare***, 247
Maui pamakani (**croftonweed**): ***Ageratina adenophora***, 21
may (*Crataegus monogyna*): Woody Plants, 454, 456
meadow
    buttercup (*Ranunculus acris*): Herbaceous Forbs, 444, 450
    deathcamas (*Zigadenus venenosus*): ***Zigadenus*** spp., 431
    hawkweed (*Hieracium caespitosum*): ***Hieracium*** spp., 209
    knapweed: ***Centaurea debeauxii***, 90, 513
    soft grass (**common velvetgrass**): ***Holcus lanatus***, 212
meadowgrass, smooth (*Poa pratensis*): Grasses & Sedges, 434, 436
meadow-pine (**horsetail, scouringrush**): ***Equisetum*** spp., 175
mealweed (*Chenopodium album*): Herbaceous Forbs, 440, 448

meansgrass (**johnsongrass**): ***Sorghum halepense***, 380
medic (*Medicago polymorpha*): Herbaceous Forbs, 442, 448
medic, barrel or bur (*Medicago polymorpha*): Herbaceous Forbs, 442, 448
Medicago spp.
    polymorpha (California burclover): Herbaceous Forbs, 442, 448
    [= *apiculata, denticulata, hispida, nigra, polycarpa*]
Mediterranean
    barley (*Hordeum marinum* ssp. *gussonianum*): ***Hordeum marinum & murinum***, 218
    grass (**Arabian mediterraneangrass**): ***Schismus*** spp., 367
    lovegrass (*Eragrostis* spp.): Grasses & Sedges, 434, 436
    mustard (**Saharan mustard**): ***Brassica tournefortii***, 64
    mustard (**shortpod mustard**): ***Hirschfeldia incana***, 211
    sage: ***Salvia aethiopis***, 356, 516
    steppegrass (*Stipa capensis*): Grasses & Sedges, 434, 436
    turnip (**Saharan mustard**): ***Brassica tournefortii***, 64
mediterraneangrass, Arabian (*Schismus arabicus*): ***Schismus*** spp., 367
mediterraneangrass, common (*Schismus barbatus*): ***Schismus*** spp., 367
medlar, neapolitan (*Crataegus monogyna*): Woody Plants, 454, 456
medusahead: ***Taeniatherum caput-medusae***, 392
Melica festucoides (***Ehrharta calycina***): ***Ehrharta*** spp., 165
Melica geniculata (***Ehrharta calycina***): ***Ehrharta*** spp., 165
melilot, white (*Melilotus albus*): Herbaceous Forbs, 442, 448
Melilotus spp.
    albus (white sweetclover): Herbaceous Forbs, 442, 448
    [= *alba, leucanthus, officinalis*]
melloncillo (*Solanum elaeagnifolium*): Herbaceous Forbs, 444, 452
Mentha daghestanica (**pennyroyal**): ***Mentha pulegium***, 275
Mentha pulegium (**pennyroyal**), 275
Menzies fiddleneck (*Amsinckia menziesii* var. *menziesii*): ***Amsinckia*** spp., 35
Mesenbryanthemum edule: ***Carpobrotus edulis***, 83
Mexican
    bamboo (**Japanese knotweed**): ***Polygonum cuspidatum*** et al., 318
    burningbush (**kochia**): ***Bassia*** spp., 54
    devil (**croftonweed**): ***Ageratina adenophora***, 21
    fan palm (*Washingtonia robusta*): ***Phoenix & Washingtonia***, 307
    -fireweed (**kochia**): ***Bassia*** spp., 54
    lovegrass (*Eragrostis* spp.): Grasses & Sedges, 434, 436
    sandbur (**puncturevine**): ***Tribulus terrestris***, 407, 517
    tobacco (**tree tobacco**): ***Nicotiana glauca***, 289
    whorled milkweed (**Mexican whorled milkweed**): ***Asclepias*** spp., 43
milfoil (**Eurasian watermilfoil**): ***Myriophyllum spicatum***, 283, 518
military grass (**downy brome**): ***Bromus diandrus, madritensis, tectorum***, 66
milk, wolf's (**leafy spurge**): ***Euphorbia*** spp., 182, 516
milkflower cotoneaster (**Parney's cotoneaster**): ***Cotoneaster*** spp., 132
milkthistle (**blessed milkthistle**): ***Silybum marianum***, 371, 515
milkthistle, blessed: ***Silybum marianum***, 371, 515
milkweed
    Mexican whorled milkweed (*Asclepias fascicularis*): ***Asclepias*** spp., 43
    narrow-leaved milkweed (*Asclepias fascicularis*): ***Asclepias*** spp., 43
    showy milkweed (*Asclepias speciosa*): ***Asclepias*** spp., 43
    whorled milkweed (**Mexican whorled milkweed**): ***Asclepias*** spp., 43
millet
    Egyptian millet (**johnsongrass**): ***Sorghum halepense***, 380
    evergreen millet (**johnsongrass**): ***Sorghum halepense***, 380
    foxtail millet (*Setaria* spp.): Grasses & Sedges, 434, 436
    millet mountain-rice (**smilograss**)*Piptatherum miliaceum*, 315
    pearl millet (*Setaria* spp.): Grasses & Sedges, 434, 436
    rice millet (**smilograss**): ***Piptatherum miliaceum***, 315
    wild millet (*Setaria* spp.): Grasses & Sedges, 434, 436
Mirabilis nyctaginea (**wild four-o'clock**), 277
mock cypress (**kochia**): ***Bassia*** spp., 54

mock orange (*Pittosporum undulatum*): Woody Plants, 454, 456
molle de Peru (**Peruvian peppertree**): *Schinus* spp., 364
moneywort (**garden loosestrife**): *Lysimachia vulgaris*, 269
monk's-head (*Taraxacum officinale*): Herbaceous Forbs, 440, 446
montbretia (*Crocosmia* x *crocosmiflora*): Herbaceous Forbs, 442, 450
Montpellier broom (**French broom**): *Genista monspessulana*, 191, 516
moon daisy (**oxeye daisy**): *Leucanthemum vulgare*, 247
morenita (**kochia**): *Bassia* spp., 54
Mormon oats (**downy brome**): *Bromus diandrus, madritensis, tectorum*, 66
morningglory
    European, field, orchard, small-flowered, or wild (**field bindweed**): *Convolvulus arvensis*, 127, 516
Moroccan mustard (**Saharan mustard**): *Brassica tournefortii*, 64
mosquitofern: *Azolla* spp., 52, 518
moss, fairy (**mosquitofern**): *Azolla* spp., 52, 518
moth mullein (*Verbascum blattaria*): *Verbascum* spp., 419
mountain nettle (*Urtica dioica*): Herbaceous Forbs, 444, 452
mountain sorrel (**red sorrel**): *Rumex acetosella*, 344
mountain-rice, millet (**smilograss**): *Piptatherum miliaceum*, 315
mouse barley (hare barley): *Hordeum marinum* & *murinum*, 218
mouse-ear hawkweed (*Hieracium pilosella*): *Hieracium* spp., 209
mousehole tree (*Myoporum laetum*): Woody Plants, 454, 456
mullein
    common mullein (*Verbascum thapsus*): *Verbascum* spp., 419
    moth mullen (*Verbascum blattaria*): *Verbascum* spp., 419
    woolly mullein (**common mullein**): *Verbascum* spp., 419
multiflora rose (*Rosa multiflora*): *Rosa* spp., 338
multiflower rose (**multiflora rose**): *Rosa* spp., 338
multiheaded thistle (**slenderflower thistle**): *Carduus* spp., 79, 513
musk mustard (*Chorispora tenella*): Herbaceous Forbs, 440, 448
musk thistle (*Carduus nutans*): *Carduus* spp., 79, 513
muskwort (*Chara* spp.): *Chara, Nitella* spp., 114
mustard
    African (**Saharan mustard**): *Brassica tournefortii*, 64
    African (*Malcolmia africana*): Herbaceous Forbs, 440, 448
    Asian (**Saharan mustard**): *Brassica tournefortii*, 64
    bead-podded (*Chorispora tenella*): Herbaceous Forbs, 440, 448
    birdsrape (*Brassica rapa*): Herbaceous Forbs, 440, 448
    black: *Brassica nigra*, 61
    blue (*Chorispora tenella*): Herbaceous Forbs, 440, 448
    common (*Brassica rapa*): Herbaceous Forbs, 440, 448
    common (**wild mustard**): *Sinapis arvensis*, 374
    desert (**London rocket**): *Sisymbrium* spp., 377
    field (*Brassica rapa*): Herbaceous Forbs, 440, 448
    field (**wild mustard**): *Sinapis arvensis*, 374
    garlic: *Alliaria petiolata*, 27
    hairy (**shortpod mustard**): *Hirschfeldia incana*, 211
    hoary (**shortpod mustard**): *Hirschfeldia incana*, 211
    Jim Hill (**tumble mustard**): *Sisymbrium* spp., 377
    Mediterranean (**Saharan mustard**): *Brassica tournefortii*, 64
    Mediterranean (**shortpod mustard**): *Hirschfeldia incana*, 211
    Moroccan (**Saharan mustard**): *Brassica tournefortii*, 64
    musk (*Chorispora tenella*): Herbaceous Forbs, 440, 448
    purple (*Chorispora tenella*): Herbaceous Forbs, 440, 448
    Saharan: *Brassica tournefortii*, 64
    shortpod: *Hirschfeldia incana*, 211
    tall (**tumble mustard**): *Sisymbrium* spp., 377
    tansy (*Descurainia sophia*): Herbaceous Forbs, 440, 448
    tenella (*Chorispora tenella*): Herbaceous Forbs, 440, 448
    tumble (*Sisymbrium altissimum*): *Sisymbrium* spp., 377
    tumbleweed (**tumble mustard**): *Sisymbrium* spp., 377
    wild: *Sinapis arvensis*, 374
myoporum (*Myoporum laetum*): Woody Plants, 454, 456
Myoporum laetum (myoporum): Woody Plants, 454, 456
***Myosotis latifolia* (broadleaf forget-me-not)**, 279

Myosotis sylvatica (**broadleaf forget-me-not**: *Myosotis latifolia*, 279
***Myriophyllum* spp.**
    aquaticum (parrotfeather), 281
        [= *brasiliense, proserpinacoides*]
    spicatum (Eurasian watermilfoil), 283, 518
myrobalan (*Prunus cerasifera*): Woody Plants, 454, 456
naiad: *Najas* spp., 286
    hollyleaf or spiny naiad (*Najas marina*): *Najas* spp., 286
    southern naiad (*Najas guadalupensis*): *Najas* spp., 286
***Najas* spp. (naiads)**
    guadalupensis (southern naiad): *Najas* spp., 286
        [= *flexilis*]
    marina (hollyleaf naiad): *Najas* spp., 286
        [= *gracilis, latifolia, major*]
naked weed (**rush skeletonweed**): *Chondrilla juncea*, 116, 515
Napa thistle (**Malta starthistle**): *Centaurea melitensis*, 97
narrowleaf or narrow-leaved
    cattail (**cattail**): *Typha* spp., 413
    dock (**curly dock**): *Rumex crispus* & *obtusifolius*, 346
    milkweed (**Mexican whorled milkweed**): *Asclepias* spp., 43
    oleaster (**Russian-olive**): *Elaeagnus angustifolia*, 169
    plantain (*Plantago lanceolata*): Herbaceous Forbs, 442, 450
neapolitan medlar (*Crataegus monogyna*): Woody Plants, 454, 456
Nebraska fern (**poison hemlock**): *Conium maculatum*, 125, 513
netseed lambsquarter (*Chenopodium berlandieri*): Herbaceous Forbs, 440, 448
nettle
    bee, blind, or dead (**henbit**, *Lamium amplexicaule*): Herbaceous Forbs, 442, 450
    silverleaf (**silverleaf nightshade**, *Solanum elaeagnifolium*): Herbaceous Forbs, 444, 452
    stinging (= American, California, coast, creek, hedge, hoary, Lyall, mountain, slender, tall) (*Urtica dioica*): Herbaceous Forbs, 444, 452
New Zealand fireweed (*Erechtites* spp.): Herbaceous Forbs, 438, 446
ngaio tree (*Myoporum laetum*): Woody Plants, 454, 456
***Nicotiana glauca* (tree tobacco)**, 289
nicotine tree (**tree tobacco**): *Nicotiana glauca*, 289
nightshade
    desert (*Solanum elaeagnifolium*): Herbaceous Forbs, 444, 452
    fetid (**black henbane**): *Hyoscyamus niger*, 226
    silverleaf (*Solanum elaeagnifolium*): Herbaceous Forbs, 444, 452
nitella (*Nitella* spp.): *Chara, Nitella* spp., 114
***Nitella* spp. (nitella)**: *Chara, Nitella* spp., 114
nodding
    foxtail (*Setaria* spp.): Grasses & Sedges, 434, 436
    plumeless thistle (**musk thistle**): *Carduus* spp., 79, 513
    smartweed (**pale smartweed**): *Polygonum persicaria* & *lapathifolium*, 320
    thistle (**musk thistle**): *Carduus* spp., 79, 513
Norta irio (*Sisymbrium irio*): *Sisymbrium* spp., 377
North Africa grass (**North African wiregrass**): *Ventenata dubia*, 417
North African wiregrass: *Ventenata dubia*, 417
Nuphar lutea (yellow pondlily): Aquatic Plants, 458, 460
    [= *luteum, polysepala*]
Nymphaea odorata (fragrant waterlily): Aquatic Plants, 458, 460
Nymphaea polysepala (*Nuphar lutea*): Aquatic Plants, 458, 460
Nymphoides nymphaeoides: ***Nymphoides peltata***, 291
***Nymphoides peltata* (yellow floatingheart)**, 291
Nymphozanthus polysepalus (*Nuphar lutea*): Aquatic Plants, 458, 460
oak, poison (**Pacific poison-oak**): *Toxicodendron diversilobum*, 403
oat or oats
    Mormon oats (**downy brome**): *Bromus diandrus, madritensis, tectorum*, 66
    oatgrass (**wild oat**): *Avena* spp., 49
    slender oat (*Avena barbata*): *Avena* spp., 49
    wheat oats (**wild oat**): *Avena* spp., 49

wild oat (*Avena fatua*): *Avena* spp., 49
oblong spurge (*Euphorbia oblongata*): *Euphorbia* spp., 182
old han schismus (**common mediterraneangrass**): *Schismus* spp., 367
old-man's-beard (± of England) (*Clematis vitalba*): Herbaceous Forbs, 444, 450
Olea europaea (olive): Woody Plants, 454, 456
oleaster (**Russian-olive**): *Elaeagnus angustifolia*, 169
oleaster, narrow-leaved (**Russian-olive**): *Elaeagnus angustifolia*, 169
olive (*Olea europaea*): Woody Plants, 454, 456
olive, Russian: *Elaeagnus angustifolia*, 169
olive, wild (**Russian-olive**): *Elaeagnus angustifolia*, 169
oneseed hawthorn (*Crataegus monogyna*): Woody Plants, 454, 456
onionweed: *Asphodelus fistulosus*, 45
Onopordum acanthium (Scotch thistle), 292, 515
orach, thorn (**fivehook bassia**): *Bassia* spp., 54
orange
    cotoneaster (*Cotoneaster franchetii*): *Cotoneaster* spp., 132
    hawkweed (*Hieracium aurantiacum*): *Hieracium* spp., 209
    mock orange (*Pittosporum undulatum*): Woody Plants, 454, 456
    paintbrush (**orange hawkweed**): *Hieracium* spp., 209
orange-eye butterflybush (*Buddleja davidii*): Woody Plants, 454, 456
orchard morningglory (**field bindweed**): *Convolvulus arvensis*, 127, 516
orchardgrass (*Dactylis glomerata*): Grasses & Sedges, 434, 436
Oryzopsis miliacea: *Piptatherum miliaceum*, 315
oval-leaf knotweed (*Polygonum arenastrum*): Herbaceous Forbs, 442, 450
Oxalis cernua: *Oxalis pes-caprae*, 295
Oxalis pes-caprae (buttercup oxalis), 295
oxalis, buttercup (**buttercup oxalis**): *Oxalis pes-caprae*, 295
oxalis, yellow-flowered (**buttercup oxalis**): *Oxalis pes-caprae*, 295
oxeye daisy: *Leucanthemum vulgare*, 247
oxeye, great (**oxeye daisy**): *Leucanthemum vulgare*, 247
oxtongue, bristly: *Picris echioides*, 313
oxygenweed: *Lagarosiphon major*, 240
ozallaik (*Peganum harmala*): Herbaceous Forbs, 444, 452
Pacific bentgrass (*Agrostis avenacea*): Grasses & Sedges, 434, 436
Pacific poison-oak: *Toxicodendron diversilobum*, 403
paintbrush, devil's (**orange hawkweed**): *Hieracium* spp., 209
paintbrush, orange (**orange hawkweed**): *Hieracium* spp., 209
pale
    madwort (*Alyssum alyssoides*): Herbaceous Forbs, 440, 448
    persicaria (**pale smartweed**): *Polygonum persicaria* & *lapathifolium*, 320
    pigeongrass (*Setaria* spp.): Grasses & Sedges, 434, 436
    smartweed (*Polygonum lapathifolium*): *Polygonum persicaria* & *lapathifolium*, 320
    paleyellow iris (**yellowflag iris**): *Iris pseudacorus*, 234
palm
    Canary Island date palm (*Phoenix canariensis*): *Phoenix* & *Washingtonia*, 307
    cotton palm (**Mexican fan palm**): *Phoenix* & *Washingtonia*, 307
    Mexican fan palm (*Washingtonia robusta*): *Phoenix* & *Washingtonia*, 307
    Washington fan palm (**Mexican fan palm**): *Phoenix* & *Washingtonia*, 307
palma christi (**castorbean**): *Ricinus communis*, 333
Palmer amaranth (*Amaranthus* spp.): Herbaceous Forbs, 438, 446
pamakani, Maui (**croftonweed**): *Ageratina adenophora*, 21
pampasgrass
    Andean, pink, or purple (**jubatagrass**): *Cortaderia* spp., 130
    common or Uruguayan (**pampasgrass**): *Cortaderia* spp., 130
    hardy (**ravennagrass**): *Tripidium ravennae*, 412
    pampasgrass (*Cortaderia selloana*): *Cortaderia* spp., 130
Panicum spp.
    dactylon: *Cynodon dactylon*, 141
    imberbe (*Setaria* spp.): Grasses & Sedges, 434, 436
    italicum (*Setaria* spp.): Grasses & Sedges, 434, 436
    laevigatum (*Setaria* spp.): Grasses & Sedges, 434, 436
    platense (*Paspalum dilatatum*): Grasses & Sedges, 434, 436
    pumilum (*Setaria* spp.): Grasses & Sedges, 434, 436
    viride (*Setaria* spp.): Grasses & Sedges, 434, 436
parlor ivy (**Cape ivy**): *Delairea odorata*, 152
Parney's cotoneaster (*Cotoneaster lacteus*): *Cotoneaster* spp., 132
parrotfeather: *Myriophyllum aquaticum*, 281
parrotfeather watermilfoil (**parrotfeather**): *Myriophyllum aquaticum*, 281
Parry's tarweed (*Hemizonia parryi*): *Hemizonia* spp., 205
parsley
    bur parsley (**hedgeparsley**): *Torilis arvensis*, 401
    hedge parsley: *Torilis arvensis*, 401
    parsley fern (**common tansy**): *Tanacetum vulgare*, 399
    poison parsley (**poison hemlock**): *Conium maculatum*, 125, 513
paspalum (*Paspalum dilatatum*): Grasses & Sedges, 434, 436
Paspalum spp.
    dactylon: *Cynodon dactylon*, 141
    dilatatum (dallisgrass): Grasses & Sedges, 434, 436
    [= *eriophorum, lanatum, ovatum, pedunculare, platense*]
Paterson's curse (**vipers bugloss**): *Echium plantagineum*, 159
pea, perennial (± sweet) (*Lathyrus latifolius*): Herbaceous Forbs, 442, 448
pea, weedy sweet (*Lathyrus latifolius*): Herbaceous Forbs, 442, 448
pearl millet (*Setaria* spp.): Grasses & Sedges, 434, 436
peavine, everlasting (*Lathyrus latifolius*): Herbaceous Forbs, 442, 448
peaweed, Austrian (**swainsonpea**): *Sphaerophysa salsula*, 388
Peganum harmala (harmal): Herbaceous Forbs, 444, 452
peganum, harmal (*Peganum harmala*): Herbaceous Forbs, 444, 452
Pennisetum spp.
    ciliare (**buffelgrass**), 297
    clandestinum (**kikuyugrass**), 299
    [= *inclusum, longistylum, rueppelii*]
    glaucum (*Setaria* spp.): Grasses & Sedges, 434, 436
    setaceum (**crimson fountaingrass**), 301
    [= *macrostachyon, rueppelianum*]
    viride (*Setaria* spp.): Grasses & Sedges, 434, 436
pennycress (*Thlaspi arvense*): Herbaceous Forbs, 440, 448
pennycress, field (*Thlaspi arvense*): Herbaceous Forbs, 440, 448
pennyroyal: *Mentha pulegium*, 275
pennyroyal, European (**pennyroyal**): *Mentha pulegium*, 275
pennywort, floating: *Hydrocotyle ranunculoides*, 224
pepper, American (**Peruvian peppertree**): *Schinus* spp., 364
pepper, false (**Peruvian peppertree**): *Schinus* spp., 364
peppercress, clasping-leaved (*Lepidium perfoliatum*): Herbaceous Forbs, 440, 448
peppercress, perennial (**perennial pepperweed**): *Lepidium latifolium*, 244
peppercress, slender perennial (**perennial pepperweed**): *Lepidium latifolium*, 244
peppergrass (*Cardaria chalepensis* or *C. draba*): *Cardaria* spp., 76
peppergrass, perennial (**hoary cress**): *Cardaria* spp., 76
peppergrass, perennial (**perennial pepperweed**): *Lepidium latifolium*, 244
pepperplant (*Capsella bursa-pastoris*): Herbaceous Forbs, 440, 448
peppertree, Brazilian (*Schinus terebinthifolius*): *Schinus* spp., 364
peppertree, Peruvian (*Schinus molle*): *Schinus* spp., 364
pepperweed
    broadleaf (broadleaved) (**perennial pepperweed**): *Lepidium latifolium*, 244
    clasping (*Lepidium perfoliatum*): Herbaceous Forbs, 440, 448
    pepperweed (**hoary cress**): *Cardaria* spp., 76
    perennial: *Lepidium latifolium*, 244
    yellowflower (*Lepidium perfoliatum*): Herbaceous Forbs, 440, 448
pepperwort, perfoliate (*Lepidium perfoliatum*): Herbaceous Forbs, 440, 448

perennial
   pea (*Lathyrus latifolius*): Herbaceous Forbs, 442, 448
   peppercress (**perennial pepperweed**): *Lepidium latifolium*, 244
   peppergrass (**hoary cress**): *Cardaria* spp., 76
   peppergrass (**perennial pepperweed**): *Lepidium latifolium*, 244
   pepperweed: *Lepidium latifolium*, 244
   pigweed (*Amaranthus* spp.): Herbaceous Forbs, 438, 446
   ryegrass (*Lolium perenne*): *Lolium* spp., 260
   slender peppercress (**perennial pepperweed**): *Lepidium latifolium*, 244
   sweet pea (*Lathyrus latifolius*): Herbaceous Forbs, 442, 448
   veldtgrass (**purple veldtgrass**): *Ehrharta* spp., 165
perfoliate pepperwort (*Lepidium perfoliatum*): Herbaceous Forbs, 440, 448
periwinkle, big (bigleaf, blue, greater, large): *Vinca major*, 423
Persian manna (**camelthorn**): *Alhagi pseudalhagi*, 25
Persicaria spp.
   lapathifolia (*Polygonum lapathifolium*): *Polygonum persicaria* & *lapathifolium*, 320
   maculata, maculosa, persicaria, ruderalis, vulgaris (***Polygonum persicaria***): *Polygonum persicaria* & *lapathifolium*, 320
   polystachya, wallichii (***Polygonum polystachyum***): *Polygonum cuspidatum* et al., 318
persicaria, pale (*Polygonum lapathifolium*): *Polygonum persicaria* & *lapathifolium*, 320
persicary (**ladysthumb**): *Polygonum persicaria* & *lapathifolium*, 320
Peruvian peppertree (*Schinus molle*): *Schinus* spp., 364
Phalaris spp.
   ammophila: *Ammophila arenaria*, 31
   aquatica (**hardinggrass**), 303
      [= *commutata, stenoptera, tuberosa*]
   arundinacea (**reed canarygrass**), 305
   maritima: *Ammophila arenaria*, 31
   setacea: *Pennisetum setaceum*, 301
Phoenix canariensis (**Canary Island date palm**): *Phoenix* & *Washingtonia*, 307
Phragmites australis (**common reed**), 309
Phragmites communis: *Phragmites australis*, 309
Phytolacca americana (**common pokeweed**), 311
Phytolacca decandra: *Phytolacca americana*, 311
pick-purse (*Capsella bursa-pastoris*): Herbaceous Forbs, 440, 448
Picris echioides (**bristly oxtongue**), 313
pigeonberry (**common pokeweed**): *Phytolacca americana*, 311
pigeon-foot (*Geranium molle*): see *Geranium* spp., 193
pigeongrass (*Setaria* spp.): Grasses & Sedges, 434, 436
pigeongrass, pale (*Setaria* spp.): Grasses & Sedges, 434, 436
pigweed (amaranth) (*Amaranthus* spp.): Herbaceous Forbs, 438, 446
pigweed (lambsquarters) (*Chenopodium album*): Herbaceous Forbs, 440, 448
Pilosella aurantiaca (*Hieracium aurantiacum*): *Hieracium* spp., 209
pin clover (**redstem filaree**): *Erodium cicutarium*, 177
pin grass (**redstem filaree**): *Erodium cicutarium*, 177
pine, athel (**athel tamarisk**): *Tamarix* spp., 395
pinegrass (**horsetail, scouringrush**): *Equisetum* spp., 175
pink
   asphodel (**onionweed**): *Asphodelus fistulosus*, 45
   hedge pink (*Saponaria officinalis*): Herbaceous Forbs, 440, 448
   pampasgrass (**jubatagrass**): *Cortaderia* spp., 130
   -weed (*Polygonum arenastrum*): Herbaceous Forbs, 442, 450
   smartweed (*Polygonum convolvulus*): Herbaceous Forbs, 444, 450
pinnate tansymustard (*Descurainia sophia*): Herbaceous Forbs, 440, 448
pin-weed (**redstem filaree**): *Erodium cicutarium*, 177
piny tarweed (**spikeweed**): *Hemizonia* spp., 205
Piptatherum miliaceum (**smilograss**), 315
Pistia stratiotes (water lettuce): 518
Pittosporum undulatum (Victorian box): Woody Plants, 454, 456

plague, devil's (**wild carrot**): *Daucus carota*, 149
Plantago lanceolata (buckhorn plantain): Herbaceous Forbs, 442, 450
plantain
   buck, buckhorn, English, lanceleaf, narrowleaf, or ribbed (*Plantago lanceolata*): Herbaceous Forbs, 442, 450
plum, cherry (*Prunus cerasifera*): Woody Plants, 454, 456
plum, purple-leafed (*Prunus cerasifera*): Woody Plants, 454, 456
plume thistle (**bull thistle**): *Cirsium vulgare*, 122, 515
plumeless thistle
   giant plumeless thistle (**musk thistle**): *Carduus* spp., 79, 513
   Italian plumeless thistle (**Italian thistle**): *Carduus* spp., 79, 513
   nodding plumeless thistle (**musk thistle**): *Carduus* spp., 79, 513
   plumeless thistle (*Carduus acanthoides*): *Carduus* spp., 79, 513
   plumeless thistle (**musk thistle**): *Carduus* spp., 79, 513
   winged plumeless thistle (**slenderflower thistle**): *Carduus* spp., 79, 513
Plymouth thistle (**Italian thistle**): *Carduus* spp., 79, 513
Poa spp.
   bulbosa (bulbous bluegrass), 316
   ciliensis (*Eragrostis* spp.): Grasses & Sedges, 434, 436
   megastachya (*Eragrostis* spp.): Grasses & Sedges, 434, 436
   pratensis (Kentucky bluegrass): Grasses & Sedges, 434, 436
      [= *agassizensis, angustifolia*]
poison
   -hemlock: *Conium maculatum*, 125, 513
   Pacific poison-oak: *Toxicodendron diversilobum*, 403
   oak (Pacific poison-oak): *Toxicodendron diversilobum*, 403
   parsley (poison hemlock): *Conium maculatum*, 125, 513
   stinkweed (poison hemlock): *Conium maculatum*, 125, 513
   western poison oak (Pacific poison-oak): *Toxicodendron diversilobum*, 403
poke
   American pokeweed (**common pokeweed**): *Phytolacca americana*, 311
   poke (**common pokeweed**): *Phytolacca americana*, 311
   pokeberry (**common pokeweed**): *Phytolacca americana*, 311
   poke sallet (**common pokeweed**): *Phytolacca americana*, 311
   pokeweed, common: *Phytolacca americana*, 311
   Virginia poke (**common pokeweed**): *Phytolacca americana*, 311
poly, grass (*Lythrum hyssopifolium*): Herbaceous Forbs, 442, 450
Polygonum spp.
   arenastrum (prostrate knotweed): Herbaceous Forbs, 438, 446
      [= *aequale, aviculare ssp. depressum, montereyense*]
   convolvulus (wild buckwheat): Herbaceous Forbs, 438, 446
   cuspidatum (**Japanese knotweed**): *P. cuspidatum* et al., 318
      [= *zuccarinii*]
   lapathifolium (**pale smartweed**): *P. persicaria* & *lapathifolium*, 320
      [= *nodosum*]
   persicaria (**ladysthumb**): *P. persicaria* & *lapathifolium*, 320
      [= *dubium, fusiforme, minus, puritanorum*]
   polystachyum (**Himalayan knotweed**): *P. cuspidatum* et al., 318
   sachalinense (**Sakhalin knotweed**): *P. cuspidatum* et al., 318
   x bohemicum (**Bohemian knotweed**): *P. cuspidatum* et al., 318
Polypogon spp.
   monspeliensis (rabbitfoot polypogon): Grasses & Sedges, 434, 436
      [= *crinitus, flavescens*]
   rabbitfoot polypogon (*P. monspeliensis*): Grasses & Sedges, 434, 436
pond lily
   Indian pond lily (*Nuphar lutea*): Aquatic Plants, 458, 460
   Rocky Mountain pond lily (*Nuphar lutea*): Aquatic Plants, 458, 460
   yellow pond lily (*Nuphar lutea*): Aquatic Plants, 458, 460
pondweed
   American (*Potamogeton nodosus*): *Potamogeton natans* & *nodosus*, 326
   broad-leafed (**floatingleaf pondweed**): *Potamogeton natans* & *nodosus*, 326

Canadian (*Elodea canadensis*): Aquatic Plants, 458, 460
closed-leaved (**leafy pondweed**): *Potamogeton & Stuckenia*, 323
common floating (**floatingleaf pondweed**): *Potamogeton natans & nodosus*, 326
crispate-leaved or crisped (**curlyleaf pondweed**): *Potamogeton & Stuckenia*, 323
curlyleaf (curled-leaved) (*Potamogeton crispus*): *Potamogeton & Stuckenia*, 323
fennel-leaf (**sago pondweed**): *Potamogeton & Stuckenia*, 323
floatingleaf (*Potamogeton natans*): *Potamogeton natans & nodosus*, 326
horned (*Zannichellia palustris*): Aquatic Plants, 458, 460
Illinois (*Potamogeton illinoensis*): *Potamogeton & Stuckenia*, 323
leafy (*Potamogeton foliosus*): *Potamogeton & Stuckenia*, 323
longleaf (long-leaved) (**American pondweed**): *Potamogeton natans & nodosus*, 326
sago (*Stuckenia pectinata*): *Potamogeton & Stuckenia*, 323
shining (**Illinois pondweed**): *Potamogeton & Stuckenia*, 323
ziz's (**Illinois pondweed**): *Potamogeton & Stuckenia*, 323
Pontederia crassipes: *Eichhornia crassipes*, 167, 518
poor man's alfalfa (**kochia**): *Bassia* spp., 54
poor man's spinach (*Chenopodium album*): Herbaceous Forbs, 440, 448
poorland flower (**oxeye daisy**): *Leucanthemum vulgare*, 247
popcorn tree (**Chinese tallowtree**): *Sapium sebiferum*, 361
***Potamogeton* spp.**
crispus (**curlyleaf pondweed**): *Potamogeton & Stuckenia*, 323
foliosus (**leafy pondweed**): *Potamogeton & Stuckenia*, 323
illinoensis (**Illinois pondweed**): *Potamogeton & Stuckenia*, 323
[= *angustifolius, lucens, zizii*]
natans (**floatingleaf pondweed**): *P. natans & nodosus*, 326
nodosus (**American pondweed**): *P. natans & nodosus*, 326
[= *americanus, lonchites*]
pectinatus (*Stuckenia pectinata*): *Potamogeton & Stuckenia*, 323
Potentilla recta (**sulfur cinquefoil**), 328
poverty sumpweed (*Iva axillaris*): Herbaceous Forbs, 438, 446
povertyweed (*Iva axillaris*): Herbaceous Forbs, 438, 446
prairie berry (*Solanum elaeagnifolium*): Herbaceous Forbs, 444, 452
prickly
broom (**gorse**): *Ulex europaeus*, 415, 516
lettuce (*Lactuca serriola*): Herbaceous Forbs, 438, 446
Russian thistle (**Russian-thistle**): *Salsola* spp., 353, 515
pride weed (*Conyza canadensis*): Herbaceous Forbs, 438, 446
primrose, water: *Ludwigia* spp., 265
primrose-willow (**creeping waterprimrose**): *Ludwigia* spp., 265
privet, glossy: *Ligustrum lucidum*, 250
prostrate knotweed (*Polygonum arenastrum*): Herbaceous Forbs, 442, 450
prostrate pigweed (*Amaranthus* spp.): Herbaceous Forbs, 438, 446
Prunus cerasifera (cherry plum): Woody Plants, 454, 456
psamma (**European beachgrass**): *Ammophila arenaria*, 31
puncture weed (**puncturevine**): *Tribulus terrestris*, 407, 517
puncturevine: *Tribulus terrestris*, 407, 517
purple
bugloss (**vipers bugloss**): *Echium plantagineum*, 159
echium (**vipers bugloss**): *Echium plantagineum*, 159
false brome (**annual false-brome**): *Brachypodium distachyon*, 57
foxglove (*Digitalis purpurea*): Herbaceous Forbs, 444, 452
-leafed plum (*Prunus cerasifera*): Woody Plants, 454, 456
loosestrife: *Lythrum salicaria*, 271, 517
lythrum (**purple loosestrife**): *Lythrum salicaria*, 271, 517
mustard (*Chorispora tenella*): Herbaceous Forbs, 440, 448
pampasgrass (**jubatagrass**): *Cortaderia* spp., 130
sesban (**red sesbania**): *Sesbania punicea*, 369
starthistle (*Centaurea calcitrapa*): *Centaurea calcitrapa & iberica*, 87, 513
-top vervain (*Verbena bonariensis*): Herbaceous Forbs, 444, 452
veldtgrass (**purple veldtgrass**): *Ehrharta* spp., 165
vipers bugloss (**vipers bugloss**): *Echium plantagineum*, 159
purse, lady's (*Capsella bursa-pastoris*): Herbaceous Forbs, 440, 448
pyracantha (*Pyracantha* spp.): Woody Plants, 454, 456
Pyracantha spp. (pyracantha): Woody Plants, 454, 456
quackgrass: *Elytrigia repens*, 172
quakinggrass, big or large (*Briza maxima*): Grasses & Sedges, 434, 436
Queen Anne's lace (**wild carrot**): *Daucus carota*, 149
Queen Mary's thistle (**Scotch thistle**): *Onopordum acanthium*, 292, 515
queendevil hawkweed (*Hieracium glomeratum*): *Hieracium* spp., 209
quickgrass (**quackgrass**): *Elytrigia repens*, 172
quickgrass, common (**bermudagrass**): *Cynodon dactylon*, 141
quitchgrass (**quackgrass**): *Elytrigia repens*, 172
rabbitbrush (*Chrysothamnus nauseosus*): Herbaceous Forbs, 438, 446
rabbitfoot polypogon, rabbitfootgrass, rabbitsfoot grass (*Polypogon monspeliensis*): Grasses & Sedges, 434, 436
radish
cultivated radish (**radish**): *Raphanus* spp., 331
jointed wild radish (**wild radish**): *Raphanus* spp., 331
radish (*Raphanus sativus*): *Raphanus* spp., 331
wild radish (*Raphanus raphanistrum* or *R. sativus*): *Raphanus* spp., 331
ragweed – common, great, or tall (*Ambrosia artemisiifolia*): Herbaceous Forbs, 438, 446
ragwort, tansy (*Senecio jacobaea*): Herbaceous Forbs, 440, 446, 515
rambler rose (**multiflora rose**): *Rosa* spp., 338
rancher's fireweed (**coast fiddleneck**): *Amsinckia* spp., 35
Ranunculus spp.
acris [= *boreanus*] (tall buttercup): Herbaceous Forbs, 444, 450
repens (creeping buttercup): Herbaceous Forbs, 444, 450
testiculatus [=*falcatus*] (bur buttercup): Herbaceous Forbs, 444, 450
rapeseed (**wild mustard**): *Sinapis arvensis*, 374
Raphanus raphanistrum (**wild radish**): *Raphanus* spp., 331
Raphanus sativus (**radish**): *Raphanus* spp., 331
rattail fescue (**rat's-tail, rat-tailed**): *Vulpia myuros*, 425
rattlebox or rattlebush (**red sesbania**): *Sesbania punicea*, 369
rattlesnake grass (*Briza maxima*): Grasses & Sedges, 434, 436
ravennagrass: *Tripidium ravennae*, 412
red
amaranth (*Amaranthus* spp.): Herbaceous Forbs, 438, 446
belvedere (**kochia**): *Bassia* spp., 54
brome (*Bromus madritensis* ssp. *rubens*): *Bromus diandrus, madritensis, tectorum*, 66
clusterberry (**Parney's cotoneaster**): *Cotoneaster* spp., 132
daisy (**orange hawkweed**): *Hieracium* spp., 209
sesbania: *Sesbania punicea*, 369
sorrel: *Rumex acetosella*, 344
starthistle (**purple starthistle**): *Centaurea calcitrapa & iberica*, 87, 513
red-ink plant (**common pokeweed**): *Phytolacca americana*, 311
redroot amaranth or pigweed (*Amaranthus* spp.): Herbaceous Forbs, 438, 446
redshanks (**ladysthumb**): *Polygonum persicaria & lapathifolium*, 320
redstem filaree or storksbill: *Erodium cicutarium*, 177
red-tailed fescue (**rattail fescue**): *Vulpia myuros*, 425
redtop (*Agrostis stolonifera*): Grasses & Sedges, 434, 436
redtop bent (*Agrostis stolonifera*): Grasses & Sedges, 434, 436
red-top sorrel (**red sorrel**): *Rumex acetosella*, 344
redweed (**common pokeweed**): *Phytolacca americana*, 311
red-weed (**red sorrel**): *Rumex acetosella*, 344
reed
bamboo reed (**giant reed**): *Arundo donax*, 40, 517
common reed: *Phragmites australis*, 309
donax reed (**giant reed**): *Arundo donax*, 40, 517
giant reed (**giant reed**): *Arundo donax*, 40, 517
reed canarygrass: *Phalaris arundinacea*, 305

reed cane (**giant reed**): *Arundo donax*, 40, 517
reed fescue (*Festuca arundinacea*): Grasses & Sedges, 434, 436
reed grass (**giant reed**): *Arundo donax*, 40, 517
   Spanish reed (**giant reed**): *Arundo donax*, 40, 517
resinbush, sweet (*Euryops multifidus*): Woody Plants, 454, 456
Rhus diversiloba: **Toxicodendron diversilobum**, 403
ribbed plantain (*Plantago lanceolata*): Herbaceous Forbs, 442, 450
ribgrass, ribwort (*Plantago lanceolata*): Herbaceous Forbs, 442, 450
rice grass (**smilograss**): *Piptatherum miliaceum*, 315
rice millet (**smilograss**): *Piptatherum miliaceum*, 315
Ricinus communis (**castorbean**), 333
ripgut brome (*Bromus diandrus*): *Bromus diandrus, madritensis, tectorum*, 66
Riverina bluebell (**vipers bugloss**): *Echium plantagineum*, 159
roadside thistle (**bull thistle**): *Cirsium vulgare*, 122, 515
robert (or Robert's) geranium (*Geranium robertianum*): Herbaceous Forbs, 442, 450; also see *Geranium* spp., 193
robin, little (*Geranium purpureum*): *Geranium* spp., 193
Robinia pseudoacacia (**black locust**), 335
   [= *pringlei*]
rocket
   dames rocket (*Hesperis matronalis*): Herbaceous Forbs, 440, 448
   London rocket (*Sisymbrium irio*): *Sisymbrium* spp., 377
   sea rocket (*Cakile maritima*): Herbaceous Forbs, 440, 448
rockspray cotoneaster (**Parney's cotoneaster**): *Cotoneaster* spp., 132
Rocky Mountain pond-lily (*Nuphar lutea*): Aquatic Plants, 458, 460
root, insane (**black henbane**): *Hyoscyamus niger*, 226
Rosa **spp.**
   canina (**dog rose**): *Rosa* spp., 338
   cathayensis (**Rosa multiflora**): *Rosa* spp., 338
   eglanteria (**sweetbriar rose**): *Rosa* spp., 338
   multiflora (**multiflora rose**): *Rosa* spp., 338
   rubiginosa (***Rosa eglanteria***): *Rosa* spp., 338
rose
   baby, Japanese, multiflower, rambler (**multiflora rose**): *Rosa* spp., 338
   dog rose (*Rosa canina*): *Rosa* spp., 338
   multiflora rose (*Rosa multiflora*): *Rosa* spp., 338
   rose clover (*Trifolium hirtum*): Herbaceous Forbs, 442, 448
   sweetbriar rose (*Rosa eglanteria*): *Rosa* spp., 338
rough catsear (**common catsear**): *Hypochaeris* spp., 230
rough cocklebur (**common cocklebur**): *Xanthium* spp., 428
rough pigweed (*Amaranthus* spp.): Herbaceous Forbs, 438, 446
roughfruit cinquefoil (**sulfur cinquefoil**): *Potentilla recta*, 328
round dock (*Malva neglecta*): Herbaceous Forbs, 442, 450
roundleaf mallow (*Malva neglecta*): Herbaceous Forbs, 442, 450
Rubus **spp.**
   armeniacus (**Himalaya blackberry**), 341
      [= *discolor, procerus*]
rue (African, Syrian, wild) (*Peganum harmala*): Herbaceous Forbs, 444, 452
ruin weed (*Peganum harmala*): Herbaceous Forbs, 444, 452
Rumex **spp.**
   acetosella (**red sorrel**), 344
      [= *angiocarpus, tenuifolius*]
   crispus (**curly dock**): *Rumex crispus & obtusifolius*, 346
   obtusifolius (**broadleaf dock**): *Rumex crispus & obtusifolius*, 346
running mallow (*Malva neglecta*): Herbaceous Forbs, 442, 450
Ruppia maritima (**widgeongrass**), 349
rush
   bog or common (*Juncus effusus*): *Juncus* spp., 238
   grass rush: *Juncus* spp., 238
   rushes: *Juncus* spp., 238
   rush skeletonweed: *Chondrilla juncea*, 116, 515
   scouring rush (**horsetail, scouringrush**): *Equisetum* spp., 175
   soft rush (*Juncus effusus*): *Juncus* spp., 238
   spreading rush (*Juncus patens*): *Juncus* spp., 238

   toad rush: *Juncus* spp., 238
Russian
   knapweed: *Acroptilon repens*, 16, 513
   -olive: *Elaeagnus angustifolia*, 169
   starthistle: *Acroptilon repens*, 16, 513
   -thistle (*Salsola tragus*): *Salsola* spp., 353, 515
   -thistle, barbwire (*Salsola paulsenii*): *Salsola* spp., 353
   tumbleweed (**Russian-thistle**): *Salsola* spp., 353, 515
rutabaga, wild (*Brassica rapa*): Herbaceous Forbs, 440, 448
rye, cereal or common (*Secale cereale*): Grasses & Sedges, 434, 436
ryegrass
   annual or Australian (**Italian ryegrass**): *Lolium* spp., 260
   English (**perennial ryegrass**): *Lolium* spp., 260
   Italian (*Lolium multiflorum*): *Lolium* spp., 260
   perennial (*Lolium perenne*): *Lolium* spp., 260
sacaline (**Sakhalin knotweed**): *Polygonum cuspidatum et al.*, 318
Saccharum ravennae: *Tripidium ravennae*, 412
safflower, woolly (**woolly distaff thistle**): *Carthamus lanatus*, 85
saffron thistle (**woolly distaff thistle**): *Carthamus lanatus*, 85
sage
   African (**Mediterranean sage**): *Salvia aethiopis*, 356, 516
   Mediterranean: *Salvia aethiopis*, 356, 516
   salt (*Iva axillaris*): Herbaceous Forbs, 438, 446
sagebrush (*Artemisia tridentata*): Woody Plants, 454, 456
sago pondweed (*Stuckenia pectinata*): Potamogeton & Stuckenia, 323
Saharan mustard: *Brassica tournefortii*, 64
sailors, blue (*Cichorium intybus*): Herbaceous Forbs, 438, 446
Saint Barnaby's thistle (**yellow starthistle**): *Centaurea solstitialis*, 100, 514
Saint James' weed (*Capsella bursa-pastoris*): Herbaceous Forbs, 440, 448
Saint Johnswort, common: *Hypericum perforatum*, 228, 516
Saint Mary's grass (**johnsongrass**): *Sorghum halepense*, 380
Saint Mary's thistle (**blessed milkthistle**): *Silybum marianum*, 371, 515
Sakhalin knotweed (*Polygonum sachalinense*): *Polygonum cuspidatum et al.*, 318
Salix spp. (**willows**), 351
sallet, poke (**common pokeweed**): *Phytolacca americana*, 311
Salsola **spp.**
   hyssopifolia (*Bassia hyssopifolia*): *Bassia* spp., 54
   paulsenii (**barbwire Russian-thistle**): *Salsola* spp., 353
   tragus (**Russian-thistle**): *Salsola* spp., 353, 515
      [= *australis, iberica, kali, pestifer, ruthenica*]
salt
   salt sage (*Iva axillaris*): Herbaceous Forbs, 438, 446
   salt tree (**athel tamarisk**): *Tamarix* spp., 395
   tamarisk salt tree (**athel tamarisk**): *Tamarix* spp., 395
saltbush
   Australian saltbush: *Atriplex semibaccata*, 47
   berry saltbush, creeping saltbush, scrambling berry saltbush (**Australian saltbush**): *Atriplex semibaccata*, 47
saltcedar (*Tamarix ramosissima*): *Tamarix* spp., 395, 517
saltcedars (*Tamarix* spp.): *Tamarix* spp., 395
salt-green (*Chenopodium berlandieri*): Herbaceous Forbs, 440, 448
saltlover (*Halogeton glomeratus*): *Halogeton*, 200, 515
saltmarsh cordgrass (**smooth cordgrass**): *Spartina* spp., 383, 517
salt-meadow cordgrass (*Spartina patens*): *Spartina* spp., 383
saltwater cordgrass (**smooth cordgrass**): *Spartina* spp., 383, 517
saltwort, common (**Russian-thistle**): *Salsola* spp., 353, 515
salvation jane (**vipers bugloss**): *Echium plantagineum*, 159
Salvia aethiopis (**Mediterranean sage**), 356, 516
Salvinia molesta (**giant salvinia**), 359
salvinia, giant: *Salvinia molesta*, 359
San Diego grass (**smilograss**): *Piptatherum miliaceum*, 315
sandalwood, false (*Myoporum laetum*): Woody Plants, 454, 456
sandbur

sandbur (annual bursage) (*Ambrosia acanthicarpa*): Herbaceous Forbs, 438, 446
    coast (coastal), common, field, grass, or mat (*Cenchrus* spp.): Grasses & Sedges, 434, 436
    Mexican (**puncturevine**): ***Tribulus terrestris***, 407, 517
    southern (*Cenchrus* spp.): Grasses & Sedges, 434, 436
    spiny (*Cenchrus* spp.): Grasses & Sedges, 434, 436
    Texas (**puncturevine**): ***Tribulus terrestris***, 407, 517
sandreed, sea (**European beachgrass**): ***Ammophila arenaria***, 31
sandspur (*Cenchrus* spp.): Grasses & Sedges, 434, 436
Sapium sebiferum (**Chinese tallowtree**), 361
Saponaria officinalis (bouncingbet): Herbaceous Forbs, 440, 448
Sarcobatus vermiculatus (greasewood): Herbaceous Forbs, 442, 448
scarlet wisteria tree (**red sesbania**): ***Sesbania punicea***, 369
Schedonorus phoenix (*Festuca arundinacea*): Grasses & Sedges, 434, 436
Schinus molle (**Peruvian peppertree**): ***Schinus*** **spp.**, 364
Schinus terebinthifolius (**Brazilian peppertree**): ***Schinus*** **spp.**, 364
Schismus **spp.** (**mediterraneangrasses**): ***Schismus*** **spp.**, 367
    Arabian schismus (**Arabian mediterraneangrass**): ***Schismus*** **spp.**, 367
    arabicus (**Arabian mediterraneangrass**): ***Schismus*** **spp.**, 367
    barbatus (**common mediterraneangrass**): ***Schismus*** **spp.**, 367
Schoenoplectus acutus (*Scirpus acutus*): Grasses & Sedges, 434, 436
Scirpus acutus (hardstem bulrush or tule): Grasses & Sedges, 434, 436
scoke (**common pokeweed**): ***Phytolacca americana***, 311
Scotch
    broom: ***Cytisus scoparius***, 147, 516
    cottonthistle (**Scotch thistle**): ***Onopordum acanthium***, 292, 515
    thistle: ***Onopordum acanthium***, 292, 515
scouringrush (common, horsetail, western): ***Equisetum*** **spp.**, 175
scourwort (*Saponaria officinalis*): Herbaceous Forbs, 440, 448
scrambling berry saltbush (**Australian saltbush**): ***Atriplex semibaccata***, 47
scutch grass (**bermudagrass**): ***Cynodon dactylon***, 141
scutch-grass (**quackgrass**): ***Elytrigia repens***, 172
sea
    burdoc (**common cocklebur**): ***Xanthium*** **spp.**, 428
    -fig (**iceplant**): ***Carpobrotus edulis***, 83
    rocket or European searocket (*Cakile maritima*): Herbaceous Forbs, 440, 448
    sandreed (**European beachgrass**): ***Ammophila arenaria***, 31
seashore vervain (*Verbena litoralis*): Herbaceous Forbs, 444, 452
seaside
    barley (**Mediterranean barley**): ***Hordeum marinum*** **&** ***murinum***, 218
    bentgrass (*Agrostis stolonifera*): Grasses & Sedges, 434, 436
    lobularia (*Lobularia maritima*): Herbaceous Forbs, 440, 448
    thistle (***Carduus tenuiflorus***): ***Carduus*** **spp.**, 79, 513
Secale cereale (common rye): Grasses & Sedges, 434, 436
sembulabula (*Cenchrus* spp.): Grasses & Sedges, 434, 436
Senecio spp.
    argutus, glomeratus, minimus: see *Erechtites* spp., Herbaceous Forbs, 440, 446
    jacobaea (tansy ragwort): Herbaceous Forbs, 438, 446
    mikanioides: ***Delairea odorata***, 152
sesban, purple (**red sesbania**): ***Sesbania punicea***, 369
Sesbania punicea (**red sesbania**), 369
Sesbania tripetii: ***Sesbania punicea***, 369
sesbania, red: ***Sesbania punicea***, 369
shavegrass (**horsetail, scouringrush**): ***Equisetum*** **spp.**, 175
sheep bur (**common cocklebur**): ***Xanthium*** **spp.**, 428
sheep sorrel (**red sorrel**): ***Rumex acetosella***, 344
sheepfoot (**birdsfoot trefoil**): ***Lotus corniculatus***, 263
shellygrass (**quackgrass**): ***Elytrigia repens***, 172
shepherd's-purse (bag, pouch) (*Capsella bursa-pastoris*): Herbaceous Forbs, 440, 448
shield-cress (*Lepidium perfoliatum*): Herbaceous Forbs, 440, 448
shining pondweed (**Illinois pondweed**): ***Potamogeton*** **&** ***Stuckenia***, 323

shore thistle (***Carduus pycnocephalus*** or ***tenuiflorus***): ***Carduus*** **spp.**, 79, 513
short whitetop (**hoary cress**): ***Cardaria*** **spp.**, 76
shortpod mustard: ***Hirschfeldia incana***, 211
showy buttercup (*Ranunculus acris*): Herbaceous Forbs, 444, 450
showy milkweed (*Asclepias speciosa*): *Asclepias* spp., 43
Siberian elm (*Ulmus pumila*): Woody Plants, 454, 456
silver thistle (**Scotch thistle**): ***Onopordum acanthium***, 292, 515
silverberry (**Russian-olive**): ***Elaeagnus angustifolia***, 169
silverleaf
    bitter apple (*Solanum elaeagnifolium*): Herbaceous Forbs, 444, 452
    cotoneaster (***Cotoneaster pannosus***): ***Cotoneaster*** **spp.**, 132
    nettle (*Solanum elaeagnifolium*): Herbaceous Forbs, 444, 452
    nightshade (*Solanum elaeagnifolium*): Herbaceous Forbs, 444, 452
Silybum marianum (**blessed milkthistle**), 371, 515
    [= *maculatum, mariae*]
Sinapis spp.
    arvensis (**wild mustard**), 374
    incana: ***Hirschfeldia incana***, 211
    nigra: ***Brassica nigra***, 61
singleseed hawthorn (*Crataegus monogyna*): Woody Plants, 454, 456
Sisymbrium spp.
    altissimum (**tumble mustard**): ***Sisymbrium*** **spp.**, 377
    irio (**London rocket**): ***Sisymbrium*** **spp.**, 377
    sophia (*Descurainia sophia*): Herbaceous Forbs, 440, 448
    tripinnatum (*Descurainia sophia*): Herbaceous Forbs, 440, 448
sixweeks fescue (**rattail fescue**): ***Vulpia myuros***, 425
sixweeks grass (**rattail fescue**): ***Vulpia myuros***, 425
six-weeks grass (**downy brome**): ***Bromus diandrus, madritensis, tectorum***, 66
skeletonweed, rush: ***Chondrilla juncea***, 116, 515
skunktail (**foxtail barley**): ***Hordeum jubatum***, 216
slender
    chess (**downy brome**): ***Bromus diandrus, madritensis, tectorum***, 66
    false-brome: ***Brachypodium sylvaticum***, 59
    nettle (*Urtica dioica*): Herbaceous Forbs, 444, 452
    oat (*Avena barbata*): *Avena* spp., 49
    perennial peppercress (**perennial pepperweed**): ***Lepidium latifolium***, 244
    pigweed (*Amaranthus* spp.): Herbaceous Forbs, 438, 446
    thistle (**Italian thistle**): ***Carduus*** **spp.**, 79, 513
    wild oat (*Avena barbata*): *Avena* spp., 49
slenderflower thistle (***Carduus tenuiflorus***): ***Carduus*** **spp.**, 79, 513
slim amaranth (*Amaranthus* spp.): Herbaceous Forbs, 438, 446
slippers, babies' (**birdsfoot trefoil**): ***Lotus corniculatus***, 263
small
    bindweed (**field bindweed**): ***Convolvulus arvensis***, 127, 516
    brassbuttons (*Cotula australis*): Herbaceous Forbs, 438, 446
    bugloss (*Anchusa arvensis*): see in *Echium plantagineum*, 159
    smutgrass (**smutgrass**): ***Sporobolus indicus***, 390
smallflower (small-flowered)
    tamarisk (*Tamarix parviflora*): *Tamarix* spp., 395, 517
    fiddleneck (**Menzies fiddleneck**): ***Amsinckia*** **spp.**, 35
    marsh elder (*Iva axillaris*): Herbaceous Forbs, 438, 446
    morningglory (**field bindweed**): ***Convolvulus arvensis***, 127, 516
smartweed
    dockleaf (**pale smartweed**): ***Polygonum persicaria*** **&** ***lapathifolium***, 320
    nodding (**pale smartweed**): ***Polygonum persicaria*** **&** ***lapathifolium***, 320
    pale (*Polygonum lapathifolium*): ***Polygonum persicaria*** **&** ***lapathifolium***, 320
    pink (*Polygonum convolvulus*): Herbaceous Forbs, 444, 450
    spotted (**ladysthumb**): ***Polygonum persicaria*** **&** ***lapathifolium***, 320
    willow (**pale smartweed**): ***Polygonum persicaria*** **&** ***lapathifolium***, 320
smilograss: ***Piptatherum miliaceum***, 315

smooth
- brome: *Bromus inermis*, 69
- catsear (*Hypochaeris glabra*): *Hypochaeris* spp., 230
- cordgrass (*Spartina alterniflora*): *Spartina* spp., 383, 517
- frogbit: *Limnobium laevigatum*, 252
  - meadowgrass (*Poa pratensis*): Grasses & Sedges, 434, 436
  - pigweed (*Amaranthus* spp.): Herbaceous Forbs, 438, 446

smotherweed (**fivehook bassia**): *Bassia* spp., 54
smutgrass: *Sporobolus indicus*, 390
smutgrass, small (**smutgrass**): *Sporobolus indicus*, 390
snakegrass (*Eragrostis* spp.): Grasses & Sedges, 434, 436
snake-grass (**horsetail, scouringrush**): *Equisetum* spp., 175
snakeroot, sticky (**croftonweed**): *Ageratina adenophora*, 21
snake-weed (**poison hemlock**): *Conium maculatum*, 125, 513
snapdragon, wild (**Dalmatian toadflax**): *Linaria dalmatica*, 254, 517
snapdragon, wild (**yellow toadflax**): *Linaria vulgaris*, 257, 517
soapwort (*Saponaria officinalis*): Herbaceous Forbs, 440, 448
soft
- brome (*Bromus hordeaceus*): *Bromus hordeaceus, japonicus*, 71
- chess (*Bromus hordeaceus*): *Bromus hordeaceus, japonicus*, 71
- grass, meadow or tufted (**common velvetgrass**): *Holcus lanatus*, 212
- rush (*Juncus effusus*): *Juncus* spp., 238

Solanum spp.
- *elaeagnifolium* (silverleaf nightshade): Herbaceous Forbs, 444, 452
  [= *dealbatum, flavidum, leprosum, obtusifolium, roemerianum, saponaceum, texense*]

soldier's button (*Cotula australis*): Herbaceous Forbs, 438, 446
Soliva tenella (*Cotula australis*): Herbaceous Forbs, 438, 446
Sophia sophia (*Descurainia sophia*): Herbaceous Forbs, 440, 448
sophia, herb (*Descurainia sophia*): Herbaceous Forbs, 440, 448
sorgho d'Alep (**johnsongrass**): *Sorghum halepense*, 380
Sorghum halepense (**johnsongrass**), 380
  [= *arundinaceum, miliaceum*]
sorgo de alepo (**johnsongrass**): *Sorghum halepense*, 380
sorrel
- cape sorrel (**buttercup oxalis**): *Oxalis pes-caprae*, 295
- common, cow, dog-eared, field, horse, mountain, red-top, or sheep sorrel (**red sorrel**): *Rumex acetosella*, 344
- red sorrel: *Rumex acetosella*, 344
- yellow sorrel (**buttercup oxalis**): *Oxalis pes-caprae*, 295

sour dock (**red sorrel**): *Rumex acetosella*, 344
sour dock (**curly dock**): *Rumex crispus & obtusifolius*, 346
sourgrass (**buttercup oxalis**): *Oxalis pes-caprae*, 295
sourgrass (**red sorrel**): *Rumex acetosella*, 344
soursob (**buttercup oxalis**): *Oxalis pes-caprae*, 295
sour-weed (**red sorrel**): *Rumex acetosella*, 344
South American
- conyza (*Conyza bonariensis*): Herbaceous Forbs, 438, 446
- spongeplant: *Limnobium laevigatum*, 252
- waterweed (**Brazilian egeria**): *Egeria densa*, 163

southern
- brassbuttons (*Cotula australis*): Herbaceous Forbs, 438, 446
- catalpa (*Catalpa bignonioides*): Woody Plants, 454, 456
- cattail (**cattail**): *Typha* spp., 413
- naiad (*Najas guadalupensis*): *Najas* spp., 286
- sandbur (*Cenchrus* spp.): Grasses & Sedges, 434, 436
- waternymph (**southern naiad**): *Najas* spp., 286

sowbane (*Chenopodium berlandieri*): Herbaceous Forbs, 440, 448
Spanish
- broom: *Spartium junceum*, 385
- reed (**giant reed**): *Arundo donax*, 40, 517
- thistle (**spiny cocklebur**): *Xanthium* spp., 428

Spartina spp. (**cordgrasses**), 383
- *alterniflora* (smooth cordgrass): *Spartina* spp., 383, 517
  [= *brasiliensis, glabra, maritima, stricta*]
- *anglica* (common cordgrass): *Spartina* spp., 383
- *densiflora* (dense-flowered cordgrass): *Spartina* spp., 383
- *patens* (salt-meadow cordgrass): *Spartina* spp., 383

Spartium junceum (Spanish broom), 385
Spartium scoparium: *Cytisus scoparius*, 147, 516
spatterdock (*Nuphar lutea*): Aquatic Plants, 458, 460
spear thistle (**bull thistle**): *Cirsium vulgare*, 122, 515
speargrass, twisted-awn (*Stipa capensis*): Grasses & Sedges, 434, 436
speedwell, water: *Veronica anagallis-aquatica*, 422
Sphaerophysa salsula (**swainsonpea**), 388
spike watermilfoil (**Eurasian watermilfoil**): *Myriophyllum spicatum*, 283, 518
spikeweed (*Hemizonia pungens*): *Hemizonia* spp., 205
spikeweed, common (**spikeweed**): *Hemizonia* spp., 205
spinach
- Australian (*Chenopodium berlandieri*): Herbaceous Forbs, 440, 448
- poor man's (*Chenopodium album*): Herbaceous Forbs, 440, 448
- wild (*Chenopodium album*): Herbaceous Forbs, 440, 448

spiny
- clotbur (**spiny cocklebur**): *Xanthium* spp., 428
- cocklebur (*Xanthium spinosum*): *Xanthium* spp., 428
- emex (*Emex spinosa*): Herbaceous Forbs, 442, 450
- naiad (**hollyleaf naiad**): *Najas* spp., 286
- sandbur (*Cenchrus* spp.): Grasses & Sedges, 434, 436
- thistle (**plumeless thistle**): *Carduus* spp., 79, 513
- three-cornered jack (*Emex spinosa*): Herbaceous Forbs, 442, 450

Spirodela spp. (**duckweed, duckmeat**): *Lemna* et al., 242
spleen amaranth (*Amaranthus* spp.): Herbaceous Forbs, 438, 446
splitgrass (**common mediterraneangrass**): *Schismus* spp., 367
spongeplant, South American: *Limnobium laevigatum*, 252
Sporobolus indicus (= *S. angustus, berterianus*) (**smutgrass**), 390
spotted
- catsear (**common catsear**): *Hypochaeris* spp., 230
- hemlock (**poison hemlock**): *Conium maculatum*, 125, 513
- knapweed: *Centaurea stoebe*, 105, 514
- knotweed or smartweed (**ladysthumb**): *Polygonum persicaria & lapathifolium*, 320
- thistle (**blessed milkthistle**): *Silybum marianum*, 371, 515

spreading
- hedgeparsley (**hedgeparsley**): *Torilis arvensis*, 401
- lovegrass (*Eragrostis* spp.): Grasses & Sedges, 434, 436
- pigweed (*Amaranthus* spp.): Herbaceous Forbs, 438, 446
- rush (*Juncus patens*): *Juncus* spp., 238

spring grass (*Anthoxanthum odoratum*): Grasses & Sedges, 434, 436
spurge
- carnation (*Euphorbia terracina*): *Euphorbia* spp., 182
- eggleaf (**oblong spurge**): *Euphorbia* spp., 182
- leafy (*Euphorbia esula*): *Euphorbia* spp., 182, 516
- oblong (*Euphorbia oblongata*): *Euphorbia* spp., 182

squarrose knapweed: *Centaurea virgata* ssp. *squarrosa*, 109, 515
squirreltail barley (**foxtail barley**): *Hordeum jubatum*, 216
squirreltailgrass (**foxtail barley**): *Hordeum jubatum*, 216
starthistle
- false (**woolly distaff thistle**): *Carthamus lanatus*, 85
- golden (**yellow starthistle**): *Centaurea solstitialis*, 100, 514
- Iberian (*Centaurea iberica*): *Centaurea calcitrapa & iberica*, 87, 514
- Malta: *Centaurea melitensis*, 97
- purple (*Centaurea calcitrapa*): *Centaurea calcitrapa & iberica*, 87, 513
- red (**purple starthistle**): *Centaurea calcitrapa & iberica*, 87, 513
- Russian: *Acroptilon repens*, 16, 513
- woolly (**woolly distaff thistle**): *Carthamus lanatus*, 85
- yellow: *Centaurea solstitialis*, 100, 514

steppegrass, Mediterranean (*Stipa capensis*): Grasses & Sedges, 434, 436
steppenraute (*Peganum harmala*): Herbaceous Forbs, 444, 452
sticktight (**houndstongue**): *Cynoglossum officinale*, 143
sticky agrimony (**croftonweed**): *Ageratina adenophora*, 21

sticky snakeroot (**croftonweed**): *Ageratina adenophora*, 21
sticky tarweed (**virgate tarweed**): *Holocarpha virgata*, 214
stiff brome (**annual false-brome**): *Brachypodium distachyon*, 57
stinging nettle (*Urtica dioica*): Herbaceous Forbs, 444, 452
stinkgrass (*Eragrostis* spp.): Grasses & Sedges, 434, 436
stinking willie (*Senecio jacobaea*): Herbaceous Forbs, 440, 446, 515
stinkweed (*Thlaspi arvense*): Herbaceous Forbs, 440, 448
stinkweed, poison (**poison hemlock**): *Conium maculatum*, 125, 513
stinkwort: *Dittrichia graveolens*, 157
Stipa capensis (Mediterranean steppegrass): Grasses & Sedges, 434, 436
  [= *retorta, tortilis*]
stock, malcolm (*Malcolmia africana*): Herbaceous Forbs, 440, 448
stonegrass (*Polygonum arenastrum*): Herbaceous Forbs, 442, 450
stonewort (***Chara, Nitella*** spp.): ***Chara, Nitella*** spp., 114
storksbill, common or redstem (**redstem filaree**): *Erodium cicutarium*, 177
Strigosella africana (*Malcolmia africana*): Herbaceous Forbs, 440, 448
strongscented grass (*Eragrostis* spp.): Grasses & Sedges, 434, 436
Stuckenia pectinata (**sago pondweed**): *Potamogeton & Stuckenia*, 323
succory (*Cichorium intybus*): Herbaceous Forbs, 438, 446
succory, gum (**rush skeletonweed**): *Chondrilla juncea*, 116, 515
sulfur (**sulphur**) **cinquefoil**: *Potentilla recta*, 328
sulphur five-fingers (**sulfur cinquefoil**): *Potentilla recta*, 328
summer cypress (**kochia**): *Bassia* spp., 54
summer lilac (*Buddleja davidii*): Woody Plants, 454, 456
summergrass (*Setaria* spp.): Grasses & Sedges, 434, 436
sumpweed, poverty (*Iva axillaris*): Herbaceous Forbs, 438, 446
sunflower, annual (common, wild) (*Helianthus annuus*): Herbaceous Forbs, 438, 446
swainsona (**swainsonpea**): *Sphaerophysa salsula*, 388
Swainsona salsula: *Sphaerophysa salsula*, 388
swainsonpea: *Sphaerophysa salsula*, 388
swainsonpea, alkali (**swainsonpea**): *Sphaerophysa salsula*, 388
swamp buttercup (*Ranunculus repens*): Herbaceous Forbs, 444, 450
sweet
  alyssum (*Alyssum alyssoides*): Herbaceous Forbs, 440, 448
  alyssum (*Lobularia maritima*): Herbaceous Forbs, 440, 448
  anise (**fennel**): *Foeniculum vulgare*, 187
  betty (*Saponaria officinalis*): Herbaceous Forbs, 440, 448
  fennel (**fennel**): *Foeniculum vulgare*, 187
  resinbush (*Euryops multifidus*): Woody Plants, 454, 456
  vernalgrass (*Anthoxanthum odoratum*): Grasses & Sedges, 434, 436
sweet pea, perennial (*Lathyrus latifolius*): Herbaceous Forbs, 442, 448
sweet pea, weedy (*Lathyrus latifolius*): Herbaceous Forbs, 442, 448
sweetbriar rose (*Rosa eglanteria*): *Rosa* spp., 338
sweetclover, white (*Melilotus albus*): Herbaceous Forbs, 442, 448
swine-bane (*Chenopodium berlandieri*): Herbaceous Forbs, 440, 448
symphonica (**black henbane**): *Hyoscyamus niger*, 226
Syria grass (**johnsongrass**): *Sorghum halepense*, 380
Syrian beancaper (*Zygophyllum fabago*): Herbaceous Forbs, 444, 452
Syrian rue (*Peganum harmala*): Herbaceous Forbs, 444, 452
tacks, tackweed (**puncturevine**): *Tribulus terrestris*, 407, 517
Taeniatherum caput-medusae (= *T. asperum, T. crinitum*) (**medusahead**), 392
tall
  buttercup or crowfoot (*Ranunculus acris*): Herbaceous Forbs, 444, 450
  fescue (*Festuca arundinacea*): Grasses & Sedges, 434, 436
  foxtail (*Setaria* spp.): Grasses & Sedges, 434, 436
  gypsophyll (**baby's breath**): *Gypsophila paniculata*, 198
  mustard (**tumble mustard**): *Sisymbrium* spp., 377
  nettle (*Urtica dioica*): Herbaceous Forbs, 444, 452
  ragweed (*Ambrosia artemisiifolia*): Herbaceous Forbs, 438, 446
  tumble-mustard (**tumble mustard**): *Sisymbrium* spp., 377
  vervain (*Verbena bonariensis*): Herbaceous Forbs, 444, 452
  whitetop (**perennial pepperweed**): *Lepidium latifolium*, 244
tallowberry, Chinese (**Chinese tallowtree**): *Sapium sebiferum*, 361

tallowtree, Chinese: *Sapium sebiferum*, 361
tamarisk (*Tamarix* spp.): *Tamarix* spp., 395
  athel tamarisk (*Tamarix aphylla*): *Tamarix* spp., 395
  Chinese tamarisk (*Tamarix chinensis*): *Tamarix* spp., 395
  early tamarisk (**smallflower tamarisk**): *Tamarix* spp., 395, 517
  five-stamen tamarisk (**saltcedar**): *Tamarix* spp., 395, 517
  French tamarisk (*Tamarix gallica*): *Tamarix* spp., 395
  late tamarisk (**saltcedar**): *Tamarix* spp., 395, 517
  smallflower tamarisk (*Tamarix parviflora*): *Tamarix* spp., 395, 517
  tamarisk salt tree (**athel tamarisk**): *Tamarix* spp., 395
Tamarix spp.
  aphylla (**athel tamarisk**): *Tamarix* spp., 395
    [= *articulata, orientalis*]
  chinensis (**Chinese tamarisk**): *Tamarix* spp., 395
  gallica (**French tamarisk**): *Tamarix* spp., 395
  parviflora (**smallflower tamarisk**): *Tamarix* spp., 395, 517
    [= *articulata, cretica, lucronensis, petteri, rubella, tetranda*]
  ramosissima (**saltcedar**): *Tamarix* spp., 395, 517
    [= *altaica, eversmannii, odessana, pallasii, pentranda*]
Tanacetum vulgare (**common tansy**), 399
tansy, common (tansy, garden tansy): *Tanacetum vulgare*, 399
tansy ragwort (*Senecio jacobaea*): Herbaceous Forbs, 440, 446, 515
tansymustard, pinnate tansymustard (*Descurainia sophia*): Herbaceous Forbs, 440, 448
taper, big (**common mullein**): *Verbascum* spp., 419
Taraxacum officinale (dandelion): Herbaceous Forbs, 440, 446
  [= *laevigatum, retroflexum, vulgare*]
tarweed
  common tarweed (**spikeweed**): *Hemizonia* spp., 205
  Fitch's tarweed (*Hemizonia fitchii*): *Hemizonia* spp., 205
  Parry's tarweed (*Hemizonia parryi*): *Hemizonia* spp., 205
  piny tarweed (**spikeweed**): *Hemizonia* spp., 205
  sticky tarweed (**virgate tarweed**): *Holocarpha virgata*, 214
  tarweed (fiddleneck) (**coast fiddleneck**): *Amsinckia* spp., 35
  virgate tarweed: *Holocarpha virgata*, 214
  yellow tarweed (**coast fiddleneck**): *Amsinckia* spp., 35
  yellowflower tarweed (**virgate tarweed**): *Holocarpha virgata*, 214
Tasmanian blue gum: *Eucalyptus globulus*, 180
tawny beardgrass (*Polypogon monspeliensis*): Grasses & Sedges, 434, 436
teasel (*Dipsacus fullonum* or *D. sativus*): *Dipsacus* spp., 154
  card (*Dipsacus fullonum* or *D. sativus*): *Dipsacus* spp., 154
  common (*Dipsacus fullonum*): *Dipsacus* spp., 154
  cultivated (**Fuller's teasel**): *Dipsacus* spp., 154
  cutleaf (*Dipsacus laciniatus*): *Dipsacus* spp., 154
  Fuller's (*Dipsacus sativus*): *Dipsacus* spp., 154
  Indian (**Fuller's teasel**): *Dipsacus* spp., 154
  wild (**common teasel**): *Dipsacus* spp., 154
tender fountaingrass (**crimson fountaingrass**): *Pennisetum setaceum*, 301
tenella mustard (*Chorispora tenella*): Herbaceous Forbs, 440, 448
testiculate buttercup (*Ranunculus testiculatus*): Herbaceous Forbs, 444, 450
Texas sandbur (**puncturevine**): *Tribulus terrestris*, 407, 517
thistle
  artichoke: *Cynara cardunculus*, 138
  asses' (**Scotch thistle**): *Onopordum acanthium*, 292, 515
  bank (**bull thistle**): *Cirsium vulgare*, 122, 515
  barbwire Russian-thistle (*Salsola paulsenii*): *Salsola* spp., 353
  bird, black, blue (**bull thistle**): *Cirsium vulgare*, 122, 515
  bristly (**plumeless thistle**): *Carduus* spp., 79, 513
  bull: *Cirsium vulgare*, 122, 515
  button bur (**bull thistle**): *Cirsium vulgare*, 122, 515
  cabbage (**blessed milkthistle**): *Silybum marianum*, 371, 515
  Canada: *Cirsium arvense*, 119, 515
  card (*Dipsacus fullonum* or *D. sativus*): *Dipsacus* spp., 154

common (**bull thistle**): *Cirsium vulgare*, 122, 515
compact-headed (**Italian thistle**): *Carduus* spp., 79, 513
corn (**Canada thistle**): *Cirsium arvense*, 119, 515
cotton (**Scotch thistle**): *Onopordum acanthium*, 292, 515
creeping (**Canada thistle**): *Cirsium arvense*, 119, 515
downy (**Scotch thistle**): *Onopordum acanthium*, 292, 515
English (*Lactuca serriola*): Herbaceous Forbs, 438, 446
false starthistle (**woolly distaff thistle**): *Carthamus lanatus*, 85
Fuller's (**bull thistle**): *Cirsium vulgare*, 122, 515
giant plumeless (*Carduus acanthoides* or *C. nutans*): *Carduus* spp., 79, 513
golden starthistle (**yellow starthistle**): *Centaurea solstitialis*, 100, 514
heraldic (**Scotch thistle**): *Onopordum acanthium*, 292, 515
holy (**blessed milkthistle**): *Silybum marianum*, 371, 515
horse (*Lactuca serriola*): Herbaceous Forbs, 438, 446
Iberian starthistle (*Centaurea iberica*): *Centaurea calcitrapa* & *iberica*, 87, 514
Italian (*Carduus pycnocephalus* or *C. tenuiflorus*): *Carduus* spp., 79, 513
Italian plumeless (**Italian thistle**): *Carduus* spp., 79, 513
lady's (**blessed milkthistle**): *Silybum marianum*, 371, 515
Malta starthistle: *Centaurea melitensis*, 97
milk (**blessed milkthistle**): *Silybum marianum*, 371, 515
multiheaded (**slenderflower thistle**): *Carduus* spp., 79, 513
musk (*Carduus nutans*): *Carduus* spp., 79, 513
Napa (**Malta starthistle**): *Centaurea melitensis*, 97
nodding or nodding plumeless (**musk thistle**): *Carduus* spp., 79, 513
plume (**bull thistle**): *Cirsium vulgare*, 122, 515
plumeless (*Carduus acanthoides* or *C. nutans*): *Carduus* spp., 79, 513
Plymouth (**Italian thistle**): *Carduus* spp., 79, 513
prickly Russian (**Russian-thistle**): *Salsola* spp., 353, 515
purple starthistle (*Centaurea calcitrapa*): *Centaurea calcitrapa* & *iberica*, 87, 513
Queen Mary's (**Scotch thistle**): *Onopordum acanthium*, 292, 515
red starthistle (**purple starthistle**): *Centaurea calcitrapa* & *iberica*, 87, 513
roadside (**bull thistle**): *Cirsium vulgare*, 122, 515
Russian starthistle: *Acroptilon repens*, 16, 513
Russian-thistle (*Salsola tragus*): *Salsola* spp., 353, 515
saffron (**woolly distaff thistle**): *Carthamus lanatus*, 85
Saint Barnaby's (**yellow starthistle**): *Centaurea solstitialis*, 100, 514
Saint Mary's (**blessed milkthistle**): *Silybum marianum*, 371, 515
Scotch: *Onopordum acanthium*, 292, 515
seaside (**slenderflower thistle**): *Carduus* spp., 79, 513
shore (*Carduus pycnocephalus* or *C. tenuiflorus*): *Carduus* spp., 79, 513
silver (**Scotch thistle**): *Onopordum acanthium*, 292, 515
slender (**Italian thistle**): *Carduus* spp., 79, 513
slenderflower (*Carduus tenuiflorus*): *Carduus* spp., 79, 513
Spanish (**spiny cocklebur**): *Xanthium* spp., 428
spear (**bull thistle**): *Cirsium vulgare*, 122, 515
spiny (**plumeless thistle**): *Carduus* spp., 79, 513
spotted (**blessed milkthistle**): *Silybum marianum*, 371, 515
Turkestan: *Acroptilon repens*, 16, 513
variegated (**blessed milkthistle**): *Silybum marianum*, 371, 515
whip (*Lactuca serriola*): Herbaceous Forbs, 438, 446
white (**blessed milkthistle**): *Silybum marianum*, 371, 515
winged (**Scotch thistle**): *Onopordum acanthium*, 292, 515
winged plumeless (**slenderflower thistle**): *Carduus* spp., 79, 513
woolly (**Scotch thistle**): *Onopordum acanthium*, 292, 515
woolly distaff: *Carthamus lanatus*, 85
woolly starthistle (**woolly distaff thistle**): *Carthamus lanatus*, 85
yellow starthistle: *Centaurea solstitialis*, 100, 514
Thlaspi arvense (field pennycress): Herbaceous Forbs, 440, 448
Thlaspi bursa-pastoris (*Capsella bursa-pastoris*): Herbaceous Forbs, 440, 448

thorn
    devil's thorn (*Emex spinosa*): Herbaceous Forbs, 442, 450
    everlasting thorn (*Pyracantha* spp.): Woody Plants, 454, 456
    fiery thorn (*Pyracantha* spp.): Woody Plants, 454, 456
    thorn broom (**gorse**): *Ulex europaeus*, 415, 516
    thorn orache (**fivehook bassia**): *Bassia* spp., 54
three-cornered jack, spiny (*Emex spinosa*): Herbaceous Forbs, 442, 450
thunderflower (**oxeye daisy**): *Leucanthemum vulgare*, 247
ticklegrass (**foxtail barley**): *Hordeum jubatum*, 216
tipton weed (**common St. Johnswort**): *Hypericum perforatum*, 228, 516
toad rush: *Juncus* spp., 238
toadflax
    broad-leaved toadflax (**Dalmatian toadflax**): *Linaria dalmatica*, 254, 517
    common toadflax (**yellow toadflax**): *Linaria vulgaris*, 257, 517
    Dalmatian toadflax: *Linaria dalmatica*, 254, 517
    toadflax (**yellow toadflax**): *Linaria vulgaris*, 257, 517
    yellow toadflax: *Linaria vulgaris*, 257, 517
tobacco, Mexican (**tree tobacco**): *Nicotiana glauca*, 289
tobacco, tree: *Nicotiana glauca*, 289
tocalote: *Centaurea melitensis*, 97
tomato weed (*Solanum elaeagnifolium*): Herbaceous Forbs, 444, 452
tongue, dog's (**houndstongue**): *Cynoglossum officinale*, 143
toothed coast fireweed (*Erechtites* spp.): Herbaceous Forbs, 438, 446
toowoomba grass (**hardinggrass**): *Phalaris aquatica*, 303
Torilis spp.
    arvensis (**hedgeparsley**), 401
        [= *africana, japonica*]
Toxicodendron spp.
    altissima: *Ailanthus altissima*, 23
    diversilobum (**Pacific poison-oak**), 403
        [= *comarophyllum, dryophyllum, isophyllum, oxycarpum, radicans, vaccarum*]
Toxicoscordion spp.: *Zigadenus* spp., 431
Trapa natans (water chestnut), 405
tree clover (*Melilotus albus*): Herbaceous Forbs, 442, 448
tree lupine (*Lupinus arboreus*): *Lupinus* spp., 267
tree tobacco: *Nicotiana glauca*, 289
tree-of-heaven: *Ailanthus altissima*, 23
trefoil, birdsfoot: *Lotus corniculatus*, 263
trefoil, broadleaf birdsfoot (**birdsfoot trefoil**): *Lotus corniculatus*, 263
Tribulus terrestris (**puncturevine**), 407, 517
Trichostema lanceolatum (vinegarweed), 410
Trifolium spp.
    hirtum (rose clover): Herbaceous Forbs, 442, 448
        [= *hispidum, oxypetasum, pictum*]
Tripidium ravennae (ravennagrass), 412
Triticum spp.
    cereale (*Secale cereale*): Grasses & Sedges, 434, 436
    cylindricum (*Aegilops cylindrica*): *Aegilops* spp., 19
    persicum (*Aegilops triuncialis*): *Aegilops* spp., 19
    repens: *Elytrigia repens*, 172
    triunciale (*Aegilops triuncialis*): *Aegilops* spp., 19
trompilla (*Solanum elaeagnifolium*): Herbaceous Forbs, 444, 452
tubercled crowfoot (*Ranunculus testiculatus*): Herbaceous Forbs, 444, 450
tuberous canarygrass (**hardinggrass**): *Phalaris aquatica*, 303
tufted lovegrass (*Eragrostis* spp.): Grasses & Sedges, 434, 436
tufted soft grass (**common velvetgrass**): *Holcus lanatus*, 212
tule (*Scirpus acutus*): Grasses & Sedges, 434, 436
tumble
    mustard (*Sisymbrium altissimum*): *Sisymbrium* spp., 377
    pigweed (*Amaranthus* spp.): Herbaceous Forbs, 438, 446
tumbleweed
    (mustard) (**tumble mustard**): *Sisymbrium* spp., 377
    (pigweed) (*Amaranthus* spp.): Herbaceous Forbs, 438, 446

(Russian thistle) (*Salsola tragus*): **Salsola** spp., 353, 515
Russian tumbleweed (**Russian-thistle**): **Salsola** spp., 353, 515
Turkestan thistle: ***Acroptilon repens***, 16, 513
turnip
    Mediterranean (**Saharan mustard**): ***Brassica tournefortii***, 64
    wild (*Brassica rapa*): Herbaceous Forbs, 440, 448
    wild (**wild radish**): ***Raphanus*** spp., 331
    wild long-fruited (**Saharan mustard**): ***Brassica tournefortii***, 64
twisted-awn speargrass (*Stipa capensis*): Grasses & Sedges, 434, 436
twitchgrass (**quackgrass**): ***Elytrigia repens***, 172
Typha **spp. (cattail)**, 413
Ulex europaeus (err. *europaea*) (**gorse**), 415, 516
Ulmus pumila (Siberian elm): Woody Plants, 454, 456
umbrellawort (**wild four-o'clock**): *Mirabilis nyctaginea*, 277
upright hedgeparsley (**hedgeparsley**): *Torilis arvensis*, 401
Urtica spp.
    dioica (stinging nettle): Herbaceous Forbs, 444, 452
        [= b*reweri, californica, cardiophylla, gracilis, holosericea, lyallii, procera, strigosissima, trachycarpa, viridis*]
Uruguay waterprimrose (*Ludwigia grandiflora*): ***Ludwigia*** **spp.**, 265
Uruguayan pampasgrass (**pampasgrass**): ***Cortaderia*** spp., 130
vanilla grass (*Anthoxanthum odoratum*): Grasses & Sedges, 434, 436
variegated thistle (**blessed milkthistle**): ***Silybum marianum***, 371, 515
varnish tree (**tree-of-heaven**): ***Ailanthus altissima***, 23
veldtgrass: ***Ehrharta*** spp., 165
    African (**purple veldtgrass**): ***Ehrharta*** spp., 165
    erect (*Ehrharta erecta*): ***Ehrharta*** spp., 165
    long-flowered (*Ehrharta longiflora*): ***Ehrharta*** spp., 165
    perennial (**purple veldtgrass**): ***Ehrharta*** spp., 165
    purple (*Ehrharta calycina*): ***Ehrharta*** spp., 165
velvet dock (or plant) (**common mullein**): ***Verbascum*** spp., 419
velvet, water (**mosquitofern**): ***Azolla*** **spp.**, 52, 518
velvetgrass, common: *Holcus lanatus*, 212
ventenata (**North African wiregrass**): ***Ventenata dubia***, 417
Ventenata dubia (**North African wiregrass**), 417
Venus-cup (*Dipsacus fullonum* or *D. sativus*): ***Dipsacus*** spp., 154
Verbascum blattaria (**moth mullein**): ***Verbascum*** **spp.**, 419
Verbascum thapsus (**common mullein**): ***Verbascum*** **spp.**, 419
Verbena spp.
    bonariensis (tall vervain): Herbaceous Forbs, 444, 452
    litoralis (seashore vervain): Herbaceous Forbs, 444, 452
        [= *bonariensis* var. *litoralis, brasiliensis, caracasana, hansenii*]
verbena, cluster-flowered (*Verbena bonariensis*): Herbaceous Forbs, 444, 452
vernalgrass, sweet (*Anthoxanthum odoratum*): Grasses & Sedges, 434, 436
Veronica anagallis-aquatica (= *V. anagallis*) (**water speedwell**), 422
vervain
    seashore (*Verbena litoralis*): Herbaceous Forbs, 444, 452
    tall (= Argentine or purple-top) (*Verbena bonariensis*): Herbaceous Forbs, 444, 452
Victorian box (*Pittosporum undulatum*): Woody Plants, 454, 456
Vinca major (= *V. pubescens*) (**big periwinkle**), 423
vinegarweed: *Trichostema lanceolatum*, 410
violet, bouquet (**purple loosestrife**): *Lythrum salicaria*, 271, 517
vipers bugloss (**blueweed**): *Echium vulgare*, 161
vipers bugloss: ***Echium plantagineum***, 159
vipers bugloss, purple (**vipers bugloss**): ***Echium plantagineum***, 159
virgate tarweed: *Holocarpha virgata*, 214
Virginia poke (**common pokeweed**): *Phytolacca americana*, 311
Vulpia megalura: ***Vulpia myuros***, 425
Vulpia myuros (**rattail fescue**), 425
wall goosefoot (*Chenopodium berlandieri*): Herbaceous Forbs, 440, 448
wand loosestrife (*Lythrum virgatum*): Herbaceous Forbs, 442, 450
Washington fan palm (***Washingtonia robusta***): **Phoenix & Washingtonia**, 307

Washingtonia spp.
    robusta (**Mexican fan palm**): **Phoenix & Washingtonia**, 307
        [= *filifera* var. *robusta, gracilis, sonorae*]
water
    chestnut: *Trapa natans*, 405
    fern (**mosquitofern**): ***Azolla*** **spp.**, 52, 518
    fringe (**yellow floatingheart**): *Nymphoides peltata*, 291
    grass (*Paspalum dilatatum*): Grasses & Sedges, 434, 436
    hyacinth: *Eichhornia crassipes*, 167, 518
    ivy (**Cape ivy**): *Delairea odorata*, 152
    lentil (*Lemna* spp.): ***Lemna, Spirodela, & Wolffia***, 242
    lettuce (*Pistia stratiotes*): 518
    speedwell: *Veronica anagallis-aquatica*, 422
    velvet (**mosquitofern**): ***Azolla*** **spp.**, 52, 518
waterbuttons, Australian (*Cotula australis*): Herbaceous Forbs, 438, 446
watergrass, large (*Paspalum dilatatum*): Grasses & Sedges, 434, 436
waterlily, fragrant (*Nymphaea odorata*): Aquatic Plants, 458, 460
waterlily, large white (*Nymphaea odorata*): Aquatic Plants, 458, 460
watermeal (***Wolffia*** spp.): ***Lemna, Spirodela, & Wolffia***, 242
watermilfoil
    Brazilian (**parrotfeather**): *Myriophyllum aquaticum*, 281
    Eurasian: *Myriophyllum spicatum*, 283, 518
    parrotfeather (**parrotfeather**): *Myriophyllum aquaticum*, 281
    spike (**Eurasian watermilfoil**): *Myriophyllum spicatum*, 283, 518
waternymph (**naiads**): *Najas* spp., 286
waternymph, southern (**southern naiad**): ***Najas*** **spp.**, 286
waterprimrose: ***Ludwigia*** **spp.**, 265
    California or creeping waterprimrose (**creeping waterprimrose**): ***Ludwigia*** spp., 265
    Uruguay waterprimrose (*Ludwigia grandiflora*): ***Ludwigia*** **spp.**, 265
waterpurslane (**waterprimrose**): ***Ludwigia*** spp., 265
watershield (*Brasenia schreberi*): Aquatic Plants, 458, 460
waterthyme (**hydrilla**): *Hydrilla verticillata*, 221, 518
waterweed
    Brazilian (**Brazilian egeria**): *Egeria densa*, 163
    broad (*Elodea canadensis*): Aquatic Plants, 458, 460
    Canadian (*Elodea canadensis*): Aquatic Plants, 458, 460
    common (**Brazilian egeria**): *Egeria densa*, 163
    dense (**Brazilian egeria**): *Egeria densa*, 163
    South American (**Brazilian egeria**): *Egeria densa*, 163
    yellow (**creeping waterprimrose**): *Ludwigia* spp., 265
watsonia (*Watsonia meriana*): Herbaceous Forbs, 442, 450
Watsonia meriana (= *bulfillifera*) (watsonia): Herbaceous Forbs, 442, 450
wax berry, white (**Chinese tallowtree**): *Sapium sebiferum*, 361
waxy mannagrass: *Glyceria declinata*, 196
waygrass (*Polygonum arenastrum*): Herbaceous Forbs, 442, 450
weavers broom (**Spanish broom**): *Spartium junceum*, 385
weedy sweet pea (*Lathyrus latifolius*): Herbaceous Forbs, 442, 448
weeping lovegrass (*Eragrostis* spp.): Grasses & Sedges, 434, 436
western poison oak (**Pacific poison-oak**): *Toxicodendron diversilobum*, 403
western scouringrush (**scouringrush**): *Equisetum* spp., 175
wheat oat (**wild oat**): ***Avena*** **spp.**, 49
wheatgrass (**quackgrass**): *Elytrigia repens*, 172
whickens (**quackgrass**): *Elytrigia repens*, 172
whin (**gorse**): *Ulex europaeus*, 415, 516
whip thistle (*Lactuca serriola*): Herbaceous Forbs, 438, 446
white
    bryony: *Bryonia alba*, 73
    daisy (**oxeye daisy**): *Leucanthemum vulgare*, 247
    goosefoot (*Chenopodium album*): Herbaceous Forbs, 440, 448
    horehound: *Marrubium vulgare*, 273
    horsenettle (*Solanum elaeagnifolium*): Herbaceous Forbs, 444, 452
    knapweed (**diffuse knapweed**): *Centaurea diffusa*, 93, 514
    melilot (*Melilotus albus*): Herbaceous Forbs, 442, 448
    pigweed (*Amaranthus* spp.): Herbaceous Forbs, 438, 446

INDEX

sweetclover (**Melilotus albus**): Herbaceous Forbs, 442, 448
thistle (**blessed milkthistle**): *Silybum marianum*, 371, 515
waterlily, large (*Nymphaea odorata*): Aquatic Plants, 458, 460
wax berry (**Chinese tallowtree**): *Sapium sebiferum*, 361
weed (*Solanum elaeagnifolium*): Herbaceous Forbs, 444, 452
whitethorn (*Crataegus monogyna*): Woody Plants, 454, 456
whitetop (*Cardaria chalepensis, C.draba, C. pubescens*): *Cardaria* spp., 76
    hairy whitetop (*Cardaria pubescens*): *Cardaria* spp., 76
    lens-podded whitetop (*Cardaria chalepensis*): *Cardaria* spp., 76
    short whitetop (**hoary cress**): *Cardaria* spp., 76
    tall whitetop (**perennial pepperweed**): *Lepidium latifolium*, 244
whiteweed (*Cardaria chalepensis, C.draba*): *Cardaria* spp., 76
whiteweed (**hoary cress**): *Cardaria* spp., 76
whiteweed (**oxeye daisy**): *Leucanthemum vulgare*, 247
whiteweed, giant (**perennial pepperweed**): *Lepidium latifolium*, 244
whorled milkweed (**Mexican whorled milkweed**): *Asclepias* spp., 43
widgeongrass: *Ruppia maritima*, 349
wild
    artichoke (**artichoke thistle**): *Cynara cardunculus*, 138
    barley (**hare barley**): *Hordeum marinum & murinum*, 218
    brier (**dog rose**): *Rosa* spp., 338
    buckwheat (*Polygonum convolvulus*): Herbaceous Forbs, 444, 450
    cane (**giant reed**): *Arundo donax*, 40, 517
    carrot: *Daucus carota*, 149
    chervil (*Anthriscus sylvestris*): *Anthriscus* spp., 39
    four-o'clock: *Mirabilis nyctaginea*, 277
    kale (**wild radish**): *Raphanus* spp., 331
    lettuce, common (*Lactuca serriola*): Herbaceous Forbs, 438, 446
    millet (*Setaria* spp.): Grasses & Sedges, 434, 436
    morningglory (**field bindweed**): *Convolvulus arvensis*, 127, 516
    mustard: *Sinapis arvensis*, 374
    oat (*Avena fatua*): *Avena* spp., 49
    oat, slender (**slender oat**): *Avena* spp., 49
    olive (**Russian-olive**): *Elaeagnus angustifolia*, 169
    radish (**radish**): *Raphanus* spp., 331
    radish (*Raphanus raphanistrum*): *Raphanus* spp., 331
    radish, jointed (**wild radish**): *Raphanus* spp., 331
    rue (*Peganum harmala*): Herbaceous Forbs, 444, 452
    rutabaga (*Brassica rapa*): Herbaceous Forbs, 440, 448
    snapdragon (**Dalmatian toadflax**): *Linaria dalmatica*, 254, 517
    snapdragon (**yellow toadflax**): *Linaria vulgaris*, 257, 517
    spinach (*Chenopodium album*): Herbaceous Forbs, 440, 448
    sunflower (*Helianthus annuus*): Herbaceous Forbs, 438, 446
    teasel (**common teasel**): *Dipsacus* spp., 154
    turnip (*Brassica rapa*): Herbaceous Forbs, 440, 448
    turnip (**wild radish**): *Raphanus* spp., 331
    turnip, long-fruited (**Saharan mustard**): *Brassica tournefortii*, 64
willie, stinking (*Senecio jacobaea*): Herbaceous Forbs, 440, 446, 515
willow smartweed (**pale smartweed**): *Polygonum persicaria & lapathifolium*, 320
willows: *Salix* spp., 351
willow-weed (*Polygonum lapathifolium* or *P. persicaria*): *Polygonum persicaria & lapathifolium*, 320
winged plumeless thistle (**slenderflower thistle**): *Carduus* spp., 79, 513
winged thistle (**Scotch thistle**): *Onopordum acanthium*, 292, 515
winter bluegrass (**bulbous bluegrass**): *Poa bulbosa*, 316
wiregrass (**bermudagrass**): *Cynodon dactylon*, 141
wiregrass, North African: *Ventenata dubia*, 417
wiregrass (prostrate knotweed) (*Polygonum arenastrum*): Herbaceous Forbs, 442, 450
wiregrass (**quackgrass**): *Elytrigia repens*, 172
wireweed (*Polygonum arenastrum*): Herbaceous Forbs, 442, 450
wireweed (**pale smartweed**): *Polygonum persicaria & lapathifolium*, 320
wisteria tree, scarlet (**red sesbania**): *Sesbania punicea*, 369

witchgrass (**quackgrass**): *Elytrigia repens*, 172
witch's brier (**dog rose**): *Rosa* spp., 338
witchweed (**Russian-thistle**): *Salsola* spp., 353, 515
woad, dyer's: *Isatis tinctoria*, 236
wokas (*Nuphar lutea*): Aquatic Plants, 458, 460
Wolffia spp. (**watermeal**): *Lemna, Spirodela, & Wolffia*, 242
wolf's milk (**leafy spurge**): *Euphorbia* spp., 182, 516
wonder plant (**castorbean**): *Ricinus communis*, 333
wood forget-me-not (**broadleaf forget-me-not**): *Myosotis latifolia*, 279
woolly
    distaff thistle: *Carthamus lanatus*, 85
    mullein (**common mullein**): *Verbascum* spp., 419
    safflower (**woolly distaff thistle**): *Carthamus lanatus*, 85
    starthistle (**woolly distaff thistle**): *Carthamus lanatus*, 85
    thistle (**Scotch thistle**): *Onopordum acanthium*, 292, 515
woolmat (**houndstongue**): *Cynoglossum officinale*, 143
wormwood, absinth (*Artemisia absinthium*): Herbaceous Forbs, 438, 446
***Xanthium* spp.**
    spinosum (**spiny cocklebur**): *Xanthium* spp., 428
    strumarium (**common cocklebur**): *Xanthium* spp., 428
        [= *californicum, calvum, campestre, canadense, chinense. macrocarpum, palustre, pennsylvanicum*]
yardgrass (*Polygonum arenastrum*): Herbaceous Forbs, 442, 450
yellow
    alyssum (*Alyssum alyssoides*): Herbaceous Forbs, 440, 448
    bristlegrass (*Setaria* spp.): Grasses & Sedges, 434, 436
    burweed (**coast fiddleneck**): *Amsinckia* spp., 35
    bush lupine (*Lupinus arboreus*): *Lupinus* spp., 267
    cockspur (**yellow starthistle**): *Centaurea solstitialis*, 100, 514
    dock (**curly dock**): *Rumex crispus & obtusifolius*, 346
    flag (**yellowflag iris**): *Iris pseudacorus*, 234
    floatingheart: *Nymphoides peltata*, 291
    forget-me-not (**coast fiddleneck**): *Amsinckia* spp., 35
    foxtail (*Setaria* spp.): Grasses & Sedges, 434, 436
    hawkweed (**meadow hawkweed**): *Hieracium* spp., 209
    locust (**black locust**): *Robinia pseudoacacia*, 335
    loosestrife (**garden loosestrife**): *Lysimachia vulgaris*, 269
    pondlily (*Nuphar lutea*): Aquatic Plants, 458, 460
    sorrel (**buttercup oxalis**): *Oxalis pes-caprae*, 295
    starthistle: *Centaurea solstitialis*, 100, 514
    tarweed (**coast fiddleneck**): *Amsinckia* spp., 35
    toadflax: *Linaria vulgaris*, 257, 517
    water-weed (**creeping waterprimrose**): *Ludwigia* spp., 265
yellowflag iris: *Iris pseudacorus*, 234
yellowflower
    pepperweed (*Lepidium perfoliatum*): Herbaceous Forbs, 440, 448
    tarweed (**virgate tarweed**): *Holocarpha virgata*, 214
yellow-flowered oxalis (**buttercup oxalis**): *Oxalis pes-caprae*, 295
yellowtuft (*Alyssum alyssoides*): Herbaceous Forbs, 440, 448
Yorkshire fog (**common velvetgrass**): *Holcus lanatus*, 212
zacate johnson (**johnsongrass**): *Sorghum halepense*, 380
zaccoto gordo (**coast fiddleneck**): *Amsinckia* spp., 35
Zannichellia palustris (horned pondweed): Aquatic Plants, 458, 460
    [=*major*]
Zantedeschia aethiopica (calla lily): Herbaceous Forbs, 438, 446
z-grass (*Zannichellia palustris*): Aquatic Plants, 458, 460
***Zigadenus* spp.**
    paniculatus (**foothill deathcamas**): *Zigadenus* spp., 431
    venenosus (**meadow deathcamas**): *Zigadenus* spp., 431
ziz's pondweed (**Illinois pondweed**): *Potamogeton & Stuckenia*, 323
zorro annual fescue (**rattail fescue**): *Vulpia myuros*, 425
zygadene: *Zigadenus* spp., 431
Zygophyllum fabago (Syrian beancaper): Herbaceous Forbs, 444, 452